Food Engineering Series

For further volumes:
http://www.springer.com/series/5996

Stavros Yanniotis • Petros Taoukis
Nikolaos G. Stoforos • Vaios T. Karathanos
Editors

Advances in Food Process Engineering Research and Applications

Springer

Editors
Stavros Yanniotis
Department of Food Science
and Human Nutrition
Agricultural University of Athens
Iera Odos 75
11855 Athens, Greece

Nikolaos G. Stoforos
Department of Food Science
and Technology
Agricultural University of Athens
Iera Odos 75
11855 Athens, Greece

Petros Taoukis
Division IV-Product and Process Development
School of Chemical Engineering
National Technical University of Athens
Iroon Polytechniou 5
15780 Zografou, Athens, Greece

Vaios T. Karathanos
Department of Nutrition and Dietetics
Harokopion University
El. Venizelou 70
17671 Kallithea, Athens, Greece

ISSN 1571-0297
ISBN 978-1-4614-7905-5 ISBN 978-1-4614-7906-2 (eBook)
DOI 10.1007/978-1-4614-7906-2
Springer New York Heidelberg Dordrecht London

Library of Congress Control Number: 2013948607

Printed on acid-free paper

Springer is part of Springer Science+Business Media (www.springer.com)

Preface

This book provides an international perspective of today's frontiers in food process engineering research and innovation. It includes selected keynote contributions to the 11th International Congress on Engineering and Food (ICEF11), which took place in Athens, Greece, 22–26 May 2011. The theme of ICEF11 was "Food Process Engineering in a Changing World." It explored the issue of how food science and engineering could contribute to the solution of vital problems in a world of increasing population and complexity, under severe constraints of limited resources of raw materials and the environment.

The 32 chapters of this book, including Chaps. 1 and 2 which introduce the theme of modern food process engineering research and innovation, are grouped according to the main themes of the Congress, namely food materials science and properties (Chaps. 3–7), advances in food process technology (Chaps. 8–16), novel food processes (Chaps. 17–21), modeling and control of food processes (Chaps. 22–25), modeling and control of food safety and quality (Chaps. 26–30), and current and future issues (Chaps. 31 and 32).

The editors wish to thank the authors for their contributions. Without their effort and expertise this book would not have been possible. We would also like to express our gratitude to Professor Emeritus George D. Saravacos, who inspired and guided the efforts of the team responsible for the successful organization of ICEF11.

Athens, Greece — Stavros Yanniotis
Athens, Greece — Petros Taoukis
Athens, Greece — Nikolaos G. Stoforos
Athens, Greece — Vaios T. Karathanos

Contents

1 **Food Process Engineering Research and Innovation in a Fast-Changing World** 1
Helmar Schubert, Heike P. Schuchmann, Robert Engel, and Kai Knoerzer

2 **Food Process Engineering Research and Innovation in a Fast Changing World: Paradigms/Case Studies** 41
Heike P. Schuchmann, Karsten Köhler, M. Azad Emin, and Helmar Schubert

Part I Food Materials Science and Properties

3 **Advances in Nanotechnology as Applied to Food Systems** 63
Jarupat Luecha, Nesli Sozer, and Jozef L. Kokini

4 **Relaxations, Glass Transition and Engineering Properties of Food Solids** 79
Yrjö H. Roos

5 **Molecular-Based Modeling and Simulation Studies of Water–Water and Water–Macromolecule Interactions in Food and Their Effects on Food Dehydration** 91
J.-C. Wang and A.I. Liapis

6 **Rheological and Structural Characteristics of Nanometre-Scale Food Protein Particle Dispersions and Gels** 111
M. Anandha Rao, Simon M. Loveday, and Harjinder Singh

7 **Transport Properties in Food Process Design** 119
M. Krokida and G.D. Saravacos

Part II Advances in Food Process Technology

8 Applying Advances in Food Process Engineering in a Changing World: The Industry Perspective 141
J. Peter Clark

9 Recent Developments in Drying Technologies for Foods 153
Sachin V. Jangam and Arun S. Mujumdar

10 Batch Coffee Roasting; Roasting Energy Use; Reducing That Use 173
Henry Schwartzberg

11 Advances and Challenges in Thermal Processing with Flexible Packages 197
Arthur Teixeira, Gaurav Ghai, and Sergio Almonacid

12 Current Knowledge in Hygienic Design: Can We Minimise Fouling and Speed Cleaning? 209
P.J. Fryer, P.T. Robbins, and I.K. Asteriadou

13 Encapsulation Systems in the Food Industry 229
Viktor Nedović, Ana Kalušević, Verica Manojlović, Tanja Petrović, and Branko Bugarski

14 Aroma Encapsulation in Powder by Spray Drying, and Fluid Bed Agglomeration and Coating 255
Turchiuli Christelle and Dumoulin Elisabeth

15 Advancements in Microbial Polysaccharide Research for Frozen Foods and Microencapsulation of Probiotics 267
Pavan Kumar Soma, Patrick D. Williams, BoKyung Moon, and Y. Martin Lo

16 Food Allergens and Processing: A Review of Recent Results 291
Milan Houska, Ivana Setinova, and Petr Kucera

Part III Novel Food Processes

17 Emerging Technologies for Targeted Food Processing 341
D. Knorr, A. Froehling, H. Jaeger, K. Reineke, O. Schlueter, and K. Schoessler

18 Nonthermal Technologies to Extend the Shelf Life of Fresh-Cut Fruits and Vegetables 375
Iryna Smetanska, Dase Hunaefi, and Gustavo V. Barbosa-Cánovas

19 Enhancing Extraction from Solid Foods and Biosuspensions by Electrical Pulsed Energy (Pulsed Electric Field, Ohmic Heating, and High-Voltage Electrical Discharge) 415
Eugène Vorobiev and Nikolai Lebovka

20 Food Structure Engineering for Nutrition, Health and Wellness 429
Stefan F.M. Kaufmann and Stefan Palzer

21 Transfer of Water and Volatiles at Interfaces: Application to Complex Food Systems 445
Andrée Voilley, Sonia Lequin, Alicia Hambleton, David Chassagne, Thomas Karbowiak, and Frédéric Debeaufort

Part IV Modeling and Control of Food Processes

22 Modeling Food Process, Quality and Safety: Frameworks and Challenges 459
Ashim Datta and Ashish Dhall

23 Mathematical Modeling of Transport Phenomena for Simulation and Optimization of Food Processing Operations 473
Ferruh Erdoğdu

24 Food Preservation Process Design 489
Dennis R. Heldman

25 Advanced Sensors, Quality Attributes, and Modeling in Food Process Control 499
Michael J. McCarthy and Kathryn L. McCarthy

Part V Modeling and Control of Food Safety and Quality

26 Predictive Modeling of Textural Quality of Almonds During Commercial Storage and Distribution 521
Li Z. Taitano and R. Paul Singh

27 Developing Next-Generation Predictive Models: Systems Biology Approach 547
D. Vercammen, E. Van Derlinden, F. Logist, and J.F. Van Impe

28 Dynamic Approach to Assessing Food Quality and Safety Characteristics: The Case of Processed Foods 567
Teresa R.S. Brandão, Maria M. Gil, Fátima A. Miller, Elsa M. Gonçalves, and Cristina L.M. Silva

29 Hyperspectral Imaging Technology: A Nondestructive Tool for Food Quality and Safety Evaluation and Inspection 581
Di Wu and Da-Wen Sun

30 Food Chain Safety Management Systems: The Impact of Good Practices 607
Raspor Peter, Ambrožič Mateja, and Jevšnik Mojca

Part VI Current and Future Issues

31 Does Biofuel Production Threaten Food Security? 629
Walter E.L. Spiess

32 Academia-Industry Interaction in Innovation: Paradigm Shifts and Avenues for the Future 645
I. Sam Saguy

Erratum E1

Index 657

Chapter 1
Food Process Engineering Research and Innovation in a Fast-Changing World

Helmar Schubert, Heike P. Schuchmann, Robert Engel, and Kai Knoerzer

1.1 Introduction

Our civilisation is currently undergoing changes at a rate higher than ever before. Information and knowledge double every 5–10 years. The global human population is still growing. While there were only three billion people living on Earth in 1960, today seven billion live on our planet, and in 2050 the population will be nine billion, most of them hoping for a better standard of living than today's. As a result of the growing population, the consumption of resources is increasing faster as well. Most of these resources, like fossil fuels, are not renewable, and furthermore, their use could be causing climate change.

Among the most significant challenges for the future are the provision of reasonably priced and sustainable supply, storage and transport of energy, clean freshwater and adequate food for all human beings. Food process engineering is involved in all three of these main challenges, directly or at least indirectly. Alternative energy sources are based on agricultural products, resulting in a new competition for farmland use for food production. Many accepted facts indicate that food will run short in the future and, therefore, food prices will increase (Schubert 2007). Global food production must increase by 70–100 % by 2050 to feed the global population (Idso 2011). Keeping in mind the decreasing amount of arable farmland (Hopp 2002) and the growing human population, new concepts will be

H. Schubert • H.P. Schuchmann (✉)
Institute of Engineering in Life Sciences, Section of Food Process Engineering, Karlsruhe Institute of Technology (KIT), D-76131 Karlsruhe, Germany
e-mail: heike.schuchmann@kit.edu

R. Engel
BASF SE, D-67117 Limburgerhof, Germany

K. Knoerzer
CSIRO Animal, Food and Health Sciences, Werribee 3030, VIC, Australia
e-mail: kai.knoerzer@csiro.au

S. Yanniotis et al. (eds.), *Advances in Food Process Engineering Research and Applications*, Food Engineering Series, DOI 10.1007/978-1-4614-7906-2_1,

required to feed the world and meet future consumer demands for food. Research and innovations in the field of food process engineering and food packaging will therefore have to focus on food security, i.e. access to sufficient, safe and nutritious food for all people at all times. A limiting factor for growing agricultural products, even today, is the sustainable supply of freshwater from precipitation. In many parts of the world, groundwater levels are falling dramatically because of intensive irrigation.

Water itself is the most important food. Freshwater shortage is a global problem, and water quality is of the utmost importance. More than two-thirds of the freshwater used on Earth is needed to produce food and animal feed. In many cases, non-renewable water resources must be tapped. That this kind of non-sustainable water supply will run dry is already on the horizon. The terms *virtual water* and *water footprint* will be explained and discussed. An adequate freshwater supply and water of high quality are prerequisites for the production of high-quality foods. Food engineers will have to seek collaboration with research and innovation on water management in the future.

Along with the problem of assuring high-quality food in large quantities for a fast growing world population, human societies will face new problems arising from the consumption of unbalanced food in unhealthy quantities, particularly in highly developed (industrialised) countries. Health-care costs explode and quality of life decreases significantly for those facing cancer or cardiovascular disease arising from the new lifestyle. Solving these problems will also require close collaboration between medical, sociological, physiological and food engineering specialists.

In face of these challenges, food process engineering offers some promising developments. Intensified research in the following areas may induce substantial improvements and technical innovations:

1. Development of adjusted processes and food products by employing the knowledge established in basic chemical engineering science using modern tools, for example computer-aided simulation, material sciences, novel measuring devices (e.g. magnetic resonance tomography) and nanotechnology. Research in these fields will lead to a better understanding of the relationship between micro- and submicro-structures and functional properties such as bioavailability or bioactivity of food compounds and enable a targeted product design.
2. Improving the process efficiency, saving energy and water, reducing waste and environmental pollution to produce high-quality food at the lowest cost and improved sustainability.
3. Ensuring food safety, for example by improving hygienic design, by providing appropriate packaging and by developing and verifying better models to ensure food safety using quantitative microbiology and new mathematical tools to improve microbial risk assessment (Peleg 2006).
4. Improved product quality control by intelligent, computer-aided automation tools, advanced monitoring and control systems and flexible manufacturing of food to manage food processes, even with the complex interacting parameters involved, and obtain personalised products.

Based on research performed in recent decades, promising novel processes such as high-pressure processing, pulsed electric field treatment or cold plasma treatment to decontaminate food and packaging surfaces are emerging as globally marketable developments. Since the 1990s the functional foods market is growing and has attracted the interest of industry. Even if nanotechnology itself will not be key in food product development, it will drive innovations in process- and product-structure-relationship understanding. Whether or not these novel processes will someday become true innovations like heat pasteurisation, cooling and freezing of food is unknown at present. True innovations in food process engineering bring changes in a big step and are, therefore, rare events. Generally, innovations in food engineering are mostly renovations or improvements of existing processes, which means taking smaller steps.

In order to remain competitive and successful, the food industry requires continuous process improvements, which demand ongoing research and development (R&D). Besides R&D in industry, research in academia and other institutions is necessary, not only to broaden basic knowledge and to create new ideas and results, but also to train, qualify and supervise students and young and open-minded researchers. Young food engineers, in collaboration with natural scientists, i.e., chemists, biologists and physicists, will become the driving force for the future of food process engineering. Interdisciplinarity will be key in solving new social problems. Thus, excellent education of enough young people is the major challenge for the future of our discipline in an ever-changing world with fast growing knowledge in science and technology.

With the help of some recent research results, the general statements mentioned earlier will be elucidated and discussed.

1.2 Water Required for Food Supply

1.2.1 Facts of a Fast Changing World

Freshwater is the most important food. For the sake of simplicity, freshwater will henceforth be referred to as water. Table 1.1 outlines some facts on our fast changing world with respect to human water demands. The world's population is growing and will do so for the foreseeable future, whereas arable land per capita is decreasing. The amount of arable land is limited by the amount of water available in a given region. Approximately 70 % of the global human water demand is associated with agricultural production (Hoekstra and Chapagain 2008; Wefer 2010; UBA 2011), 20 % with industry, including power stations, and 10 % with domestic use (Hüttl and Bens 2012). Thus, water is the limiting factor and most important resource in the production of food. Global water consumption (Table 1.1) is increasing and will reach the lower limit of the available renewable water resources by 2025 according to Zehnder (2002).

Table 1.1 Some facts about the fast changing world

Year	1950	1975	2000	2025
World population (10^9) (UN 2011)	2.5	4.1	6.1	8
Global arable land (m^2/capita) (Hopp 2002)	5,100	3,400	2,100	1,500
Global water[a] consumption (L/(capita · day)) (Zehnder 2002[b])	1,500	2,100	2,700	3,300
Global water consumption (km^3/year) (Zehnder 2002[b])	1,400	3,000	6,000	9,000[c]

[a]Water: here always freshwater
[b]Estimated values
[c]Lower limit of available renewable water resources per year according to Zehnder (2002)

In order to asses the water supply to the Earth's population, regional water availability is more important as an indicator than global water availability. Some regions have an excess of water, some have a satisfactory or sufficient water supply, and some suffer from water stress (1,700–1,000 m^3/(capita · year)), water shortage (<1,000–500 m^3/(capita · year)), and water scarcity (<500 m^3/(capita · year)) (Falkenmark and Widstrand 1992). Finally, the temporal dimension of water availability must be considered. Dry periods during the growing season have an impact on the yield per hectare in agriculture and may be compensated by additional irrigation.

Regions unable to carry out sufficient farming on their own (e.g., due to water shortages), therefore, depend on importing raw materials for food and feed from abroad. The lion's share of water utilised by humans is needed for food production because plants require huge amounts of water from seed to harvest. As will be shown later, approximately 1,000–10,000 kg of water is needed to produce 1 kg of foodstuffs. In other words, it is either possible to transport 1 kg of food or to transport the 1,000-fold to 10,000-fold mass of water to an arid region that will then be able to produce the raw food materials on its own. These thoughts lay the foundation for the concept of virtual water explained in Sect. 1.2.3.

1.2.2 The Earth's Water

According to the global water cycle, water evaporates from the oceans (71 % of the Earth's surface) and land surface (29 % of the Earth's surface), condenses in the atmosphere through the formation of clouds and yields precipitation. Thus, water is not only permanently distributed across the Earth's land surface, it is also purified by a process similar to that in a water distillation plant. By means of evaporation and, subsequently, condensation, the oceans' salt water is converted into freshwater.

While the oceans contain an almost unlimited amount of salt water, only 2.5 % of the water on Earth is available as freshwater (in this paper simply water). Only a small share of about 0.13 % of the Earth's water (Shiklomanov 1993; Zehnder 2002) is stored in a way that allows for immediate access and use by the flora and fauna, including humans, outside the oceans. In order to guarantee sustainable utilisation, the amount of water stored must remain constant over a long period. This means that precipitation is the only renewable resource available for sustainable water management.

It is important to point out that water is usually contaminated or heated, which may induce thermal pollution, but not fully consumed or used up, as is the case with, for example, fuels and electric energy. Although the purification of water can be highly complex and, therefore, expensive, no water will be lost from a global point of view. The law of mass conservation for water

$$P + I - R - E - \Delta S = 0, \tag{1.1}$$

where P = mass of precipitation, I = mass of inflow from upstream regions ($I = 0$ if the balance scope is a river basin), R = mass of water runoff, E = mass of evaporated water (including transpiration) and ΔS = changes in stored mass of water (positive when stored mass increases), is valid for a certain period of time and, from a regional perspective, for a limited balance scope (control volume). However, we need to remember that the amount of evaporated water is no longer available within the control volume, whereas the water outflow may be repeatedly used and purified before finally leaving the system boundaries. As a consequence, it seems fair to refer to the amount of evaporated water as water consumption since – at the time of evaporation – this amount is no longer available to the control volume. As the evaporated water will return to the Earth's surface as rainfall at some point in time, at least a certain share (depending on the control volume's size) of this evaporated water may be re-added to the respective region. This secondary effect will not be addressed in this work. (Natural) evaporation is made up of evaporation (E) and transpiration (T) and is referred to as evapotranspiration:

$$ET = E + T. \tag{1.2}$$

Evaporation refers to the evaporation from moist solid surfaces such as soil (soil evaporation), plant surfaces moistened by precipitation (interception) or other water surfaces without biological or physiological processes. Transpiration is defined as the evaporation from plants by means of biological-physiological processes.

Compared to other kinds of water use, evaporation is of particular importance. This is also stressed in the following scenario: if in a certain region (control volume) water is used for manufacturing non-agriculturally produced goods or for rendering services, this water will be available for further use after sufficient purification and will finally end up in the balance scope's sewerage. For the production of agricultural goods, however, water is primarily needed for evapotranspiration purposes – and due to its evaporation, this water will no longer be available to the region. In the case of large-scale irrigation of dry soils, the mass of evaporating water E, according to Eq. 1.1, reduces the water runoff R as long as P and I remain constant and ΔS disappears on a long-term average (sustainability) basis. On the whole, a large-scale cultivation of arid regions that utilises additional irrigation for agricultural purposes may eventually have an impact on the global water cycle.

1.2.3 Virtual Water and Water Footprint Concepts

In a recent paper (Schubert 2011) the concepts of virtual water and water footprints are described in detail and a modified concept of virtual water is proposed. In the following section a short excerpt of this paper is presented.

1.2.3.1 Definitions

The virtual water V of a product (e.g., good, electrical power, service) is the volume of water used to produce a product in the various steps of the production chain. The virtual water content v of a good is the virtual water volume per kilogram of product. Virtual water also contains the actual amount of water that exists in a certain product, particularly since this water was also necessary for the production of this good. The actual water content is usually negligible compared to v.

In London in 1994, Allen was the first to introduce the concept of virtual water (Allen 1994; World Water Council 2004). It was based on analyses by Israeli water experts who found that it seemed to make more sense for their arid country to import water-intensive goods than to cultivate (let alone export) such products themselves (World Water Council 2004). Import here mainly refers to commodities such as cereals. High-quality and, thus, costly agricultural raw materials may be produced locally as long as proportional water costs remain sufficiently low.

Allen had initially developed the concept of virtual water as a metaphor – but then attempted to apply it to illustrate the actual amount of water utilised by humans. Moreover, the concept allows us to track the flows of virtual water since trade with goods whose production requires large amounts of water can be understood as a virtual transfer of water. This kind of transfer is referred to as virtual water trade (Hoekstra 2003; World Water Council 2004). As soon as a water-intensive agricultural product is exported, the exporting region will lack the amount of water required for evapotranspiration, whereas the importing region will save this amount of water.

Water footprint *WF* is defined by the total volume of water that is required per time unit for a person or a particular group of people. It comprises both the directly required volume of water and the specific indirect (i.e., virtual) amount of water for manufacturing goods or providing services that are required by that particular person or group. The water footprint is typically expressed in annualised terms. It is also customary to refer to cities, regions, states or companies where goods are manufactured or services provided. The water footprint concept was coined by Hoekstra (Hoekstra and Hung 2002).

A comprehensive collection of data has been worked out by the UNESCO-IHE Institute for Water Education in Delft (cf. Hoekstra and Chapagain 2008). In 2008, the Water Footprint Network (WFN 2008) was established; it provides extensive research on the issues of virtual water and water footprint.

As for global trade, the water footprint of nations provides another variable. For example, Germany's water footprint has been measured at 125 km^3 water/year $\approx$ 4,300 L of water/(capita · day) (Hoekstra 2008). Other assessments (Sonnenberg et al. 2009) provide estimated values of 160 km^3 water/year $\approx$ 5,000 L of water/(capita · day). The global water footprint for agricultural products has been determined as being 7,500 km^3/year $\approx$ 3,000 L of water/(capita · day) (Hoekstra and Chapagain 2007). Further, it should be recalled that, depending on the applied definition of water footprint, different dimensions will arise, such as cubic meters of water/(person · year), cubic meters of water/(enterprise · year), cubic meters of water/(region · year), or cubic meters of water/(nation · year).

In recent research, the author of the water footprint concept suggested applying it to both products and services (Hoekstra 2008). A product's water footprint is defined by the volume of water required to manufacture a certain mass of the product at the actual place of manufacture. Usually, the water footprint will therefore be indicated in cubic meters of water per kilogram of product. As Hoekstra notes, this definition is in line with the definition of virtual water content. However, it always refers to the place of manufacture to allow for a regional reference point. The same applies to services offered.

1.2.3.2 Application of Concepts

Although at first glance the concept of virtual water appears simple, it has proven difficult to practically apply it. Firstly, it is necessary to make a distinction between services and products. In terms of services, it remains an open question as to what extent water demand should also consider the virtual water needs of the person who renders this service. In the case of products, we have to consider the type of water use. When the form of utilisation of water is the sole source of its contamination, reutilisation will be possible as soon as the water has been purified. Water demand at a certain production location can be reduced substantially by means of appropriate recirculation with integrated waste-water treatment. Water that evaporates into the atmosphere will, however, be lost for the respective system and can, thus, be regarded as consumed in a regional sense (cf. Sect. 1.2.2). Hence, the majority of water demand that is required for household purposes, for manufacturing industrial products, or the treatment and processing of raw materials into food products needs to be assessed in a different way from the amount of water needed for evapotranspiration to generate agricultural raw materials.

Another difficulty arises from the question of how to consider the type and quality of water. Experts have come to classify water into green, blue, and grey (as explained in Sect. 1.2.3.3) and, thus, at least to some degree, have started to take into account the origin, type and quality of water. What is still missing is a suitable approach to assessing virtual water capable of taking into account water quality and water management in a satisfactory way. In Sect. 1.2.3.5 the large differences between sustainable and unsustainable water withdrawal in a given region will be elucidated.

Finally, the amount of water needed for the production of goods may largely depend on the production site (Hoekstra 2008). The product may either be a primary (e.g., a freshly harvested coffee bean), intermediate (roasted coffee bean), or final product (ready-to-drink coffee).

For crop plants, virtual water is mainly determined by evaporation occurring during the time span between sowing and harvest. For plants such as rice, the period of field preparation immediately prior to sowing or planting will be added (Chapagain and Hoekstra 2010). Above all, evapotranspiration is responsible for determining the amount of virtual water contained in the final product. Compared to the product-related amount of water needed for evapotranspiration, the water required for treatment and processing amounts to only a few percent in the case of modern enterprises in the food industry (Brabeck-Letmathe 2008). Evapotranspiration depends primarily on factors such as the type of plant and its water demand, water availability, meteorological data at a given location and the form of land management. In the case of artificial irrigation, water management, as well as the type and efficiency of irrigation, will have an essential impact on defining the amount of virtual water per harvested product.

It is important to precisely define the product to which virtual water is to be assigned in each particular case. If the harvested material is to be used for different purposes, the virtual water should be assigned proportionally to the respective subquantities. For cereals we would have to differentiate between grains (or the flour produced thereof) on the one hand and straw as well as husks (which may be used for the production of energy) on the other. The single subquantities may be assigned in terms of either weight or value. Research has not always taken into account the set of problems connected to these issues. However, we should consider them as the content of virtual water will vary by a factor of 2 or more according to the assignments mentioned earlier.

In most cases, the virtual water content of animal products is significantly higher than the virtual water content of crop products. Animals need to be fed and watered throughout their lives. In terms of animal housing, additional water will be required for cleaning purposes. For agricultural production, the predominantly vegetarian feed requires large amounts of water that will be allocated to the animal products as virtual water. The total amount of real and virtual water required by the animal and the water required for processing are proportionately allocated to the diverse animal products (meat and sausage products, eggs, milk, cheese, among others). This proportional allocation may also occur in terms of weight or, more frequently, by market value. For ready-to-eat animal products, research data (Mekonnen and Hoekstra 2010) suggest an estimated 3,000–15,000 L/(kg product) of virtual water content, compared to approximately 1,000–3,000 L/(kg product) for ready-to-use crop products. As the virtual water content is significantly higher for cotton, coffee beans and cocoa beans, these plants clearly constitute an exception. Values for other selected products are displayed in Tables 1.2 and 1.3.

The following reference values are applicable to ready-to-eat food (Brabeck-Letmathe 2008):

Table 1.2 Virtual water content for selected food products

	Virtual water content in litres per kilogram of product	
Product	For different countries	Global average
Sugar cane	100–200	170
Maize	400–1,900	900
Milk (cow)	650–2,400	1,000
Wheat	620–2,400	1,300
Soybean	1,100–4,100	1,800
Rice	1,000–4,600	2,900
Chicken meat	2,200–7,700	3,900
Pork	2,200–7,000	4,900
Beef	11,000–38,000	15,500
Coffee (roasted)	5,800–33,500	21,000

Source: Hoekstra and Chapagain (2007)

Table 1.3 Water demand (approximately virtual water) for selected products

Product	Virtual water (litres)	Virtual water content (litres/kg product)
1 cup of tea (250 mL)	35	140
1 glass of beer (250 mL)	75	300
1 glass of wine (250 mL)	240	1,000
1 cup of coffee (125 mL)	140	1,100
1 slice of bread (30 g)	40	1,300
1 slice of bread with 10 g cheese	90	2,300
1 egg (40 g)	135	3,400
1 hamburger (150 g)	2,400	16,000
1 cotton t-shirt (250 g)	2,000	8,000
1 microchip (2 g)	32	16,000

Source: Hoekstra and Chapagain (2007)

Approximately 10 L of water for 1 kcal (4.2 kJ) of meat is required, compared to 1 L of water for 1 kcal of vegetable food.

Accordingly, a global average of approximately 3,000 L of water per adult person per day is therefore required for foodstuff (the energy supply recommended per day of an adult person varies from 1,600–3,300 kcal and is on average approximately 2,500 kcal $\approx$ 10,000 kJ). Numerous parameters, simplifications and various understandings can considerably affect the estimated amount of virtual water content. In general, there is a lack of information on the reliability of the estimated values published.

1.2.3.3 Blue, Green and Grey Water

Blue water refers to groundwater and surface water. Blue water is readily available to humans. It may be collected and transported and is, for example, used in agriculture for artificial irrigation. In contrast, green water is capillary water in the soil or stored in plants. It results from precipitation and is utilised for local farming and forestry. As a long-term global average, 65 % is available as green water, while 35 % is available as blue water (Zehnder 2002). However, huge regional and (due to drought periods and floods) temporal variations occur.

As mentioned earlier, water is usually not consumed. Instead, the use of water usually results in contamination or – in the case of cooling water (direct cooling) – warming. Only the water that evaporates into the atmosphere can be said to be consumed as it is usually no longer available to a given region. In contrast, contaminated blue water may be purified in sewage treatment plants and, therefore, reused once or several times.

Contaminated water is frequently referred to as grey water. In many countries, it is utilised in agriculture after (or without) purification. The share of grey water is defined differently in the virtual water concepts (Hoekstra and Chapagain 2008). In general, the production of goods involves contamination of water. As a measure of contamination, Hoekstra chose the volume of water that would be needed to dilute the contaminated water to an extent just about sufficient to reach a tolerable standard concentration c_{max} of undesired substances (Hoekstra and Chapagain 2008). The dilution water related to the product mass is defined as the share of grey water:

$$v_{grey} = m_s / \left(m_p \cdot c_{max}\right). \tag{1.3}$$

In this equation, m_s indicates the mass of undesired substances contained in the water per year (kg/year), while m_p refers to the mass of the product manufactured per year. Together with the share of the utilised, product-related green (v_g) and blue (v_b) water, the total content of virtual water can be summarised by the following equation:

$$v = v_g + v_b + v_{grey}. \tag{1.4}$$

Correspondingly, the following equation illustrates the total water footprint

$$WF = WF_g + WF_b + WF_{grey}. \tag{1.5}$$

In general, the grey water footprint and the grey virtual water content are to be regarded as notional variables that serve as a guide for water quality and as such do not always need to be provided as real amounts of water. In contrast, green (in the case of agriculture) and blue virtual water do have to exist as real amounts since they are used for production purposes. For these reasons, it is problematic to add the share of grey water in Eqs. 1.4 and 1.5 in order to determine the total share of virtual water or the virtual water footprint. The determination of a tolerable standard concentration c_{max} of contamination should similarly be regarded as problematic. Therefore, the dilution approach is problematic in terms of agricultural products. Nonetheless, it may be acknowledged that this approach provided a preliminary means of assessing water quality in terms of virtual water content and the water footprint.

The share of utilised green water (v_g or WF_g) is unproblematic for both the environment and the farmer. Green water is supplied immediately through precipitation and in general results in no (or very little) opportunity costs.

The share of utilised blue water may, however, be problematic. Sustainable water management becomes possible in a region for which the water influx is sufficiently high (due to direct precipitation or due to an inflow from upstream regions) on a long-term average basis when compared with the total amount of evaporation. However, as soon as the amount of blue water that is withdrawn or released into the atmosphere (by means of evaporation) exceeds the amount of water that can be re-added on a long-term average basis, a deficit emerges. As a consequence, water management will cease to be sustainable. Many regions around the world often resort to artificial irrigation, thereby utilising water in an unsustainable manner (Hahn 2009). In some cases, this has resulted in a dramatic lowering of the groundwater level or shrinkages of inland waters, such as in the Aral Sea (Giese et al. 1998). This is why special attention has to be paid to the virtual blue water content and the blue water footprint. What is the maximum share of blue water that may be utilised in a certain region without endangering sustainable management? Unfortunately, research data to date have failed to answer this crucial question.

1.2.3.4 Assessing the Concepts

1.2.3.4.1 Trade in Goods as an Indicator of Virtual Water Flow

Trade in goods necessitates their transport. However, this obviously does not involve the transportation of the sometimes tremendous amount of virtual water required for manufacturing these commodities. The transportation of virtual water between regions and nations is defined as the virtual water flow. The amount of virtual water exported by a country is referred to as the virtual water export. This value indicates the water that was required at the place of manufacture for the production of exported goods or the provision of exported services. The virtual water import refers to the virtual water imported by a country due to an import of products or services from abroad. For a specified period, it is possible to devise a balance for virtual water flow. The virtual water balance of a certain country is defined as positive (net import) if more water has been imported than exported. In the reverse case (net export), one speaks of a country's negative virtual water balance for a specified period. These concepts have been described extensively in the academic literature (cf. Hoekstra and Chapagain 2008). The Americas (North, South and Central) and Australia are the regions with the biggest net export of virtual water worldwide. In contrast, Europe (particularly western Europe) and Central Asia and South Asia are the regions with the highest net import of virtual water.

As data of the virtual water flow provided by different authors indicate, the values tend to fluctuate enormously depending on various estimates and are sometimes inconsistent (cf. Schubert 2011). Furthermore, from the viewpoint of economic trade theory, the concept of virtual water flow has been discussed (Gawel and Bernsen 2011).

1.2.3.4.2 The Water Footprint

As outlined in Sect. 1.2.3.1, the water footprint is a concept that is immediately comprehensible – the amount of water used per time measured by a certain entity. *Entity* may refer to an individual or a group of individuals as well as to companies, regions or nations. However, the water footprint may also be defined as the utilised amount of water per amount of a certain product. In this case, the concept is identical to the virtual water content (Sect. 1.2.3.1). The water footprint manages to avoid the term virtual, though, which is often misinterpreted by the public. The particular meaning is specified by respective dimensions, such as cubic meters of water/(person · year), cubic meters of water/(region · year), or cubic meters of water/(kg product). If, for instance, a consumer buys an imported product, he will leave a footprint in the exporting country according to the amount of water that was required for this good's production in the exporting country. The poorer the water management and the less effective the production in the exporting country is, the higher the water footprint per product mass will be.

The water footprint concept is also founded on the idea that it elucidates the responsibility of both producers and consumers for water demand (Hoekstra and Chapagain 2008) with the aim of reducing it. The water footprint concept should allow producers to compare their own water demand with the water demand of their competitors in a way which encourages greater efficiency. However, it has been shown (Schubert 2011) that the concept of the water footprint will be of little use in reducing water wastage and water consumption. Although consumers are not able to exert a direct influence here, the virtual water and the water footprint concepts may, however, help to raise awareness of the high water consumption of all of us, particularly in highly developed countries.

Our eating habits – and especially the amount of meat consumed – have an impact on our personal water footprint. A rough estimate indicates that the annual water demand per capita required for nutritional purposes will be 350 m^3 (about 1,000 L of water per day) for vegetarians (Schubert 2007). For a mixed diet with a 20 % share (in relation to energy) of meat and meat products, it will amount to 1,000 m^3 of water (2,700 L of water per day). As the indicated values are minimum estimated values, we can expect the average water consumption per capita to be slightly higher (Brabeck-Letmathe 2008). As the example illustrates, the personal water footprint resulting from a mixed diet is two to three times higher than for a vegetarian diet. As such, consumers' eating habits have a substantial influence on water consumption. However, with some exceptions, experience shows that providing information like this does not lead to significant changes in people's eating habits. It is, therefore, doubtful whether the water footprint concept can help to influence consumers' decisions. It is not enough, however, to pose the general question about where our water really comes from (as is done by many brochures). It is crucial to ask what *kind* of water we are talking about. This will be addressed in the following discussion.

Green virtual water originates from renewable precipitation and, therefore, entails no – or only little – opportunity costs. As a consequence, this share of virtual water flows (or global water footprints) caused by international trade could not be critical in terms of an overexploitation of water resources in the exporting country. Recent estimates indicate that less than 10 % of water resources are to be regarded as particularly critical for the production of agricultural raw materials (Schubert 2011). The current conceptualisations of virtual water and water footprint do not provide information on the share of non-sustainable water in virtual water flows or in a product's level of virtual water, nor does it help to determine those regions with an overexploitation of the resource water. Given the fact that it is not necessary to purchase water at cost-covering market prices in many regions of the globe due to subsidies, this information would, however, be useful not only for producers but also consumers. Producers could improve water management, while consumers could acquire information about the goods and the source countries that are associated with an overexploitation of local water resources. The present conceptualisations of virtual water and water footprint are unable to do this. As will be explained in the next section, it is therefore more useful to employ modified versions of these concepts.

1.2.3.5 Modified Concept of Virtual Water and Water Footprint

As explained in a recent paper (Schubert 2011), the concept of virtual water has frequently been used in an attempt to influence consumers and reduce their consumption of imported products, which require large volumes of water during their production. If merely green water were needed for the agricultural production of such goods, these efforts would not be justifiable, as rainwater entails no or only minimal opportunity costs. In contrast, however, the utilisation of rainwater for agricultural products creates a benefit for the exporting country that would usually not exist without this agriculture. Coffee plants cultivated in climatically favourable regions with sufficient precipitation are a good example of a sensible use of green water for comparatively expensive export goods. Even if artificial irrigation is necessary, it does not have to be detrimental as long as sustainable water management can be warranted in this region. The current concepts of virtual water and the water footprint do not provide answers to these related questions. They may even lead to misguided attempts to influence consumers and could impair the export potential of economically poor developing countries. Furthermore, the water footprint concept does not allow for an adequate assessment of production conditions.

Similarly, the current virtual water and water footprint concepts have shown to be of little use concerning questions about appropriate and sustainable forms of land use, effective water management, avoidance of water wastage and more general questions concerning the sustainable use of water. However, differentiating between green and blue water footprints is not sufficient, as blue water may also have a renewable share that is available for sustainable use.

With reference to what was indicated previously, the author (Schubert 2011) suggests complementing and modifying the concept of virtual water as follows: only the share of water required for the production of goods or the provision of services that has been unsustainably taken from (or contaminated to a certain extent at) the site of production or the place where the service is provided is to be defined as virtual water. The share of water utilised on site in a sustainable fashion to produce the respective good or to provide the particular service shall not be taken into account. The modified concept, thus, refers to the unsustainably utilised share of water, which, therefore, will henceforth be referred to as unsustainable virtual water or unsustainable water footprint. The main difficulty with the concepts lies in the quantification of sustainable water use, which cannot be determined by a single indicator (Hoekstra and Chapagain 2008). Nevertheless, we attempt to take small steps in order to move closer to our objective. For the time being, water pollution will be excluded.

Green water used by agriculture that is consumed by means of evapotranspiration is to be regarded as sustainably utilised and, thus, will remain unconsidered. Blue water in a producing control system (e.g., an agricultural holding, a company, a free trade, a region or a nation) may be used in a sustainable way as well. To achieve sustainable water management, it is a basic prerequisite that – as a long-term average – the amount of water withdrawn from a certain system (control volume) due to evaporation and water run-off must not exceed the amount of water added by inflow and precipitation. In other words, the amount of water stored in a certain system has to remain constant on a long-term average – and, for instance, the groundwater level must not decline over time. An adequate amount of water run-off is required to avoid a concentration of undesired substances in water beyond certain limit values. If the basic requirements mentioned are not met, the water resource is not seen to be managed in a sustainable manner. In this case, the evaporated blue water must be considered as virtual water (just as in the previous concept) or unsustainable virtual water.

As the interpretation of available data on virtual water and the water footprint of harvested products demonstrates, there is a plethora of data on green, blue, and (in more recent publications) grey virtual water. However, no data are available as to the volume of unsustainable blue virtual water. The author, therefore, argues for collating (or building upon) the existing data. As a first step, the water balance of a respective region may be used to observe whether the amount of water stored in this region remains constant on a long-term average basis and, thus, sustainability is possible, or whether it decreases instead, leading to water stress and unsustainable water management. Once a region has been identified where the water withdrawn due to evaporation and water run-off exceeds the amount of water added by inflow and precipitation, the evaporated share of blue water in the total amount of blue water utilised for irrigation purposes is referred to as unsustainable virtual water.

Additional information on grey water according to the definition by Hoekstra (Sect. 1.2.3.3) can be useful to characterise the degree of water contamination in agriculture caused for example by fertilisers or inappropriate use of plant protection products.

Unfortunately, there are no reliable data about the share of unsustainably withdrawn blue water. A rough estimate indicates that the unsustainable global blue water footprint amounts to approximately 500 km^3/year for harvested products (Schubert 2011). This corresponds to 7 % of the global water footprint for agricultural raw products. The unsustainable water footprint for internationally traded agricultural raw products is about 200 km^3/year, equalling approximately 3 % of the global water footprint for agricultural raw products. However, these data are rough estimates only; hence, reliable data should be collected. As a first step, it appears appropriate to confine the focus to the clear-cut cases of unsustainable water management in agriculture.

Although the generally estimated unsustainable global water footprint of 500 km^3/year for agricultural products (or 200 km^3/year for internationally traded agricultural raw products) is small compared to the total water footprint, it still amounts to a huge volume of water. The reduction of the water required by agriculture to the critical share of water suggested here is designed to allow us to highlight water problems in a more targeted way than is presently possible. The proposed concept provides a first step for counteracting this development by documenting the cases and regions where water resources are utilised in an unsustainable fashion. It helps to identify, for example, the goods produced in regions that require more water than is actually available in the long term.

The modified concept complements and extends the previous concept and has two main advantages:

1. Producers and producing regions can see if they withdraw water in an unsustainable manner through the production of goods (and if so, how much). They are, thus, also given an indication of the water pollution they cause.
2. Consumers and the importing regions receive information on the production processes which are based on an unsustainable use of water resources and learn about the extent of this water demand.

For both consumers and producers, the modified concept identifies the sectors, regions and goods that are responsible for an overexploitation of water resources and helps to quantify this overexploitation.

1.2.4 Water as a Scarce Resource

Water is a scarce resource in both temporal and regional terms. It is already apparent that many areas suffer from serious water shortage or are threatened by water crisis (Hahn 2009). In many regions non-renewable water resources have to be tapped, and this kind of unsustainable water supply will run dry. If no action is taken, the availability of water will deteriorate and may have disastrous impacts on both landscapes and human beings (Giese et al. 1998). Today, more than 900 million people lack access to safe water, and more than one-third of humanity does not even have access to basic sanitary services (UNNC 2010). Such data indicate that apart from energy provision, the provision of water of adequate quality is one of the biggest global challenges.

Sustainable water management for food production is largely based upon sustainable agriculture. Processing companies in the food industry depend on high-quality agricultural raw products that cannot be produced permanently without high-quality water or under conditions of sustainability (cf. Schubert 2007). For these reasons, the Sustainable Agriculture Initiative (SAI) platform was established (Jöhr 2003). Founded by the biggest European companies in the food industry, 18 more companies have decided to join the platform as members (SAI Platform 2003). The globally operating companies within the food industry have recognised the important role played by a sustainable water supply for agriculture guaranteeing high-quality agricultural raw products. For this reason, effective water management and a sustainable use of water are evidently the top priorities of the biggest company in the food industry (Nestlé 2007). The provision of water of adequate quality is, therefore, the key to feeding people and will increasingly become a limitation on the provision of agricultural products (Schubert 2007).

To avoid running dry in the future, more research, innovation and interdisciplinary collaboration is required to:

- Improve water management in agriculture, e.g., efficient irrigation, and in the food industry, e.g. water recycling, water quality;
- Develop novel waste-water treatment, for instance, innovative, low cost membrane technology (e.g., forward osmosis); and
- Develop improved seawater desalination using solar power.

Furthermore, it would be helpful if subsidies for artificial irrigation were discarded and reasonably priced water were used; no subsidies were provided for biofuels since, for example, 1 L of biodiesel from soybean or rapeseed requires 14,000 L of water (Gerbens-Leenes et al. 2009); and consumers changed their eating habits as the personal water footprint resulting from a mixed diet with 20 % share of meat is two to three times higher than for a vegetarian diet (cf. Sect. 1.2.3.4.2).

1.2.5 Global Problem: Malnutrition

In the majority of cases malnutrition means the consumption of unbalanced food in insufficient or in excessive quantities. On the one hand, more than 900 million people, i.e., approximately 14 % of the world's population in developing countries suffer from hunger (FAO 2010), although on a global scale enough food is produced to feed everyone. The main reasons for this are poverty and increasing food prices. On the other hand, 1.5 billion adults, or approximately 20 % of the world's population, mainly in highly developed countries, are overweight or obese (WHO 2011). Health problems and exploding health-care costs are the main consequences of overweight and obesity.

Food engineers may assist in solving some of these problems by developing healthy diets (e.g., improving the bioavailability of bioactive substances, reducing the fat, sugar and salt content of processed food); improving process efficiency and sustainability, for example by saving energy and water and reducing waste; ensuring food safety by, for example, using quantitative microbiology and new mathematical tools (Peleg 2006); and improving product quality control using advanced monitoring and control systems or flexible manufacturing systems.

The main challenge for the future of food engineering is providing an excellent education for enough young people. Excellent education also means attracting the best students and imparting not only technological knowledge but also natural sciences and humanities because multidisciplinarity will be key in solving most current and future problems.

In the following sections, recent research results from our laboratory will be presented as examples to elucidate some of the statements and conclusions mentioned previously.

1.3 Examples of Promising Developments in Food Process Engineering

1.3.1 Development of Adjusted Processes and Food Products Using Knowledge in Basic Chemical Engineering and Modern Tools

1.3.1.1 Formulation of Oil-in-Water Emulsions Containing Poorly Soluble Bioactive Components

The formulation of functional foods with the health benefits of naturally occurring bioactive components has been of great interest for many years. With the positive effects of many substances already established, the effect of bioactive components, such as carotenoids, phytosterols or anthocyanins, greatly depends on their formulation (Ax 2004; Mattson et al. 1982; Engel 2007; Frank 2012). For many substances, however, formulation proves problematic due to their poor stability in food matrices or due to their lipophilic character and low solubility in food or pharmaceutical components. A further challenge in formulating hydrophilic bioactive components is achieving microbial stability.

Although solutions of bioactive components usually exhibit the greatest bioavailability or dose response, the low solubility or poor stability of either the solution or the bioactive components themselves would require the uptake of large quantities of the solution. In the case of lipophilic bioactive components,

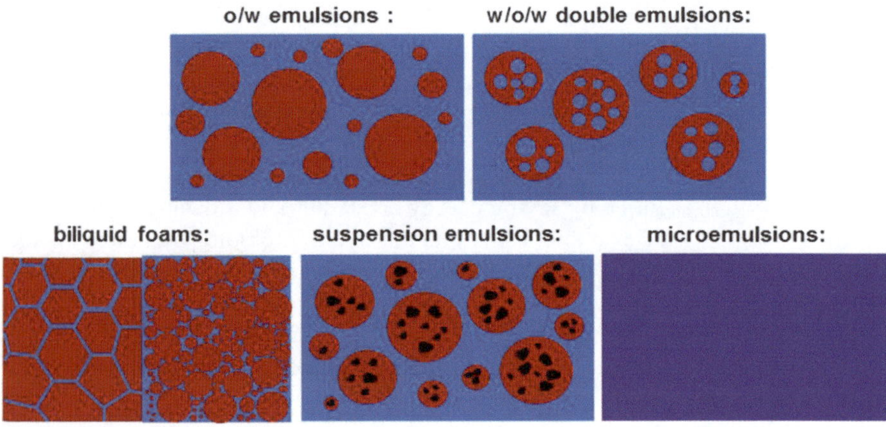

Fig. 1.1 Schematic illustration of various disperse systems for formulation of bioactive components

mostly triglycerides would have to be used as solvent, which would therefore be taken up in large quantities in order to reach a sufficient daily dosage of the bioactive component.

For purely lipophilic bioactive components, formulation in the disperse phase of oil-in-water (o/w) emulsions is a good means for formulation in solution (Ax 2004). If the emulsion droplets are sufficiently small, lipophilic substances like carotenoids can be supersaturated in the disperse oil phase due to the lack of seeds for crystallisation in most of the large number of smallest droplets. In this way, the amount of triglycerides associated with the appropriate dose is greatly reduced (Bunnell et al. 1985). Furthermore, by formulation in an o/w emulsion, the lipophilic substance is transferred into a water-dispersible system that can be applied to almost all kinds of food. Several emulsion-based formulation systems, partly accessible in combination with innovative production processes, provide further solutions to the aforementioned challenges. Besides conventional o/w miniemulsions, Fig. 1.1 depicts schematic illustrations of alternative carrier systems for formulations of both lipophilic and hydrophilic bioactive components.

Water-in-oil-in-water (w/o/w) double emulsions can, for instance, greatly improve the stability of hydrophilic bioactive components like anthocyanins if these are in the inner water phase (Frank 2012).

If sufficient concentrations of lipophilic bioactive components cannot be achieved in conventional o/w emulsions or a certain texture of the product is required, highly concentrated o/w emulsions – biliquid foams – can be the formulation of choice. Although representing an o/w emulsion with the aforementioned opportunity for stabilising supersaturated oily solutions, these systems can, for example, exhibit the texture of a water-in-oil (w/o) spread and a yield stress.

Still, for many lipophilic bioactive components, insolubility in food compounds or instability in solution can be an obstacle for formulation in the aforementioned

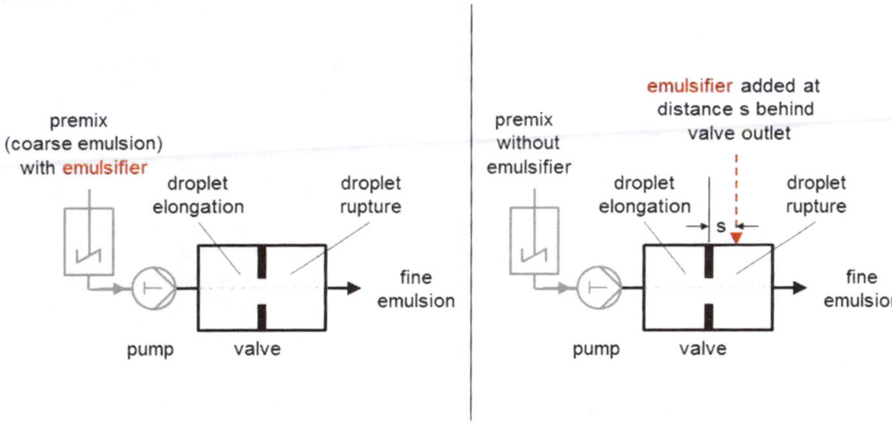

Fig. 1.2 Process of homogenisation with emulsifier present in premix or added in zone of droplet disruption at distance *s* behind orifice (Source: Kempa 2009)

systems. In this case, suspension emulsions with the bioactive component finely dispersed in the oil phase could be an option. Compared to lipophilic dispersions, these systems could also prevent ripening of the bioactive component particles, which usually drastically reduces bioavailability due to the formation of large crystals.

In addition, microemulsions, which are thermodynamically stable, offer new opportunities for the formulation of various difficult-to-formulate bioactive components. However, only a few food-grade microemulsion systems are known to date, and their stability during ingestion and until reaching the site of uptake or the site of effect of the bioactive component is questionable. In this section we will focus on thermodynamically unstable emulsions that must be kinetically stabilised in an adequate manner (cf. Schubert 2005).

Emulsion Formation. One of the conventional routes to miniemulsions with submicron droplets, which are required, for example, for supersaturation of bioactive components, is high-pressure homogenisation. Here, a coarse pre-emulsion is forced through an orifice by the homogenisation pressure. Due to the elongational flow before the orifice, the droplet is elongated and then, right behind the orifice, the deformed droplet is disrupted into smaller droplets by the turbulent flow (Fig. 1.2, left side).

Recent results from investigations into the effect of an emulsifier on the deformation of pre-emulsion droplets have revealed that the emulsifier does not facilitate droplet deformation in any way that supports droplet disruption: emulsions with the same final composition were produced in a high-pressure homogeniser (Kempa 2009). In one emulsion, the emulsifier was already present in the pre-emulsion, whereas in the other, the emulsifier was added at a certain distance *s* behind the orifice, as depicted in Fig. 1.2, right side.

Despite the different modes of emulsifier addition, Fig. 1.3 shows that the resulting droplet sizes in both emulsions are the same for both modes of emulsifier

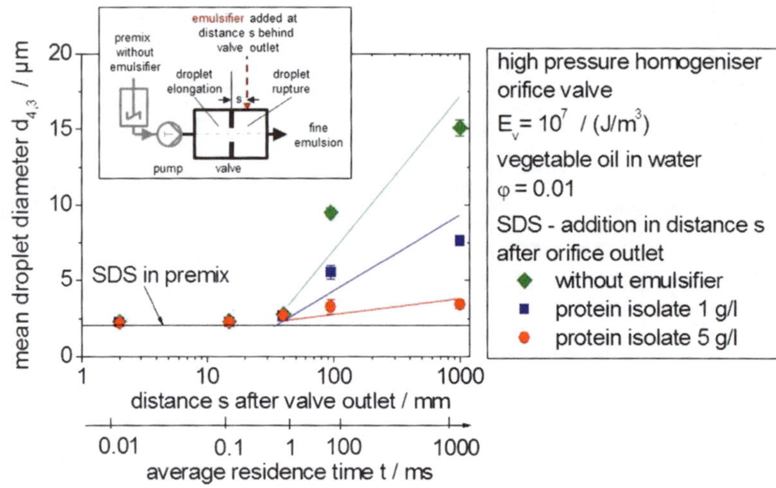

Fig. 1.3 Mean droplet size versus emulsifier addition in premix or at various distances *s* behind orifice valve for two emulsifier concentrations and without emulsifier as reference (Source: Kempa 2009)

addition, provided the emulsifier is added early enough behind the orifice to prevent agglomeration and coalescence of the newly formed droplets. The general rule is that emulsifiers do not improve droplet break-up by reducing the interfacial tension but may prevent droplet coalescence immediately after droplet disruption due to the different time scales for breakage and for stabilising the newly created interface (Kempa 2009). These findings allow the use of emulsifiers which were formerly deemed unsuitable for high pressure homogenisers, such as certain proteins which could be of great use in formulating bioactive components in high concentration and at high bioavailability.

For the formation of emulsions, for instance, with ingredients sensitive to mechanical stress, membrane emulsification is often suggested as the method of choice (Vladisavljevic et al. 2005). However, achieving sufficient disperse phase fractions and the required small droplet sizes requires long processing times in conventional direct membrane emulsification where all of the disperse phase has to flow through the membrane. An alternative process is premix membrane emulsification, where the entire emulsion premix is pushed through a membrane at a comparatively low pressure of up to approximately 10 bar. The process, which has a high transmembrane flux, can be repeated several times in order to achieve very narrow droplet size distributions without harming the ingredients of the product (Badolato Bönisch 2009). Single-use filter membranes with a suitable membrane support structure can be used as long as they are wettable with the continuous phase of the desired emulsion type. However, if for example, an o/w emulsion is pushed through a hydrophobic membrane, a phase inversion can take place in parallel to the reduction of droplet size, as depicted in Fig. 1.4 (Suzuki 2001). The high

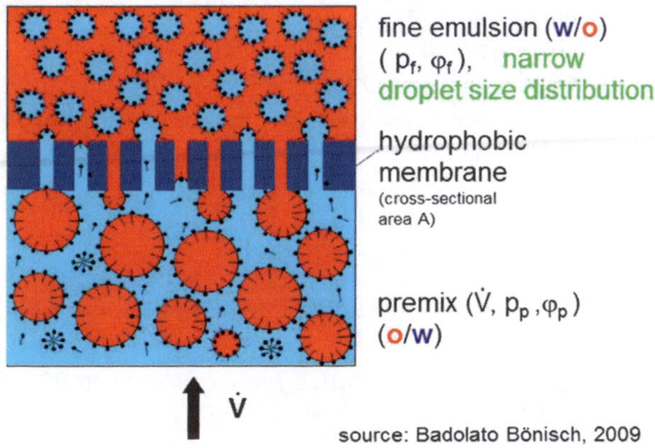

Fig. 1.4 Schematic illustration of premix membrane emulsification with phase inversion; flow from *bottom* to *top* (Source: Badolato Bönisch 2009)

transmembrane flux, combined with low pressure drop and simple construction of the membranes, makes the process of premix membrane emulsification very suitable for the formulation of bioactive components in emulsion systems.

Formulation of Bioactive Components in Stable Supersaturated Solutions. A prerequisite for the formulation of a bioactive component in a stable, supersaturated solution is its dissolution in a suitable solvent. Usually, for food applications, the selection of an appropriate solvent is one of the key aspects in the successful formulation of bioactive components. For the formulation of lipophilic bioactive components in food products, triglycerides represent a capable group of solvents that can be used without restrictions. For most bioactive components, the solubility in, for example, vegetable oil greatly increases with rising temperature. Therefore, in a first step, the dissolution of sufficient concentrations of a bioactive component can be achieved by dissolving it at the highest possible temperature. The temperature used in this step can be limited by either the stability of the bioactive component or the stability of the solvent. However, the stability of triglycerides compared to that of many bioactive components for food applications is better in most cases, allowing for dissolution temperatures in excess of 200 °C. Above that, the influence of residence time at the elevated temperature must be considered. A high temperature–short time regime is the method of choice in one of the examples given below, as it is in food processing in general.

The second step towards a formulation with a stable, supersaturated solution of a bioactive component is the transfer of the hot solution to the disperse phase of an o/w emulsion. By dispersing the solution, supersaturated by cooling down to ambient temperature, into submicron droplets, the supersaturated state can be conserved, allowing for sufficient shelf life of the formulation, as explained earlier. The key aspect of the novel process for formulation in supersaturated solution is to handle the recrystallisation kinetics of the bioactive component appropriately. Whereas for some substances recrystallisation is kinetically hindered or slow enough to allow for cooling down the solution, transferring it to a pre-emulsion and then homogenising it at room temperature, other bioactive components recrystallise immediately upon supersaturation. In the latter case, the process steps of emulsion formation and homogenisation must be carried out at the temperature of dissolution, provided the stability of both the solvent and the bioactive component makes this possible. The process for the formulation of two bioactive components according to the concept developed previously is described in two examples in the following sections.

1.3.1.1.1 Formulation of Carotenoids

Properties and Nutritional Effects of Carotenoids. Carotenoids can be found in many fruits and vegetables, as well as in crustaceans and certain fish. Several epidemiologic studies have shown their potential to reduce the risk of certain diseases, such as various types of cancer or cardiovascular diseases (cf. Ax 2004). Carotenoids are insoluble in water, and their solubility in oil is very poor – approximately 1 g/L at room temperature. This explains their low bioavailability from unprocessed food, where they usually do not occur in solution. Their bioavailability greatly depends on their physical state, with crystalline carotenoids having the lowest bioavailability.

Incorporation of carotenoids into highly disperse droplets of an o/w emulsion can yield a formulation with high bioavailability of the carotenoids, which is water-dispersible at the same time and, thus, can be easily applied to almost any food product. In this formulation, the carotenoids have to be dissolved in the oil phase. For the studies presented in what follows, the carotenoids β-carotene, astaxanthin and lycopene were used.

To achieve a higher concentration of dissolved carotenoids in the formulation, the dissolution step was carried out at temperatures between 140 °C and 210 °C, with subsequent supersaturation as described earlier. Due to the temperature sensitivity of carotenoids, the heat treatment was limited to processing times below 4 s. Although 100 °C for β-carotene and 140 °C for lycopene would be sufficient to reach a concentration of 3 g/L in vegetable oil, a slightly higher dissolution temperature should be applied in order to avoid any crystals remaining (Ax 2004).

Process for Formulation of Carotenoids. The whole process for the formulation of the carotenoids in a supersaturated state in the disperse phase of o/w emulsions is depicted in Fig. 1.5 (Ax 2004; Schweikert and Kolter 1997). In order to avoid influences of minor lipid components in vegetable oil, a refined medium-chain triglyceride (Bergabest MCT-oil, Schumann und Sohn, Karlsruhe, Germany) was used as solvent. The disperse-phase fraction of the emulsion was set to 20 %. The carotenoids (BASF SE, Ludwigshafen, Germany) had a purity of 95 % and were present as the all-trans isomer. Tween-20 (polyoxyethylene sorbitan fatty acid ester) was used as emulsifier, at a concentration of 1 % with respect to the total formulation.

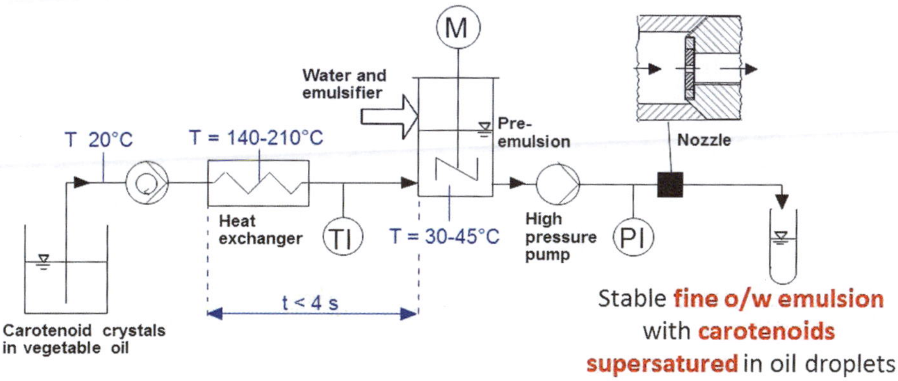

Fig. 1.5 Schematic illustration of process for supersaturation of carotenoids in the oil droplets of an o/w emulsion (Source: Ax 2004)

The oil phase containing the dissolved carotenoids was produced by first preparing a suspension of carotenoid crystals in the oil. In order to guarantee for short dissolution times, this suspension was then subjected to milling to reduce the particle size to below 20 μm. The resulting suspension was then heated to the desired dissolution temperature by pumping it through a heat exchanger. After the dissolution step, the hot lipid solution was immediately subjected to emulsification in the watery Tween-20 solution using a rotor-stator device (Ultraturrax, IKA, Staufen, Germany), resulting in a Sauter diameter of the oil droplets of between 2 and 10 μm, depending on the duration of dispersal. This pre-emulsion with carotenoid supersaturated in the oil droplets was then high-pressure homogenised to achieve the desired final product of a submicron o/w emulsion. In the experimental setup, an orifice valve with 80 μm in diameter was used to reach a high homogenisation pressure even at low volumetric flow. The droplet size distribution was determined by laser diffraction spectrometry with polarisation intensity differential scattering (PIDS) technology in a Coulter LS230 (Beckman Coulter, Brea, CA, USA) (Ax 2004).

Assessment of Carotenoid Bioavailability. The bioavailability of carotenoids from this novel formulation was investigated using a cell culture test with HT 29- or Caco-2-colon carcinoma cells. The cellular uptake of carotenoids in this model system serves as an indicator for the bioavailability of the carotenoids in vivo in human colon epithelia cells. The cellular uptake with respect to the Sauter diameter of the emulsion droplets is depicted in Fig. 1.6, which shows an increased uptake for decreasing diameters, especially for astaxanthin and β-carotene. However, the bioavailability of carotenoids also greatly depends on the emulsifier system used for the formulation, as can be seen from Fig. 1.7. The reference in this picture is a formulation of carotenoids in liposomes, which are known for the high bioavailability of substances incorporated into their double layer, but it is still significantly surpassed by the formulation in an o/w emulsion with a certain emulsifier system.

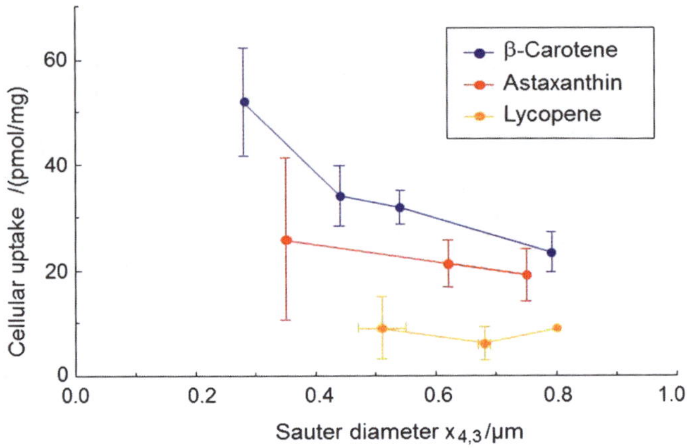

Fig. 1.6 Cellular uptake of carotenoids versus droplet size of o/w emulsion carrying carotenoid supersaturated in oil phase (Source: Ax 2004)

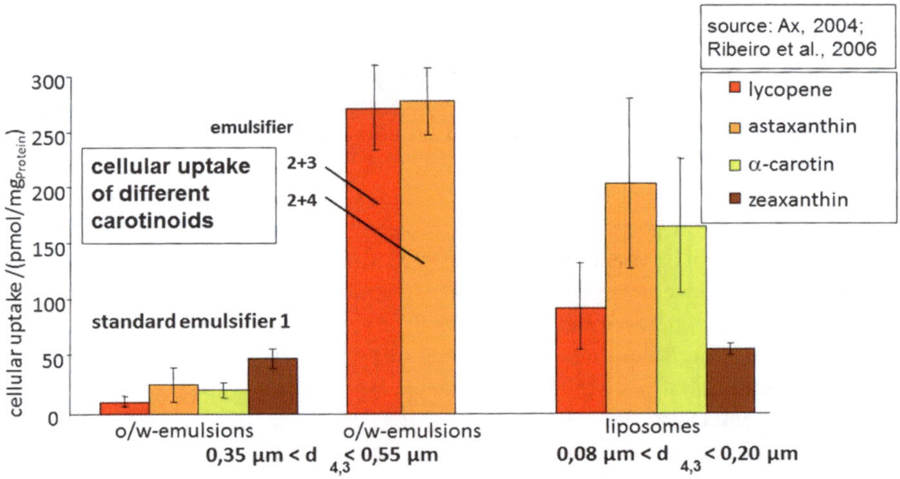

Fig. 1.7 Cellular uptake of carotenoids from o/w emulsions stabilised with standard emulsifier or with emulsifier system and from liposomes as reference (Source: Ax 2004; Ribeiro et al. 2006)

These results clearly show that the formulation of bioactive components, especially lipophilic ones, in emulsion systems can be highly beneficial in terms of both bioavailability and achievable concentration in a given amount of triglyceride. With the novel formulation process described previously, these systems are also much more easily accessible with food-grade ingredients than, for example, a formulation in liposomes (Ax 2004).

1.3.1.1.2 Formulation of Phytosterols

Among the positive results of formulating poorly soluble lipophilic bioactive components mentioned earlier, other bioactive components of plant origin should also be incorporated into an emulsion formulation as a supersaturated solution. However, their interface activity and different crystallisation kinetics may require some adaptations of the process. How these challenges can be mastered will be described in the following section on the formulation of phytosterols as a supersaturated solution in an o/w emulsion.

Properties and Nutritional Effects of Phytosterols. Phytosterols are present in all foods of plant origin, especially seeds and oils, in concentrations of up to 5 % as free phytosterols, as phytosterol fatty acid esters, or glycosidically bound. Different phytosterols, such as β-sitosterol, stigmasterol or campesterol, are distinguishable from the structurally comparable cholesterol and from each other by corresponding variations of the side chain or its additional methyl groups.

Phytosterols are well known for their cholesterol-lowering and anti-carcinogenic effects, which became evident in both animal and human studies (Pollak 1953; Awad and Fink 2000), as well as their potential to reduce the number and size of gallstones and the risk for colon cancer (Pollak 1985). Negative side effects, such as hormone-like action, were not observed (Ling and Jones 1995; Jones et al. 1997). Depending on the dose and formulation, phytosterols not only lower the total serum-cholesterol concentration but also the ratio of the concentrations of low-density lipoprotein to high-density lipoprotein-bound cholesterol in serum (Mattson et al. 1982; Miettinen et al. 1995).

The solubility of phytosterols in water is negligibly low. Their solubility in triglycerides at room temperature reaches approximately 2–3 % but increases to approximately 35 % at 100 °C. However, due to rapid crystallisation, supersaturated solutions are not stable. Considering a recommended daily intake of approximately 2 g, the solubility in triglycerides is far too low for the incorporation of sufficient amounts in functional foods. Above that, phytosterols' molecular structure makes them absorb to interfaces between the lipophilic solution and an aqueous phase in a way that yields crystallisation in a polymorphic form with an even lower solubility in an organic solvent or oil (Jandacek et al. 1977; Christiansen et al. 2002; Bar et al. 1984). This interface crystallisation, which can be seen in the insert of Fig. 1.8, does not allow for their formulation in o/w emulsions in line with the concepts described earlier.

Therefore, formulations of phytosterols in state-of-the-art functional foods rely on phytosterol fatty acid esters. By esterifying phytosterols with fatty acids, their solubility in oil increases approximately tenfold and allows for incorporation of the required quantities into foods with a sufficient content of oil or fat. However, the daily intake of phytosterols required to obtain a certain decrease in the serum-cholesterol level highly depends on the formulation. Between 10 and 20 g/day of crystalline phytosterols is required for a serum-cholesterol reduction of approximately 10 %, whereas only 2–3 g/day of esterified phytosterol are necessary to

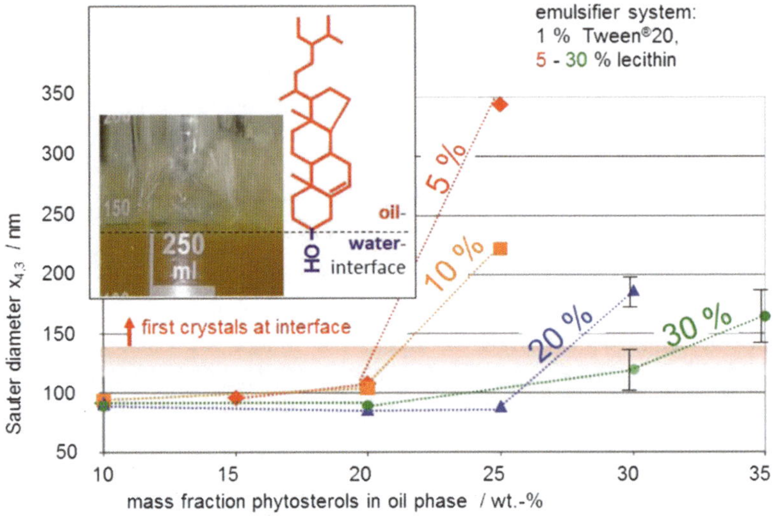

Fig. 1.8 Sauter diameter in o/w emulsions carrying phytosterols in disperse oil phase for various concentrations of lecithin as oil-soluble emulsifier. *Insert*: phytosterol crystals of less soluble modification forming at oil–water interface due to interface crystallisation and molecular structure of β-sitosterol (Source: Engel 2007)

achieve a comparable effect. Data from the literature suggest that the same effect can be achieved with an even smaller amount of solved free phytosterols (Jones et al. 1997; Mattson et al. 1982). Therefore, it is useful to supersaturate phytosterols in the oil phase of o/w emulsions, overcoming the problem of interface crystallisation. Since the site of action of phytosterols is the small intestine, emulsions loaded with phytosterols must also reach this site unchanged.

This was achieved by applying an emulsifier system of one oil-soluble and one water-soluble emulsifier in order to reach the highest possible dose response and good applicability in food systems. Therefore, especially the passage through the stomach is critical. In order to find a formulation suitable to meet these demands, emulsion systems capable of stabilising phytosterols against interface crystallisation were subjected to stability tests in an in vitro stomach model.

Stabilisation of Poorly Soluble, Lipophilic and Surface Active Substances in O/W Emulsions. In order to prevent interface crystallisation of phytosterols in o/w emulsions, an emulsifier system of one oil-soluble and one water-soluble emulsifier can be applied (Engel 2007). Phytosterols are prevented from reaching the interface and crystallising at it, provided the emulsifier molecules reach the interface more quickly and absorb to it better than the phytosterol molecules present in the oil phase. Besides being food-grade, the emulsifiers are required to lie in a certain range of hydrophilic-lipophilic-balance (HLB) values: one emulsifier should be oil soluble, however, with an HLB value as high as possible in order to firmly attach to the oil–water interface from the side of the oil phase. The second emulsifier should be water soluble and have an HLB value as low as possible in order to do the same from the watery side. Beyond that, the emulsifier molecules should be sufficiently small for rapid diffusion to the interface. To compare as many emulsifiers as possible by their HLB values, the extended concept for HLB values was applied (Engel 2007).

Trying to fulfil the requirements stated earlier, experiments with various food-grade emulsifiers were carried out with the aim of stabilising as high concentrations of phytosterols in the dispersed phase as possible. In order to achieve high phytosterol concentrations, the formation and homogenisation of the emulsions were carried out at a temperature of approximately 95 °C. Only when the emulsion droplets were sufficiently small to allow for supersaturation of the phytosterols, namely after high-pressure homogenisation at 1,000 bar in a Microfluidizer (Microfluidics, Newton, MA, USA), were the emulsions cooled down to room temperature. They were then analysed with respect to their particle size distribution by laser diffraction and PIDS technology (cf. Sect. 1.3.1.1) and with respect to the phytosterol crystals that had formed upon first contact of the oil and water phase in the formation of the pre-emulsion by light microscopy. The resulting Sauter diameters are displayed in Fig. 1.8, where a sharp increase in the Sauter diameter indicates the crystallisation of phytosterols upon first contact with the aqueous phase.

Furthermore, taking into account the requirement of stability at high temperatures, only a small selection of emulsifiers remained. Studying emulsifiers of the sucrose fatty acid esters, experimental data revealed that indeed oil-soluble emulsifiers with high HLB values could stabilise phytosterols against interface crystallisation far better than those with lower HLB values of a similar group. For the water-soluble emulsifiers, among others, the group of Tween emulsifiers (cf. Sect. 1.3.1.1) was investigated in terms of their ability to stabilise phytosterols against interface crystallisation. Here, experimental data show that the size of the emulsifier molecule and, thus, the time required for diffusion to the interface are even more important than the HLB value of the emulsifier.

The study of stabilisation of phytosterols in the disperse phase of o/w emulsion shows that interface-active lipophilic substances can be formulated in these systems provided an appropriate emulsifier system is applied which prevents the substances reaching too high concentrations at the interface. Thus, phytosterols can be approximately ten-fold supersaturated in the dispersed phase without the occurrence of interface crystallisation. As long as no crystallisation occurs, the influence of concentration of phytosterols in the disperse phase on the droplet size distribution in the emulsion is detectable, although only to a very small extent. Depending on the emulsifier system applied, droplet sizes either increase very little or decrease significantly due to the phytosterols' acting as co-emulsifier. Concerning the effect of the emulsifier system on emulsion stability, most emulsifier systems can stabilise an emulsion over a shelf life exceeding 60 days without any change in particle size distribution.

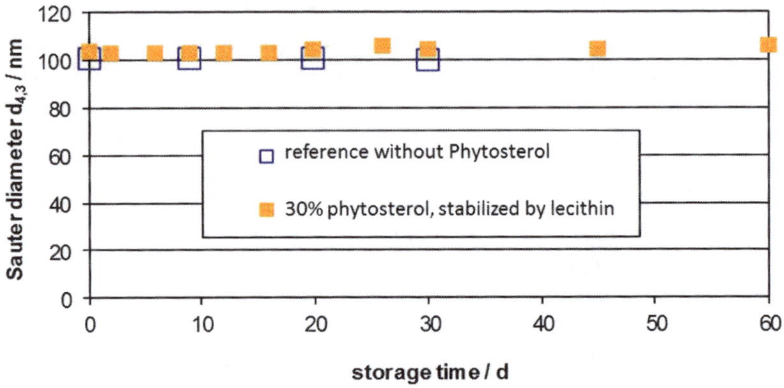

Fig. 1.9 Sauter diameter in phytosterol-loaded o/w emulsions over a period of 60 days and a phytosterol-free reference emulsion (Source: Engel 2007)

Investigation of Stability of Phytosterol-Loaded O/W Emulsions During Passage Through Stomach. To achieve the greatest possible reduction in, for example, serum-cholesterol levels, phytosterols have to reach their site of activity in the state with the best dose response, i.e., as a supersaturated solution in the oil droplets of an o/w emulsion. However, this requires that the emulsion does not break or be otherwise destabilised during storage and during the passage of the stomach since the site of activity of the phytosterols is the small intestine. Figure 1.9 shows that no change in droplet size was observable in emulsions loaded with phytosterols for 60 days.

To find out whether the formulations described previously are suitable to deliver free, dissolved phytosterols to the intestine, their stability under stomach conditions were investigated in an in vitro model stomach system in cooperation with the Institute for Process Engineering of the Max-Rubner-Institute (MRI) in Karlsruhe, Germany.

In a stirred, jacketed fermenter, stomach conditions were simulated with regard to temperature, dilution, electrolytes, stomach enzymes, agitation, residence time and pH.

Particle size distributions of the emulsions were determined by laser diffraction and PIDS technology (cf. Sect. 1.3.1.1) before and after incubation.

Depending on the emulsifier system, phytosterol-loaded emulsions did not undergo any changes during incubation with stomach conditions. The particle size distributions in the emulsion before and after incubation displayed in Fig. 1.10 were similar within the limits of precision of the measurement. Some combinations of emulsifiers, however, were susceptible to stomach conditions and could not prevent flocculation, leading to the emulsion's complete breaking.

Phytosterol Formulation as Basis for Further Work. With phytosterols formulated in o/w emulsions, there are formulations at hand for detailed investigations into the

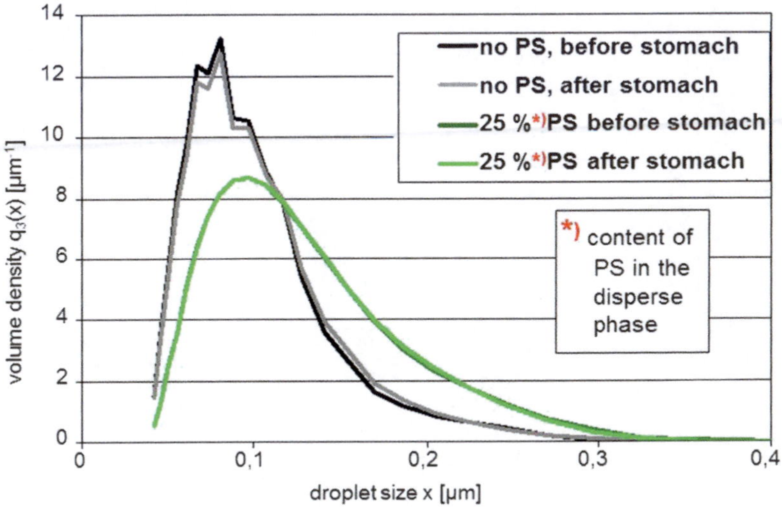

Fig. 1.10 Volume density distribution of phytosterol-loaded o/w emulsion before and after passage of model stomach and of phytosterol-free reference emulsion (Source: Engel 2007)

dose–response relationship of free phytosterols in a non-crystalline state formulated in a water-dispersible carrier system.

Beyond that, the example of phytosterols as a group of poorly soluble, lipophilic substances exhibiting interface activity clearly shows that, despite these properties, these substances can be incorporated into water-dispersible food-grade systems, especially o/w emulsions, at high supersaturation for formulation in functional foods or for pharmaceutical application.

1.3.1.1.3 Formulation of Intrinsically Microbial Stable Emulsions

Besides the physicochemical stability of an emulsion system and the state of bioactive components formulated therein, biological or microbial stability is of great importance for sufficient shelf life, food safety and, therefore, consumer health. A novel approach to achieving microbial stability via the food structure has been developed for w/o emulsions.

Provided that pathogenic microorganisms and those responsible for food spoilage require water for proliferation, they must be of a hydrophilic nature. Thus, in an emulsion system, they are bound to the aqueous phase due to their wetting behaviour. To generate a food structure that prohibits microbial proliferation, w/o emulsions with different droplet sizes in a range of 4–21 µm were prepared with lysogeny broth (LB) as aqueous phase and inoculated with *E. coli* K12. Pure LB of the same inoculation was used as a reference. The emulsions and the reference were incubated for several hours. Samples were taken on an hourly basis and the load of microorganisms determined. The results are depicted in Fig. 1.11 (Badolato Bönisch 2009).

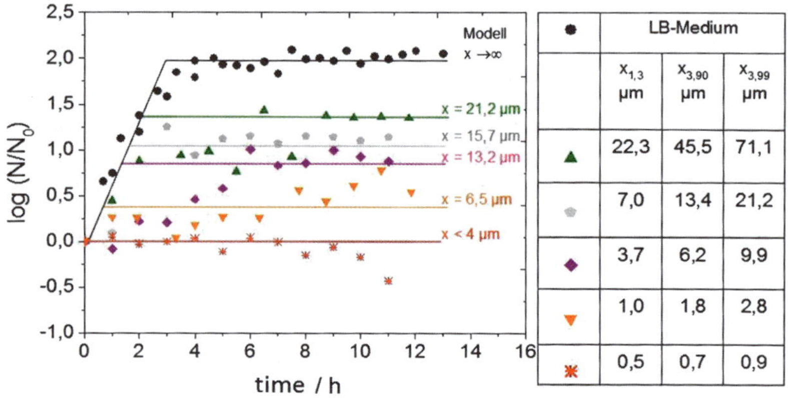

Fig. 1.11 Proliferation of microorganisms in watery phase (lysogeny broth) of a w/o emulsion over a period of 16 h for various droplet diameters (Source: Badolato Bönisch 2009)

The growth curve of the reference clearly shows the microbial growth in the log phase and the following stationary phase specific to the boundary conditions, such as the availability of nutrients and number concentration of microorganisms. Compared to this reference, the growth behaviour of the microorganisms in the emulsion systems is different both in terms of duration of the log phase as well as the number of microorganisms in the stationary phase. The duration of the log phase decreases with decreasing droplet diameter, as does the number of microorganisms in the stable stationary phase. The fact that microbial growth can be observed in emulsions with larger droplets shows that neither the ingredients of the emulsion nor its preparation has any considerable negative impact on microbial growth. At a droplet diameter below 4 µm, no microbial growth can be observed. This result can be explained by the size of the droplets, which approaches the size of the microorganisms themselves, which leads to a lack of water, nutrients and space required for the proliferation of microorganisms in the individual droplet. Therefore, this emulsion can be considered intrinsically microbially stable (Badolato Bönisch 2009).

These results show the immense influence of food structure on the microbial stability of food and the opportunity to produce microbially stable food without the need for additional preservatives.

1.3.1.2 Multiphysics Simulation of Microwave Processing and Utilisation of Magnetic Resonance Imaging

By heating volumetrically, microwave processes have several advantages over conventional heating processes. The main advantage is the increase in process rates, hence, improvement in the quality of microwave-heated, dried, pasteurised or sterilised products. However, to ensure product safety and satisfy regulatory requirements, temperature distributions must be as uniform as possible. To design a process providing the required temperature uniformity is challenging because a number of microwave-inherent factors inevitably lead to the development of hot and cold spots. Assessing the extent of the temperature non-uniformity is essential for the process design. Measuring transient, spatially resolved three-dimensional temperature distributions, however, is not feasible with common, established temperature measurement systems, such as thermocouples, fibre-optic probes or infrared cameras (Knoerzer 2006).

To address these challenges and allow for a convenient way to visualise the temperature evolution in a microwave-treated product, a new simulation approach was developed that is capable of predicting transient temperature profiles of arbitrarily shaped products in three dimensions subjected to microwave treatment. To validate the simulations, a magnetic resonance imaging (MRI) tomograph was used, together with a new temperature measurement protocol, enabling the visualisation and quantification of three-dimensional temperature distributions over time in a microwave process.

1.3.1.2.1 The Microwave System

A microwave device was developed with Gigatherm (Grub, Switzerland) to fit into an MRI tomograph (Avance 200 SWB tomograph, maximum sample diameter 64 mm, magnetic flux density 4.7 T; Bruker Biospin MRI, Ettlingen, Germany). By introducing microwave power (at 2,450 MHz) directly into the tomograph where the sample to be heated is located, the design avoids temperature equilibration between microwave heating and measuring of the temperature distribution. For a detailed description of the system the reader is referred to Knoerzer (2006).

This system, together with the MRI tomograph and an appropriate pulse protocol, allows for measuring three-dimensional temperature distributions during microwave heating in real time.

1.3.1.2.2 The Model

Geometry. The geometry of the model microwave cavity is a cylinder ($h = 300$ mm, $d = 84$ mm).

The model has been designed in full three dimensions to be a good approximation of the actual microwave chamber utilised in the experiments.

The samples used were predominantly simple shapes (cylinders, spheres) of a model food with known temperature-dependent thermo-physical properties (Knoerzer et al. 2004). In addition, a numerical method was developed to allow the introduction of models of real food structures based on a three-dimensional spin-density measurement technique (Knoerzer 2006), as described in the section "*Computational Methods*".

Governing Equations. Microwaves, being electromagnetic waves, are described by Maxwell's equations:

$$\nabla \cdot \vec{D} = \rho, \tag{1.6}$$

$$\nabla \cdot \vec{B} = 0, \tag{1.7}$$

$$\nabla \times \vec{E} = -\frac{\partial \vec{B}}{\partial t}, \tag{1.8}$$

$$\nabla \times \vec{H} = \vec{j} + \frac{\partial \vec{D}}{\partial t}, \tag{1.9}$$

where $\vec{D}$ is the electric flux density (As/m^2), ρ is the charge density (As/m^3), $\vec{B}$ is the magnetic flux density (Vs/Am), $\vec{E}$ is the electric field (V/m), t is the time (s), $\vec{H}$ is the magnetic field (A/m) and $\vec{j}$ is the electric current density (A/m^2).

The interaction of electromagnetism with matter is expressed by the material equations or constitutive relations (Eqs. 1.10, 1.11 and 1.12), where the permittivity or dielectric constant ε (the interaction of non-conducting matter with an electric field), the conductivity σ and the permeability μ (the interaction with a magnetic field) describe the material effect on the electric flux, magnetic flux and current density:

$$\vec{D} = \varepsilon_0\, \varepsilon_r \cdot \vec{E}, \tag{1.10}$$

$$\vec{B} = \mu_0\, \mu_r \cdot \vec{H}, \tag{1.11}$$

$$\vec{j} = \sigma \cdot \vec{E}. \tag{1.12}$$

The dissipation of electromagnetic energy, leading to a temperature increase in the microwave-treated product, can be expressed as

$$P_V = \frac{P}{V} = 2 \cdot \pi \cdot f \cdot \varepsilon_0 \cdot \varepsilon''_r \cdot E^2, \tag{1.13}$$

with P_V the volumetrically dissipated power (W/m^3), as a function of microwave frequency f, dielectric properties (ε_0, ε'') and the electric field E.

Knowing the dissipated power P_V as the heat source, the spatially resolved, transient temperature distribution can be calculated by the energy conservation equation (Fourier equation):

$$\rho C_p \frac{\partial T}{\partial t} - \nabla \cdot (k \cdot \nabla T) = P_v. \tag{1.14}$$

This enables the coupling of electromagnetism and heat transfer.

More information on the governing equations and boundary conditions applied in the model can be found in Knoerzer et al. (2008).

Process Conditions. In the uncontrolled microwave heating simulation of model food cylinders, the microwave power was set to constant values of $P = 19$ W and $P = 23.1$ W between $t = 200$ s and $t = 650$ s and $P = 0$ at all other times. In the feedback-controlled simulation, the microwave power was set to 37 W as long as the temperature in the hottest spot was below the predefined maximum temperature 70 °C and set to $P = 0$ when this temperature was exceeded. A stop condition was triggered when the temperature in the coldest spot was equal to or greater than the predefined target temperature (60 °C) for a certain time.

At the start of the process, both sample and surrounding air were in thermal equilibrium at $T_0 = 25$ °C. In the feedback-controlled simulation the initial temperature of sample and surrounding air was set to $T_0 = 25$ °C.

Material Properties. The thermo-physical properties, i.e. loss factor ε'', permittivity ε', thermal conductivity λ, density ρ, heat capacity C_P and their variation with water content and temperature, were determined experimentally for a model food, developed for microwave heating research, and fitted with third-order polynomials (Knoerzer et al. 2004). These equations were implemented in the subdomain settings of the simulation software used. In the case of the chicken wing heating, the thermo-physical properties of fat, meat and bone material were taken from the literature.

Computational Methods. The partial differential equations describing the electromagnetic part of the model were solved with the finite-difference time domain (FDTD) method using QuickWave-3D (QWED, Warsaw, Poland). The finite-element method was chosen for the energy balance (heat transfer) part and solved using COMSOL Multiphysics (COMSOL, Stockholm, Sweden). The interface between both software packages was programmed as a graphical user interface (GUI) in MATLAB 7.4 (Mathworks, Natick, MA, USA). The cross-software communication was enabled by transforming the different output formats into a universal matrix format.

A MATLAB 7.4 code was used to transform the structure data of real foods as determined by MRI into a format readable by the simulation software packages and to allocate the corresponding material properties. Different materials yield different intensities in MRI spin density measurements. Thus, it is possible by multithreshold setting to subdivide a full three-dimensional MRI image (three-dimensional matrix) into submatrices corresponding to the different materials and allocate the respective thermo-physical properties at the specific locations to these materials. In the case of the simulated heating of a chicken wing, the system was divided into meat, fat and bone material.

An additional code was developed in MATLAB 7.4 to allow for a feedback-controlled simulation. This code was able to identify and observe the hottest and coldest spots in the simulated heating scenario, virtually turning the power off after

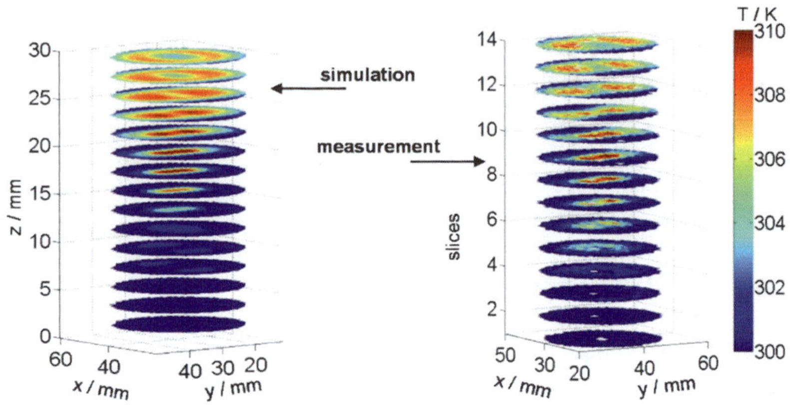

Fig. 1.12 Visual comparison between simulated (*left*) and measured (*right*) heating of model food cylinder (at a discrete time) (Source: Knoerzer et al. 2008)

a predefined maximum temperature in the hot spot is reached and turning it back on after falling below that value according to the description in the "*Process Conditions*" subsection. The simulation was programmed to stop after a predefined temperature in the coldest spot is reached and held for a defined time.

Model Validation. Validation was performed by visually comparing simulated and measured three-dimensional temperature distribution results. Average temperature profiles at measured points (hot and cold spots) were compared with profiles provided by the model at the same locations. A MATLAB routine was developed to plot average measured temperatures within axial cross sections at specific points in time versus simulated ones. The routine was used to perform a statistical analysis to calculate the coefficient of determination, R^2.

The sophisticated method of temperature mapping using MRI was used to measure temperatures in three dimensions throughout the heated samples. Due to the complexity of this approach and the space limitations, the reader is referred to Knoerzer et al. (2009) for a detailed description of this measurement technique and the underlying physical principles.

1.3.1.2.3 Results and Discussion

Comparison of Simulated and Measured Heating Profile of a Model Food Cylinder. As mentioned earlier, both the simulation and measurement methodologies yield transient three-dimensional temperature profiles throughout the microwave-treated samples, allowing point-for-point comparison, at various time steps throughout the process, for model validation.

Visual (Qualitative) Comparison of Three-Dimensional Temperature Distributions. Figure 1.12 shows both simulated and measured three-dimensional temperature distributions in a microwave-heated model food cylinder at a time of 250 s and a microwave power of 19 W.

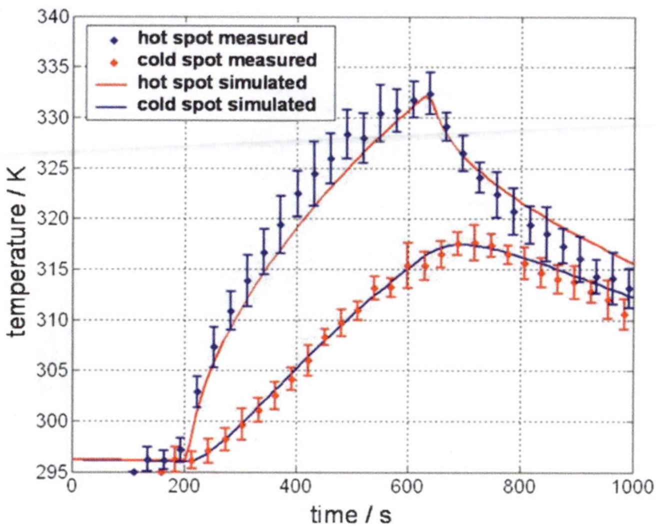

Fig. 1.13 Comparison of heating curves in hot and cold spots in a microwave-heated cylindrical model food sample (microwave power $P = 19$ W) (Source: Knoerzer et al. 2008)

An excellent agreement of simulation and measurement was found in all investigated process scenarios.

Quantitative Model Validation. Figure 1.13 shows temperature profiles in a hot and a cold spot of the heated model food cylinder at a microwave power of 19 W. As can be seen, simulation and measurement agreed well.

Simulated Heating of a Chicken Wing. The simulated heating of a "real" product, a chicken wing, was performed by introducing the product structure into the model through the multi-threshold MRI spin density measurement described earlier, allocating structure and material properties for fat, meat and bone components. The microwave power in this example was set to 25 W for 600 s. Figure 1.14 shows the temperature distribution in the chicken wing after heating. It is obvious that uncontrolled process conditions led to the development of unacceptable non-uniform temperature distributions caused by a number of interacting microwave inherent factors, described in detail in the literature (Ringle and Donaldson 1975; Knoerzer 2006; Wäppling-Raaholt and Ohlsson 2005; Ryynänen et al. 2004; Risman 1992; Sinell 1986; Ohlsson and Bengtsson 2001; Buffler 1993).

Controlled Microwave Heating of a Model Food Cylinder. In simulated microwave heating scenarios, temperature distributions are known at any point in time. This fact allows a feedback control loop to be simulated on the basis of developing temperatures to ensure uniform heating. Such a feedback control loop is difficult to accomplish in an actual microwave process, particularly since the temperature data are required at any point in the product to ensure that both the hot and cold spots

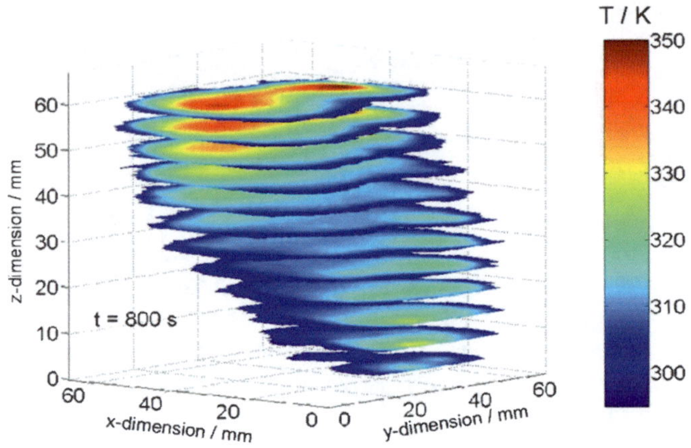

Fig. 1.14 Three-dimensional temperature distribution in a chicken wing after microwave treatment of 600 s (power was switched on after 200 s) as calculated using new simulation procedure (Source: Knoerzer et al. 2008)

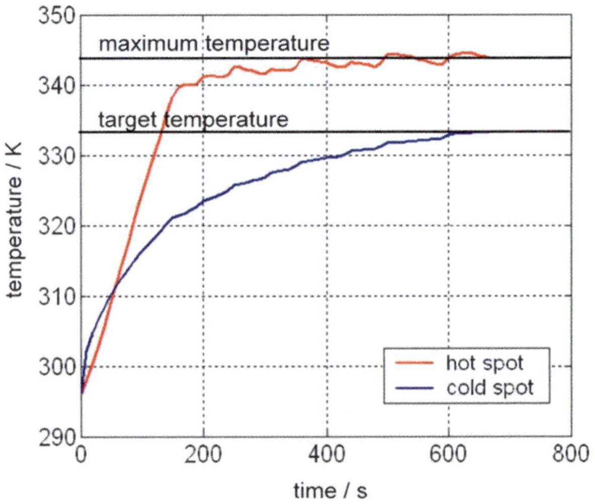

Fig. 1.15 Temperature profile in hot and cold spots in feedback-controlled heating scenario, with combined microwave/hot air treatment (Source: Knoerzer et al. 2008)

(which will be in different locations depending upon the process and product; also, locations will change during processing) are captured.

Figure 1.15 shows the heating curves for the hottest and the coldest spots in a microwave-heated cylindrical model food sample with assisted heating through elevated air temperature at product target temperature level. The microwave power was set to $P = 37$ W as long as the temperature in the hot spot was below the maximum temperature and to $P = 0$ as soon as this temperature was exceeded. The maximum temperature to be reached in the hot spot was set to 70 °C (343.15 K) and the minimum temperature to be reached and held for 30 s was set to 60 °C (333.15 K).

The output of the feedback-controlled simulation is a microwave power pulse program that can be incorporated into the control unit of a microwave oven and thus allows for repeating the simulated scenario in an actual process by controlling the microwave power.

It must be noted that the temperature values used in this study are representative for any temperature level and range that might be required to ensure a minimal time-temperature treatment, yielding, for example, a microbiologically safe product in a pasteurisation process. The model was designed as a platform that is easily adjustable to different microwave system configurations, food products and required process conditions.

Models of the type described in this section can be used both to design and optimise processes and to demonstrate the safety of the process for regulatory bodies.

References

Allen JA (1994) Overall perspectives on countries and regions. In: Rogers P, Lydon P (eds) Water in the Arabic World – perspectives and prognoses. Harvard University Press, Cambridge

Awad AB, Fink CS (2000) Phytosterols as anticancer dietary components: evidence and mechanism of action. J Nutr 130:2127–2130

Ax K (2004) Emulsionen und Liposomen als Trägersysteme für Carotinoide. Dissertation Institut für Lebensmittelverfahrenstechnik, Universität Karlsruhe (19.12.2003), Shaker, Aachen

Badolato Bönisch G (2009) Untersuchung zur mikrobiologischen Stabilisierung von Emulsionen. Dissertation Institut für Bio- und Lebensmitteltechnik, Bereich I: Lebensmittelverfahrenstechnik, Universität Karlsruhe

Bar LK, Garti N, Sarig S, Bar R (1984) Solubilities of cholesterol, sitosterol, and cholesteryl acetate in polar organic solvents. J Chem Eng Data 29(4):440–443

Brabeck-Letmathe P. (2008) Energie und Wasser. Presentatation at the HWZ Hochschule für Wirtschaft, Zürich, 11.07.2008

Buffler CR (1993) Microwave cooking and processing. Engineering fundamentals for the food scientist. AVI Book, New York

Bunnell RH, Drisoll W, Bauernfeind JC (1985) Coloring water-base food with β-carotene. Food Technol 12:536–541

Chapagain AK, Hoekstra AY (2010) The blue, green and grey water footprint of rice from production and consumption perspectives, vol 40, Value of water research report series. UNESCO-IHE, Delft

Christiansen LI, Rantanen JT, von Bonsdorff AK, Karjalainen MA, Yliruusi JK (2002) A novel method of producing a microcrystalline b-sitosterol suspension in oil. Eur J Pharm Sci 15:261–269

Engel R (2007) Flüssige, wasserdispergierbare Phytosterol-Formulierungen zur Senkung des Serum-Chlosterolspiegels. Dissertation Institut für Lebensmittelverfahrenstechnik, Universität Karlsruhe

Falkenmark M, Widstrand C (1992) Population and water resources: a delicate balance, Population bulletin 47:3. Population Reference Bureau, Washington, DC

FAO (2010) 925 million in chronic hunger worldwide. Food and Agriculture Organization of the United Nations (FAO) Media Centre, Rome, 14 September 2010

Frank K (2012) Formulieren von Anthocyanen in Doppelemulsionen. Dissertation Institut für Bio- und Lebensmitteltechnik, Bereich I: Lebensmittelverfahrenstechnik, Karlsruher Institut für Technologie (KIT)

Gawel E, Bernsen K (2011) What is wrong with virtual water trading? UFZ-Diskussionspapiere 1/2011, Helmholtz-Zentrum für Umweltforschung (UFZ). http://hdl.handle.net/10419/45252

Gerbens-Leenes W, Hoekstra AY, van der Meer TH (2009) The water footprint of bioenergy. Proc Natl Acad Sci USA 106(25):10219–10223

Giese E, Bahro G, Betke D (1998) Umweltzerstörungen in Trockengebieten Zentralasiens (West- und Ost-Turkestan). Ursachen, Auswirkungen, Maßnahmen. Franz Steiner, Stuttgart

Hahn HH (2009) Sitzen wir bald auf dem Trockenen? In: Boßler G, Strobel S (eds) Wassernotstand, Herrenalber forum Band 59. Evangelische Akademie Baden, Karlsruhe, pp 36–55. ISBN 978-3-89674-560-6

Hoekstra AY (ed) (2003) Virtual water trade. Proceedings of the international expert meeting on virtual water trade. IHE, Delft, February 2003

Hoekstra AY (2008) The water footprint of food. In: Förare J (ed) Water for food, The Swedish research council for environment agriculture sciences and spatial planning. Formas, Stockholm, pp 49–60

Hoekstra AY, Chapagain AK (2007) Water footprint of nations: water use by people as a function of their consumption pattern. Water Resour Manag 21(1):35–48

Hoekstra AY, Chapagain AK (2008) Globalization of water. Sharing the planet's freshwater resources. Blackwell, Oxford

Hoekstra AY, Hung PQ (2002) Virtual water trade: a quantification of virtual water flows between nations in relation to international crop trade, Value of water research report series no. 11. IHE, Delft

Hopp V (2002) Wasser – Wasserkrise. Bedeutung für Natur, Menschen und Technik. CIT plus, Praxismagazin der Chemie Ingenieur Technik (CIT) 5:4–9

Hüttl RF, Bens O (eds) (2012) Georessource Wasser – Herausforderung Globaler Wandel. Acatech Studie/Springer, Heidelberg

Idso CD (2011) Estimates of global food production in the year 2050. Will we produce enough to adequately feed the world? Center for the Study of Carbon Dioxide and Global Change. www.co2science.org

Jandacek RJ, Webb MR, Mattson FH (1977) Effect of an aqueous phase on the solubility of cholesterol in an oil phase. J Lipid Res 18:203–210

Jöhr H (2003) Lösungsansätze in der global agierenden Lebensmittelwirtschaft. In: Girnau M et al (eds) Nachhaltige Agrar- und Ernährungswirtschaft. Erich Schmidt, Berlin

Jones PJH, MacDougal DE, Ntanios FY, Vanstone CA (1997) Dietary phytosterols as cholesterol-lowering agents in humans. Can J Physiol Pharmacol 75:217–227

Kempa LM (2009) Untersuchungen zum Einfluss des Emulgators auf die Kurzzeitstabilität feindisperser Emulsionen. Dissertation Institut für Bio- und Lebensmitteltechnik, Bereich I: Lebensmittelverfahrenstechnik, Karlsruher Institut für Technologie (KIT)

Knoerzer K (2006) Simulation von Mikrowellenprozessen und Validierung mittels bildgebender magnetischer Resonanz. Dissertation Institut für Lebensmittelverfahrens-technik, Universität Karlsruhe

Knoerzer K, Regier M, Erle U, Schubert H (2004) Development of a model food for microwave processing and the prediction of its physical properties. J Microw Power Electromagn Energy 39(3–4):167–177

Knoerzer K, Regier M, Schubert H (2008) A computational model for calculating temperature distributions in microwave food applications. Innov Food Sci Emerg Technol 9(3):374–384

Knoerzer K, Regier M, Hardy EH, Schuchmann HP, Schubert H (2009) Simultaneous microwave heating and three-dimensional MRI temperature mapping. Innov Food Sci Emerg Technol 10 (4):537–544

Ling WH, Jones PJH (1995) Minireview: dietary phytosterols: a review of metabolism, benefits and side effects. Life Sci 57(3):195–206

Mattson FH, Grundy SM, Crouse JR (1982) Optimizing the effect of plant sterols on cholesterol absorption in man. Am J Clin Nutr 35:697–700

Mekonnen MM, Hoekstra AY (2010) The green, blue and grey water footprint of farm animals and animal products. Value of Water Research Series No. 48 UNESCO-IHE, Delft, The Netherlands

Miettinen TA, Puska P, Gylling H, Vanhanen HT, Vartiainen E (1995) Reduction of serum cholesterol with sitostanol-ester margarine in a mildly hypercholesterolemic population. N Engl J Med 333(20):1308–1312

Nestlé (ed) (2007) Der Nestlé-Bericht zum Wassermanagement. Nestlé S.A., Avenue Nestlé 55, CH-1800 Vevey

Ohlsson T, Bengtsson NE (2001) Microwave technology and foods. Adv Food Nutr Res 43:65–140

Peleg M (2006) Advanced quantitative microbiology for foods and biosystems: models for predicting growth and inactivation. CRC Taylor & Francis, Boca Raton/London/New York

SAI Platform (2003) Sustainable Agriculture Initiative Platform, Sustainable agriculture information. www.saiplatform.org

Pollak OJ (1953) Reduction of blood cholesterol in man. Circulation 7:702–706

Pollak OJ (1985) Effect of plant sterols on serum lipids and atheriosclerosis. Pharmacol Ther 31:177–208

Ribeiro HS, Guerrero JMM, Briviba K, Rechkemmer G, Schuchmann HP, Schubert H (2006) Cellular uptake of carotenoid-loaded oil-in-water emulsions in colon carcinoma cells in vitro. J Agric Food Chem 54(25):9366–9369

Ringle EC, Donaldson DB (1975) Measuring electric field distribution in a microwave oven. Food Technol 29(12):46–54

Risman PO (1992) Metal in the microwave oven. Microwave World 13(1):28–33

Ryynänen S, Risman PO, Ohlsson T (2004) Hamburger composition and microwave heating uniformity. J Food Sci 69(7):M187–M196

Schubert H (ed) (2005) Emulgiertechnik – Grundlagen, Verfahren und Anwendungen. B. Behr's, Hamburg

Schubert H (2007) Lebensmittelwirtschaft. Materialien Nr. 17 der Interdisziplinären Arbeitsgruppe "Zukunftsorientierte Nutzung ländlicher Räume – Landinnovation –" der Berlin-Brandenburgischen Akademie der Wissenschaften, Jägerstr. 22/23, D-10227 Berlin

Schubert H (2011) The virtual water and the water footprint concepts, acatech materialien Nr. 14. acatech, München/Germany

Schweikert L, Kolter K (1997) Stabile, zur parenteralen Verabreichung geeignete Carotinoid-Emulsionen, EP 0800824 A1

Shiklomanov IA (1993) World water resources. In: Gleick PH (ed) Water in crisis. Oxford University Press, New York/Oxford

Sinell HJ (1986) Der Einfluss der Mikrowellenbehandlung auf Mikroorganismen im Vergleich zur konventionellen Hitzebehandlung, DFG-Abschlussbericht Si/55 24–1

Sonnenberg A, Chapagain AK, Geiger M, August D (2009) Der Wasser-Fußabdruck Deutschlands. Woher stammt das Wasser, das in unseren Lebensmitteln steckt? Springer, Berlin/Heidelberg/New York

Suzuki K (2001) Membrane emulsification method and its application. J Jpn Soc Food Sci Technol 48(3):169–175

UBA – Umweltbundesamt (2011) Daten zur Umwelt:Wassernutzung in der Landwirtschaft. UBA-Homepage www.umweltbundesamt.de/daten-zur-umwelt, Dessau-Roßlau, aktualisiert Januar 2011

UN – United Nations (2011) World population prospects: the 2010 revision. United Nations Population Division, New York, May 2011

UNNC – United Nations News Centre (2010) Assembly declares access to clean water and sanitation is a human right. United Nations News Centre, New York, 28 July 2010

Vladisavljevic GT, Shimizu M, Nakashima T (2005) Preparation of uniform multiple emulsions using multi-stage premix membrane emulsification technique. In: Schubert (2005):433–468

Wäppling-Raaholt B, Ohlsson T (2005) Improving the heating uniformity in microwave processing. In: Schubert H, Regier M (eds) The microwave processing of foods. Woodhead, Cambridge

Wefer G (ed) (2010) Dynamische Erde – Zukunftsaufgaben der Geowissenschaften. Strategieschrift der DFG-Senatskommission für die Geowissenschaftliche Gemeinschaftsforschung, Online Ausgabe

WFN (2008) Water Footprint Network, initiated by Arjen Y. Hoekstra and conducted by the University of Twente, The Netherlands. www.waterfootprint.org

WHO (2011) Obesity and overweight. World Health Organization (WHO) Media Centre, fact sheet No. 311, March 2011

World Water Council (ed) (2004) E-conference synthesis: virtual water trade – concious choices, March 2004, ISBN: 92-95017-10-2 Wold Wide Fund For Nature – WWF Deutschland, Frankfurt am Main. www.wwf.de

Zehnder AJB (2002) Wasserressourcen und Bevölkerungsentwicklung. Nova Acta Leopold NF 85 (323):300–418

Chapter 2
Food Process Engineering Research and Innovation in a Fast Changing World: Paradigms/Case Studies

Heike P. Schuchmann, Karsten Köhler, M. Azad Emin, and Helmar Schubert

2.1 Introduction

Facing the challenges of a fast changing world (Schubert et al. PART I), food process engineering offers some promising developments. Part II concentrates on insights gained from model-based simulation offering an improved understanding of underlying mechanisms and – based on this – technical innovations. Two examples from ongoing research will emphasize the need to use basic chemical engineering tools in food process engineering research, as highlighted in Part I:

1. Development of adjusted processes and food products by utilizing the knowledge established in basic chemical engineering using modern tools, such as numerical process simulation, online process analysis, novel measuring devices, microstructuring apparatus engineering, and nanostructuring technologies. Research in these fields will lead to a better understanding of the relationship between microstructures and submicron structures and functional properties, such as consumer-oriented sensory or texture properties or bioavailability/bioactivity of food components, and will therefore enable a targeted product design.
2. Improving the process efficiency, and therefore saving energy, water, and material, and reducing waste and environmental pollution to achieve high-quality food products under conditions of environmental and economic sustainability.

These examples show the principal approach using chemical engineering tools in the design of food products and processes. It therefore complements Chap. 3 (*Some Examples of Promising Developments in Food Process Engineering*) in Part I of this book. Specific emphasis is given to material characterization, modern

H.P. Schuchmann (✉) • K. Köhler • M.A. Emin • H. Schubert
Section of Food Process Engineering, Karlsruhe Institute of Technology (KIT), Institute of Engineering in Life Sciences, D – 76131 Karlsruhe, Germany
e-mail: heike.schuchmann@kit.edu

S. Yanniotis et al. (eds.), *Advances in Food Process Engineering Research and Applications*, Food Engineering Series, DOI 10.1007/978-1-4614-7906-2_2,

equipment design techniques, and numerical simulation tools. The examples cover the design of new homogenization valves for emulsification and dispersion applications (such as dairy homogenization) and the design of food extruders for the stabilization and encapsulation of lipophilic bioactive components, as utilized for the manufacture of functional foods. Foods that may be produced by the selected processes are dairy and emulsion-based functional drinks, breakfast cereals, snacks, and starch-based encapsulation systems for aromas or functional triglycerides. The innovative technology highlighted in the first example allows for reducing energy consumption significantly. In the case of its application in dairy homogenization, energy consumption may be reduced by as much as 90 %. Therefore, modern process engineering tools also support food industry in their sustainability efforts.

In this chapter, only the basic approaches are summarized. Details can be found in the relevant publications (as cited in the text) and Ph.D. dissertation works of Köhler (2010) and Emin (2013). Such approaches require collaboration with research groups having specific competences in the respective research fields required. The authors want to take this opportunity to express their specific thanks to their collaboration partners: Andreas Hensel and Klaus Schubert of the Institute of Micro Process Engineering at KIT/Campus Nord, Bernhard Hochstein and Norbert Willenbacher from the Institute of Applied Mechanics at KIT/CS, and Atze Jan van der Goot and Remko Boom from the Institute of Food and Bio Process Engineering at Wageningen University.

2.2 Enhancing Process Efficiency to Achieve High-Quality Food Products Under Conditions of Improved Economic and Environmental Sustainability

2.2.1 Adapted Homogenization Orifices for Reducing Energy and Improving Product Quality in Homogenization and Emulsification Processes

2.2.1.1 Challenges

High-pressure homogenizers were developed approximately 100 years ago (Gaulin 1899) during the period of industrialization. The basic idea of combining a high-pressure pump and a disruption system, like a valve, was presented at the Exposition Universelle, a world fair held in Paris, and endures to this day. However, there is ongoing technical improvement, particularly in high-pressure pumps and valves, motivated by either daily application problems or new product challenges.

Current technology allows working at volume streams of up to 50,000 L/h and pressures of up to 10,000 bar. Homogenization pressures in industrial applications today are in the range of 50–2,000 bar. Mainly piston pumps serve as high-pressure

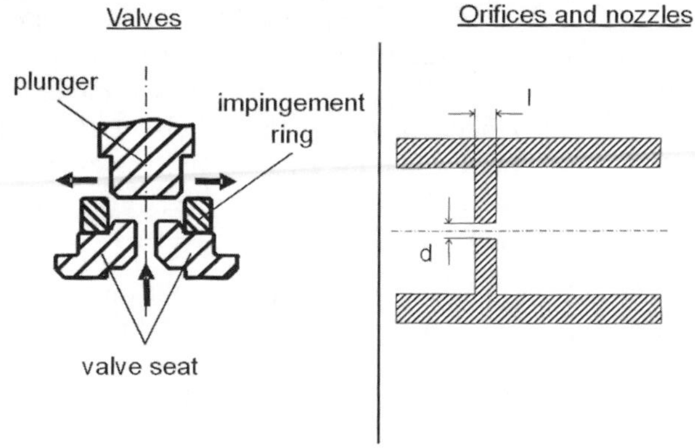

Fig. 2.1 Disruption systems: flat valve (*left*) (APV, 2010, Types of SPX APV valves, personal communication) and schematic drawing of a simple orifice as published by (Stang 1998; Freudig 2004; Aguilar et al. 2008) (*right*)

pumps. On a bench scale, commonly a single-piston pump delivers the volume stream, or rather the pressure, while up to eight piston pumps are found in production plants. The single-piston pump has the disadvantage that the pressure and the volume stream vary dramatically over the processing time, which results in a pulsation of the stresses on the product. Transient stresses act on droplets, and therefore products are heterogeneous and product properties are difficult to control. To reduce the pulsation, several pistons are combined in a phase-shifted manner. Valves are used to control the different pistons, which generally do not influence the quality of the emulsion.

2.2.1.2 Conventional Technical Solution

Droplet disruption and disaggregation of agglomerates large occur in the disruption system of a high-pressure homogenizer. Here, the pressure built up by the pump is released over a very short distance, resulting in specific flow conditions being responsible for droplet disruption. The disruption systems currently available on the market can be divided into valves and nozzles.

Valves. Valves, also known as radial diffusors, are common high-pressure disruption systems. The fundamental idea is to reduce the flow diameter with a valve plunger that is pushed to a valve seat forming a small gap (Fig. 2.1, left). The fluid enters the valve via a central hole in the valve seat, pumped by the high-pressure pump. It is then deflected by 90° and flows through a radial gap between the valve plunger and the valve seat. Often flat valves are fitted with an impingement ring to deflect the fluid a second time before it leaves through a drain hole. The impingement ring forms a defined outlet cross section and protects the valve housing against damage through abrasion and cavitation effects.

In valve systems like that shown in Fig. 2.1 (left), the pressure can be easily adjusted independent of the volume stream. This can be realized by varying the plunger's position and, therefore, the gap between valve seat and plunger. Further, the pressure can be adjusted automatically to the volume stream using a spring to pressure-load the plunger. Flow conditions within flat valves have been studied extensively (Phipps 1974; Kiefer and Treiber 1975; Treiber and Kiefer 1976; Freudig 2004; Innings and Tragardh 2005).

Orifices and Nozzles. An orifice presents the technically simplest way to increase pressure and transfer this energy into high flow velocities (Fig. 2.1, right) (EN ISO 5167-1 2003). In contrast to valves, simple orifices are constructed with no movable parts. This has the advantage that orifices are easy to make. At a constant viscosity of the emulsion the homogenizing pressure can be adjusted by the volume stream, the orifice hole diameter, and the cross-section area. Increasing the volume stream at a target pressure loss requires a numbering up of the orifices as done, for example, in Bayer 1997, 2001). Numbering up is limited only by the requirement that the minimum distance between the holes must be on the order of ten times the bore diameter (Aguilar et al. 2008). To ensure a constant homogenizing pressure even for fluctuating volume flow rates, the number of orifices must also be automatically adapted.

The Bayer AG company [now Bayer Technology Services GmbH (BTS)] was among the first to patent and commercialize orifices for high-pressure homogenization applications (Bayer 1991). Basic research on the flow conditions in orifices of circular cross section and their effect on droplet disruption was first published by Stang and Schubert in Karlsruhe (Stang 1998). This was followed by extensive research in several groups, also resulting in several patents (Grace 1982; Walstra 1983; Cook and Lagace 1985; Stone et al. 1986; Armbruster 1990; Muschiolik et al. 1995; Stang 1998; Penth 2000; Kolb 2001; Tesch 2002; Floury et al. 2004; Freudig 2004; Aguilar et al. 2008). Typical dimensions of an orifice are bore diameters of $d = 0.1$–1 mm and thickness of $l = 0.4$ to several mm. Today the well-known Microfluidizer also uses orifices of a specific design (so-called Y- or Z-chamber).

Conventional Dairy Homogenization Process. In conventional dairy processing, raw milk is separated prior to homogenization into a low-fat phase (0.03–0.3 vol.-% fat, skim milk) and a fat-enriched phase (30–42 vol.-% fat, cream) using a separator. In conventional full-stream homogenization, milk is then standardized to the final product fat content by mixing these two phases in various ratios. The mixture is then homogenized (Fig. 2.6, left). During the first energy crisis in Europe (1973), the dairy industry started partial-stream homogenization processes. Here, cream is diluted with skim milk to a fat content of 15–17 vol.-%, then homogenized, followed by standardizing again to the target fat concentration of, for example, 3.5 vol.-% in full cream milk. This reduces the homogenization energy required, as the less continuous phase must be compressed to the homogenization pressure (Kessler 2002).

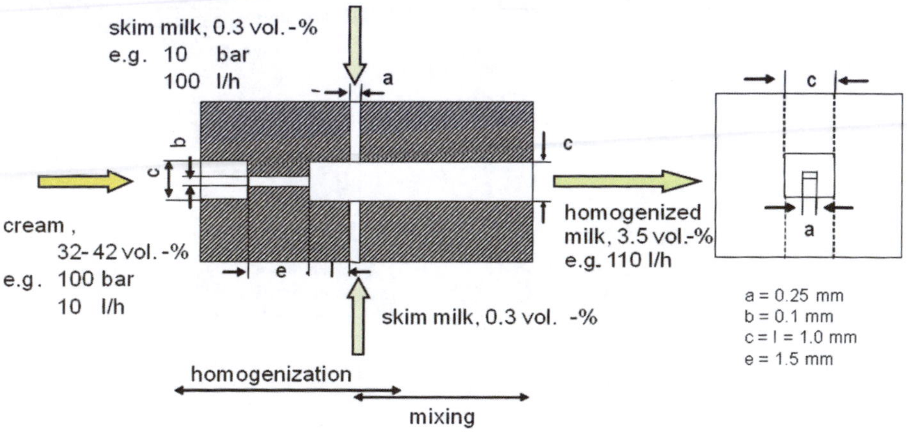

Fig. 2.2 Principal design of a combined homogenization nozzle and T-shaped micro-mixer: simultaneous emulsification and mixing (SEM) nozzle (Köhler et al. 2007)

This two-step remixing process, interrupted by high-pressure homogenization, is required as aggregations of fat globules (casein bridging) are found in homogenized cream above 13 vol.-% of fat (Kessler 2002; Köhler et al. 2008). The negative impact of aggregation can be compensated by an increased homogenizing pressure up to fat contents of 17 vol.-%. At fat contents higher than that, the process is mainly controlled by the coalescence and aggregation of the fat globules, leading to unsatisfactory homogenization results (see Fig. 2.3, Δ). Coalescence of newly formed fat droplets is observed until adsorbing dairy proteins have stabilized the droplets. In the stabilization of milk fat globules, a secondary droplet membrane is built up by adsorbing casein micelles, submicelles, lactoalbumins, and lactoglobulins (Walstra and Oortwijn 1982; Dalgleish et al. 1996). Because adsorbed casein micelles tend to adsorb at more than one fat globule at a time and strongly interact, bridges between fat globules are formed at increased fat contents, resulting in high aggregation rates. These fat-globule aggregates can be partially destroyed in a second homogenizing step (Darling and Butcher 1978), as is done in conventional technical processes. However, with increasing fat-globule concentrations, coalescence and aggregation rates increase simultaneously (Ogden et al. 1976; Walstra 1999; Köhler et al. 2007), limiting partial homogenization to 17 vol.-% of fat. Thus the energy reduction potential cannot be fully exploited.

Simultaneous Emulsification and Mixing: The Idea. In conventional dairy processing lines, the mixer for the standardization is located several meters behind the homogenizer. There, aggregation by casein bridging is fully completed. To limit aggregation, droplets must be separated from each other within milliseconds after their disruption. For this purpose, a novel homogenization orifice was developed: the main idea behind this is to combine the homogenization and standardization process step in one process unit, which is referred to as a *simultaneous emulsification and mixing (SEM) unit*. SEM nozzles (Fig. 2.2) allow for the dilution of fat globules directly after being generated. For dilution, skim milk is used; thus, emulsifier molecules (here:

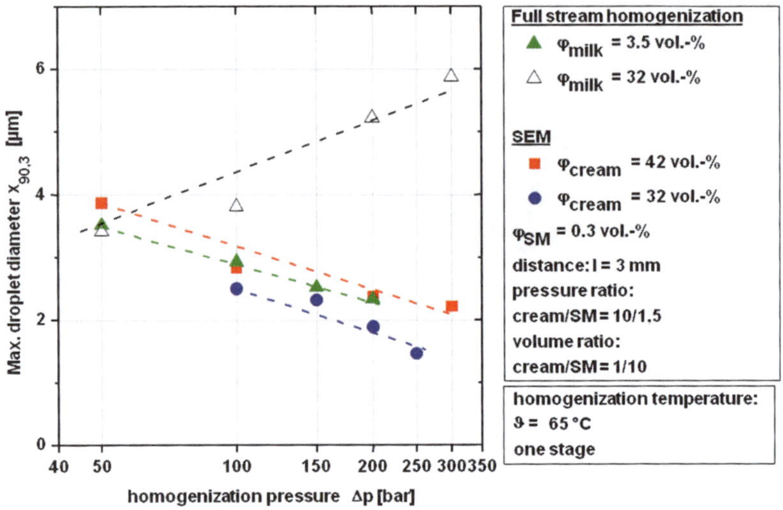

Fig. 2.3 Influence of fat content of homogenized cream and homogenization pressure on characteristic maximum droplet diameter $x_{90,3}$ (90 % quantile of volume distribution) in full stream and partial homogenization processing. When SEM nozzles are used, cream of 32 and 42 vol.-% fat were mixed with skim milk (φ φ = 0.3 vol.-%) within the nozzle. In conventional full-stream homogenization milk (volume fat content φ φ = 3.5 vol.-%) and cream (φ φ = 32 vol.-%) were homogenized as full stream (Köhler et al. 2008)

dairy proteins) are additionally mixed into the cream at the critical process stage (droplet breakup) with high mixing intensity. As the main stream, cream is pumped through the nozzle orifice at homogenizing pressure, with a fat content determined by the separator located in the dairy process line prior to the homogenizer, usually ranging from 32 to 42 vol.-%. Skim milk runs as a mixing stream at pressures on the order of 0.1–20 % of the homogenizing pressure.

As shown in Fig. 2.3, it was proven that the homogenization of cream and its direct standardization within one process unit (SEM-nozzle) results in fat-globule sizes comparable to a full-stream homogenization process. A comparison of SEM partial homogenization results at 32 and 42 vol.-% fat to those of conventional full-stream homogenization at 3.5 vol.-% fat shows that product quality is fully maintained.

2.2.1.3 Design of SEM Nozzles Based on Modern Process Engineering Tools: Numerical Process Simulation and Microprocessing Apparatus Design

For the design and technical layout, knowledge of the fluid dynamics in this device is essential. Local flow velocities were predicted by Fluent (Ansys, Canonsburg, PA, USA), and through postprocessing, local stresses acting on fat droplets were evaluated. Figure 2.4 shows the results of the impact of the distance between the

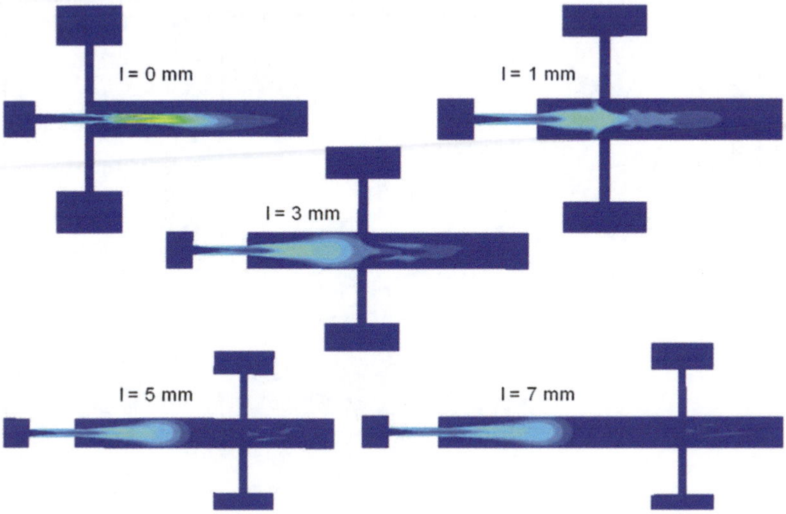

Fig. 2.4 Distribution and intensity of specific turbulent kinetic energy κ in SEM nozzle. Parameters varied: distance between homogenization orifice outlet and lateral injection ($l = 0$, 1, 3, 5, and 7 mm)

homogenization orifice outlet and lateral injection on local flow and the resulting specific turbulent kinetic energy.

Depending on the point of injection of high volumes of skim milk into the main stream (which in fact corresponds to 10 % of the volume only), the distribution and intensity of the specific turbulent kinetic energy κ in the orifice outlet is affected. For a SEM-nozzle geometry with lateral injection at a distance of $l = 0$ mm (directly at the orifice outlet), the turbulent kinetic energy κ dissipation is concentrated in the flow center (Fig. 2.4). However, injection at larger distances changes the areas and intensities of energy dissipation significantly. For an injection at a distance of $l = 1$ mm, two areas of turbulent kinetic energy dissipation κ are found, one in front and a second one (which is less pronounced) behind the mixing inlet point. If the distance is larger ($l = 3$ mm), then the turbulent energy dissipation area is initially compressed in the region directly after the nozzle outlet. Increasing the distances (here: from 5 to 7 mm) no longer affects the flow conditions in the region of droplet disruption.

In conclusion, the injection point locality influences both the mixing quality and the disruptive stresses in the flow: their impact is counteracting. While a distance of $l = 0$ mm should be optimal whenever mixing dominates the overall result, higher distances will be better whenever disruption must proceed undisturbed. The compression of the flow resulting from the injection of high volumes can even be positive in terms of creating an internal counterpressure.

These simulation results could be indirectly validated by experiments. For this, SEM nozzles with injection points at distances of $l = 0$, 1, 3, 5, and 7 mm (Fig. 2.2)

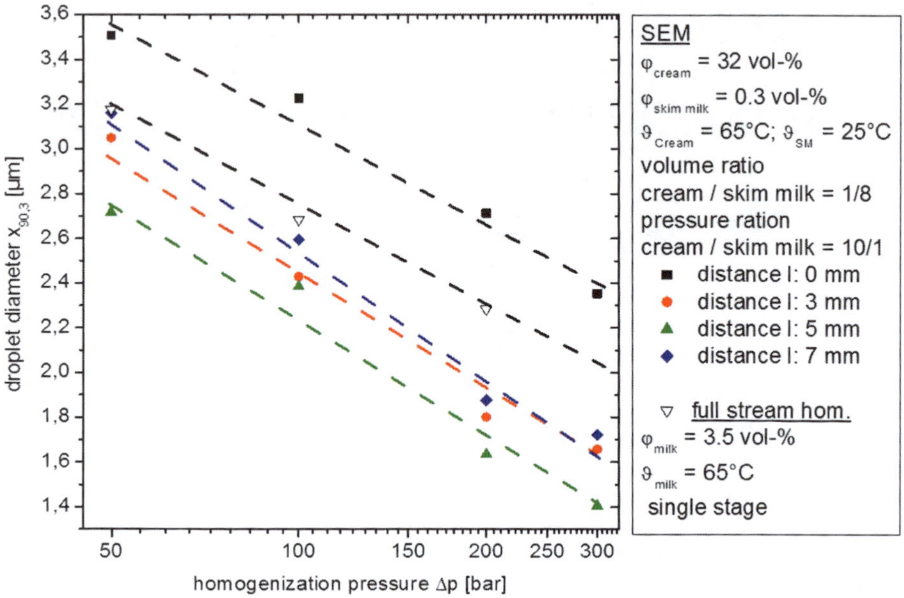

Fig. 2.5 Homogenization results (maximum droplet diameter $x_{90,3}$ of volume distribution) for partial homogenization in novel SEM nozzle by changing distance between nozzle outlet and mixing point at typical homogenization pressures of 50–300 bar (Köhler et al. 2009)

were built. Cream of 32 vol.-% milk fat was homogenized at 100 bars as the main stream, and skim milk of 0.3 vol.-% was injected laterally at different distances. The mixing ratio was adapted to produce full cream milk (3.5 vol.-% fat). As shown in Fig. 2.5, distances that are too large (here: 7 mm) negatively affect the mixing quality and, thus, allow fat-globule coalescence and aggregation. The smallest fat-globule sizes and, therefore, the best disruption and stabilization of fat droplets were found for skim milk injection at a distance of 5 mm. Smaller distances resulted in decreased fat-globule disruption, while greater distances resulted in increased globule aggregation and coalescence.

2.2.1.4 Technical Solution and Implementation, Process Integration

Replacing the conventional homogenization valve by a SEM nozzle in a dairy homogenization process line allows for eliminating the mixing unit. No technical construction must be carried out on the plant except for adjusting the pipes (Fig. 2.6, right). Standardization is done directly in the homogenization step. Thus, SEM-nozzle technology allows for process integration by eliminating processing units, combining processing steps, and simplifying processing lines.

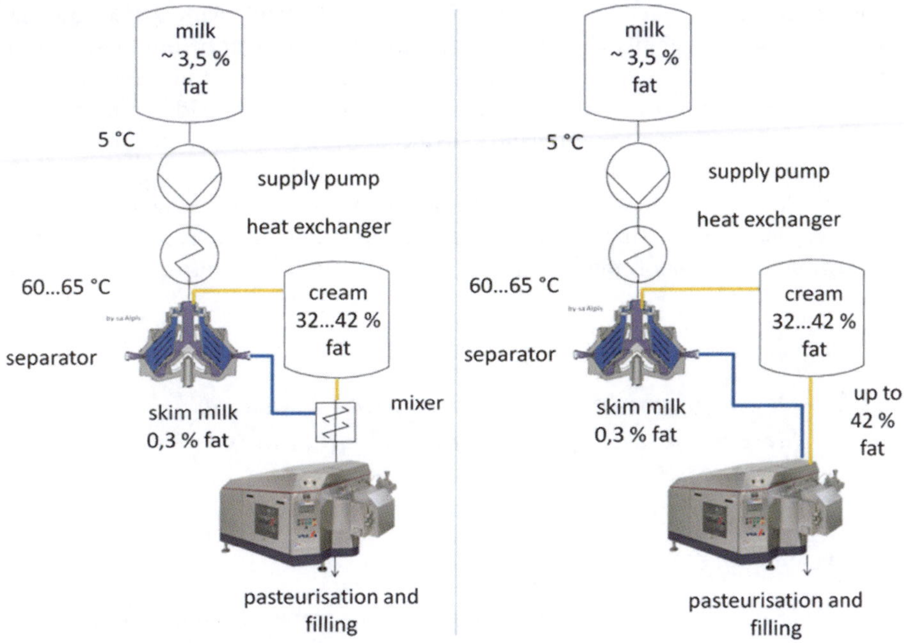

Fig. 2.6 Flow chart of conventional full stream (*left*) and novel partial homogenization process (*right*)

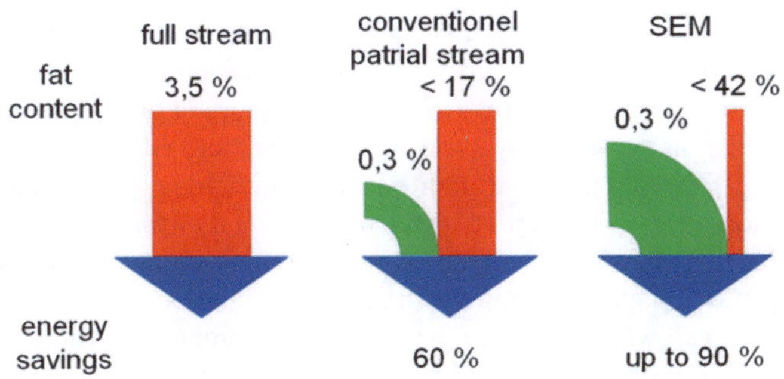

Fig. 2.7 Comparison of conventional dairy homogenization and novel SEM technology with regard to energy savings

2.2.1.5 Impact on Process Efficiency and Product Quality

Compared to conventional dairy homogenization process lines, the new SEM process requires only 10 % of the homogenization energy input (with regard to full stream processing) and only 60 % of the energy (with regard to conventional partial homogenization processing), as illustrated in Fig. 2.7. This results in considerable energy and cost savings in dairy processing with no detrimental effects on product quality. In addition, two mixing units can be eliminated from the process line, resulting in less investment, cleaning, and maintenance costs. It could also be demonstrated that maintenance costs due to abrasion on homogenization nozzles are also significantly reduced by using SEM nozzles (Hecht et al. 2012).

Thus, this example demonstrates how modern process engineering tools, such as computational fluid dynamics (CFD) simulation and microprocess-engineered processing units, support the food industry in its sustainability efforts.

2.2.2 Development of Adapted Extrusion Processes for Encapsulating Functional Components in Starch-Based Matrices

2.2.2.1 Challenges

The encapsulation of functional hydrophobic components into starch-based matrices via extrusion processing is a promising area that has attracted growing interest by the food industry as well as by the chemical, pharmaceutical, and medical industries (Ellis et al. 1998; Mano et al. 2003; Shogren et al. 1993; Yilmaz et al. 2001). For the encapsulation of lipophilic bioactive components, a lipophilic carrier is required. In foods, triglycerides serve as nontoxic, food-grade carriers. Triglyceride droplets, especially small ones (e.g., <1 μm) are known to encapsulate high concentrations of bioactive substances and increase their bioavailability (Horn and Rieger 2001; Ribeiro et al. 2006). To control important product properties, such as the color intensity, stability, and release characteristics, of the bioactive substances, the dispersed phase's (oil droplets') morphology (droplet concentration, size distribution, charge, interfacial characteristics, and physical state) is crucial in designing delivery systems with specific functional performances suitable for different types of lipophilic components and food matrices. Thus, understanding droplet breakup behavior in thermoplasticized, starch-based matrices under extrusion conditions is essential in designing encapsulation processes. This in turn is a well-known problem in basic chemical engineering.

In laminar flow, as mainly found in extruders, shear and elongational stresses are responsible for droplet deformation and breakup (Cheng and Manas-Zloczower 1997) and the molecular status of starch biopolymers representing the encapsulating matrix material (van den Einde et al. 2004). The resulting droplet sizes, in addition, depend on the viscosity ratio between the dispersed phase (lipophilic components) and the continuous phase (thermoplasticized starch matrix under extrusion conditions), which are temperature and shear-rate dependent (Taylor 1934a). Hence, knowledge of temperature, elongational, and shear profiles inside the extruder channels is the key to controlling the resulting matrix and

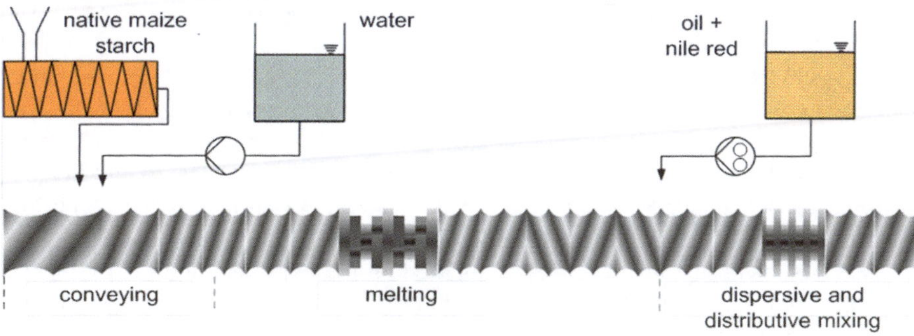

Fig. 2.8 Schematic illustration of experimental setup, screw geometry, and mixing zone

emulsion characteristics and, therefore, the bioavailability and stability of the lipophilic bioactive ingredients encapsulated within the oil droplets and starch matrix. However, the complex geometry of extruder channels leads to a wide and nonuniform, transient distribution of the stresses in the flow field. The measurement of the exposed stresses in extruders is a very difficult task, if it is even at all feasible.

CFD simulation of flow characteristics and mixing ability inside extruders is, therefore, a promising tool for improving our understanding of the underlying mechanisms and representation of the influence of process parameters.

2.2.2.2 Experimental Data: Extrusion Trials and Droplet Size Measurements

The extrusion trials were conducted using a corotating, high-speed, twin-screw extruder, ZSK 26 Mc (Coperion, Stuttgart, Germany), consisting of seven barrels with a length-to-diameter ratio of 29. In the first part of the extruder, native maize starch was mixed with water, which was then plasticized under thermal and mechanical stresses generated by screw rotation. The medium-chain triglyceride oil was initially mixed with Nile red to make them visible under confocal laser scanning microscopy (CLSM). The oil was then incorporated into the plasticized starch matrix at the end of the extruder. In this zone, the extruder barrel was completely filled due to the pressure generated at the extrusion die, which was verified by inline mounted pressure sensors. An illustration of the experimental setup, screw geometry, and mixing zone can be seen in Fig. 2.8.

To obtain the size distribution of the dispersed oil droplets in the extruded starch samples, a CLSM was used. The sections cut out from every sample were placed on a glass slide and covered with a cover slide sealed with bidistilled water. The sections were analyzed using a confocal microscope (Zeiss LSM 510 Meta, Oberkochen, Germany). The images obtained from CLSM were further analyzed using image-processing software (Image Pro Plus 6.0, Media Cybernetics, Silver Spring, MD, USA) to determine the droplet size distribution. Area-weighted cumulative size distribution was chosen for plotting.

2.2.2.3 Material Characterization: Online Rheological Measurements

Rheological characteristics of the plasticized starch must be determined under extrusion conditions, which cannot be realized in conventional laboratory rheological equipment. As the biopolymer molecules of starch (particularly amylopectins) are degraded when subjected to thermomechanical stresses, as found in cook extrusion processes, their rheological properties depend on processing conditions and, thus, must be determined online. For this purpose, a multistep slit-die rheometer was designed, manufactured, and mounted on the extruder for all experiments. Details are discussed elsewhere (Horvat et al. 2013). The rheological properties of the starch matrices under extrusion process conditions can be calculated from the pressure drop of the flow in the slit-die, which is measured using flush-mounted transducers. Various shear rates necessary to obtain a viscosity curve were generated by increasing the cross section of the slit-die without any disturbance on the extrusion process parameters, thereby keeping the thermomechanical history of the product constant. Van den Einde et al. (2004) and Liu et al. (2010) showed that the transformation of starch molecules under mechanical stresses depends only on the maximum stress applied. For the screw geometry selected for this study, it was expected that the maximum stresses had been generated at the inverse screw elements. Therefore, it was assumed that the molecular structure, and thus the viscosity of the starch, would not change further at the end of the extruder (mixing zone) and could, therefore, be measured by the online slit-die rheometer.

2.2.2.4 Numerical Simulation

The flow and corresponding mixing characteristics (e.g., shear rate, shear stress, mixing index) of the plasticized starch matrix in the mixing region of the extruder were predicted by the CFD code ANSYS POLYFLOW (Ansys, Canonsburg, PA, USA), which is a finite-element code dedicated to highly viscous flows. The flow in a corotating, twin-screw extruder is three-dimensional and unsteady due to moving and intermeshing parts (two rotating intermeshing screws in a nonmoving barrel). To simplify the setup of such a computational model and to avoid the use of remeshing algorithms, the mesh superimposition technique implemented by Avalosse (1996) in the aforementioned software package was used. Since the dispersive and distributive mixing of the oil phase into the plasticized starch matrix takes place at the very end of the screws (Fig. 2.8), only this part was selected as the computational domain. Flow through the mixing elements (45° kneading blocks) is assumed to be isothermal in accordance with negligible temperature changes in the material in this region. The rheological data of plasticized starch required for the simulation were measured with the multistep online slit-die rheometer (Sect. 2.2.2.3) and fitted to the Bird-Carreau viscosity model for calculating the apparent viscosity η_{app}:

$$\eta_{app} = \eta_0 \left(1 + \left(\lambda \dot{\gamma}_{app}\right)^2\right)^{\frac{n-1}{2}} \tag{2.1}$$

where η_0 is the zero shear viscosity, λ is a fitting parameter (called natural time by the authors), η_{app} is the apparent shear rate, and n is the power law index. Since the flow rate of the starch is relatively low (11 kg/h), creeping flow is assumed and gravity effects are neglected. After investigating the influence of the mesh design and interpolation method on the pressure profile along the extruder, 503,000 mesh elements, enriched minielement velocity, and linear pressure for mixed interpolations were selected (results not shown). To obtain a good spatial resolution, a time-dependent simulation was conducted with 40 time steps for half a turn of the screws. To calculate the shear stresses experienced by the droplets in the extruder channel, virtual particles were launched at the same time in the flow domain. The marker particles are massless, volumeless, and noninteracting with each other. Under these assumptions, the particles can be located by integrating the velocity vectors with respect to time. Initially, these particles are randomly distributed in an inlet section of the flow domain where the triglyceride oil is injected in experiments and their trajectories between the inlet and outlet are calculated from the velocity profiles. Along the trajectory, stresses experienced by each particle are predicted. To obtain statistically good results, 1,500 particles were tracked.

2.2.2.5 Evaluation of Dispersive Mixing Efficiency

One of the most important parameters used to describe the droplet behavior under flow is a dimensionless number called the capillary number (Ca) (Taylor 1932, 1934b). The capillary number represents the ratio between the deforming viscous forces and the shape-conserving interfacial forces:

$$Ca = \tau / (2\sigma / D) \tag{2.2}$$

where τ is the shear stress, σ the interfacial tension between the droplet and matrix phase, and D the diameter of the droplet. The capillary number corresponding to the critical value at which the droplet becomes unstable and breaks is called the critical capillary number (Ca_{crit}). The critical capillary number depends on the viscosity ratio (p), which is a dimensionless parameter describing the ratio between the viscosities of the dispersed phase and the continuous phase. This dependency was established and validated for Newtonian systems (Bentley and Leal 1986; Debruijn 1991; Grace 1982). The authors found that droplet breakup in a simple shear flow occurred most favorably in an intermediate range of $0.1 < p < 1$. At lower p-values – as found for the systems investigated here – Ca_{crit} steadily increases with decreasing p-values. No breakup is possible at $p > 4$. An empirical fit to Grace's data at $p < 0.1$ was given by Hinch and Acrivos (1980):

Fig. 2.9 Influence of screw speed on maximum shear stress experienced by droplets during their flow in mixing zone of extruder

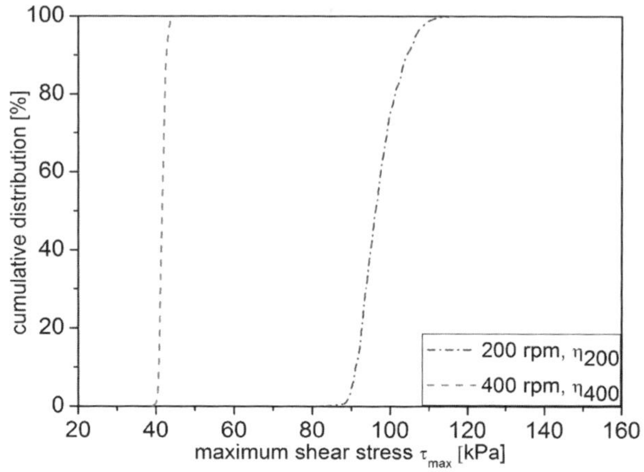

$$Ca_{crit} = 0.17 \cdot p^{-0.55} \tag{2.3}$$

For large capillary numbers (Ca/Ca_{crit}) and for viscosity ratios $p < 1$, droplets are stretched into long, slender filaments (Grace 1982). As a consequence, small disturbances present at the interface of the drop filaments grow and finally result in the disintegration of the thread into very small droplets (Grace 1982). The deformation time also plays an important role in the breakup of the droplets, which was reported to decrease with increasing capillary ratios (Ca/Ca_{crit}) (Grace 1982). Therefore, not only the capillary number or shear stress but also the ratio of the capillary number to the system-specific critical capillary number (Ca/Ca_{crit}) should be taken into account in order to evaluate the dispersive mixing in an extrusion process.

To evaluate the final droplet size distribution, it is essential to know the initial droplet size distribution and the stresses experienced by the droplets with their exposure time. Few data have been published on this question (e.g., Deroussel et al. 2001; Huneault et al. 1995; Potente et al. 2001). It remains a very challenging task regarding the rheological complexity of a starch-based matrix and the wide distribution of stresses owing to the complex geometry of a twin-screw extruder. Based on the physically well-grounded assumption that the maximum stresses within an extruder channel are mainly generated at the screw tips, where the flow is dominated by a simple shear flow, these values were taken from the simulation and used to evaluate the efficiency of dispersive mixing within the extruder under different process conditions.

Figure 2.9 depicts the maximum shear stresses experienced by droplets during their flow in the extruder channel for two different screw speeds (200 and 400 rpm). Since the dispersive mixing ability of the flow was intended to be evaluated, the droplet diameter was kept constant and arbitrarily set to 10 μm. It was expected that

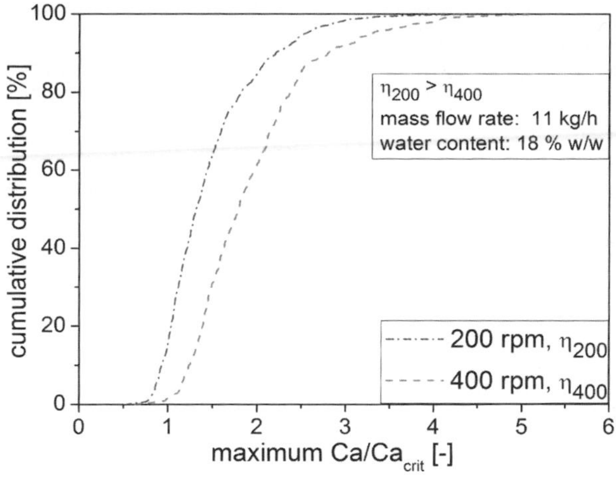

Fig. 2.10 Influence of screw speed on maximum capillary ratios experienced by droplets during their flow in mixing zone of extruder (Emin et al. 2011)

increasing screw speed would lead to an increase in shear stresses since this increased the shear rate. However, there was a negative correlation between shear stress and the viscosity of the plasticized starch matrix since higher shear stresses lead to a higher degree of starch degradation (van den Einde et al. 2003). This resulted in a decreased shear viscosity and, therefore, in lower maximum stresses at the mixing zone of the extruder. Here, it is worth mentioning that the starch was exposed to maximum shear stresses at the inverse screw element region located before the simulated mixing region. Increasing screw speeds decrease the maximum shear stresses and maximum capillary number experienced by the particles tracked in the simulated flow domain due to a dramatic decrease in starch viscosity at higher screw speeds.

Although the shear stresses are responsible for droplet breakup, the capillary ratio (Ca/Ca_{crit}) must be considered to evaluate the dispersive mixing efficiency as discussed earlier. Therefore, the capillary ratio experienced by each droplet is monitored during particle tracking, and the maximum values are used to form a cumulative distribution for the 1,500 particles tracked. The results are depicted in Fig. 2.10 for screw speeds of 200 and 400 rpm. The viscosity of the plasticized starch, showing shear thinning flow behavior, has a strong influence on the obtained shear stress profile in the extruder. From Eq. 2.3, it can be deduced that the matrix viscosity, strongly influencing the viscosity ratio, also determines the critical capillary number. Therefore, the viscosity of the plasticized starch has counteracting effects on the droplet breakup, which makes the evaluation of the dispersive mixing in plasticized starch very complicated. The computation of the capillary ratio, which includes both the counteracting effects of the starch viscosity, however, makes it possible to qualitatively evaluate the effect of process parameters on droplet breakup in dispersive mixing.

According to the results depicted in Fig. 2.10, it is expected that smaller droplet sizes would be obtained in plasticized starch matrices extruded at higher screw

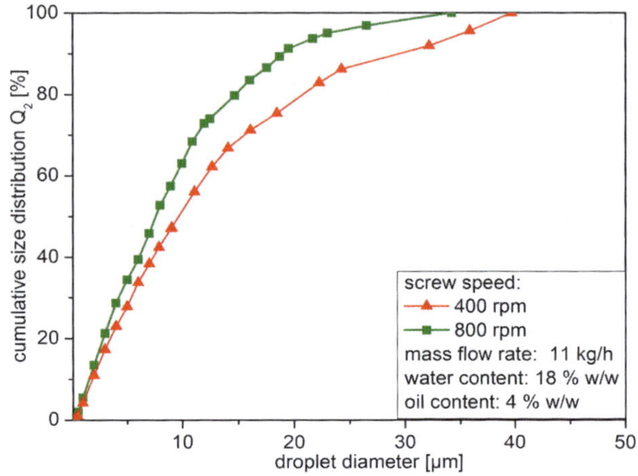

Fig. 2.11 Cumulative droplet size distributions obtained from extrusion experiments

speeds, regardless of the fact that the droplets are exposed to lower maximum shear stresses (Fig. 2.9).

To validate this simulation result, experiments were carried out at different screw speeds. Oil was added to the mixing zone of the extruder, and the resulting droplet size distributions were measured. The results are shown in Fig. 2.11, which confirms the computational results: increasing the screw speed (here from 400 to 800 rpm) led to smaller droplet sizes. Results for screw speeds of 200 rpm are not plotted because the oil was dispersed poorly; in fact, it dripped out of the starch matrix when leaving the extruder, which also confirmed the results.

2.2.2.6 Concluding Remarks

Evaluation of droplet breakup through dispersive mixing of oil in rheologically complex plasticized starch matrices within an extruder is a very challenging task because several parameters, such as the complex rheological behavior of the matrix, the initial droplet size distribution, and the nature and strength of the flow with the time of exposure, need to be known. Experimentally assessing these parameters is practically impossible. Basic chemical engineering approaches using modern numerical tools, however, can provide invaluable insights into the mechanisms in such processes. Flow simulation offers access to local data on flow velocities and the resulting shear stresses, capillary numbers, and capillary ratio, the latter being parameters of relevance for droplet breakup in laminar flows. Thus, the influence of process parameters on the efficiency of dispersive mixing in such a system can be investigated reasonably. Therefore, the methods and models developed can be used for process parameter adaptation and optimization, which was shown in this example for the objective of optimizing dispersive mixing processes, even when these are governed by complex flows, as can be found in the homogenization process of oil in plasticized starch matrices in an extruder.

References

Aguilar FA, Köhler K, Schubert H, Schuchmann HP (2008) Herstellen von Emulsionen in einfachen und modifizierten Lochblenden: Einfluss der Geometrie auf die Effizienz der Zerkleinerung und Folgen für die Maßstabsvergrößerung. Chemie Ingenieur Technik 80(5):607–613

Armbruster H (1990) Untersuchungen zum kontinuierlichen Emulgierprozeß in Kolloidmühlen unter Berücksichtigung spezifischer Emulgatoreigenschaften und der Strömungsverhältnisse im Dispergierspalt. Dissertation, Universität Karlsruhe (TH)

Avalosse T (1996) Numerical simulation of distributive mixing in 3-D flows. Macromol Symp 112:91

Bayer AG (1991) Preparation of pharmaceutical or cosmetic dispersions. Patent US4996004

Bayer AG (1997) Verfahren und Vorrichtung zur Herstellung einer parenteralen Arzneistoffzubereitung. Patent WO9717946

Bayer AG (2001) Dispersion nozzle with variable throughput. Patent WO 01/05517

Bentley B, Leal L (1986) An experimental investigation of drop deformation and break-up in steady, two-dimensional linear flows. J Fluid Mech 167(1):241–283, Cambridge University Press

Cheng H, Manas-Zloczower I (1997) Study of mixing efficiency in kneading discs of co-rotating twin-screw extruders. Polym Eng Sci 37(6):1082–1090, Wiley

Cook EJ, Lagace AP (1985) Apparatus for forming emulsions, Biotechnology Development. Patent 4 533 254

Dalgleish DG, Tosh SM, West S (1996) Beyond homogenization: the formation of very small emulsion droplets during the processing of milk by a Microfluidizer. Neth Milk Dairy J 50(2):135–148

Darling DF, Butcher DW (1978) Milk-fat globule membrane in homogenized cream. J Dairy Res 45(2):197–208

Debruijn RA (1991) Deformation and breakup of drops in simple shear flows. Doctoral Thesis. Eindhoven University, The Netherlands

Deroussel P, Khakhar DV, Ottino JM (2001) Mixing of viscous immiscible liquids. Part 1: computational models for strong – weak and continuous ow systems. Chem Eng Sci 56:5511–5529

Ellis RP, Cochrane MP, Dale MFB, Duffus CM, Lynn A, Morrison IM, Prentice RDM (1998) Starch production and industrial use. J Sci Food Agric 77(3):289–311, Wiley Online Library. doi:10.1002/(SICI)1097-0010(199807)77:3<289::AID-JSFA38>3.3.CO;2-4

Emin MA (2013) Dispersive mixing of oil in plasticized starch by extrusion processing to design functional foods, Verlag Dr. Hut, München, ISBN 978-3-8439-1112-2

Emin MA, Köhler K, Schlender M, Schuchmann HP (2011) Characterization of mixing in food extrusion and emulsification processes by using CFD. In: Nagel WE et al (eds) High performance computing in science and engineering, vol 10. Springer, Heidelberg pp 443–462

EN ISO 5167-1 (2003) Durchflussmessung von Fluiden mit Drosselgeräten in voll durchströmten Leitungen mit Kreisquerschnitt Teil 1: Allgemeine Grundlagen und Anforderungen (ISO 5167-1:2003); Deutsche Fassung EN ISO 5167-1:2003

Floury J, Bellettre J, Legrand J, Desrumaux A (2004) Analysis of a new type of high pressure homogeniser. A study of the flow pattern. Chem Eng Sci 59(4):843–853

Freudig B (2004) Herstellen von Emulsionen und Homogenisieren von Milch in modifizierten Lochblenden. Dissertation, Universität Karlsruhe (TH), 3-8322-3147-1

Gaulin A (1899) Appareil et Procédé pour la Stabilisation du Lait. Patentnr.: Brecet nr. 295596

Grace HP (1982) Dispersion phenomena in high-viscosity immiscible fluid systems and application of static mixers as dispersion devices in such systems. Chem Eng Commun 14(3–6):225–277

Hecht LL, Schlender M, Köhler K, Schuchmann HP (2012) Abrasion in high-pressure homogenization orifices: a new method to quantify the impact of particle loaded fluids. Wear 289:138–144

Hinch EJ, Acrivos A (1980) Long slender drops in a simple shear flow. J Fluid Mech 98(2):305. doi:10.1017/S0022112080000171

Horn D, Rieger J (2001) Organic nanoparticles in the aqueous phase—theory, experiment, and use. Angew Chem Int Ed 40(23):4330–4361, Wiley Online Library

Horvat M, Emin MA, Hochstein B, Willenbacher N, Schuchmann HP (2013) A multiple-step slit-die rheometer for rheological characterization of extruded starch melts. J Food Eng 116(2): 2398–2403

Huneault M, Shi Z, Utracki LA (1995) Development of polymer blend morphology during compounding in a twin-screw extruder. Part IV: a new computational model with coalescence. Polym Eng Sci 35(1):115–127

Innings F, Tragardh C (2005) Visualization of the drop deformation and break-up process in a high pressure homogenizer. Chem Eng Technol 28(8):882–891

Kessler HG (2002) Food and bio process engineering – dairy technology. Verlag A. Kessler, München

Kiefer P, Treiber A (1975) Prall und Stoß als Zerkleinerungsmechanismen bei der Hochdruck-Homogenisation von O/W-Emulsionen. CIT 47(13):573

Köhler K (2010) Simultanes Emulgieren und Mischen, Logos Verlag, Berlin, ISBN 978-3-8325-2716-7

Köhler K, Aguilar FA, Hensel A, Schubert K, Schubert H, Schuchmann HP (2007) Design of a microstructured system for homogenization of dairy products with high fat content. Chem Eng Technol 30(11):1590–1595. doi:10.1002/ceat.200700266

Köhler K, Aguilar FA, Hensel A, Schubert K, Schubert H, Schuchmann HP (2008) Design of a microstructured system for the homogenization of dairy products at high fat content part II: influence of process parameters. Chem Eng Technol 31(12):1863–1868

Köhler K, Aguilar FA, Hensel A, Schubert H, Schuchmann HP (2009) Design of a microstructured system for the homogenization of dairy producs at high fat content- part III: influence of geometric parameters. Chem Eng Technol 32(7):1120–1126

Kolb GE (2001) Zur Emulsionsherstellung in Blendensystemen. Dissertation, Universität Bremen, 3-8265-9204-2

Liu W-C, Halley PJ, Gilbert RG (2010) Mechanism of degradation of starch, a highly branched polymer, during extrusion. Macromolecules 43(6):2855–2864. doi:10.1021/ma100067x

Mano JF, Koniarova D, Reis RL (2003) Thermal properties of thermoplastic starch/synthetic polymer blends with potential biomedical applicability. J Mater Sci Mater Med 14(2):127–135

Muschiolik G, Roeder R-T, Lengfeld K (1995) Druckhomogenisator. Patentnr. DE 195 30 247

Ogden LV, Walstra P, Morris HA (1976) Homogenization-induced clustering of fat globules in cream and model systems. J Dairy Sci 59(10):1727–1737

Penth B (2000) Method and device for carring out chemical and physical processes. Patentnr. WO/2000/061275

Phipps LW (1974) Cavitation and separated flow in a simple homogenizing valve and their influence on the break-up of fat globules in milk. J Dairy Res 41(3):339–347

Potente H, Bastian M, Bergemann K, Senge M, Scheel G, Winkelmann T (2001) Morphology of polymer blends in the melting section of co-rotating twin screw extruders. Polym Eng Sci 41(2):222–231. doi:10.1002/pen.10723

Ribeiro HS, Guerrero JMM, Briviba K, Rechkemmer G, Schuchmann HP, Schubert H (2006) Cellular uptake of carotenoid-loaded oil-in-water emulsions in colon carcinoma cells in vitro. J Agric Food Chem 54(25):9366–9369. doi:10.1021/jf062409z

Shogren RL, Fanta GF, Doane WM (1993) Development of starch-based plastics – a reexamination of selected polymer systems in historical perspective. Starch – Stärke 45(8):276–280. doi:10.1002/star.19930450806

Stang M (1998) Zerkleinern und Stabilisieren von Tropfen beim mechanischen Emulgieren. Dissertation, Universität Karlsruhe (TH). doi:10.1002/cite.330700917

Stone HA, Bentley BJ, Leal LG (1986) An experimental-study of transient effects in the break-up of viscous drops. J Fluid Mech 173:131–158

Taylor GI (1932) The viscosity of a fluid containing small drops of another fluid. Proc R Soc Lond Ser A 138(834):41–48

Taylor GI (1934a) The formation of emulsions in definable fields of flow. Proc R Soc Lond Ser A 146(858):501–523

Taylor GI (1934b) The formation of emulsions in definable fields of flow. Proc R Soc A Math Phys Eng Sci 146(858):501–523. doi:10.1098/rspa.1934.0169

Tesch S (2002) Charakterisieren mechanischer Emulgierverfahren: Herstellen und Stabilisieren von Tropfen als Teilschritte beim Formulieren von Emulsionen. Dissertation, Universität Karlsruhe (TH)

Treiber A, Kiefer P (1976) Kavitation und Turbulenz als Zerkleinerungsmechanismen bei der Homogenisation von O/W-Emulsionen. Chemie Ingenieur Technik 48(3):259

van den Einde RM, van der Goot AJ, Boom RM (2003) Understanding molecular weight reduction of starch during heating-shearing processes. J Food Sci 68(8):2396–2404. doi:10.1111/j.1365-2621.2003.tb07036x

van den Einde RM, Akkermans C, van der Goot AJ, Boom RM (2004) Molecular breakdown of corn starch by thermal and mechanical effects. Carbohydr Polym 56(4):415–422. doi:10.1016/j.carbpol.2004.03.006

Walstra P (1983) Formation of emulsions. In: Becher P (ed) Encyclopedia of emulsion technology, vol 1. Marcel Dekker, New York, pp 57–128

Walstra P (1999) Casein sub-micelles: do they exist? Int Dairy J 9(3–6):189–192

Walstra P, Oortwijn H (1982) The membranes of recombined fat globules. 3. Mode of formation. Neth Milk Dairy J 36(2):103–113

Yilmaz G, Jongboom ROJ, Feil H, Hennink WE (2001) Encapsulation of sunflower oil in starch matrices via extrusion: effect of the interfacial properties and processing conditions on the formation of dispersed phase morphologies. Carbohydr Polym 45(4):403–410. doi:10.1016/S0144-8617(00)00264-2, Elsevier

Part I
Food Materials Science and Properties

Chapter 3
Advances in Nanotechnology as Applied to Food Systems

Jarupat Luecha, Nesli Sozer, and Jozef L. Kokini

3.1 Introduction

The worldwide human population reached seven billion in the year 2011 and continues to grow (Bloom 2011). This leads to several important global challenges with respect to the availability, manufacturability, preservation, safety, sustainability, vulnerability, and wholesomeness of foods. Overcoming these challenges will necessitate the development of new and advanced technologies. Nanotechnology is one such toolbox that might offer solutions and opportunities. Nanotechnology is defined as the ability to control and restructure matter at the atomic or molecular levels, allowing the manipulation of dissimilar properties and phenomena at those scales to create new functionalities, leading to new properties and new devices that might make food manufacturing more economical and efficient (Roco 2011). This novel technology has been emerging in a broad range of disciplines, particularly with applications from electronics to biomedicine (Paul and Robeson 2008; Kirby 2011). Although the applications of nanotechnology in food and agriculture are new when compared to their use in electronics, biomedicine, and pharmaceuticals, advances in nanotechnology such, as DNA microarrays, microelectromechanical systems, microfluidics, and

J. Luecha
Food Technology Program, Mahidol University, Kanchanaburi Campus, 199 Moo 9, Tambon Loom Soom, Amphur Sai Yok, Kanchanaburi 71150, Thailand

N. Sozer
VTT Technical Research Centre of Finland, P.O. Box 1000, FI-02044 VTT, Espoo, Finland

J.L. Kokini (✉)
Department of Food Science and Human Nutrition, Illinois Agricultural Experiment Station, College of Agricultural, Consumer and Environmental Sciences, University of Illinois Urbana-Champaign, 228 Mumford Hall, 1301 W. Gregory Drive, Urbana, IL 61801, USA

Micro and Nanotechnology Laboratory, University of Illinois Urbana-Champaign, 208 N. Wright Street, Urbana, IL 61801, USA
e-mail: kokini@illinois.edu

S. Yanniotis et al. (eds.), *Advances in Food Process Engineering Research and Applications*, Food Engineering Series, DOI 10.1007/978-1-4614-7906-2_3,

nanocomposites are leading to applications focused on food technology that are unique and empowering (Sozer and Kokini 2009). The use of nanoencapsulation, nanotubes, nanoemulsions, nanoparticles, and nanosensors enables food scientists to successfully deliver nutrients and nutraceuticals and detect pathogens and contaminants in a more efficient way, develop eco-friendly packaging materials that have good mechanical and barrier properties, develop miniaturized analytical tools for various applications, and develop imaging tools to characterize structure and chemistry better than ever.

This chapter presents examples from the recent findings of nanotechnology research in the food science and technology area. We focus on the current nanotechnology applications in food science ranging from functional foods, delivery systems, food packaging, and food safety.

3.2 Nanostructured Materials and Their Applications in Food

Nanostructured materials are materials that have at least one dimension at the nanometer scale and are fabricated via one of two approaches, a bottom-up approach or a top-down approach (Sozer and Kokini 2009). Nanostructured materials can be classified into several types based on shape, origin, or application. Several materials have been used for the fabrication of nanostructured materials, but only food biopolymers will be covered in this chapter.

3.2.1 Engineering of Food Nanoparticles

Many researchers in the nanotechnology field are interested in engineering nanoparticles (Chaudhry et al. 2010). Starch as an abundant natural polymer has been used as a substrate for the fabrication of nanoparticles. Generally, starch granules are approximately 30 μm in diameter; thus the only way to obtain nanoparticles from starch is by size reduction, a top-down approach. Preparation techniques include acid hydrolysis and mechanical treatment. Starch nanoparticles can be potentially used as nanofillers in polymer nanocomposites for packaging applications, adhesives, food additives, and drug carriers (LeCorre et al. 2010). Starch nanoparticles were successfully fabricated by reactive extrusion with glyoxal as a cross-linking agent. The particle size varied with time and temperature applied in the extrusion process, with the smallest particle being 160 nm. In suspension form, starch nanoparticles exhibited greater viscosity than starch with larger particles. The viscosity of starch nanoparticle suspension tended to be stable over a wide range of shear rates compared to the viscosity of native starch suspensions (Song et al. 2011).

Bio-based nanoparticles can also be formed via a bottom-up approach as a self-assembling process. Chitosan nanoparticles were formed with modified lecithin. Nanoparticles were obtained from vigorously mixing the chitosan polymer and modified lecithin at a specific ratio and pH, resulting in positively charged nanoparticles with diameters ranging between 123 and 250 nm, enabling encapsulation of hydrophilic compounds (Chuah et al. 2009).

Phytoglycogen, a carbohydrate-based polymer from maize mutant *sugary-1 (su1)*, has been reported to form high-density solid nanoparticles (Huang and Yao 2011). Phytoglycogen was hydrolyzed with β-amylose to obtain intact outer-layer nanoparticles and then spontaneously subjected to succinate or an octynyl succinate substitution to obtain negatively charged nanoparticles with particle sizes ranging from 30 to 100 nm (Bi et al. 2011). These nanoparticles were used as carriers for an antimicrobial agent, nisin, and the inhibition of *Listeria monocytogeneses* by nisin-loaded phytoglycogen nanoparticles was monitored. Based on the deep-well model, nisin-loaded phytoglycogen nanoparticles retained nisin activity for a longer period of time as a result of the electrostatic and hydrophobic interactions of nisin and the surface of phytoglycogen-derived nanoparticles (Bi et al. 2011).

3.2.2 Food Protein and Synthetic Nanotubes

Due to their great surface area, nanotubes have become increasingly attractive for several applications. Most of the research on nanotubes focuses on the use of carbon-based nanotubes, while some bio-based nanotubes are gaining increased attention and importance. What follows are the research topics related to the applications of nanotubes in food science and technology.

Nanotubes can be formed from bio-based materials via a self-assembly process. Understanding of the conditions that controls the self-assembly process of proteins allows manipulation of proteins' functionalities. The self-assembly process of proteins is considered a bottom-up approach in the formation of nanostructures (Whitesides and Boncheva 2002). Specific proteins or polypeptides can self-assemble into protein nanotubes. The fabrication of partially hydrolyzed α-lactalbumin nanotubes has been investigated extensively (Graveland-Bikker and de Kruif 2006; Graveland-Bikker et al. 2006, 2009). Under certain hydrolysis conditions of α-lactalbumin using the enzyme protease from *Bacillus licheniformis* in the presence of di- or trivalent ions, hydrolyzed α-lactalbumins assemble into hollow protein nanotubes about 20 nm in the outer diameter, 8.7 nm in the inner cavity, and up to several micrometers in length (Graveland-Bikker and de Kruif 2006; Graveland-Bikker et al. 2009). Structural characterization showed that the protein nanotubes were assembled from the arrangement of dimeric building blocks in such a way that the β-pleated sheets stacked perpendicular to the longitudinal axis (Graveland-Bikker et al. 2009). These nanotubes, made entirely from milk proteins, were stable and fairly rigid and could potentially be used as nanocarriers for delivering drugs, nutrients, and bioactive compounds and in controlled-release applications. Moreover, at low concentrations, protein nanotubes can form a transparent strong gel that is reversible, thereby enabling novel applications, including structural protein-based functional foods (Graveland-Bikker and de Kruif 2006; Loveday et al. 2009).

Aside from self-assembly, hollow cylindrical tubes can be synthesized via layer-by-layer assembly of a protein using polycarbonate nanoporous membranes (pores 400 nm in diameter) as templates. The nanotubes, comprised of six layers of human serum albumin and poly-l-arginine, were dispersible in water. By adding a functional layer inside the nanotubes, the functional nanoparticles with particle sizes smaller than the opening end of the nanotubes were trapped inside the hollow cavity. These findings enable the delivery of functional food-, enzyme-, and virus-containing nanoparticles using biologically friendly vehicles as protein-based nanotubes for food and biomedical applications (Komatsu 2012).

For synthetic nanotubes, the study of the toxicity of single-walled carbon nanotubes (SWNTs) to bacterial cells showed that the direct exposure of *Escherichia coli* K12 to precipitate SWNTs caused cell membrane damage, resulting in inactivation of the bacterium (Kang et al. 2007). Since carbon nanotubes are toxic to cells, their application in food technology is limited to antimicrobial applications. Carbon nanotubes are not the only class of nanotubes that are attractive to food scientists. Halloysite, a clay mineral with a hollow nanotube structure approximately 10–150 nm in diameter, has found application in the areas of biosorbents for heavy metal ions and dyes and as nanofillers in biopolymer matrices. These nanotubes were functionalized with amylose, resulting in an increase in dispersion of the halloysite in a hydrophilic substrate, enabling compatibility with several kinds of hydrophilic polymers (Chang et al. 2011).

3.2.3 Nanodots or Quantum Dots for Imaging of Food Structures

Semiconductor nanocrystals or quantum dots (QDs) have been deployed as markers for biological processes because they are more photostable and brighter than commonly used fluorescent dyes. Besides their excellent optical properties, the ability of QDs to change color depending on their size enables imaging of several biologically targeted molecules at the same time. QD immunolabeling has proven effective for the detection of several pathogenic bacteria beneficial in food safety considerations (Zhao et al. 2009). QDs conjugated with antibodies of pathogenic bacteria, including *Salmonella typhimurium*, *Shigella flexneri*, and *E. coli* O157:H7, were able to selectively bind and facilitate the fluorescence imaging of individual bacteria from a mixture of pathogenic bacteria in a very short time (Fig. 3.1), resulting in cost-effective and rapid food pathogen detection devices (Zhao et al. 2009). QD-based microarrays on 96-well plates were also developed for rapid detection and quantification of *E. coli* O157:H7 with a high sensitivity of approximately 10 cfu/mL without sample enrichment (Sanvicens et al. 2011).

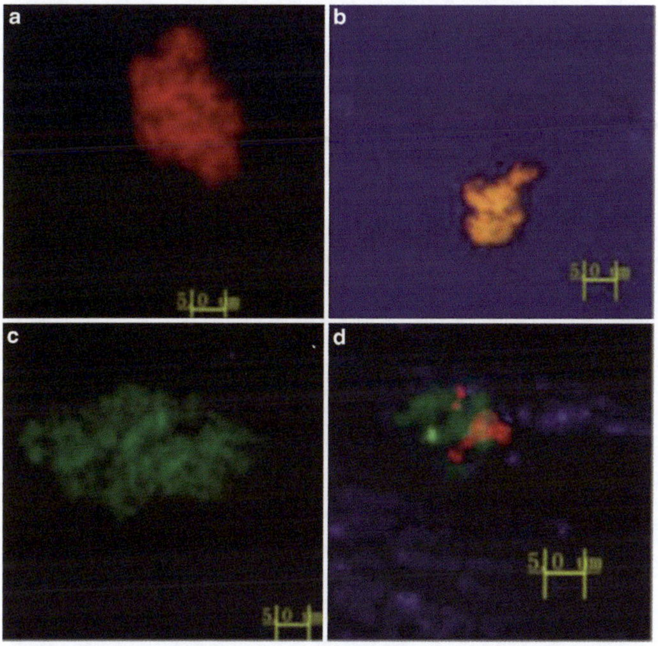

Fig. 3.1 Fluorescent microscopic images of *S. thyphimurium* (**a**), *S. flexneri* (**b**), *E. coli* O157:H7 (**c**), and a mixture of the three (**d**), all conjugated with quantum dots (Reproduced from Zhao et al. 2009 with permission from ACS Publications)

QDs were also used in developing cost-effective enzyme-immobilization techniques. The fabrication of synthetic composite nanofiber QD mixtures by electrospinning showed an ability to promote enzyme immobilization, resulting in a decrease in enzyme activity loss during multiple cycles. The QDs were believed to induce enzyme immobilization in the nanofiber patch (Hwang et al. 2011).

Recently, our laboratory has developed a procedure for visualizing gluten protein distribution in flat bread using QDs. QDs, owing to their unique optical and chemical properties, are excellent imaging probes for long-term, high-quality imaging with the ability to conjugate multiple targets. QDs have several applications in the biomedical field. However, the concept is new for food science research. Before imaging, QDs need to be functionalized to be water soluble (organic capping) and stable (inorganic shell and core), and they must contain biological tags, too, in order to selectively and covalently bind to polymers within a food matrix (Fig. 3.2). The size of the inorganic core determines the color; the inorganic shell improves brightness, whereas the organic capping makes QDs water soluble and stable in buffers. Various covalent or noncovalent binding strategies can be applied to conjugate –COOH, $-NH_2$, or –SH groups present on QD surfaces to the functional groups present on the biomolecules. We developed a detailed protocol for the use of QDs in imaging a gluten matrix by taking advantage of

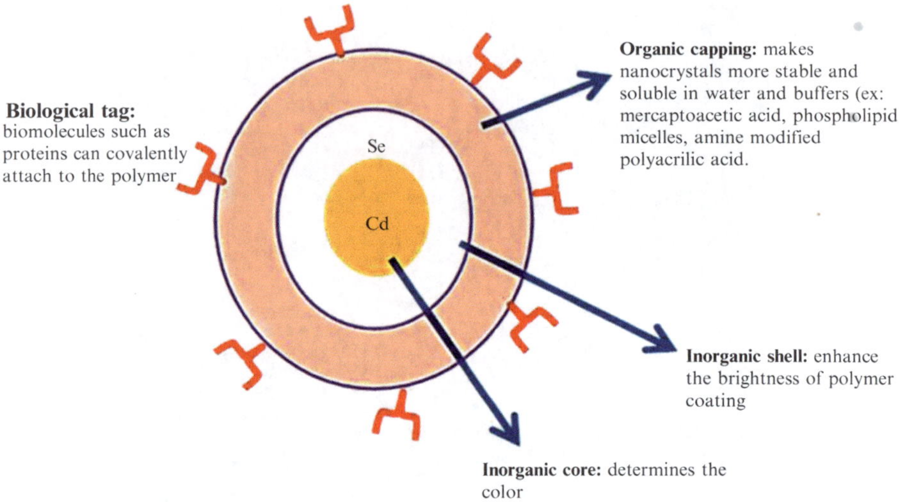

Fig. 3.2 Multifunctional quantum dots can be used to target and image complex systems simultaneously (Sozer and Kokini, unpublished data)

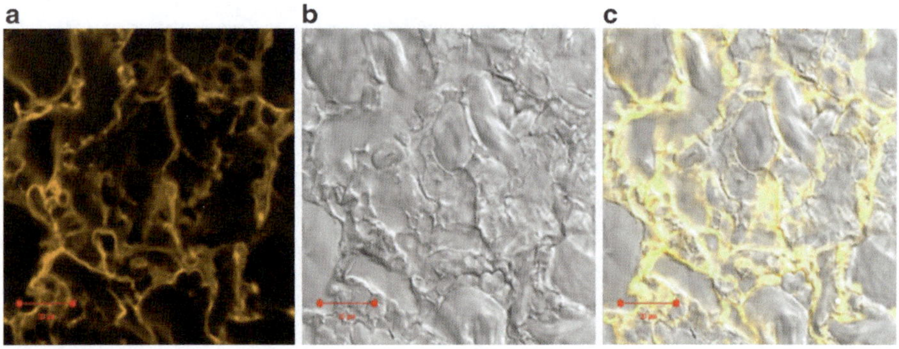

Fig. 3.3 Bread sample labeled with quantum dots: (**a**) fluorescence image, (**b**) differential interference contrast (DIC) image, (**c**) merged fluorescence-DIC image. Combination of fluorescence signals and DIC images used to locate protein network and starch granules (Sozer and Kokini, unpublished data)

bioconjugation principles and rules of immunohistochemistry (Sozer et al. 2012). In our study we used water-soluble CdSe/ZnS core/shell QDs containing carboxyl-terminated groups and covalently bound them to gluten in flat bread samples by EDC cross-linking agent (1-Ethyl-3-[3-dimethylaminopropyl] carbodiimide hydrochloride) (Fig. 3.3). The ongoing research may lead to a better understanding of protein distribution and its morphology within food matrices during and after food processing (Sozer et al. 2012).

3.2.4 Design of Biopolymer Nanocomposite Structures for Green Sustainable Packaging Applications

Polymer-nanoclay nanocomposite technology has been proven to improve the mechanical, gas barrier, and thermal properties of several synthetic polymers and biopolymers for packaging applications. The enhancement is a result of well-dispersed and uniform distribution of 1-nm-thick single layer of nanoclay sheets in polymer matrices (Alexandre and Dubois 2000; Paul and Robeson 2008). This technique has been used extensively in several polymer systems in several commercially available packaging products such as plastic wraps and bottles for carbonated beverages (Maul 2005; Lan 2009). The most widely used nanomaterials for polymer nanocomposite packaging applications are nanoclays and silver nanoparticles.

Nanocomposite techniques have been used with several biobased polymers of food origin such as wheat gluten, soy protein, and zein (Tunc et al. 2007; Angellier-Coussy et al. 2008; Kumar et al. 2010; Luecha et al. 2010). The film fabrication techniques used for the dispersion of nanoclays and fabrication play an important role in the resulting nanocomposite structures that will lead to improvements in physical properties. The types of nanoclay, plasticizer, and other modification reagents are also important. For example, the impact of nanoclay concentration was more pronounced in zein films made by an extrusion blowing technique than by a solvent casting technique. The incorporation of nanoclay into zein films has shown that adequate preparation methods, together with the right chemical affinity between zein and nanoclay, would result in improvements in physical properties (Luecha et al. 2010). Nanocomposite chitosan films with nanoclays combined with silver nanoparticles were found to improve both mechanical and barrier properties as well as antimicrobial activity. These intercalated nanocomposite films were able to inhibit the growth of *Staphylococcus aureus*, *Leuconostoc monocytogenes*, *Salmonella typhimurium*, and *E. coli*, thus creating biodegradable antimicrobial films with improved physical properties for packaging and coating applications (Rhim et al. 2006).

3.3 Functional Foods and Delivery Systems

3.3.1 Nanoencapsulation

Controlled release of active compounds can be achieved by nanoencapsulation in nanosized spheres consisting of micelle, liposome, nanoemulsion, and biopolymeric nanoparticles. A micellar encapsulation technique was used to encapsulate beta carotene in chitosan micelles. It was reported that these nanosized particles were resistant to a physiological environment, which prolonged the life

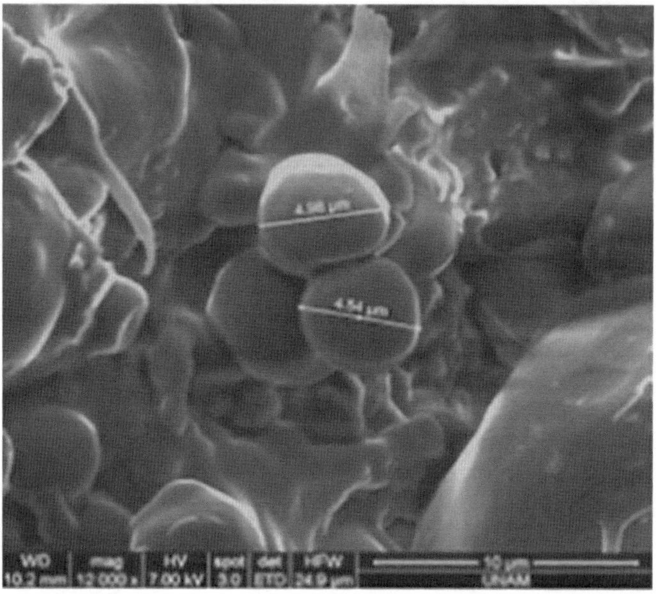

Fig. 3.4 Scanning electron microscopic image of bread crumb with addition of 10 % encapsulated omega-3 complex as spheres approximately 4 μm in diameter (Modified from Gokmen et al. 2011 with permission from Elsevier)

span of active compounds in the bloodstream, resulting in an increase in bioavailability (Huang et al. 2009). The micellar encapsulation technique was employed to produce hydrophobically modified starch micelles for curcumin – an anticancer bioactive compound – encapsulation (Yu and Huang 2010). The encapsulated water-insoluble curcumin showed better water solubility and interaction with cancer cells. Encapsulation of hydrophobic bioactive compounds with starch micelles improved water solubility and can be used for a variety of functional food systems.

High-amylose corn starch nanoparticles were used to nanoencapsulate omega-3 fatty acids added to a bread product. First, the flax seed oil was encapsulated in the starch nanoparticle using a high-pressure homogenizer, followed by freeze drying to obtain a free-flowing nanoencapsulated powder. This powder was then added to the bread formulation at 10 % by weight. The nanoencapsulated particles remained intact inside the bread crumb (Fig. 3.4). It was also shown that the addition of nanoencapsulated powder decreased bread volume expansion during baking and increased the hardness of the bread slightly. However, nanoencapsulation helped reduce lipid oxidation and the formation of thermal oxidation-derived compounds, protecting the omega-3 fatty acid, which could potentially be used for functional food applications (Gokmen et al. 2011).

Zein nanoparticles coated with carboxymethyl chitosan were prepared for encapsulation of vitamin D3. The carboxymethyl chitosan coating layer protected the nanoparticles from photodegradation and improved the controlled release of vitamin D3 in the stomach and the intestinal environment (Luo et al. 2012).

3.3.2 Nanoemulsions

Nanoemulsions, are among the promising colloidal delivery systems that have been employed for the fabrication of several functional ingredients used in food and beverage products. Sucrose monopalmitate, a biologically friendly and biodegradable surfactant, was used for the fabrication of lemon oil nanoemulsion following blending, heating, and homogenization steps. The nanoemulsion was stable in the range of refrigeration temperatures up to room temperatures and started to coalesce at higher temperatures. The nanoemulsion started to agglomerate at high salt concentrations. With a simple fabrication technique, nanoemulsion-containing nontoxic surfactants can potentially play an important role in functional beverages (Rao and McClements 2011).

Thymoquinone, an anticancer and antioxidant phytochemical compound, was loaded to poly(D-L-lactide-co-glycolide) (PLGA) nanoparticles fabricated by an emulsion solvent evaporation technique using anionic molecular micelle surfactants. The thymoquinone nanoparticles were monodisperse with an average particle size of less than 200 nm. The release rate of the thymoquinone nanoparticles was approximately 20 % lower than the free thymoquinone. Moreover, the antioxidant activity and the in vitro cancer cell inhibitory test properties were prolonged through encapsulation by nanoparticles, showing that PLGA nanoparticles are a good candidate for drug and phytochemical delivery systems (Ganea et al. 2010). In addition, the PLGA nanoparticles fabricated via nanoprecipitation using a surfactant made from vitamins C and E were quickly absorbed into the cytoplasm (Astete et al. 2011). To enhance the bioavailability of active compounds encapsulated in nanoparticles for targeted delivery systems, the inclusion of magnetic nanoparticles in PLGA nanoparticles enabled directly targeting these nanoparticles to any organs or any part of the body (Astete et al. 2007).

3.4 Microfluidic Devices for Food Safety and Food Analysis

Microfluidic devices are widely used in the design of both bioanalytical and diagnostic microdevices or chips. Miniaturized chips have several advantages, including that they only consume a small amount of sample and reagents and reduce the requirements for laboratory space, labor, and expertise. The applications of microfluidic devices include mixing and creating nanoemulsions and the detection of contaminants, toxicants, and pathogens in foods. Recently, the demand for environmentally friendly polymers, such as naturally derived polymers, has increased significantly. Zein, a biodegradable polymer of agricultural origin with

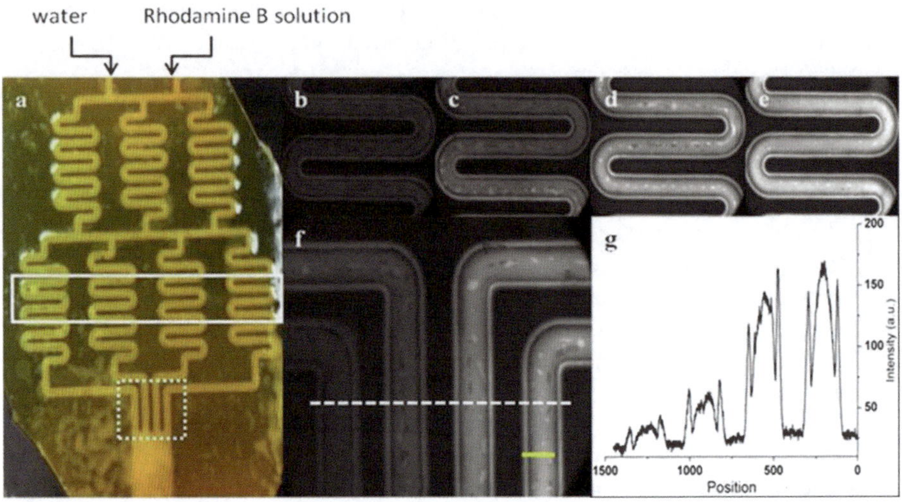

Fig. 3.5 A zein-glass microfluidic concentration gradient generator with two levels of serpentine channels for mixing two different chemical fluids introduced from two inlets (**a**). Fluorescence microscopic images of a second level of serpentine channels (*solid box* in Fig. 3.5a) enclose an increasing concentration of rhodamine B solution (**b–e**). Final concentrations of rhodamine B solution at the exit region of microfluidic mixer (*dashed box* in Fig. 3.5a) (**f**). Fluorescence intensity profile across exit region along *dashed line* in Fig. 3.5f (**g**). A scale bar is 500 μm (Reprinted from Luecha et al. 2011 with permission from RSC Publishing)

interesting properties, was successfully used in the design and fabrication of zein microfluidic devices (Luecha et al. 2011). The zein-based microdevices were successfully fabricated using soft lithography with solvent vapor deposition bonding to create closed channels. We demonstrated that zein microfluidic devices could potentially be used as a mixing component in integrated microfluidic unit operations and for prestoring reagents in microfluidic devices (Fig. 3.5).

3.4.1 Detection of Chemicals, Toxins, and Contaminants

The botulinum toxins are deadly neurotoxins that could be used for bioterrorism. Their rapid detection is very critical for food security. Several attempts have been made to develop microfluidic sensor chips capable of rapid detection of botulinum toxins instead of conventional mouse assays that are time consuming and labor intensive. The botulinum toxin type A was detected using a fluorophore-tagged peptide conjugated onto silica beads embedded in polydimethyl siloxane (PDMS)-based microfluidic channels. The fluorophore-tagged peptide was recognized and cleaved by the toxins, thereby releasing some fluorophore fragments into the buffer solution. The design of this channel directed the solution containing the fragments to another chamber, which allowed quick solvent evaporation, resulting in an

Fig. 3.6 (**a**) The macroscopic image of a microfluidic device that are smaller than a penny. (**b**) Setup of microfluidic device coupled with dielectrophoresis. (**c**) Schematic of series of reagent conjugation for capturing *Listeria monocytogenese* (Reprinted from Koo et al. 2009 with permission from ACS Publications)

increase in the fluorescence intensity. This microfluidic device had a detection sensitivity of approximately 10 pg from 1 mL of buffer solution for the botulinum toxin type A, with a detection time of approximately 30 min (Frisk et al. 2008).

Food adulteration concerns have increased dramatically in several countries around the world. PDMS-based microfluidic devices comprised of four reaction reservoirs were developed for the detection of formaldehyde in food samples. The chemical reaction between formaldehyde and substrates occurring in the reaction reservoirs resulted in violet light absorbed product in which it was measured and recorded using an inverted microscope. These microfluidic devices required just 1 min reaction time with a detection limit of 5 g/kg of food (Weng et al. 2009).

Melamine has been found in tainted milk products. Disposable microfluidic electrophoresis platforms coupled with UV detection units were fabricated on polymethylmetacrylate (PMMA) board. The milk sample was introduced into the microdevices without presample preparation. The detection limit of melamine was 0.23 μg/L of samples in 75 s detection time with no interference from the protein content in the milk (Zhai et al. 2010).

3.4.2 Detection of Microorganisms and Pathogens in Foods

The real-time detection of pathogens in foods has become increasingly important for food safety. Microfluidic biosensor platforms were used to detect *Listeria monocytogenese*. By conjugating on the channel surface a series of reagents with one that was a Listeria receptor protein, the Listeria was captured inside the microfluidic devices with the enhanced localization of bioreceptor using dielectrophoresis (Fig. 3.6). This microfluidic sensor chip was capable of real-time detection of pathogens and specific targeting of only one microorganism in the pool of pathogens (Koo et al. 2009).

Microfluidic platforms for *E. coli* O157:H7 detection integrated several units, such as the immobilization of the *E. coli* inside the microfluidic channels, enzymatic amplification, and optical detection (Li and Su 2005). UV-visible

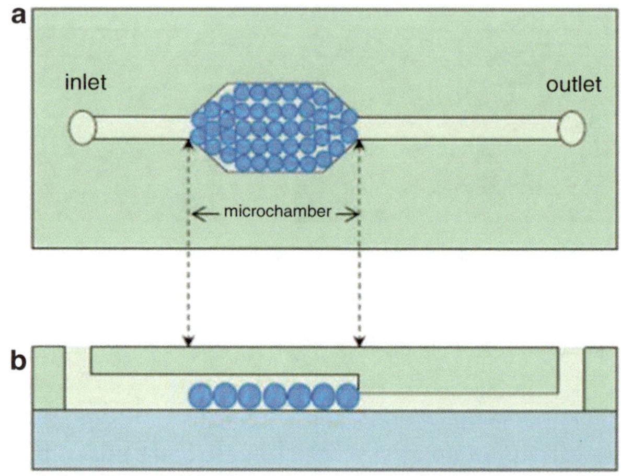

Fig. 3.7 Schematic of microfluidic chip containing glass beads in microchamber (**a**) top view, (**b**) cross-sectional view (Reproduced from Guan et al. 2010 with permission from Springer Science + Business Media, LLC)

spectroscopy was used for measurement of the enzymatic product that was later correlated back to the concentration of the *E. coli*. This method was able to detect *E. coli* in a concentration range of 10–100 cfu/mL in 2 h. Another method utilized microfluidic devices to detect *E. coli* food samples made from a combination of a glass substrate and PDMS cover, as shown in Fig. 3.7 (Guan et al. 2010). These integrated microfluidic platforms consisted of an immunocapture unit using antibody-modified glass beads, an ATP bioluminescence unit using a commercial cell viability assay, the transportation of reagents and sample units using pumps, and a detection unit with a photodiode and a camera. The technology used in the integrated microfluidic devices yielded highly accurate, rapid, sensitive, and reproducible detection of *E. coli* without sample preparation. These reliable microfluidic devices had a detection limit ranging from 10^1 to 10^5 of viable bacteria/μL sample with a short detection time of approximately 20 min (Guan et al. 2010).

3.5 Concluding Remarks

Nanotechnology is having a noticeable impact on food technology. Advanced nanostructure fabrication methods allow the formation of nanoparticles and nanotubes from polymers of food origin, enabling novel edible materials that can be used in several applications. Quantum dots assist chemical and biological material detection by their powerful imaging ability. Nanoemulsion and nanoencapsulation technologies are good tools for producing high-bioavailability-encapsulated bioactive compounds. The use of nanoparticles, such as carbon nanotubes and nanoclays, results in many useful functionalities. Miniaturized microfluidic systems enable portable and disposable microchips for rapid detection and diagnosis. Nanotechnology-enabled food science and technology research will continue to advance to provide sustainable solutions for food-related issues and challenges in the fast growing world.

References

Alexandre M, Dubois P (2000) Polymer-layered silicate nanocomposites: preparation, properties and uses of a new class of materials. Mater Sci Eng: R: Rep 28(1–2):1–63

Angellier-Coussy H, Torres-Giner S, Morel M, Gontard N, Gostaldi T (2008) Functional properties of thermoformed wheat gluten/montmorillonite materials with respect to formulation and processing conditions. J Appl Polym Sci 107(1):487–496

Astete CE, Kumar CSSR, Sabliov CM (2007) Size control of poly(d,l-lactide-co-glycolide) and poly(d,l-lactide-co-glycolide)-magnetite nanoparticles synthesized by emulsion evaporation technique. Colloids Surf A Physicochem Eng Asp 299(1–3):209–216

Astete CE, Dolliver DD, Whaley MM, Khachatryan LL, Sabliov CM (2011) Antioxidant poly(d,l-lactide-co-glycolide) acid nanoparticles made with α-tocopheral-ascorbic acid surfactant. ACS Nano 5(12):9313–9325

Bi L, Yang L, Narsimhan G, Bhunia AK, Yao Y (2011) Designing carbohydrate nanoparticles for prolonged efficacy of antimicrobial peptide. J Control Release 150(2):150–156

Bloom DE (2011) 7 billion and counting. Science 333(6042):562–569

Chang PR, Yanfang X, Dongliang W, Ma X (2011) Amylose wrapped halloysite nanotubes. Carbohydr Polym 84(4):1426–1429

Chaudhry Q, Watkins R, Castle L (2010) Nanotechnologies in the food arena: new opportunities, new applications, new concerns. In: Chaudhry Q, Castle L, Watkins R (eds) Nanotechnologies in foods, vol 4, RSC nanoscience & nanotechnology. RSC, Cambridge, UK

Chuah AM, Kuroiwa T, Ichikawa S, Kobayachi I, Nakajima M (2009) Formation of biocompatible nanoparticles via the self-assembly of chitosan and modified lecithin. J Food Sci 74(1):N1–N8

Frisk MA, Berthier E, Tepp WH, Johnson EA, Beebe DJ (2008) Bead-based microfluidic toxin sensor integrating evaporative signal amplification. Lab Chip 11(8):1793–1800

Ganea GM, Fakayode SO, Losso JN, van Nostrum CF, Sabliov CM, Warner IM (2010) Delivery of phytochemical thymoquinone using molecular micelle modified poly (D, L lactide-co-glycolide) (PLGA) nanoparticles. Nanotechnology 21(28):285104

Gokmen V, Mogol BA, Lumaca RB, Fogliano V, Kaplun C, Shimoni E (2011) Development of functional bread containing nanoencapsulated omega-3 fatty acids. J Food Eng 105(4):585–591

Graveland-Bikker JF, de Kruif CG (2006) Unique milk protein based nanotubes: Food and nanotechnology meet. Food Sci Technol 17(5):196–203

Graveland-Bikker JF, Schaap IA, Schmidt CF, de Kruif CG (2006) Structural and mechanical study of a self-assembling protein nanotube. Nano Lett 6(4):616–621

Graveland-Bikker JF, Koning RI, Koerten HK, Geels RBJ, Heeren RMA, de Kruif CG (2009) Structural characterization of α-lactalbumin nanotubes. Soft matter 5(10):2020–2026

Guan X, Zhang H, Bi YN, Zhang L, Hao D (2010) Rapid detection of pathogens using antibody-coated microbeads with bioluminescence in microfluidic devices. Biomed Microdevices 12(4):683–691

Huang Y, Yao Y (2011) Particulate structure of phytoglycogen nanoparticles probed using amyloglucosidase. Carbohydr Polym 83(4):1665–1671

Huang YP, Yu HL, Guo L, Huang QR (2009) Preparation, characterization, and applications of octanoyl-chitosan-polyethylene glycol monomethyl ether amphiphile. The 238th ACS National meeting, Washington, DC, 16–20 Aug 2009. AGFD220

Hwang ET, Tatavarty R, Lee H, Kim J, Gu MB (2011) Shape reformable polymeric nanofibers entrapped with QDs as a scaffold for enzyme stabilization. J Mater Chem 21(14):5215–5218

Kang S, Pinault M, Pfefferle LD, Elimelech M (2007) Single-walled carbon nanotubes exhibit strong antimicrobial activity. Langmuir 23(17):8670–8673

Kirby CJ (2011) Nanotechnology in the food sector. In: Brennan JG, Grandison AS (eds) Food processing handbook, 2nd edn. Wiley-VCH Verlag GmbH & Co. KGaA, Weinheim

Komatsu T (2012) Protein-based nanotubes for biomedical applications. Nanoscale 4(6):1910–1918

Koo OK, Liu YS, Shuaib S, Bhattacharya S, Ladisch MR, Bashir R, Bhunia A (2009) Targeted capture of pathogenic bacteria using a mammalian cell receptor coupled with dielectrophoresis on a biochip. Anal Chem 81(8):3094–3101

Kumar P, Sandeep KP, Alavi S, Truong VD, Gorga RE (2010) Effect of type and content of modified montmorillonite on the structure and properties of Bio-nanocomposite films based on soy protein isolate and montmorillonite. J Food Sci 75(5):N46–N56

Lan T (2009) Nanocomposite materials for packaging films applications. Symposium on nanomaterials for flexible packaging presentation. PLACE flexible packaging summit, Columbus. http://www.tappi.org/content/events/09PLACESY/Symp_Papers/lan.pdf

Le Corre D, Bras J, Dufrene A (2010) Starch nanoparticles: a review. Biomacromolecules 10(5):1139–1153

Li Y, Su XL (2005) Microfluidic-based optical biosensing method for rapid detection of *Escherichia Coli* O157:H7. J Rapid Method Autom Microbiol 14(1):96–109

Loveday SM, Rao MA, Creamer LK, Singh H (2009) Factors affecting rheological characteristics of fibril gels: the case of β-lactoglobulin and α-lactalbumin. J Food Sci 74(3):R47–R55

Luecha J, Sozer N, Kokini JL (2010) Synthesis and properties of corn zein/montmorillonite nanocomposite films. J Mater Sci 45(13):3529–3537

Luecha J, Hsiao A, Brodsky S, Liu GL, Kokini JL (2011) Green microfluidic devices made from corn proteins. Lab Chip 11(20):3419–3425

Luo Y, Teng Z, Wang Q (2012) Development of zein nanoparticles coated with carboxymethyl chitosan for encapsulation and controlled release of vitamin D3. J Agric Food Chem 60(3):836–843

Maul P (2005) Barrier enhancement using additive. PIRA international conference, Brussel. http://www.nanocor.com/tech_papers/BARRIER%20ENHANCEMENT%20USING%20ADDITIVES%20110605.pdf

Paul DR, Robeson LM (2008) Polymer nanotechnology: nanocomposites. Polymer 49:3187–3204

Rao J, McClements DJ (2011) Food-grade microemulsions, nanoemulsions and emulsions: fabrication from sucrose monopalmitate and lemon oil. Food Hydrocoll 25(6):1413–1423

Rhim JW, Hong SI, Park HM, Ng PKW (2006) Preparation and characterization of chitosan-based nanocomposite films with antimicrobial activity. J Agric Food Chem 54(16):5814–5822

Roco MC (2011) The long view of nanotechnology development: the national nanotechnology initiative at 10 years. Nanotechnology research directions for societal needs in 2020. Sci Policy Rep 1:1–28

Sanvicens N, Pascual N, Fernandez-Arquelles MT, Adrian J, Costa-Fernandez JM, Sanchez-Baeza F, Sanz-Medel A, Marco MP (2011) Quantum dot-based array for sensitive detection of *Escherichia Coli*. Anal Bioanal Chem 399(8):2755–2762

Song D, Yonathan ST, Deng Y (2011) Starch nanoparticle formation via reactive extrusion and related mechanism study. Carbohydr Polym 85(1):208–214

Sozer N, Kokini JL (2009) Nanotechnology and its applications in the food sector. Trends Biotechnol 72(2):82–89

Sozer N, Sivaguru M, Kokini JL (2012) Use of quantum nanodot crystals as imaging probes for cereal proteins. J Food Eng (submitted)

Tunc S, Angellier H, Cahyana Y, Chalier P, Gontard N, Gastaldi E (2007) Functional properties of wheat gluten/montmorillonite nanocomposite films processed by casting. J Membr Sci 289 (1–2):159–168

Weng X, Chan HC, Jiang H, Li D (2009) Rapid detection of formaldehyde concentration in food on a polydimethylsiloxane (PDMS) microfluidic chip. Food Chem 114(3):1079–1082

Whitesides GM, Boncheva M (2002) Beyond molecules: Self-assembly of mesoscopic and macroscopic components. Proc Natl Acad Sci USA 99(8):4769–4774

Yu H, Huang Q (2010) Enhanced in vitro anti-cancer activity of curcumin encapsulated in hydrophobically modified starch. Food Chem 119(2):669–674

Zhai C, Qiang W, Sheng J, Lei J, Ju H (2010) Pretreatment-free fast ultraviolet detection of melamine in milk products with a disposable microfluidic device. J Chromatogr A 1217(5):785–789

Zhao Y, Yi M, Chao Q, Jie N, Ge Y, Shen H (2009) Simultaneous detection of multifood-borne pathogenic bacteria based on functionalized quantum dots coupled with immunomagnetic separation in food samples. J Agric Food Chem 57(2):517–524

Chapter 4
Relaxations, Glass Transition and Engineering Properties of Food Solids

Yrjö H. Roos

4.1 Introduction

The engineering properties of solids in food materials are highly dependent on their physical state, i.e. amorphous non-crystalline, crystalline or liquid. Variations of food material properties and states may occur as a result of changes in external thermodynamic conditions, such as pressure and temperature, and within materials because of changes in plasticiser or solvent (water) contents. The food polymer science approach introduced by Levine and Slade (1986) has been successful in explaining the time-dependent characteristics of amorphous food components and cryostabilisation in the manufacture of frozen foods. However, complex food systems are composed of numerous miscible, partially miscible, immiscible, partially crystalline and partially amorphous components. This makes the understanding and control of the properties of individual food systems in various processing and storage conditions very challenging and different from the behaviour of synthetic polymers and somewhat similar pharmaceutical preparations with more definite characteristics.

Glass-transition data have been published for numerous food components, primarily carbohydrates (Roos 1993) and proteins (Aguilera et al. 1993), as well as food solids, such as milk (Jouppila et al. 1997) and apples (Bai et al. 2001), which typically include carbohydrates, proteins, lipids and minor components in complex and often cellular, local and specific structures. The glass-transition data of complex food systems with highly heterogeneous microstructures can be problematic as most glass transitions measured result from those of single or miscible components. More importantly, water as a plasticiser can be differently distributed within various components. Typical glass transition measurements use differential scanning calorimetry (DSC), which gives global glass transitions for food solids but

Y.H. Roos (✉)
School of Food and Nutritional Sciences, University College Cork, Cork, Ireland
e-mail: yrjo.roos@ucc.ie

S. Yanniotis et al. (eds.), *Advances in Food Process Engineering Research and Applications*, Food Engineering Series, DOI 10.1007/978-1-4614-7906-2_4,

cannot measure changes in individual components responsible for thermal properties. The mechanical properties of food systems have been measured by dynamic mechanical analysis (DMA) and related to dielectric properties measured by dielectric analysis (DEA) (Moates et al. 2001). These measurements give information on dielectric or mechanical relaxations alpha-relaxation shows relaxation times above the glass transition. Glass-transition measurements using DSC primarily determine the glass-transition temperature, T_g, its temperature range and the magnitude of the change in heat capacity over the glass transition (ΔC_p). DSC may also be used to determine enthalpy relaxations associated with a glass transition and to show the dynamic effects of molecular packaging and aging on the material response to plasticisation at a glass transition. Enthalpy relaxations reflect changes in molecular mobility around the glass transition and their effects on the translational diffusion of the non-crystalline phase components above the glass transition. Hence, rates of diffusion-controlled reactions involving glass components may show increasing rates above the T_g. A typical diffusion-controlled property above the glass transition is the crystallisation of the non-crystalline components, such as lactose in milk-based powders. It has also been recognised that the flow properties and stickiness of powders are affected by the glass transition (Bhandari and Howes 1999). Glass-transition measurements emphasise information about the temperature range over which dramatic changes in material properties may occur, but no information on the extent of changes in material characteristics in specific conditions, e.g., during food processing and storage, can be obtained.

The DMA and DEA measurements of material properties show relaxation times above the glass transition. The relaxation time at the onset temperature of the change in heat capacity measured by DSC is 100 s and corresponds to the viscosity of 10^{12} Pa s typical of the solid, glassy state of materials (Roos 1995; Angell 1991). Furthermore, it is well known that the glass transition and material properties related to the glass transition are time-dependent (Roos 1995), i.e., rates of various changes above the T_g may correlate with the structural relaxation times. The changes in relaxation times above the glass transition according to the Williams-Landel-Ferry relationship was emphasised by Levine and Slade (1986), and the 'fragility' concept developed by Angell (1991) has aimed at showing the structural relaxation properties of 'strong' and 'fragile' glass formers above the glass transition. Our approach has been to develop knowledge on the dielectric and mechanical properties, and the corresponding structural relaxation times, of food systems at temperatures around and above the calorimetric glass transition. The present review will address the glass transition, relaxations and engineering properties of food components and real food systems taking into account the complexity of food systems and the effects of various components and their miscibility on food processing and storage stability.

4.2 Relaxation Times and Fluidness

The glass transition is an important property of food solids in such processes as dehydration, extrusion and freezing and helps to understand the properties of confectionary, edible films and frozen foods, for example. More importantly, several delivery systems for sensitive food components and pharmaceuticals use non-crystalline solids as protective matrices, or the active components need to remain non-crystalline for rapid release and uptake. It is also important to note that reactions that control the stability of biological materials require the presence of a non-crystalline state of reactants and may therefore be controlled by the physical state and plasticisation of the substances. This often means understanding the properties of food solids at low or intermediate water contents where water is a strong plasticiser (Levine and Slade 1986). Non-crystalline solids form stable glassy states, and vitrification is often a prerequisite for the success of stabilisation of dehydrated and frozen materials. Thermal and water plasticisation that results in a glass transition and induce translational molecular mobility causes structural 'fluidness' as structural relaxation times decrease at and above the glass transition. Fluid, fluidness and fluidity are used here as terms referring to the characteristics of glass-forming materials above their glass transitions and associated material-specific changes in structural relaxation times in their super-cooled liquid states. This is somewhat different from the 'fragility' concept developed by Angell (1991), which is defined as the steepness of the T_g-scaled Arrhenius plot near the glass transition or sharpness of the glass transition (Angell 2002). Such fragility and material fluidness show similarities, but fluidness refers clearly to a super-cooled liquid state while non-crystalline, solid glasses are generally mechanically fragile without translational mobility or fluidness. Fluidness can also relate to various important properties of glass-forming food materials, including reactivity, collapse, caking, stickiness and various other phenomena which occur above the glass transition. It also takes into account the non-Arrhenius behaviour and temperature dependence of the activation energy of super-cooled liquids at and above the glass transition.

Glass transition is a universal property like other changes in the state of materials, i.e. it refers to the non-equilibrium liquid and solid forms of the super-cooled state typical of inorganic glass formers, synthetic polymers and sugars (Angell 2002), and it is a property of materials existing below their equilibrium melting temperature with no well-defined structure. The glass transition in all materials shows complexity due to simultaneous rapid changes in kinetic and thermodynamic processes. Therefore, individual materials may show an indefinite number of molecular arrangements and glass structures with varying levels of molecular packing and order. The extent of molecular packing and order is a result of the rate of molecular 'freezing' towards the glassy state during cooling or annealing or aging processes at conditions allowing slow molecular arrangements in the vicinity of the glass transition. In food systems, the molecular arrangements may also change as a result of fluctuations in temperature, water content and water activity, and it may vary within the food microstructure. The non-equilibrium properties of glass-forming materials underpin, for example, the enthalpy and volume relaxations occurring around the glass transition as well as aging processes below the glass transition.

Vitrification of materials at the glass transition differs significantly from melting and crystallisation, which are well-defined thermodynamic changes in systems with high chemical purity, such as metals. Glass transition has no latent heat and is a property of pure materials and of mixtures of two or more miscible food components. A glass transition in food systems at a well-defined water content can be a specific, complex property of numerous food components. However, most glass-transition data available are limited to binary mixtures of single food components and water or food solids with poor compositional characterisation. The glassy state and properties of super-cooled liquids above the glass transition have been of great interest across materials science because of the fundamental effects of the glass transition on the physicochemical properties of materials, and particularly the effects of glass transition on the material fluidness above the glass transition. The non-equilibrium characteristics of the super-cooled liquid and glassy states of materials underscore the importance of understanding the time-dependent nature of glass-forming food solids. The relative rates of change in the glassy and super-cooled liquid states are related to structural relaxation times that indicate molecular mobility and its variations at different temperatures, pressures and levels of water plasticisation. Such relaxation times may be derived from measurements of dielectric and rheological properties, such as complex moduli and dielectric loss and permittivity. Relative relaxation times may also use measurements of viscosity and other molecular-mobility-dependent parameters.

Williams et al. (1955) pointed out a general problem in the use of a common reference temperature, such as 25 °C, to describe changes in glass-forming material properties. They found that mechanical and dielectric relaxation times above the glass transition followed a simple, universal relationship when relaxation times of various glass-forming materials were compared. A common, arbitrarily chosen reference temperature of approximately 50 K above the glass-transition temperature, T_g, was used as a reference temperature, T_s. Although the reference temperature $T_g = T_s$ could be used, T_s was preferred because of the difficulties of measurements of relaxation times at the glass transition, the problems of varying T_g values caused by residual solvents and plasticisers, and thermal history effects on the T_g measurements. The Williams–Landel–Ferry (WLF) relationship (4.1) (Williams et al. 1955) was derived from the general behaviour of most inorganic and organic glass formers that showed a similar temperature dependence of relaxation times and viscosity over a temperature range of T_g to T_g + 100 K (Fig. 4.1). This was different from the parent Vogel–Tammann–Fulcher (VTF) relationship (4.2), which used a reference temperature, T_o, well below the T_g (Williams et al. 1955).

It is important to note that the VTF relationship has a form that is similar to that of the Arrhenius equation, but the VTF model uses a T_0 as a reference temperature instead of 0 K and allows determination of apparent activation energy. Also, the WLF equation can be written in the form of the VTF relationship, as shown by

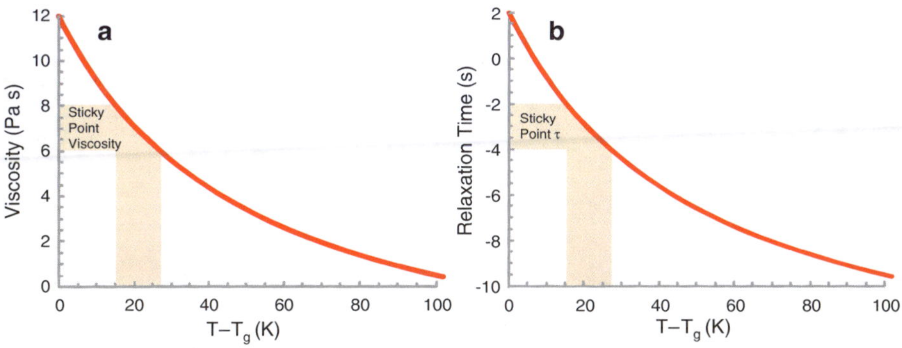

Fig. 4.1 Viscosity (**a**) and relaxation time, τ (e.g. dielectric relaxation) (**b**) above the glass transition temperature, T_g, predicted by the Williams–Landel–Ferry (WLF) relationship with the universal constants $C_1 = -17.44$ and $C_2 = 51.6$ K

Eqs. 4.1–4.3, but T_0 is replaced by T_s or T_g and the constant B is replaced by a $T - T_s$ or $T - T_g$-dependent parameter that includes the constants C_1 and C_2. An assumption of $B = 2.303\ C_1C_2$ has been suggested (Angell 1997), but this assumption is valid only at large $T - T_g$ values of >100 °C, where deviation from the Arrhenius temperature dependence of structural relaxation times is minimal. The activation energy in the WLF equation is highly temperature and material dependent (Williams et al. 1955) and follows the deviation of the temperature dependence of structural relaxation times from the Arrhenius behaviour. As discussed by Angell (1997), the deviation from the Arrhenius behaviour in the WLF equation is measured by the C_2 constant. Although the problems of using universal reference temperatures are well known, as described by Williams et al. (1955), the fragility concept introduced by Angell (1991) uses individual T_g values on the absolute temperature scale and plots of relaxation times against T_g/T. Such plotting uses 0 K as a reference temperature for all materials and identifies the T_g at $T_g/T = 1$ for each material independent of its specific numeric value on the absolute temperature scale. This shows the logarithmic values of the structural relaxation times against T_g-multiplied activation energies (log τ vs. $T_g \times 1/T$), indicating that the fragility is T_g value dependent and that materials with a similar fluidness but different T_g also have different fragilities. Therefore, materials with the same activation energy in the vicinity of T_g also have different fragilities if their T_g values differ (Fig. 4.2). The fragility concept has successfully pointed out differences in the structural relaxation times of various materials above the T_g, but it does not seem to measure the fluidness of biological materials close to the glass transition, which is important for understanding glass-former properties in the food and pharmaceutical industries.

The use of viscosity and relaxation time models assumes that the viscosity of the super-cooled liquid state approaches 10^{12} Pa s at the glass transition and the dielectric relaxation time becomes approximately 100 s at the onset of the calorimetric glass transition (Angell 1991). The WLF constants, C_1 and C_2, may be

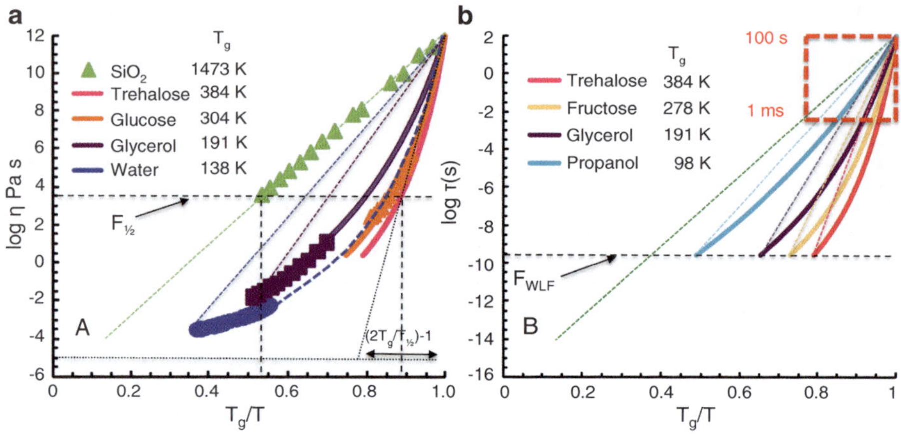

Fig. 4.2 Viscosity, η (**a**) and relaxation time, τ, as predicted by the Williams–Landel–Ferry (WLF) relationship (**4.1**) (**b**) above the glass transition temperature, T_g, shown in 'T_g-scaled' Arrhenius plots using the abscissa of T_g/T according to Angell (1991) with experimental viscosity data for glucose (Parks and Gilkey 1929), glycerol (Segur and Oberstar 1951) and water (Hallett 1963). In accordance with Angell (1991), fragility is shown as the $F_{½} = (1-T_g/T)$ fragility or 'steepness index', m = $[m_{min}(1 + F½)/(1-F½)]$, where F_{min} is 16 for τ_{min} and 17 for η_{min}. F_{WLF} refers to the WLF-predicted τ at $T-T_g$ = 100 K. The apparent differences in $F_{½}$ and F_{WLF} shown result only from differences in the individual T_g values of the substances. Plotting of the data against viscosity against $T-T_g$ gives a single curve (Fig. 4.1)

assigned 'universal' values of −17.44 and 51.6, respectively, when $T_s = T_g$ is used as the reference temperature (Williams et al. 1955). Although the WLF relationship may fit to viscosity and relaxation time data above the glass transition, it is obvious that changes around the glass transition occur gradually showing an upward concavity when plotted against $T-T_g$, as shown by Peleg (1996). This problem, however, is not apparent, and downward concavity is obtained when the data are plotted against $1/T-T_g$ (4.4). In such plots, η_g approaches the high viscosity of 10^{12} Pa s for the glassy state when $T-T_g$ approaches 0 K. When viscosity or relaxation time data are shown isothermally against water activity or water content, the Fermi model proposed by Peleg (1996) may be preferred because of its simplicity and parameters may be used to measure the broadness and extent of the changes in stiffness at the glass transition:

$$\log a_T = \log\left(\frac{\tau}{\tau_s}\right) = \log\left(\frac{\eta}{\eta_s}\right) = \frac{-C_1(T-T_s)}{C_2+(T-T_s)}, \tag{4.1}$$

$$\eta = \eta_0 e^{B/(T-T_0)}, \tag{4.2}$$

$$\ln \eta = \ln \eta_0 + B\frac{1}{T-T_0}, \tag{4.3}$$

$$\log \eta = -\left(C_1 - \log \eta_g\right) + \left(\frac{C_1}{\frac{1}{C_2} + \frac{1}{T-T_g}}\right) \frac{1}{T - T_g}, \tag{4.4}$$

where a_T is the ratio of the relaxation times, τ and τ_s, or viscosities, η and η_s, at temperature T and a reference temperature, T_s, respectively, and C_1, C_2, and B are constants. T_g refers to the onset temperature of glass transition in a differential scanning calorimetry heating scan.

The use of the glass-transition-anchored reference temperature showing the effects of the glass transition of various substances to their structural relaxation times is important for comparisons of various glass-forming materials as mobility-related properties change above the glass transition. The changes appear very different if T_g/T plotting instead of a specific, individual glass-transition-related reference temperature is used for the various glass-forming materials. The WLF equation is a simple approach because it makes it possible to plot relaxation times (or viscosity) against $(T - T_g)$. As shown by Eq. 4.4 and pointed out by Angell (1997), the WLF constant C_1 is a 'scaling parameter' defining the number of logarithmic decades for the change in relaxation time or viscosity above the glass transition. For a relaxation time of 100 s at the onset of glass transition, this means that the relaxation time approaches 10^{-C_1-2} s as $T - T_g$ approaches infinity, and the use of $C_1 = 16$ was suggested by Angell (1997). Alternatively, the viscosity approaches 10^{-C_1-12} when $T - T_g$ approaches infinity, with $C_1 = 17$. The C_2 of the WLF relationship can be taken as the 'fluidness' parameter, which becomes different depending on the changes in structural relaxation times above the material-specific T_g.

The classification of glass-forming materials as 'strong' and 'fragile', as developed by Angell (1991), aims at using a fragility parameter (slope), m, as a measure of the deviation from the Arrhenius temperature dependence above at their respective glass-transition temperature. However, as shown in Fig. 4.2, the fragility parameter fails to compare glass-forming materials if they show differences in T_g. Also, the same substance can have different fragilities depending on the level of plasticisation simply due to the artefact of differences in T_g values. Strong liquids are those following the Arrhenius relationship, e.g. SiO_2, while fragility increases with increasing deviation from the linearity of relaxation times against the reciprocal temperature (Fig. 4.2). Fragility may also be derived from the parameter D of the modified VTF relationship (4.5) as $F_{VTF} = 1/D$, which varies from 0 to 1. Fragility has also been defined as $F_{1/2}$ fragility (Angell 2002), as shown in Fig. 4.2 for viscosity:

$$\tau = \tau_0 e^{DT_0/(T-T_0)} \tag{4.5}$$

The fragility approach with experimental data shows that SiO_2 is a strong glass former with Arrhenius behaviour above the T_g. Organic glass formers and water appear to be highly fragile glass formers. Although the fragility indexes in Fig. 4.2

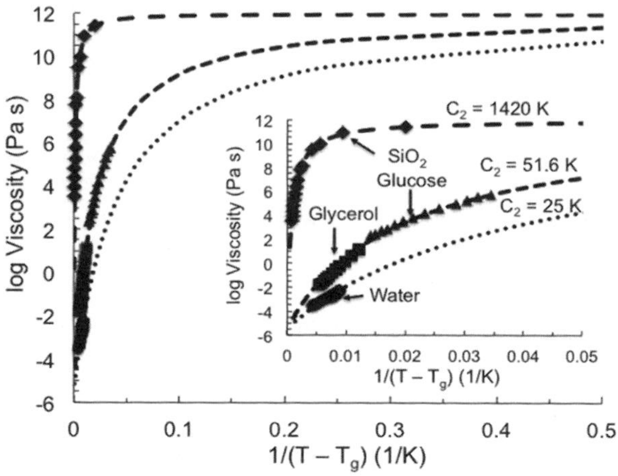

Fig. 4.3 Fluidness plot for common glass-forming materials showing experimental viscosity data of Fig. **4.2** for SiO_2, glucose, glycerol and water in their liquid and super-cooled liquid states. The constant $C_1 = 17.44$ of the WLF equation was used as the 'scaling' parameter and C_2 was used as the fluidness parameter

for glycerol, glucose and trehalose are different, they show similar changes in viscosity and relaxation times above the T_g and should be noted as being equally fluid (Fig. 4.3). The difference in the apparent fragility of these materials, for example, is a serious problem and a limitation of the fragility approach, which results from the differences in the individual T_g values and the T_g/T scaling. Water shows the highest fluidness, and water-plasticised food components may be assumed to increase in fluidness with increasing water plasticisation. Despite these limitations of the fragility concept, it appears quite obvious that the fluidness of food materials increases with increasing water content. Alternatively, as suggested by the 'Arrhenius or VTF'-type WLF relationship of Eq. 4.4, a plot of log τ against $1/T{-}T_g$ shows differences in the structural relaxation times of glass-forming materials around and above their glass transitions. The relationships also indicate that a decreasing C_2 in Eq. 4.4 increases the activation energy and the temperature dependence of the viscosity or structural relaxation times around the glass transition. This is particularly important over the temperature range of 10–20 K typical of the glass transition.

DEA and DMA studies of food materials have shown a significant decrease in relaxation times around the glass transition when derived from the respective frequency-dependent dielectric loss or loss modulus α-relaxation temperatures. Measurements of the relaxation times and effects of food composition on the relaxation properties of amorphous components in complex foods is fundamental for understanding food properties in processes and storage at high solid contents or low temperatures. Our studies have shown that the T_g value of a glass former and particularly the presence of other components and water in food systems are more significant factors affecting food fluidness and properties than the fragility of a single glass former, as defined by Angell (1991).

4.3 Relaxation Times in Food Systems

According to Angell (1991, 2002) relaxation times, such as dielectric relaxation times and calorimetric relaxations at the onset T_g, are typically 100 s and assumed to approach 10^{-14} s at high temperatures (Fig. 4.2). The structural relaxation times of food systems decrease to 10^{-3} s at 20–30 °C above the T_g. This corresponds to a decrease in viscosity from 10^{12} to 10^5 Pa s, which agrees with critical viscosities for collapse in freeze-drying (Bellows and King 1973) and stickiness in spray-drying (Downton et al. 1982), as well-known examples.

The WLF relationship has been useful in relating the viscosities of super-cooled liquids to their stickiness (Downton et al. 1982) and times to crystallisation of amorphous sugars (Roos and Karel 1990). The assumption made is that the universal WLF constants apply and that the viscosity of the glass-forming liquids above the T_g follows the WLF relationship, as shown in Fig. 4.1. The data in Fig. 4.1 show stickiness at a surface viscosity of 10^6 to 10^8 Pa s, corresponding to a dielectric relaxation time of 10^{-4} to 10^{-2} s and a surface contact time of 1–10 s at the 'sticky point' (Downton et al. 1982). These values were shown to apply to a non-crystalline 7:1 mixture of sucrose and fructose (Downton et al. 1982). The sticky points were found at approximately 20 °C above the onset temperature of the glass transition, T_g, measured by DSC (Roos and Karel 1990). Several other studies have confirmed the relationships of the glass transition and stickiness of amorphous solids (Boonyai et al. 2004).

Dielectric and mechanical analyses of food systems allow for the determination of relaxation times at the α-relaxation, which can be related to the stickiness characteristics of food powders as well as other mechanical changes in food solids, such as collapse of structure or viscosity and diffusion. Dielectric analyses of fructose and glucose show that the α-relaxation time decreases from 10^2 s at the onset of the calorimetric T_g to 10^{-2} to 10^{-3} s at the sticky point at $T-T_g$ of 20 °C.

The crystallisation of amorphous sugars occurs time-dependently above the glass transition. Dielectric α-relaxations suggest that an α-relaxation time of 1–2 s at 10 °C above the onset T_g corresponds to a time to crystallisation of amorphous lactose (Roos and Karel 1992) of 10 days, and α-relaxation times of approximately 10^{-2} s correspond to crystallisation within 30 h.

The relaxation time data show that there is a significant increase in the mobility of carbohydrate components in food materials within the glass-transition temperature range measured by DSC. The most important range to be considered in using the T_g data of sugar-containing food materials in relating glass transition to processing characteristics and storage stability is within the $T-T_g$ of 20 °C, but it might be different for complex food matrices.

4.4 Dielectric and Mechanical Relaxations of Food Solids

Amorphous food components have been studied for glass transitions, dielectric relaxations, mechanical relaxations, spectroscopic properties and various other characteristics showing changes in molecular mobility at and around the glass transition. There are, however, very few studies on the properties of food solids with varying carbohydrate and protein compositions (Fig. 4.4). Our recent approach in studies of the stickiness properties of food solids has been to use materials, particularly dairy-based systems, with various carbohydrate and protein compositions (Silalai and Roos 2010, 2011a, b). These materials have been studied for their sticky points and dielectric and mechanical relaxations at various water activities, a_w. The results have revealed significant differences in the glass transition and relaxation behaviour of systems containing low molecular weight sugars with maltodextrins or proteins. We have shown that carbohydrates and proteins may form phase-separated regions in food systems, e.g. food powders. In a carbohydrate–protein system, the

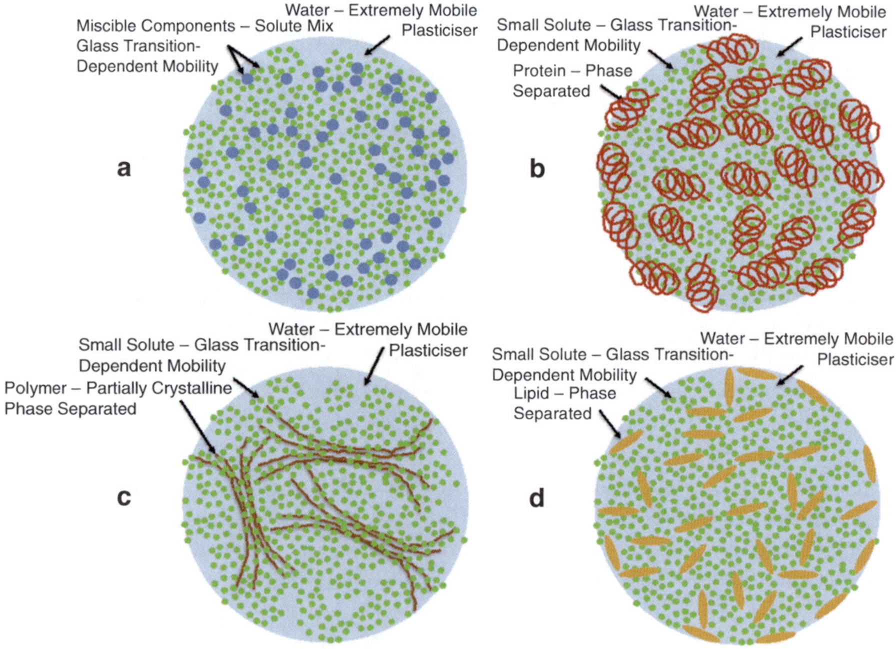

Fig. 4.4 Food components with high solid contents, e.g. in food powders. Miscible components, such as sugars, maltodextrins and water, form amorphous structures showing glass formation according to the component properties (**a**); carbohydrate and protein systems are plasticised by water but show phase separation of protein, and the glass-forming properties are determined primarily by the carbohydrate phase (**b**); high molecular weight components, such as starch, may show partial crystallinity and phase separation from a continuous, low molecular weight, glass-forming carbohydrate matrix (**c**); a lipid phase is phase-separated from a continuous, glass-forming carbohydrate phase (**d**)

carbohydrate phase typically showed an almost protein-content-independent glass transition in a DSC study, and the glass transition approached that of the carbohydrate at increasing water contents (Silalai and Roos 2010). The sticky point was found for skim milk–milk protein solids systems to occur at increasing temperatures above the T_g with rising protein content (Silalai and Roos 2010). The results also showed that although the T_g was at an almost constant, a_w-dependent temperature, stickiness developed at the higher temperature the higher was the protein content leading to a larger temperature difference between the sticky point and glass transition (Silalai and Roos 2011a). The α-relaxation time corresponding to the sticky point decreased with increasing a_w, but there were increases in the temperature difference of the sticky point to T_g with increasing a_w. This could be related to a_w-dependent interactions of the components in water-plasticised systems and availability of the carbohydrate phase for the formation of liquid bridges at particle surfaces (Silalai and Roos 2011a).

Skim milk–maltodextrin solid systems showed quite different properties from those of skim milk–protein systems (Silalai and Roos 2011b). Skim milk–maltodextrin solid systems formed a carbohydrate-rich phase with a maltodextrin-content-dependent T_g (described in Fig. 4.3a, b). The sticky point occurred at approximately 20 °C above the maltodextrin-content- and molecular-size-dependent T_g. These studies showed that stickiness in carbohydrate-protein powders is affected by the carbohydrate, and the sticky point is higher the higher the protein content. In skim milk–maltodextrin solid systems, a higher maltodextrin content gave a higher T_g and the sticky point was at a higher temperature but at an approximately constant $T-T_g$. Similar differences in glass-forming properties and behaviour-affecting characteristics of various other food systems, as shown in Fig. 4.4, may be expected.

4.5 Conclusion

Understanding glass-transition-related relaxations and their coupling with the engineering properties of food materials is essential for the design of complex food and nutrient delivery systems. The macroscopic glass-transition behaviour of food systems may often be misleading in the prediction of characteristics of food components and their storage stability because relaxation times determined for mixtures of carbohydrates and proteins vary and need to be interpreted carefully. The fragility concept, because of its limitations, cannot explain the glass-forming properties of food systems, but studies of relaxations around and above the glass transition give new information on fluidness that can advance innovations in food formulation by mapping the engineering properties of food components and their mixes and the engineering of novel nutrient delivery systems.

References

Aguilera JM, Levi G, Karel M (1993) Effect of water content on the glass transition and caking of fish protein hydrolyzates. Biotechnol Prog 9:651–654

Angell CA (1991) Thermodynamic aspects of the glass transition in liquids and plastic crystals. Pure Appl Chem 63:1387–1392

Angell CA (1997) Why C1 = 16–17 in the WLF equation is physical – and the fragility of polymers. Polymer 38:6261–6266

Angell CA (2002) Liquid fragility and the glass transition in water and aqueous solutions. Chem Rev 102:2627–2650

Bai Y, Rahman MS, Perera CO, Smith B, Melton LD (2001) State diagram of apple slices: glass transition and freezing curves. Food Res Int 34:89–95

Bellows RJ, King CJ (1973) Product collapse during freeze drying of liquid foods. AIChE Symp Ser 69(132):33–41

Bhandari B, Howes T (1999) Implication of glass transition for the drying and stability of dried foods. J Food Eng 40:71–79

Boonyai P, Bhandari B, Howes T (2004) Stickiness measurement techniques for food powders: a review. Powder Technol 145:34–46

Downton DP, Flores-Luna JL, King CJ (1982) Mechanism of stickiness in hygroscopic, amorphous powders. Ind Eng Chem Fundam 21:447–451

Hallett J (1963) The temperature dependence of the viscosity of supercooled water. Proc Phys Soc 82:1046–1050

Jouppila K, Kansikas J, Roos YH (1997) Glass transition, water plasticization, and lactose crystallization in skim milk powder. J Dairy Sci 80:3152–3160

Levine H, Slade L (1986) A polymer physico-chemical approach to the study of commercial starch hydrolysis products (SHPs). Carbohydr Polym 6:213–244

Moates GK, Noel TR, Parker R, Ring SG (2001) Dynamic mechanical and dielectric characterisation of amylose-glycerol films. Carbohydr Polym 44:247–253

Parks GS, Gilkey WA (1929) Studies on glass IV. Some viscosity data on liquid glucose and glucose-glycerol solutions. J Phys Chem 33:1428–1437

Peleg M (1996) On modeling changes in food and biosolids at and around their glass transition temperature range. Crit Rev Food Sci Nutr 36:49–67

Roos Y (1993) Melting and glass transitions of low molecular weight carbohydrates. Carbohydr Res 238:39–48

Roos YH (1995) Phase transitions in foods. Academic, San Diego

Roos Y, Karel M (1990) Differential scanning calorimetry study of phase transitions affecting the quality of dehydrated materials. Biotechnol Prog 6:159–163

Roos Y, Karel M (1992) Crystallization of amorphous lactose. J Food Sci 57:775–777

Segur JB, Oberstar HE (1951) Viscosity of glycerol and its aqueous solutions. Ind Eng Chem 43:2117–2120

Silalai N, Roos YH (2010) Roles of water and solids composition in the control of glass transition and stickiness of milk powders. J Food Sci 75:E285–E296

Silalai N, Roos YH (2011a) Coupling of dielectric and mechanical relaxations with glass transition and stickiness of milk solids. J Food Eng 104(3):445–454

Silalai N, Roos YH (2011b) Mechanical α-relaxations and stickiness of maltodextrin-milk solids systems around glass transition. J Sci Food Agric 91(14):2529–2536

Williams ML, Landel RF, Ferry JD (1955) The temperature dependence of relaxation mechanisms in amorphous polymers and other glass-forming liquids. J Am Chem Soc 77:3701–3707

Chapter 5
Molecular-Based Modeling and Simulation Studies of Water–Water and Water–Macromolecule Interactions in Food and Their Effects on Food Dehydration

J.-C. Wang and A.I. Liapis

5.1 Introduction

From a fundamental perspective, various food materials consist mainly of biopolymeric molecules that form structures with seams and pores in which other food components are physically or chemically accommodated. Water (moisture) is a major component of food materials and can act as a solvent, medium for heat and mass transfer, or participant in chemical or biological reactions in food systems. Water is thus a major factor (Franks 1985; Barbosa-Canoras et al. 2007) in determining the properties of foods in the various stages of processing and packaging (Maroulis and Saravacos 2003; Barbosa-Canoras et al. 2007). Since 1953 food scientists and engineers have employed the concept of equilibrium relative humidity (ERH) or, as it is most often called, water activity, a_w, to construct phenomenological relations for food materials so that they could study as a function of a_w (1) moisture adsorption/desorption, (2) water migration and textural changes, (3) glass transition, (4) microbial growth, (5) degradation rates of chemical and enzymatic reactions, and (6) powder state changes (Chen and Mujumdar 2006; Jayaraman and Das Gupta 2006; Labuza 2011).

From a fundamental point of view, water owes its ubiquity in foods and its effects on food structures and processing to its indispensable and unique interactions with the macromolecules and dissolved solute species of foods. These water interactions are constantly present in foods and play a very important role in determining the structure of the food material at the beginning of a given process and during the processing of the food. They also determine the distributions of water molecules in the food material and in different energetic states, which cause the water molecules to exhibit values of a_w unlike that of the pure bulk water

J.-C. Wang • A.I. Liapis (✉)
Department of Chemical and Biochemical Engineering, Missouri University of Science and Technology, 400 West 11th Street, Rolla, MO 65409-1230, USA
e-mail: ail@mst.edu

S. Yanniotis et al. (eds.), *Advances in Food Process Engineering Research and Applications*, Food Engineering Series, DOI 10.1007/978-1-4614-7906-2_5,

under the same temperature and to be categorized as liquid-like, weakly bound, strongly bound, and nonfreezable in the evolving structure of the food material during processing. Although the phenomenological variable of water activity, a_w, could then be used in continuum material and energy balance models to study the behavior of the food material during processing and packaging, it cannot provide information about the force fields and relative importance of the different interactions in the food material, nor can it provide information with respect to the changes in the physicochemical properties of water and the porous structures of the food during processing or packaging.

Food dehydration (drying) refers to water removal from food and represents a very important process in the preparation and preservation of foods (Maroulis and Saravacos 2003). It is a process that requires the supply of energy to enable water molecules to break loose from their interactions and escape from the porous structures of the foods. The process of food dehydration and the preservation of dehydrated foods can thus be directly related to the interactions of water molecules with the macromolecules of food because the water–macromolecule and the water–water interactions dictate the removal of water molecules during food dehydration, the structures of foods before, during, and after dehydration, and the dehydration strategy and efficiency by which the supplied heat is utilized. However, the water activity variable, a_w, cannot provide the relationship of the evolution of the structural changes of the food material with the changes of the water–macromolecule and water–water interactions that occur during dehydration. This aspect of food science and engineering remains largely unexplored due mainly to the limitations of current experimental techniques and theoretical methods.

Molecular dynamics (MD) is a modeling approach (Allen and Tildesley 1987; Rapaport 1997) that employs molecular-based interaction potentials to construct a model system that provides a satisfactory representation of the real system and allows various properties to be studied at a molecular level. Over the years MD has been demonstrated to be a methodology capable of investigating systems with complex geometries and molecules as well as studying structural and dynamic properties that include the transport and interaction of molecules at interfaces (e.g., adsorption). For multiscale studies, MD is capable of characterizing microscopic, mesoscopic, and even macroscopic porous media, identifying responsible mechanisms and evaluating their relative importance, which can make significant contributions to the formulation of appropriate constitutive equations for use in macroscopic systems described by continuum mechanics models (Riccardi et al. 2009c). These capabilities can be employed to complement both experimental and theoretical approaches and to conceive and design new processes. When properly implemented (Riccardi et al. 2008, 2009a, b, 2010a; Wang and Liapis 2011a, b, c), MD can also provide the capability to probe and obtain detailed information at the molecular level and scientific understanding with regard to the water–macromolecule and water–water interactions in foods. In this respect, our recent MD modeling and simulation studies, which have constructed and employed atomistic and coarse-grain models to investigate the behavior and structure of porous polymeric adsorbent media formed by dextran chains (Riccardi et al. 2009a, b; Wang and Liapis 2011a), can be considered to have demonstrated the efficacy of this new modeling approach for food science and engineering. In this work, we employ MD modeling and simulations to elucidate the water–water and water–macromolecule interactions in model food materials and examine the dependence of such interactions on the evolving pore structures of foods during dehydration.

5.2 System Formulation and Simulation Models and Methods

Polysaccharides are one of the main ingredients in many foods and are also biopolymers of great importance with applications in many fields. Amylose is one of the two major components of starch and is a linear polysaccharide that is made up of D-glucose monomers connected together via α-D(1 $\rightarrow$ 4) bonds (Fig. 5.1a). Molecular model systems constructed with multiple amylose chains that form realistic porous structures and are filled with different water contents can thus be considered physicochemically appropriate for representing food materials and their dynamic stages of dehydration for relevant fundamental studies. Although computer modeling studies of saccharide molecules have been attempted, they were mostly concerned with conformations of single simple saccharide molecules and usually without water (French and Brady 1990). For the purpose of this work, which involves MD simulations and large computational time and length scales, coarse-grain models are considered a sensible choice and an M3B model (Molinero and Goddard 2004) developed for amylose is employed. In the M3B model, each glucose monomer as depicted in Fig. 5.1b is represented by three beads denoted as B1, B4, and B6 and located at the positions of the C1, C4, and C6 carbons. A unique one-to-one correspondence exists between the M3B configurations and the corresponding full atomistic descriptions, which allows structural mapping matrices to be used to convert between the two modeling approaches. In general, the interactions between the M3B beads of the amylose chains can be expressed as a sum of valence and nonbonded contributions, $U_t = U_v + U_{nb}$, where U_t, U_v, and U_{nb} represent the total interaction energy, the valence potential energy for bonds within the same amylose chain, and the nonbonded potential energy between different amylose chains, respectively. In this work, the SHAKE algorithm (Allen and Tildesley 1987; Rapaport 1997) is used to constrain the bond lengths to equilibrium values (Molinero and Goddard 2004) between consecutive M3B beads. As a result, the valence potential energy, U_v, has two contributions: three-body interactions associated with the bending angle between three connected beads, U_b, and four-body interactions associated with the torsion (dihedral) angle between four connected beads, U_{tor}. The angle bending motion is treated by the following harmonic potential, which is a rather standard approach:

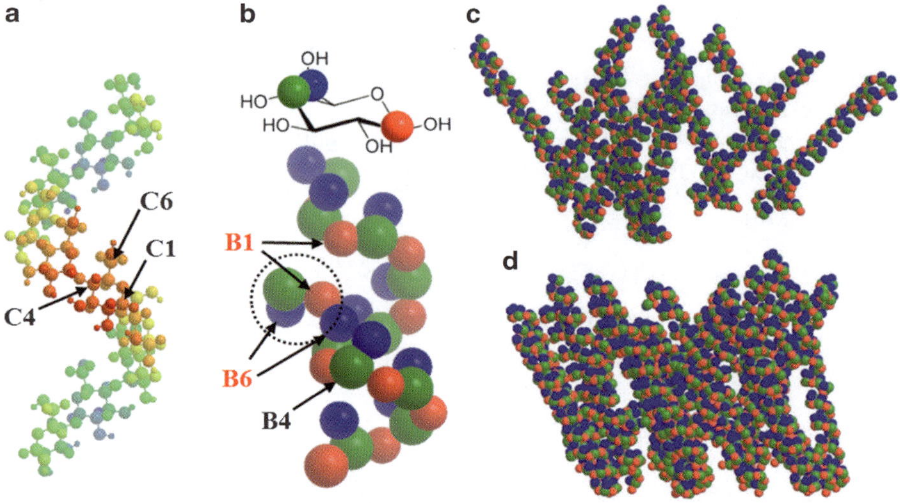

Fig. 5.1 (**a**) Atomistic representation of α-D(1 $\rightarrow$ 4)-linked amylose chain. (**b**) M3B coarse-grain representations of glucose monomer and amylose chain with side branch encircled and linked to main chain via a (1 $\rightarrow$ 6) bond. Side views of initial food porous structures formed by (**c**) 10 amylose chains and (**d**) 20 amylose chains

$$U_{\mathrm{b}}(\theta) = \frac{1}{2} K_{\theta} (\theta_i - \theta_0)^2, \tag{5.1}$$

where θ_i and θ_0 denote the instantaneous and equilibrium values of a bending angle, respectively, and K_θ is a coarse-grain bending force constant. For the four-body valence interactions, each coarse-grain torsion is considered to be represented by a sum of shifted dihedral functions:

$$U_{\mathrm{tor}}(\phi) = \sum\nolimits_{j=1}^{3} \frac{1}{2} B_j \left[1 + \cos(j\phi - \phi_j^0)\right], \tag{5.2}$$

where the coarse-grain torsional angle $\phi = \phi_{inkl}$ denotes the angle between the planes formed by *ink* and *nkl* (*i,n,k,l* are B1, B4, and B6) beads, and each term has a different integral periodicity j, barrier B_j, and phase ϕ_j^0. The non-bonded like and unlike interactions between the amylose M3B beads and the water molecules are considered to be described by the Morse potential,

$$U_{\mathrm{nb}}(r_{ij}) = D_{0,ij} \left\{ \left[\exp\left(-0.5\alpha_{ij} \left(\frac{r_{ij}}{R_{0,ij}} - 1 \right) \right) - 1 \right]^2 - 1 \right\}, \tag{5.3}$$

where r_{ij} is the distance between a pair of M3B beads and/or water molecules, $R_{0,ij}$ is the distance of minimum energy $D_{0,ij}$, and α_{ij} is the steepness parameter of the interaction potential. The mixing rule used for unlike pair interactions is $D_{0,ij} = \sqrt{D_{0,i}D_{0,j}}$, $R_{0,ij} = \sqrt{R_{0,i}R_{0,j}}$, and $\alpha_{ij} = \frac{1}{2}(\alpha_i + \alpha_j)$. The M3B parameters were fitted to reproduce the results from a combination of atomistic simulations and experimental data for the gas phase and amorphous bulk phase of sugars over a wide range of pressures, including such quantities as helical structures, excluded volume interactions, and the distribution of torsional configurations about the inter-monomeric bonds (Molinero and Goddard 2004). The M3B model has also coarse-grained each water molecule to a spherical particle with a proper size (3.77 Å) and interaction strength that are not meant to represent the characteristics of a single water molecule but rather effective values for reproducing the density and interaction energetics of water, which are actually determined by the hydrogen bonds among water molecules. It is worth noting here that similar and realistic porous structures in the extender layers of dextran, a α-D(1 $\rightarrow$ 6)-linked polysaccharide, were obtained using a modified M3B model and the corresponding full atomistic model based on the mapping matrix in our previous studies (Riccardi et al. 2009a, b, 2010a; Wang and Liapis 2011a).

Amylose molecules are composed of 40 monomers per main polymeric chain, of which 5 % of the glucose monomers have side branches that are one monomer long and linked to randomly selected B6 sites because B4 sites have already been used for inter-monomeric bonding. For illustration and comparison, the atomistic and M3B representations of an amylose chain are shown in Fig. 5.1a, b, respectively. The amylose chains in our MD simulation systems are immobilized on a surface and completely or partially immersed in water. Based on our previous studies (Riccardi et al. 2008, 2009a, b, 2010a), the following nonflat surface is suitable for immobilizing amylose chains to create realistic pore structures:

$$U_{is}(x,y,z) = 2\pi\varepsilon_{is}\left(\frac{\sigma_{is}}{R_{0s}}\right)^2 \times \left[\frac{2}{5}\left(\frac{\sigma_{is}}{z_s}\right)^{10} - \left(\frac{\sigma_{is}}{z_s}\right)^4 - \frac{\sqrt{2}}{3\left(\frac{R_{0s}}{\sigma_{is}}\right)\left(\frac{z_s}{\sigma_{is}} + \frac{0.61}{\sqrt{2}}\frac{R_{0s}}{\sigma_{is}}\right)^3}\right], \tag{5.3a}$$

$$z_s = z - a\left[2 + \cos\left(\frac{2\pi}{b}x\right) + \cos\left(\frac{2\pi}{b}y\right)\right], \tag{5.3b}$$

where U_{is} is the interaction potential energy between the nonflat surface and a particle i at (x,y,z) whose size and interaction strength are denoted by σ_i and ε_i. The interaction parameters are determined by the mixing rules, $\sigma_{is} = (\sigma_{is} + R_{0s})/2$ and $\varepsilon_{is} = \sqrt{\varepsilon_i \varepsilon_{0s}}$, where R_{0s} and ε_{0s} represent the nearest distance and the interaction strength, respectively, between the pseudo-monomers adopted as the building blocks of the model surface (Riccardi et al. 2008, 2009a, b, 2010a). The values of a, b, R_{0s}, and ε_{0s} used in this work are the same as in our previous studies (Riccardi et al. 2008, 2009a, b, 2010a). Specifically, 16 trigonometric waves with $a = 4.688$Å as the amplitude parameter and $b = 25$ Å as the wavelength parameter are arranged in a square pattern on a surface whose linear dimensions are 100×100 Å in the lateral directions where periodic boundary conditions are also applied.

Two density cases with 10 and 20 amylose chains immobilized at randomly selected surface sites are studied, and their initial conditions are shown in Fig. 5.1c, d, where the initial orientations of the chain molecular axes are perpendicular locally to the immobilization sites on the sinusoidal surface. Here it should be noted that the B1, B4, and B6 beads in Fig. 5.1b–d, as well as in other simulation snapshots included in this work, are consistently colored in red, green, and blue, respectively. The first two monomers of each chain are held rigid throughout each MD simulation in order to prevent the amylose chains from collapsing onto the surface to result in diminished pore structures. With such a modeling strategy, the use of different values for the sinusoidal parameters a and b can provide an indirect control of the resultant pore structures. The simulations in this work are carried out using the leap frog integration algorithm together with a Gaussian thermostat method for controlling the temperature at 298 K. Before dehydration, the model food systems are filled with water to the highest Z positions (perpendicular to the surface with $Z = 0$ corresponding to the lowest points of the nonflat surface) of the amylose chains and extra water molecules are removed from the top at the beginning of system equilibration. To represent different stages of food dehydration, different amounts of water molecules are removed based on their Z positions from the top and the systems are re-equilibrated before data are collected and analyzed.

5.3 Results and Discussion

Before dehydration, the model food systems with 10 and 20 immobilized amylose chains are hydrated to the highest Z positions of the amylose chains and for simplicity can be labeled as having 100 % water content in the pore spaces. Since the main factor accounting for the differences between the model systems is the density of the food macromolecules, the differences in the energetics of the water molecules examined can be attributed with certainty to the effects of pore structures that result from different numbers of amylose chains occupying the same physical space. It is thus beneficial to characterize the pore structures of the model systems first. Shown in Fig. 5.2 are the top-down and side views of the porous structures formed by 10 and 20 amylose chains, respectively, with 100 % water content, but no water molecule is included for visual clarity. It is apparent that other than the first two monomers, the amylose chains exhibit significant structural flexibility in their responses to the interactions among themselves and with water. In general, the porous structure with 20 amylose chains that represent food systems with a higher density of polysaccharide molecules can be seen in Fig. 5.2 to have smaller pores.

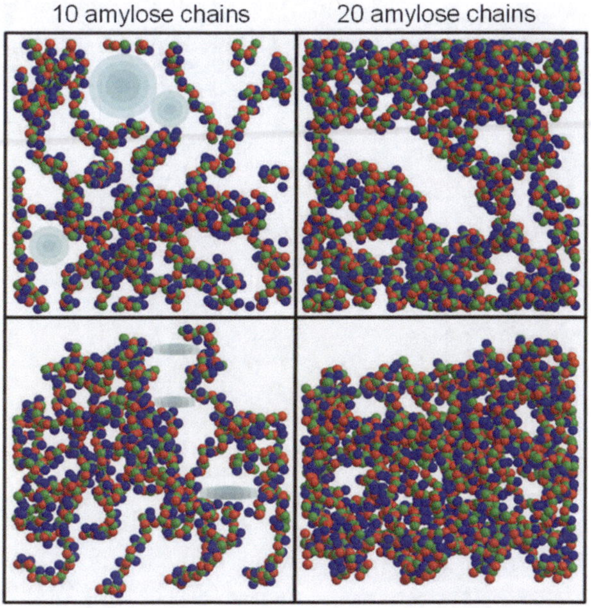

Fig. 5.2 Simulation snapshots of porous structures formed by 10 and 20 amylose chains with 100 % water content. *Top* and *bottom* panels: top-down and side views, respectively

To provide a more detailed and quantitative picture, the number of pore openings of different sizes as a function of distance from the nonflat surface is quantitatively characterized using a two-level lattice representation (Zhang et al. 2005; Riccardi et al. 2008, 2009a, b, 2010a). In this approach, the volume element of each system is first divided into a lattice of bigger square prisms whose lateral dimensions along the lateral x and y directions are 5 Å. The volume of each square prism is then divided into a lattice of smaller square prisms whose lateral dimensions are 1 Å. As schematically shown in Fig. 5.2, cylindrical disks with varying radii and a fixed thickness equivalent to the effective molecular diameter of water (3.77 Å) are tested at each of the lattice points in the two-level lattice representation. By knowing the coordinates of the M3B beads from the MD simulations and by excluding their radii from their distances to the centers of the cylindrical disks, we can rigorously and efficiently locate and assess the pore openings three-dimensionally. The results are shown in Fig. 5.3, where the smallest Z position corresponds to the highest point of the nonflat surface and the pores are grouped into different size categories to facilitate comparison. It can be seen from Fig. 5.3 that increased density of food macromolecules generally results in more pores and reduces the sizes of larger pores. As expected, the largest pores tend to develop from the outer regions near the water–vapor interfaces or above because there the chain monomers and segments are less constrained structurally and energetically. Judging by the significant levels of criss-cross among curves that represent pores of different size groups, the model food porous structures can be described as having reasonably good pore connectivity (Meyers and Liapis 1999; Zhang et al. 2005; Riccardi et al. 2008, 2009a, b, 2010a, b).

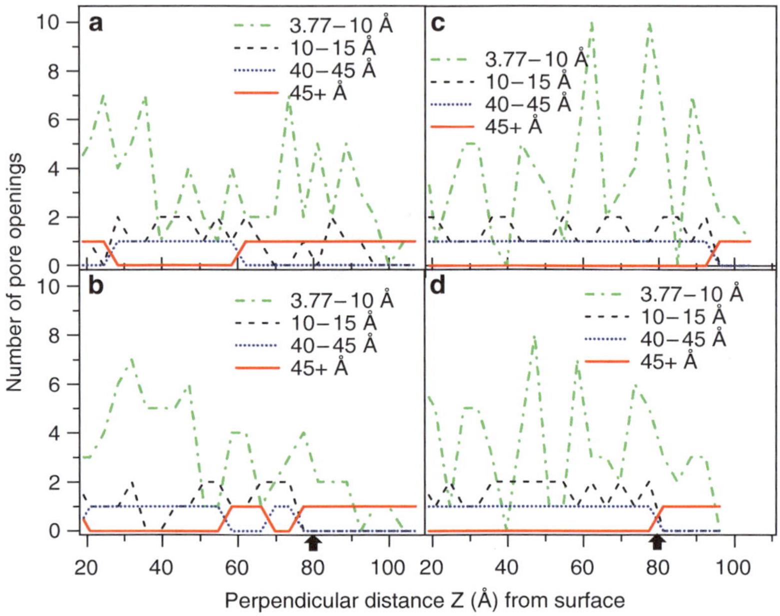

Fig. 5.3 Distributions of number of pore openings in porous structures formed by (**a**) 10 amylose chains and 100 % water content, (**b**) 10 amylose chains and 60 % water content, (**c**) 20 amylose chains and 100 % water content, and (**d**) 20 amylose chains and 60 % water content. The *arrows* in (**b**) and (**d**) indicate the position of the topmost water molecule

The energetics of water molecules in this work are analyzed mainly as a function of their perpendicular position (Z) from the surface and of their distance (r) to the nearest amylose chain. In terms of meaning and insight, the former provides a cross-sectional average of the water interaction energetics, while the latter manifests the effects of food macromolecules on the interactions of water molecules under their direct influence. Furthermore, to minimize possible complications caused by the water interactions with the nonflat surface, only water molecules above Z = 31.7 Å (corresponding to the highest point of the nonflat surface plus 2.5 times the water–water interaction equilibrium distance) are included in the analysis of the energetics. To establish a reference system for comparison, a model system with a single immobilized amylose chain immersed in a sufficient amount of water to develop a bulk-like region above the amylose chain is also studied. Such a reference system can be considered as representing systems with very low densities of food macromolecules and very large pores. The results of the analysis of the energetics are presented in Fig. 5.4, where negative numbers are used in order to reflect the attractive nature of the interactions and k_BT, equivalent to 2.478 kJ/mol, is used to facilitate the comparison between the interaction energies and the thermal energy. It can be readily seen in Fig. 5.4a, d that the coarse-grained model employed in this work produces an average potential energy of approximately 24 k_BT per water molecule for water molecules sufficiently far from the amylose chain. This energy

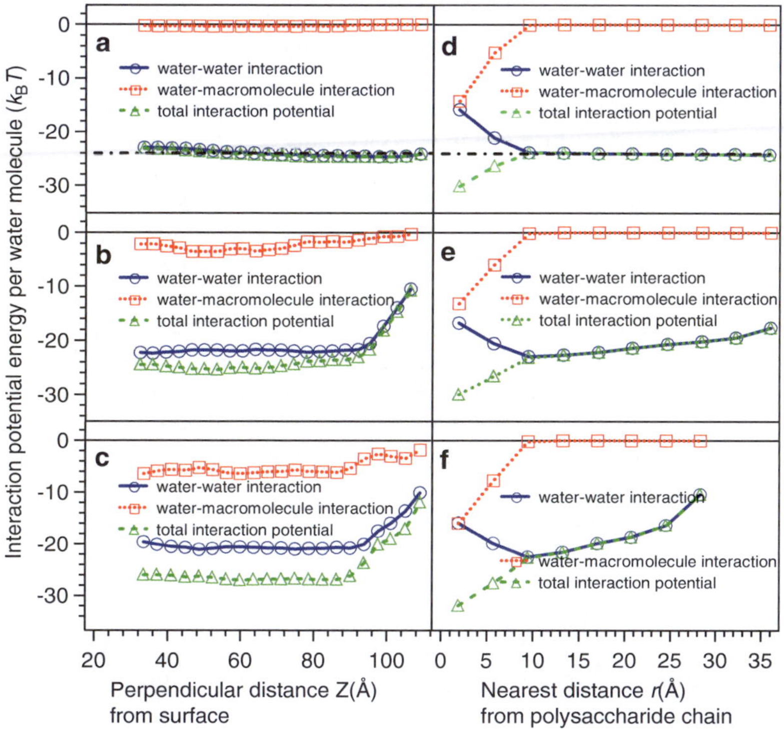

Fig. 5.4 Energetics of water molecules at different perpendicular positions (Z) from surface and different radial distances (r) from nearest amylose chain in systems with one amylose chain (panels **a** and **d**), 10 amylose chains (panels **b** and **e**), and 20 amylose chains (panels **c** and **f**). Numbers on ordinate axis are multiples of k_BT (equivalent to 2.478 kJ/mol)

can be taken to be the water–water interaction energy in large pores and approximates $\Delta U_{\rm deh}$, the internal energy change of water molecules entering the vapor phase from large pores. It also suggests a dehydration enthalpy of $\Delta H_{\rm deh} = 3{,}166 \times 10^3$ J/kgH_2O, which is approximately 11.5 % larger than the experimental value of $2{,}840 \times 10^3$ J/kgH_2O determined by Sheehan and Liapis (1998). Water molecules close to the amylose chain can be seen in Fig. 5.4d to have reduced water–water interactions, which is attributable to reduced coordination numbers of water molecules within the hydration interaction range in the proximity of the macromolecule. However, the energetic gain in the water–macromolecule interactions significantly outweighs the partial loss of the water–water interactions, leading to much stronger overall interactions and strongly bound water molecules adjacent to the macromolecule. In such systems that correspond to low densities of food macromolecules, the number of strongly bound water molecules only amounts to a small fraction and is thus unable to significantly alter the overall water interaction energetics.

Compared with the energetics of water molecules in very large pores shown in Fig. 5.4a, it is apparent from Fig. 5.4b, c that the porous structures formed by 10 and

20 amylose chains have reduced the average water–water interactions by approximately 2–3 k_BT per water molecule across all interior sections, and the level of reduction is more significant with smaller pores in the higher-density system. This effect of pore structures is caused by the reduction of water–water hydration numbers due to the presence of the polysaccharide chains. However, the amylose chains provide new water–macromolecule interactions in return that make the water overall potential energy stronger than that of water in the bulk phase or in very large pores. Different energetic states of water molecules in food porous structures under the influence of food macromolecules can be further elucidated by the distribution of water energetics as a function of the distance to the nearest food macromolecule (Fig. 5.4d–f). Again, reduction of water–water interactions and strengthening of water–macromolecule and overall interactions can be easily observed, and their levels appear to increase with the density of food macromolecules or, equivalently, smaller pores. It is worth pointing out here that, while the interaction energetics of strongly bound water molecules individually do not change much in Fig. 5.4d, e, it is the significantly increased number of strongly bound water molecules due to a higher density of food macromolecules that results in the differences observed in Fig. 5.4a, b. From the point of view that these strongly bound water molecules are trapped in deep energetic wells and subject to impedance caused by nearby macromolecules that have different structures, they can be understood as being not easily rearranged into configurations that resemble ice structures and, thus, very likely to remain nonfreezable in food under normal conditions. Following this perspective and as suggested by the energetic data presented earlier, water molecules sufficiently far from food macromolecules are either weakly bound to the macromolecules or in liquid-like states. The energetic analyses presented earlier can be used to explain why food dehydration tends to be more difficult and more energy consuming than simple water evaporation. Additionally, the water interaction energetics in Fig. 5.4 also reflect certain characteristics of the amylose pore structures. The fact that the interaction energetics shown in Fig. 5.4b, c exhibit relatively minor variations inside the porous structures below the interfaces suggests that the amylose porous structures have less varying pore size distributions in the Z direction, which is consistent with the distributions of the number of pore openings shown in Fig. 5.3.

Defined as a ratio of the partial pressure of water vapor in equilibrium with a water-containing food over the equilibrium vapor pressure of pure water at the same temperature, water activity, a_w, is a macroscopic thermodynamic property whose value in foods falls between 0 (completely dry food) and 1. High water activities mean water molecules are not only freer to escape the food structures but also more available for water–water interactions, microbial growth, and chemical and enzymatic reactions, which also implies decreased food stability. During food dehydration, as water content decreases, water activity also decreases and exhibits very nonlinear relationships (Marinos-Kouris and Maroulis 2006; Bhandari and Adhikari 2008). Empirically, strongly bound water refers to the last small amount of water molecules that remain trapped until very low water vapor pressures or, equivalently, very low a_w values are realized. Based on the results and discussion presented earlier and in Fig. 5.4, strong water–macromolecule interactions underlie the strongly bound water in foods and hence are a cause of the lowering of the water activity a_w in foods. From this point of view, the decreased pore sizes resulting from higher food macromolecular densities can have a twofold effect that depresses a_w: an increased capillary effect (e.g., Barbosa-Canoras et al. 2007, p. 116) and enhanced water–macromolecule interactions, as shown in this work. Since the former cannot completely explain the significant a_w changes observed during water adsorption and desorption and between different food systems, the latter must play a very important role in determining a_w and be included in our pursuit of an adequate understanding of food science and engineering.

To complement the insight and understanding from the energetics and analyses on a per-water-molecule basis, quantitative changes in the amounts of water molecules due to changes in porous structures are examined. As the number of amylose chains increases from 10 to 20 within the same simulation box, the total number of water molecules filling the pore structures understandably decreases significantly, but the number of strongly bound water molecules within the first hydration shell from the amylose chains increases by approximately 90 % and within the second hydration shell by approximately 45 %. Furthermore, a body of experimental studies and data (King 1971; Liapis and Bruttini 2006) has also clearly shown that food macromolecules have very low thermal conductivities and in effect represent excellent insulators. Based on these facts and the number and the energetics of water molecules discussed earlier, we could thus conclude that it is more energy consuming and more difficult to dehydrate food systems with higher polysaccharide densities. Additionally, it could also be argued that the physicochemical nature of food macromolecules plays a very important role in determining water activity, a_w, and the dynamic stages of interaction energetics and food structure during food dehydration because it dictates not only the thermal conductivity of foods but also the water–macromolecule interactions by rendering a different affinity for water and a different compatibility of the resultant porous structures with water configurational structures (Wang and Liapis 2011c).

In this work, we study food dehydration with three dehydrated stages of the amylose-based model systems where 20, 40, and 60 % water removal are prepared and labeled as having 80, 60, and 40 % water content, respectively. The water molecules are removed from the top of the amylose porous structures based on their perpendicular Z positions, and the systems are then re-equilibrated to represent dehydrated stages. Since the fundamental structural and interaction elements are in essence intact, such a strategy could be considered rational and capable of uncovering relevant insights. As examples and also as representations of visual evidence, simulation snapshots of the initial conditions and the equilibrated porous structures with 60 % water content are shown in Figs. 5.5 and 5.6. It can be seen that as the upper parts of the amylose chains become dehydrated without support from the water–macromolecule and water–water interactions, they are energetically driven to form compact conformations locally, and some would even bend downward so as to retain interactions with water. Consequently, the structural changes in the amylose porous structures in response to food dehydration are in such a

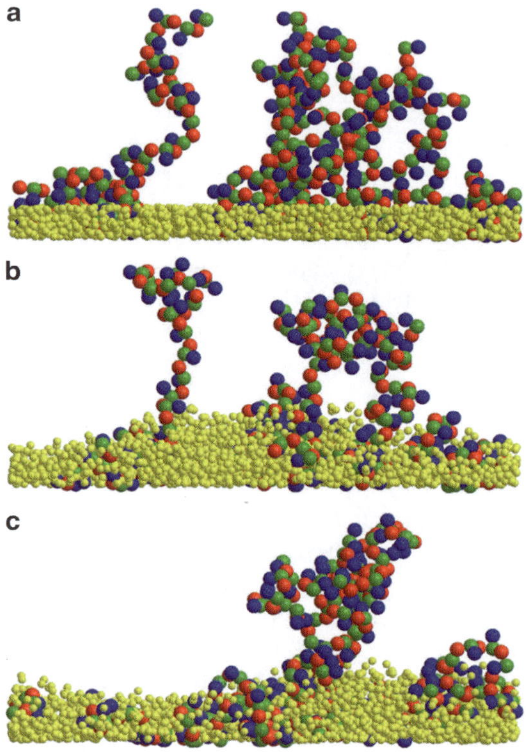

Fig. 5.5 Upper regions ($Z > 63.2$ Å) of (**a**) x-z side view of initial condition, (**b**) x-z and (**c**) y-z side views of equilibrated porous structures formed by 10 amylose chains and with 60 % water content, where water molecules are reduced in size and in *yellow* for visual clarity

direction that the main body of the porous structures becomes shorter. It is worth noting here that the formation of local compact conformations strengthens the stiffness of the chain molecules and makes it possible for the structures to be stable or sufficiently metastable without support from water–water and water–macromolecule interactions. More importantly and interestingly, water molecules also rearrange themselves and develop from planar interfaces in the initial conditions into nonplanar interfaces that resemble the curvatures of concave menisci, in particular in the space of relatively large pores. In general, underlying the formation of a meniscus is the capillary effect that results from a combination of surface tension (caused by cohesion within the liquid or more fundamentally attractive liquid–liquid intermolecular interactions) and adhesion between the liquid and surrounding pore surface (caused by attractive liquid–pore surface intermolecular interactions). A concave meniscus occurs when the liquid molecules have a stronger attraction with the pore surface than with each other or, equivalently, when the liquid molecules can acquire stronger overall interactions by moving from the bulk liquid phase to being adjacent to and interacting with the pore surface. Consistent with this mechanistic perspective of the capillary effect, the energetic analyses examined in this work and reported in Fig. 5.4 find that strong water–macromolecule interactions provide the driving force for capillary action and the anchor for

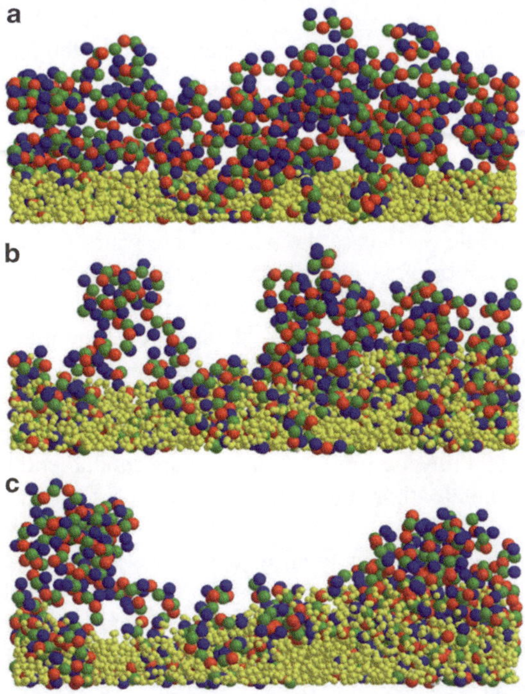

Fig. 5.6 Upper regions ($Z > 56.9$ Å) of (**a**) x-z side view of initial condition, (**b**) x-z and (**c**) y-z side views of equilibrated porous structures formed by 20 amylose chains and with 60 % water content, where water molecules are reduced in size and in *yellow* for visual clarity

meniscus formation. Further evidence comes from a comparison of the heights of the interfaces and the number of water molecules in the same upper regions shown in Figs. 5.5 and 5.6. The equilibrated systems can be seen to have their interfaces driven to higher Z positions and have pulled in certain amounts of water from the interiors below to the regions under consideration, which in Fig. 5.5 is 427 water molecules, or 2.6 % of the remaining water molecules, and in Fig. 5.6 is 132 water molecules, or 0.97 % of the remaining water molecules. As a result, the water–water interaction potential and the water–macromolecule interaction potential employed in this work could be considered appropriate and provide our MD modeling and simulation approach with an adequate capability to generate and investigate various levels of capillary effect in molecular detail in food systems where food molecules have a wide range of densities and can form continuous or noncontinuous pore surfaces.

Based on the results and discussion presented earlier, the employed interaction potentials and model systems in this work are able to capture the essence of the physics of the food materials under consideration and make the findings and insights obtained from this work relevant to the actual food systems and food processing involving water removal. Because of their strong water–macromolecule interactions and because of the presence of chain macromolecules in close proximity, the strongly bound water molecules that underlie the formation of concave menisci can be understood to be configurationally hindered, and hence their dehydration would be accompanied by larger entropy changes as compared to those of weakly bound water located far from the food macromolecules or in the middle of the pores. However, because the last amounts of water to be dehydrated from foods are the strongly bound water molecules adjacent to the macromolecules, it could be concluded that the enthalpic effect is more important than the entropic effect during food dehydration. From this perspective, water removal during food dehydration could be understood as starting from the largest pores and from the middle regions of the pores where water molecules have weaker interactions. As a result of these phenomena and tendencies, the water–vapor interfaces in foods and during food dehydration could be expected to be nonplanar inside individual pores as well as across lateral sections of a food material, and the water dehydration rate would not be uniform but would depend on the distributions of pore sizes. From this point of view, relevant macroscopic continuum models that take into account the occurrence and effects of nonplanar water–vapor interfaces (Sheehan and Liapis 1998) during food dehydration are fundamentally more sound and practically more accurate than those that assume planar interfaces, which can lead to skewed parameter values and inaccurate interpretations. Furthermore, it is perhaps more factual and beneficial to regard water activity, a_w, not as a macroscopic uniform variable but as a microscopic or local variable on the basis of the persistent existence of different pore sizes and water–macromolecule interactions during food dehydration. Although it is still doubtful if such locally varying water activity can be measured by any current experimental technique, it seems plausible to extend the molecular-based modeling and simulation approach employed in this work to study water activity, a_w, as a function of different local environments existing or emerging in food systems during food dehydration.

As discussed earlier, dehydration reduces the amount of water molecules remaining in food and further allows more space for the remaining water molecules to be accommodated in the structure. Both responses weaken the effects of excluded volumes and intermolecular interactions of water and generally lead to decreased numbers of pore openings and reduced pore sizes and volumes in porous food structures, in particular in the regions above the water–vapor interfaces where the food macromolecules are exposed and could rely only on internal or external macromolecule–macromolecule interactions. The changes in pore sizes due to dehydration can be seen in Fig. 5.3 and in Fig. 5.7, where the distances from water to the nearest amylose chains gradually become shorter as more water is removed. Figure 5.7 also shows that lower-density polysaccharide systems consistently have larger pores than their high-density counterparts during dehydration. More importantly, a comparison of Figs. 5.3 and 5.7 indicates that in food systems with a relatively high density of food macromolecules and smaller pores, further reduction in pore sizes by dehydration increases the water–macromolecule interactions while slightly decreasing the water–water interactions. Therefore, while the overall thermal conductivity of the food material decreases during dehydration as the removal of water proceeds and because the thermal conductivity of water is much greater than that of food macromolecules, the results of this work show that food dehydration can further strengthen the interaction energies and

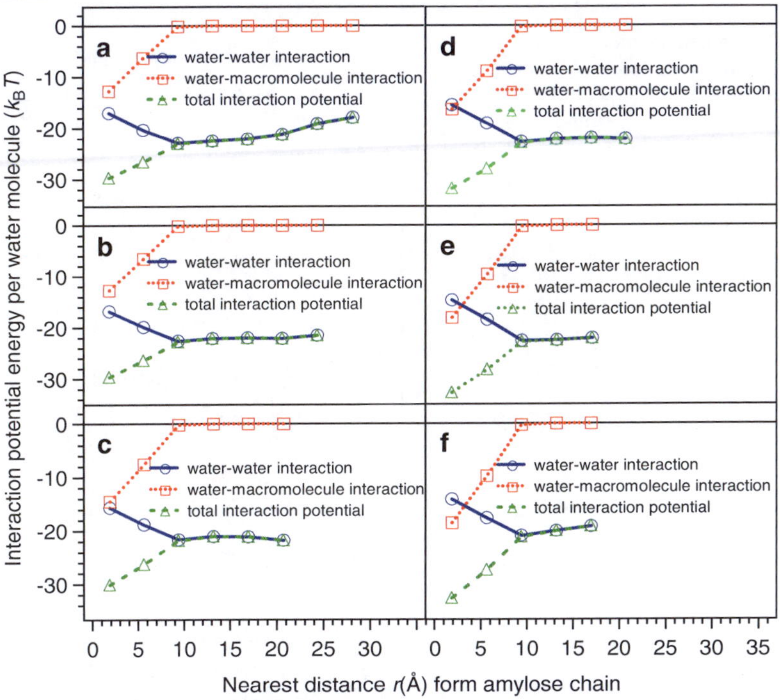

Fig. 5.7 Energetics of water molecules at different distances from nearest amylose chain in systems with (**a**) 10 amylose chains and 80 % water content, (**b**) 10 amylose chains and 60 % water content, (**c**) 10 amylose chains and 40 % water content, (**d**) 20 amylose chains and 80 % water content, (**e**) 20 amylose chains and 60 % water content, and (**f**) 20 amylose chains and 40 % water content. Numbers on ordinate axis are multiples of k_BT (equivalent to 2.478 kJ/mol)

lower the a_w values of the remaining water molecules. Consequently, the difficulty of removing water molecules from the deeper regions of food materials would increase as dehydration progresses.

Since the pore structures of food materials can vary with type, depth, and dehydration time, and because these variations significantly affect the energetics of water–water and water–macromolecule interactions, it could be beneficial to dynamically adjust the heat supply during food dehydration. Relevant to this topic in our current work is the average interaction potential energy of water molecules in food structures at different dehydration stages. In general, an upward or downward trend of this average potential energy indicates a higher or lower amount of heat flux needed for dehydration at similar rates. The average interaction potential energy per water molecule calculated for the model food systems with 100, 80, 60, or 40 % water content is presented in Fig. 5.8a. Not surprisingly, the results confirm the previous findings discussed earlier that (1) systems with higher densities of food chain molecules result in greater water interaction potential energies and require more heat to dehydrate than those with lower chain densities

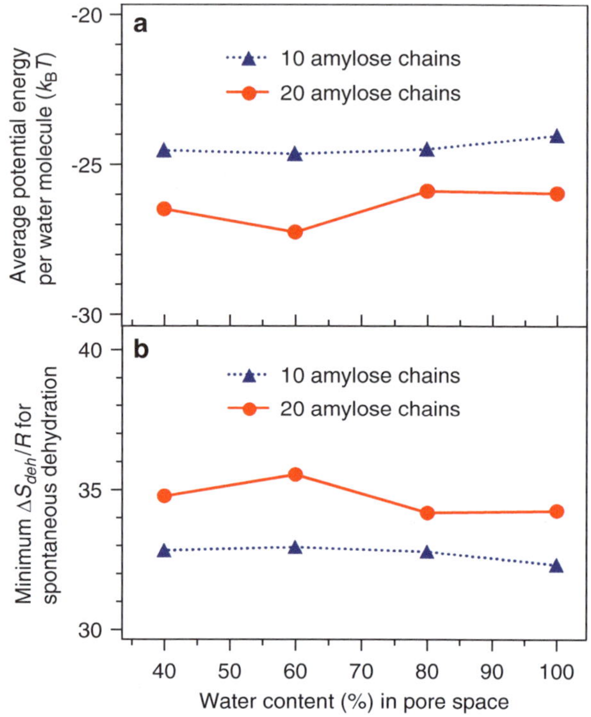

Fig. 5.8 (**a**) Average interaction potential energy per water molecule where numbers on ordinate axis are multiples of k_BT (equivalent to 2.478 kJ/mol) and (**b**) average minimum entropy requirement for spontaneous dehydration under various water contents in pore space of model food systems where numbers on ordinate axis are multiples of R (equivalent to 8.314 J/mol-K)

and (2) decreased water content during dehydration also gives rise to enhanced water interactions and requires more heat to dehydrate the remaining water. Here it is important to emphasize that the water interaction energies shown in Fig. 5.8a are averaged over all water molecules in the pore spaces, including strongly bound, weakly bound, and liquid-like water molecules and, thus, should not be used as the only or major type of energy criterion for assessing food dehydration because it is highly possible, as suggested by the results of this work, that the rate of water removal will not be uniform but will depend strongly on the distributions of pore sizes and locations of water molecules with respect to the food macromolecules in all food systems. Because of the chain flexibility and strong water–macromolecule interactions, it could be further inferred that (i) a certain residual amount of water molecules could be trapped in small pores or constrained by narrow bottlenecks that would not be removed by typical dehydration means; this especially concerns small pores having low initial water content, and (ii) rehydration might not return dehydrated foods to their original pore structures and original water content. These aspects could be studied using the system formulation and simulation models employed in this work.

If we consider the Gibbs free energy, ΔG_{deh}, for dehydration where $\Delta G_{\mathrm{deh}} \approx \Delta H_{\mathrm{deh}} - T\Delta S_{\mathrm{deh}} \approx \Delta U_{\mathrm{deh}} + RT - T\Delta S_{\mathrm{deh}}$ and ideal gas behavior for low-pressure water vapor, we could estimate the average minimum entropy requirement, ΔS_{deh}, for dehydration to occur spontaneously ($\Delta G_{\mathrm{deh}} \leq 0$). The results of this thermodynamic analysis based on our model food systems and simulation data are plotted in Fig. 5.8b, where we can see that dehydration (I) experiences greater entropy change when the density of food molecules is higher and (II) is generally accompanied by increased minimum entropy requirements as water content decreases. These trends of ΔS_{deh} are consistent with the previous results and discussion but could also depend on the physicochemical nature of the food macromolecules (Wang and Liapis 2011c). Although this type of thermodynamic analysis is both insightful and beneficial for providing an averaged entropic overview of the dehydration of various food systems, it is worth noting that the local entropy requirements across individual pores and sections of food systems can vary with the distributions of pore sizes and distances to the pore wall formed by food macromolecules.

5.4 Conclusions

A MD modeling and simulations approach was rationally developed and used to construct and study amylose-based porous layers that could be considered good physicochemical representations of food materials. The findings from our MD studies indicate that the presence of food macromolecules lowers water–water interactions for nearby water molecules by reducing their hydration numbers and disturbing their configurations, but it also provides new water–macromolecule interactions that could significantly outweigh the partial loss of water–water interactions to make the water molecules adjacent to the macromolecules strongly bound and very likely nonfreezable and having low values of water activity a_w. As a result, water molecules in the pore space of porous food systems are not in the same energetic and configurational states and, thus, most likely do not have uniform distributions of water activity a_w, water removal rate, and dehydration energy and entropy requirements within individual pores and across lateral sections. These effects of pore structures were found to be greater in food systems with higher densities of food macromolecules and smaller pores.

The model food porous materials were found to have good pore connectivity for water molecules. During dehydration, water content decreased and the remaining water molecules were found to be able to develop seemingly concave menisci in large pores and nonplanar interfaces across lateral sections of the foods. The dehydrated food macromolecules either adopted compact conformations locally for enhanced self support or bent downward to retain contact with water, both causing the main body of the porous layers to decrease in size. In general, dehydration also reduces pore sizes and the number of pore openings, increases water–macromolecule interactions, and decreases the value of the overall thermal conductivity, so that more energy (heat), longer times, and/or greater temperature gradients are needed to further dehydrate the porous materials. In addition, the minimum entropy requirement for food dehydration was found to be greater in food systems with higher densities of food macromolecules and lower water content.

However, based on the results and discussion presented in this work, dehydration of food materials can be reasonably expected to be driven more by an enthalpic effect than by an entropic effect and to start from the largest pores and from the middle regions of individual pores where the values of the local water activity are higher. The approaches used in this work, including system formulation, interaction potential models, and energetic and entropic analyses, could also be adopted in the construction and subsequent study of many other aspects of food engineering and science, including rehydration of dehydrated foods and food materials with different numbers and types of solutes and monomers per polymer chain.

References

Allen MP, Tildesley DJ (1987) Computer simulation of liquids. Clarendon Press, Oxford, UK

Barbosa-Canoras GV, Fontana AJ Jr, Schmidt SJ, Labuza TP (eds) (2007) Water activity in foods: fundamentals and applications. Blackwell Publishing, Oxford, UK

Bhandari BR, Adhikari BP (2008) Water activity in food processing and preservation. In: Chen XD, Mujumdar AS (eds) Drying technologies in food processing. Wiley-Blackwell, Singapore, pp 55–86

Chen XD, Mujumdar AS (2006) Drying of herbal medicines and tea. In: Mujumdar AS (ed) Handbook of industrial drying. CRC Press, Boca Raton, pp 635–646

Franks F (1985) Biophysics and biochemistry at low temperatures. Cambridge University Press, Cambridge, UK

French AD, Brady JW (eds) (1990) Computer modeling of carbohydrate molecules, ACS symposium series 430, American Chemical Society

Jayaraman KS, Das Gupta DK (2006) Drying of fruits and vegetables. In: Mujumdar AS (ed) Handbook of industrial drying. CRC Press, Boca Raton, pp 605–633

King CJ (1971) Freeze-drying of foods. CRC Press, Boca Raton

Labuza TP (2011) Water relations in food: paradigm shifts to supplant "cook and look". In: Taoukis PS, Stoforos NG, Karathanos VT, Saravacos GD (eds) Proceedings of the 11th international congress on engineering and food (ICEF11), vol I. p 75

Liapis AI, Bruttini R (2006) Freeze drying. In: Mujumdar AS (ed) Handbook of industrial drying. CRC Press, Boca Raton, pp 257–283

Marinos-Kouris D, Maroulis ZB (2006) Transport properties in the drying of solids. In: Mujumdar AS (ed) Handbook of industrial drying. CRC Press, Boca Raton, pp 81–119

Maroulis ZB, Saravacos GD (2003) Food process design. Marcel Dekker, New York

Meyers JJ, Liapis AI (1999) Network modeling of the convective flow and diffusion of molecules adsorbing in monoliths and in porous particles packed in a chromatographic column. J Chromatogr A 852:3–23

Molinero V, Goddard WA III (2004) M3B: a coarse grain force field for the simulation of oligosaccharides and their water mixtures. J Phys Chem B 108:1414–1427

Rapaport DC (1997) The art of molecular dynamic simulation. Cambridge University Press, New York

Riccardi E, Wang J-C, Liapis AI (2008) Rational surface design for molecular dynamics simulations of porous polymer adsorbent media. J Phys Chem B 112:7478–7488

Riccardi E, Wang J-C, Liapis AI (2009a) Porous polymer adsorbent media constructed by molecular dynamics modeling and simulations: the immobilization of charged ligands and their effect on pore structure and local nonelectroneutrality. J Phys Chem B 113:2317–2327

Riccardi E, Wang J-C, Liapis AI (2009b) The design by molecular dynamics modeling and simulations of porous polymer adsorbent media immobilized on the throughpore surfaces of polymeric monoliths. J Chromatogr Sci 47:459–466

Riccardi E, Wang J-C, Liapis AI (2009c) Protein adsorption in porous polymeric adsorbent particles: a multi-scale modeling study on inner radial humps in the concentration profiles of adsorbed protein induced by non-uniform ligand density distributions. J Sep Sci 32:3084–3098

Riccardi E, Wang J-C, Liapis AI (2010a) A molecular dynamics study on the transport of a charged biomolecule in a polymeric adsorbent medium and its adsorption onto a charged ligand. J Chem Phys 133:084904-1–084904-12

Riccardi E, Wang J-C, Liapis AI (2010b) Effects on the dynamic utilization of the adsorptive capacity of chromatographic columns induced by non-uniform ligand density distributions. J Sep Sci 33:2749–2756

Sheehan P, Liapis AI (1998) Modeling of primary and secondary drying stages of the freeze drying of pharmaceuticals in vials: numerical results obtained from the solution of a dynamic and spatially multi-dimensional lyophilization model for different operational policies. Biotechnol Bioeng 60:712–728

Wang J-C, Liapis AI (2011a) Design of polymeric porous adsorbent media and the dynamic behavior of transport and adsorption of bioactive molecules in such media. Chem Ing Tech 83:152–165

Wang J.-C, Liapis AI (2011b) Water-macromolecule interactions in food dehydration and the effects of pore structures on such interactions. In: Taoukis PS, Stoforos NG, Karathanos VT, Saravacos GD (eds) Proceedings of the 11th international congress on engineering and food (ICEF11), vol I. pp 77–78

Wang J-C, Liapis AI (2011c) Water-water and water-macromolecule interactions in food dehydration and the effects of the pore structures of food on the energetics of the interactions. J Food Eng 110:514–524

Zhang X, Wang JC, Lacki KM, Liapis AI (2005) Construction by molecular dynamics modeling and simulations of the porous structures formed by dextran polymer chains attached on the surface of the pores of a base matrix: characterization of porous structures. J Phys Chem B 109:21028–21039

Chapter 6
Rheological and Structural Characteristics of Nanometre-Scale Food Protein Particle Dispersions and Gels

M. Anandha Rao, Simon M. Loveday, and Harjinder Singh

6.1 Introduction

Self-assembly of molecules occurs in many biological materials. One well-known example of biological self-assembly is the folding of proteins into their compact three-dimensional structures. However, a wide variety of pathological conditions arise when proteins fail to fold or to remain folded correctly (Dobson 2003). The deposition of protein aggregates with fibrillar structures has been implicated in amyloidosis, a generic term for a subset of protein misfolding diseases (e.g. Creutzfeldt-Jakob) and in amyloid-related disorders (e.g. Huntington's disease). Typically, the fibrils are long, unbranched and often twisted structures that are 1–100 nm in diameter. Amyloid fibrils formed from proteins almost exclusively involve β-sheets whose strands run perpendicular to the fibril axis and an irreversible process (Dobson 2003).

Many peptides and proteins, including many globular food proteins that are used as thickening and gelling agents or emulsifiers, form amyloid-like structures when heated in vitro at a low pH (~2), low ionic strength and high temperature (~85 °C). The fibrils possess mechanical properties that are desirable to create useful structures (Gosal et al. 2004a). Because of the high ratio of length (micrometre scale) to diameter (nanometre scale), fibril dispersions can form meso-phases such as nematic phases, providing opportunities for the development of new protein-based functional foods (Sagis et al. 2004).

The two major globular proteins in milk and whey are β-lactoglobulin (β-lg) and α-lactalbumin (α-la). They have been used to create nanometre-scale fibrils whose dispersions and gels have different rheological characteristics. Gelation of globular

M.A. Rao (✉)
Cornell University, Geneva, NY 14456, USA
e-mail: mar2@cornell.edu

S.M. Loveday • H. Singh
Riddet Institute, Massey University, Palmerston North, New Zealand

S. Yanniotis et al. (eds.), *Advances in Food Process Engineering Research and Applications*, Food Engineering Series, DOI 10.1007/978-1-4614-7906-2_6,

whey proteins can be induced by heating alone – heat-set gels – or by heating and cooling, and then adding salts or lowering the pH (chemically or biologically) at ambient temperature –cold-set gels. In heat-set gels, only part of the protein is aggregated and available for the building of a protein network at the point where the microstructure is formed (Verheul and Roefs 1998). In contrast, in cold-set gelation, all the protein is present in aggregated form and is available to participate in the protein network from the beginning of gelation (Alting et al. 2003).

Measured rheological responses occur at the macroscopic level, but they are affected by the properties at the molecular and microscopic levels. A major challenge is to establish links between the macroscopic rheological properties of a food with its structure and microstructure. Loveday et al. (2011a) reviewed the formation and characterisation of nanometre-scale food protein particles and their gels.

Dynamic rheological experiments provide suitable means for monitoring the gelation process of many biopolymers and characterizing the resulting gels (Rao 2007). In these experiments, a sinusoidal oscillating stress or strain with a frequency ω is applied to the material and the phase difference between the oscillating stress and strain and the amplitude ratio are measured. The following rheological parameters are obtained: the storage modulus G' (Pa), the loss modulus G'' (Pa) and the loss tangent ($\tan\delta = G''/G'$). For strain values within the linear range of deformation, G' and G'' are independent of strain.

Creep-compliance studies conducted in the linear viscoelastic range also provide valuable information on the viscoelastic behaviour of gels (Rao 2007). In a creep experiment, an undeformed sample is suddenly subjected to a constant shearing stress, σ. The strain, γ, will increase with time and will approach a steady state, when the strain rate is constant. The data are analysed in terms of creep compliance, defined by the relation

$$J(t,\ \sigma) = \gamma(t)/\sigma.$$

A typical creep-compliance versus time curve can be divided into three principal regions (Rao 2007): a region of instantaneous compliance in which the bonds between the different structural units are stretched elastically, a time-dependent retarded elastic region in which the bonds break and reform, but not all of them break and reform at the same rate, and a linear region of Newtonian compliance in which the units flow past one another as a result of rupture of the bonds. The creep-compliance experimental data on a viscoelastic gel can be described by mechanical models, made up of retarded moduli and viscosities, to obtain insight into its structure.

Here, the structural parameters of nanometre-scale food protein fibrils, especially those of β-lg, and particles are outlined. In addition, the rheological characteristics of their dispersions and gels are discussed.

6.2 Materials and Methods

The involvement of β-lg and α-la in nanometre-scale aggregates and the functionality of cold-set gels produced from them were studied by Rabiey and Britten (2009). Solutions with a total protein concentration of 8 % w/v at pH 7.5 containing mixtures of β-lg and α-la in different ratios were heated at 80 °C for 20 min. The heated solutions of nanometre-scale aggregates were cooled to room temperature, diluted to 5.5 % w/v and acidified with glucono-δ-lactone (GDL) to pH 5.0, causing gelation.

Fibrils of β-lg were produced by either incubation in the solvents water–ethanol and water–trifluoroethanol or by heating at pH 2, 75 °C and 80 °C (Gosal et al. 2004a). The morphology of the fibrils was characterized by two scales: the persistence length, l_p, and the contour length L_c. The persistence length, defined as $l_p = \kappa/(k_B T)$ (κ- bending stiffness of a filament or fibril, N m; k_B – Boltzmann's constant, 1.38×10^{-23}, N m K^{-1}), is the typical length at which thermal fluctuations begin to bend the polymer in different directions; it is used to characterize the flexibility or rigidity of a filament (Loveday et al. 2011a). The contour length, L_c, of a filament is its length at maximum extension. A filament is considered to be flexible when $l_p << L_c$ and rigid when the opposite holds ($l_p >> L_c$); many biological filaments are in a third intermediate category: semi-flexible filaments with l_p and L_c of comparable magnitudes.

Partial hydrolysis of α-la with a serine proteinase from *Bacillus licheniformis* was shown to form hollow tubes approximately 20 nm in diameter and several micrometres in length (Ipsen et al. 2001; Ipsen and Otte 2007). The minimum concentration of α-la necessary to form nanometre-diameter tubes was 3.0 % w/v at 50 °C, 75 mM Tris buffer, pH 7.5, 2 mol Ca^{2+}/mol α-la.

6.3 Results and Discussion

Rheology of β-lg and α-la aggregates. Increasing the proportion of α-la compared to β-lg produced smaller particles with more negative ζ-potential, and these remained smaller than 100 nm at pH 5.5. However, aggregates made with 40 % or more β-lg began to further aggregate into larger particles at the same pH (Rabiey and Britten 2009). The particle size (nm) decreased linearly with an increasing proportion of α-la (%) at pH 7.5:

$$\text{Particle size (nm)} = 41.3 - 0.13 \times (\alpha\text{-la, } \%); R^2 = 0.983$$

Creep-compliance rheological tests were conducted on gels that were created in situ in a parallel plate geometry with a 1 mm gap; these gels had protein particles greater than 1,000 nm. The creep-compliance data, $J(t)$ versus time, were fitted to a modified Burger model:

$$J(t) = \frac{1}{G_1} + \left[\frac{t}{G_2(t+\tau)}\right] + \frac{t}{\eta},$$

where G_I is the instantaneous modulus, G_2 is the retarded elastic modulus, η is the viscosity associated with viscous flow, and τ is the retardation time. The values of G_I and G_2 decreased linearly with an increasing proportion of α-la (%). However, values of η decreased exponentially and those of τ varied over the narrow range of 22.0 to 24.4 s. The relationships between the Burger model parameters and the proportion of α-la (%) were as follows:

$$G_1 = 6.58 - 0.05 \times (\%\ \alpha\text{-la}); R^2 = 0.953,$$

$$G_2 = 12.86 - 0.12 \times (\%\ \alpha\text{-la}); R^2 = 0.954,$$

$$\eta = 6253.2 \exp(-0.021 \times (\%\ \alpha\text{-la})); R^2 = 0.985.$$

Increasing the proportion of α-la in the polymer dispersions resulted in more turbid gels that were characterised by an open microstructure (Rabiey and Britten 2009). Further, the gel with 100 % α-la was too soft to be characterised by the creep recovery test.

Structure of β-lg fibrils. Often, β-lg fibrils are created at pH 2 and low ionic strength by heating at 80 °C. Loveday et al. (2010) observed that when β-lg adjusted to pH 1.6–2.4 was heated at 80 °C, the morphology of the resulting fibrils was little affected by pH, but fibrils formed faster at pH < 2. Thioflavin T (ThT) fluorescence increased sigmoidally with heating time; fluorescence changed little for the first few hours (lag phase), then increased rapidly (growth phase) and finally plateaued (stationary phase). Both NaCl and $CaCl_2$ increased the growth rate to an approximately equal extent on an equimolar basis, but $CaCl_2$ also decreased the lag time substantially, whereas NaCl had little effect on lag time.

Without added salts, the fibrils were semi-flexible, 5–10 nm in diameter and up to 5 μm long (Fig. 6.1). The long semi-flexible fibrils were entangled with others in loose networks approximately 5–10 μm wide. In addition to long semi-flexible fibrils, a small number of spherical aggregates with diameter 15–25 nm were seen.

With the addition of NaCl up to 50 mM, the morphology of the fibrils was virtually unchanged, although at 50 mM fibrils were often "wavy" in appearance. There were also short, tightly curled fibrils similar to the worm-like fibrils produced when β-lg is incubated with solvents (Gosal et al. 2004a). At higher NaCl concentrations, such as at 100 mM NaCl (Fig. 6.2), the two types of fibrils again coexisted, and the proportion of worm-like fibrils increased with increasing NaCl concentration.

From transmission electron microscope images of the fibrils, the calculated persistence lengths of the fibrils were found to decrease with the addition of NaCl (Loveday et al. 2011a, 2010).

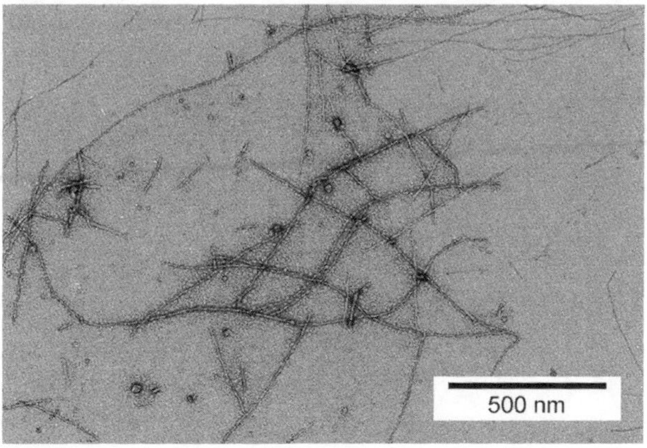

Fig. 6.1 Transmission electron microscope image of fibrils obtained by heating 1 % (w/v) β-lactoglobulin at pH = 2.0, 80 °C for 6 h

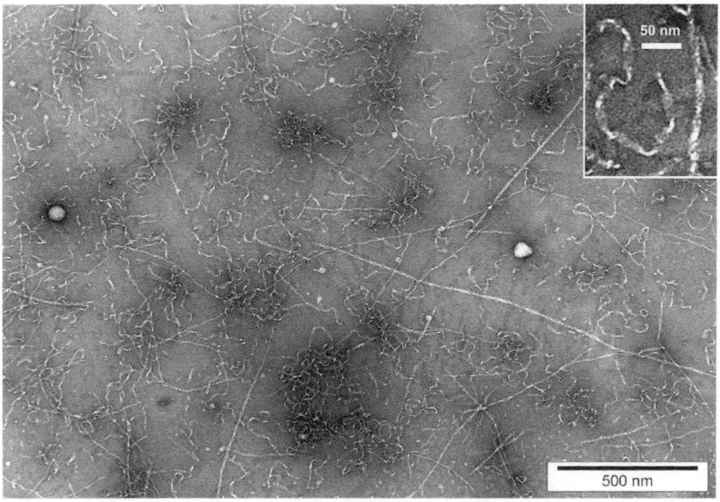

Fig. 6.2 Transmission electron microscope image of fibrils obtained by heating 1 % (w/v) β-lactoglobulin with 100 mM NaCl, pH = 2.0, 80 °C for 6 h

Other investigations into the effect of heating temperature (75–120 °C) (Loveday et al. 2011b) and interactions between NaCl, $CaCl_2$, temperature and pH (Loveday et al. 2011c) have shown that the rate of self-assembly, and to a lesser extent the morphology of fibrils, are quite sensitive to environmental conditions.

Rheology of β-lg fibrils. Rheological properties of gels from β-lg fibrils that were produced by incubation in the solvents water–ethanol and water–trifluoroethanol and by heating at pH 2, 75 °C and 80 °C were studied (Gosal et al. 2004b, c). Useful information on the β-lg fibril gels – gelation time, t_c, critical gel concentration, c_0,

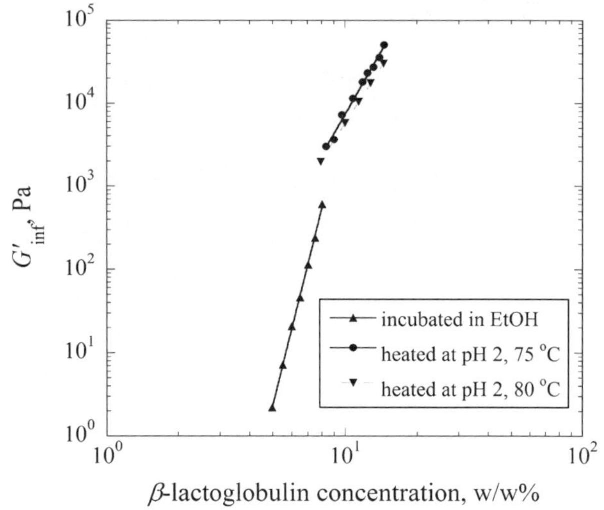

Fig. 6.3 Values of G'_{inf} of β-lactoglobulin gels created from fibrils obtained by incubation in water-ethanol and those by heating at pH 2 at 75 °C and 80 °C (Gosal et al. 2004b, c)

and the equilibrium value of the storage modulus, G', at long gelation times, G'_{inf} – can be determined from experimental rheological data. Values of t_c and G'_{inf} can be obtained using the equation

$$G' \approx G'_{inf} \exp(-B/t),$$

where B is a constant. The values of G'_{inf} of β-lg gels created from fibrils obtained by incubation in water–ethanol and those by heating at pH 2 at 75 °C and 80 °C (Gosal et al. 2004b, c) are shown in Fig. 6.3. It can be seen in Fig. 6.3 that the values of G'_{inf} for the gels from fibrils obtained by heating β-lg at pH 2 at either 75 °C or 80 °C were higher than those of the ethanol-incubated β-lg fibril gels. Insights into the nature of the gels were obtained by applying the well-known cascade and percolation theories. The differences in the values of G'_{inf} of the gels are likely due to differences in the characteristics of the fibrils: l_p, L_c, and filament density n (filaments m^{-3}).

Data on these variables were not available for fibrils made by incubation in solvents, but l_p and L_c were reported for heat-induced fibrils in some studies (Loveday et al. 2010). We recently investigated the effect of $CaCl_2$ on G'_{inf} for fibrils made by heating 10 % w/w whey protein isolate at pH 2 and 80 °C (Loveday et al. 2011c). Adding 40–80 mM $CaCl_2$ decreased l_p (Loveday et al. 2010) but increased G'_{inf} (Loveday et al. 2011c). The increased fibril flexibility (lower l_p) resulting from the presence of $CaCl_2$ apparently allowed fibrils to entangle better, forming a stronger network.

Structure and rheology of α-la fibrils. Partial hydrolysis of α-la with a serine proteinase from *B. licheniformis* resulted in fibrils that are tubes approximately 20 nm in diameter (Ipsen et al. 2001; Ipsen and Otte 2007). Such a fibril gel from a

10.0 % w/v α-la solution had a higher modulus than a heat-set gel from a 10 % w/w β-lg solution, pH 2.5; further, the gelation times of α-la fibrils were longer than those of β-lg fibrils. It was suggested that one reason for the higher modulus might be the greater stiffness of the α-la fibrils (Loveday et al. 2009).

6.4 Conclusion

Cold-set gels were prepared from mixtures of the proteins β-lg and α-la. Increasing the proportion of α-la compared to β-lg resulted in smaller particles with more negative ζ-potential; the particles remained smaller than 100 nm at pH 5.5. However, aggregates made of 40 % or more β-lg aggregated into larger particles at the same pH. The particle size (nm) decreased linearly with increasing proportion of α-la (%) at pH 7.5. The parameters of the Burger model – instantaneous modulus G_I, retarded elastic modulus, G_2, and viscosity η – decreased with an increasing proportion of α-la (Rabiey and Britten 2009).

Dispersions and gels with different rheological properties can be created using fibrils from various food proteins. The protein β-lg has been the subject of many studies. Heat-induced conversion of β-lg into fibrils occurs at low pH and low ionic strength. Fibrils from whey protein isolate, bovine serum albumin, ovalbumin, soy glycinin, conglycinin, soy protein isolate and κ-casein have also been studied (Loveday et al. 2011a). However, more studies are needed to understand the effects of pH and salts on fibril formation. Systematic studies should be conducted to obtain data on the persistence and contour lengths of fibrils, as well as on fibril density (or fibril content). In addition, more data on the rheological properties of the fibril gels and dispersions created using a wide range of protein concentrations are needed.

To derive an expression for the gel modulus (Palmer and Boyce 2008), the force-extension behaviour of the individual filaments is captured using an applicable constitutive relationship. That relationship is then used in conjunction with a network model to obtain the three-dimensional multiaxial stress–strain behaviour of the network. The F-actin filament modulus (Palmer and Boyce 2008) was shown to be a function of the material properties l_p, L_c, filament density n (filaments m^{-3}), and the initial filament end-to-end length r_0. In principle, r_0 and L_c are measurable from micrographs, and l_p is measurable from single-molecule bending. The filament density, n (filaments m^{-3}), is defined as $n = \rho_L/L_c$, where the polymer length density (μm m^{-3}), ρ_L, is estimated from the experimental monomer concentration, the linear polymer density (Da μm^{-1}), and the molecular mass of each monomer (Da $monomer^{-1}$). Both the linear polymer density and the molecular mass are defined *a priori*. Thus the persistence and the contour lengths play important roles in the magnitude of the gel modulus.

References

Alting AC, Hamer RJ, de Kruif CG, Visschers RW (2003) Cold-set globular protein gels: interactions, structure and rheology as a function of protein concentration. J Agric Food Chem 51:3150–3156

Dobson CM (2003) Protein folding and misfolding. Nature 426:884–890

Gosal WJ, Clark AH, Ross-Murphy SB (2004a) Fibrillar β-lactoglobulin gels: part 1. Fibril formation and structure. Biomacromolecules 5(6):2408–2419

Gosal WJ, Clark AH, Ross-Murphy SB (2004b) Fibrillar β-lactoglobulin gels: part 2. Dynamic mechanical characterization of heat-set systems. Biomacromolecules 5(6):2420–2429

Gosal WJ, Clark AH, Ross-Murphy SB (2004c) Fibrillar β-lactoglobulin gels: part 3. Dynamic mechanical characterization of solvent-induced systems. Biomacromolecules 5(6):2430–2438

Ipsen R, Otte J (2007) Self-assembly of partially hydrolysed α-lactalbumin. Biotechnol Adv 25:602–605

Ipsen R, Otte J, Qvist KB (2001) Molecular self-assembly of partially hydrolysed α-lactalbumin resulting in strong gels with a novel microstructure. J Dairy Res 68:277–286

Loveday SM, Rao MA, Creamer LK, Singh H (2009) Factors affecting rheological characteristics of fibril gels: the case of β-lactoglobulin and α-lactalbumin. J Food Sci 74:R47–R55

Loveday SM, Wang XL, Anema SG, Rao MA, Creamer LK, Singh H (2010) Tuning the properties of β-lactoglobulin nanofibrils with pH, NaCl and $CaCl_2$. Int Dairy J 20:571–579

Loveday SM, Rao MA, Singh H (2011a) Food protein nanoparticles: formation, properties and applications. In: Bhandari B, Roos Y (eds) Food materials science and engineering. Blackwell Publishing, Oxford, UK, pp 263–294

Loveday SM, Wang XL, Rao MA, Anema SG, Singh H (2011) β-lactoglobulin nanofibrils: effect of temperature on fibril formation kinetics, fibril morphology, and the rheological properties of fibril dispersions. Food Hydrocoll 27:242–249

Loveday SM, Wang XL, Anema SG, Rao MA, Singh H (2011b) Effect of pH, NaCl, $CaCl_2$ and temperature on self-assembly of β-lactoglobulin into nanofibrils: a central composite design study. J Agric Food Chem 59:8467–8474

Loveday SM, Su J, Rao MA, Anema SG, Singh H (2011c) Effect of calcium on the morphology and functionality of whey protein nanofibrils. Biomacromolecules 12:3780–3788

Palmer JS, Boyce MC (2008) Constitutive modeling of the stress–strain behavior of F-actin filament networks. Acta Biomater 4:597–612

Rabiey L, Britten M (2009) Effect of protein composition on the rheological properties of acid-induced whey protein gels. Food Hydrocoll 23:973–979

Rao MA (2007) Rheology of fluid and semisolid foods: principles and applications, 2nd edn. Springer, New York

Sagis LMC, Veerman C, van der Linden E (2004) Mesoscopic properties of semiflexible amyloid fibrils. Langmuir 20:924–927

Verheul M, Roefs SPFM (1998) Structure of whey protein gels, studied by permeability, scanning electron microscopy and rheology. Food Hydrocoll 12:17–24

Chapter 7
Transport Properties in Food Process Design

M. Krokida and G.D. Saravacos

7.1 Introduction

Adequate, nutritious, and safe foods at affordable prices are needed for a growing world population. The design and operation of food processes and food processing plants require material and energy (heat) balances on the process flowsheet. The sizing of food processing equipment is based on momentum (flow), heat, and mass transfer rate equations (Maroulis and Saravacos 2003; Saravacos and Maroulis 2011).

Material balances include overall and partial (component) balances throughout the whole process. In most food processing operations, heat is the main energy quantity considered in preliminary energy balances. The thermal properties required are specific heat and enthalpy changes, which do not change very much in a given material, and they can be taken from the literature (Rao et al. 2005).

Transfer equations require reliable values of the basic transport properties of foods, such as viscosity, thermal conductivity, and mass diffusivity, and approximate values of heat and mass transfer coefficients. Transport properties are affected strongly by the composition and physical structure of the food product, such as apparent density and porosity (Saravacos and Maroulis 2001).

Most liquid foods are non-Newtonian fluids whose apparent viscosity is affected strongly by the size and concentration of the dissolved or suspended particles. The effect of temperature on the rheology of fluids depends on the composition and concentration. The viscosity of concentrated solutions of Newtonian fluids, such as sugars, decreases sharply as the temperature is increased, while the apparent viscosity of non-Newtonian fluids is affected slightly (Rao 2007).

M. Krokida (✉) • G.D. Saravacos
National Technical University, 15780 Athens, Greece
e-mail: krok@chemeng.ntua.gr

S. Yanniotis et al. (eds.), *Advances in Food Process Engineering Research and Applications*, Food Engineering Series, DOI 10.1007/978-1-4614-7906-2_7,

The thermal conductivity and mass diffusivity in complex fluid and solid foods cannot be predicted theoretically. Experimental values of these properties are needed under specified conditions.

The thermal conductivity of solids decreases at higher porosities, while the opposite effect is observed in mass diffusivity. Mass diffusivity increases strongly with the temperature in low-porosity solids, while a smaller effect is observed in porous materials.

Heat and mass transfer coefficients are needed in food process design. Experimental and empirical values from the literature are normally used.

7.2 Transport Properties of Gases and Liquids

The transport properties of gases and liquids are used extensively in the design and analysis of various chemical engineering operations. The properties of simple gases and liquids can be evaluated and correlated using the principles of molecular dynamics and thermodynamics (Reid et al. 1987; Saravacos and Maroulis 2011).

Estimations of the transport properties of gases and liquids can be facilitated using momentum, heat, and mass transfer analogies, which relate heat and mass transfer coefficients (Saravacos and Maroulis 2001).

The transport of momentum, heat, and mass in one direction (x) is expressed by the following analogous transport equations (shear develops in the perpendicular y direction):

$$\tau = -\eta\ (du/dy), \tag{7.1}$$

$$(q/A) = -\lambda\ (dT/dx), \tag{7.2}$$

$$(J/A) = -D\ (dC/dx), \tag{7.3}$$

where τ (Pa/m^2) is the shear stress and η [(Pa s) or kg/(m s)] is the viscosity, q/A (W/m^2) is the heat flux, λ [W/(m K)] is the thermal conductivity, J/A (kg/m^2 s) is the mass flux, D (m^2/s) is the mass diffusivity, u (m/s) is the velocity, T (K) is the temperature, and C (kg/m^3) is the concentration of the species being transported.

The transport properties (η, λ, D) of real (nonideal) gases can be estimated from molecular dynamics equations as functions of temperature, pressure, molecular weight, and the molecular parameter collision diameter and collision integrals (Saravacos and Maroulis 2001).

These equations predict that the viscosity (η) and the thermal conductivity (λ) of gases will increase with the square root of temperature ($T^{0.5}$), while the mass diffusivity will increase with $T^{1.5}$.

At constant temperature (T) the mass diffusivity (D) of gases and vapors is inversely proportional to pressure (P), i.e., the product (P D) is constant.

Table 7.1 Viscosity (η) and thermal conductivity (λ) of gases and liquids (25 °C)

Material	η, mPa s	λ, W/(m K)
Gases		
Air	0.017	0.025
Oxygen	0.018	0.020
Nitrogen	0.018	0.026
Carbon dioxide	0.015	0.016
Water vapor	0.010	0.020
Liquids		
Water	0.90	0.62
Ethanol	1.04	0.15

The Wilke–Chang equation (Saravacos and Maroulis 2011) shows that the mass diffusivity (D) is inversely proportional to the viscosity (η) and directly proportional to the temperature (T) of a solution, i.e., $(D\,\eta)/T = \text{constant}$.

The mass diffusivity (D) of a particle of radius (r) in a liquid of viscosity (η) and temperature (T) is given by the Stokes–Einstein equation

$$D = (k_B T)/(6\pi\,\eta\,r), \tag{7.4}$$

where $k_B = 1.38 \times 10^{-23}$ J/(molecule K) is the Boltzmann constant.

The effect of temperature (T) on the viscosity of liquids is expressed by the Arrhenius equation

$$\eta = \eta_0 \exp(E_a/RT), \tag{7.5}$$

where E_a is the activation energy and η_0 is a constant.

A similar Arrhenius equation is used to describe the effect of temperature on the mass diffusivity (D) in liquids:

$$D = D_0 \exp(-E_a/RT). \tag{7.6}$$

It should be noted that viscosity decreases with temperature, while the opposite effect is observed with mass diffusivity ($E_a > 0$).

The typical values of transport properties of gases and liquids, useful in food process engineering calculations, are given in Tables 7.1 and 7.2.

The ratio of viscosities of a material in the liquid (L) to the gas (G) phase is equal to the ratio of the thermal conductivities in the respective phases, varying by approximately 100 times, i.e., $(\eta_L/\eta_G) = (\lambda_L/\lambda_G) = 10 - 10^3$.

The ratio of the mass diffusivities of a molecule in the liquid (L) to the gas (G) phase is approximately $(D_L/D_G) = 1/10^4$.

Table 7.2 Mass diffusivity of gases (D_{AB}) in air and water (25 °C)

Diffusing material (A)	In air (B)[a], $\times$ 10^{-5} m^2/s	In water (B), $\times$ 10^{-9} m^2/s
Oxygen	1.7	1.7
Nitrogen	1.8	1.9
Carbon dioxide	1.9	2.0
Ethanol	1.3	1.3
Water	2.0	1.1
NaCl	–	1.2

[a]Atmospheric pressure

7.3 Food Structure and Transport Properties

Food structure at the molecular, microscopic, and macroscopic levels is important not only for the study and evaluation of food texture and quality but also for the analysis and correlation of the transport properties of foods, such as viscosity, thermal conductivity/diffusivity, and mass diffusivity. At the macroscopic level, the structural (macroscopic) properties, such as density, porosity, shrinkage, particle size, and shape, can be defined as quantitative parameters of physical meaning for the characterization of structural changes of foods during processing and storage. These properties are strongly related to the transport properties of solid and semisolid food materials. Food materials are classified into two categories:

- Continuous solids, in which shrinkage and porosity develop when water is removed;
- Particulate or granular materials, such as starch granules, in which porosity is a dependent variable that can be controlled, e.g., by compression or vibration.

Density, shrinkage, and porosity are the most common structural properties, and they are included in the mass-and volume-related properties that constitute one of the main groups of mechanical properties, along with rheological, morphological, acoustic, and surface properties (Rahman and McCarthy 1999).

A detailed terminology for the most common mass- and volume-related properties has been presented by Michailidis et al. (2009a). Measurement techniques used to determine the density, shrinkage, and porosity of food materials are given by Michailidis et al. (2009b).

Table 7.3 shows the measurement methods for density, porosity, and shrinkage of food materials.

A large amount of information regarding the transport properties of food materials has been published in the literature, including the properties of fruits, vegetables, milk, juices, ice cream, fish/meat products, model foods, fried products, extrudates, and more. In addition to models proposed for specific materials, effort has been made to develop generic models on a theoretical basis that would apply to all materials using parameters that had a physical meaning. Zogzas et al. (1994), Krokida and Maroulis (1997), and Saravacos and Maroulis (2001) developed a semiempirical model to predict structural properties, such as apparent and material

Table 7.3 Measurement techniques for density, porosity, and shrinkage

Estimated property	Method	Measured property	Application
Apparent density			
	Geometric dimension	Mass and characteristic dimensions (for boundary volume estimation)	Regularly shaped solids
	Buoyant force	Weight in air and buoyant force (sample must be coated)	Irregularly shaped solids
	Liquid displacement	Mass and displacement liquid volume (sample must be coated)	Irregularly shaped coated solids
	Gas pycnometer	Mass and displacement gas volume (sample must be coated)	Irregularly shaped coated solids
	Solid displacement	Mass and displacement solid volume	Irregularly shaped solids
Material density			
	Liquid displacement	Mass and displacement liquid volume	Ground solids; can be used for the measurement of density as a function of temperature
	Pycnometer method	Mass and displacement gas volume	Ground solids
	Mercury porosimetry	Pressure versus mercury intrusion and extrusion volume	Determination of pore volume and pore size distribution of ground solids
	Gas adsorption	Volume adsorbed versus relative pressure	Determination of pore volume and pore size distribution of ground solids
Particle density			
	Liquid displacement	Mass and displacement liquid volume	Irregularly shaped solids
	Gas pycnometer	Mass and displacement gas volume	Irregularly shaped solids
Bulk density			
	Bulk density	Mass and container volume	Liquids or materials packed or staked in bulk into a known volume container

(continued)

Table 7.3 (continued)

Estimated property	Method	Measured property	Application
Apparent shrinkage			
	Geometric dimension	Characteristic dimensions (for boundary volume)	Regularly shaped solids
	Liquid displacement	Displacement liquid volume (sample coated)	Irregularly shaped coated solids
	Gas pycnometer	Displacement gas volume (sample must be coated)	Irregularly shaped coated solids
	Solid displacement	Displacement solid volume	Irregularly shaped solids
Porosity			
	Direct method	Bulk volume and volume of compacted material	Soft solids
	Optical microscopic method	Microscopic view of a random section of the porous material	Solids presenting sectional porosity same as bulk porosity
	X-ray microtomography	Optical image of porous material	Solids; can be used in visualization and measurement of microstructural features
	Density method[a]	–	Solids with known densities
	Fragile agglomerates method	Mass and optical sizing of individual agglomerates	Fragile agglomerates ("instant powders")

[a]If densities are unknown, pore volume or bulk volume should be measured using techniques for volume measurement. The volume of regularly shaped particles can also be measured with methods used for irregularly shaped solids but the geometric dimension method is more accurate in this case. Powders are classified as irregularly shaped solids

density, shrinkage, and porosity. Material density (ρ_m), apparent density (ρ_a), and porosity (ε) are given respectively by the following equations:

$$\rho_m = \frac{1 + M_w}{\frac{1}{\rho_s} + \frac{M_w}{\rho_w}}, \tag{7.7}$$

$$\rho_a = \frac{1 + M_w}{\frac{1}{\rho_{a_i}} + \beta' \frac{M_w}{\rho_w}}, \tag{7.8}$$

$$\varepsilon = 1 - \frac{\rho_a}{\rho_m}, \tag{7.9}$$

where ρ_s is the actual density of dry solids, ρ_w is the actual density of the enclosed water, M_w is the material moisture content, and β' is the volume-shrinkage coefficient.

The preceding model was further expanded by Boukouvalas et al. (2006) to incorporate the effect of temperature on the material density of food materials. The dry solid material density ρ_s is related to temperature through the linear equation

$$\rho_s = s_0 + s_1 T, \tag{7.10}$$

where s_0 and s_1 are parameters. Water density is also considered temperature-dependent and can be calculated through a second-degree polynomial:

$$\rho_w = w_0 + w_1 T + w_2 T^2, \tag{7.11}$$

where T is the material temperature, and the constants of the preceding equation are $w_0 = 9.97 \times 10^2$, $w_1 = 3.14 \times 10^{-3}$, and $w_2 = -3.76 \times 10^{-3}$ (Maroulis and Saravacos 2003).

In fried products, when oil is considered as another phase, the foregoing equations are transformed into

$$\upsilon = \upsilon_i + \beta' \left(\frac{M_w}{\rho_w} + \frac{Y}{\rho_L} \right), \tag{7.12}$$

$$\rho_a = \frac{1 + M_w + Y}{\frac{1}{\rho_{a_i}} + \beta' \left(\frac{M_w}{\rho_w} + \frac{Y}{\rho_L} \right)}, \tag{7.13}$$

where ρ_L is the oil density and Y the oil content.

Krokida and Maroulis (1997) and Krokida et al. (2001) applied the previous model to the study of the structural properties of fruits and vegetables and their correlation with the material moisture content. The effect of a drying method on the apparent density of banana, apple, carrot, and potato at various levels of moisture content is presented in the following figure, using a large set of experimental measurements.

Parameters affecting structural properties include, among others, the maturity of the initial material (e.g., fruit) and the parameters of the process taking place. In dried fruits and vegetables, the parameters of the drying process are of utmost importance (e.g., temperature and humidity in air drying, pressure and power in microwave drying, oil type and temperature in frying). Pretreatments such as blanching or osmotic drying also influence the final structure. In extrusion cooking structural properties are influenced by both raw material characteristics and extrusion parameters, such as barrel temperature, screw rotational speed, and feed rate (or equivalent residence time or apparent velocity of the food material in the extruder barrel) (Fig. 7.1).

Lazou et al. (2007) examined the influences of process conditions and material characteristics on the porosity of corn–legume-based extrudates. Four different types of legumes – chickpea, Mexican bean, white bean, and lentils – were used to form mixtures with corn flour in a ratio ranging from 10 to 90 % corn/legume, and the following simple power model was developed:

$$\varepsilon = \varepsilon_0 \cdot \left(\frac{T}{T_r}\right)^{n_T} \cdot \left(\frac{\tau}{\tau_r}\right)^{n_\tau} \cdot \left(\frac{X_c}{X_{cr}}\right)^{n_X} \cdot \left(\frac{C}{C_r}\right)^{n_C} \cdot \left(\frac{N}{N_r}\right)^{n_N}, \tag{7.14}$$

where T is the extrusion temperature, τ is the residence time, N is the screw speed, X is the feed moisture content, and C is the ratio of the materials (kg corn/kg legume).

Oikonomopoulou and Krokida studied the effect of extrusion parameters on the porosity of corn/freeze-dried carrot mixtures.

Figure 7.2 presents the porosity of extrudates as a function of mean residence time, feed moisture content, extrusion temperature, and corn-to-legume or carrot ratio (CP: corn/chickpea, CM: corn/mexican bean, CB: corn/white bean, CL: corn/lentils, carrot: corn/freeze-dried carrot).

In recent years, research on structural properties has aimed at the study of pore formation and distribution rather than the measurement of an average value of a property for the whole food material. Advanced methods, such as light microscopy, transmission electron microscopy, scanning electron microscopy (SEM), scanning probe microscopy, X-ray microscopy, light scattering, magnetic resonance imaging (MRI), and spectroscopy, are employed to reveal the microstructure of the material and the pore distribution as it influenced by the applied treatment. Oikonomopoulou et al. (2011) studied the structural properties of freeze-dried rice as affected by process conditions. The alterations of porosity as a function of boiling period and freeze-drying conditions were visualized with SEM photographs and scanning microscopy. An increase in the boiling time led to changes in the microstructure of cooked rice kernels. At higher boiling periods both porosity and average pore

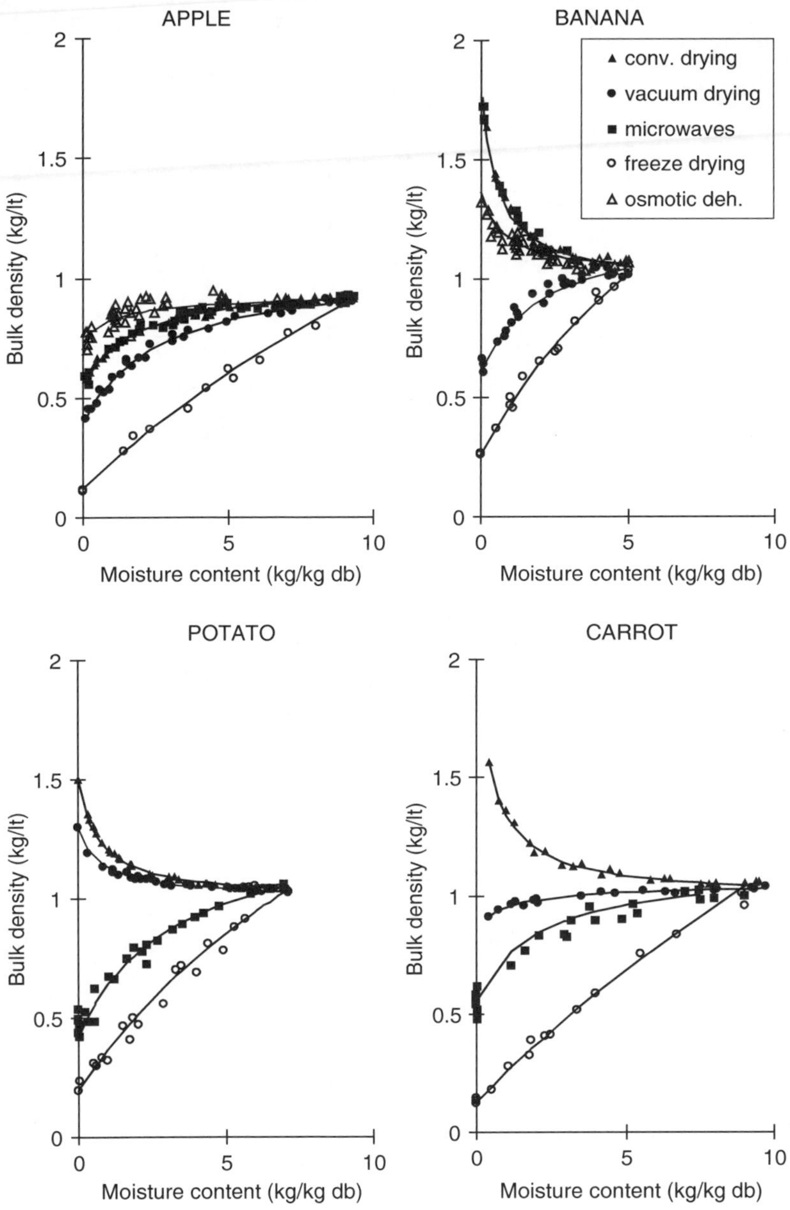

Fig. 7.1 Variation of bulk density with material moisture content for various drying methods during dehydration

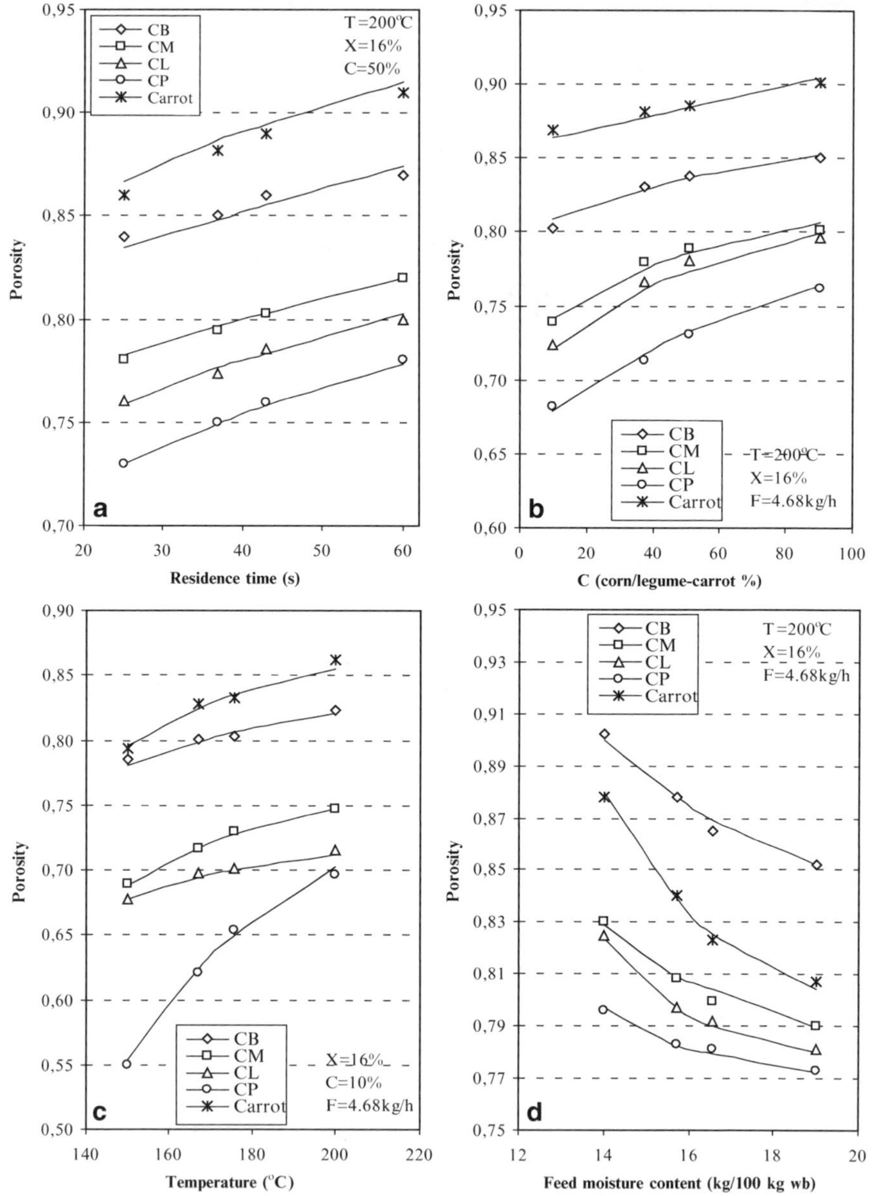

Fig. 7.2 Comparative figures of porosity for extrudates: (**a**) porosity as a function of residence time, (**b**) porosity as a function of corn/legume ratio-carrot, (**c**) porosity as a function of temperature, (**d**) porosity as a function of feed moisture content

size seemed to increase. Porosity depends on water uptake and is higher at longer boiling times where water uptake is higher. In addition, as far as freeze-drying pressure is concerned, as the absolute pressure in the freeze-drying chamber increases, the porosity decreases, compared to processing at low pressures.

7.4 Rheological Properties of Foods

The rheological properties of fluid and semifluid food products are very important in food process design. The viscosity of Newtonian fluids, or the apparent viscosity (rheological constants) of non-Newtonian fluids, is used in calculations involving fluid flow, mixing, heating, cooling, and evaporation processes.

The Reynolds number (Re), characterizing the type of flow, is used extensively:

$$\mathrm{Re} = (\mathrm{d\,u}\,\rho)/\eta = (\mathrm{dm})/(\mathrm{A}\,\eta), \tag{7.15}$$

where d is the pipe diameter, u is the fluid velocity, ρ is the fluid density, m is the fluid flow rate, A is the pipe cross-sectional area, and η is the (apparent) viscosity of the fluid.

The viscosity of Newtonian fluids is defined by the Poiseuille equation for laminar fluid flow:

$$(\mathrm{d}\Delta \mathrm{p})/4\mathrm{L} = \eta(8\mathrm{u}/\mathrm{d}) \quad \text{or} \quad \tau = \eta\,\gamma, \tag{7.16}$$

where d and L are the diameter and length of the pipe, respectively, u is the velocity, τ is the shear stress, and γ is the shear rate.

For non-Newtonian fluids, the Poiseuille equation becomes

$$(\mathrm{d}\Delta \mathrm{p})/4\mathrm{L} = \mathrm{K}[(3\mathrm{n}+1)/4\mathrm{n}]^2(8\mathrm{u}/\mathrm{d})^{\mathrm{n}} \text{ or } \tau = \mathrm{K}\,\gamma^{\mathrm{n}}, \tag{7.17}$$

where K, the flow consistency coefficient, and n, the flow behavior index, are characteristic rheological constants.

The Reynolds number (Re) of non-Newtonian fluids can be calculated from the basic Eq. 7.15, replacing the Newtonian viscosity (η) by the apparent viscosity (η_α):

$$\eta_\alpha = \mathrm{K}\,\gamma^{\mathrm{n}-1}. \tag{7.18}$$

Most non-Newtonian fluid foods are pseudoplastic ($n < 1$), and their apparent viscosity decreases at higher shear rates (velocities or agitation speeds).

The viscosity of Newtonian fluid foods varies widely from approximately 1 mPa s (water) to approximately 10 Pa s (honey) (Saravacos and Maroulis 2001). The activation energy for viscosity increases sharply as the concentration

Table 7.4 Viscosity (η) and activation energy (E_a) of Newtonian fluid foods

Material	η (25°C), mPa s	E_a, kJ/mol
Water	0.9	14
Sucrose solution		
15 %	1.5	20
65 %	40	60
Vegetable oil	50	48
Honey 70 %	10×10^3	65

Table 7.5 Flow consistency coefficients (K) and flow behavior indices (n) of pseudoplastic fluid foods

Material	K (25°C), mPa s^n	n, –
Orange juice		
Concentrate 65 °Brix	0.4	0.76
Tomato paste 35 °Brix	120	0.30
Applesauce	26	0.28
Mayonnaise	6.3	0.50
Mustard	18.5	0.40

of sugar solution is increased. Table 7.4 gives typical viscosities and activation energies of fluid foods.

Clarified (depectinized) fruit juices are nearly Newtonian fluids with viscosity and activation energy close to the corresponding sugar solution. At very high sugar concentrations, the clarified juices tend to become non-Newtonian pseudoplastic ($n < 1$).

Fluid foods, containing macromolecules and suspended particles, are usually preudoplastic non-Newtonian fluids with large K values and smaller than 1 (n) indices. Table 7.5 shows the typical rheological constants of some important fluid foods.

The activation energy (E_a) for the flow of pseudoplastic fluid foods is much lower than for a Newtonian solution of the same concentration. Concentrated particle suspensions, such as tomato concentrates and apple sauce, have a lower E_a (15–25 kJ/mol) than fruit juice concentrates (approximately 40 kJ/mol). This means that temperature has a relatively small effect on pseudoplastics.

7.5 Heat Transport Properties

The heat transport properties that are of importance in food process design are the thermal conductivity (λ) and the thermal diffusivity (α) of the food product.

The thermal conductivity (λ) of solid foods is affected strongly by the moisture and the physical structure of the material, especially its apparent density and porosity. This is due to the large difference in thermal conductivity between the pure solid and the air (or gas) contained in the food material.

Table 7.6 Typical thermal conductivities (λ) of unfrozen foods (25 °C)

Material	% Moisture	λ, W/(m K)
Water	100	0.620
Sucrose solution	85	0.570
Potato	82	0.510
Tomato concentrate	65	0.550
Beef	74	0.450
Fish	80	0.460
Milk	85	0.530
Edible oil	–	0.180
Dried fruit	18	0.240

Table 7.7 Empirical values of thermal conductivity (λ) and diffusivity (α)

Food material	λ, W/(m K)	α, m^2/s
1. Moisture > 30–40 %	$\lambda_1 = 0.40-0.58$	$\alpha_1 = 1.4 \times 10^{-7}$
2. Frozen foods	$\lambda_2 = 2.5\ \lambda_1$	$\alpha_2 = 3\alpha_1$
3. Dry food	$\lambda_3 = 0.1\lambda_1$	$\alpha_3 = 0.8\alpha_1$
4. Fats and oils	$\lambda_3 = (0.25-0.50)\lambda_1$	$\alpha_3 = 0.6\alpha_1$

The thermal diffusivity (α) is normally estimated from the thermal conductivity (λ), using the equation

$$\alpha = \lambda/(\rho C_p), \tag{7.19}$$

where ρ is the density and C_p is the specific heat of the food material.

In estimations of the thermal diffusivity (α) of food materials, the following typical values of density (ρ) and specific heat (C_p) are used:

Water: $\rho = 1{,}000\ kg/m^3$, $C_p = 4.18$ kJ/(kg K);
Ice: $\rho = 917\ kg/m^3$, $C_p = 2.10$ kJ/(kg K);
Nonfrozen foods: $\rho = 1{,}150\ kg/m^3$, $C_p = 3.00$ kJ/(kg K);
Edible oils: $\rho = 920\ kg/m^3$, $C_p = 1.70$ kJ/(kg K).

The thermal diffusivity of nonfrozen food products does not change much, varying in a range of 1.0×10^{-7} to $1.5 \times 10^{-7}\ m^2/s$.

Table 7.6 shows typical values of λ of nonfrozen foods

The thermal conductivity of frozen foods below the freezing point (T < 0 °C) is higher than in the nonfrozen product [1.0–1.5 W/(m K)].

For design purposes, the following approximate values of thermal conductivity (λ) can be used (Table 7.7).

Rahman (1995) presented a comprehensive review of heat transfer correlations that applies in most of food processes. Very low heat transfer coefficients are found in air processes under natural convection.

7.6 Mass Transport Properties

Solid foods have a heterogeneous structure, and they may interact with the diffusing compounds; thus, the diffusivity of small molecules in solids is much lower than in liquids, affecting the rates of various physical and chemical processes involving mass transfer.

The basic mass transport property in food materials is the mass diffusivity (D), defined by the Fick equation of unsteady transport in the x direction:

$$\delta C/\delta t = D(\delta^2 C/\delta t^2), \tag{7.20}$$

where C is the concentration of the diffusing component in the solid, x the diffusion direction, and t the time.

The Fick diffusion equation is similar to the Fourier heat conduction equation. The diffusivity (D) is an overall mass transport property that includes molecular diffusion and transport by various flow mechanisms.

Solutions of the unsteady-state (transient) diffusion equation for constant diffusivity are available in graphical form for the basic shapes of slab, infinite cylinder, and sphere, similar to the solutions of the Fourier equation for unsteady-state heat conduction.

If the resistance to mass transfer at the surface of a solid is significant, compared to the resistance inside the solid, then it must be taken into consideration. For this purpose the Biot number (Bi) is used, which for mass transfer is defined by the equation

$$Bi = (k_c L)/D, \tag{7.21}$$

where k_c is the interphase mass transfer coefficient, L the solid half-thickness, and D the mass diffusivity.

The mass diffusivity increases significantly as the temperature of the material is raised, following the Arrhenius Eq. 7.6. The activation energy (E_a) depends strongly on the physical structure of the food material, with higher values obtained in nonporous solids.

The mass diffusivity is estimated using various methods, such as drying rate, sorption rate, and penetration of diffusing component.

Food biopolymers, such as starch and proteins, reduce D due to the complex networks and sorption of the small molecules on the polymer. The mass diffusivity may change during drying due to the formation of pores.

Of particular importance to food processing is the moisture diffusivity, which may control various physical and chemical changes in foods.

Table 7.8 shows typical moisture diffusivities (D) and the energy of activation for diffusion of various food systems (Saravacos and Maroulis 2001).

The mass diffusivity (D) of solutes (small molecules) of food components is important in various food processes and in food storage. The value of D depends on

Table 7.8 Typical moisture diffusivities (D) and energy of activation for diffusion (E_a) of various food systems

Food material	Moisture %	D, $\times 10^{-10}$ m^2/s	E_a, kJ/mol
Apple	50	2.0	60
Potato	30	5.0	45
Freeze-dried fruit	5	50.0	15
Bread	30	2.0	40
Pasta	13	0.3	40
Meat	60	1.0	35
Fish	40	0.5	35

Table 7.9 Typical values of diffusivity (D) of solutes in food materials

Solute	Food material	D, $\times 10^{-10}$ m^2/s
Ethanol	Water solution	8.4
Acetic acid	Water solution	12.1
Glucose/fructose	Water solution	7.0
Sucrose	Water solution	5.4
Sodium chloride	Water solution	12.0
Lactoglobulin	Water solution	0.7
Hemoglobin	Water solution	0.7
Coffee solubles	Coffee/water	1.0
Cottonseed oil	Cottonseed/hexane	0.27

the size of the molecule of the diffusing component, the physical structure of the food system, and the temperature.

Table 7.9 shows typical values of D of solutes in common food systems.

7.7 Heat and Mass Transfer Coefficients

Design and engineering analysis of food process operations and equipment requires heat and mass transfer coefficients, which are used in transfer rate equations. Theoretical prediction of the transfer coefficients is not possible, and empirical values are used for the appropriate transfer equation.

Empirical dimensionless equations for a given operation can provide approximate transfer coefficients that are satisfactory for preliminary design calculations.

Heat transfer operations have been studied and applied more extensively than mass transfer, and more reliable heat transfer data are available in the literature.

7.7.1 Heat Transfer Coefficients

The design of heat transfer operations and equipment is based on heat transfer coefficients, which are obtained from the literature or from experimental data on industrial or pilot plant equipment.

Heat transfer controls several food process operations, such as heating, thermal processing, refrigeration/freezing, and evaporation. High heat transfer coefficients are important in process economics and food product quality.

Heat exchangers of various types are used in the heating and cooling of various fluid foods. In addition to thermal efficiency, food heat exchangers must meet strict sanitary and food safety requirements.

The convection heat transfer coefficient (h) is defined by the equation

$$q = h\,\Delta T, \tag{7.22}$$

where q is the heat transfer rate and ΔT is the temperature difference.

For complex systems, the overall heat transfer coefficient (U) is used:

$$q = U\Delta T, \tag{7.23}$$

where the temperature difference (ΔT) is the overall driving force for heat transfer and involves the entire system.

The overall heat transfer coefficient is calculated from the heat transfer coefficients and the thermal conductivities of the components of the system. In preliminary food process design, empirical values of U from the literature are normally used.

The coefficient varies widely according to the heat transfer system, for example, h = 10 W/(m^2 K) for air heating/cooling to h = 10,000 W/(m^2 K) for condensing water vapors.

Empirical dimensionless equations can be used to estimate h in a given heat transfer system. They are mainly correlations of the following dimensionless numbers: Reynolds, $Re = (d\,u\,\rho)/\eta$; Nusselt $Nu = (h\,d)/\lambda$; and Prandtl $Pr = (C_p\,\eta)/\lambda$.

Regression analysis of the data in the literature resulted in the empirical equation of the heat transfer factor (j_H), (Saravacos and Maroulis 2001):

$$j_H = \left[h/(u\,\rho C_p)\right] Pr^{2/3} \quad \text{heat transfer factor}, \tag{7.24}$$

where u is the fluid velocity, ρ the density, C_p the specific heat, $Pr = \eta/(\rho\,\alpha)$ the Prandtl number, η the viscosity, and $\alpha = \lambda/(\rho\,C_p)$ the thermal diffusivity (λ = thermal conductivity).

In heat transfer calculations, an average heat transfer factor can be used:

$$j_H = 0.344\ Re^{-0.433}, \tag{7.25}$$

where $Re = d\,u\,\rho/\eta$ is the Reynolds number.

Table 7.10 gives typical overall heat transfer coefficients that are useful in food process design.

Table 7.10 Typical overall heat transfer coefficients useful in food process design

Heat transfer system	U, W/(m^2 K)
Air heating	10
Water heating	1,000
Oil heating	300
Pulp heating	500
Evaporation	2,000
Condensation of water vapors	5,000

7.7.2 *Mass Transfer Operations*

Mass transfer operations involve the transfer of various components within a phase or between phases by molecular diffusion and natural or forced convection. Mass is transferred by concentration or partial pressure gradients. Typical mass transfer processes of importance in food engineering include drying (food dehydration), extraction (oil refining), distillation (recovery of ethanol and volatiles), and gas absorption (carbonation of food liquids).

Mass transfer rate equations require data on phase equilibria and mass transfer coefficients.

7.7.3 *Phase Equilibria*

Phase equilibria, such as vapor–liquid (distillation), gas–liquid (absorption), and liquid–liquid (extraction), are important in calculations involving mass transfer between phases.

Vapor–liquid equilibria in a binary mixture (A, B) are expressed by the equation

$$y = (\alpha_{AB} x)/[1 + (\alpha_{AB} - 1)], \tag{7.26}$$

where y and x are the mole fractions of volatile component A in the vapor and liquid phases, respectively, and α_{AB} is the relative volatility of A to B.

Gas–liquid equilibria are expressed by the equation

$$p = H\ x, \tag{7.27}$$

where p is the partial pressure of the gas, x the mole fraction of the dissolved gas, and 1/H the solubility of the gas.

Liquid–liquid equilibria are expressed by the equation

$$y = K\ x, \tag{7.28}$$

where y and x are the concentrations of component A in the extract and the residue, respectively, and K is the partition coefficient of A between extract and residue.

7.7.4 Mass Transfer Coefficients

The mass transfer of a component between two phases involves two steps – from phase I to the interphase and from the interphase to phase II. For a gas–liquid operation (absorption), the transfer rate of component A (J_A) is given by the equation

$$J_A = k_y(y - y_i) = k_x(x_i - x), \qquad (7.29)$$

where y and x are the bulk concentrations of the component in the gas and liquid phases, respectively, y_i and x_i are the concentrations at the interphase, and k_y and k_x are the mass transfer coefficients in the gas and liquid phases.

The units of the mass transfer coefficients depend on the concentration units. Since the concentration at the interphase cannot be measured easily, overall mass transfer coefficients (K_y and K_x) are used according to the transfer equation

$$J_A = K_y(y - y_e) = K_x(x_e - x), \qquad (7.30)$$

where y_e and x_e are the equilibrium concentrations of the component in the gas and liquid phases, respectively.

In mass transfer operations of interest to food processing, such as evaporation and drying, the driving force is either the partial pressure difference (p–p_e) in Pa or concentration difference (C_e–C) in kg/m^3. Thus, the transfer rate is expressed in kg/(m^2 s) and is given by the equation

$$J_A = K_G(p - p_e) = K_C(C_e - C). \qquad (7.31)$$

The transfer coefficients are given as K_G [kg/(m^2 s Pa)] and K_C (m/s).

The two coefficients are related as $K_G = K_C/(RT)$.

The mass transfer coefficients depend on the geometry of the system, the fluid velocity, and the thermophysical properties of the fluid phases.

The prediction of mass transfer coefficients is analogous to the prediction of heat transfer coefficients (h). Empirical correlations of the data in the literature are used; they contain the dimensionless numbers Sherwood ($Sh = K_C d/D$), Schmidt [$Sc = \eta/(\rho\ D)$], and Reynolds [$Re = (d\ u\ \rho)/\eta$], where D is the mass diffusivity of the transferred component (Saravacos and Maroulis 2001).

The mass transfer factor (j_M) is defined by the equation

$$j_M = (K_C/u)\ Sc^{2/3}. \qquad (7.32)$$

A regression analysis of mass transfer data of the literature yielded the following approximate equation of the mass transfer factor (j_M)

$$j_M = 1.11 Re^{-0.54}, \qquad (7.33)$$

where $Re = (d\ u\ \rho)/\eta$ is the Reynolds number of the flow system.

References

Boukouvalas CJ, Krokida MK, Maroulis ZB, Marinos-Kouris D (2006) Effect of material moisture content and temperature on the true density of foods. Int J Food Prop 9(1):109–125

Brodkey RS, Hershey HC (1988) Transport phenomena. A unified approach. McGraw-Hill, New York

Gekas V (1992) Transport properties of foods and biological materials. CRC Press, New York

Krokida MK, Maroulis ZB (1997) Effect of drying method on shrinkage and porosity. Drying Technol 15(10):2441–2458

Krokida MK, Maroulis ZB, Rahman MS (2001) A structural generic model to predict the effective thermal conductivity of granular materials. Drying Technol 19(9):2277–2290

Lazou AE, Michailidis PA, Thymi S, Krokida MK, Bisharat GI (2007) Structural properties of corn-legume based extrudates as a function of processing conditions and raw material characteristics. Int J Food Prop 10(4):721–738

Maroulis ZB, Saravacos GD (2003) Food process design. Marcel Dekker, New York

Michailidis PA, Krokida MK, Rahman MS (2009a) Data and models of density, shrinkage and porosity in food properties handbook, 2ed

Michailidis PA, Krokida MK, Bisharat GI, Marinos-Kouris D, Rahman MS (2009b) Measurement of density, shrinkage and porosity in food properties handbook, 2ed

Oikonomopoulou VP, Krokida MK, Karathanos VT (2011) Structural properties of freeze-dried rice. J Food Eng 107(2011):326–333

Rahman MS (1995) Food properties handbook. CRC Press, New York

Rahman MS, McCarthy OJ (1999) A classification of food properties. Int J Food Prop 2(2):93–99

Rao MA, Rizvi SSH, Datta AE (2005) Engineering properties of foods, 3rd edn. Taylor and Francis, New York

Reid RC, Prausnitz JM, Poling BE (1987) The physical properties of gases and liquids, 4th edn. McGraw-Hill, New York

Saravacos GD, Maroulis ZB (2002) Transport properties of foods. Marcel Dekker, New York

Saravacos GD, Maroulis ZB (2011) Food process engineering operations. CRC Press, New York

Zogzas NP, Maroulis ZB, Marinos-Kouris D (1994) Densities, shrinkage and porosity of some vegetables during air drying. Drying Technol 12(7):1653–1666

Part II
Advances in Food Process Technology

Chapter 8
Applying Advances in Food Process Engineering in a Changing World: The Industry Perspective

J. Peter Clark

8.1 Introduction

This chapter is a synthesis of two presentations at the International Congress on Engineering and Food (ICEF11) in Athens, Greece, in May 2011. The two presentations are "Food Process Engineering in a Changing World – The Industry Perspective," presented as one of two opening lectures, and "Applying Advances in Food Process Technology in Industry," which opened the session on Innovation in traditional processing.

Opinions and observations on the commercialization of food process innovations expressed here are based on over 40 years of experience involving many different technologies in many parts of the world. To provide an appropriate context, some of those technologies and locations are listed below.

Technologies:

- Freeze drying
- Membrane concentration
- Cooking extrusion
- Plant leaf protein
- Dihydrochalcone sweetener
- Retort pouch foods
- "Aseptic" fresh sausage
- 2 liter PET soft drink bottle
- High-fructose corn syrup replacing sugar in baked goods
- Acidified foods
- Blend in canned vitamin mix
- Automated material handling, cereal, and ice cream
- Reduced water consumption for ice cream

J.P. Clark (✉)
644 Linden Avenue, Oak Park, IL 60302, USA
e-mail: Jpc3@att.net

S. Yanniotis et al. (eds.), *Advances in Food Process Engineering Research and Applications*, Food Engineering Series, DOI 10.1007/978-1-4614-7906-2_8,

- Cold extraction for baby food
- Aseptic juices

International experiences have included the following locations:

- North America
- Egypt
- Moldova
- Albania
- Malawi
- Haiti
- India
- Honduras
- China
- Europe

The point is that these observations have at least some international perspective and cover a range of products, technologies, and processes.

8.2 Issues in Commercialization

First, we consider some examples of technologies that have succeeded, some that have failed to catch on, and some for which the conclusion is unclear. Then we will discuss what lessons might be learned.

8.2.1 Successes

Here are some clearly successful food technologies. Why are they successful?

8.2.1.1 Canning

Canning in metal cans, glass jars and plastic, or, more recently, paperboard composite containers is undeniably ultrasafe, with a commercial record of billions of units with scarcely any human fatalities over more than 100 years. Canning involves heating hermetically sealed containers to at least 121 °C for low-acid foods, which requires a pressure vessel or retort. High-acid foods are sterilized at lower temperatures after filling or may be hot filled, sealed, then cooled. Compared with other preservation techniques canning is relatively inexpensive and the quality of meats, fruits, and vegetables is adequate for most consumers. Products are shelf stable and ready to eat. There is a huge infrastructure in place all around the world, and the equipment is rugged and relatively easy to understand and use. Despite its long history, there is still active research conducted in the field with resulting innovations such as hydrostatic retorts, two-piece cans, rotary cookers, rotary retorts, shaking retorts, and computer controls.

8.2.1.2 Aseptic

Aseptic processing and packaging involves sterilizing foods and containers separately then filling the sterile containers in a sterile environment. Because foods are cooled before filling, containers do not need the strength to resist high temperatures and pressures, as do containers used in canning. This makes packaging less expensive – lightweight plastics, paperboard laminates, and large polymer bags, for instance. Containers are sterilized in a variety of ways: with chemicals (hydrogen peroxide) for paperboard, by irradiation for bags, and by the heat of thermoforming for plastic tubs. Very large tanks are sterilized by dilute solutions of iodoforms displaced with sterile nitrogen. The sterile environment is created by steam or chemical sterilization and maintained by sterile air filtration.

Foods are sterilized by heat in scraped surface heat exchangers or tubular heat exchangers and cooled with similar equipment. Recently, microwave heating has been investigated as a fast heating method for pumpable foods.

Foods with particulates require special care in sterilization to ensure that the center temperatures reach the minimum required for the type of food.

The flavor and texture of aseptically processed foods may be superior to that of post-package-processed foods, but the big advantage of aseptic processing is lower-cost packaging. Capital cost is higher than that for equivalent canning capacity because filling rates are generally slower for aseptics and the equipment controls are more complex.

8.2.1.3 Membrane Separations

Membrane separations include ultrafiltration and reverse osmosis, in which fluids are forced through porous polymer or inorganic membranes. Ultrafiltration removes large molecules, such as proteins, while permitting water and smaller molecules to pass. Reverse osmosis membranes permit only water to pass through, retaining all soluble material. Compared with evaporation, in which heat is applied to remove water as vapor, membrane concentration is less expensive because the energy to generate fluid pressure is less than that required to vaporize water. Because many flavors are heat sensitive, the quality of concentrates prepared by membranes is superior to those prepared by evaporation. Controlling the porosity and permeability of membranes permits difficult separations and fractionations, such as isolating various functional milk proteins from cheese whey (the liquid left when milk solids and fats are precipitated to make cheese). New membranes and process configurations continue to be developed.

8.2.1.4 Individual Quick Freezing

Individual quick freezing (IQF) is a relatively inexpensive preservation process in which single pieces of fruit, vegetable, or proteins (shrimp, fish filets, chicken chunks) are frozen, often by exposure to very cold cryogenic gases or by mechanically cooled air. This can be contrasted with freezing in bulk blocks, which is still done for some foods. IQF preserves quality, extends a seasonal supply, and is versatile, in that the same equipment can be used for many products.

8.2.1.5 Centralized Processing

Meats, poultry, and some field crops are processed in large, centralized plants, as distinguished from smaller, more localized facilities. This is primarily a business practice rather than a technology and is driven by economies of scale and reduced labor costs, often achieved by mechanization and automation. There is an increased risk of foodborne illness because more material is at risk for cross contamination, but there is also the opportunity to apply such technologies as modified atmosphere (MAP) and vacuum packaging, in which oxygen levels in shelf-ready packages are reduced to extend shelf life.

8.2.2 Limited Success

One lesson is that innovations take time to catch on in the food industry. Thus, those technologies with limited success also are relatively young.

8.2.2.1 Irradiation

Irradiation sterilizes foods without cooking or heating, thereby preserving fresh flavor, color, and texture. Irradiation uses gamma rays from radioactive isotopes such as cobalt 60, electron beams, or X-rays by aiming electron beams at metal targets. The various forms of energy differ in their penetration depth and the complexity of the equipment. Cobalt 60 poses hazards to workers and so must be kept behind shields or under water. Electron beams have limited penetration but are adequate for treating packaging material and smaller packages of food. X-rays have greater penetration depth but add complexity to the equipment.

Irradiation can address the growing issue of pathogens in foods such as ground beef, which arises in part from centralized processing and packaging. It is not very expensive compared to other preservation processes and can provide some unique benefits. On the other hand, there exists a perception on the part of consumers that irradiation is harmful. This perception is not well founded, but it is widely held and inhibits the greater adoption of irradiation.

8.2.2.2 High Hydrostatic Pressure

High hydrostatic pressure processing (HPP) involves applying pressures of 60,000 pounds per square inch (psi) or more to foods in flexible packages for relatively short time periods. Vegetative cells are killed. Due to adiabatic heating, the food temperature rises by approximately 20 °C. Additional heat may be applied in some cases to combine the effects of high pressure and heat.

HPP does not appear to inactivate enzymes, which is often an additional objective in thermal processing. So far, HPP has seen limited use, specifically for guacamole, ready-to-eat (RTE) cooked meat, and some fruit juices. The equipment tends to be small because the pressure vessel required has thick walls. Due to the cycling of pressure and metal fatigue, the equipment may have limited a service life, adding to costs.

8.2.2.3 Microwave and Radio Frequency Heating

Microwaves and radio frequencies are forms of electromagnetic energy that interact with water molecules and other food components to heat material from the inside as compared with other forms of heat transfer in which energy is supplied from the surface. The result can be very fast and selective heating. However, some desirable reactions in cooking, such as browning, may not occur. Microwaves have been applied in food processing to finish baking crackers and biscuits and in tempering frozen blocks of food. Unfortunately, ice is a poor absorber of microwave energy, so tempering or thawing must be carefully controlled to avoid uneven heating.

The power tubes used to generate microwave energy are expensive and prone to failure. Radio frequency heating has some advantages over microwaves but is not as well developed.

8.2.2.4 Retort Pouch Foods

Retort pouch foods were developed by Natick Labs of the U. S. Department of Defense as an improved military ration, replacing small, single-serve metal cans. The retort pouch is a flexible container made of laminated polymer films with an aluminum foil barrier layer to protect against light and oxygen. The critical development was a sealant layer of polypropylene that could withstand the elevated temperatures of pressure cooking. The narrow profile of the package was also presumed to provide improved flavor and texture quality compared with a cylindrical can of the same volume due to reduced overcooking.

The package was commercialized in the USA by ITT Continental Baking Company and, later, by Kraft Foods to deliver shelf-stable, RTE meat containing entrees. In the case of Continental Baking, this commercialization effort involved introducing to the public a new food category (shelf-stable entrees), a new brand (Continental Kitchens), and a new package (the retort pouch). Any one of these is a daunting challenge; it should not be surprising that the venture failed. A second line of products, Deli On Your Shelf, seemed promising because it had a lower price, a name that communicated better and novel products that invited trial. However, the venture folded before that line could prove itself. Kraft's venture also folded.

Retort pouch foods are still popular in Europe and Japan, where there are fewer frozen and refrigerated options, refrigerators are typically smaller (making shelf-stable foods more desirable), and food items that lend themselves to pouches are common (sauces, stews, curries).

In the USA, retort pouches are used as originally intended for military and emergency rations, but also for solid items such as tuna and cooked chicken. They have been used for pet foods, baby foods, and weight loss diet foods, but those volumes are relatively small.

8.2.2.5 Freeze Drying

Freeze drying involves sublimation (conversion of solid to vapor) of water from frozen ice in food to water vapor, leaving behind a porous solid close in shape to the starting piece of food. Rehydration for use, if desired, is more effective than for conventionally dried foods, which typically shrink and harden. Flavor is better than foods dried in other ways because the process occurs at low temperatures.

Freeze drying is usually a batch process occurring in vacuum chambers and taking a long time because of the low temperature. This makes for high capital and operating costs. As a result, freeze drying for foods is used, but mostly for relatively valuable products such as fruits for cereals, herbs, and meals for backpackers and emergency supplies.

8.2.2.6 Dry Pasteurization

Dry pasteurization is applied to materials such as nuts, spices, and teas that are harmed by moisture and yet are vulnerable to contamination, due in part to the places and conditions in which they are grown. Microbes are more resistant to heat when dry than when wet, so it is useful to provide moisture when using heat to kill harmful microbes. However, this must be done carefully with dry materials to prevent staling, clumping of powders, and loss of desirable texture.

At least four proprietary procedures exist for delivering this process. Some use electrically heated surfaces of conveying equipment to heat powders in a steam atmosphere, while others use chambers to first expose packaged powders to steam and then to vacuum. In general, the idea is to moisten the surface of the nuts or powders, heat quickly, then just as quickly dry the surface and cool the solids. The processes have typically been developed in Europe and slowly introduced to US markets.

8.2.3 *Examples of Little or No Success So Far*

In addition to retort pouch foods in the USA (unlike other places in the world), there are interesting process technologies that have had enough time to enjoy at least limited success but have not. It is instructive to consider these as well.

8.2.3.1 Freeze Concentration

Freeze concentration involves turning water in a solution, such as fruit or vegetable juice, into ice crystals and then physically removing them to produce a concentrated solution of soluble solids. The ice crystals are pure water. The same process can be used to generate pure water from brine, seawater, or waste streams. The flavor of the concentrate from food liquids is superior to that from thermal evaporation because only low temperatures are used in freeze concentration.

However, the costs of freeze concentration are high, even though energy requirements are theoretically lower than for evaporation because the heat from fusion is only 144 BTU/lb of water compared with 1,000 BTU/lb for vaporization. One reason is that the cost of energy as heat is generally lower than that of the same amount as refrigeration. Further, evaporators can be made multieffect, reducing the amount of energy required roughly in proportion to the number of effects (at the cost of increased equipment, of course). Thus, a seven-effect evaporator has about the same energy demand as a freeze concentrator of the same water removal capacity.

A more significant issue is the challenge of yield, in which losses of valuable solubles occur due to incomplete separation of ice from concentrate. As the concentrate increases in solids, it becomes more viscous. The low temperature aggravates this phenomenon. The viscous concentrate clings to the ice crystals.

One solution is to melt some ice to make liquid water, which washes the surface of the ice, but this dilutes the concentrate. If ice crystals grow too fast, they can occlude concentrate (capture the thick solution within the crystal). If there are too many nuclei (centers of crystal growth), there is too much surface to wash. In some systems, crystals ripen by being agitated and slightly warmed, so small ones dissolve and others grow.

All these issues make freeze concentration complex and more expensive than competing processes, such as membrane separation or well-run evaporators with essence recovery (condensation of flavor vapors for addition to concentrate).

8.2.3.2 Pulsed Electric Fields

Pulsed electric fields (PEFs) involve passing fluids through a device in which they are exposed to an alternating electric field between electrodes. The field strength is typically very high, so clearances are small. The electronic pulse generator is similar to that used in radar systems. The fields are presumed to kill bacteria by perforating cell membranes and allowing the contents to leak out. PEF has the same effect on fruit and vegetable cells and so can affect the yield of juice and the extraction of cellular substances, such as sugar. By killing the hydrophilic (water-loving) cells in sewage, PEF can improve sludge thickening in waste treatment plants.

However, so far, PEF has proven difficult to scale up because of the narrow clearances required and the high power levels. The electrodes erode and need relatively frequent replacement. PEF does not appear to impact enzymes. There are no commercial suppliers of the equipment at present.

One interesting potential variant of PEF is the use of somewhat lower power fields to make some foods digestible very quickly – a novel form of cooking. This is still in the early stages of development.

8.2.3.3 Meat Analogs

Meat analogs are products typically made from soy protein isolates, though conceivably other sources are possible, that mimic the texture of meats such as chicken, beef, or pork. Flavors and colors are added to spun fibers and fabricated into chunks, strips, or other forms. Less sophisticated meat extenders are made by cooking extrusion of soy flours or concentrates to make granules. Alone, they are not mistaken for meat, but, added to ground meat, they can be acceptable and inexpensive additives that are similar in nutrient content.

At one time, meat analogs were going to solve world hunger by replacing inefficient animal production with plants while satisfying our cultural bias favoring meat. This has not happened yet. Meat extenders are routinely used in inexpensive pizza toppings, chilis, emergency feeding programs, and other places that are impacted more by economics than choice. Meat analogs, as originally conceived, satisfy some niche markets among strict vegetarians. So far, they have been too expensive and poor in organoleptic quality for general commercial success. These are solvable issues, in principle, but they haven’t been solved yet.

8.2.3.4 Ready or Chilled Meals

Ready or chilled meals are refrigerated, complete, restaurant-quality, preplated meals centrally prepared but eaten at home, perhaps after heating briefly in a microwave oven. Several attempts to manufacture such items in the USA on a commercial scale have failed, even with large, sophisticated firms behind the effort. Modified atmosphere packaging (MAP) was applied with expensive domed packages, selling at kiosks at commuter train stations.

Issues included such short shelf lives that distribution was limited to a small area around a manufacturing site and high cost due to the manual labor involved. Much more successful are sales of RTE meal components, such as roasted chickens, salads, desserts, soups, and side dishes. These are simpler to cook and package. They often can be prepared in stores rather than more remote factories. If prepared locally, they need not be packaged for long-distance shipping. At least in the USA, the major foods centrally prepared and refrigerated are used in food service for restaurants, schools, and institutions.

8.2.3.5 Fads

One last category of less successful food technologies are those that support fads. Examples include soft-center cookies, mini cookies, comic-themed cereals, and, perhaps, organic foods. Each of these, while relatively short-lived, nonetheless offers some lessons about realistic commercialization.

Soft-center cookies used two different batters, with different amounts of sweetener to affect texture in order to produce a cookie with a brittle exterior and soft interior. Proprietary technology was claimed, but it was disputed, and the effect was relatively easy to reproduce or imitate. As a result, the product category was a short-lived novelty, and any premium value was quickly eroded. The technology was not very robust, and any benefit to the consumer was fleeting.

Mini cookies posed a similar challenge, with even less technology. These were simply small versions of existing products. Manufacturing was a challenge because equipment, such as sandwiching machines, had to be modified. Rather than develop and pay for specialized packaging machines, manual packing was used, seriously impacting margins. Versions of mini cookies and crackers are still seen in stores, but more often now they are packaged in bags rather than trays – a much less expensive approach.

Comic-themed cereals are intended to be short-lived products, usually based on a simple corn or oat flour extruded shape that is sugar coated and dyed to appeal to children. The shapes are changed to resemble some licensed cartoon character. This involves machining and installing new extruder dies and then learning how to operate the extruders and dryers for the new pieces. One wonders if all that effort is worth the cost.

Perhaps more controversially, are organic foods a fad? They are said to be one of the fastest growing categories, albeit from a small base. There are organic counterparts to nearly every food type, some sold in organic departments of grocery stores or in standalone stores, while others are sold side by side with conventional foods. Even the definition of organic is debated, with some relying on USDA certification (in the USA) and others looking to independent certification or other definitions.

The largest single category of organic foods is dairy, followed by meats and produce. There are relatively few organic processed foods, in part because of the challenge of securing all the necessary ingredients. There is an impact on manufacturing, including in relation to performing cleaning and sanitation, avoiding cross contamination, segregating and tracking ingredients, and reformulating certain products without preservatives and the usual processing aids.

A significant point is whether manufacturing organic foods is worth the bother. Do they confer a long-term benefit to the consumer? Will the interest and growth last?

8.3 Principles of Commercialization

The principles of successful commercialization are as follows:

- Economics
- Market need
- Right team
- Timing
- Robust technology

8.3.1 Economics

No technology can be commercialized if it does not offer significant benefits to the consumer. This includes being able to sell a product at an attractive price that still provides an acceptable margin to the manufacturer. Product costs are impacted by both capital and operating costs. Both of these cost categories are influenced by broad trends, such as minimizing energy consumption, waste, and water use. Typically, conservation and sustainability appear to have short-term consequences of increasing capital costs. Conventional investment analysis may make such investments unattractive. However, the long-term benefits can be significant, though difficult to quantify. Enlightened management takes a long view and recognizes that the firm and its employees are citizens of the world and need to share responsibility for the future of the environment.

8.3.2 Market Need

Is the market real or perceived? Is it a niche market or broad based? The markets for foods in the USA and Europe are relatively flat because population growth is flat. This suggests a focus on the developing world, with serious consequences for commercialization of innovations. There are stable trends that define markets: food safety, concern for health, management of chronic disease, and allowable ingredients. Food cultures are intertwining, for example, salsa, Greek yogurt, and curries have become popular outside of their places of origin. Finally, if there is a real market need, can it be effectively communicated?

8.3.3 Right Team

The right team for successful commercialization of innovation is multidisciplinary and multifunctional. This means that it includes engineers, food scientists, marketers, financial managers, operations, logistics – all the elements needed to take a concept from idea to store shelf. The team also needs an executive sponsor capable of making decisions that commit the firm financially. There are multiple phases to a typical commercialization project, and the appropriate leadership skills may change with the phase. This means that team leadership may change and adapt appropriately. At the same time, coherency and continuity are critical, so membership in the team should be stable, if possible. Each change in team membership causes loss of time and accumulated experience.

8.3.4 Timing

The timing of a commercialization project needs to be favorable in the sense that the market is ready, there is sufficient time, and the resources are available. The team cannot always control these factors, but it must be aware of them. Development efforts are often underestimated. Milestones are established early and often are immovable. Examples are launch dates for advertising, lead times for equipment, printing, and ingredients, and budget deadlines. Nonetheless, critical decisions are often delayed, compressing time severely.

There are lessons to be learned from past failures, including how to motivate decision makers.

8.3.5 Robust Technology

Robust technology is appropriate for the location where it is to be used. It is also compatible with existing operations (if any) and is comprehensible or easy to understand. Especially in the developing world, these requirements translate into a few concepts:

- Keep it simple
- Combine steps if possible (e.g., premixes)

- Build on indigenous materials and culture as appropriate
- Keep it inexpensive
- Use local materials if possible
- Reduce waste
- Respect culture
- Address nutrition and food safety

8.4 Conclusion

The adoption of innovations in the food industry can be slow because the industry is conservative, margins are low, adequate processes and infrastructure are already in place (in the developed world), and there is an overriding and appropriate concern for food safety. On the other hand, cost is more important than quality, so any development that offers cost reduction with no impact on safety will be of interest. The most successful innovations address real needs, including developing concerns about the impact of food on long-term health.

There is great opportunity in the developing world because processed foods actually accelerate development by releasing labor, especially that of women, to the economy. Women in Africa spend long hours preparing food when they could be acquiring an education or performing more rewarding work.

Finally, innovations in food processing must contribute to sustainability by reducing consumption of energy, water, and other resources while utilizing more scarce agricultural materials.

Chapter 9
Recent Developments in Drying Technologies for Foods

Sachin V. Jangam and Arun S. Mujumdar

9.1 Introduction

It is well known that several hundred designs of drying equipment are in commercial production due to the diverse physical forms of wet feedstocks to be dried to different specifications and at different production capacities. In fact, even a cursory overview of the published literature on drying research and development (R&D), which has exploded over the last three decades, shows that in particular the number of dryer types that have been proposed and examined at the laboratory scale – and in a few cases at the pilot scale – runs into several hundred. Some of the conventional dryers, such as fluidized beds or rotary dryers, have scores of variants, some of which can be considered innovative due to their nonconventional features. Of course, what was deemed innovative three decades ago cannot be considered innovative any longer. Nevertheless, many of the so-called innovative dryer concepts have not yet made it into industrial practice in a big way. In this chapter we will consider a select set of developments in drying technologies that are not commonly found in industrial practice but that do offer some key advantages, which warrants their serious consideration. This review is biased in favor of ideas that have been published in archival literature or presented at conferences. There are remarkably few patents on industrial dryers, probably because of the high cost of patenting and limited return on investment in protecting the intellectual property (IP). Many of the ideas presented openly in the literature are undoubtedly used by some industries dispersed around the world, but it is not possible to determine who is applying which innovation in

S.V. Jangam (✉)
Department of Chemical & Biomolecular Engineering, National University of Singapore, 4 Engineering Drive 4, Blk E5 #02–09, Singapore 117585, Singapore
e-mail: chejsv@nus.edu.sg

A.S. Mujumdar
Department of Chemical & Biomolecular Engineering, Hong Kong University of Science and Technology, Clear Water Bay, Kowloon, Hong Kong

S. Yanniotis et al. (eds.), *Advances in Food Process Engineering Research and Applications*, Food Engineering Series, DOI 10.1007/978-1-4614-7906-2_9,

practice except by chance. Readers should be aware of this inherent limitation of any such overview of drying technologies.

As with all innovations, many of the recently developed drying systems are incremental in nature, that is, they enhance performance of a well-known drying system marginally but in a cost-effective manner at low risk and low incremental cost to the user. Often they are simply combinations of well-established technologies that take advantage of each individual technology. For example, when drying a solid with both external and internal moisture, one may employ a dryer type that is particularly effective in removing surface moisture and follow it with another type that cost-effectively removes internal moisture that requires a longer residence time. This leads to the well-known spray dryer, fluid-bed, two-stage dryers for, for example, milk and detergents. A conventional spray dryer cannot provide the residence time needed for full drying, so one can only surface-dry droplets in the spray chamber and then transfer the semiwet particles into a fluid bed or a vibrated bed dryer to take out the internal moisture that requires a longer dwell time. This is incremental innovation. Radical innovations in drying are very few and rarely implemented due to the high risk in scale-up, high capital costs, and lack of prior experience. As the cost of energy rises and carbon taxes are imposed around the world, there may be more incentive to taking the necessary risks and going for radical innovations in decades to come. In this chapter we will limit ourselves to some recently developed drying concepts and examine their advantages and limitations. Much R&D needs to be done in most cases along with proper life cycle analysis (LCA) before selecting new drying technologies. It is important to state at the outset that new or innovative drying technologies do not automatically imply superior dryers for a given product because a well-designed and operated conventional dryer may in fact be better than a novel one.

Kudra and Mujumdar (2009) have provided a detailed overview of recent drying technologies. For an in-depth analysis of most conventional dryers the reader us referred to the Mujumdar (2007a), Jaya and Sokhansanj (2007) and a recent review article (Jangam 2011a). In this overview we will not examine conventional dryers in any depth. It should be pointed out that often, although not in all cases, drying of foods may be preceded by some form of pretreatment step, for example, osmosis, mechanical, or chemical treatment, to enhance drying rates or modify a quality parameter. Sometimes a field such as ultrasonics or microwaves may be applied to speed up drying. Blanching is often applied to disinfect foods or inactivate enzymes. We will not consider these processes in this overview. There is adequate literature on these aspects that the reader can access readily.

9.2 Need for New Developments in Drying

Mujumdar (2007a) and Kudra and Mujumdar (2009) have identified many reasons for the development of new dryers, although a massive number of them are already discussed in the literature. They are:

- Higher thermal efficiency,
- Better control of product quality,

- Better control of dryer and drying systems,
- Lowering the overall cost (capital and operating),
- Higher capacity than existing drying systems,
- Sustainability in overall process,
- Drying new products such as functional foods,
- Reducing the risk of fire, explosion, and toxic hazards.

Indeed, almost every type of traditional dryer has numerous variants that possess some unique characteristics. It is highly unusual to have a variant that only possesses advantages and no limitations. Thus, the final selection is always a compromise based on the needs of the product specification. Selection of dryers is a complex processes; it becomes increasingly complex as the variety of dryers keeps increasing. It is worthwhile to note that the inherently nonlinear nature of drying at both macroscopic and microscopic levels makes scale-up of dryers a challenge. This is one of the reasons that one finds few truly innovative technologies brought into practice. This is true of drying as well.

In this chapter we will examine a selection of dryer types for drying particulates and slurry-form materials, which constitute most of the foods that are dried industrially. The overview is not all-inclusive. We have selected the following topics for further discussion.

- Heat pump drying
- Superheated steam drying
- Impinging stream drying
- Intermittent batch drying
- Variable pressure drop drying
- Contact-sorption drying
- Spray drying advances
- Hybrid drying techniques
- Pulse combustion drying
- Sustainability in food drying
- Mathematical techniques

9.3 Recent developments in food drying

9.3.1 Heat-pump-assisted drying

Although heat-pump-assisted drying is not new, it can be considered a form of nonconventional drying since it is still not commonly employed in industry. It has great potential in improving energy efficiency and product quality. A heat-pump-assisted dryer (HPD) consists of two main components, a heat pump system (mechanical/chemical or other type) and a convective dryer. The basic principle of a closed-loop HPD is described in Fig. 9.1, where a heat pump system is coupled

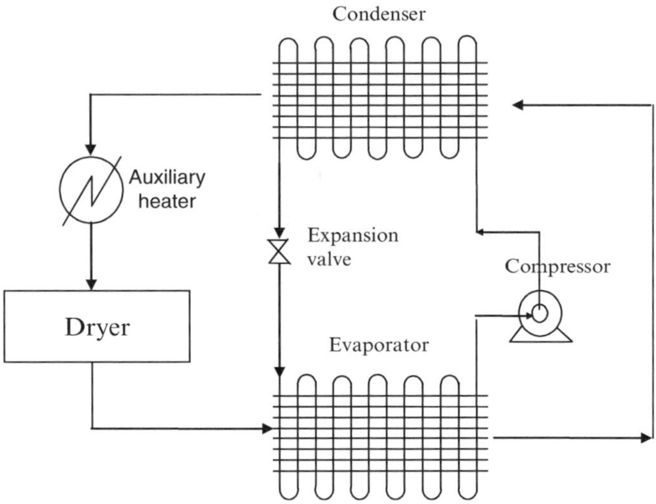

Fig. 9.1 Basic configuration of heat pump dryer with auxiliary heater

with a convective dryer (Islam and Mujumdar 2008a, b; Alves-Filho and Efremov 2009; Jangam and Mujumdar 2011).

The most important feature of HPD is the use of low-temperature, dehumidified air as the convective drying medium. The heat pump recovers the sensible and latent heats by condensing moisture from the drying air. The recovered heat is recycled back to the dryer through heating of the dehumidified drying air; hence, the energy efficiency is increased substantially as a result of the recovery of heat, which otherwise is lost to the atmosphere in conventional dryers. The possibility of a wide range of operating conditions (air temperature, humidity) can be used to make a better quality product. The closed system of a heat pump is very useful for the retention of volatile components of the foods to be dried, which otherwise are lost using the common convective dryers (Islam and Mujumdar 2008a, b).

The HPD system also has several limitations, such as maintenance, higher capital cost, and negative environmental impacts. Hence, there is tremendous scope for improving the performance of HPD in several ways, such as by using multistaging of the mechanical heat pump cycle, cascade heat pumps, or heat pipes to improve the coefficient of performance (COP) – a measure of the efficiency of a heat pump system. Other systems, such as absorption refrigeration and chemical heat pumps, have also been used recently to enhance performance. Of these, the chemical heat pump (CHP) system has a good potential for high-temperature applications (Ogura and Mujumdar 2000). The CHP stores or absorbs thermal energy in the exhaust from convective dryers, solar energy, or geothermal energy in the form of chemical energy via endothermic reactions in specially designed reactors and rejects heat at a desired level via exothermic reactions (Wongsuwan et al. 2001). CHPs operate by using thermal energy from the heat source and release no gases that may have a negative environmental impact. Advances in CHPs involve mainly developing new chemical reactions for wider temperature range applications and less environmental impact. CHP systems have also been used along with solar energy (Daghigh et al. 2010).

One advantage of a CHP over a mechanical heat pump is its eco-friendly nature. CHPs operate by using thermal energy from the heat source and release no gases that may have a negative environmental impact, which is why it is environmentally friendly. There is scope for R&D to enhance, for example, reactor performance, reactor design, and new materials for chemical heat pumps. Recently, Jangam (2011) reported on several challenges and R&D opportunities in CHP drying.

Several heating modes have also been used along with convective heating mainly to enhance the drying rate during final stages. This includes microwave, infrared, Radio Frequency (RF), and solar energy individually or simultaneously. Solar-assisted heat pump dryers have tremendous potential to make the HPD system more sustainable. Solar heat can be used for auxiliary heating either by direct heat transfer to the dehumidified air or using the stored solar energy in a phase-change material (Daghigh et al. 2010). On the other hand, solar energy can also be used to evaporate the refrigerant in an evaporator if insufficient energy can be extracted from the air. Hawlader et al. (2008) reported such a complex solar-assisted heat pump drying system that uses solar energy for both heating the refrigerant in an evaporator as well as heating the drying air. Daghigh et al. (2010) reviewed the application of solar-assisted heat pump drying technique to agricultural and marine products. They listed a number of applications of various solar-assisted HPD system layouts.

Mujumdar (1991) first proposed the use of intermittency in HPD some 20 years ago. This can save considerably on operating and capital costs since it allows the use of a lower-capacity heat pump or a single heat pump to service several drying chambers. This strategy is currently under active R&D at several laboratories. Furthermore, it is possible to use a smaller heat pump to service two or more drying chambers in cyclical mode, which may dry the same or different products in different chambers (Mujumdar 2006; Chua et al. 2002). Islam and Mujumdar (2008a) proposed a complex HPD system that makes use of multimode heat inputs and multiple drying chambers. The interested reader can refer to those researchers' book chapters and research articles (Islam et al. 2003; Islam and Mujumdar 2008a; Jangam and Mujumdar 2011).

Atmospheric temperature freeze drying can be accomplished using a heat pump. Vacuum freeze drying is very expensive as a result of using a highly efficient vacuum system and refrigeration system for very long drying cycles. However, freeze drying can also be achieved at atmospheric pressure if controlled drying conditions are attained. Atmospheric freeze drying (AFD) can be achieved using, for example, a heat pump system or a vortex tube. A heat pump system can be efficiently used to carry out AFD using air at a very low temperature, below the freezing point (generally between -3 and -10 °C). The process is very similar to the common HPD, except that the air is cooled to a very low temperature before being used to dry the frozen products (Claussen et al. 2007a). The drying rates are very sluggish; however, excellent product quality can be achieved at lower cost compared to conventional vacuum drying. Interested readers can refer to a detailed review by Claussen et al. (2007a), which presents a detailed concept of AFD, applications in various sectors including foods, recent developments, and challenges. A continuous production line is a key factor in the development of such a technique, and the first continuous plant for drying of vegetables was started in 2005 in Hungary, designed by a Norwegian company. It is worth noting that the technique has been successfully applied to various food products, such as green peas, red chili, sweet corn, and others (Alves-Filho and Strommen 1996; Strommen et al. 2004; Claussen et al. 2007a, b; Alves-Filho et al. 2007). In the case of vortex tube AFD, compressed air is passed through a well-known mechanical device known as a vortex tube. The air is then split into two streams of high- and low-temperature air. The lower-temperature air stream from the vortex tube is then used for AFD. The use of this technique is not feasible in terms of cost-effectiveness for high throughput.

9.3.2 Superheated Steam Drying

In superheated steam drying (SSD), drying air in a convective dryer is replaced with superheated steam. Superheated steam can be used in different types of convection dryers. Figure 9.2 shows the simple concept of SSD. Superheated steam at a certain pressure enters the drying chamber and removes the moisture from the wet solids; the exhaust from the dryer is also superheated steam, although at a lower specific enthalpy. Part of the steam can be recycled back after compression, and the excess can be either used directly or purged out of the system. Figure 9.3 shows the possibilities of various convective dryers that can use superheated steam as a drying medium. Superheated steam is an attractive drying medium for some products that yield better product quality in the absence of oxygen. The net energy consumption of SSD can be minimized only if the exhaust steam coming out of the dryer is utilized elsewhere in the plant and not "charged" to the dryer (Mujumdar 1990, 2006; Kumar and Mujumdar 1990; Shibata 2011). To achieve efficient energy recovery from the exhaust steam, it is necessary to have a leak-proof system to avoid air leakage. The use of superheated steam eliminates the risk of fire and explosion hazards as a result of an absence of oxygen.

Typically SSD can also give higher drying rates in both constant and falling rate periods under certain conditions. The higher drying rates in constant rate periods are possible only when the temperature used is above the inversion temperature. During the constant rate period, where hot air and superheated steam are used at the same operating temperature, the driving force (difference in the surface temperature and drying gas temperature) is higher in air drying. This is because the solid temperature is equal to the wet bulb temperature in the case of air drying, while in SSD, it equals the saturation temperature of steam at the operating pressure. However, the heat transfer coefficient in air drying is lower than that in SSD

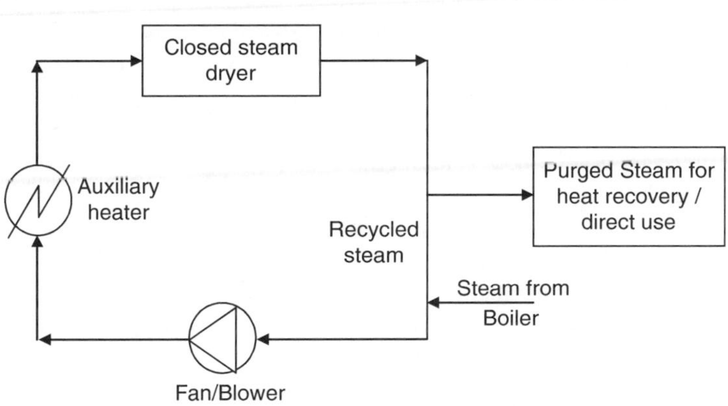

Fig. 9.2 Typical scheme of superheated steam drying system

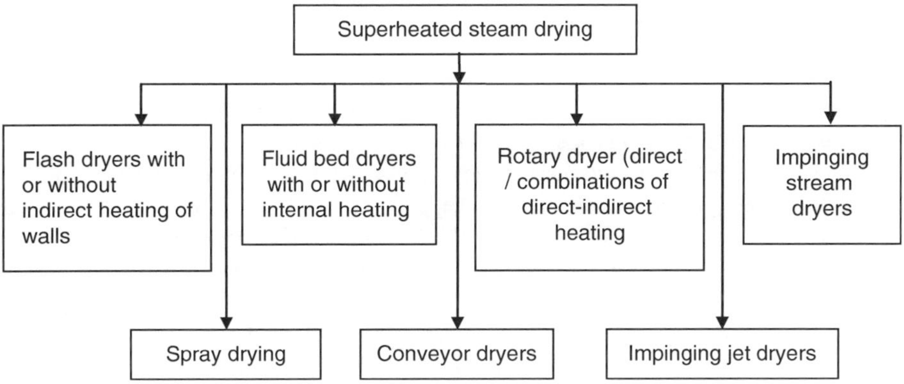

Fig. 9.3 Types of superheated steam dryers

because of the better thermal properties of steam. Hence there exists a certain temperature called the inversion temperature above which the drying rates in SSD are higher than air drying in a constant rate period. However, the reason for the higher drying rates in falling rate periods is different. The surface temperature is higher in SSD, and hence the water mobility is higher in falling rate periods, which results in faster drying. This can, of course, adversely affect product quality.

The quality of superheated-steam-dried products tends to be better than that from conventional hot air dryers because the surface does not collapse. SSD also allows for the pasteurization, sterilization, and deodorization of food products. This is particularly important for food products that require a high standard of hygienic processing. A closed SSD system enables recovery of odor, dust, or other hazardous components. The desirable organic compounds can be recovered and fed back to the dried product. Mujumdar has discussed the principles, advantages, limitations, and diverse applications of SSD technologies in a number of papers and books (Mujumdar 1990, 1992, 2007b; Kumar and Mujumdar 1990), including the

Handbook of Industrial Drying (Mujumdar 2006). However, SSD has several limitations such as the complexity of the process, which requires that the system be made leakproof to avoid leakage of atmospheric air into the system, which will result in problems during condensation and compression if the steam is recycled. SSD cannot handle products that melt or undergo glass transition unless low pressures are employed. At low pressure the low density and thermal conductivity of superheated steam lowers the heat transfer coefficient and makes the process very slow.

A number of products, such as fruits and vegetables, have been successfully dried in low-pressure, superheated steam dryers (Hasibuan and Daud 2009; Speckhahn et al. 2010; Pronyk et al. 2010). Since heat transfer for drying is still by convection, the drying rates are very low at low steam pressures. Although one can obtain good quality at low pressures, the process is still not popular due to the large equipment size as a result of the low drying rates. It may be necessary to include supplementary heat sources, for example, microwaves, radiation, or conduction, to speed up the drying rates at low steam pressures. As energy costs rise and legislation on carbon emissions becomes more stringent, SSD will become more attractive.

SSD entails considerable capital costs, along with the high cost of energy recovery, which makes the operation less feasible for higher throughput. The use of more efficient compressors for the recycling of steam will increase the use of this technology for various applications. The exhaust heat also needs to be utilized either to preheat the product or supply the heat indirectly to the solids in order to enhance overall efficiency. Cleaning of steam for recirculation or compression is another difficulty associated with SSD. Currently, SSD has very special applications; hence more work needs to be carried out at the laboratory and pilot scales to make wide use of SSD. Capital costs are higher for SSD, which has also slowed down its adoption for industrial applications.

9.3.3 Impinging Stream Drying

Impinging stream dryers are dryers in which the evaporation of moisture from the solids or liquid droplets occurs in an impingement zone, which develops as a result of a collision between two opposing gas streams out of which at least one stream has wet dispersed particles to be dried. The impingement zone offers very high heat, mass, and momentum transfer, which results in the rapid removal of surface water. A detailed discussion of the dryer concept can be found elsewhere (Kudra and Mujumdar 2009). These dryers are mainly innovative alternatives to flash dryers to dry particulate solids of very high moisture content (Kudra and Mujumdar 1989; Kudra et al. 1995). Other advantages of impinging dryers include their smaller footprints and high robustness due to the absence of moving parts. However, the design of such a system is very important, particularly the feeding arrangement and the design of the impinging pipes, which affect the value of the volumetric heat transfer coefficient and, in turn, the water evaporation rate. Studies on impinging stream drying (ISD) are still rather limited, and no major industrial applications have been reported.

Recently, Choicharoen et al. (2010) carried out a performance evaluation of impinging dryers with okara as a model material and concluded that ISD gives very high volumetric heat transfer coefficients, which can be a significant limitation. The ISD designs proposed in Kudra and Mujumdar (1989, 2009) have been, examined in detail by Nimmol and Devahastin (2010) and Choicharoen et al. (2010). However, there are some 30 other variants that need to be examined via modeling and by experimentation. There is much scope for innovation here, but few efforts have been reported to date. It is worth pointing out some limitations of lab studies. The values of volumetric heat/mass transfer coefficients are sensitive to the characteristic values of particle diameter and dryer volume, where the heat/mass transfer processes actually occur. Very short drying times for some materials, where internal heat/mass transfer rates should govern the kinetics, can possibly be due to microcracks in the material due to mechanical collisions in the impingement zone, which give easy pathways for the moisture to diffuse out. This can affect product quality, although the drying times are very short. Scale-up of impinging stream dryers is therefore a major problem.

9.3.4 Intermittent Batch Drying

By varying the airflow rate, temperature, humidity, or operating pressure individually or in tandem, the operating condition of a drying process can be monitored to optimize the energy input and maintain product quality. The objective is to obtain high energy efficiency without subjecting the product to temperatures beyond its permissible temperature limit while maintaining a high moisture removal rate. There are two ways to apply intermittent heat input. The first one is to subject the drying materials to intermittent heat input or time-varying flow of the drying medium or use cyclically varying operating pressure in the drying chamber. The main purpose is to allow internal moisture to migrate to the material surface during the nonactive phase of drying, often termed the tempering period. Intermittent drying consists of two distinctive drying periods, active drying and nonactive drying. During active drying, heat input is applied by the drying medium, while during the nonactive drying period, heat input or flow of the drying medium is stopped or reduced. The two distinctive periods are carried out in an intermittent mode. Since water content at the surface is increased during the tempering period, the drying rate during subsequent active drying period is increased. However, since the rate of drying is very slow during the passive period the overall drying time is increased somewhat but it is offset by the reduction in energy consumed and improved product quality. Islam et al. (2003) have carried out a numerical study of the time-varying heat input in convective dryers and suggested various routes to enhance drying rates.

The second intermittent drying strategy is to apply stepwise changes in operating conditions in order to minimize the total energy requirement. The drying rate toward the end of the process is controlled by internal diffusion where the external factors have limited effect on the drying kinetics. As such, one way to reduce energy loss is to gradually reduce the heat input to the materials during the drying process. One can also vary the mode of heat input (e.g., convection, conduction, radiation, or microwave/radio-frequency heating). Multiple heat inputs can be used to remove both surface and internal moisture simultaneously.

Intermittent drying can be applied to any direct dryer or batch dryer such as, for example, a tray dryer, convective dryer, conveyor dryer, fluidized-bed dryer, and spouted-bed dryer. Mujumdar (1991) identified and proposed for the first time the use of multiple modes of variable levels of heat input, simultaneous or consecutive, as well as cyclical variations in velocity or operating pressure as technologies of the future for batch and continuous HPD processes. Using multiple modes of heat input, it is possible to speed up drying kinetics without adversely affecting the quality of the dried products. Dryers such as rotary, spouted-bed and multicylinder paper dryers are all inherently intermittent since heat is supplied intermittently due to the inherent operational mode of the dryer, although none of the operating variables, such as flow rate, temperature, or pressure, is altered with time. They are no longer termed intermittent since the on and off times of heat input cannot be altered independently of the other operating variables.

9.3.5 Variable Pressure Drop Drying

Variable pressure drop drying – sometimes called swell drying – is carried out either by the so-called controlled instantaneous pressure drop method (Déetente Instantanée Contrôlée, "DIC") or the cyclic pressure drop method [Déshydratation par Détentes Successives, (DDS)] (Maache-Rezzoug et al. 2002; Cong et al. 2008). The DIC method, proposed by Allaf and coworkers, consists of three stages. It starts with a vacuum created in a drying chamber, followed by injection of steam to the product for some time, followed by a sudden reduction of pressure. Because of the sudden change in pressure, vaporization of moisture takes place, accompanied by swelling of the material, resulting in a product of high porosity. This also increases the specific surface area and reduces the diffusion resistance there by increasing the drying rate in final stages. However, in a successive decompression DDS process, a thermo-sensitive product undergoes a series of cycles, during which it is placed at a certain pressure for a defined time and then subjected to an instantaneous decompression in a vacuum (Maache-Rezzoug et al. 2002; Haddad et al. 2004). The product is maintained under a vacuum for a defined time before the following cycle begins. Each decompression step results in partial water removal by autovaporization. The major advantages of these technique are a reduction in drying time and highly porous dried products with no shrinkage. These techniques of instantaneous and cyclic pressure drop have been successfully applied to the industrial drying of various food products including grains, fruits, medicinal herbs, vegetables, and fish (Maache-Rezzoug et al. 2002; Cong et al. 2008; Abdulla et al. 2009; Abdulla et al. 2010). This process has been in commercial use since 1993, mainly in France.

9.3.6 *Contact-Sorption Drying*

Contact-sorption drying is a combination of two drying techniques, contact drying, in which a wet material is contacted with a heated surface causing the removal of the moisture as a result of heat exchange, then sorption drying, in which moisture is transferred from wet solids to the sorbent particles. The efficiency of drying can be enhanced by use of adsorbents such as zeolites, which have a strong attraction for water. A typical contact-sorption drying technique involves good mixing of wet solid particles with the sorbent particles to achieve good heat and mass transfer between the two and then separation of the two media. The sorbent particles are regenerated and returned to the dryer. Typical inert sorbent particles (also called as carrier) used for this purpose include molecular sieves, zeolites, activated carbon, and silica gel. Typical applications of contact-sorption drying include the drying of various grains and pieces of fruit. The fruit pieces can also be dried in the presence of sugar granules, which absorb moisture from the wet solids and result in a sweeter taste. Kudra and Mujumdar (2009) discuss in more detail the contact-sorption drying system. Rahman and Mujumdar (2008) proposed AFD for foods in a vibro-fluidized bed dryer of adsorbent particles. This resulted in a reduced drying time and enhanced product quality. These adsorbents can also be used for the dehumidification of air to be used for drying. The system consist of two beds of sorbent particles, one in desiccation mode (to remove water from the drying air) and the other in regeneration mode. As noted earlier, adsorbents must be edible, stable, and easy to regenerate for the drying of foods. The overall system is rather complex and hence not commonly used in practice.

9.3.7 *Spray Drying*

Spray drying (SD) is a well-known and very commonly used industrial technique for large-scale drying of heat-sensitive liquid, slurry, or pasty food material into powder form. The particular objective in SD is to achieve rapid removal of liquid without damaging product quality. SD has a number of applications in the food sector including, for example, the drying of fruit juices, instant coffee, whey, and milk products and the encapsulation of essential oils and volatiles and other extracts. Although SD has the advantage of being able to dry rapidly without increasing the product surface temperature very much, the most known limitation of spray dryers is their very low energy efficiency because high-temperature drying air is used. Wall deposition is another common problem in SD. Hence, the most important things to consider in the improvement of SD design are energy efficiency, modifying the drying chamber design to reduce wall deposition, and improving the atomization technique to simulate particles of expected sizes and shapes.

The pressure nozzle and rotating disc atomizer are the most commonly used atomization techniques in SD depending on the scale of operation. However, recent trends and demands for specially engineered particles of a particular size and size distribution and atomization of highly viscous fluids, which are otherwise difficult to atomize using conventional techniques, has resulted in the development of novel atomization techniques, such as ultrasonic atomizers and multiple fluid channel atomizers. Walzel (2011) has reviewed various spraying methods and their effect on the quality and morphology of the spray-dried product.

Recently, numerous research articles have been published on improving spray dryer performance, both experimentally and using advanced computational tools. Cal and Sollohub (2010) reported a useful review of spray drying in general. They discussed various aspects including different atomizers and their selection, the design of drying chambers, various air-droplet contact systems, and other aspects. Kuriakose and Anandaramakrishnan (2010) presented an extensive review of the use of computational fluid dynamics (CFD) in SD of food products. They summarized the numerous approaches used by researchers to study, for example, air flow patterns, particle trajectories, particle residence time distribution, and chamber design to reduce wall depositions, along with the key findings. They also summarized novel approaches to the simulation of SD, including a drying kinetic model, population balance model, and the reaction engineering approach (REA) used by Li and Chen (2006) in some of their recent work. It was concluded that computational tools such as CFD can be effectively used for better design and control of industrial spray dryers. However, the researchers also pointed out various challenges for further CFD work in SD, such as the modeling of particle interactions, modeling of turbulence, agglomeration, wall depositions, and thermal degradation of particles.

9.3.8 Hybrid Drying Techniques

Using a single drying technique does not allow much flexibility of operating conditions for most of the drying, with the heat pump dryer being, perhaps, the sole exception. For this reason hybrid drying techniques have become more popular because of the flexibility they give to drying operations. Hybrid drying systems may include either the use of more than one dryer for drying a particular product (multistage drying), the use of more than one mode of heat transfer, various methods of heat transfer, or multiprocessing dryers. Multistage drying is a widely accepted method of enhancing drying performance. Figure 9.4 describes various

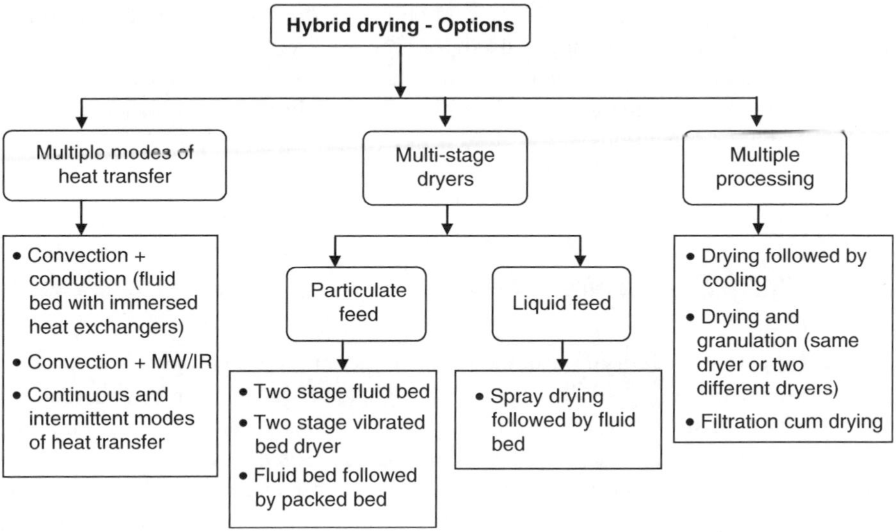

Fig. 9.4 Possible hybrid drying systems for food drying

multistage drying options for various feed properties. For particulate drying, either variants of fluid bed or fluid bed with some other technique can be used in series to achieve faster drying. However, for liquid feedstock, generally SD is followed by a fluid-bed dryer to reduce the moisture content to an acceptable level, which is not possible using a spray dryer alone (Kudra and Mujumdar 2009). Infrared drying is also useful for the removal of the final traces of moisture at a faster rate. These techniques are typically combined with other drying methods to overcome the limitations of uneven heating resulting from focusing, corner and edge heating, inhomogeneous electromagnetic fields, and irregular shape and nonuniform composition of material. However, the startup costs for these techniques are relatively high and sophisticated mechanical and electronic components are required (Zhang et al. 2006). Microwave vacuum drying has been shown to produce dried products with improved texture and color. Microwave fields allow for volumetric heating whereby heat is transferred to the inner core of a material without the need for a temperature gradient, even in the initial stage of drying. A combination of microwave and vacuum drying results in improved color and texture of dried products over air-dried products. The reduction of drying times in microwave assisted drying is beneficial for color, porosity, aroma, shrinkage, and rehydration. Microwave-assisted freeze drying is another recently developed hybrid technique for various food products. The use of microwave freeze drying results in much faster drying rates with very high product quality. A recent review by Zhang et al. (2010) explains in detail the advances in microwave-assisted drying of vegetables, fruits, and aquatic products with a focus on drying kinetics and product quality. The method has already been applied for numerous food products to prepare newfangled snack foods. Woo and Mujumdar (2010) reviewed the application of electric and magnetic fields in freezing and their effect on freeze drying. Spray freeze drying is another hybrid drying technique used to prepare highly porous particles. The techniques involve spraying of liquid in a liquid nitrogen atmosphere and then freeze drying the frozen droplets. However, this technique is more common in pharmaceutical applications.

9.3.9 Pulse Combustion Drying

Pulse combustion (PC) drying is another drying technology with good potential as the reported work suggests it offers very high heat transfer rates with reduced emission (Zbicinski 2002; Wu and Mujumdar 2006; Kudra 2008; Kudra and Mujumdar 2009). Nevertheless, the technique has some issues including noise and scale-up issues, which need to be sorted out before any progress is made. The term pulse combustion essentially means an intermittent (pulse) combustion of gaseous, liquid, or solid fuels. Such a periodic combustion of fuel generates intense pressure and velocity and, to a certain extent, temperature waves propagated from a combustion chamber that can be used for different purposes such as drying. The mechanism of PC is very complex and discussed in detail elsewhere (Kudra 2008; Kudra and Mujumdar 2009).

In general, PC drying can be achieved either within the gas jet coming out of the combustor or in a separate drying chamber. PC drying technology can utilize one or multiple pulse combustors to produce high-temperature and high-velocity pulsating jets. Short drying time, high energy efficiency, improved product quality, and environmentally friendly operation are noted as the key advantages of this drying technology (Kudra and Mujumdar 2009). The drying of liquid material within the gas jet from the combustor is another way of carry out PC drying. Such a technique is especially useful for biomaterials. The technique mainly involves the atomization of liquid using the sound energy generated by the combustor (Zbicinski et al. 2000). The liquid or pasty material is introduced into a gas stream coming out of the combustor entering at the top of the drying chamber. The particle size achieved using such a system is very fine, with a residence time of milliseconds in the high-temperature zone. This allows drying of even highly temperature-sensitive products. PC spray dryers can handle corrosive products because there are no wear parts in the atomization system. This system has been used for a number of food products such as yeast, spices, fibers, and eggs. Kudra and Mujumdar (2009) have given a detailed comparison of conventional and PC spray drying based on various characteristics such as atomization, heat transfer rates, particle properties, and overall efficiency. Considerable R&D is essential at each scale to make this technique acceptable for a wider range of products.

9.3.10 Enhancing the Energy Efficiency of Industrial Dryers for Foods

As stated earlier, drying is a highly energy-intensive unit operation and competes with distillation (Baker 2005; Mujumdar 2006). Food dehydration especially uses considerable energy for some reasons such as temperature sensitivity, which limits the drying temperature and, hence, leads to longer drying times. Another reason for the high energy use involved in food dehydration is that the drying of food products is essentially diffusion controlled, which results in longer drying times. Baker (2005) has summarized various reasons for the poor efficiency of industrial dryers in general and gives some su ggestions as to how to enhance dryer efficiency. Baker states that legislation, monitoring of dryers, and the intensification of the drying process are three interrelated factors that affect energy consumption. Some possible methods of achieving energy efficient drying systems are periodic auditing, process monitoring, use of waste heat, renewable energy, recovery of energy from dryer exhaust, and the recirculation of the drying medium, which can result in greater energy efficiency. Recently, Kudra et al. (2009) proposed a simple Excel-based calculation tool to examine the energy performance of convective dryers. The calculated values of specific energy consumption and energy efficiency can then be used to compare with the ideal adiabatic dryers to identify the scope for improvement. HPD – as discussed earlier – is another useful way of increasing energy efficiency. Recently, self-heat recuperative drying has been reported for dryers using various types of drying media (Fushimi et al. 2011). In this technique, the drying gas is essentially compressed and recycled back for drying as well as partial conductive heating of drying solids. This system has been reported to give very high energy efficiency. However, the process seems unfeasible for drying media other than superheated steam because condensation and compression (two main steps) are difficult and very expensive.

9.3.11 Sustainability in Industrial Drying of Foods

The amount of energy used is directly proportional to the emission of greenhouse gases. The high amount of energy required for drying is mainly supplied by fossil fuels, which have high global warming potential. Recently, companies have demonstrated an awareness of the issue and a desire to make the processes sustainable – mainly by reducing greenhouse gas emissions and overall energy consumption. In the previous section, some ideas were presented about how to minimize energy consumption; however, using renewable sources of energy has tremendous potential to achieve the objectives of sustainability. Renewable energy sources, for example, solar and wind, should be used frequently to minimize fossil fuel usage as current concerns over potential energy shortage and global climate change will likely result in legislative action.

A solar drying system, particularly for agro and marine products, is viable and has proven to be very useful, especially in developing countries where labor costs are low and the cost of fossil fuel energy is very high. In the future, larger systems should be designed that utilize solar thermal, photovoltaic panels combined with wind power. The notable limitation of common renewable energy sources is their intermittent availability; however, this limitation can be overcome by applying the new ideas to thermal and electrical energy storage. Biomass is also used in some places to replace oil and gas as a source of backup heating in the absence of solar/wind energy. There have been few reports on the storage of thermal energy using water pools, pebble beds, and in phase-change materials that could be connected to the use of intermittent energy sources like solar and wind energy. However, much more R&D is needed at the systems level to make this concept commercially viable. The main part of the R&D is needed on the fabrication of an efficient solar collection system, better air circulation in solar dryers for uniform product quality, use of appropriate biomass as a supplementary option, and use of wind energy for drying (either for heating or blowing of drying air) wherever possible. Since all renewable energy sources are intermittent, it is necessary to provide backup heat or heat storage during the day (for solar energy) using possibly phase change material (PCM) heat store exchangers.

It is worth mentioning that for any new dryer to be used for some application, a life cycle assessment is essential to check its feasibility, although it may provide much better quality compared to conventional dryers. The life cycle assessment will check a number of important issues, such as emissions, safety, energy consumption, and total cost.

9.3.12 Use of Mathematical Modeling

As discussed earlier, the drying of foods is a complex process as one must take into account numerous factors that will determine the choice of dryer. It is extremely difficult and expensive, not to mention potentially time consuming, to experimentally test complex drying systems. Recently, CFD has emerged as a promising tool to evaluate and improve the performance of several unit operations. The drying of foods also requires the development of new and innovative techniques to enhance product quality and establish a cost-effective and sustainable process. With advances in mathematical tools and improved computational power, CFD has been found to be very useful for predicting the drying phenomenon in various industrial dryers (Jamaleddine and Ray 2010). Jamaleddine and Ray (2010) carried out a very comprehensive survey of CFD techniques applied to diverse problems in industrial drying. They reported that CFD solutions have been used in drying to optimize, retrofit, develop equipment and processing strategies, and replace expensive and time-consuming experimentation. Mujumdar and Wu (2008) highlighted the need for a cost-effective solution that could push innovation and creativity in drying. This is easily attainable using highly efficient tools such as CFD. Some of the key advantages of CFD in the drying sector are its ability to provide information on the comparison of different geometries and its use as a powerful tool for troubleshooting purposes, including the evaluation of the effect of various parameters even in complex geometries (Mujumdar and Wu 2008). Response surface methodology is another tool that can be used in experimental design and in the optimization of process parameters (Mujumdar and Wu 2008). This tool can be used to relate several input variables and response (output) variables using regression analysis. The significant parameters can be identified based on the relation between input and output parameters. There are number of research articles on the drying of food products making use of response surface methodology for the optimization of operating parameters for various drying systems (Jangam and Thorat 2010; Han et al. 2010; Mestry et al. 2011).

Computational methods can be profitably used to simulate drying in complex heterogeneous materials. However, it is always difficult to make or describe a real structure using these methods. Perre (2005) reported a new tool that provides a finite-element description of a real structure of materials at the microscopic, anatomical, and cellular levels. It involves the development of a finite-element mesh based on a digital image of complex heterogeneous and anisotropic products. Although it was essentially developed for the study of the cellular structure of wood, the technique can be used for various food products and it can give a more realistic structure of a material. Devahastin and Niamnuy (2010) have reviewed the use of mathematical modeling for food quality.

Recently developed software such as Simprosys developed by Simprotek (http://www.simprotek.com/) can be a very useful tool for evaluating various opportunities to study and improve a drying system as a whole. Such a software can be effectively used to assess various options for energy savings and sustainability such as using waste heat and indirect heating, recycling dryer exhaust, recycling part of the dried material, and utilizing renewable energy for drying. This software is based on heat and mass balances and can be used to solve/optimize complicated flow sheets in a faster way. Simprosys can also handle all pre- and postdrying operations such as, for example, heater, fan, filter, cyclone, and scrubber/condenser (Gong and Mujumdar 2010). Various dryers and drying systems can be compared based on the foot print and energy consumption, for example, in a faster manner (Gong and Mujumdar 2010). This software also has a database for different drying gas-solvent systems, which allows the use of Simprosys for drying processes that involve the removal of solvents other than water as well the use of different drying media such as N_2. Further, the components of renewable energy could be added to study their use to make the process cost-effective.

9.4 Closing Remarks

This chapter covered some recently developed drying technologies with the potential for industrial applications. Since no reports were found on actual applications and industrial experience with new technologies, it is difficult to make definitive recommendations. It seems that heat-pump-assisted drying, superheated steam drying at lower pressures, hybrid drying technologies, and in-batch drying use of heat input that matches the energy needs dictated by instantaneous drying kinetics will emerge as some of the more popular drying technologies. For special applications the use of a microwave field with/without vacuum presents some advantages. Since vacuum freeze drying is an extremely slow and, hence, expensive process, for highly heat-sensitive materials microwave-assisted freeze drying presents certain advantages. The use of renewable energy such as solar and wind will be limited to smaller-scale agricultural or farm-level applications in the near future.

References

Abdulla G, Belghit A, Allaf K (2009) Impact of instant controlled pressure drop treatment on moisture adsorption isotherm of cork granules. Drying Technol 27(2):237–247

Abdulla G, Belghi A, Allaf K (2010) Impact of the instant controlled pressure drop treatment on hot air drying of cork granules. Drying Technol 28(2):180–185

Alves-Filh O, Eikevik T, Mulet A, Garau C, Rossello C (2007) Kinetics and mass transfer during atmospheric freeze drying of Red pepper. Drying Technol 25(7–8):1155–1161

Alves-Filho O, Efremov G (2009) Heat pump drying: experimentation, theory, design and practice. CRC Press, Boca Raton

Alves-Filho O, Strommen I (1996) The application of heat pump in drying of biomaterials. Drying Technol 14(9):2061–2090

Baker CGJ (2005) Energy efficient dryer operation – an update on developments. Drying Technol 23(9–11):2071–2087

Cal K, Sollohub K (2010) Spray drying technique. I: hardware and process parameters. J Pharm Sci 99(2):575–586

Choicharoen K, Devahastin S, Soponronnarit S (2010) Performance and energy consumption of an impinging stream dryer for high-moisture particulate materials. Drying Technol 28 (1–3):20–29

Chua KJ, Chou SK, Hawlader MNA, Ho JC, Mujumdar AS (2002) On the study of time-varying temperature drying – effect on drying kinetics and product quality. Drying Technol 20:1579–1610

Claussen IC, Ustad TS, Strømmen I, Walde PM (2007a) Atmospheric freeze drying – a review. Drying Technol 25(6):947–957

Claussen IC, Andresen T, Eikevik TA, Strommen I (2007b) Atmospheric freeze drying-modeling and simulation of a tunnel dryer. Drying Technol 25(12):1959–1965

Cong DT, Al Haddad M, Rezzoug Z, Lefevre L, Allaf K (2008) Dehydration by successive pressure drops for drying paddy rice treated by instant controlled pressure drop. Drying Technol 26(4):443–451

Daghigh R, Ruslan MH, Sulaiman MY, Sopian K (2010) Review of solar assisted heat pump drying systems for agricultural and marine products. Renew Sustain Energy Rev 14:2564–2579

Devahastin S, Niamnuy C (2010) Modelling quality changes of fruits and vegetables during drying: a review. Int J Food Sci Technol 45:1755–1767

Fushimi C, Kansha Y, Muhammad A, Mochidzuk K, Kaneko S, Tsutsumi A, Matsumoto K, Yokohama K, Kosaka K, Kawamoto N, Oura K, Yamaguchi Y, Kinoshita M (2011) Novel drying process based on self-heat recuperation technology. Drying Technol 29(1):105–110

Gong ZX, Mujumdar AS (2010) Simulation of drying nonaqueous systems – an application of simprosys software. Drying Technol 28(1–3):111–115

Haddad J, Juhe F, Louka N, Allaf KA (2004) Study of dehydration of fish using successive pressure drops (DDS) and controlled instantaneous pressure drop (DIC). Drying Technol 22 (3):457–478

Han Q-H, Yin L-J, Li S-J, Yang B-N, Ma J-W (2010) Optimization of process parameters for microwave vacuum drying of apple slices using response surface method. Drying Technol 28 (4):523–532

Hasibuan R, Daud WRW (2009) Quality changes of superheated steam–dried fibers from oil palm empty fruit bunches. Drying Technol 27(2):194–200

Hawlader MNA, Rahman SMA, Jahangeer KA (2008) Performance of evaporator-collector and Air collector in solar assisted heat pump dryer. Energy Convers Manage 49:1612–1619

Islam MR, Mujumdar AS (2008a) Heat pump-assisted drying. In: Chen XD, Mujumdar AS (eds) Drying technologies in food processing. Blackwell Publishing, West Sussex

Islam MR, Mujumdar AS (2008b) Heat pump-assisted drying. In: Mujumdar AS (ed) Guide to industrial drying. Three S Colors Publication, Mumbai

Islam MR, Ho JC, Mujumdar AS (2003) Convective drying with time-varying heat input: simulation results. Drying Technol 21(7):1333–1356

Jamaleddine TJ, Ray MB (2010) Application of computational fluid dynamics for simulation of drying processes: a review. Drying Technol 28(1–3):120–154

Jangam SV (2011a) An overview of recent developments and some R&D challenges related to drying of foods. Drying Technol 29(12):1343–1357

Jangam SV (2011b) Chemical heat pumps to enhance drying efficiency-some thoughts for new R&D opportunities. Drying Technol 29(6):610–611

Jangam SV, Mujumdar AS (2011) Heat pump assisted drying technology – overview with focus on energy, environment and product quality. In: Tsotsas E, Mujumdar AS (eds) Modern drying technology, vol 4. Wiley Interscience, UK

Jangam SV, Thorat BN (2010) Optimization of spray drying of ginger extract. Drying Technol 28 (12):1426–1434

Jaya DS, Sokhansanj S (2007) Drying of foodstuffs. In: Mujumdar AS (ed) Handbook of industrial drying, 3rd edn. CRC Press, Boca Raton, pp 521–546

Kudra T (2008) Pulse-combustion drying: status and potentials. Drying Technol 26 (12):1409–1420

Kudra T, Mujumdar AS (1989) Impinging stream dryers for particles and pastes. Drying Technol 7:219–266

Kudra T, Mujumdar AS (2009) Advanced drying technologies, 2nd edn. CRC Press, Boca Raton

Kudra T, Mujumdar AS, Meltser V (1995) Impinging stream dryers. In: Mujumdar AS (ed) Handbook of industrial drying, 2nd edn. Marcel Dekker Incl., New York

Kudra T, Platon R, Navarri P (2009) Excel-based tool to analyze the energy performance of convective dryers. Drying Technol 27(10–12):1302–1308

Kumar P, Mujumdar AS (1990) Superheated steam drying – a state of the art survey. In: Mujumdar AS (ed) Drying of solids. Meerut, Sarita Prakashan

Kuriakose R, Anandharamakrishnan C (2010) Computational fluid dynamics (CFD) applications in spray drying of food products. Trends Food Sci Technol 21:383–398

Lin SXQ, Chen XD (2006) A model for drying of an aqueous lactose droplet using the reaction engineering approach. Drying Technol 24(11):1329–1334

Maache-Rezzoug Z, Rezzoug SA, Allaf K (2002) Development of a new drying process – dehydration by cyclical pressure drops (D.D.S.): application to the collagen gel. Drying Technol 20(1):109–129

Mestry AP, Mujumdar AS, Thorat BN (2011) Optimization of spray drying of an innovative functional food: fermented mixed juice of carrot and watermelon. Drying Technol 29 (10):1121–1131

Mujumdar AS (1990) Superheated steam drying: principles practice and potential for use of electricity. Canadian Electrical Association, Montreal. Report No. 817, U 671

Mujumdar AS (1991) Drying technologies of the future. Drying Technol 9:325–347

Mujumdar AS (1992) Superheated steam drying of paper: principles, status and potential. In: Mujumdar AS (ed) Drying of solids. International Science Publisher, New York
Mujumdar AS (2006) Some recent developments in drying technologies appropriate for post-harvest processing. Int J Postharvest Technol Innov 1:76–92
Mujumdar AS (2007a) Handbook of industrial drying, 3rd edn. CRC Press, Boca Raton
Mujumdar AS (2007b) Superheated steam drying. In: Mujumdar AS (ed) Handbook of industrial drying, 3rd edn. CRC Press, Boca Raton
Mujumdar AS, Wu Z (2008) Thermal drying technologies-Cost-effective innovation aided by mathematical modeling approach. Drying Technol 26:146–154
Nimmol C, Devahastin S (2010) Evaluation of performance and energy consumption of an impinging stream dryer for paddy. Appl Therm Eng 30(14–15):2204–2212
Ogura H, Mujumdar AS (2000) Proposal for a novel chemical heat pump dryer. Drying Technol 18 (4):1033–1053
Perre P (2005) Meshpore: a software able to apply image based meshing techniques to anisotropic and heterogeneous porous media. Drying Technol 23(9–11):1993–2006
Pronyk C, Cenkowski S, Muir WE (2010) Drying kinetics of instant Asian noodles processed in superheated steam. Drying Technol 28(2):304–314
Rahman SMA, Mujumdar AS (2008) A novel atmospheric freeze-drying system using a vibro-fluidized Bed with adsorbent. Drying Technol 26(4):393–403
Shibata H (2011) Fundamentals of superheated steam drying. Kyushu University Press, Fukuoka
Speckhahn A, Srzednicki G, Desai DK (2010) Drying of beef in superheated steam. Drying Technol 28(9):1072–1082
Strommen I, Eikevik T, Alves Filho O, Syverud K (2004) Heat pump drying of sulphate and sulphite cellulose. In: Proceedings of the 14th international drying symposium (IDS 2004), vol B. Sao Paulo, pp 1225–1232
Walzel P (2011) Influence of the spray method on product quality and morphology in spray drying. Chem Eng Technol 34(7):1039–1048
Wongsuwan W, Kumar S, Neveu P, Meunier F (2001) A review of chemical heat pump technology and applications. Appl Therm Eng 21:1489–1519
Woo MW, Mujumdar AS (2010) Effects of electric and magnetic field on freezing and possible relevance in freeze drying. Drying Technol 28(4):433–443
Wu Z, Mujumdar AS (2006) R&D needs and opportunities in pulse combustion and pulse combustion drying. Drying Technol 24(11):1521–1523
Zbicinski I (2002) Equipment, technology, perspectives and modeling of pulse combustion drying. Chem Eng J 86(1–2):33–46
Zbicinski I, Smucerowicz I, Strumillo C, Crowe C (2000) Application of pulse combustion technology in spray drying process. Braz J Chem Eng 17(4):441–450
Zhang M, Tang J, Mujumdar AS, Wang S (2006) Trends in microwave related drying of fruits and vegetable. Trends Food Sci Technol 17:524–534
Zhang M, Jiang H, Lim RX (2010) Recent developments in microwave-assisted drying of vegetables, fruits, and aquatic products – drying kinetics and quality considerations. Drying Technol 28(11):1307–1316

Chapter 10
Batch Coffee Roasting; Roasting Energy Use; Reducing That Use

Henry Schwartzberg

10.1 Introduction

On an annual basis, 6.7 billion kg of coffee beans are roasted worldwide, mostly in batch roasters. Near the end of roasts, roasting coffee emits volatile organic compounds (VOCs) and carbon monoxide (CO), which in industrialized countries must be oxidized in afterburners. Roasting with afterburning uses roughly 11.2×10^{12} kJ fuel energy/year and causes emission of roughly 7.0×10^{8} kg CO_2/year. This paper describes these processes, factors affecting energy use for them, and presents and evaluates ways to reduce such use.

Figure 10.1 depicts the basic parts of a batch roaster with afterburner. Roman italic symbols indicate stream temperatures and the stream or stream component masses that enter or leave each part during a roast. Stream mass-flow rates vary during a roast. Total stream masses are shown and will be used for mass and enthalpy balances for entire roasts. Roaster gas contains air partly depleted in oxygen, combustion products, and CO_2, H_2O, VOCs, and CO emitted by roasting coffee. Mass $(G{-}F_R)$ of roaster gas is heated from temperature T_{GO} to T_{GI} by mixing it with a mass F_R of combustion products (whose temperature is T_F) in the roaster furnace and is then drawn by the blower into the roasting chamber (chamber for short), where it transfers heat to coffee beans. T_{GI} usually ranges from 260 to 520 °C. Prolonged use of $T_{GI} > 520$ °C causes beans tips to burn. T_{GI} is controlled by either stepwise or continuous automatic adjustment of the burner-gas flow rate, F'_R. Normally the combustion-air/fuel ratio used is automatically kept constant, so T_F normally remains constant.

H. Schwartzberg (✉)
Food Science Department, University of Massachusetts, Amherst, MA, United States
e-mail: schwartzberg@foodsci.umass.edu

S. Yanniotis et al. (eds.), *Advances in Food Process Engineering Research and Applications*, Food Engineering Series, DOI 10.1007/978-1-4614-7906-2_10,

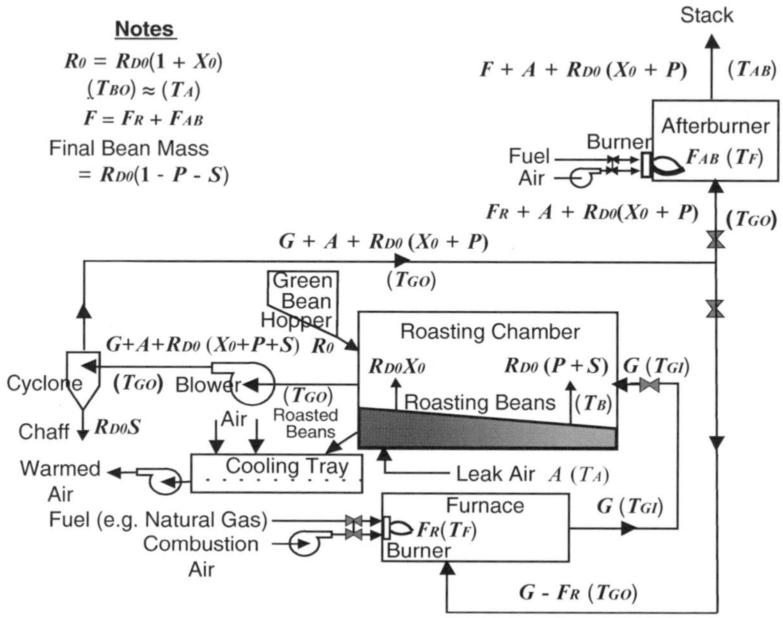

Fig. 10.1 Components, flow layout, and stream masses and temperatures for a gas-fired batch coffee roaster in which roaster gas recirculates and that is equipped with an afterburner

10.1.1 Green-Bean Load

The green coffee bean batch weight is R_O. Its dry weight is R_{DO}. Its moisture content weighs $R_{DO}X_O$, and its initial temperature is T_{BO}. Approximately 75 % of commercially roasted beans are of Arabica varieties (e.g., Colombians, Santos, Costa Ricans, Kenyans) and approximately 25 % are of Robusta varieties (e.g., beans from Vietnam, Uganda, and Ivory Coast). Robustas are of lower quality but are less costly. With minor exceptions, beans are roughly hemiellipsoidal and have a groove in their flatter surface. Green bean lots are graded in size to prevent uneven roasting. Arabica beans in the large size range are roughly 10.3 mm long, 7.3 mm wide, and 4 mm high. Dimensions of Arabica beans in the smallest commercial-size range are roughly three-quarters as large. The length/width ratios for Robustas are smaller than for Arabicas. A typical load for a medium-size batch roaster, 120 kg of green beans, would contain roughly 700,000 beans if large Arabicas are used, or roughly 1,570,000 beans if Arabicas of the smallest commercial size are used. Batches for very large roasters contain roughly four times as many Arabica beans.

10.1.2 Roasting-Induced Changes

Chamber elements or pneumatic transport bring about bean–gas contact and continually mix freshly heated beans with the rest of the batch, thereby improving the uniformity of heating. The bean temperature, T_B, rises as roasting proceeds. Moisture vaporizes internally and diffuses out of the beans, but not fast enough to prevent steam pressure buildup within bean cells. The beans soften, and the internal pressure causes them to expand (Geiger 2004). The expanding beans shed the mass, $R_{DO}S$, of the enveloping parchment (chaff). Moisture loss ultimately causes beans to harden, setting the expansion (Geiger 2004). Reactions within cells and in bean cell walls produce more than 800 identified roasting-reaction products and many intermediates. A small fraction of the products have desirable flavors or aromas prized by users of roasted coffee. Pigmented reaction products cause bean color, originally light green or tan, to turn brown and progressively darken (Mondry 2004; Schenker 2000). Reaction-generated water vapor, CO_2 (Geiger 2004; Schenker 2000; Jansen 2006), CO (Jansen 2006; Valdovinos-Tijerino 2005), and VOCs (Jansen 2006; Dorfner 2004) escape from beans, causing dry matter mass loss, $R_{DO}P$. Roasted beans retain some reaction-formed water, but virtually all the water originally in unroasted beans evaporates by the time the beans reach their end-of-roast temperature, T_{BF}, at which time the mass of the beans is equal to $R_{DO}(1 - P - S)$.

Static pressures are kept negative in most roasting chambers to prevent the escape of gas, vapor, or smoke. Therefore, mass A of fresh air leaks into the chamber during a roast.

Gas leaves the chamber at $T_{GO} \cdot T_{GO}$ is substantially lower than T_{GI} because roaster gas has transferred heat to the beans, but also because roaster gas has mixed with in-leaking air and gas and vapor emitted by the beans. The mass $G + A + R_{DO}(X_O + P + S)$ of gas and entrained solids leaves the chamber. The entering mass is G. The exiting gas passes through the blower and then the cyclone, which removes and discharges the mass, $R_{DO}S$, of entrained chaff. Then the gas splits into (1) a larger stream, $(G - F_R)$, which passes through the furnace and then, accompanied by the mass, F_R, of added burner gas, recycles to the chamber; and (2) a smaller stream, $[F_R + A + R_{DO}(X_O + P)]$, which passes through the afterburner.

10.1.3 VOC Oxidation: Afterburners

Roasting beans emit VOCs near the end of roasting. VOC levels in gas leaving the roasting chamber rise rapidly and, at the end of a medium roast, reach roughly 4,400 ppm (Jansen 2006). The part of that gas that passes through the afterburner is heated there to T_{AB} by admixing the mass, F_{AB}, of combustion products (again at temperature T_F) from a burner. If T_{AB} is high enough and the gas-residence time in the afterburner is adequately long, enough VOCs will oxidize in the afterburner to yield afterburner-discharge gas containing less than 50 ppm VOC. The afterburner-discharge gas then flows out to the atmosphere via a stack.

The gas stream that recycles also contains VOCs. Some of that VOC is destroyed by oxidization as the recycle stream passes through the roaster furnace. The fraction of entering VOCs destroyed in the furnace depends on T_{GI} and how long the heated gas resides there. At roughly normal operating conditions, recycled gas leaving the furnace contains 300 ppm VOC (Rothfus 1986). The VOC not destroyed reenters the chamber, where it is joined by freshly created VOCs.

Catalytic afterburners for coffee roasters operate at $T_{AB} \geq 400$ °C and provide 0.05 to 0.1 s of gas residence time. The catalyst must be replaced every 3 or 4 years and should be used with a prefilter to prevent particulate matter from blinding it. A supplementary blower is often required to overcome pressure drop across the catalyst bed and prefilter. Thermal afterburners are cheaper but require $T_{AB} \geq 760$ °C and 0.4 s of gas residence time. To satisfy safety regulations, VOC levels in feeds for both thermal and catalytic afterburner should be less than 25 % of their lower explosive limit (LEL). The LEL for typical combustible gases ranges from around 10,000 to 30,000 ppm. Peak VOC levels as high as 10,000 ppm, i.e., potentially explosive levels, have been reported (European Commission 2006), very likely for a roast where T_{BF} was very high, e.g., ≥ 255 °C. Afterburner T_{AB} must be increased to provide less than 50 ppm VOC output when afterburner-input VOC levels are higher than normal. The mass of gas that leaves the afterburner and stack is equal to $F + A + R_{DO}(X_O + P)$, where $F = F_R + F_{AB}$.

Regenerative thermal oxidizers (RTOs) use regenerative heat exchange to heat afterburner feeds from T_{GO} to appropriate T_{AB} (900–1,000 °C) and then cool gas discharged from the RTO almost to T_{GO}. Thus they require little or no heat input or use of flame-based combustion after startup (Schmidt 2008). RTO should only be used for feed VOC levels less than 10 % of their LEL. RTOs also require frequent switching of flow direction through the regenerator bed or beds, may experience fouling problems, and, at present, are rarely used for coffee roasters.

10.1.4 NO_X and CO

Nitrogen oxides (NO_X) are generated during fuel combustion in the burners for the roaster furnace and afterburner. The higher the burner temperature, T_F, the greater the NO_X level. Thus afterburners destroy VOC levels but create added NO_X. NO_X emission levels for coffee roasters are higher for darker roasts (Schmidt 2008), i.e., roasts producing large amounts of VOC. NO_X levels can be reduced through the use of low NO_X burners, i.e., burners where combustion products recycle into the combustion zone. Because little fuel combustion is needed to operate RTOs, NO_X emission was markedly reduced when they were used (Schmidt 2008). European Union (EU) regulations limit emission of NO_X to 100 mg/Nm^3. NO_X levels from most roasters exceed this limit, particularly when thermal afterburning is used. New catalyst formulations can reduce NO_X levels to approximately 52 mg/Nm^3.

Roasting beans emit CO. Many intermediates, including CO, form as VOCs oxidize. Therefore, significant amounts of CO may be produced when VOCs in recycled roaster gas partially oxidize as that gas passes through the roaster furnace.

CO can also form due to incomplete VOC oxidation in an afterburner, particularly if there has been loss of catalyst activity, residence time is insufficient, flow channeling occurs, or T_{AB} is too low. EU regulations limit CO emission levels to 100 mg/Nm^3. CO emission levels from some roasters are significantly higher (Schmidt 2008).

10.1.5 Recycling

The flow layout in Fig. 10.1 is for an R roaster, i.e., one in which most gas leaving the chamber recycles to the chamber. In older R roasters, all chamber-discharge gas is heated to T_{GI} in the furnace before splitting into recycled gas and afterburner feed. Then if $T_{GI} > T_{AB}$, the afterburner feed is not heated further. Most R roasters use catalytic afterburners (Schmidt 2008). Component sequences for R roasters may differ from that shown in Fig. 10.1. The cyclone may come before the blower (Clarke 1979), or added blowers may be used to overcome high local pressure drops. In spouted-bed roasters, the blower is located just before the chamber (Clarke 1979), and static pressures are positive in the chamber, which is well sealed. Then $A = 0$. A is positive for 95 % of roasters. Green-bean-load weights, R_0, for R roasters are normally $\leq$ 500 kg and $\geq$ 60 kg.

10.1.6 Single-Pass Operation

Single-pass (SP) roasters are usually used when R_0 is less than 60 kg. Gas passes through SP roasters only once. Often, entering fresh air mixes with flames from three lance burners and the mixed gas then flows through the roasting chamber, blower, cyclone, afterburner, and stack. Toward the end of a roast, first one burner and then another are turned off, thereby reducing T_{GI} in steps. Thermal afterburners are usually used with small roasters, i.e., SP roasters (Schmidt 2008).

10.1.7 Mass-Flow Rates

Adding a prime ($'$) to a symbol for a stream or component mass converts that symbol into one for the corresponding mass-flow rate or one for a mass rate of change with respect to time. Thus G' indicates the gas mass-flow rate entering the roaster chamber, F'_R the mass rate of furnace burner gas output, and $R_{D0}X'$ and $R_{D0}P'$ the respective mass rates of moisture evaporation and reaction product loss from beans. The heat capacity of burner gas and roaster gas are roughly equal. Therefore, a rough enthalpy balance around the furnace yields

$$\frac{F'_R}{G'} = \frac{(T_{GI} - T_{GO})}{(T_F - T_{GO})}. \tag{10.1}$$

Since T_{GO} progressively rises during roasting, F'_R/G' is smallest at the end of roasts.

10.1.8 Ending Roasts

In industrial roasters, bean temperatures are measured with sturdy sheathed probes. The resulting measured bean temperatures, T_{BM}, deviate from the corresponding actual T_B (Schwartzberg 2006a). Often, roasts are ended when the T_{BM} reaches a set value corresponding to a desired T_{BF}. If the T_{BF} is less than 190 °C, then flavor and aroma development will be inadequate. In most R roasters, a controlled volume of quench water is sprayed on the beans when a set end-of-roast T_{BM} is reached. Most of the water evaporates, thereby cooling the beans rapidly and stopping the roast. Then the beans are dumped into the cooling tray, where they form a shallow bed, which is stirred and through which ambient air is drawn, further cooling the beans. Beans at the air-outflow side of the cooler bed often still remain quite hot when those at the inflow side are already cool. Pneumatically agitated coolers that provide good bean mixing and more uniform bean cooling are used with some R roasters. Bean quenching is sometimes carried out in these coolers, rather than in the roasting chamber itself. Air cooling alone, without water quenching, is usually done for SP roasters.

10.1.9 Bean Temperature Profiles

The outcome of coffee roasting reactions depends on the initial reactant concentrations, which in turn depend on the bean variety and source. The outcome of these reactions also depends on how T_B varies with time, t, i.e., the "T_B profile." Figure 10.2 shows a T_B profile and corresponding T_{BM} profiles obtained with two properly positioned bean-temperature probes of different diameter during a "regular" roast, i.e., one where roasting time, t_R, ranges from 480 to 720 s. Properly positioned probes are closely surrounded by moving roasting beans so that hot gas flow does not affect the T_{BM} (Schwartzberg 2006a). A T_{BM} affected by hot gas flow tends to lie above T_B. Probes are contacted by hot gas alone when empty roasting chambers are preheated before green beans are loaded. Thus the T_{BM} is initially much higher than the T_B. After beans are added, the T_{BM} first falls toward the T_B, which is rising. Then, for properly positioned probes, the T_{BM} passes through a minimum as it reaches the T_B and starts rising. Thereafter, both the T_B and the T_{BM} rise, with the T_B above the T_{BM}. Thermometric lag ($T_B - T_{BM}$) increases with probe size. Thus, once the T_{BM} has passed through its minimum, the T_{BM} tends to lie farther below the T_B for the larger-diameter probe than for the smaller one

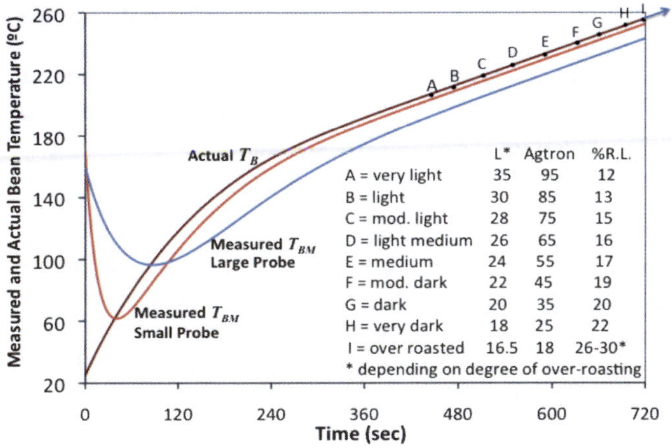

Fig. 10.2 Actual and measured bean temperature profiles for a regular roast. Degrees of roast darkness and total roasting loss obtained for different T_{BF}; Agtron and C.I.E. L* reflectance-based roast colors for those T_{BF}; roasting loss, Agtron, roast darkness, and T_B values for points A to I based on Mondry (2004); L* for those points estimated using data by (Schenker 2000)

(Schwartzberg 2006a). ($T_B - T_{BM}$) tends to be proportional to dT_B/dt. Therefore, it is difficult to correctly compare T_{BM} profiles and T_{BF} values from roasters with different probe arrangements or to compare T_{BF} for fast and slow industrial roasts. Useful T_B data have been obtained for fluidized-bed roasts by implanting very fine thermocouples in coffee beans (Geiger 2004; Schenker 2000; Eggers and von Blittersdorff 2005). T_B gradients within beans are large when dT_B/dt is large (Eggers and von Blittersdorff 2005). T_B profiles and T_{BF} can be calculated from properly measured T_{BM} profiles (Schwartzberg 2006a). For individual roasters, control of such T_{BM}-based profiles provides T_B profile control. T_{BM} profiles are often roughly adjusted by the use of one or two step reductions in T_{GI} or G'. Control systems developed by Eichner continuously regulate T_{GI} and G' for some roasters and thereby provide T_{BM} profiles that consistently and continuously conform to "learned" T_{BM} profiles that yielded desirable roasted coffee.

10.1.10 Roast Darkness

Points A–I in Fig. 10.2 indicate roast darkness and the corresponding T_B for a regular roast. The T_B values were obtained by correcting data for roast darkness versus T_{BM} (Mondry 2004) for thermometric lag. A indicates a very light roast, B a light roast, E a medium roast, G a dark roast, and I overroasting, i.e., nearly black beans. The inset table in Fig. 10.2 lists total as-is roasting loss and instrumentally measured "roast color" values for these points (Mondry 2004).

Roast color values are widely used to characterize degree of roast. Agtron colors are based on the reflectance of narrow-spectrum near-IR light from ground, roasted beans and are often used by specialty roasters. Other near-IR-based roast-color meters are also used in industry (Mondry 2004). C.I.E. (Commission Internationale de l'Eclairage) roast colors known as L* values are based on the reflectance of broad-spectrum visible light and have been used in academic research on roasting (Geiger 2004; Schenker 2000; Baggenstoss 2008). The L* values listed in Fig. 10.2 are rough estimates based on measured L* for coffee bought from Swiss retailers (Schenker 2000). Hunter L values, a slightly different roast color measure, also based on the reflectance of broad-spectrum visible light, are used by Folgers, the largest U.S. roaster of coffee, and are cited in process descriptions in patents obtained by that company. In the end-of-roast color range, L values are roughly 74 % as high as L* values. As t_R and T_{BF} increase, bean darkness increases, roast color decreases, and the coffee's flavor character changes. Very darkly roasted beans tend to be oilier or become oily after roasting (Geiger 2004) and do not keep as well as lighter roasted beans.

10.1.11 Aroma Concentration Histories and Profiles

Coffee flavor is strongly influenced by absolute and relative concentrations of coffee aroma compounds within beans. Concentrations of some aroma compounds peak when roast colors are light, i.e., roughly between points B and E on Fig. 10.2, and then decline markedly as roasts become darker (Schenker 2000; Baggenstoss 2008; Gianturco 1967). Concentrations of other aromas have broader peaks, which are reached at medium to dark roasts and decline less as roasts become darker (Schenker 2000; Baggenstoss 2008). The concentration of an aroma compound decreases when the sum of its rates of decomposition and escape as a VOC exceeds its rate of formation. Some aroma concentrations rise without declining even for overroasting. Maximum concentrations for individual aroma compounds are affected by the T_B profile used and range from 0.25 to $>$ 300 mg/kg green bean dry matter (Baggenstoss 2008).

Roasted coffee's flavor character depends on the T_B profile used as well as on roast color. ETH researchers (Schenker 2000; Baggenstoss 2008) used different T_B profiles (measured with fine thermocouples implanted in beans) to roast identical green coffees to the *same L* roast color*. The distribution of coffee aroma concentrations differed when the T_B profile differed. For the *same roast color*, faster roasts required a higher T_{BF} (Geiger 2004; Schenker 2000), e.g., 245 °C for $t_R = 180$ s compared to 230 °C for $t_R = 300$ s and 212 °C for $t_R = 720$ s.

10.1.12 Emission of Reaction Products

Figure 10.3 shows how emissions of P and its gas and vapor components: reaction-H_2O, CO_2, CO, and VOC each expressed in terms of kilograms of gas or vapor emitted per kilogram of dry green beans increase as the T_B increases during a regular roast. The P were estimated by subtracting an estimated initial as-is bean

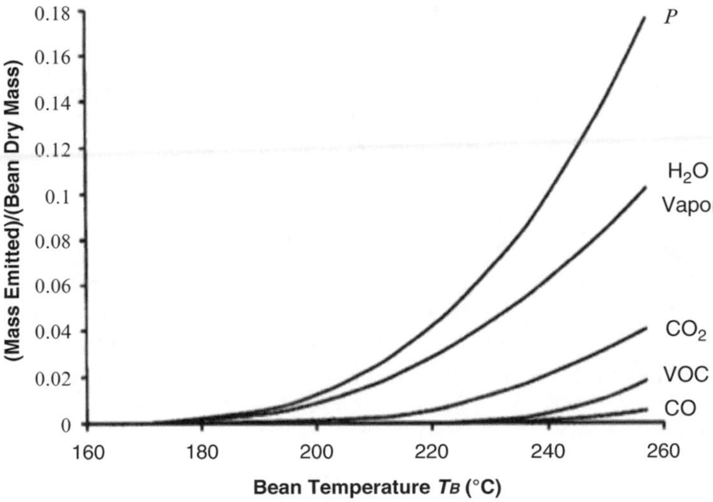

Fig. 10.3 Cumulative kilogram of gas and vapor products of roasting reactions emitted per kilogram of dry coffee bean feed versus T_B. The total cumulative kilogram of all such products emitted per kilogram of dry coffee bean feed is P

moisture content of 0.1 and an estimated as-is chaff loss of 0.01 from total as-is roasting loss values for T_{BM} (Mondry 2004) corresponding to the plotted T_B. Other data (Geiger 2004; Jansen 2006; Valdovinos-Tijerino 2005; Dorfner 2004) were used in plotting the curves for cumulative emission of reaction-H_2O, CO_2, VOC, and CO. Reaction-H_2O emission starts at $T_B \approx 160$ °C for low-speed, low-T_{BF} roasts and at $T_B \approx 200$ °C for high-speed, high-T_{BF} roasts (Geiger 2004). CO_2 emission starts at around 190 °C (Geiger 2004). CO emission starts at around 220 °C (Valdovinos-Tijerino 2005). All emission rates increase very rapidly as the T_B rises after emission starts. The main emitted reaction products are water vapor (58–70 %) and CO_2, (10–27 %), with the CO_2/H_2O ratio increasing with degree of roast (Geiger 2004).

VOC-loss data (Jansen 2006) were used to calculate the cumulative kilogram of VOCs emitted per kilogram of dry green beans. VOC concentrations were measured by online mass spectrometry for several types of roasts (Dorfner 2004). The onset of VOC emission and shape of the cumulative VOC emission curve were based on Dorfner's data for a medium roast where *convective heating* of beans predominated. VOCs were emitted mainly during the last 22 % of the roast. In contrast, when Dorfner roasted beans by *conductive heating* in a sample roaster, VOCs were emitted during more than 70 % of the roast. VOCs were also emitted very early in a roast where Dorfner used strong preheating (initial $T_{BM} = 236$ °C). Thus conductive heating and locally high bean temperatures near the start of roasts favor early emission of VOCs. VOCs are acidic, and their acidity tends to increase almost until the very end of roasts (Dutra et al. 2001). Acetic acid is a major VOC constituent (Jansen 2006).

10.1.13 Explosion Hazards

VOC and CO emission is most rapid near the end of roasts. Purging of chamber discharge gas, VOCs, and CO via the afterburner and stack is also lowest then. Thus, near the end of roasts, VOC and CO concentrations in roasting chambers can reach temporarily ignitable or even explosive levels. This can be prevented if F'_R is kept large enough at that time to provide adequate VOC and CO purging near the end of roasts. Increasing the combustion-air/fuel ratio will increase F'_R. Thus, combustion-air/fuel ratios are often set large enough to keep $(A' + F'_R)/G' \geq 1/3$ at the end of roasts. The T_F decreases as the combustion-air/fuel ratio increases. Since the combustion-air/fuel ratio normally remains constant, the T_F will be lower for the whole roast. Dilution of VOCs and CO in roasting chambers increases as G/R_{DO} increases. Thus, when G/R_{DO} is substantially greater than 3, final $(A' + F'_R)/G' < 1/3$ can be used.

10.2 Heat Transfer in Roasting Chambers

Roasting beans fall across, are thrown across, or entrain in hot gas flowing through the roasting chamber, or encounter hot gas flowing through spaces in a moving bed of beans. The gas transfers heat to contacted beans. The beans mix, and more recently or more strongly heated beans transfer heat to less recently or less strongly heated beans. Therefore bean-to-bean variation in the T_B is small. Equation 10.2 describes bean heating when the rate of loss of heat from the chamber to the environment is negligibly small, a reasonable approximation for well-designed roasters (Schwartzberg 2002):

$$R_D C_B \frac{dT_B}{dt} = EG'C_G(T_{GI} - T_B) + R_D(H_{EV}X' + Q'_{REAC}) - MC_M\frac{dT_M}{dt}. \quad (10.2)$$

R_D is the current dry mass of beans, C_G the mean gas heat capacity. X', the rate of change of bean dry-basis moisture content, is negative. C_B is the beans' dry-basis heat capacity. H_{EV}, the latent heat of evaporation of moisture from beans, is approximately equal to 2,760 kJ/kg.

$E \equiv \Delta T_G/(T_{GI} - T_B)$ is the heat-transfer efficiency. ΔT_G is the heat-transfer-induced gas-temperature drop in the roasting chamber. In-leaking air mixes with roaster gas leaving the chamber. Therefore, $T_{GO} < (T_{GI} - \Delta T_G)$. If gas flow is reasonably pluglike,

$$E = 1 - f_C \exp\left[-\frac{UA_{BE}}{f_C G' C_G}\right]. \quad (10.3)$$

where U is the gas-bean heat-transfer coefficient, A_{BE} the bean surface heat-transfer area, and f_C the fraction of gas that actually contacts beans. If complete back-mixing of gas occurs,

$$E = \frac{(UA_{BE})/(G'C_G)}{1 + (UA_{BE})/(G'C_G)}. \tag{10.4}$$

U tends to be proportional to $(G'_{REL})^{0.5}$, where G'_{REL}, the mass velocity of roaster gas relative to beans, is the vector sum of G' plus mass-velocity components due to mechanically and gravitationally imparted bean motion. Therefore, U is somewhat less than proportional to $(G')^{0.5}$, and E decreases as G' increases.

10.2.1 Exothermic Heating

Q'_{REAC} is the rate of heat generation by roasting reactions. In some studies, differential scanning calorimetry (DSC) measurements made in conditions where evaporation of reaction water and initial bean moisture were suppressed showed Q'_{REAC} to be positive (Raemy 1981; Raemy and Lambelet 1982). This reaction environment was abnormal because bean moisture and reaction-generated water progressively evaporate and also provide evaporative cooling during roasting (Schwartzberg 2006b). Q'_{REAC} was positive only at high T_B in DSC tests where evaporation was not suppressed (Eggers 2004). Bean heat generation can cause fires in unmixed beans when the T_B is high. Nevertheless, T_B profiles for both experimental (Schenker 2000; Geiger 2004; Baggenstoss 2008) and industrial roasts do not usually appear to be noticeably affected by exothermic heating.

10.2.2 Bean-Hardware Heat Exchange

M and C_M are respectively the mass and heat capacity of roaster chamber parts, and dT_M/dt is their rate of temperature change. Preheated roaster hardware heats beans at the start of roasts. Near the end of roasts, beans may transfer heat to roaster hardware. The net effect of $MC_M(dT_M/dt)$ on T_B rise is usually small when convective transfer of heat from gas to beans predominates, as it does in most large modern roasters.

10.2.3 dT_B/dt Variation

The main factors affecting dT_B/dt are G', E, $(T_{GI} - T_B)$, and X'. $(T_{GI} - T_B)$ progressively decreases as the T_B rises. Therefore, dT_B/dt progressively decreases. dT_B/dt decreases more for T_{GI} close to T_{BF} than for T_{GI} well above T_{BF}. When $-X'$

is large, evaporative cooling depresses dT_B/dt markedly. Thus, T_B profiles are significantly lower for large X_0 than for small X_0 (Schenker 2000). G' is proportional to gas density in the blower, which in commercial roasters (except spouted bed roasters) is, in turn, inversely proportional to T_{GO} in degrees K. T_{GO} progressively rises as T_B rises. Thus G' progressively decreases as roasting progresses, thereby augmenting the dT_B/dt decrease (Schwartzberg 2004). R_D reductions due to dry-matter loss tend to increase dT_B/dt. DSC data (Singh et al. 1997), when analyzed, indicate that C_B is the sum of a moisture-dependent term, $C_{BM}X = 5.0X$, and a dry-matter-based, T_B-dependent term, $C_{BD} = 1.099 + 0.007T_B$. Thus, C_B changes affect dT_B/dt in a complex way.

10.2.4 Roasting Time

Integration of Eq. 10.1 with small terms eliminated yields the following approximation for roast time, t_R, in seconds when T is in degrees Celsius and all other variables are in SI form:

$$t_R \approx \frac{\bar{R}_D\bar{C}_B}{E\bar{G}'C_G}\left[\ln\left(\frac{T_{GI}-T_{BO}}{T_{GI}-T_{BF}}\right)+\frac{\Delta H_{EV}X_0}{\bar{C}_B(T_{GI}-\bar{T}_B)}\right]$$
$$\approx \frac{0.51\bar{R}_D}{\bar{E}\bar{G}'}\left[\ln\left(\frac{T_{GI}-T_{BO}}{T_{GI}-T_{BF}}\right)+\frac{1250X_0}{\bar{C}_B(T_{GI}-160)}\right]. \tag{10.5}$$

Terms with bars over them are mean effective values. Increasing G'/R_D or T_{GI} reduces t_R.

Roasts with t_R between 180 and 300 s are now widely used in the USA. These high-speed roasts are obtained mainly by increasing G'/R_0. High-speed roasted coffees have bulk densities up to 34 % lower than that of regular roast coffee (Small and Horrell 1993) and contain more available soluble matter (Mahlmann and Schecter 1985; MacAllister and Spotholz 1964; Price et al. 1991), more sucrose, more 3-caffeolquinic acid and aroma (Price et al. 1991), and more residual water (Rothfus 1986; MacAllister and Spotholz 1964). The soluble-matter increase has been attributed to the hydrolytic breakdown of polymeric cell wall carbohydrates (Brandlein et al. 1988), most likely arabinogalactans. Even shorter t_R have been used, but the use of $t_R < 180$ s is uncommon nowadays. In Europe, t_R between 270 and 390 s are frequently used. High-speed roasting usually is carried out in newer kinds of roasters that can or must use high G', e.g., batch spouted-bed roasters, rotating-scoop roasters, rotating-bowl roasters, or continuous roasters where jets of gas cause local spouting in moving beds of beans (Clarke 1979). $G = \bar{G}'(t_R)$. If one multiplies both the left- and right-hand sides of Eq. 10.5 by $\bar{G}'$, then it can be seen that G is inversely proportional to E. Because E decreases as $\bar{G}'$ increases, higher G/R_0, e.g., $G/R_0 > 10$, are used for high-speed roasts than for regular roasts, where G/R_O usually ranges from 0.9 to 3.3. $\bar{G}'/R_O$ for high-speed roasts is often more than 33 times greater than $\bar{G}'/R_O$ for regular roasts. Therefore, blower energy use per kilogram of bean feed is much greater for high-speed roasts than for regular roasts.

10.3 Computation Methods

Overall mass balances were based on the mass flows shown in Fig. 10.1. Masses R_O, F, and A enter the roaster. Mass $R_{DO}S$ leaves as chaff and $[F + A + R_{DO}(X_O + P)]$ leaves as stack gas. The mass of roasted beans in the chamber as a roast ends is equal to $R_{DO}(1 - P - S)$.

10.3.1 Enthalpy Balances

Overall enthalpy balances for regular roasts were used to compute Q_{SR}, the kJ thermal energy used to roast a kilogram of coffee feed. T_A was used as a reference temperature. Stream or stream-component masses divided by R_{DO} and multiplied by the relevant heat capacity were integrated between T_A and the stream temperature, thereby providing specific stream enthalpies: Q_B for roasted beans, Q_G for stack gas and Q_C for chaff, all on a kJ/kg dry-bean basis. Exothermic heating of beans and net bean-hardware heat transfer were assumed to be negligible. It was also assumed that (a) $H_{EV} = 2{,}760$ kJ/kg if the beans' initial water content evaporated at 160 $_0$C; (b) T_P, the average temperature at which dry-matter loss $R_{DO}(P)$ occurs, is $(T_{BF} + 200\ °\mathrm{C})/2$; (c) chaff was heated to 200 °C before leaving the system; and (d) both T_A and T_{BO}, the initial T_B, are 20 °C.

Stack-gas temperatures are somewhat greater than the T_{AB} because the oxidation of VOCs and CO in an afterburner generates heat. The enthalpy contribution due to that temperature rise is exactly cancelled by a corresponding combustion energy contribution assignable to incoming green coffee. Therefore, both the combustion energy contribution and the stack gas temperature rise due to VOC and CO oxidation were neglected in carrying out enthalpy balances for the system, i.e., T_{AB} was used as the discharge temperature of stack gas and the combustion energy contribution for green coffee was assumed to be zero.

From the overall roaster enthalpy balance, $Q_{SR} = (Q_B + Q_G + Q_C)/(1 + X_O)$. Q_G is the sum of four terms: (a) $Q_F = FC_G(T_{AB} - T_A)/R_{DO}$; (b) $Q_A = (A/R_{DO})C_G(T_{AB} - T_A)$; (c) $Q_X = X_O[C_{BM}(160 - T_A) + H_{EV} + C_{WV}(T_{AB} - 160)]$; and (d) $Q_P = P\{1.099(T_P - T_A) + 0.0035[(T_P)^2 - (T_A)^2] + C_G(T_{AB} - T_P)\}$.

Q_{SR} is also equal to $(FC_G)(T_F - T_A)/[R_{DO}(1 + X_O)]$. Equating the two expressions for Q_{SR}, one can solve for $F/[R_{DO}(1 + X_O)]$, which was used as follows to evaluate Q_{SR}:

$$Q_{SR} = \frac{(T_F - T_A)(Q_B + Q_A + Q_X + Q_P + Q_C)}{(T_F - T_{AB})(1 + X_0)}. \quad (10.6)$$

$Q_B = (1 - P - S)\{1.099(T_{BF} - T_A) + 0.0035[(T_{BF})^2 - (T_A)^2]\}$. $C_G = 1.12$ kJ/(kg °C). $C_{WV} = 2.09$ kJ/(kg °C). $Q_C = S\{1.099(200 - T_A) + 0.0035[(200)^2 - (T_A)^2]\}$. Figure 10.3 was used to determine the P for the chosen T_{BF}. It was assumed that $S = 0.01$ kg chaff/kg dry bean.

10.3.2 Other Heat Input

Gas heating and afterburning usually continue during quenching, bean loading and unloading, and preheating. The kJ energy/kg bean feed used during these periods is

$$Q_{ADD} = \frac{(T_F - T_A)[C_G(A_Q + A_D + A_P)(T_{AB} - T_A) + W_E C_{WV}(T_{AB} - 100) + Q_M]}{(T_F - T_{AB})R_{D0}(1 + X_0)}. \quad (10.7)$$

The kilogram of air in-leakage during quenching, bean discharge from the chamber, and preheating and bean loading, i.e., A_Q, A_D, and A_P, respectively, vary with roaster type, quench time (15–30 s), preheat and bean charge time (30–90 s), and bean discharge time (5–40 s) and the sizes of the bean charge and discharge ports. $(A_Q/R_0) \approx 0.005$, $(A_D/R_0) \approx 0.041$, and $(A_P/R_0) \approx 0.027$ are reasonable values for rotating-drum roasters. (Q_M/R_0), the heat transferred to roaster hardware during preheating, was assumed to be approximately equal to 45 kJ/kg green beans, a reasonable value for large, rotating-drum roasters (W_E/R_{D0}), the kilograms of quench water evaporated per kilogram of dry green beans, $= C_{BD}(T_{BF} - T_{BC})/\Delta H_Q$. T_{BC}, the T_B after quenching, was assumed to be 120 °C. $\Delta H_Q \approx 2{,}602$ kJ/kg, the enthalpy rise for water between 20 °C liquid and 100 °C steam. The total kJ thermal energy provided by roaster fuel per kilogram of green coffee beans, Q_{TOT}, equals $Q_{SR} + Q_{ADD}$.

10.4 Results

An Excel spreadsheet program that can be used for wide ranges of feed and operating conditions was set up to compute Q_{SR} and Q_{TOT}. Reference Q_{SR} and Q_{TOT} were computed for a medium-darkness regular roast in an R roaster with catalytic afterburner, where $T_{AB} = 400$ °C. The reference feed and operating conditions and Q_{SR} and Q_{TOT} are listed in boldface in the first row of Table 10.1. $A/R_{D0} = 0.136$. T_F was set at 1,260 °C to keep $(A' + F'_R)/G' \geq 1/3$ at the end of roasts. The Q_{TOT} computed for the reference roast is 3.3 % lower than Q_{TOT}

Table 10.1 Q_{SP} and Q_{TOT} for various roaster-feed, roaster-operating, and afterburning conditions

T_{BF} (°C)	P (kg dry matter/kg dry beans)	T_F (°C)	T_{AB} (°C)	X_O (moisture dry basis)	A/R_{DO} (kg air/kg dry beans)	Q_{SR} (kJ/kg dry beans)	Q_{TOT} (kJ/kg dry beans)
236[a]	**0.068**[a]	**1,260**	**400**[b]	**0.111**	**0.136**	**1,417**	**1,621**
214[c]	**0.022**[c]	1,260	400[b]	0.111	**0.109**	**1,308**	**1,491**
255[d]	0.160	1,260	**480**[b]	0.111	**0.162**	**1,710**	**1,999**
236[a]	0.068	1,260	400[b]	**0.133**	0.136	**1,500**	**1,705**
236[a]	0.068	1,260	400[b]	**0.075**	0.136	**1,273**	**1,477**
236[a]	0.068	1,260	400[b]	0.111	**0.068**	**1,379**	**1,584**
236[a]	0.068	**1,530**[e]	400[b]	0.111	0.136	**1,313**	**1,502**
236[a]	0.068	1,260	**480**[f]	0.111	0.136	**1,615**	**1,878**
236[a]	0.068	1,260	**760**[g]	0.111	0.136	**2,807**	**3,426**
236[a,h]	0.068	1,260	**760**[g]	0.111	**0.733**[h]	**3,004**	**3,266**
236[a,h]	0.068	1,260	**400**[b]	0.111	**0.733**[h]	**1,747**	**1,856**

[a]Medium roast
[b]Catalytic afterburner fed directly by roasting chamber
[c]Very light roast
[d]Very dark roast
[e]Higher T_F (CO_2 and H_2O start to slightly decompose into CO, H_2 and O_2 at $T_F > 1{,}530$ °C)
[f]Catalytic afterburner fed from roaster furnace
[g]Thermal afterburner
[h]SP roaster

reported for similar industrial roasts (Rothfus 1986). The Q_{SR} computed for the reference roast is 13 % higher than Q_{SR} measured for a somewhat lighter roast in a modern R roaster with catalytic afterburner (Dorfner 2004).

10.4.1 Effects of Operating and Feed Conditions

Q_{SR} and Q_{TOT} were also computed for R-roaster-based roasts with significantly different feed or operating conditions. The changed condition or conditions and corresponding Q_{SR} and Q_{TOT} are shown in boldface in other rows in Table 10.1. For very light roasts, $T_{BF} = 214$ °C, $P = 0.022$, and, since t_R is smaller, $A/R_{DO} = 0.109$. Therefore, Q_{SR} and Q_{TOT} decrease. For very dark roasts, $T_{BF} = 255$ °C, $A/R_{DO} = 0.162$, $P = 0.160$, and the VOC load is very high. Therefore, T_{AB} was set at 480 °C instead of 400 °C. Thus Q_{SR} and Q_{TOT} are markedly higher than for the reference roast. Q_{SR} and Q_{TOT} are (a) greater for high bean moisture content ($X_O = 0.133$), (b) lower for low moisture content ($X_O = 0.075$), (c) decreased by reducing air in-leakage ($A/R_{DO} = 0.068$), (d) lower for higher burner temperatures ($T_F = 1530$ °C), and (e) larger when the catalytic afterburner feed is heated in the roaster furnace rather than in the afterburner itself (row 8 in Table 10.1, where $T_{AB} = T_{GI} = 480$ °C instead of 400 °C). Q_{TOT} is roughly twice as great for thermal afterburning ($T_{AB} = 760$ °C) as for standard catalytic afterburning ($T_{AB} = 400$ °C).

10.4.2 SP Roasting

A/R_{DO} is markedly greater (0.733 in Table 10.1) for SP roasters than for R roasters. Thus Q_{SR} is greater, particularly if thermal afterburning is used (row 10 in Table 10.1, where $T_{AB} = 760$ °C and $Q_{SR} = 3{,}004$ kJ/kg). Line 11 in Table 10.1, where $A/R_{DO} = 0.733$ and $T_{AB} = 400$ °C, shows that Q_{SR} and Q_{TOT} for SP roasters would be greatly reduced if catalytic afterburning were used. The required A/R_{DO} for SP roasting depends on E and T_{GI} and tends to increase if either E or T_{GI} decreases. Thus, the Q_{SR} computed for SP roasters would be even greater if E or T_{GI} were lower than currently assumed. Since quenching is not used for SP roasters, both (A_Q/R_O) and $(W_E/R_O) = 0$. Thus, Q_{ADD} is significantly smaller for SP roasters than for R roasters. Thus, when thermal afterburning is used in both cases, the computed Q_{TOT} for an SP roaster (line 10 in Table 10.1) is smaller than the corresponding Q_{TOT} for an R roaster (line 9 in Table 10.1). As shown later, additional thermal energy is required to oxidize VOCs in bean-cooling air when SP roasting is used. That additional thermal energy exceeds the Q_{ADD} reduction caused by not using quenching.

10.4.3 Effects of E, G′, and T_{GI} in R Roasters

E, G', and T_{GI} affect t_R for R roasters, but, along with t_R, they do not directly affect enthalpy balances and Q_{SR} and Q_{TOT} for such roasters. Both A/R_{DO} and heat loss to surroundings from an R roaster are proportional to t_R. Thus, for R roasters, lower E, G', or T_{GI} indirectly causes slight increases in Q_{SR} and Q_{TOT}.

10.4.4 High-Speed Roasting

Even though higher T_{BF} are employed when high-speed roasting is used, P decreases by 0.005–0.015 (Geiger 2004; Schenker 2000; Baggenstoss 2008). A/R_{DO} also decreases. These combined effects cause Q_{TOT} to decrease. The Q_{TOT} decrease for high-speed roasting is obtained using higher G' provided by added electrical energy for the blower. The amount of thermal energy used to generate that added electrical energy exceeds the computed savings in Q_{TOT}.

10.4.5 Combined Effects of Operating and Feed Condition Changes

Other than using catalytic instead of thermal afterburning or avoiding very dark roasting, Q_{TOT} savings from individual process changes are less than 10 %.

Combining them provides greater, but slightly less-than-additive, savings. For example, low X_O, low A/R_{DO}, and high T_F, when combined, reduce Q_{TOT} by 17.4 %. Energy savings per kilogram of *roasted* coffee are greater than those per kilogram of *green* coffee. The previously cited combined green-coffee-based Q_{TOT} savings of 17.4 % is a 20.3 % savings when reckoned per kilogram of *roasted* coffee. Reducing X_O and, where possible, reducing P by using lighter roasts can provide attractive thermal-energy and product-yield savings. It has been claimed that predrying of beans (1) reduces roast color gradients within beans, particularly when high-speed roasting is used (Kirkpatrick et al. 1992; Jensen et al. 1994); (2) prevents the formation of rubbery tastes in dark-roasted coffees, particularly Robustas (Jensen et al. 1994); and (3) reduces P when combined with a modified T_B profile (Adler and Feldbrugge 1971).

10.4.6 Dealing with Very Dark Roasting

Besides using more energy, very dark roasting reduces roasted-coffee yield, destroys desirable flavor and aroma notes, increases fouling, and reduces afterburner catalyst life. It has been claimed that blending 1–20 % of darkly roasted coffee made from predried Robusta beans with 80–99 % of normally roasted coffee will provide coffee that has a dark brew color, a high flavor strength/rubbery taste ratio, and good flavor/strength balance (Jensen et al. 1994). Similar blends might satisfy those who prefer dark-roast coffee, better utilize light and medium roast flavor and aroma contributions, increase roasted coffee yield, and save energy.

10.4.7 VOC in Bean-Cooler Air

Water-quenched freshly roasted coffee beans continue to emit VOCs while being air-cooled. In some locales, air that has passed through the cooling beans must be sent through an afterburner to oxidize the VOC. Afterburning should be used only during the brief time at the start of cooling when VOC emission into cooling air is significant (e.g., for roughly the first 15 s of a 210–240 s air-cooling period for regular roasts in R roasters). If beans are cooled in a moderately stratified fashion, limiting afterburning of discharged cooler air to the first 15 s of discharge would require roughly 50 kJ/kg green beans compared to the roughly 810 kJ/kg that would be used if all discharged cooler air passed through the afterburner. Thermal stratification during bean cooling causes different bean strata to cool at different rates but saves energy when cooler air must be subjected to afterburning. To minimize the adverse consequences of bean-cooling-time unevenness, enough quenching should be used prior to air cooling to virtually stop all roasting reactions.

If prior water quenching is not used, as usually occurs with SP roasters, VOC emission during air cooling lasts longer and more energy must be expended for afterburning during such cooling. This causes total thermal-energy expenditures for both SP roasting and cooling to be markedly greater the Q_{TOT} listed for SP roasting alone in lines 10 and 11 of Table 10.1. Beans enter the air cooler at temperature T_{BF} rather than around 120 °C, and roasting reactions continue during air cooling. To minimize nonuniform roasting due to thermal stratification during air cooling of roasted beans, air coolers for SP roasters provide quite shallow beds of beans.

In some roasting systems, beans are air-cooled in a spouted bed and thoroughly mix during cooling. In such cases, limiting afterburning of discharged cooler air to the first 15 s of discharge would require roughly 175 kJ/kg green beans; 2,740 kJ/kg green beans would be used if all discharged cooler air were subjected to afterburning. Additional energy would be required for afterburning of cooler-discharge air if, as sometimes occurs, roasted beans are water quenched in the spouted bed near the start of air cooling.

10.4.8 Reducing Stack-Gas Heat Outflow

Nearly 75 % of heat input to roasters goes out the stack. The latent heat of the water vapor in stack gas cannot be readily recovered, but much of the sensible heat content of stack gas can be transferred to streams inside or entering the roasting system. This will reduce the stack gas's outflow temperature (henceforth called T_{OUT}) and stack-gas mass outflow, thereby reducing Q_{SR} and Q_{TOT} significantly. T_{OUT} will vary during roasts. The reduced Q_{SR} and Q_{TOT} obtained from using such heat exchange were computed by substituting effective-average T_{OUT} for T_{AB} in the aforementioned Excel program.

10.4.8.1 Heat-Exchange System

Countercurrent heat transfer from stack gas to (a) afterburner feed, then (b) chamber-discharge gas, then (c) the combustion air used in the afterburner and roaster furnace can markedly reduce effective-average T_{OUT}.

$CR \equiv (m_H{}'c_H)/(m_S{}'c_S)$, where $m_H{}'$ is the mass-flow rate of the heated gas, c_H is its effective heat capacity, and $m_S{}'$ and c_S are respectively the mass-flow rate and effective heat capacity of the stack gas. Differences between c_H and c_S are small for the pairs of streams involved in each proposed heat-exchange step; and $CR \approx m_H{}'/m_S{}'$ for the step. T_{OUT} after a heat-exchange step cannot be lower than the corresponding inlet temperature of the heated stream. If $CR < 1$, then ΔT_{OUT} for a step $= (CR)\varepsilon(\Delta T_{IN})$, where ΔT_{IN} is the difference between the inlet temperatures of the streams involved. ε, the effectiveness of heat-transfer in the exchanger involved, is a function of CR and NTU, the number of heat-transfer units provided by the exchanger. $NTU = U_E A_E/(m_H{}'c_H)$. U_E and A_E are respectively the overall heat-transfer coefficient and heat-transfer area provided by the exchanger. CR, NTU, ε, and ΔT_{IN} for each of the aforementioned heat-exchange steps vary during roasting.

Afterburner feed is heated in the first heat exchanger, and $m_S{}'$ exceeds the $m_H{}'$ by F'_{AB}. Consequently, $CR < 1$. The heating of afterburner feed acts to reduce F'_{AB} and thereby raise CR, but the T_{OUT} from the exchanger will be markedly greater than the T_{GO}.

CR is substantially greater than 1.0 in the second heat exchanger, where stack-gas transfers heat to chamber-discharge gas. The T_{OUT} for stack gas from that exchanger (also the T_{IN} for stack gas entering the third exchanger) will be fairly close to the T_{GO}. The T_{GO} progressively increases during roasting. Therefore, the T_{IN} for stack gas entering the third exchanger progressively increases.

Stack gas transfers heat to combustion air for the afterburner and roaster furnace in the third exchanger. (ΔT_{IN}), CR, and NTU for the exchanger vary markedly during roasting. The CR and NTU variations cause changes in ε. For example, if ε is 0.80 at the roast's start, it will be 0.95 at its end and lie between those values at intermediate times. During the initial 80 % of a reference roast, $P' \approx 0$. Thus, for the third exchanger, $CR \approx F'/[F' + A' + R_{DO}X']$, ≈ 0.76 initially, then drops to 0.72 as F' decreases and X' peaks, and increases to 0.8 as X' decreases. Near the end of roasts, $X' \approx 0$, P' increases sharply, F' is small and steadily decreases, $CR \approx F'/[F' + A' + R_{DO}P']$ and decreases rapidly from 0.8 to 0.42.

If the T_{GO} varies normally and CR and ε vary as previously indicated, then the T_{OUT} for stack gas leaving the third exchanger for roasts at reference conditions will initially be approximately equal to 85 °C, remain around 110 °C for most of the roast, and then rapidly rise to roughly 200 °C as the roast ends. If the T_{OUT} for time intervals are multiplied by the fraction of stack-gas mass outflow occurring in that interval, one obtains an effective-average T_{OUT} of 120 °C, which in turn causes the Q_{TOT} to decrease to 1,005 kJ/kg green beans (a 38 % reduction). Moisture will condense on heat-transfer surfaces in the third-exchanger at T_{OUT} below the stack-gas dew point but will later completely evaporate when the T_{OUT} rises above 100 °C.

A higher effective-average T_{OUT} and smaller percentage Q_{TOT} reductions will be obtained for darker roasts. If heat exchange was less efficient and the T_{OUT} averaged 200 °C, the Q_{TOT} would be reduced to 1,148 kJ/kg green beans for roasts at reference conditions (a 29 % reduction). If stack gas only transferred heat to the afterburner feed, the T_{OUT} would average 300 °C and the Q_{TOT} for roasts at reference conditions would be reduced to 1,360 kJ/kg green beans (a 16 % reduction).

Transferring heat to the combustion air for burners will raise the T_F and thus may consequently increase NO_X production in those burners (Donley and Lewandowski 1996). Therefore, it may be preferable to replace the third heat exchanger with a step in which partly cooled stack gas transfers heat to the green-bean feed for the roaster. A mini rotating-scoop roaster has been used to provide similar preheating of green beans (Schmidt 2008).

10.4.8.2 Compact Heat Exchange

Roasters are designed to minimize duct length and consequent loss of heat from ducts. Therefore, heat-exchange systems employed to recover energy used for roasting must be compact. A compact heat exchanger and afterburner developed for SP roasters (Finken et al. 1998) might be adapted to recover heat in R roasters. Heat-transfer fluids or heat pipes might be used to exchange heat compactly in the third or second stage of the proposed three-stage heat-exchange scheme, i.e., after stack gas has cooled enough to prevent thermal degradation of the media used.

10.4.8.3 Fouling

Concentrations of VOCs in chamber-discharge gas tend to be inversely proportional to G'/R_{DO}. E tends to increase and the T_{GO} consequently tends to decrease as G'/R_{DO} decreases. Leak-air inflow also reduces the T_{GO} more when G'/R_{DO} is low, and T_{GO} reductions due to heat loss are greater. Therefore, high-boiling point VOCs may condense and foul surfaces in the first and second heat exchangers when G'/R_{DO} is small. Such fouling will be particularly bad for very dark roasting. Condensed VOCs initially are oily but harden over time. The heat transferred from stack gas may raise chamber-discharge-gas and exchanger-surface temperatures enough to prevent such fouling or cause condensed VOCs to evaporate while still oily. The heat-transfer-induced rise in chamber-discharge-gas temperature increases as G'/R_{DO} decreases.

Since combustion air does not contain condensable VOCs, heating that air alone would not cause fouling but would cause a heat-exchange pinch that would limit Q_{TOT} savings.

10.4.8.4 Proper Purging of VOCs and CO

Q_{TOT} reductions decrease F_R and stack outflow and purging of VOCs and CO from the roasting chamber. Thus, large Q_{TOT} reductions could cause VOC or CO levels in the chamber to become high enough to cause puffing (repeated nonexplosive ignitions) or even an explosion near the ends of roasts, problems that are most likely to occur when G/R_{DO} is small or for very dark roasts. Automatically adding extra burner air to increase F'_R enough to provide adequate purging when monitored VOC or CO concentrations approach dangerous levels can prevent these problems. The extra burner air would be added through an added burner-air line and a valve that is opened only when VOC or CO levels start to become high. Though the T_F would drop while the extra burner air is added, T_{GI}–based feedback control would prevent the T_{GI} from changing. Adding extra burner air near the end of roasts will also diminish the normal third-exchanger CR drop for that period. A higher T_F can be used early in roasts, when VOC and CO levels are low, if the burner-air/fuel ratio is adjusted upward enough at the end of roasts.

10.4.9 Afterburner Bypassing

VOC levels are acceptably low, and roughly 90 % of heat use occurs during the first 75–80 % of regular roasts. Q_{TOT} could be significantly reduced if most stack gas bypassed the afterburner during this period. Burner fuel and air and a small portion, e.g., 10 %, of stack gas would still be sent through the afterburner at that time to keep it at operating temperature. All stack gas would go through the afterburner when VOC levels started rising. F'_R, the main contributor to normal stack-gas outflow, decreases markedly toward the end of roasting, and the T_{GO} is highest then. Thus, afterburner feed rates and F'_{AB} will be low when the afterburner is fully online. For roasts in reference conditions, this arrangement would reduce the average T_{OUT} to 280 °C and Q_{TOT} to 1,314 kJ/kg green beans (a 19 % reduction). If E is high, the percentage reduction in Q_{TOT} would be greater.

Emitted VOC levels were only reduced to 150 ppm when intermittent afterburning was used for SP roasters, (Rothfus 1986). VOC output levels of 50 ppm should be readily attainable for R roasters if (a) excessive roaster preheating is avoided and (b) online VOC monitoring or a critical T_{BM} is used to both indicate when VOC levels are about to become excessive and initiate full afterburning. T_{BM} profiles could also be slightly adjusted to limit VOC and CO production time. A more substantial T_{BM} profile adjustment might even limit VOC production to the last 10 % of roasting and save more energy. Afterburner bypassing would be less effective for roasts where G'/R_{D0} is high and the T_{GI} is only moderately greater than the T_{BF}. In such cases, the T_B and T_{BM} profiles level off very markedly during the middle and late stages of roasting and VOC emission starts earlier.

10.4.10 Combined Energy-Saving Methods

Stack-gas heat exchange and afterburner bypassing cannot be effectively combined without losing much of the energy saving provided by heat exchange. However, each of these methods can be effectively combined with other energy-saving steps, e.g., reducing X_0 and A/R_{D0}, raising the T_F, or using lighter roasts. The percentage of Q_{TOT} savings for the added steps would only apply to energy not already saved, but total Q_{TOT} savings could be as high as 49 % if stack-gas heat exchange is combined with these added steps, and as high as 33 % if afterburner bypassing is combined with these steps. In the first case, as much as 57 % of the energy input per kilogram of *roasted* coffee could be saved, and in the second case, as much as 38 %.

10.5 Conclusions

Wherever possible, existing thermal afterburners should be replaced by catalytic afterburners. Catalytic afterburner feeds for R roasters should be heated in the afterburner itself rather than in the roaster furnace. Energy consumption for coffee roasters with afterburners can be reduced by up to 38 % by transferring heat from roaster stack gas first to afterburner feed gas, then to roaster gas, and, finally, to combustion air or to green beans. The feasibility of this method for reducing roasting energy depends on whether fouling of surfaces in the first two heat exchangers by condensable VOC can be minimized or prevented. Largely bypassing the afterburner when VOC concentrations are low (an approach not affected by fouling) can save up to 19 % of current roasting energy use. Additional energy can be saved by also reducing air in-leakage, using higher T_F, storing green coffee beans at conditions that reduce their moisture content, and replacing very dark roasts with lighter roasts admixed with small amounts of a very dark roast.

Extra roaster-furnace burner air should be automatically and temporarily provided if and when the use of energy saving methods causes VOC or CO levels to temporarily become dangerously high inside roasters.

Air that has been used to cool roasted beans should be subjected to afterburning only when its VOC content is excessive.

Acknowledgment Joachim Eichner, President of Praxis International, provided information about industrial roasting practices that proved useful in preparing this paper.

References

Adler IL, Feldbrugge AH (1971) Method of roasting coffee, United States patent, US 3,589,912

Baggenstoss J (2008) Coffee roasting and quenching technology – Formation and stability of aroma compounds. ETH dissertation no. 17696, Swiss Federal Institute of Technology, Zurich

Brandlein LS, Schecter SM, Mahlmann JP (1988) Coffee roasting method. United States patent, US 4,737,376

Clarke RJ (1979) Roasting and grinding. In: Clarke RJ, Macrae R (eds) Coffee, vol 2, Technology. Elsevier Applied Science, Barking, pp 73–107

Donley E, Lewandowski D (1996) Optimized design and oxidizing parameters for minimizing emissions during VOC thermal oxidation. Met Finish 94(11):52–58

Dorfner R (2004) Online Techniken zur Echtzeit Analyse von Fluchtigen Verbindungen in Kaffee Rostgasen. Doctoral thesis, Technischen Universität, Munchen

Dutra ER, Oliveira LS, Franca AS, Afonso RJCF (2001) A preliminary study on the feasibility of using the composition of coffee roasting exhaust gas for the determination of the degree of roast. J Food Eng 47(3):241–246

Eggers R (2004) Heat and mass transfer during roasting – new process developments. Paper presented at the 20th international conference on coffee science, Bangalore, 11–15 Oct 2004

Eggers R, von Blittersdorff M (2005) Temperature field during the roasting and cooling of coffee beans, Seminar 77 (presented Eurotherm, Parma, 20–23 June 2005)

European Commission (2006) Roasting air emissions. In: integrated pollution prevention and control reference document on best available techniques in the food, drink and milk industries. p 139 http//www.ftp.jrc.es/eippcb/doc/wt_bref_0806.pdf

Finken H, Jansen G, Naves E (1998) Arrangement for roasting vegetable bulk material, such as coffee beans. United States patent, US 5,718,164

Geiger R (2004) Development of coffee bean structure during roasting – investigations on resistance and driving forces. ETH dissertation no. 15430, Swiss Federal Institute of Technology, Zurich

Gianturco M (1967) Coffee flavor. In: Schultz HW, Day EA, Libbey EM (eds) The chemistry and physiology of flavor. AVI Publishing Westport, Connecticut, pp 431–449

Jansen GA (2006) Coffee roasting, magic, art, science: physical changes and chemical reactions. SV Corporate Media GmbH, Munich

Jensen MR, Kirkpatrick SJ, Leppla JK (1994) High yield roasted coffee with balanced flavor. United States patent, US 5,322,703

Kirkpatrick SJ, Bertagna RW, Gutwein RW (1992) Process for making reduced density coffee. United States patent, US 5,160,757

MacAllister RV, Spotholz CH (1964) Coffee roasting. United States patent, US 3,122,439

Mahlmann JP, Schecter SM (1985) Controlled coffee roasting. United States patent, US 4,501,761

Mondry VA (2004) Measuring roast color: can the various popular systems be cross-referenced. Paper presented at the annual conference Specialty Coffee Association of America, Atlanta, 23–26 Apr 2004

Price SE, Kussin RP, Fruhling RJ, Harris MB (1991) Ultrafast roasted coffee. United States patent, US 4,988,590

Raemy A (1981) Differential thermal analysis and heat flow calorimetry of coffee and chicory products. Thermochim Acta 43(2):229–236

Raemy A, Lambelet PA (1982) Calorimetric study of self heating in coffee beans and chicory. J Food Technol 17(4):452–460

Rothfus B (1986) Roasting technology. In: Coffee consumption. Gordian-Max Rieck GmbH, Hamburg, pp 112–173

Schenker S (2000) Investigations on the hot air roasting of coffee beans. Doctoral dissertation ETH no. 13620, Swiss Federal Institute of Technology, Zurich

Schmidt K (2008) Mitigating the environmental impact of coffee roasting. Paper presented at the National Coffee Association of America 7th annual convention, Aventura, 6–8 Mar 2008

Schwartzberg H (2002) Modeling bean heating during batch roasting of coffee beans. In: Welte-Chanes J, Barbosa-Canovas G, Aguilera J (eds) Engineering of food for the 21st century. CRC Press, Boca Raton, pp 871–890

Schwartzberg H (2004) Effects of the characteristic behavior of roaster blowers on bean heating in batch roasters. Paper presented at the 9th international congress on engineering and food, Montpellier, 7–11 Apr 2004

Schwartzberg H (2006a) Improving industrial measurement of the temperature of roasting coffee beans. Paper presented at the 21st international conference on coffee science, Montpellier, 11–15 Sept 2006

Schwartzberg H (2006b) Modeling exothermic heat generation during the roasting of coffee. Paper presented at the 21st international conference on coffee science, Montpellier, 11–15 Sept 2006

Singh P, Singh R, Bhamidipati S, Singh S, Barone P (1997) Thermophysical properties of fresh and roasted coffee powders. J Food Process Eng 20(1):31–50

Small LE, Horrell RS (1993) High yield coffee technology. In proceedings of the 15th international conference on coffee science, Montpellier, 6–11 June 1993, pp 719–726

Valdovinos-Tijerino B (2005) Etude de la Torréfaction du Café: Modélisation et Développement des Outils Pour Maitriser la Qualité du Produit en Ligne. Thése de Doctorat. Université Paris VI, Paris

Chapter 11
Advances and Challenges in Thermal Processing with Flexible Packages

Arthur Teixeira, Gaurav Ghai, and Sergio Almonacid

11.1 Introduction

The never-ending quest for improved quality retention at low cost with minimum risk to public health has challenged the food engineering community to seek ways to avoid unnecessary overprocessing and design process conditions with much greater precision. This quest has brought about the development of improved methods for calculating optimum retort process conditions of time and temperature, as well as the introduction of automation and robotics in materials handling operations on the cook room floor in major food plants. These developments were summarized in a presentation and proceedings paper from the 2008 ICEF-10 Congress in Chile (Teixeira 2011).

More recently, the food industry has seen retortable flexible pouches and trays gaining popularity in the marketplace. The limited strength of these flexible packages has posed engineering challenges because they lack the strength of metal cans and glass jars. Internal pressure develops inside the flexible packages during thermal processing as a result of internal water vapor pressure and the expansion of entrapped gases with increasing temperature. This internal pressure will become greater than the saturation pressure of the steam in the retort and could easily cause packages to burst, as seen in Fig. 11.1, unless counterbalanced with carefully controlled overriding air pressure (Ghai et al. 2011). This has driven food

A. Teixeira (✉)
University of Florida, Gainesville, FL, USA
e-mail: atex@ufl.edu

G. Ghai
Food and Drug Administration, College Park, MD, USA

S. Almonacid
Universidad Tecnica Federico Santa Maria, Valparaiso, Chile

S. Yanniotis et al. (eds.), *Advances in Food Process Engineering Research and Applications*, Food Engineering Series, DOI 10.1007/978-1-4614-7906-2_11,

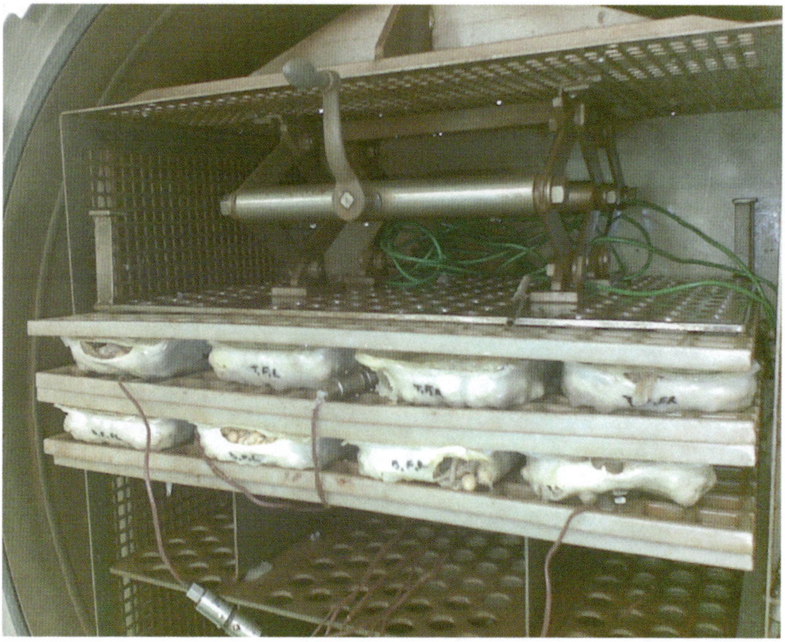

Fig. 11.1 Flexible retortable trays having burst as a result of insufficient overriding retort pressure during pilot plant retort experiment

equipment manufacturers to introduce new types of retort systems specially designed to provide more accurate and closer control of overriding retort pressure during processing.

A common practice in industry is the use of online deflection detectors that can be attached to flexible packages within the retort (Ellab, Centennial, CO, USA). When the packages swell, due to higher package internal pressures, these detectors send signals to an online control system to increase the retort pressure in order to maintain a minimal differential across the package. These detectors play an important role in establishing the external pressure profile required, but their employment is time consuming and labor intensive and must be repeated for each new process.

Profiles of overriding air pressure required during the retorting of foods in flexible packages are largely determined by trial and error or real-time online detecting devices as described earlier. However, various approaches have been used to develop the mathematical relationships between internal pressure buildup and process conditions with varying degrees of success. Several mathematical equations have been proposed for calculating the pressure of water vapor over a surface of liquid water (Vomel 2006). The International Association for the Properties of Water and Steam (IAPWS) recommended an equation for this purpose (Wagner and Pruss 2002) that was adopted for use in this study, along with the ideal gas law for noncondensable gases. In the case of sucrose solutions, Raoult's law was employed to predict the vapor pressure of water. The objective of this work was to develop a methodology to experimentally measure internal pressure profiles during the retorting of simple food systems and a mathematical model for predicting these pressure profiles. Model development was based on the fundamental principles of thermodynamics, as described by Ghai et al. (2011).

11.2 Materials and Methods

11.2.1 Design and Fabrication of Test Module for Head Space Pressure Measurement

A strong rigid container module capable of withstanding a high internal temperature and pressure with the least possible volumetric and thermal expansion was designed and fabricated for this work. Such a module was needed to capture the pressure profile generated in a flexible package that would otherwise cause volumetric expansion of the package. A thick rigid aluminum test vessel (module) in the size and shape of a cylindrical food can was fabricated and instrumented as shown in Fig. 11.2 (top). The lid of the module was of equal thickness and fabricated to fit tightly over the casing and tightened against an O-ring. Three separate holes were drilled through the lid of the module to accommodate pressure and temperature measurement probes and for pulling a vacuum, as illustrated in the schematic in Fig. 11.2 (bottom).

Internal pressures and temperatures were measured with precalibrated wireless data loggers possessing both pressure and temperature sensors, as shown in Fig. 11.3 (top). These data loggers also contained an internal data acquisition and memory system to record and store data from each retort run, which was downloaded to a data receiving station, shown at the bottom of Fig. 11.3 (Ellab, Centennial, CO, USA). The instrumented module was filled with the model food systems and subjected to a retort process to obtain experimental data against which model predictions could be compared.

11.2.2 Design of Experiments

A pilot scale retort, Stock Pilot Rotor PR-900-0 BV-VA (Stock America, Grafton, WI, USA) was equipped with independent temperature and pressure control systems. Retort experiments were conducted under a still cook in full water immersion in accordance with a preprogrammed process cycle. One data logger was carefully lowered into the bottom of the module, and a known amount of product was added while maintaining a consistent measured headspace volume. Filling and sealing of the module were carried out under ambient conditions, and the initial temperature of the product was recorded. A wired thermocouple was mounted through the fitting on the module lid and fastened tightly. Air was withdrawn using the vacuum apparatus, and the initial internal pressure was set at

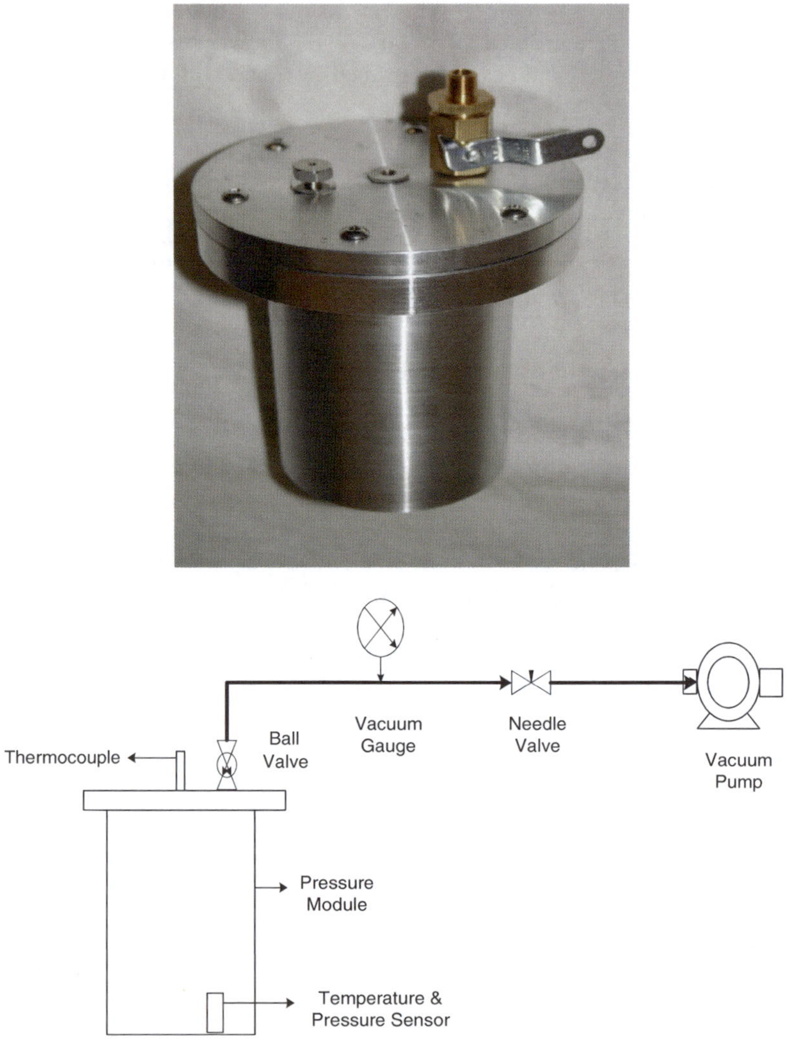

Fig. 11.2 Rigid pressure-tight module assembly (*top*) and system schematic for achieving initial partial vacuum in rigid pressure-tight module (*bottom*)

10 in. Hg absolute, typical in the canned food industry. The module was then placed in the center of the retort, and a second data logger was planted at a secure location elsewhere inside the retort. In addition, two external wired thermocouples were placed, one right next to the module and the other closer to the mercury in a glass (MIG) thermometer well inside the retort. The module was held tightly using a cage clamp system and then exposed to the respective process cycles with holding times of 30–90 min, depending on the product; and average come-up and cool-down

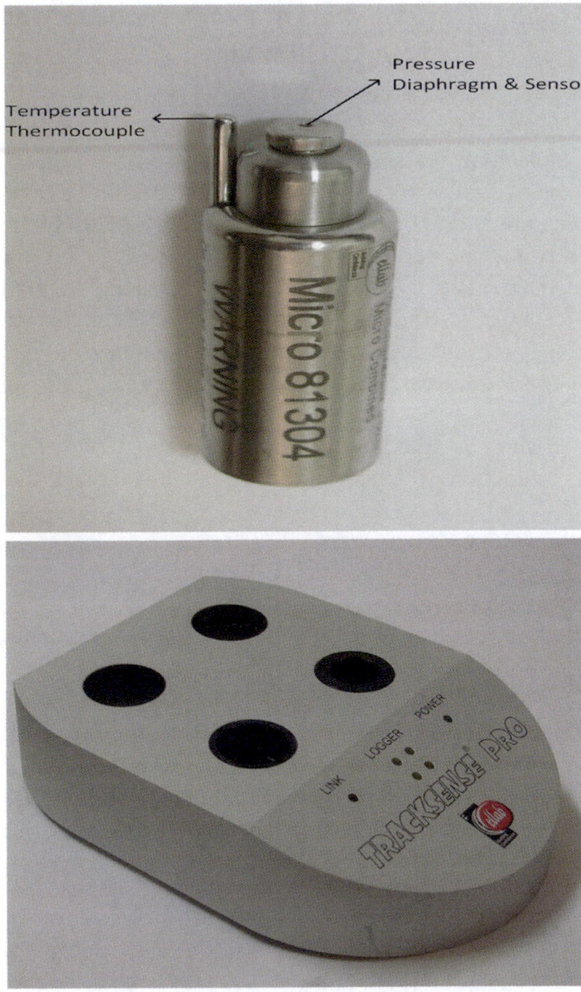

Fig. 11.3 Wireless instrumentation for measuring temperature and pressure, showing data logger (*top*) and logger reading station for downloading data files (*bottom*)

times of 10–12 min and 8–10 min, respectively. Three replicate experimental runs were carried out with each of four simple food systems:

- Distilled water,
- 10 % sucrose solution,
- Green beans in water, and
- Sweet peas in water.

Details of a sample preparation can be found in Ghai et al. (2011).

11.2.3 Development of Mathematical Model for Predicting Headspace Pressure

A mathematical model to predict the internal pressure of plain distilled water as a simple food system can be expressed as the sum of the pressure exerted by the vapors of the water itself (water vapor pressure) as a function of temperature and the pressure exerted by the gas phase (noncondensable gasses) as a function of initial and final temperatures. An equation derived by Wagner and Pruss (2002) was recommended by the International Association for the Properties of Water and Steam (IAPWS) in 1995 and adopted for use in this study for the case of pure water, along with the ideal gas law for noncondensable gases. In the case of saline and sucrose solutions, Raoult's law was employed to predict the vapor pressure of water.

Using the equation recommended by IAPWS, the vapor pressure of pure water as a function of temperature was calculated using the following expression:

$$\begin{aligned}\mathrm{Log}_e\left(P_w/22.064^*10^6\right) = 647.096/T^*\ (&-7.85951783^*v + 1.84408259^*v^{1.5}\\ &- 11.7866497^*v^3 + 22.6807411^*v^{3.5}\\ &- 15.9618719^*v^4 + 1.80122502^*v^{7.5}),\end{aligned} \tag{11.1}$$

with T (temperature) in K, P_w (water vapor pressure) in Pa, and $v = 1-T/647.1$ where the value 647.1 K is the triple point of water.

The pressure exerted by the gaseous phase can be calculated using the equation for an ideal gas undergoing a reversible adiabatic (isentropic) process:

$$(PV = nRT),\ \text{in which } PV^\gamma = \text{constant},\ \text{and } P^{\gamma-1}T^{-\gamma} = \text{constant}, \tag{11.2}$$

where P is pressure, V is specific or molar volume, T is temperature, and γ is the heat capacity ratio, adiabatic index, or isentropic expansion factor:

$$\gamma = Cp/Cv = (\alpha + 1)/\alpha. \tag{11.3}$$

Cp is the specific heat at constant pressure, Cv the specific heat at constant volume, and α the number of degrees of freedom divided by 2 (3/2 for a monatomic gas and 5/2 for a diatomic gas).

Thus, for a monatomic ideal gas $\gamma = 5/3$, for a diatomic gas (such as nitrogen and oxygen, the main components of air) $\gamma = 7/5$, and for water vapor $\gamma = 1.32$ (Van Wylen and Sonntag 1965; Lange 1970; White 1986).

However, the thermal process experienced by the module in this work was neither adiabatic nor isentropic. Entropy had to be transferred from the retort heating medium to the module. Therefore, the approach taken here is only a first approximation.

Using the preceding pressure-temperature equation for an isentropic process taking place inside the module, the relationship between pressure and temperature at two different conditions was expressed as follows:

$$P_1{}^{\gamma-1}/T_1{}^{\gamma} = P_2{}^{\gamma-1}/T_2{}^{\gamma}. \tag{11.4}$$

Since the majority of the headspace was filled with water vapor, due to vacuum withdrawal, a value of 1.32 for γ was suitably chosen. This gave the expression needed to calculate the final pressure exerted by the gaseous phase as a function of the initial pressure and the initial and final temperatures to be

$$P_2 = P_1{}^{*}(T_2/T_1)^{4.1}. \tag{11.5}$$

Pressures exerted by the liquid and gaseous phases were calculated separately using the preceding equations and added later to obtain the final predicted internal pressure profile of the pure water food system.

11.3 Results and Discussion

Experimental data from the retort process with pure distilled water are presented in Fig. 11.4. Profiles of measured retort temperature and pressure over time can be seen superimposed on the profiles of the internal product (distilled water) temperature and pressure over time. It can be seen that the pressure inside the module gradually increased as the temperature of product kept on increasing, until it reached the desired process temperature. Profiles for the module and the retort show that the pressure differential across the module during the hold period was approximately 23 kPa. This pressure differential, while acting on a flexible package, such as a plastic tray with approximately 300 in.^{2} of surface area, would generate a force greater than 650 N attempting to strain the package and burst it. A higher overriding retort pressure would be needed to avoid any undue strain on the package.

A comparison of model-predicted internal pressure profiles with those measured experimentally for all four simple food systems (pure water, 10 % sucrose solution, green beans in water, and sweet peas in water) is shown in Fig. 11.5a–d. Note that the subfigures a–d in Fig. 11.5 are in order of increasing compositional complexity of the model food systems. Likewise, the difference between the measured and predicted pressures at steady state (constant retort temperature) increases with the complexity of the food systems. In the case of pure distilled water in Fig. 11.5a, there is fairly close agreement between measured and predicted pressures, showing a difference (error in model prediction) of only 10 kPa. Such good agreement in this case is not surprising since the system of pure water with noncondensable gases satisfies the assumptions upon which the model was based. However, it is

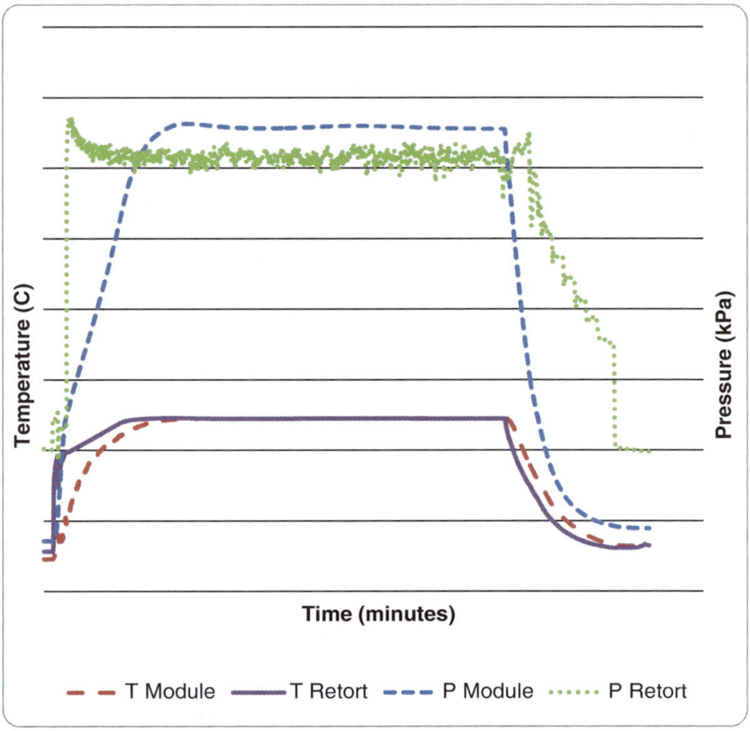

Fig. 11.4 Experimentally measured temperature and pressure profiles within retort chamber and within rigid pressure-tight module filled with distilled water undergoing thermal processing

interesting to note that the difference between measured and predicted pressures increased to approximately 15 kPa for the 10 % sucrose solution, then to 20 kPa for green beans in water, and reached up to 30 kPa for the sweet peas in water.

These results suggest that the model can perform reasonably well for the simple case of pure distilled water but is not adequate for predicting pressure with sufficient accuracy in the case of a more complex food system, such as sweet peas in water. Sweet peas contain soluble sugars and starches that will impart noticeably lower vapor pressures to the system as compared with pure water, and the first-generation model developed here was incapable of taking such factors into account.

Many factors come into play when it comes to complex food systems, where thermodynamic relationships are difficult to characterize. Salts and other compounds dissociate into ions and other functional groups when dissolved in water. These individual groups interact with one another while in solution and affect the colligative properties of the solution, such as vapor pressure. An alternative approach to this modeling effort could be based on the work of Fredenslund et al. (1975), who introduced the Generalized UNIversal Functional Activity Coefficient (UNIFAC). This approach has proven to be a fast and reliable tool for predicting liquid-phase activity coefficients. Use of the UNIFAC models for

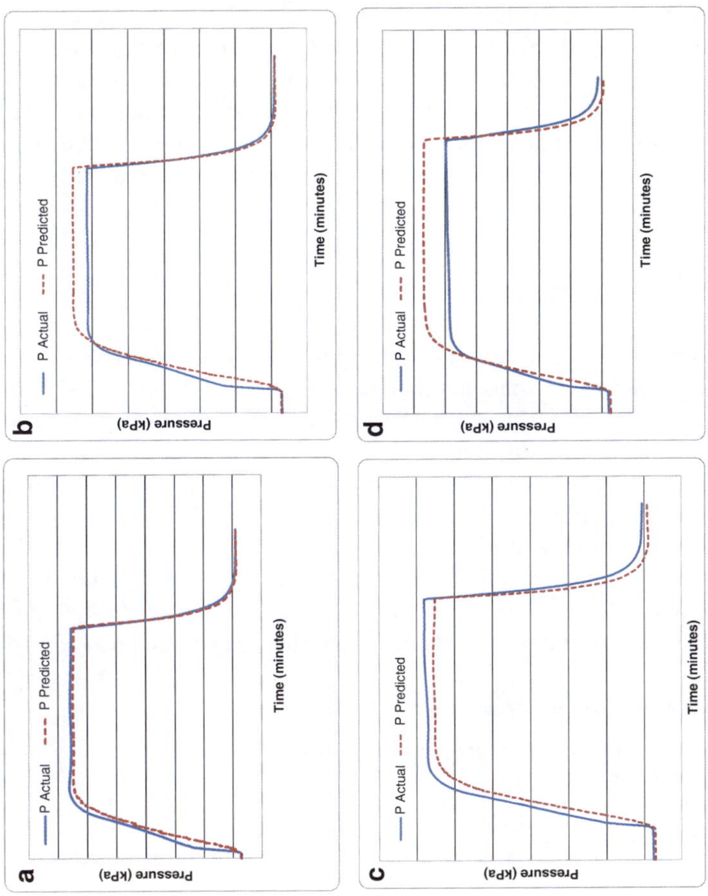

Fig. 11.5 Comparison of model-predicted internal pressure profiles with experimentally measured pressure profiles for (**a**) distilled water, (**b**) 10 % sucrose in distilled water, (**c**) green beans in distilled water, and (**d**) sweet peas in distilled water

predicting headspace pressures in a rigid module undergoing a retort process is the subject of companion research under way by colleagues at the Universidad Tecnica Federico Santa Maria in Valparaiso, Chile. The work reported here was undertaken in parallel with the work in Chile as a collaborative effort in an attempt to evaluate alternative approaches.

Results thus far pertain to observations of the measured and predicted pressure profiles during steady state when the retort temperature remained constant. Considerable disagreement can be seen during the early come-up period. That is likely due to experimental error caused by the existence of a nonuniform temperature distribution within the module during the early stages of heating. The sensing probe on the data logger was always resting at the bottom of the module, which is likely a cold spot during natural convection, while temperatures in the headspace would have been higher, producing greater pressure than that predicted by the model in response to the cold spot temperature. This can be easily remedied by relocating and affixing the data logger to keep the sensing probe well into the headspace.

11.4 Conclusions

- A strong, rigid, pressure-tight module proved to be an effective tool for experimentally determining the internal pressure profiles of various food systems during retorting and can be used to determine optimum retort pressure profiles to minimize pressure differentials across flexible packages.
- The relative simplicity of the model developed in this study (IAPWS equation for water vapor pressure, ideal gas law for noncondensable gas pressure, and Raoult's laws for dissolved solutes) renders it inadequate for application to food products with complex composition and chemistry. Improved models are needed that would likely require the inclusion of factors related to the chemical and physical composition of food products and should be the subject of further study.

Acknowledgments The authors are grateful to:

- Ellab, Centennial, CO, USA, for loan of the wireless data logger system throughout the duration of the project;
- Chilean Foundation for Development (FONDECYT Project 1090689) for its support of collaboration; and
- Florida Agricultural Experiment Station at the University of Florida's Institute of Food and Agricultural Sciences, Gainesville, FL, USA.

References

Fredenslund A, Jones RL, Prausnitz JM (1975) Group-contribution estimation of activity coefficients in non-ideal liquid mixtures. AIChE J 21:1086–1099

Ghai G, Teixeira AA, Welt BA, Goodrich-Schneider R, Yang W, Almonacid S (2011) Measuring and predicting head space pressure during retorting of thermally processed foods. J Food Sci 76(3):E298–E308

Lange NA (1970) Lange's handbook of chemistry, 10th edn. McGraw-Hill, New York
Teixeira AA (2011) Innovations in thermal treatment of food. Chapter 11. In: Aguilera JM (ed) Food engineering interfaces, Food engineering series. Springer Science and Business Media, LLC, New York, pp 247–259
Van Wylen GJ, Sonntag RE (1965) Fundamentals of classical thermodynamics. Wiley, New York
Vomel H (2006) Saturation vapor pressure formulations: discussion of saturation vapor pressure formulations. CIRES, University of Colorado, Boulder
Wagner W, Pruss A (2002) The IAPWS formulation 1995 for the thermodynamic properties of ordinary water substance for general and scientific use. J Phys Chem Ref Data 31:387–535
White FM (1986) Fluid mechanics, 4th edn. McGraw-Hill, New York

Chapter 12
Current Knowledge in Hygienic Design: Can We Minimise Fouling and Speed Cleaning?

P.J. Fryer, P.T. Robbins, and I.K. Asteriadou

12.1 Introduction

Within a manufacturing environment many different types of equipment are used to produce a product, including tanks, pumps, valves and pipe work. Any surface with which a material come into contact during its production will have to be cleaned to an appropriately safe and hygienic standard to enable the use of the equipment for further production. Fouling of food processing equipment is rapid and may result from many processes; in addition, in many plants there is a need to change between products as one process line is used to make many different Foods. Cleaning in place (CIP) is typically used to clean the inside of a process plant. The hygienic design of processes is a complex subject, and a number of guidelines have been produced by organisations such as the European Hygienic Engineering and Design Group (EHEDG). The cleaning of a process plant is time and resource consuming within a manufacturing operation and contributes significantly to the environmental footprint of the plant (Eide et al. 2003). Efforts to minimise the use of time, water, CIP chemicals and energy can improve both the efficiency of a plant and its environmental impact.

12.1.1 Fouling Processes

Fouling – the formation of unwanted surface deposit on process equipment – is a significant problem in process industries. In food processing, fouling lowers the heat transfer efficiency in heat exchangers, raises the pressure drop throughout

P.J. Fryer (✉) • P.T. Robbins • I.K. Asteriadou
School of Chemical Engineering, University of Birmingham, Birmingham, UK
e-mail: p.j.fryer@bham.ac.uk

S. Yanniotis et al. (eds.), *Advances in Food Process Engineering Research and Applications*, Food Engineering Series, DOI 10.1007/978-1-4614-7906-2_12,

the plant and endangers safe operation by giving bacteria places to grow. Deposits form due to the adhesion of species to the surface followed by cohesion between elements of the material; the result is a surface layer that resists removal. Fouling research has formed an important part of heat transfer research for many years (Kern and Seaton 1959), and since its description as 'the last unresolved problem in heat transfer' (Taborek et al. 1972), many papers and conference proceedings have discussed advances in the area [for recent conference proceedings see Müller-Steinhagen et al. (2011)]. Many food components are thermally unstable. As a result, during thermal processing, the formation of fouling deposits from foods is a severe problem.

A range of organic and mineral materials can be deposited on process surfaces. Epstein (1981) developed a general classification of fouling mechanisms, and many of these can be identified in foods, including (1) *reaction fouling*, where fouling results from reactions such as protein denaturation (Changani et al. 1997); for example, the whey protein b-lactoglobulin is deposited preferentially from a significant fraction of milk pasteuriser deposits; (2) *biological fouling,* from adhesion of single organisms to the growth of thick biofilms (Boulange-Petermann 1996); and (3) *crystallisation or precipitation fouling*, where a dissolved component of a fluid is deposited, such as calcium carbonate from boiling water or calcium phosphate from milk at UHT temperatures. A subset is *solidification fouling:* solidification of a fluid or components of it (Fernandez-Torres et al. 2001) such as ice from water or starch from a food fluid.

Fouling from foods may result from more than one mechanism, for example, proteins and calcium phosphate interact to create milk fouling (Simmons et al. 2007) and particles caused by the reaction of proteins in a bulk are transported and deposited on a wall. Fouling profiles can also be significantly affected by small changes in fluid composition: for example, adding calcium phosphate to whey protein concentrate (WPC) changes the rate and extent of fouling and the deposit composition (Rosmaninho et al. 2008). Predictive models for fouling are rare both because of the complexity of the processes that take place and because minor variations in composition yield significant changes in the rate and extent of fouling, so it may be difficult to know enough detail to generate an accurate model at the plant rather than the laboratory scale. Several researchers (e.g. De Jong et al. 2002a, b; Schutyser et al. 2008; Petermeier et al. 2002; Benning et al. 2003) have proposed methods for predicting fouling and reducing the severity of the problem; some of these methods use indeterminate models to take account of the variation found industry-wide. Integration of fouling and cleaning models in the food industry has not been reported.

12.1.2 Cleaning Processes

It is usually not possible to run food lines for longer than 24 h. Cleaning is needed because of the degradation of performance caused by fouling, because of possible damage to products caused by microbial growth at the plant, and because of the need to change over between different products to avoid cross contamination when many products are made on the same line. Cleaning takes several hours and thus forms a large fraction of total production time – the greatest cost may be loss of the potential product that cannot be made as the plant is being cleaned (Goode et al. 2010). Minimising cleaning costs and environmental effects is critical for industry.

The focus of the food industry is on cleaning efficiency, as cleaning can take up to 25 % of operational time. Cleaning processes are often highly automated, with CIP technologies commonplace (such as Tamime 2008). Designing a protocol for a given situation is still, however, semi-empirical (Fryer et al. 2006). Practical cleaning protocols are commonly overdesigned, but there is little incentive to experiment with safety margins, which might endanger safe production. However, as the costs of utilities and of waste disposal increase, minimising the environmental impact of processes takes on greater importance. Cleaning can generate large volumes of dirty water of very high or low pH, forming a significant part of the effluent load of a process. To make the process more environmentally sustainable, a better understanding of cleaning is needed that can be used both to design better cleaning processes and to minimise the cleaning and effluent costs. Determination of the endpoint of cleaning is critical – whilst sensors for fouling are relatively common (such as Wallhäußer et al. 2011) because the effects of a thick layer are relatively easy to measure, measuring the endpoint of cleaning, which corresponds to the removal of a last piece of fouling, is much more difficult.

12.1.3 Design to Reduce Fouling

In many cases in food processing the reactions that give rise to fouling are inevitable parts of the heating process to pasteurise or sterilise the material. A number of possible solutions exist:

- Employ designs which remove or minimise fouling, such as the scraped surface heat exchanger in which any deposit is removed by the scraper. This is an expensive alternative to conventional plate heat exchangers, however.
- Remove the heated surface on which fouling reactions will be fastest, for example, ohmic heating, in which an electric current is passed through the fluid and heating occurs by resistance heating, has been considered as an option to minimise fouling because it has no hot surfaces (such as Ghnimi et al. 2008, 2009).
- Use a non-thermal process such as high-pressure processing or UV pasteurisation (Simmons et al. 2012). The economics of these are complex because these processes may be more expensive than conventional heating.

In practice, there is usually no economic alternative to using conventional heating. Fouling can be reduced by conventional hygienic design and good manufacturing practice, with protocols and standards developed by organisations such as EHEDG, which include keeping velocities high, minimising the roughness of the steel surface and minimising dead spaces such as T-junctions. Process lines must be cleaned regularly – the need is to develop operating and cleaning protocols which minimise the problem.

12.2 Prototype Cleaning Map

Significant problems in cleaning practice (and research) including the following:

- The difficulty of comparing cleaning of different fouling deposits. For example there is currently no reliable way to predict the time required to clean one deposit based on that of another;
- The lack of effective on-line measurement methods to identify the cleaning end-point;
- Scale-up: if the cleaning mechanism is not understood, then scaling up lab or pilot plant data to production is basically empirical. Model deposits can give different results in real systems; for example using whey proteins to simulate pasteuriser behaviour (Christian et al. 2002) gives different fouling patterns to milk.

The practical consequences of this are that the best way of cleaning each type of soil has been developed largely independently and CIP protocols are established and implemented without effective measurement of the end-point. The long-term aim of cleaning research is to find ways to predict cleaning behaviour that can be used to design equipment, for example by incorporating cleanability into computational design codes (e.g. Jensen et al. 2005; Asteriadou et al. 2007). To do this requires better understanding of cleaning processes (Wilson 2005).

12.2.1 Classification of Cleaning Problems

We recently (Fryer and Asteriadou 2009) tried to classify the cleaning problems of the food and personal care industries by the severity of problems and the mechanisms of removal. Figure 12.1 shows cleaning problems classified by the *type of fouling deposit* and the *type of cleaning agent needed*. The chemical and mechanical properties of the deposit control the type of chemical needed. The fouling deposits most difficult to clean are as follows.

Type 1: Cleaning of highly viscous or visco-elastic or viscoplastic fluids by water
In some cases, fouling forms basically by solidification, such as for viscous foods, starch-based sauces and confectionery fluids such as chocolates or crème fillings.

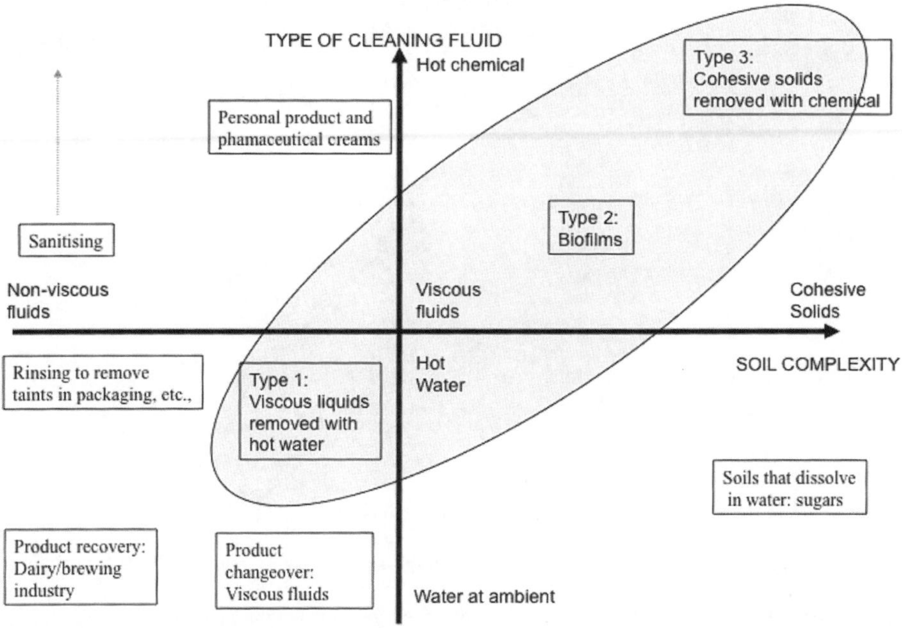

Fig. 12.1 Prototype cleaning map [from Fryer and Asteriadou (2009)]

At the end of a run, thick layers of product are left in tanks and may fill pipe work. These materials can, however, be removed by the action of water alone. Product recovery can be important here – for example, emptying filled pipe work back into the product tank both minimises waste and increases profitability.

Type 2: Cleaning and killing of biofilms
Although it is possible to remove some biofilms with water (Blel et al. 2009a, b), normal practice is to use biocide, which changes the removal characteristics of the deposit (Burfoot and Middleton 2009).

Type 3: Cleaning of solid deposits by chemical action
Here solids are formed by deposition of the components of the process fluid. The nature of the deposit will determine the type of cleaning chemical used, such as sodium hydroxide to remove organic films and acids to remove mineral scales.

CIP involves fluid flow across a deposit, and thus both fluid mechanical and chemical forces act on the deposit. Cleaning must overcome both the cohesive forces that bind elements of deposits together and the adhesion forces between deposit and surface. This may be done by fluid action alone or by the combined effects of fluid and chemicals.

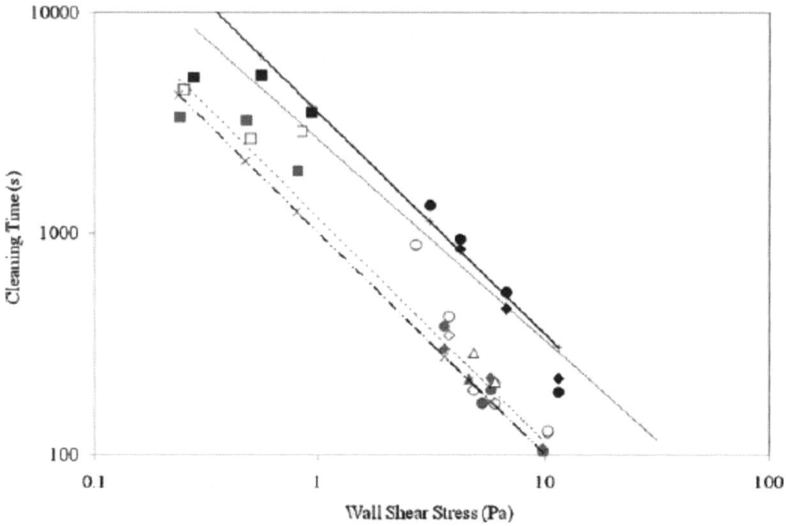

Fig. 12.2 Cleaning time versus wall shear stress for lab-scale (*squares*) and pilot plant experiments (0.3 m length = diamonds, 1 m length = circles, 2 m length = triangles), in a range of temperatures (20 °C = black, 40 °C = white, 50 °C = grey data points) [from Cole et al. (2010)]

12.2.2 Mechanical Fluid Removal

Type 1 deposits can be removed by fluid action alone. The rheology of the deposit and the rheological and flow properties (such as the shear force applied to the deposit) of the cleaning fluid will thus be the factors of removal. It should be possible to predict the cleaning time from the fluid dynamics of the flows [such as by the models developed by Sahu et al. (2007)]. However, only limited data on the displacement of one fluid by another with significantly different viscosities are available (e.g. Grassi et al. 2008). Henningsson et al. (2007) describe rinsing yoghurt out of a tube using water and used a volume-of-fluid computational fluid dynamics (CFD) model to simulate rinsing; as expected, the final stages of cleaning, with a thin layer on the wall, were most difficult to measure and to model.

Cole et al. (2010) show for the removal of toothpaste from pipe work (a problem similar to the removal of viscous foods) that cleaning time is inversely proportional to shear stress and is not a function of the tube length. Figure 12.2 shows data for two length scales – lab and pilot plant – showing similar behaviour.

Figure 12.3a shows the weight of toothpaste remaining in a pipe after water has been flowing through it for various lengths of time. Initially the pipe is full. Palabiyik et al. (2014) show that there are three stages of removal:

(1) A short *core cleaning* stage (similar in length to the fluid residence time in a pipe; here <10 s and thus not seen in the figure) before water breaks through,

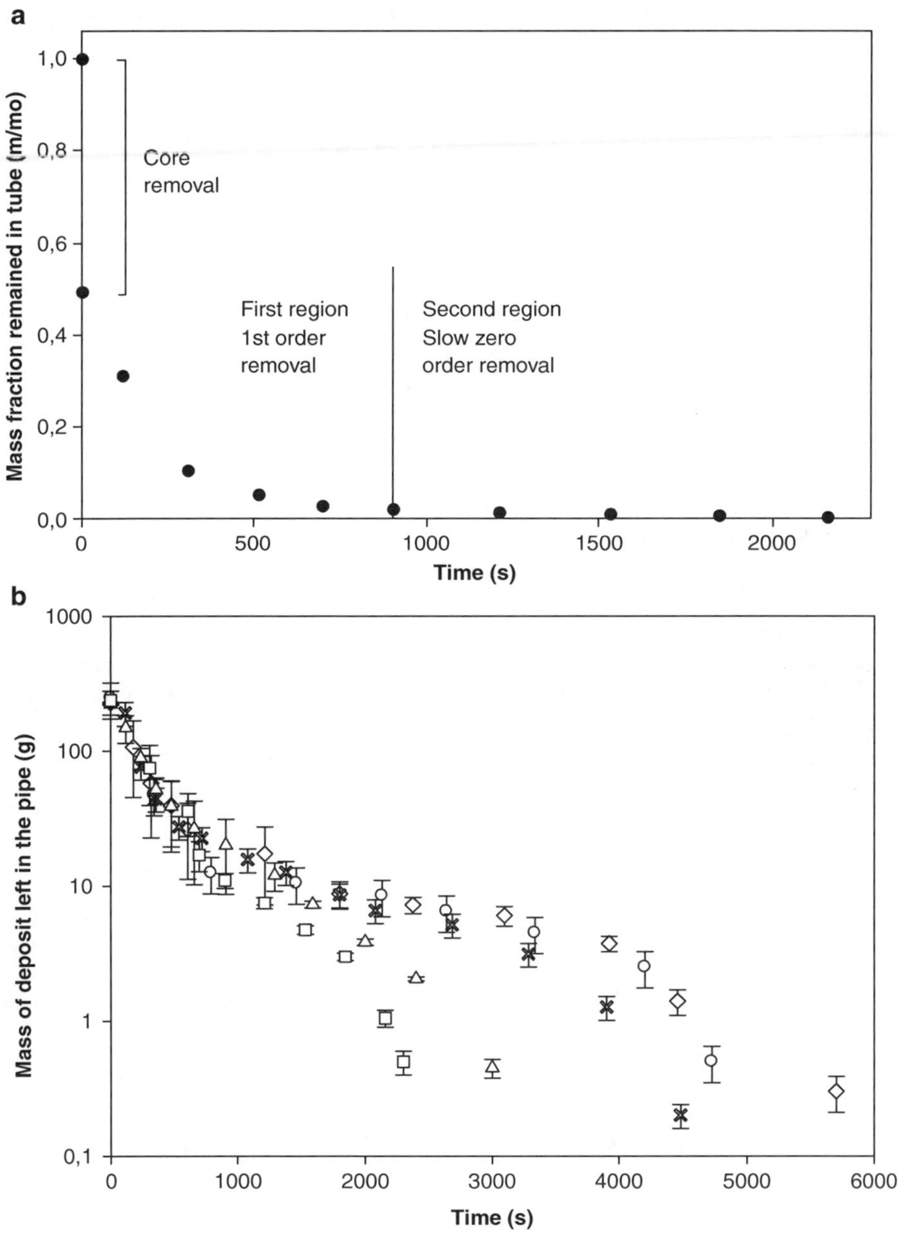

Fig. 12.3 (**a**) Graph of weight fraction of toothpaste in a 1 m pipe during cleaning with water at 0.55 m/s and 15 °C, showing three stages of cleaning process; a very short core removal process followed by film removal, followed by slow removal of a surface patches; (**b**) cleaning profiles of thin film of toothpaste at same Reynolds number of 12100 at lab scale. Cleaning is performed at 15 °C, 0.55 m/s (*squares*), 20 °C, 0.48 m/s (*triangles*), 30 °C, 0.38 m/s (*crosses*), 40 °C, 0.31 m/s (*circles*) and 50 °C, 0.26 m/s (*diamonds*)

which leaves a film of material on the wall; in Fig. 12.2 about half of the product is removed in this stage.
(2) A *film removal* stage in which the loss of material follows exponential behaviour; about 90 % of the remaining deposit is removed in this stage.
(3) A final *patch removal* stage controlled by the slow removal of patches of deposit from the wall, although only the last remaining percentage of the material is removed in this stage.

The need is to maximise the product that may be recoverable in the first two stages and then minimise the cleaning time to remove the final film.

Complex effects are observed. Figure 12.3b shows the effects of different velocities, but the same empty-tube Reynolds number of 12100. It can be seen that in the *film removal* stage (0–1,000 s) the process is first order, and the Reynolds number appears to be capable of describing the behaviour, i.e. data for different conditions lie on the same line. However, in the *patch removal* stage, the Reynolds number does not correlate well with the data, and the data scale with velocity; the slowest condition corresponds to the slowest cleaning velocity. Significant changes in overall cleaning time (at least a factor of 2) can be made by changing the velocity.

It is clear why this type of cleaning is not a function of length – the first stage of core removal is over in a few seconds and the final stages are those of removal from islands of deposit, where the surface shear stress will be more significant. More work needs to be done to understand these effects and optimise the processes of cleaning and product recovery in these systems. Most recently, Taghavi et al. (2012) presented and correlated data for laminar flows, but the introduction of turbulence in real cleaning flows makes modelling more complex.

12.2.3 Diffusion–Reaction Removal

For both Types 2 and 3 deposits, chemicals are needed for removal. The cleaning process will thus involve diffusion of the chemical to or into the deposit and a reaction that transforms it into a removable form. There are three different mechanisms:

- *Dissolution:* for example the dissolving of mineral salts in acid, proteins in alkali or sugars in water.
- *Bulk cohesive failure:* removal through breakdown of cohesive forces between parts of the deposit;
- *Surface adhesive failure:* removal by breakdown of adhesive forces that bind deposit to the surface.

In some cases one mechanism will predominate; for example, some starch films exhibit removal by adhesive failure between the surface and the deposit. In practice, different mechanisms may be in control under different conditions. Liu et al. (2002,

2006a, b, c) have used micromanipulation probes to deform and remove a range of deposits, including milk and starch, and identified cohesive and adhesive failure modes as a function of variables such as deposit thickness and interfacial energy. Cohesive failure occurs preferentially at high temperatures and chemical concentrations, i.e. where reaction rates are high. The same types of effects have been identified using other types of probes, such as dynamic gauging (e.g. Gu et al. 2011).

Concerning fluid shear, cleaning will depend on the following factors:

- *Mass transfer* of chemical to the deposit-fluid interface or of removed material;
- *Diffusion* of active species through fouling deposit; Mercade-Prieto et al. (2007, 2008) give estimates for whey protein gels;
- *The rate of reaction*; little information is easily accessible, although Yoo et al. (2007) give qualitative data on the effect of alkali on whey protein gels and show that the effects of temperature predominate;
- *The effect of the reaction* on the deposits and surface. In protein cleaning, reactions with alkali can produce an expanded deposit that is easier to remove, but for some deposits (e.g. Bird and Fryer 1991) excessive concentration of alkali gives a shrunken and gelled cohesive deposit that is difficult to clean.

Most mechanistic studies on diffusion–reaction films used milk or milk simulants because of the commercial importance of milk processing. Milk protein swells on contact with alkali (except at very high pH) into a form that can be removed. Protein behaviour was studied by Mercade-Prieto and Chen (2006); Mercade-Prieto et al. (2007), who investigated whey protein and b-lactoglobulin gels and solutions at high pH and related protein breakdown to a number of effects seen in cleaning, providing a mechanistic explanation for the observation of Bird and Fryer (1991). Mercade-Prieto et al. (2007) also show that dissolution depends on the aggregation structure, i.e. on the fouling conditions. However, full kinetic models for cleaning of these systems are not yet available.

Gillham et al. (1999, 2000) studied alkali-based cleaning of whey protein deposits on stainless steel and found that different mechanisms controlled the cleaning at different stages; removal of the last layers was largely controlled by shear, with flow reversal significantly increasing removal. The effect of flow pulsing has been studied by Blel et al. (2009a, b) and Augustin et al. (2010), who use CFD to predict surface shear stress.

The effect of swelling during cleaning on process plants can be seen in Fig. 12.4, which shows pressure drop and heat transfer coefficient changes in a section of a plate heat exchanger through which water and alkali solutions are passed [see also Christian and Fryer (2006)]. It can be seen that:

- Pressure drop increases rapidly on contact with alkali, showing the effect of swelling, which is to reduce the flow channel in the exchanger and thus increase the pressure drop; if fouling is excessive, then this swelling can cause plate exchanger gaskets to rupture.

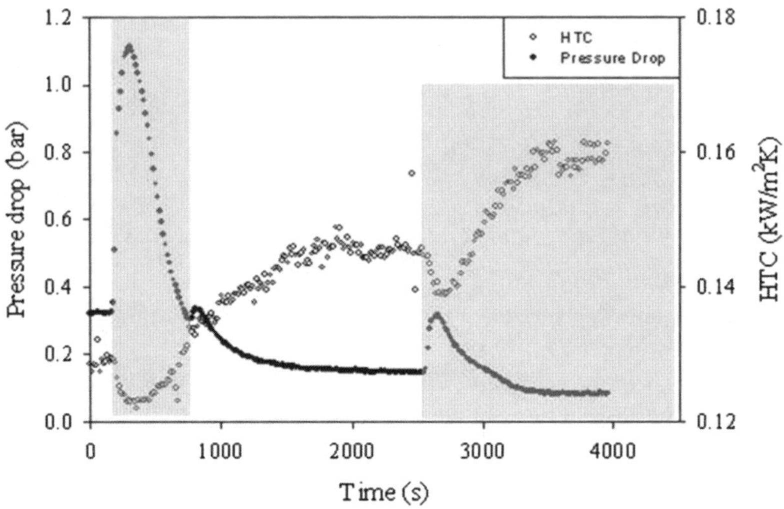

Fig. 12.4 Comparison of heat transfer coefficient and pressure drop in a section of plate heat exchanger during chemical exposure experiments to remove protein-based deposit. A cleaning chemical concentration of 0.1 wt% was used. *Shaded area*: chemical flow through UHT section [from Christian (2003); see also Christian and Fryer (2006)]

- After a short time the pressure drop begins to decrease and the heat transfer coefficient begins to increase due to removal of the swollen film by the fluid. There are thus two processes going on, the swelling reaction and the removal.
- However, when water replaces alkali, the rate of cleaning slows – there is a small initial *increase* in the pressure drop, as the alkali within the fouling layer continues to cause swelling, but the removal reaction slows.
- Both the pressure drop and the heat transfer coefficient reach an asymptote towards the end of the water cleaning, suggesting that the surface might be clean – but when alkali flow is resumed, another swelling peak is observed, followed by a further decrease in the pressure drop towards a true clean state.

It is more complex to correlate or model this behaviour than that of Type 1 as the processes under way involve both reactions and fluid mechanics. The data suggest that the cleaning rate increases with increasing temperature and flow rate, as would be expected, and that as the temperature increases, the effects of the reaction increase, so that different processes may be in control under different conditions. For example, Fig. 12.5a shows data for the cleaning of brewery yeast films (Goode et al. 2010) showing the effects of both temperature and flow velocity; the greatest effect of velocity is at low temperatures, where reaction effects will be least pronounced. The effect of fouling conditions, temperature and flow rate for WPC films is summarised in Fig. 12.5b (Christian 2003), which again shows that as the

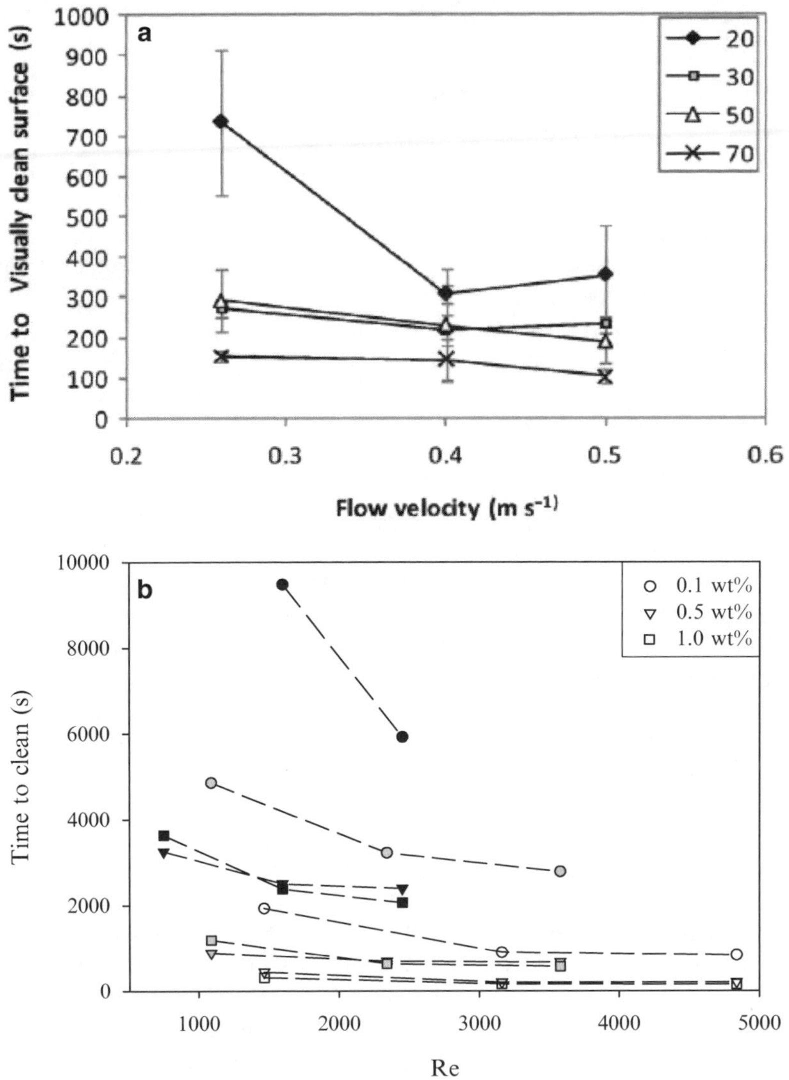

Fig. 12.5 Cleaning times for (**a**) yeast film deposited above fermentation vessels using 2 wt% chemical at temperatures at 20–70 °C and flow velocities of 0.26–0.5 ms^{-1} [from Goode et al. (2010)]; (**b**) whey protein films as a function of the concentration of protein used and flow temperature; *black*: 30 °C, *grey*: 50 °C and *open*: 70 °C (Christian 2003; Christian and Fryer 2006)

temperature increases, the effect of velocity increases and that the effect of increasing the WPC concentration is to increase the cleaning time.

The optimisation problem has to do with deciding whether the time gained by cleaning at a high temperature or high chemical concentration is justified by the cost.

12.3 Modification of Surfaces

12.3.1 An Antifouling Surface?

The balance between adhesive and cohesive effects can be changed by changing the heat transfer surface. It is well known that bacterial adhesion can be minimised by changing the surface energy (Baier 1980), and it is possible to control marine biofouling by surface modification (Rosenhahn et al. 2008) and to reduce mineral fouling (Zhao et al. 2005).

Amongst many other studies, a large European Union (EU) project (Rosmaninho et al. 2007, 2008; Rosmaninho and Melo 2008) investigated the effects of different surfaces on milk deposit adhesion and found that it was possible to change the rate and extent of fouling and cleaning.

Changing the surface energy will change the adhesion between deposit and surface, but not the cohesive forces between elements of the deposit. The forces involved in removing deposits from different surfaces have been studied using both dynamic gauging and micromanipulation probes (Liu et al. 2006a; Saikhwan et al. 2006). The authors found that adhesive failure occurred preferentially over a range of surface energies (22–28 mN/m) where the surface binding was least strong (the data of Fig. 12.6 are representative) in that there is a minimum adhesive force which is reflected either in reduced fouling or a faster rate of cleaning.

The search for a surface to prevent fouling is a very active subject of research in a variety of areas (Yu et al. 2011), and these surfaces are being used in applications such as membranes (e.g. Sui et al. 2012). Techniques developed from nanotechnology are increasingly being used to develop a wide range of different surface films using tailored deposition, for example research on highly hydrophobic surfaces (Roach et al. 2008) has applications such as self-cleaning solar cells (Park et al. 2011). It is likely that some of these will have application in food processing by making it possible either to resist fouling or minimise cleaning time. For example, Kananeh et al. (2010) showed 80 % reductions in cleaning time for dairy deposits for electropolished surfaces and surfaces coated in polyurethane. The problem of making non-fouling food contact surfaces is very complex because of the variety of species that are contained in foods; it is unlikely that a surface that resists one material would be able to prevent fouling from another. Any surface must, however, be cheap enough to make and stable enough to survive industrial use, but once such surfaces are developed, the fouling and cleaning problem will diminish significantly.

12.3.2 Use of Nanoscale Measurements of Adhesion and Cleaning

One issue will be the testing of candidate surfaces without having to do it on an industrial scale. It would, for example, be very useful to carry out experiments on novel surfaces on a small scale because such experiments can be very expensive.

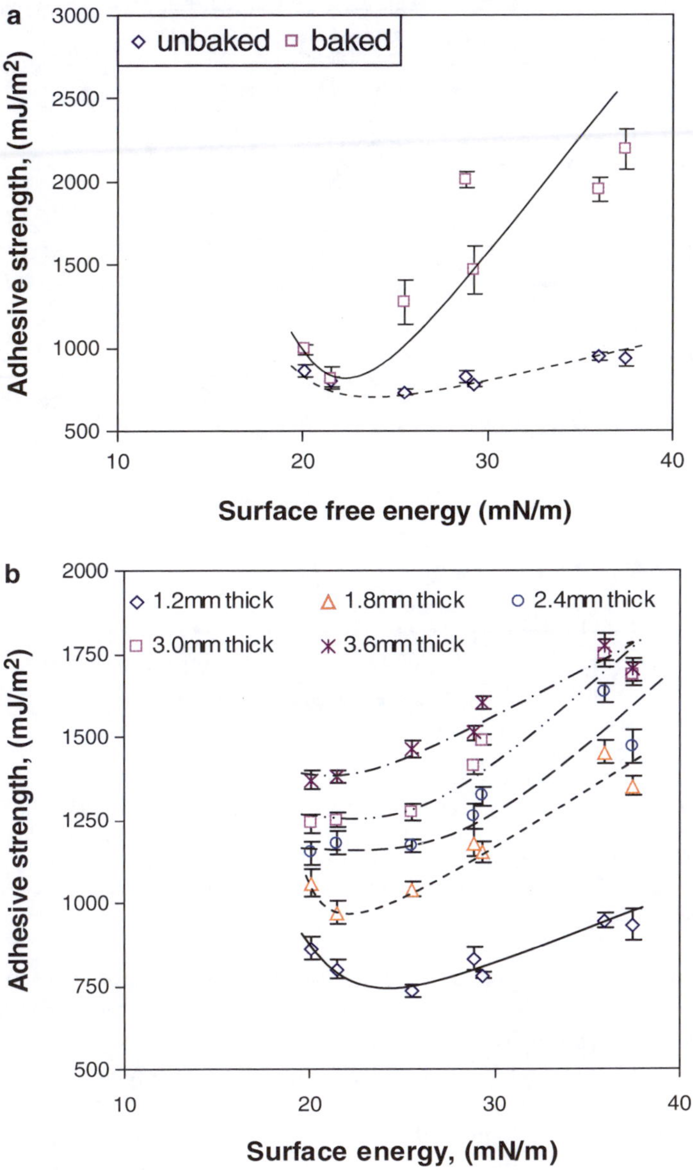

Fig. 12.6 Variation of apparent adhesive strength with surface free energy for (**a**) baked and unbaked tomato pastes and (**b**) unbaked tomato paste samples of different thicknesses

We have recently studied (Akhtar et al. 2010) interactions at the nanoscale, using atomic force microscopy (AFM) to study how surfaces and deposits interact, and at the mesoscale, using micromanipulation probes and bench-top cleaning rigs to study cleaning at the micron to millimetre level. At the nanoscale, results were

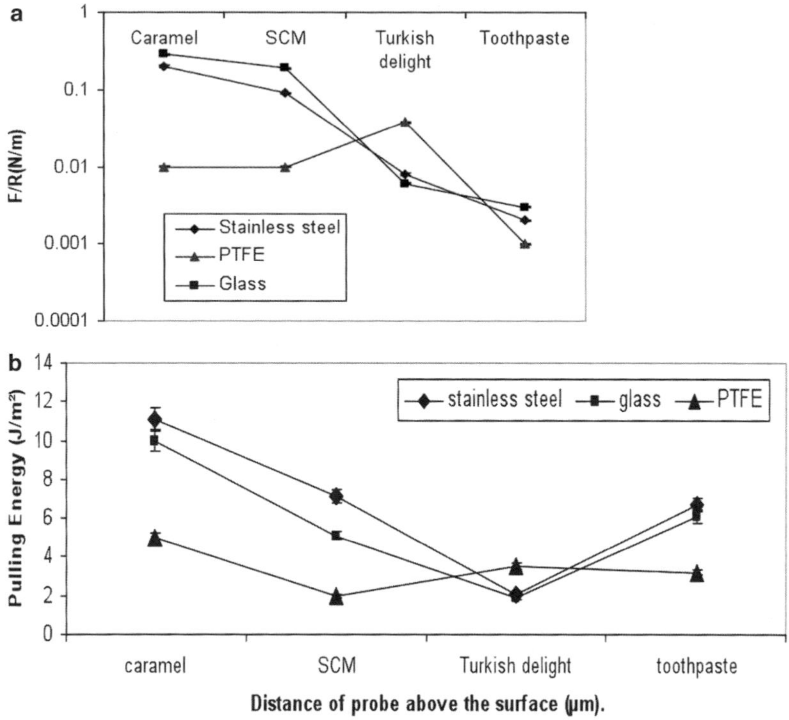

Fig. 12.7 (**a**) AFM force measurements. Stainless steel, glass and TCTFPS micro particles were immersed in deposits and then retracted. Deposits were spread out between 50 and 60 μm thick on a glass slide. Data show results from five different contact positions on caramel deposit. The approach speed for all experiments was 3 μm/s, and then a 5 s pause on deposit and 0.25 μm/s retract. Data are averages of more than 25 readings. (**b**) Summary of micromanipulation experiments at 20 μm cut height of caramel, SCM, Turkish delight and toothpaste on glass, stainless steel and TCTFPS (Akhtar et al. 2010)

obtained using AFM, in which a probe attached to micro particles of different materials was put in contact with surfaces coated with different types of possible fouling substrate: toothpaste and three confectionary soils, sweetened condensed milk (SCM), caramel and 'Turkish delight' (an agar-based material). At the meso-scale, micromanipulation was used to pull deposits from a 25 mm coupon, and the force per unit area required to clean the surface was measured. Surfaces tested included glass, stainless steel and vapour-deposited trichloro(3,3,3-trifluoropropyl) silane (henceforth referred to as TCTFPS), which has surface properties close to those of polytetrafluoroethylene.

Figure 12.7a summarises adhesion data, which vary over two orders of magnitude; the largest values are observed with caramel and SCM, with smaller values for Turkish delight and toothpaste. Overall:

- The glass microparticles show the highest adhesive behaviour with caramel, SCM and toothpaste.

- The greatest adhesion between the TCTFPS-coated microparticles and the surface is seen for Turkish delight, although the forces are much weaker than for the other two systems.

Figure 12.7b shows data sets for the micromanipulation experiments, plotting the results to remove the deposit for a cut height of 20 μm, i.e. where the cut is close to the surface, and where the force measured will be most related to the surface-deposit interactions. In three out of the four cases, stainless steel has the highest measured force, whilst for Turkish delight, the strongest force between deposit and surface is found for the TCTFPS.

Both methods can be used to get an idea of interfacial forces. The clear difference between Fig. 12.7a, b is that a log scale is needed to visualise the variation in the AFM force values seen for the toothpaste and TCTFPS data, whilst the micromanipulation data do not show the same variation, with values between 2 and 12 J/m^2. The micromanipulation measurement records the total force required to remove the deposit from the surface, which will involve both surface-deposit interactions and some measure of the rheology of the deposits and how they deform. AFM is a purer measurement and measures the surface-deposit interactions; this suggests that the magnitude of the surface interactions varies by a much larger amount than is shown by the micromanipulation measurements.

The data for the two systems compare well:

- For Turkish delight, both AFM and micromanipulation record that the fluorinated surface shows the greatest adhesion; this agreement shows that both techniques can be used to assess adhesion forces between substrates and soils.
- Data for stainless steel and glass are more closely matched, but both are greater than fluorinated surfaces for caramel, SCM and toothpaste.

The experiments suggest that micromanipulation data can provide information related to the detail shown in the AFM. That the result for Turkish delight is the same for both systems suggests that they are measuring similar effects at different length scales.

AFM is difficult to use, and the experimental set-up requires parameters such as the right size particle and cantilever stiffness. Many of the surfaces being developed are, however, first made at the nanometre scale; these results suggest that it would be possible to predict mesoscale behaviour from atomic-scale measurement, and thus this may act as a test of possible new surfaces.

12.4 Conclusion: Future Work Needed

The problems of fouling and cleaning were reviewed briefly. As the importance of making food processes sustainable increases, losses due to cleaning will become more significant. Fouling is well understood in comparison to cleaning, especially in the dairy industry. Cleaning processes have been classified by deposit severity and the corresponding cleaning mechanism. The food industry is very cautious and slow to change, but the advances in tailored surfaces being made in nanoscience suggest that it might soon be possible to design genuinely non-stick – or easy-to-clean – surfaces. Whether these will be commercially viable will depend on the costs of the cleaning process; more work needs to be done to establish both the operating costs in terms of the equipment and chemicals needed and the opportunity cost – what increases in production could be achieved or what new products could be made if cleaning were eliminated.

Much of the basic engineering of cleaning has been solved; CIP processes can be developed for all food formulations. The basics are well known – increasing the flow rate increases the cleaning rate – but design rules are still needed.

- For Type 1, *fluid mechanical removal*, the need is to maximise product recovery and minimise cleaning time. Shear stress is a critical parameter; but because this is a purely fluid mechanical problem, it should be possible to simulate the cleaning process or to predict the cleaning time of one fluid based on that of another, knowing only the rheological and interfacial properties of the deposit and the surface and of the cleaning fluids. The problem is analogous to annular flow in gas–liquid systems (in which generally a core of low-viscosity air displaces a wall film of water), and the same kind of mathematics can be used to develop computational models.
- For Types 2 and 3, the need is to develop efficient models for *diffusion–reaction removal* which can predict cleaning times for various materials; this will require diffusion and kinetic data. Precise data will be difficult to find; however, some progress has been made recently with relatively simple models of how fouling deposits age (e.g. Coletti et al. 2010).

The final stage will be to incorporate both kinds of model into design codes for pieces of equipment and for process plants, and thus to enable inherently cleanable equipment to be developed at the design stage, i.e. building the understanding into codes developed by groups such as the EHEDG.

Acknowledgments The Birmingham work discussed in this paper includes results from the ZEAL project TP//ZEE/6/1/21191, which involves Alfa Laval, Cadbury, Ecolab, Newcastle University, Scottish & Newcastle, GEA Process Engineering, Unilever UK Central Resources, Imperial College of Science Technology and Medicine, GlaxoSmithKline, Bruker Optics and the University of Birmingham. The project is co-funded by the Technology Strategy Board's Collaborative Research and Development programme, following open competition. For more information visit http://www.innovateuk.org.

References

Akhtar N, Bowen J, Asteriadou K, Robbins PT, Zhang Z, Fryer PJ (2010) Matching the nano- to the meso- scale: measuring deposit surface interactions with atomic force microscopy and micromanipulation. Food Bioprod Process 88:341–348

Asteriadou K, Hasting T, Bird M, Melrose J (2007) Predicting cleaning of equipment using computational fluid dynamics. J Food Process Eng 30(1):88–105

Augustin W, Fuchs T, Föste H, Schöler M, Majschak J-P, Scholl S (2010) Pulsed flow for enhanced cleaning in food processing. Food Bioprod Process 88:384–391

Baier RE (1980) Adsorption of microorganisms to surfaces. Wiley, New York, Substrate Influences on Adhesion of Microorganisms and Their Resultant New Surface Properties

Benning R, Petermeier H, Delgado A, Hinrichs J, Kulozik U, Becker T (2003) Process design for improved fouling behaviour in dairy heat exchangers using a hybrid modelling approach. Food Bioprod Process 81(C3):266–274

Bird MR, Fryer PJ (1991) An experimental study of the cleaning of surfaces fouled by whey proteins. Food Bioprod Process 69:13–21

Blel W, Le Gentil-Lelievre C, Benezech T, Legrand J, Legentilhomme P (2009a) Application of turbulent pulsating flows to the bacterial removal during a cleaning in place procedure. Part 1: experimental analysis of wall shear stress in a cylindrical pipe. J Food Eng 90(4):422–432

Blel W, Legentilhomme P, Benezech T, Legrand J, Le Gentil-Lelievre C (2009b) Application of turbulent pulsating flows to the bacterial removal during a cleaning in place procedure. Part 2: effects on cleaning efficiency. J Food Eng 90:433–440

Boulange-Petermann L (1996) Processes of bioadhesion on stainless steel surfaces and cleanability: a review with special reference to the food industry. Biofouling 10(4):275–300

Burfoot D, Middleton K (2009) Effects of operating conditions of high pressure washing on the removal of biofilms from stainless steel surfaces. J Food Eng 90(3):350–357

Changani SD, Belmarbeiny MT, Fryer PJ (1997) Engineering and chemical factors associated with fouling and cleaning in milk processing. Exp Therm Fluid Sci 14(4):392–406

Christian GK (2003) Removal of food deposits by fluid flow. Ph.D. thesis, University of Birmingham

Christian GK, Fryer PJ (2006) The effect of pulsing cleaning chemicals on the cleaning of whey protein deposits. Food Bioprod Process 84:320–328

Christian GK, Changani SD, Fryer PJ (2002) The effect of adding minerals on fouling from whey protein concentrate: development of a model fouling fluid for a plate heat exchanger. Food Bioprod Process 80(4):231–239

Cole PA, Asteriadou K, Robbins PT, Owen EG, Montague GA, Fryer PJ (2010) Comparison of cleaning of toothpaste from surfaces and pilot scale pipework. Food Bioprod Process 88:392–400

Coletti F, Ishiyama EM, Paterson WR, Wilson DI, Macchietto S (2010) Impact of deposit aging and surface roughness on thermal fouling: distributed model. AIChE J 56:3257–3273

De Jong P, Te Giffel MC, Kiezebrink EA (2002a) Prediction of the adherence, growth and release of microorganisms in production chains. Int J Food Microbiol 74(1–2):13–25

De Jong P, Te Giffel MC, Straatsma H, Vissers MMM (2002b) Reduction of fouling and contamination by predictive kinetic models. Int Dairy J 12(2–3):285–292

Eide MH, Homleid JP, Mattsson B (2003) Life cycle assessment (LCA) of cleaning-in-place processes in dairies. Lebensm Wiss Technol 36:303–314

Epstein N (1981) Thinking about heat transfer fouling: a 5×5 matrix. Heat Transf Eng 4(1):43–56

Fernandez-Torres MJ, Fitzgerald AM, Paterson WR, Wilson DI (2001) A theoretical study of freezing fouling: limiting behaviour based on a heat and mass transfer analysis. Chem Eng Process 40(4):335–344

Fryer PJ, Asteriadou K (2009) A prototype cleaning map: a classification of industrial cleaning processes. Trends Food Sci Technol 20(6–7):255–262

Fryer PJ, Christian GK, Liu W (2006) How hygiene happens; the physics and chemistry of cleaning. Int J Dairy Technol 59:76–84

Ghnimi S, Flach-Malaspina N, Dresch M, Delaplace G, Maingonnat JF (2008) Design and performance evaluation of an ohmic heating unit for thermal processing of highly viscous liquids. Chem Eng Res Des 86:626–632

Ghnimi S, Zaid I, Maingonnat JF, Delaplace G (2009) Axial temperature profile of ohmically heated fluid jet: analytical model and experimental validation. Chem Eng Sci 64:3188–3196

Gillham CR, Fryer PJ, Hasting APM, Wilson DI (1999) Cleaning-in-place of whey protein fouling deposits: mechanisms controlling cleaning. Food Bioprod Process 77(C2):127–136

Gillham CR, Fryer PJ, Hasting APM, Wilson DI (2000) Enhanced cleaning of whey protein soils using pulsed flows. J Food Eng 46(3):199–209

Goode KR, Asteriadou K, Fryer PJ, Robbins PT, Picksley P (2010) Characterising the cleaning mechanisms of yeast and the implications for cleaning in place (CIP). Food Bioprod Process 88:365–374

Grassi B, Strazza D, Poesio P (2008) Experimental validation of theoretical models in two-phase high-viscosity ratio liquid-liquid flows in horizontal and slightly inclined pipes. Int J Multiphase Flow 34(10):950–965

Gu T, Albert F, Augustin W, Chew YMJ, Mayer M, Paterson WR, Scholl S, Sheikh I, Wang K, Wilson DI (2011) Application of fluid dynamic gauging to annular test apparatuses for studying fouling and cleaning. Exp Therm Fluid Sci 35:509–520

Henningsson M, Regner M, Ostergren K, Tragardh C, Dejmek P (2007) CFD simulation and ERT visualization of the displacement of yoghurt by water on industrial scale. J Food Eng 80 (1):166–175

Jensen BBB, Friis A, Benezech T, Legentilhomme P, Lelievre CL (2005) Local wall shear stress variations predicted by computational fluid dynamics for hygienic design. Food Bioprod Process 83(C1):53–60

Kananeh AB, Scharnbeck E, Kuck UD, Rabiger N (2010) Reduction of milk fouling inside gasketed plate heat exchanger using nano-coatings. Food Bioprod Process 88(4):349–356

Kern DQ, Seaton RE (1959) A theoretical analysis of thermal surface fouling. Br Chem Eng 4:258–262

Liu W, Christian GK, Zhang Z, Fryer PJ (2002) Development and use of a micromanipulation technique for measuring the force required to disrupt and remove fouling deposits. Food Bioprod Process 80(C4):286–291

Liu W, Fryer PJ, Zhang Z, Zhao Q, Liu Y (2006a) Identification of cohesive and adhesive effects in the cleaning of food fouling deposits. Innov Food Sci Emerg Technol 7(4):263–269

Liu W, Christian GK, Zhang Z, Fryer PJ (2006b) Direct measurement of the force required to disrupt and remove fouling deposits of whey protein concentrate. Int Dairy J 16(2):164–172

Liu W, Zhang Z, Fryer PJ (2006c) Identification and modelling of different removal modes in the cleaning of a model food deposit. Chem Eng Sci 61(22):7528–7534

Mercade-Prieto R, Chen XD (2006) Dissolution of whey protein concentrate gels in alkali. AIChE J 52(2):792–803

Mercade-Prieto R, Falconer RJ, Paterson WR, Wilson DI (2007) Swelling and dissolution of beta-lactoglobulin gels in alkali. Biomacromolecules 8(2):469–476

Mercade-Prieto R, Paterson WR, Chen XD, Wilson DI (2008) Diffusion of NaOH into a protein gel. Chem Eng Sci 63(10):2763–2772

Müller-Steinhagen H, Malayeri RR, Watkinson AP (2011) at www.heatexchanger-fouling.com

Palabiyik I, Olunloyo B, Fryer PJ, Robbins PT (2014) Flow regimes in the emptying of pipes filled with viscoelastic material. Chem Engng Res Des. (in press)

Park Y-B, Im H, Im M, Choi Y-K (2011) Self-cleaning effect of highly water-repellent microshell structures for solar cell applications. J Mater Chem 21:633–636

Petermeier H, Benning R, Delgado A, Kulozik U, Hinrichs J, Becker T (2002) Hybrid model of the fouling process in tubular heat exchangers for the dairy industry. J Food Eng 55(1):9–17

Roach P, Shirtcliffe NJ, Newton MI (2008) Progess in superhydrophobic surface development. Soft Matter 4:224–240

Rosenhahn A, Ederth T, Pettitt ME (2008) Advanced nanostructures for the control of biofouling: the Fp6 Eu integrated project ambio. Biointerphases 3(1):IR1–IR5

Rosmaninho R, Melo LF (2008) Protein-calcium phosphate interactions in fouling of modified stainless-steel surfaces by simulated milk. Int Dairy J 18(1):72–80

Rosmaninho R, Santos O, Nylander T, Paulsson M, Beuf M, Benezech T, Yiantsios S, Andritsos N, Karabelas A, Rizzo G, Müller-Steinhagen H, Melo LF (2007) Modified stainless steel surfaces targeted to reduce fouling – evaluation of fouling by milk components. J Food Eng 80:1176–1187

Rosmaninho R, Rizzo G, Müller-Steinhagen H, Melo LF (2008) Deposition from a milk mineral solution on novel heat transfer surfaces under turbulent flow conditions. J Food Eng 85:29–41
Sahu KC, Valluri P, Spelt PDM, Matar OK (2007) Linear instability of pressure-driven channel flow of a newtonian and a herschel-bulkley fluid. Phys Fluids 19(12):11–122101
Saikhwan P, Geddert T, Augustin W, Scholl S, Paterson WR, Wilson DI (2006) Effect of surface treatment on cleaning of a model food soil. Surf Coat Technol 201(3–4):943–951
Schutyser MI, Straatsma J, Keijzer PM, Verschueren M, De Jong P (2008) A new web-based modelling tool (Websim-Milq) aimed at optimisation of thermal treatments in the dairy industry. Int J Food Microbiol 128(1):153–157
Simmons MJH, Jayaraman P, Fryer PJ (2007) The effect of temperature and shear rate upon the aggregation of whey protein and its implications for milk fouling. J Food Eng 79(2):517–528
Simmons MJH, Alberini F, Tsoligkas AN, Gargiuli J, Parker DJ, Fryer PJ, Robinson S (2012) Development of a hydrodynamic model for the UV-C treatment of turbid food fluids in a novel 'SurePure turbulator™' swirl-tube reactor. Innov Food Sci Emerg Tech 14:122–134
Sui Y, Gao X, Wang Z, Gao C (2012) Antifouling and antibacterial improvement of surface-functionalized poly(vinylidene fluoride) membrane prepared via dihydroxyphenylalanine-initiated atom transfer radical graft polymerizations. J Membr Sci 394:107–119
Taborek J, Palen JW, Aoki T, Ritter RB, Knudsen JG (1972) Fouling – major unresolved problem in heat transfer. Chem Eng Prog 68(2):59–67
Taghavi SM, Alba K, Moyers-Gonzalez M, Frigaard IA (2012) Incomplete fluid–fluid displacement of yield stress fluids in near-horizontal pipes: experiments and theory. J Non-Newton Fluid Mech 167–168:59–74
Tamime AV (2008) Cleaning-in-place: dairy, food and beverage operations. Wiley-Blackwell, London
Wallhäußer E, Hussein WB, Hussein MA, Hinrichs J, Becker TM (2011) On the usage of acoustic properties combined with an artificial neural network – a new approach of determining presence of dairy fouling. J Food Eng 103(4):449–456
Wilson DI (2005) Challenges in cleaning: recent developments and future prospects. Heat Transf Eng 26(1):51–59
Yoo JY, Chen XD, Mercade-Prieto R, Wilson DI (2007) Dissolving heat-induced protein gel cubes in alkaline solutions under natural and forced convection conditions. J Food Eng 79 (4):1315–1321
Yu Q, Zhang Y, Wang H, Brash J, Chen H (2011) Anti-fouling bioactive surfaces. Acta Biomater 7:1550–1557
Zhao Q, Liu Y, Wang C, Wang S, Muller-Steinhagen H (2005) Effect of surface free energy on the adhesion of biofouling and crystalline fouling. Chem Eng Sci 60(17):4858–4865

Chapter 13
Encapsulation Systems in the Food Industry

Viktor Nedović, Ana Kalušević, Verica Manojlović, Tanja Petrović, and Branko Bugarski

13.1 Introduction

Encapsulation may be defined as a process to entrap one substance (active agent) within another substance (wall material). Also, it may be defined as a technology in which a physical barrier is applied to protect the bioactive components against any undesirable environmental condition. Applications of such technology have increased in the food industry since encapsulated materials can be protected from moisture, heat or other extreme conditions, thereby enhancing their stability and maintaining viability.

The encapsulated substance, except an active agent, can be called the core, fill, active, internal or payload phase. The substance that is encapsulating is often called the coating, membrane, shell, capsule, carrier material, external phase or matrix (Wandrey et al. 2010; Fang and Bhandari 2010). The entrapped or coated material is usually a liquid but can be a solid or a gas. Actives include, for example, flavours, antimicrobial agents, nutraceutical and therapeutic actives, vitamins, minerals, antioxidants, colours, acids, alkalis, buffers, sweeteners, nutrients, cross-linking agents, chemical leavening agents, enzymes and cells (Brownlie 2007). Encapsulates are usually in the form of particles with diameters ranging from a few nanometres to a few millimetres (Wandrey et al. 2010).

Encapsulation was originally introduced in the area of biotechnology to make production processes more efficient. Nowadays there is a growing interest in bioactive target release in the body via emerging encapsulation technologies that implement food-grade encapsulants. However, it should be emphasised that the

V. Nedović (✉) • A. Kalušević • T. Petrović
Department of Food Technology and Biochemistry, Faculty of Agriculture, University of Belgrade, Belgrade, Serbia
e-mail: vnedovic@agrif.bg.ac.rs

V. Manojlović • B. Bugarski
Department of Chemical Engineering, Faculty of Technology and Metallurgy, University of Belgrade, Belgrade, Serbia

S. Yanniotis et al. (eds.), *Advances in Food Process Engineering Research and Applications*, Food Engineering Series, DOI 10.1007/978-1-4614-7906-2_13,

application of carrier materials that can be used for encapsulation in the food industry is limited to some extent (e.g. proteins, carbohydrates, fats and food-grade emulsifiers). This chapter will review the encapsulation systems that can be applied considering the materials that are usable in the food industry.

13.2 Why Use Encapsulation Technologies?

In the food industry, encapsulation of actives can be applied for a variety of reasons. Protection, controlled delivery, taste masking and easy handling are among the most frequently listed in the literature.

13.2.1 Protection of Actives

Protection of food/feed ingredients is a primary reason for encapsulation. Encapsulation aims to preserve the stability of an active compound during processing, storage and consumption. Protection can be achieved from a variety of influences that might cause the loss of functionality of an ingredient or degradation of an ingredient before it has time to act. One of the problems here is interactions between substances, which may affect the product's acceptable shelf life, such as in the case of functional compounds in food products. Functional compounds, such as vitamins, minerals and polyphenols, carry various potential health benefits. However, producers of functional foods face many difficulties because most functional compounds are sensitive to various factors and characterised by rapid inactivation (Wandrey et al. 2010; Fang and Bhandari 2010). These compounds are usually highly susceptible to environmental, processing or gastrointestinal conditions; encapsulating them makes it possible to slow down the degradation processes (e.g. oxidation or hydrolysis) or prevent degradation (caused by heat, light, moisture) until the active compounds are delivered to the desired sites (McClements and Lesmes 2009). In this way, the bioactive component can be kept and release the compounds full functionality. Therefore, encapsulation has been used as an approach to effectively prevent undesirable influences on the colour, flavour, texture, and sensory and to preserve characteristics of final products.

13.2.2 Controlled Delivery

Encapsulation is a useful tool to provide delivery of bioactive molecules (e.g. antioxidants, minerals, vitamins, phytosterols) and living cells (e.g. probiotics) at a desired place or within an appropriate time (Wandrey et al. 2010; Fang and Bhandari 2010). In this way, the proper delay of release of active compounds for the right sensory or functional benefits is achieved. Actually, there are two distinct types of release, delayed and sustained. The first one refers to a delayed release mechanism of actives to a desired location. Examples of delayed release are numerous: encapsulated probiotic bacteria are protected from gastric acidity, and release occurs in the small intestine; flavour is released upon microwave heating of ready-to-eat meals; the release of encapsulated sodium bicarbonate happens upon baking. The second type is sustained release – a mechanism designed to maintain a constant concentration of actives at the target site. Examples of this category include encapsulated flavours and sweeteners for chewing gum so that the rate of their release is reduced to maintain a desired flavour effect throughout the time of chewing.

13.2.3 Taste Masking

Many ingredients used in the food industry are unpalatable, and their taste needs to be masked until they reach the taste sensors in the mouth. Examples of these ingredients include polyphenols, vitamins and minerals, which are added to improve the nutritional properties of the food product. This is also the reason for encapsulation in the feed industry, where higher concentrations of such ingredients are employed in animal diet.

13.2.4 Easy Handling

Encapsulation technology can be used to allow easier handling of the core material during production of the food products. Handling can be simplified by giving a uniform position of the core, promoting simple mixing and converting a liquid to a solid form (e.g. in the case of plant extracts and essential oils) (Brownlie 2007; Wilson and Shah 2007).

13.3 Materials Used for Encapsulation

13.3.1 Criteria for Selection of Encapsulation Materials

Materials used for the design of the protective shell of encapsulates must be food-grade, biodegradable and able to form a barrier between the internal phase and its surroundings. Many substances may be used to coat or encapsulate solids, liquids or gases of different types and properties. However, the regulations for food additives are very rigid, and these substances must be certified for food applications as generally recognised as safe (GRAS) materials. Actually, the whole food process should be designed to meet the safety requirements of governmental agencies such as the European Food Safety Authority (EFSA) or Food and Drug Administration (FDA) in the USA (Wandrey et al. 2010).

Before selecting an encapsulation material, it is important to consider some criteria such as functionality that the encapsulate should provide to the final product, potential restrictions for the coating material, concentration of encapsulates, required type of release and stability requirements. Besides these criteria, cost constraints are a key factor for selecting the most appropriate material. Biopolymers represent the majority of materials used for encapsulation in the food sector. In addition, materials used for encapsulation should be natural, provide maximal protection of the active material against environmental conditions, hold actives within the capsule structure during processing or storage under various conditions, not interact with the encapsulated material, possess good rheological characteristics at high concentrations if so desired and have easy workability during encapsulation. Before the final choice is made, the physical and chemical characteristics of the materials should be taken into account, especially behaviour under conditions present in food formulations (Wandrey et al. 2010).

13.3.2 Commonly Used Materials

13.3.2.1 Polysaccharides

Among all materials, polysaccharides are the most widely used for encapsulation in food applications. The most commonly used polysaccharides are starch and their derivates (e.g. amylose, amylopectin, dextrins, maltodextrins, polydextrose), as well as cellulose and their derivatives. Carbohydrates are outstanding contenders for encapsulation applications since they form an integral part of many food systems, form shapes in a wide range of sizes and have desirable physicochemical properties such as solubility, melting, phase change and others. Finally, but no less important, they are cost-effective (Brownlie 2007). Carbohydrates such as starch and cyclodextrins have good capability to absorb volatiles from the environment. The unique helical structure of the amylose molecule makes starch a very efficient vehicle for encapsulating molecules like lipids, flavours and others (Conde-Petit et al. 2006). Also, the carbohydrates inulin and trehalose provide additional benefits: inulin is a prebiotic ingredient that can improve survival of probiotic bacteria while trehalose serves as a support nutrient for yeasts. Ethyl cellulose is a good material to encapsulate, for example, water soluble vitamins because it is water soluble itself, and as the shell width increases, the water permeability of the actives (e.g. vitamin) is reduced (Brownlie 2007; Shahidi and Han 1993).

Plant exudates and extracts – gum arabic, gum tragacanth, gum karaya, mesquite gum, galactomannans, pectins and soluble soybean polysaccharides – have been employed as well. Marine extracts such as carrageenans and alginate are also present in foods. Gum arabic is a commonly used capsule material due to its emulsification characteristics, viscosity and solubility, but a major disadvantage is its high costs. Alginates and hydrocolloids are extracted from kelp and react with calcium ions to form a stable gel. For example, they can be used to entrap flavour oils at ambient temperatures. Increased aroma retention was found using alginate beads during baking of crackers (de Roos 2003). Microbial and animal polysaccharides like dextran, chitosan, xanthan and gellan have been used, too. Chitosan is a membrane material which possesses excellent biodegradable and biocompatible characteristics, a unique polymeric cationic characteristic and gel-forming, film-forming and heat-shrinking properties. Because of these and many other features, chitosan has been extensively examined mainly in the pharmaceutical industry for its potential use in the development of drug delivery systems (Wandrey et al. 2010; Shahidi and Han 1993; Park et al. 2005; Khor and Lim 2003).

13.3.2.2 Proteins

Apart from natural and modified polysaccharides, proteins are also appropriate materials for encapsulation. Milk and whey proteins such as caseins, gelatine and gluten are some of the most commonly used protein wall materials. These are able to form stable emulsions with volatile flavour components, but their low solubility in cold water, potential to react with carbonyls and high costs are limiting factors in their application (Shahidi and Han 1993).

13.3.2.3 Lipids

Lipid materials used for food applications are fatty acids and fatty alcohols, waxes (beeswax, carnauba wax, candelilla wax), glycerides and phospholipids. Natural waxes with a melting point between 60 °C and 80 °C are used as suitable materials for the encapsulation of aromas. Important advantages of these materials include their superior ease of handling, stability and safety. Natural waxes, such as insect waxes and bees wax, and plant waxes, like candelilla and carnauba wax, are permitted additives in the European Union (Milanovic et al. 2010; Donhow and Fennema 1993).

13.3.2.4 Polymers

In addition to those mentioned earlier, other materials are employed for encapsulation in the food industry, such as food-grade polymers, for example polyvinylpyrrolidone, paraffin, polypropylene, polyvinylacetate, polystyrene and polybutadiene (Wandrey et al. 2010; Brownlie 2007).

13.4 Encapsulation Techniques

A number of techniques are available for the encapsulation of food compounds. Since encapsulating compounds are very often in liquid form, many technologies are based on drying. Various techniques like spray-drying, spray-bed-drying, fluidised bed coating, spray-chilling, spray-cooling or melt injection are used to encapsulate active agents (Gibbs et al. 1999; Porzio 2007). In what follows, these techniques are described in more detail.

13.4.1 Spray-Drying

Spray-drying is one of the oldest and most widely used encapsulation techniques in the food industry. The principle of spray-drying is to dissolve a core in a dispersion of a matrix material, forming an emulsion or a suspension. The matrix should have a high solubility in water because only water-based dispersions are applied in spray-drying. Commonly used matrix materials are hydrophilic carbohydrate molecules, while hydrophilic and hydrophobic active molecules can be used as the core materials (Vos et al. 2010). Spray-drying technology offers a high production rate and relatively low operating costs, and resulting powders are stable and easily applied. These are the main reasons for the broad application of spray-drying in industrial settings. Also, most evaporation takes place in milliseconds to a few seconds, which is well suited for heat-sensitive products such as proteins, enzymes, flavours, living cells and others. The size of spray-dried particles/powders is between 10 and 100 μm (Zuidam and Heinrich 2010). Smaller sized particles are more desirable from the standpoint of sensorial and textural characteristics of the final product. Although spray-dryers are widespread in the food industry, the technique has several disadvantages such as complexity of the equipment, non-uniform conditions in the drying chamber and difficult control of particle size. Spray-drying is typically used for the preparation of dry, stable food additives and flavours (Desai and Park 2005). The technique has been applied in the treatment of a wide range of bioactive food components such as vitamins, minerals, flavours, pigments, unsaturated oils, enzymes and living cells. According to the literature, approximately 80–90 % of encapsulates are spray-dried, while the others are mostly prepared by spray-chilling, freeze-drying, melt extrusion and melt injection (Milanovic et al. 2010; Porzio 2007).

13.4.2 Spray-Chilling and Spray-Cooling

Spray-chilling and spray-cooling are technologies used to produce lipid-coated active agents. Which one will be applied depends on the melting point of lipids.

In the case of spray-chilling, the melting temperature is in a range of 34–42 °C, while for spray-cooling it is higher. The principle of these technologies is contrary to that of spray-drying. The agent is dissolved in lipids, which are present as dry particles, or present as aqueous emulsion. Spray-cooling is a technology with the potential to achieve high yields and can be run in batch or continuous processing modes. So far, it has been used for dry products to conserve minerals, flavours, proteins and enzymes (Vos et al. 2010). In the case of spray-chilling, particles are kept at a low temperature in a set-up similar to that of fluidised bed spray granulation (Zuidam and Shimoni 2010; Gouin 2004).

13.4.3 Vacuum- and Freeze-Drying

Vacuum- and freeze-drying are very similar drying processes, but the former is faster and cheaper because it operates at a temperature above the freezing point of the solvent. Freeze-drying has been proposed as an alternative for spray-drying heat-sensitive bioactive food components (Bilensoy and Hincal 2009). The major disadvantages of freeze-drying are the high energy input and long processing time. In addition, during processing a barrier with an open porous structure between the active agent and its surroundings is formed; this high-porosity wall offers poor protection when prolonged release of an active is required (Zuidam and Shimoni 2010).

13.4.4 Fluidised Bed Coating

Fluidised bed coating is an encapsulation technique in which a coating is applied onto powder particles in a batch processor or a continuous set-up. The powder particles are suspended by an air stream at a specific temperature and sprayed with an atomised coating material (Fig. 13.1). The choice of coating material is greater than for traditional spray-drying. The coating material might be an aqueous solution of cellulose or starch derivatives, proteins or gums, but also an emulsifier. In general, the core is always solid. This technique may be applied to give a second coating to spray-dried products or to products with a sensitive core, such as oils. Also, fluidised bed coating has been used to create an additional layer of molecules for targeted release in the gut (Vos et al. 2010; Dewettinck and Huyghebaert 1999).

13.4.5 Melt Injection and Melt Extrusion

Melt injection and melt extrusion are processes to encapsulate an active agent in a carbohydrate melt. In melt injection, the melt is pressed through a filter and then

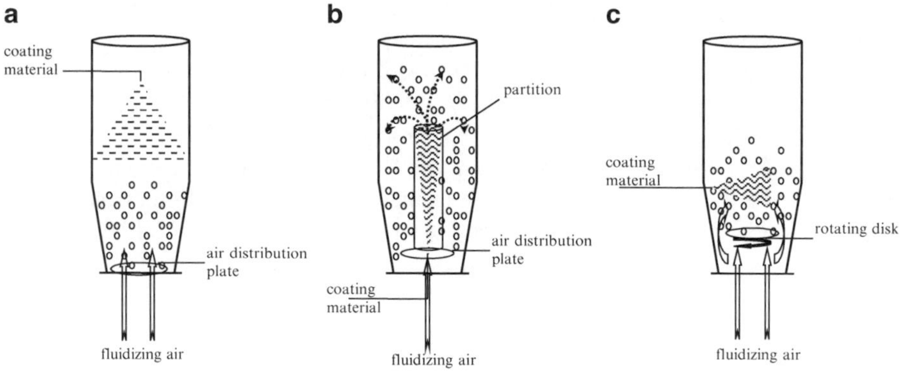

Fig. 13.1 Spray-coating techniques for the microencapsulation of cells: (**a**) fluidised bed top spray coating; (**b**) fluidised bed bottom spray-coating with a Wurster device; (**c**) fluidised bed tangential spray coating (Manojlovic et al. 2010a)

quenched by a dehydrating solvent such as isopropanol or liquid nitrogen. The active is encapsulated owing to the hardening of a wall material in contact with the solvent (Porzio 2004). Particles made by this process are water-soluble with sizes ranging from 200 to 2,000 μm. Melt extrusion is a process similar to melt injection, with few differences between the two: in the case of melt extrusion, the extrudates are not surface washed. In addition, melt injection utilises screws in a vertical position, while in melt extrusion the screws are in the horizontal position.

13.4.6 Extrusion Methods

Extrusion methods are based on the dispersing of an aqueous solution of polymer (usually 0.6–3 wt% sodium alginate) and an active ingredient into droplets dropping into a gelling bath (in the case of alginate, the gelling bath is 0.05–1.5M calcium chloride solution). The dripping tool can simply be a pipette, syringe, vibrating nozzle, spraying nozzle, jet cutter or atomizing disk (Fig. 13.2) (Wandrey et al. 2010). Among all extrusion technologies, JetCutter has been found to be the best technology for large-scale/industrial applications (Prüsse et al. 2008). Electrostatic extrusion has advantages over other extrusion technologies and is an especially effective technique for the production of very small particles down to 50 μm. The method is based on the use of electrostatic forces to stir the liquid filament at the tip of a capillary/needle and to form a charged stream of small droplets. Nedovic et al. (2001a) investigated electrostatic droplet generation as a technology for the production of alginate microbeads loaded with yeast. The obtained results demonstrated that this technology could be used for the controlled production of small microbeads highly loaded with yeast cells. The production of uniform small-diameter beads (in a range of 250 μm to 2.0 mm) depended on the applied

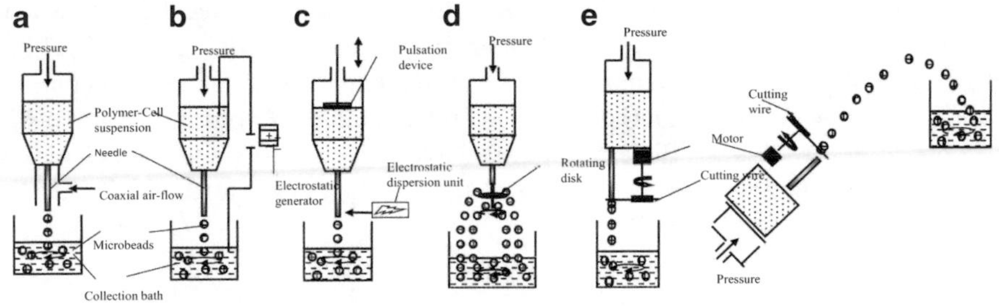

Fig. 13.2 Extrusion techniques: (**a**) coaxial air flow; (**b**) electrostatic extrusion; (**c**) vibration technique; (**d**) atomisation technique; (**e**) JetCutter with normal mode and soft-landing mode (modified from 24, the copyright holder: Institute of Chemistry, Slovak Academy of Sciences)

conditions. Although exposed to high electrostatic potentials (up to 8 kV) during immobilisation process, cells do not lose their viability, as confirmed by experiments on the fermentation activity of immobilised brewing yeast cells (Nedovic et al. 2001a).

13.4.7 Emulsification

Another frequently used encapsulation technology is emulsification. There are several types of emulsions: water/oil emulsions and oil/water emulsions; an alternative to those two is a water/oil/water double emulsion. The size of microcapsules depends on the reactor design and mixer geometry and can be controlled by varying the agitation speed and water/oil ratio (Kailasapathy 2009). An oil-in-water emulsion can be dried by different drying methods such as spray- or freeze-drying to produce a powder. Such dried emulsions have been used as encapsulates or an instant formulation for numerous food products (Zuidam and Shimoni 2010). Thus, by an easy and simple procedure it is possible to control the formation of micelles, different shapes of glycerides (hexagonal, cubic, lamellar) that can envelope one or more bioactive molecules. This technology is already widely applied for the controlled release of aromas and flavours (Augustin and Hemar 2009).

13.4.8 Coacervation

Coacervation is a modified emulsification technology. A complex is formed when a solution of an active component is mixed with a matrix molecule. By changing the ion concentration, the type of matrix or ratio of the active component and the matrix molecule, it is possible to produce capsules with different characteristics and sizes. This immobilisation/encapsulation technology is mostly applied for bioactive food molecules like flavours, citrus and vegetable oils, vitamin A and some water-soluble bioactive molecules rather than for bioactive living cells (Gibbs et al. 1999; Augustin and Hemar 2009). Gelatine, gum acacia and polysaccharides are often used as coating materials (Wilson and Shah 2007; Gibbs et al. 1999). This encapsulation process is efficient but expensive and limited by the deficiency of suitable encapsulating materials approved for food application (Shahidi and Han 1993).

13.4.9 Molecular Inclusion

Molecular inclusion in cyclodextrins and liposomal vesicles provides some specific features of bioactives; however, these techniques are more expensive and, therefore, less widely used. Cyclodextrins have a lipophilic inner pocket of approximately 5–8 Å, in which an active molecule of the right size can be reversibly entrapped in an aqueous environment. The size of the lipophilic interior of the capsule can be changed by varying the number of glucose units of the cyclodextrin molecules (Bilensoy and Hincal 2009; Harada et al. 2009). Molecular inclusion is used to increase the solubility of lipophilic molecules and to protect molecules from inactivation or degradation. However, the small size of the ring-forming hole limits its loading capacity. Liposomes are spherical particles with ranging in size from 30 nm to several microns. The mechanism for the creation of liposomes is basically the hydrophilic-hydrophobic interactions between polar lipids (mostly phospholipids) and water molecules. Although expensive, it has been used to preserve aromas in the food industry (Vos et al. 2010).

13.5 Examples of Microenapsulates in Food Products

There are several reasons that one would employ an encapsulation technology, and this chapter has already reviewed most of them. In brief, this technology may, for example, provide barriers between sensitive bioactive materials and the environment, allowing taste and aroma differentiation, mask unpleasant tastes or odours, stabilise food ingredients or increase their bioavailability, extend product shelf life or provide a controlled and sustained release. Currently, there has been a significant increase in the number of products that contain bioactive components with health-promoting or disease-preventing effects. Many ingredients and food-fortifying compounds have been subjected to encapsulation, such as vitamins, aroma, micronutrients, essential and fish oils, peptides, pigments and living cells, etc. (Gibbs et al. 1999; Vos et al. 2010). Some examples of microencapsulates will be discussed in this section.

13.5.1 Probiotic Microencapsulates

Probiotics are highly sensitive to variations in pH, mechanical stress, transport conditions and digestive enzymes in the stomach. Probiotic refers to live microorganisms which when administered in adequate amounts confer a health benefit on the host (FAO/WHO 2006). These living cells need to survive the food process, storage, and food intake before they can be useful. In recent years, probiotics have been the driving force in the design of functional foods, especially in the form of fermented dairy products and dietary supplements. To produce health benefits, probiotic strains should be present in a viable form at a suitable level during food processing, storage and intake and maintain high viability throughout the gastrointestinal tract. Despite the importance of these beneficial microorganisms, many investigations have shown their poor viability, especially for bifidobacteria in fermented products (Medina and Jordano 1994; Shah et al. 1995). The survival of probiotic strains in dairy products is affected by a range of factors including low pH, the presence of hydrogen peroxide and oxygen toxicity (oxygen permeation through packaging), storage temperature, stability in dried or frozen form and compatibility with traditional starter culture during fermentation (Shah 2000). The introduction of microencapsulation techniques for the protection of sensitive probiotic cultures has resulted in a greatly enhanced viability of these microorganisms in food products and target delivery in the gastrointestinal tract (Manojlovic et al. 2010b). Additional benefits for application of encapsulated probiotics are their easier incorporation into the food system, persistent characteristics, greater viability and stability during storage. The choice of encapsulation system is always crucial since it must be efficient and easily incorporated into food without interfering with the texture and taste of the food. To select the appropriate encapsulation techniques for probiotic strains, two main aspects should be taken into account: their size (generally between 1 and 5 μm diameter), which excludes the possibility of using nanotechnologies, and the fact that they must be in a viable form. One of the ‘gentle’ approaches to encapsulating probiotics is based on capsule formation by the entrapment of probiotics within a polymeric matrix using extrusion or emulsion techniques. The commonly used supporting materials for probiotics are kappa-carrageenan, gellan, agarose, gelatin, alginate, chitosan, xanthan, starch and locust bean gum (Zuidam and Shimoni 2010; Petrovic et al. 2007). The size of the beads should be less than 100 μm to avoid the negative sensorial influence of capsules in food products (Truelstrup-Hansen et al. 2002). Although alginate is frequently used for the entrapment of probiotics by extrusion techniques, it has undesirable properties such as susceptibility to and degradation by acid. Nevertheless, to overcome this problem, it is recommended that alginate be mixed with other polymer compounds. Another strategy is to coat the capsules with another compound or to use different additives (Krasaekoopt et al. 2003). In addition, it was reported that a mixed gel of gellan and xanthan gums had better technological properties for microencapsulation by extrusion technique than alginate (McMaster and Kokott 2005). Encapsulation by emulsion results in

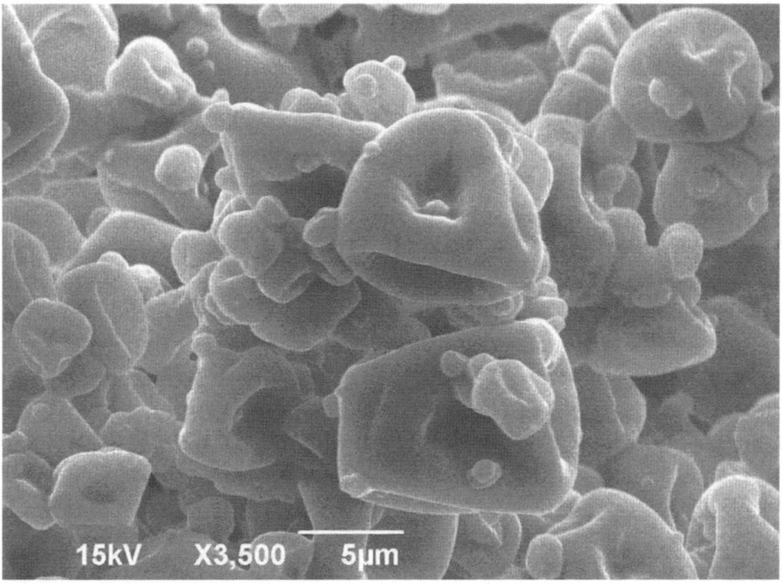

Fig. 13.3 Scanning electron micrograph of spray-dried *L. plantarum* BBA in skim milk powders

small-diameter beads, and it has been used to scale up applications (Burgain et al. 2011). However, residual oil in capsules may not be suitable for applications in low-fat functional food.

The other convenient techniques for encapsulation of probiotics are drying (fluidised bed, spray and freeze) and spray-coating techniques. Most probiotic strains do not survive well the high temperatures and dehydration during the spray-drying process. Loss of viability is principally caused by cytoplasmatic membrane damage, although the cell wall, ribosomes and DNA are also affected at higher temperatures (Teixeira et al. 1997). The commonly used supporting materials for spray-drying of probiotic bacteria are skim milk powder, whey protein, starch, gelatin, non-fat dry milk solids, maltodextrin, gum arabic, gum acacia, sugars and prebiotics (Manojlovic et al. 2010b). The microparticles of potential probiotic strains *Lactobacillus plantarum* BBA, isolated from prematurely born infants, in 20 % skim milk is shown in Fig. 13.3. Lian et al. (2002) observed that the survival of bifidobacteria after spray-drying varied with strains and was highly dependent on the type of carrier and outlet-air temperatures employed during spray-drying. The characteristics of carriers, such as thermal conductivity and porosity, may be determinative for the survival of spray-dried probiotics.

It was found that microencapsulation in alginate microparticles coated with high molecular weight chitosan improved the survival of acid-sensitive *L. bulgaricus* in simulated gastric juice (Lee et al. 2004). Probiotic bacteria encapsulated by different techniques could be applied in different non-dairy food matrices like mayonnaise, soft foods, biscuits, sausages, orange and apple juices and dry beverages (Burgain et al. 2011). Maillard and Landuyt (2008) investigated the possibility of using the probiotic cells encapsulated by spray-coating in chocolate. They showed that the survival of probiotics in chocolate was increased by up to three times at the small-intestine level when compared to a dairy matrix. Nowadays the health benefits associated with probiotics are well documented. Therefore, the probiotic market has a secure future because of increasing consumer demands for food with health-promoting effects.

13.5.2 Aroma Encapsulates

Aroma usually contains a mixture of volatile and odorous organic molecules. In addition, flavours are usually expensive, and therefore food manufacturers are generally concerned with the preservation of aromatic additives (Milanovic et al. 2010). Thanks to encapsulation, a food compound such as aroma is covered with a protective wall material and protected against evaporation, chemical reactions (such as flavour–flavour interactions, light-induced reactions, oxidation) or migration in a food (de Roos 2003; Madene et al. 2006). Flavour encapsulation can be accomplished by a variety of methods: spray-drying, spray-chilling or spray-cooling, spray bed drying and others. Examples of carrier material used for the encapsulation of flavours by spray-drying are mono- and disaccharides, maltodextrin, corn syrup solids, modified starches, gum arabic, larch gum, milk or soy proteins, hydrolysed gelatin and their various combinations. Natural waxes also appear to be promising for the encapsulation of aroma agents as they provide long-term retention of compounds (Mellema et al. 2006). Carnauba wax microparticles encapsulating ethyl-vanilline (Fig. 13.4) were produced by a melt dispersion technique (Milanovic et al. 2010). Spray-chilling is a convenient technology to produce lipid particles with aroma (McClements and Lesmes 2009). However, the most commonly used processes on an industrial level are spray-drying and extrusion (Wandrey et al. 2010). Encapsulation of flavours by extrusion in glassy carbohydrate matrices has been used for volatile and unstable flavours. Since flavour is one of the most important characteristics of food, the ultimate goal of encapsulation is to control aroma release and to improve stability during processing and consumption of the final product. In general, aroma is released from food before and after eating, which depends on the aroma features and physical state of the matrix.

13.5.3 Encapsulation of Phenolic Compounds

Polyphenols and other compounds that show high antioxidant activities produce mostly unpleasant feelings during eating, such as bitter taste and astringency. The effectiveness of polyphenols depends on preserving their stability, bioactivity and bioavailability. Plant polyphenols are usually consumed from herbal extracts. The

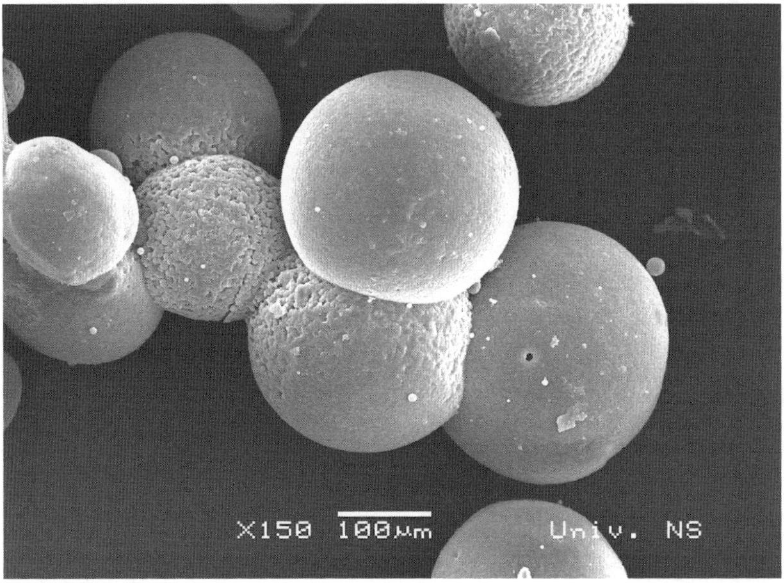

Fig. 13.4 Scanning electron micrograph of carnauba wax microparticles encapsulating ethyl-vanilline

most common method for preparing herbal extracts is percolation followed by lyophilisation. Among other techniques, spray- and vacuum-drying have also been employed with good results (Schimidt 1997). The utilisation of encapsulated polyphenols, instead of free compounds, can effectively alleviate some deficiencies (Bell 2001). The unpleasant taste of the most phenolic compounds is one of the factors which limit their application at higher concentrations.

Conversion of liquid plant extracts into a powder is another reason for encapsulation. Another problem is that only a small proportion of polyphenols remains available following oral administration due to insufficient gastric residence time, low permeability and solubility within the gut. In addition, their instability under conditions encountered in food processing and storage (temperature, oxygen, light) or in the gastrointestinal tract (pH, enzymes, presence of other nutrients) limit the activity and potential health benefits of components like polyphenols (Bell 2001). From the literature, it is clear that the utilisation of encapsulated polyphenols instead of free compounds can lead to improvements in both the stability and bioavailability of compounds in vivo and in vitro. Spray-drying has been succesfully applied for the encapsulation of plant extracts containing polyphenolic compounds such as olive leaf extract, β-carotene, *Amaranthus* betacyanin extracts, *Ilex paraguariensis* extracts and others (Kosaraju et al. 2006; Cai and Corke 2000; Desobry et al. 1997a; Deladino et al. 2008). However, spray-drying usually employs high temperatures, around 100 °C under the best circumstances when all measures are taken to prevent overheating of the particles. One of the alternatives is to apply an extrusion technique (Belscak-Cvitanovic et al. 2011). Figure 13.5 shows

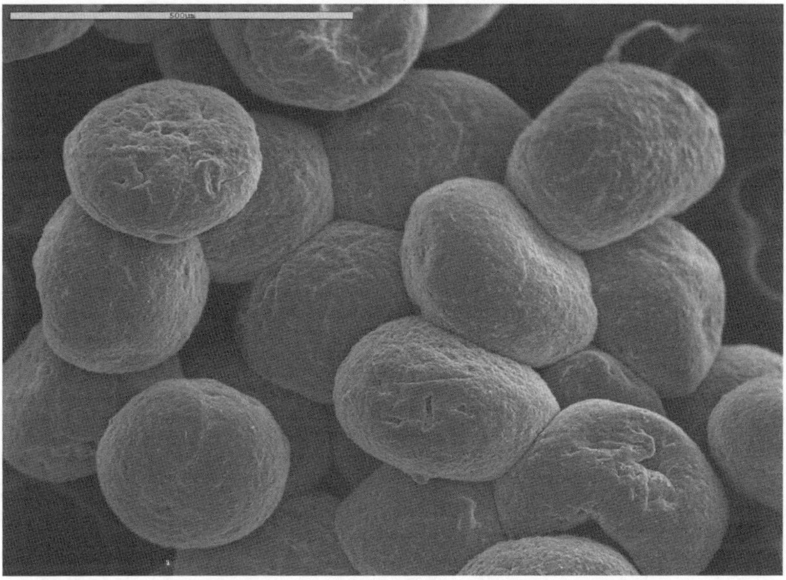

Fig. 13.5 Scanning electron micrograph of lyophilised alginate-inulin microbeads encapsulating *Thymus serpyllum* L.

alginate-inulin microparticles which are produced by electrostatic extrusion followed by lyophilisation.

Future research on polyphenol encapsulation is likely to focus on aspects of delivery and the potential use of co-encapsulation methodologies, where two or more bioactive ingredients can be immobilised together simultaneously to provide synergistic activity between them. An example of such a system was reported recently in which polyphenolic compounds from medicinal plant extracts were encapsulated together with vitamin C to achieve synergistic antioxidant activity (Belscak-Cvitanovic et al. 2011). It can be foreseen that, with a deep understanding of the health benefits of polyphenols and new strategies for stabilisation of fragile nutraceuticals, encapsulated polyphenols will play an important role in increasing the efficacy of functional foods.

13.5.4 Encapsulation of Essential Oils

Essential oils are slightly soluble in water and transfer their odour and taste to water. These oils contain terpenes, phenols, alcohols, aldehydes, esters, ketones and other compounds and have a wide spectrum of biological activities, including antimicrobial activity against a wide range of food-borne pathogens. The encapsulation of essential oils into different nanospheres has been used as a controlled release vehicle with site-specific delivery properties to maximise their antimicrobial activity (Parris et al. 2005).

13.5.5 Encapsulation of Vitamins

The main problem with water- and lipid-soluble vitamins is their stability, which can be improved by encapsulation. Most vitamins have an unpleasant taste that must be masked. Various techniques (such as spray-drying, spray-cooling, spray-chilling and fluidised bed coating) have produced satisfactory results. Spray-chilling and fluidised bed coating have been used for the encapsulation of water-soluble vitamins, while spray-drying of emulsions is generally employed for the encapsulation of lipid-soluble vitamins. Each process should be optimised by a proper selection of the wall material, active-to-wall ratio, temperature, and additives until the final product is as stable as possible. Encapsulated vitamins are used as antioxidants and for increasing the nutritional value of food. Such coated vitamins are desirable in foods and usually applied in baked goods, cereal bars, biscuits, crackers, dry soup, pet food and dairy products (Brownlie 2007; Wilson and Shah 2007).

13.5.6 Encapsulation of Fish Oils

Encapsulation can also be employed to prevent off-flavour of fish oil by preventing contact between light, oxygen or metal ions with fish oil (Garg et al. 2006). The exclusion of oxygen and trapping of volatile off-flavours might be possible by the entrapment of fish oil in a glassy state or in amorphous materials. The inclusion of oxygen from air during the encapsulation process should be avoided as much as possible. Another reason to encapsulate fish oils is to convert liquid oil into a powder. This simplifies handling during incorporation into food products. However, a disadvantage of many encapsulated fish oils is their storage instability in aqueous food products since they easily disintegrate or dissolve. For this reason the insolubility of encapsulates in aqueous food products is an important requirement. Another requirement is the insolubility of encapsulates in gastrointestinal fluids. That is, the aim of encapsulation is to preserve the bioavailablity of fish oils upon consumption. The most frequently used technology for encapsulation of fish oil is spray-drying. In addition, processes such as melt injection, extrusion and coacervation have also been used often. Proteins, sugars, gums, pectin, maltodextrin, glucose syrup, modified cellulose or starch can be applied as wall materials (Wandrey et al. 2010; Beindorff and Zuidam 2010).

13.5.7 Encapsulation of Carotenoids

Carotenoids (β-carotene, lycopene, astaxanthin, lutein and zeaxanthin) are sensitive to heat, oxidation and light due to their unsaturated chemical structures. Consequently, it is very important to prevent the degradation of carotenoids (isomerisation or oxidation), which decreases the nutritional properties, coloration and quality of final products. Since carotenoids are very expensive, some cost-effective techniques of encapsulation are desirable. Spray-, drum-, and freeze-drying have been applied, and these techniques provide particle sizes of approximately 30, 105, and 80 μm, respectively (Desobry et al. 1997b). Liposomes and cyclodextrins are also able to serve as host molecules to encapsulate carotenoids (Shahidi and Han 1993). Maltodextrins were found to be an effective carrier in protecting and improving the solubility and bioavailability of carotenoids (Ribeiro et al. 2010; Beatus et al. 1985; Higuera-Ciapara et al. 2002). Gellan, mesquite gum and gum arabic have been used as wall materials.

Astaxanthin, reddish-orange pigment, belongs to carotenoids. It is a highly unsaturated molecule with powerful antioxidant properties and thus can easily be degraded by thermal or oxidative processes during production and storage. Production of undesirable flavour compounds may result from the oxidation of astaxanthin. Therefore, this compound is often handled as stabilised emulsions or microcapsules (instead of in crystalline form). For example, synthetic pigments were encapsulated in a chitosan matrix cross-linked with glutaraldehyde using the method of multiple emulsion/solvent evaporation (Higuera-Ciapara et al. 2002).

13.5.8 Encapsulation of Colouring Agents

Colouring agents containing natural or synthetic substances are commonly used as additives in the food industry. A well-known problem is that colouring agents tend to shift within the food product or into the environment of the product. Typical examples of food products containing colouring agents are cakes and desserts with jelly fruit filling. In addition to these, it happens in different meat products, e.g. surimi. The migration of colouring agents into non-coloured layers results in highly undesirable appearances. For these products colouring agents were encapsulated within a mixture of carbohydrate-based wall materials (Higuera-Ciapara et al. 2002). Some authors encapsulated various colouring agents in partially hydrogenated vegetable oil to incorporate them into edible coating for snacks, chicken legs, fish and similar products (Shahidi and Pegg 1995).

13.6 Immobilisation of Cells and Enzymes in Food Processing

Encapsulation may be used to immobilise cells or enzymes in food processing applications, such as fermentation processes and metabolite production processes. There is an increasing demand to find suitable solutions that provide high productivity and, at the same time, satisfy demands for adequate quality of the final food products.

Adsorption, gel entrapment and covalent binding are the accepted methods of immobilisation used in various bioprocesses. Immobilisation of microbiological cells by entrapment within natural or synthetic polymers or by adsorption onto solid (in)organic carrier materials (Figs. 13.6 and 13.7) is a research area that has been attracting increased interest. Among materials for cell immobilisation, natural polysaccharides such as alginate, chitosan, pectate and carrageenan, proteins like gelatine and collagen and synthetic polymers like polyvinyl alcohol are the most popular. Natural gelling polysaccharides represent an emerging group due to their advantage of being non-toxic, biocompatible and cheap (Popplewell and Porzio 2001). They can be gelled into hydrophilic matrices under mild conditions, allowing cell entrapment with minimal loss of viability. As a result, very high biomass loadings can be achieved. Gels are mostly used in the form of spherical beads with diameters ranging from approximately 0.3 to 5 mm. However,

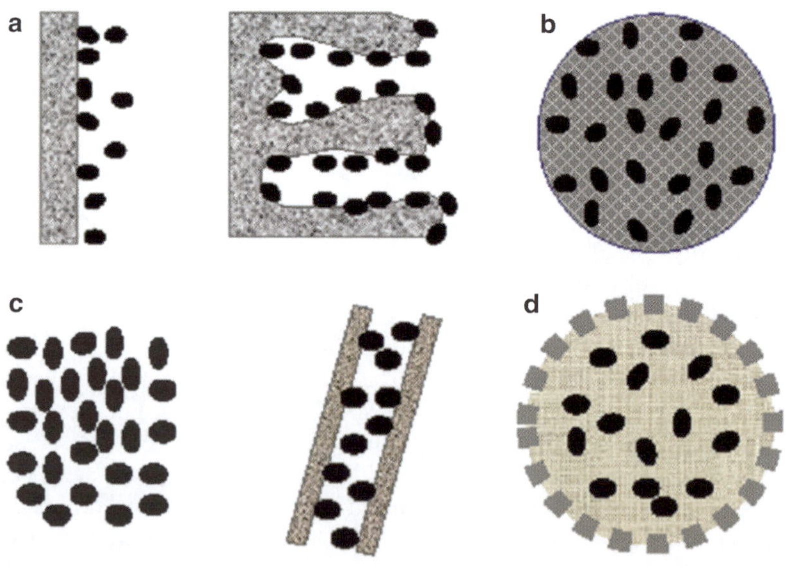

Fig. 13.6 Methods for cell immobilisation: (**a**) adsorption to a preformed carrier; (**b**) entrapment within a matrix; (**c**) self-aggregation; (**d**) containment behind a barrier [modified from Nedovic et al. (2001b)]

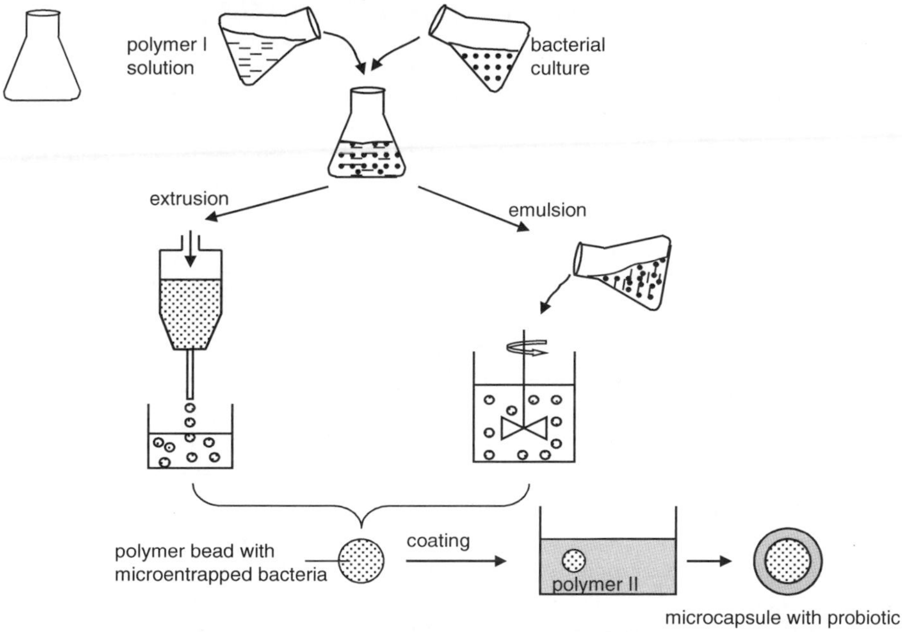

Fig. 13.7 Gel–particle technologies for the microencapsulation of cells [modified from Manojlovic et al. (2010a), Krasaekoopt et al. (2003) and Champagne and Fustier (2007)]

mechanical stability is an important disadvantage of gels. It has often been noticed that the gel structure is destroyed due to the growth of the cells and not because of intensive carbon dioxide production.

13.6.1 Features of Immobilisation Technology

The immobilisation of cells provides ease of biomass separation and recovery, lower risk of microbial contamination, better use of equipment and, as a consequence of these and other benefits, higher productivity and efficiency. Immobilised cell technology offers numerous other potential advantages compared to free cell systems, such as higher cell densities and cell loads, shorter reaction times, smaller bioreactor sizes that lower capital costs, reuse of the same biocatalysts for prolonged periods of time, development of continuous processes which may be performed beyond the nominal washout rate, better substrate utilisation, reduced risk of microbial contamination, simplified process design, constant product quality and protection of cells, and faster fermentation rates. However, the two most important disadvantages are the complexity of production process and cost constraints.

13.6.2 Immobilised Cells in Fermentation Processes

To meet the increasing demand for alcohol-free beer, several methods have been developed including alcohol removal from the product or limited fermentation of wort. In the second case, production is much better when immobilised cells are used (Nedovic et al. 2001b). In other processes like primary beer fermentation and wine and cider fermentation, immobilisation technology is still under scrutiny at the lab or pilot levels. These processes are significantly more complex and have various side reactions important for flavour formation and final beverage quality. Currently, the major challenge for successful application of immobilised yeast cell technology on an industrial scale is yeast physiology control and fine-tuning of flavour formation during fermentation processes (Nedovic et al. 2010).

13.6.2.1 Beer Production by Immobilised Biocatalysts

Beer production with immobilised yeast has been the subject of extensive research for more than 30 years. Traditional beer fermentation systems use freely suspended yeast cells to ferment wort in an unstirred batch reactor. The traditional primary fermentation for lager beer takes about 1 week with secondary fermentation a few weeks. Treatment at a higher fermentation temperature and with a selected specific yeast strain allows the production of lager beer in 12–15 days. Immobilised cell technology is able to produce lager beer in a much shorter period, usually 1–3 days, and the ultimate goal is the production of beer with satisfactory final quality in that period of time (Verbelen et al. 2010; Willaert and Nedovic 2006). Nowadays, as a result of detailed and comprehensive research, immobilised yeast technology is a well established technology for beer maturation and alcohol-free and low-alcohol beer production. However, the situation is more complex in primary fermentation, and this process is still under examination on the lab and pilot levels (Nedovic et al. 2001b; Willaert and Nedovic 2006). An important issue of this research area is the production of beer by immobilised yeast in continuous culture with the same properties as traditionally manufactured beers. Achieving satisfactory sensory characteristics in such a short time is a major difficulty. In some cases, cell proliferation and metabolic activity is limited, and this can result in deficient free amino nitrogen consumption and therefore an unbalanced flavour profile of the final beer. Therefore, different approaches to the adaptation of immobilised systems were investigated in order to obtain the proper final beer quality. The researchers aim to explore new bioreactor designs and new materials for cell immobilisation and to find solutions for changing cell physiology and metabolism. Another direction would be to use genetically modified yeast strains to develop the capability to produce larger or smaller amounts of one or more flavour compounds.

13.6.2.2 Wine and Cider Production by Immobilised Biocatalysts

The well-known transformation/fermentation of grape must and apples by microbial activity results in the production of wine and cider, respectively. In these processes, various compounds affect the sensory properties of wine and cider (Kourkoutas et al. 2010). The use of immobilised yeast cells in wine production has been explored with a view towards reducing labour requirements, simplifying time-consuming procedures and, thereby, reducing costs. To be convenient in wine production, a method must be economical, easily performed in industrial conditions and not cause oxidation and contamination of wine (Diviès et al. 1994). In wine- or cider-making, the main objective is to achieve an adequate product quality. Microbial cell immobilisation can improve the efficiency of malolactic fermentation, the capacity for cell recycling, cell stability and viability, and an improvement in product quality (Kourkoutas et al. 2004). The use of immobilised cells in wine and cider production offers other advantages such as simplified systems for removing microbial cells from batch processes, greater tolerance toward inhibitory substances, smaller-scale fermentation facilities and possibilities for using a variety of microbial strains including genetically modified organisms. Immobilisation in various materials aims to increase the tolerance of malolactic bacteria to changes in composition of fermentation medium and to speed up the process. In cider production, research has focused on simultaneous alcoholic and malolactic fermentation by the co-immobilisation of two different species or by the same microorganism, often using genetic modification. However, some potential disadvantages must also be mentioned, such as cell overgrowth, which increases the turbidity of the fermented beverage, mechanical stability of the matrix used to immobilise microbial cells, and loss of activity from prolonged operation. The selection of a suitable carrier and bioreactor system is a challenge, and many factors should be taken into account, such as product quality, safety and stability, investments, operating costs and legality (Kourkoutas et al. 2010).

13.6.2.3 Dairy and Meat Fermentations by Immobilised Biocatalysts

Besides in beverage production processes, immobilised cell/enzyme technology has been used in dairy and meat fermentation and in enzymatic processes. Immobilisation may be very useful in improving the stability of probiotics and protective cultures in fermented foods. Immobilised cell technology can provide protection of cells during fermentation and drying, protection against bacteriophage attack, inhibition of undesirable flora, enhanced survival of cells after heating and freezing, improved stability of cells during storage and acceleration of flavour development. In fermented meat and milk, the main microorganisms used are lactic acid bacteria. Basically, cells are micro-entrapped into gel particles and added to the growth medium. Extrusion and emulsion techniques are commonly used for immobilisation of lactic cultures in gels (Fig. 13.7). A specific feature is that most of the biomass is located on the surface of gel beads, principally because of mass-transfer limitations of substrates and fermentation products. Cells are therefore released from the beads into the surrounding medium. This feature is desirable when a system is used for continuous inoculation of milk such as in the dairy industry.

However, there are also some disadvantages in biomass production. The two most important ones are higher investment costs and lower-than-expected yields. Another important disadvantage is contamination of the released cells by bacteriophages, which can happen when milk used for cheese production is insufficiently heat treated by pasteurisation; then bacteriophages from the raw milk can survive and contaminate the bioreactor (Champagne et al. 1993; Macedo et al. 1999).

13.7 Conclusions

Encapsulation of food ingredients is not a new concept and is widely applied in the food industry. However, the rapid growth in the variety and quantity of functional food products on the market continuously forces companies and, consequently, researchers to investigate and open up new opportunities for the use of encapsulation technologies in the food industry. It is safe to assume that in the future encapsulated bioactives will play a significant role in increasing the efficacy of functional foods. With advanced strategies for the stabilisation of food ingredients and the development of new approaches, it will be possible to improve the nutritional properties and health benefits of food compounds.

Immobilisation of cells has successfully achieved industrial applications in several food sectors, such as the beverage industry (beer and wine production). The main advantages of immobilised cell technology are the high productivity and efficiency of continuous fermentation. On the other hand, a number of difficulties remain unresolved. Therefore, current and future research should focus on overcoming the existing complexities, which are mainly connected with modified immobilised cell metabolism. In that respect, some of the major challenges and, at the same time, fields of investigation are bioreactor design, carrier materials, immobilisation techniques and process conditions.

References

Augustin MA, Hemar Y (2009) Nano- and micro-structured assemblies for encapsulation of food ingredients. Chem Soc Rev 38:902–912

Beatus Y, Raziel A, Rosenberg M, Kopelman IJ (1985) Spray-drying microencapsulation of paprika oleoresin. Lebensm Wiss Technol 18:28–34

Beindorff CM, Zuidam NJ (2010) Microencapsulation of fish oil. In: Zuidam NJ, Nedovic VA (eds) Encapsulation technologies for food active ingredients and food processing. Springer, Dordrecht, pp 161–187

Bell LN (2001) Stability testing of nutraceuticals and functional foods. In: Wildman REC (ed) Handbook of nutraceuticals and functional foods. CRC Press, New York, pp 501–516

Belscak-Cvitanovic A, Stojanovic R, Manojlovic V et al (2011) Encapsulation of polyphenolic antioxidants from medicinal plant extracts in alginate-chitosan system enhanced with ascorbic acid by electrostatic extrusion. Food Res Int 44(4):1094–1101

Bilensoy E, Hincal AA (2009) Recent advances and future directions in amphiphilic cyclodextrin nanoparticles. Expert Opin Drug Deliv 6:1161–1173

Brownlie K (2007) Marketing perspective in encapsulation technologies in food applications. In: Lakkis JM (ed) Encapsulation and controlled release technologies in food systems. Blackwell, Iowa, pp 213–235

Burgain J, Gaiani C, Linder M, Scher J (2011) Encapsulation of probiotic living cells: from laboratory to industrial application. J Food Eng 104(4):467–483

Cai YZ, Corke H (2000) Production and properties of spray-dried Amaranthus betacyanin pigments. J Food Sci 65(6):1248–1252

Champagne CP, Fustier P (2007) Microencapsulation for the improved delivery of bioactive compounds into foods. Curr Opin Biotechnol 18:184–190

Champagne CP, Girard F, Rodrigue N (1993) Production of concentrated suspensions of thermophilic lactic acid bacteria in calcium alginate beads. Int Dairy J 3(3):257–275

Conde-Petit B, Escher F, Nuessli J (2006) Structural features of starch-flavor complexation in food model systems. Trends Food Sci Technol 17(5):227–235

de Roos KB (2003) Effect of texture and microstructure on flavour retention and release. Int Dairy J 13:593–605

Deladino L, Anbinder PS, Navarro AS, Martino MN (2008) Encapsulation of natural antioxidants extracted from Ilex paraguariensis. Carbohyd Polym 71:126–34

Desai KGH, Park HJ (2005) Recent developments in microencapsulation of food ingredients. Drying Technol 23:1361–1394

Desobry SS, Netto FM, Labuza TP (1997) Comparison of spray-drying, drum-drying, and freeze-drying for β-carotene encapsulation and preservation. J Food Sci 62(6):1158–1162

Dewettinck K, Huyghebaert A (1999) Fluidized bed coating in food technology. Trends Food Sci Technol 10:163–168

Diviès C, Cachon R, Cavin J-F, Prévost H (1994) Theme 4: immobilized cell technology in wine production. Crit Rev Biotechnol 14:135–153

Donhow IG, Fennema O (1993) Water vapour and oxygen permeability of wax films. J Am Oil Chem Soc 70:867–873

Fang Z, Bhandari B (2010) Encapsulation of polyphenols – a review. Trends Food Sci Technol 21 (10):510–523

FAO/WHO (2006) Probiotics in food. Health and nutritionl properties and guidelines for evalution. FAO food and nutritional paper no 85. ISBN:92-5-105513-0

Garg ML, Wood LG, Singh H, Moughan PJ (2006) Means of delivering recommended levels of long chain n-3 polyunsaturated fatty acids in human diets. J Food Sci 71(5):R66–R71

Gibbs BF, Kermasha S, Alli I, Mulligan CN (1999) Encapsulation in the food industry: a review. Int J Food Sci Nutr 50:213–224

Gouin S (2004) Microencapulation: industrial appraisal of existing technologies and trends. Trends Food Sci Technol 15:330–347

Harada A, Takashima Y, Yamaguchi H (2009) Cyclodextrin-based supramolecular polymers. Chem Soc Rev 38:875–882

Higuera-Ciapara I, Felix-Valenzuela L, Goycoolea FM, Argüelles-Monal W (2002) Microencapsulation of astaxanthin in a chitosan matrix. Carbohydr Polym 56(1):41–45

Kailasapathy K (2009) Encapsulation technologies for functional foods and nutraceutical product development. CAB Rev Perspect Agric Vet Sci Nutr Nat Resour 4(6):1–19

Khor E, Lim LY (2003) Implantable applications of chitin and chitosan. Biomaterials 24:2339–2349

Kosaraju SL, Dath L, Lawrence A (2006) Preparation and characterisation of chitosan microspheres for antioxidant delivery. Carbohyd Polym 64:163–167

Kourkoutas Y, Bekatorou A, Banat IM, Marchant R, Koutinas AA (2004) Immobilization technologies and support materials suitable in alcohol beverages production: a review. Food Microbiol 21(4):377–397

Kourkoutas Y, Manojlovic V, Nedovic VA (2010) Immobilization of microbial cells for alcoholic and malolactic fermentation of wine and cider. In: Zuidam NJ, Nedovic VA (eds) Encapsulation technologies for food active ingredients and food processing. Springer, Dordrecht, pp 327–344

Krasaekoopt W, Bhandari B, Deeth H (2003) Evaluation of encapsulation techniques of probiotics for yoghurt. Int Dairy J 13(1):3–13

Lee JS, Cha DS, Park HJ (2004) Survival of freeze-dried *Lactobacillus bulgaricus* KFRI 673 in chitosan-coated calcium alginate microparticles. J Agric Food Chem 52:7300–7305

Lian WC, Hsiao HC, Chou CC (2002) Survival of bifidobacteria after spray drying. Int J Food Microbiol 74:79–86

Macedo MG, Champagne CP, Vuillemard JC, Lacroix C (1999) Establishment of bacteriophages in an immobilized cells system used for continuous inoculation of lactococci. Int Dairy J 9(7):437–445

Madene A, Jacquot M, Scher J, Desobry S (2006) Aroma encapsulation and controlled release – a review. Int J Food Sci Technol 41:1–21

Maillard M, Landuyt A (2008) Chocolate: an ideal carrier for probiotics. Agro Food Ind Hi-Tec 19(3):13–15

Manojlovic V, Bugarski B, Nedovic V (2010) Immobilised cells. In: Flickinger MF (ed) Encyclopedia of industrial biotechnology: bioprocess, bioseparation, and cell technology. Wiley, Hoboken, pp 1–18

Manojlovic V, Nedovic V, Kailasapathy K, Zuidam NJ (2010) Encapsulation of probiotics for use in food products. In: Zuidam NJ, Nedovic VA (eds) Encapsulation technologies for food active ingredients and food processing. Springer, Dordrecht, p 269

McClements D, Lesmes U (2009) Structure-function relationships to guide rational design and fabrication of particulate food delivery systems. Trends Food Sci Technol 20(10):448–457

McMaster LD, Kokott SA (2005) Micro-encapsulation of *Bifidobacterium lactis* for incorporation into soft foods. World J Microbiol Biotechnol 21:723–728

Medina LM, Jordano J (1994) Survival of constitutive microflora in commercially fermented milk containing bifidobacteria during refrigerated storage. J Food Prot 56:731–733

Mellema M, Van Benthum AJ, Boer B, Von Harras J, Visser A (2006) Wax encapsulation of water-soluble compounds for application in foods. J Microencapsul 23(7):729–740

Milanovic J, Manojlovic V, Levic S, Rajic N, Nedovic V, Bugarski B (2010) Microencapsulation of flavors in carnauba wax. Sensors 10(1):901–912

Nedovic VA, Obradovic B, Leskosek-Cukalovic I, Trifunovic O, Pesic R, Bugarski B (2001a) Electrostatic generation of alginate microbeads loaded with brewing yeast. Process Biochem 37:17–22

Nedovic VA, Obradovic B, Leskosek-Cukalovic I, Vunjak-Novakovic G (2001b) Immobilized yeast bioreactor systems for brewing – recent achievements. In: Hofman M, Thonart P (eds) Focus in biotechnolgy series engineering and manufacturing for biotechnology. Kluwer Academic, Dordrecht, pp 277–292

Nedovic V, Manojlovic V, Leskosek-Cukalovic I, Bugarski B, Willaert R (2010) State of the art in immobilized/encapsulated cell technology in fermentation processes. In: Aguilera JM, Simpson R, Welti Chanes J, Bermudez-Aguirre D, Barbosa-Canovas G (eds) Food engineering interfaces, Food engineering series. Springer, New York, pp 119–47

Park JH, Ye M, Park K (2005) Biodegradable polymers for microencapsulation of drugs. Molecules 10:146–161

Parris N, Cooke PH, Hicks KB (2005) Encapsulation of essential oils in zein nanospherical particles. J Agric Food Chem 53:4788–4792

Petrovic T, Nedovic V, Dimitrijevic-Brankovic S, Bugarski B, Lacroix C (2007) Protection of probiotic microorganisms by microencapsulation. CI&CEQ 13(3):169–174

Popplewell LM, Porzio MA (2001) Fat-coated encapsulation compositions and method for preparing the same. US Patent no 6,245,366

Porzio M (2004) Flavor encapsulation: a convergence of science and art. Food Technol 58 (7):40–47

Porzio MA (2007) Flavor delivery and product development. Food Technol 61(1):22–29

Prüsse U, Bilancetti L, Bucko M et al (2008) Comparison of different technologies for alginate beads production. Chem Pap 62(4):364–374

Ribeiro HS, Schuchmann HP, Engel R, Walz E, Briviba K (2010) Encapsulation of carotenoids. In: Zuidam NJ, Nedovic VA (eds) Encapsulation technologies for food active ingredients and food processing. Springer, Dordrecht, pp 211–253

Schimidt PC (1997) Technological aspects of the development and production of plant extracts. Pharm Ind 59:69

Shah NP (2000) Probiotic bacteria: selective enumeration and survival in dairy foods. J Dairy Sci 83:894–907

Shah N, Lankaputhra WEV, Britz ML, Kyle WSA (1995) Survival of *Lactobacillus acidophilus* and *Bifidobacterium bifidum* in commercial yoghurt during refrigerated storage. Int Dairy J 5 (5):515–521

Shahidi F, Han XQ (1993) Encapsulation of food ingredients. Crit Rev Food Sci Nutr 33 (6):501–547

Shahidi F, Pegg RB (1995) Stabilized cooked cured-meat pigment. US Patent no 5,425,956

Teixeira P, Castro H, Mohácsi-Farkas C, Kirby R (1997) Identification of sites of injury in *Lactobacillus bulgaricus* during heat stress. J Appl Microbiol 83:219–226

Truelstrup-Hansen L, Allan-Wojotas PM, Jin YL, Paulson AT (2002) Survival of Ca-alginate microencapsulated Bifidobacterium spp. in milk and simulated gastrointestinal conditions. Food Microbiol 19(1):35–45

Verbelen PJ, Nedovic VA, Manojlovic V et al (2010) Bioprocess intensification of beer fermentation using immobilised cells. In: Zuidam NJ, Nedovic VA (eds) Encapsulation technologies for food active ingredients and food processing. Springer, Dordrecht, pp 303–326

Vos P, Faas MM, Spasojevic M, Sikkema J (2010) Review: encapsulation for preservation of functionality and targeted delivery of bioactive food components. Int Dairy J 20(4):292–302

Wandrey C, Bartkowiak A, Harding SE (2010) Materials for encapsulation. In: Zuidam NJ, Nedovic VA (eds) Encapsulation technologies for food active ingredients and food processing. Springer, Dordrecht, pp 31–100

Willaert R, Nedovic VA (2006) Primary beer fermentation by immobilised yeast – a review on flavour formation and control strategies. J Chem Technol Biotechnol 81:1353–1367

Wilson N, Shah NP (2007) Microencapsulation of vitamins. ASEAN Food J 14(1):1–14

Zuidam NJ, Heinrich J (2010) Encapsulation of aroma. In: Zuidam NJ, Nedovic VA (eds) Encapsulation technologies for food active ingredients and food processing. Springer, Dordrecht, pp 127–160

Zuidam NJ, Shimoni E (2010) Overview of microencapsulates for use in food products or processes and methods to make them. In: Zuidam NJ, Nedovic VA (eds) Encapsulation technologies for food active ingredients and food processing. Springer, Dordrecht, pp 3–31

Chapter 14
Aroma Encapsulation in Powder by Spray Drying, and Fluid Bed Agglomeration and Coating

Turchiuli Christelle and Dumoulin Elisabeth

14.1 Introduction

Aromas in food products bring typical characteristics called taste and flavour. Aroma compositions (called aroma) are complex mixtures of different molecules, more or less stable with temperature, light and oxygen and soluble or not in water. They are often very strong if pure, which means they must be diluted in a neutral support. They must often be protected against the environment until they are needed for use in specific conditions, where their final release must be controlled.

In this study the objective was to replace liquid aroma by aroma powder in sweets and chewing gum. These products must deliver the right balance of flavour in specific conditions of temperature and humidity with a persistent intensity. The global perception during chewing will depend on the composition in various molecules and on their release when moist, neutral saliva reaches the active component. These properties may be built using flavour-containing microcapsules that are able to release core material progressively through mechanical stress and diffusion/dissolution.

Microencapsulation has become an attractive approach to converting liquid food flavourings into a dry and free-flowing powder form that is easy to handle and to incorporate into a dry system (Risch and Reineccius 1995; Gibbs 1999; Madene 2006; Vandamme 2007). The powder properties depend closely on the initial formulation and on the production process: fraction of aroma in the matrix; integrity of shell; practical conditions of the initial formulation such as viscosity, aroma dispersion and required temperature (Gouin 2004). The dry final powder

T. Christelle
AgroParisTech, UMR1145 Ingénierie Procédés Aliments, F-91300 Massy, France

Univ Paris-Sud, F-91405 Orsay, France

D. Elisabeth (✉)
AgroParisTech, UMR1145 Ingénierie Procédés Aliments, F-91300 Massy, France
e-mail: elisabeth.dumoulin@agroparistech.fr

S. Yanniotis et al. (eds.), *Advances in Food Process Engineering Research and Applications*, Food Engineering Series, DOI 10.1007/978-1-4614-7906-2_14,

grains (microns to millimetres) will be made of aroma dispersed in a wall material, with a controlled size distribution. Different sizes may give different time releases. Microcapsules must ensure the protection of aroma against oxygen and moisture, especially during storage. The main encapsulation techniques, which vary in cost, are spray-drying, spray-cooling, drum-drying, extrusion and coacervation. A fluidised-bed process is used to modify the solid particle properties such as size, solubility and stability by agglomeration and additional coating, keeping in mind the difficulty of coating very thin particles (<100 μm) (Gouin 2004; Finney et al. 2002; Guignon et al. 2002; Fuchs et al. 2006; Jimenez et al. 2006).

Spray-drying remains the dominant method to produce a dry aroma emulsion (Jafari et al. 2008). This continuous process uses the spraying of emulsions in small drops with high surface exchange for heat and mass transfer when in contact with hot air. Solvent evaporation from initial drops is very fast and the temperature reached by the product will be low, consequently minimising the alteration and loss of volatile aroma components able to modify the final aroma profile in powder. Volatile losses may occur mainly during the early stages of formation and drying of drops, before the formation of a dry surface layer, retaining flavour molecules (Rosenberg et al. 1990; King 1995). Besides the composition of wall materials, the main loss factors for aroma molecules are related to molecular weight, relative volatility, polarity type and the ratio of aroma to wall. Optimal conditions will depend on the spraying mode and drop fast initial drying in relation to feed emulsion properties such as high total solid content and small drop size (~1 μm) (Reineccius 2004; Soottitantawat et al. 2005; Jafari et al. 2008).

The nature and quantity of food-grade shell materials are important for an optimal dosage and for providing good protection during storage against moisture and oxygen, ensuring a long shelf life. They will contribute, even to a slight degree, to the composition, texture and taste of the final product. The mechanisms of release, mainly disruption and diffusion, will be considered with the global final product composition and structure. In the case of powders added to a formulated paste, the solid particles must be limited in size so as not to be perceived in the mouth.

The properties of wall materials differ by molecular weight and stability with temperature, including phase transitions. They may be hydrophilic or hydrophobic, and some may have emulsifying properties. The main materials are polysaccharides, celluloses, proteins, gums, fats, waxes, glyceride fatty esters and modified starches, alone or in combination. Studies have been reported mainly on pilot or laboratory equipment for model oil molecules (Finney et al. 2002; Soottitantawat et al. 2005).

Encapsulation efficiency is usually characterised by the percentage of oil encapsulated, the fraction on the surface of particles and stability against oxidation during storage with variable atmosphere humidity and time (Senoussi et al. 1995; McNamee et al. 2001; Krishnan et al. 2005). For example, the rate of release and oxidation of encapsulated d-limonene in carbohydrates was investigated with structural changes and a glass transition temperature of the capsule matrixes in relation to relative humidity (Soottitantawat et al. 2005). In parallel, single drops drying (in constant conditions) were studied to understand the drying mechanisms (Hetch and King 2000).

To control the release of aroma, multi-stage encapsulation was proposed with different aroma molecules distributed between the core (solid, paste, liquid) and the solid external layer, which is often made of sugars and polyols. In this layer, aroma is usually protected to avoid aroma and water migration using various encapsulation methods and supports. This provides two levels of release (Merrit et al. 1985; Pearl et al. 1993; Cho and Park 2002; Clark and Shen 2004; Lakkis 2007; Jeon et al. 2008). For example, an oil-in-water emulsion with coffee aroma was sprayed on coffee powder to form a dry coating layer enriched in aroma, with release in hot water (Buffo et al. 2002).

Spray-dried powders have a small diameter, in a range of 10–30 μm, resulting in poor properties for reconstitution and handling. The aggregation of these small particles into large porous agglomerates yields powders with improved properties such as wettability and dispersibility in liquid. Agglomeration is possible by creating contacts and links between particles. The main parameters are the surface state of particles, the contact duration and intensity, and the adhesion mechanisms to form strong bridges. Those bridges may be made of the same material as the basic powders or of a different material. In the case of flavour powders with wall materials capable of being rehydrated and forming inter-particle bridges, the fluid-bed process will be favoured using wet growth agglomeration. Water or aqueous binder solution is sprayed on the surface of the moving fluidised particles. If moist, sticky particles collide, then liquid bridges will form at the points of contact, with the bridges being subsequently dried (*agglomeration*). If the sprayed solution dries on the particle surface before collision, then the surface will be coated progressively with binder deposit (*coating*). In agglomeration, it is important to avoid the collapse of carriers; this phenomenon is influenced by the glass transition temperature, water activity, process temperature and spraying rate (Liu and Rushmore 1996).

The present study focused on the preparation of dry emulsion powders with several shapes and structures, maintaining a constant chemical composition for carriers and aroma (20 % w/w). Powders were first prepared by spray-drying, then agglomerated in a fluid bed with different binder solutions, and finally coated with a dry aroma emulsion. The powders obtained with the different processes were compared for their physicochemical properties and evaluated by incorporation into a chewing gum base paste.

14.2 Preparation of Powders

14.2.1 Products

The support matrix was made of maltodextrin (MD) (Glucidex DE12, Roquette, Lestrem, France) and acacia gum (AG) (Instant Gum AA, CNI/Nexira, Rouen, France), with bulk densities of 0.47 and 0.3 g.cm^{-3} respectively. Both soluble in water, they were used with a weight ratio MD/AG equal to 3/2 (Bhandari et al. 1992).

The pure aroma molecules were as follows: T, top note, ester, water soluble; C, middle note, aldehyde; F, end note, lactone (Fruitaflor, Versailles, France), all with a boiling point greater than130 °C and a molecular weight on the order of T $<$ M $<$ F. M1 was a mix: (T) 96.5 %, (C) 0.93 %, (F) 2.5 % (% w/w) representing ~30 % of the real aroma composition M2.

14.2.2 Emulsion MD/AG/M1 (or M2) + Water (40 % w/w)

Aroma was added to the carrier aqueous solution MD/AG (3/2) (30 °C; stored for 12 h at 4 °C). At ambient temperature, the aroma (pure, M1, M2), representing 20 % of the total solid content (MD/AG: aroma = 4:1), was incorporated under mechanical agitation. The final emulsion was obtained by agitation with Polytron (PT-3000, Kinematica, Eschbach, Germany); (4,000 rpm, 25 min). The emulsion size was less than 5 μm. Volumes of 4.5 L were prepared for each spray-drying trial. The emulsion AG/M2 (1.6/1, 67.5 % water) was prepared using the same procedure.

The analyses performed on the emulsions and powders and the techniques used are described in Table 14.1.

14.2.3 Spray-Drying

The flavour emulsion was spray-dried in a co-current spray dryer (Niro Minor, Niro, Soborg, Denmark; 0.5 m^3, h = 1.2 m) equipped with a centrifugal wheel atomiser (Fig. 14.1). The process conditions were as follows: air inlet temperatures: 140 °C, 160 °C, with constant feed rate of respectively 46, 28 g/min; rotational speed of atomiser: 25,000 rpm (5 bars); air flow rate: 110 kg/h. Temperatures for inlet/outlet air, for liquid feed and final powder (<50 °C) were measured. The powder was separated from air in a cyclone and stored either in plastic bags or glass containers. The process powder yield was defined as the ratio between the total collected powder and the theoretical powder quantity from the sprayed emulsion.

Table 14.1 Analyses and techniques

Analyses	Techniques
Viscosity of emulsion	Viscosimeter co-axial cylinders (Rheomat 180, Lamy, Champagne au Mont d'Or, France)
Size	Laser granulometer (MS2000, Malvern, Orsay, France): wet mode for emulsion (initial or reconstituted); dry mode (air) for powders (<100 μm)
Size distribution	Sieving for powders >100 μm (14 sieves; diameter 53 mm; 100–3,150 μm; 5 g)
Water content of powder	Oven 105 °C–24 h
Water activity/sorption isotherm	Aw-meter (Thermoconstanter, Novasina, Genève, Suisse; GBX FA-st lab)
True, packed, bulk density, porosity of powder	Air pycnometer (Accupyc 1330, Micromeritics, Verneuil, France)
	Graduated test tube with/without tapping
Wettability of powder	Contact time with water surface before disappearance (5 g, 100 ml, 20 °C)
Observation of solid particles	Optical microscopy (BX60, Olympus, Rungis, France) & SEM (JSM-5200, Jeol, Croissy, France)
Chromatographic aroma analysis	Hewlett Packard 5890, capillary column + IFD (after extraction)
Powder = > emulsion/ coalescence	Solution with same concentration DM/water, and microscopy observation
Sensory evaluation	Chewing gum paste with 0.6 % w/w aroma

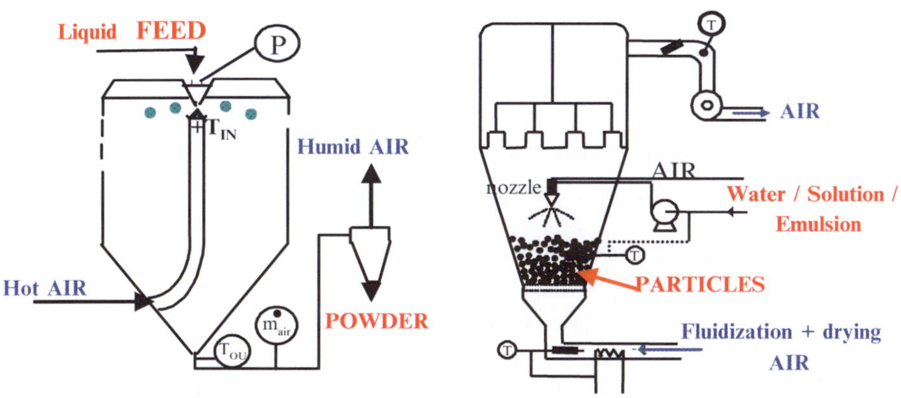

Fig. 14.1 Spray dryer Niro Minor and fluidised bed Uni Glatt

14.2.4 Fluidised-Bed Agglomeration and Coating

In a batch fluidiser (Uni-Glatt, Weimar, Germany) solid particles (~500 g) were fluidised by hot air (70 °C) and moistened on the surface by spraying a binder solution to finally obtain, after drying, either larger agglomerates or coated

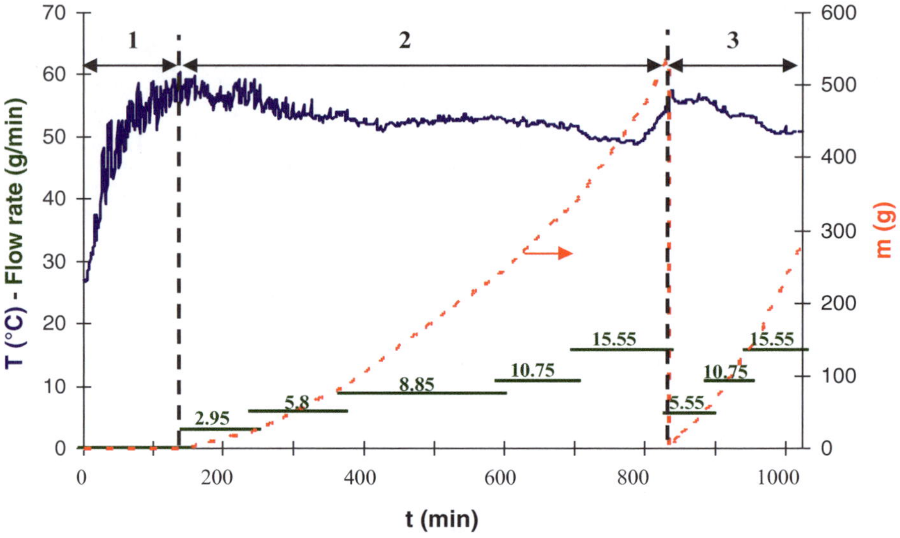

Fig. 14.2 Evolution of bed temperature, sprayed-liquid flow rate and mass of liquid sprayed during trials 4.1 and 4.2 (1. heating, 2. agglomeration, 3. coating)

individual particles (Fig. 14.1). The air bed temperature was maintained between 47 °C and 60 °C. All the parameters were adapted to prevent any collapse of the bed. After initial stabilisation of temperatures, the sprayed-liquid flow rate was progressively increased (i.e. 2.9 to 16 g/min) (Fig. 14.2). The air flow rate was adjusted to compensate for the increased weight of particles (i.e. 80–199 kg/h) keeping a constant bed height under the nozzle. The duration was fixed by liquid quantity to spray or particle size to reach. The process powder yield was the ratio between the final powder mass and the theoretical mass of solids.

14.3 Powder Properties

An aroma emulsion was first spray-dried in thin powder. Then agglomerates were produced in a fluidised bed in three ways by top spraying binder: water (1) or aroma/carrier/water emulsion (2) on fluidised atomised aroma powders; or acacia gum/aroma/water emulsion on fluidised maltodextrin powders (3) (see Fig. 14.3).

Coating was carried out by spraying aroma/carrier/water emulsion on agglomerated particles (>200 μm), with two spraying modes, top or bottom (with internal Wurster tube). Coating is a slow process with a layer built progressively with successive deposits of emulsion drops and drying.

All these conditions finally led to flavour powders that maintained the same total chemical composition, but with different structures, the aroma being present inside the particles (atomised, agglomeration with water) or in the bridges of agglomerates and in the coating layer (Fig. 14.3).

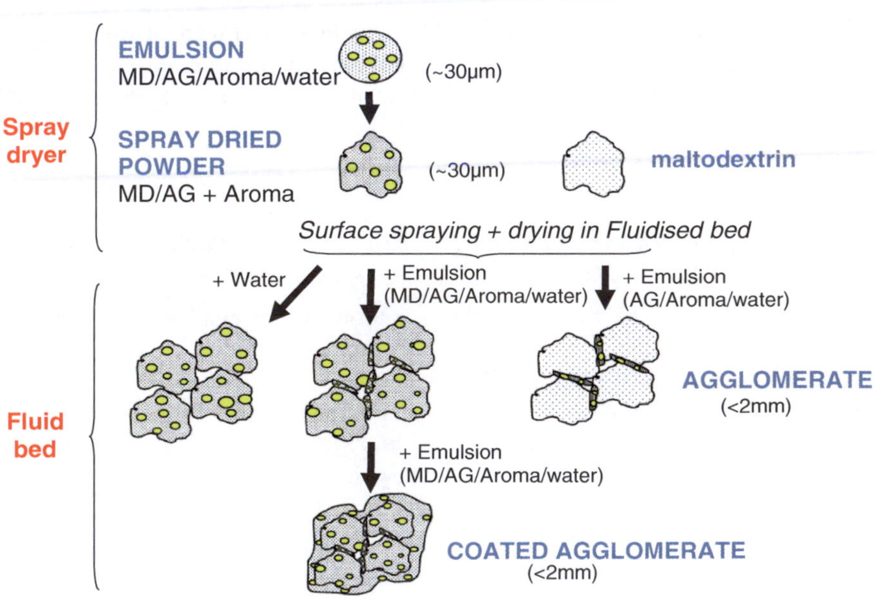

Fig. 14.3 Production of spray-dried powder, agglomerates and coated agglomerates

14.3.1 Spray-Dried Aroma Powders

Spray-drying conditions were determined in preliminary trials using emulsions with the pure aromas alone and mixed. Then the emulsion MD/AG/M1 was spray-dried with two inlet air temperatures, 160 °C and 140 °C, at a constant air flow rate, leading to powders SD160 and SD140. The liquid feed flow rate (respectively 45.8 and 27.6 g/min) was adapted to keep the outlet temperature between 70 and 90 °C, to enable collection of the powder without sticking. The powder yield was greater than 75 %. Drying was more efficient at 160 °C than at 140 °C and faster (respectively 109 and 232 min to spray-dry 4,600 g emulsion). For the lower inlet air temperature the initial drying of the drop was less efficient, leading to slower drying in the chamber. Compared to SD140, SD160 powder corresponded to a lower mean powder water content (3.8 and 4.5 g/100 g DM), water activity (0.1 and 0.13) and bulk density (0.49 and 0.51 g/cm^3). Similar results were observed in the literature (Finney et al. 2002).

In both cases, the mean particle size was low, 20–30 µm, with a narrow size distribution, and powder wettability was bad (>15 min). The reconstituted emulsions showed no coalescence of aroma drops.

Total aroma content in powders was 14 % (SD140) and 16.7 % (SD160), corresponding to a retention of 70 % and 83 % respectively. The relative composition in T, C, F was preserved, with a slight decrease of F. During sensory tests the three molecules were perceived distinctly, strongly for the SD160 powder and satisfactorily for SD140.

One trial with the emulsion MD/AG/M2 (M2 real aroma) with a larger piece of equipment (factor 6) was carried out in similar conditions (air 160 °C). The resulting powder yielded encouraging results concerning aroma retention and perception.

14.3.2 Agglomerated Spray-Dried Powders in Fluidised Bed

The final objectives of the study were to obtain aroma encapsulation not only inside the atomised particles but also more accessible on their surface such as in a coating layer. Before coating in the fluidised bed it was necessary to increase the atomised powder size by agglomeration. For this agglomeration the particles surfaces need to be sticky to induce the solid bridges. Two kinds of spray on the surfaces were adopted: either water; or the binder emulsion (with aroma) to get the aroma into the solid interparticule bridges (Fig. 14.3) (Fuchs et al. 2006; Turchiuli et al. 2005).

The powder MD/AG/M2 obtained with the larger spray dryer was agglomerated by spraying water on the fluidised particles. The T, C, F aroma content was decreased (5.3 % instead of 6.4 %) maintaining a good relative proportion of T, C, F. Air flow and temperature in the fluidised bed during the agglomeration process probably contributed to stripping of part of the oil remaining on the surface. The size distribution was very large, and the powder properties that were similar to those of the other trials will be described further.

In another trial, fluidised maltodextrin powder (240 g) was agglomerated by spraying the aqueous emulsion AG/M2 (800 g). The conditions were applied to obtain a final powder with the same composition MD/AG/aroma (20 % w/w aroma in powder, with 6.4 % of T, C, F for M2). The size increase was rapid up to 160 μm and then slowed down, leading to a narrow final size distribution of around 450 μm. The final concentration in the T, C, F molecules was low (1.4 instead of 6.4 % with M2), but the proportion of each molecule was preserved. In that case the maltodextrin was the support and the whole aroma was associated with acacia gum in the bridges between particles, with insufficient protection during the long coating process (93 min), which may explain losses.

Powders SD140 and SD160 were agglomerated by spraying MD/AG/M1 emulsion on 500 g of fluidised SD powder. In that case aroma was distributed both inside grains and in bridges.

For the agglomeration of SD160 leading to SD160A, the powder yield was good, 88 %. The size distribution was large, with a median diameter d_{50} of 608 μm: size up to 1,800 μm and 46 % less than 450 μm. After spraying 340 g emulsion on 500 g powder (58 min) the aroma concentration in the final powder was 17.6 % with a good ratio of T, M, F.

For SD140 agglomeration, two trials (SD140A, SD140B) were compared by spraying different quantities of emulsion. With a lower quantity sprayed (450 g, 53 min), the powder size distribution of SD140A was bimodal: 133 μm (14.6 % with $d < 200$ μm) and 486 μm (85 % with $d > 200$ μm), with a high mean powder water content (7.3 g/100 g DM) and yield of 85 %. Some coalescence of aroma was observed in the reconstituted emulsion.

For SD140B with more emulsion sprayed (600 g, 70 min) with slower feed flow rate increase, the size distribution was larger up to 2,000 μm. A high fraction (36 %) less than 200 μm might have been due to an absence of agglomeration of small particles or the appearance of new particles by the breaking of some agglomerates. A lower mean powder water content of 5.5 g/100 g DM was observed, with a high powder yield of 93 %. The total aroma content was 16 % instead of 17 % for SD140A, with losses probably due to longer heating.

For the three agglomerated powders SD160A, SD140A and SD140B, the mean preservation of aroma molecules T and C was 80 %, and 60 % for the end note F. More specifically, preservation was better for shorter processes and losses were higher for fine particles. The sensory perception was strong for SD160A and with separation of the three notes for the others, SD140A being globally too strong and SD140B too short in persistence.

All the agglomerated powders had a low bulk density (0.40 $kg.m^{-3}$) and improved wettability (<7 min) due to their porous structure compared to the atomised powder. The wettability was not equal for all the particle size fractions, the large agglomerates being more rapidly wetted.

For atomised and agglomerated powders, the aspect of the surface was not smooth, with numerous asperities.

14.3.3 Coated Agglomerates

The three types of agglomerates, SD160A, SD140A and SD140B, were fluidised for coating by top spraying the emulsion MD/AG/M1/water (no modification of global composition). The sprayed quantity of emulsion was calculated to cover regularly the surface of particles with a coating layer of 10 μm. No significant size increase was expected.

For SD160A (346 g emulsion sprayed on 484 g powder for 32 min) the coating yield was 97 %. The powder size distribution gave one peak with d_{50} of 750 μm (disappearance of fine particles 20 % < 315 μm), showing some agglomeration occurring together with coating.

For powders SD140A and SD140B only agglomerates larger than 200 μm were coated. For SD140A, even with a slow pulverisation rate (<9 g/min on 400 g powder) it was necessary to stop to clean the nozzle. Total spraying time was 62 min. The resulting size was high (d_{50} = 1,090 μm) corresponding to agglomeration with some coating.

For SD140B two modes of coating were tested by spraying in the top mode (similar to SD140A) and in the bottom mode (particles circulation through internal Wurster tube). In both cases the initial mass load of agglomerates to coat was reduced to 150 g with 160 g emulsion sprayed (384 g on 400 g powder for 140A). In the top mode, the final powder had a narrow monomodal size distribution

($d_{50} = 572$ µm). In that case, the initial larger agglomerates had disappeared. For the bottom mode we collected a thin homogeneous powder ($d_{50} = 452$ µm), but large dry agglomerates were stuck on the inner tube.

The coated powders SD140A and SD140B had a regular shape and smooth surface, characteristic of coated particles. Wettability was less than 4 min and water content less than 6 g/100g DM. During these coating trials the total aroma content was decreased to 15 % and 10 % for the bottom mode. For the tested conditions, the coating process led to more aroma losses than during spray-drying and agglomeration, probably because of the prolonged heating of the dry thin aroma emulsion deposit on fluidised particles. To improve this behaviour, it would be necessary to modify the composition of the coating and the aroma.

14.4 Conclusions

In this study, aroma powders were produced using three processes – spray-drying, fluidised-bed agglomeration of solid particles and then coating. The carriers for aroma protection were maltodextrin and acacia gum. The studied aroma composition was composed of three types of molecules, with the proportions as in a real aroma formula.

The atomised powders (140–160 °C) and their agglomeration with an aroma emulsion gave a global positive answer with respect to the production of aroma powder (~20 % w/w) and global taste and intensity in chewing gum paste compared to the use of liquid aroma. The agglomeration process led to a large distribution of powder sizes in the studied conditions. Optimised conditions could be found for controlled aroma release. No coalescence of aroma oil was observed during the various processes (spraying, heating, drying), and the final total aroma content was greater than 17 % for all powders.

The coating process must be improved by considering a more stable composition for the coating layer, leading to a modified total composition. More generally, the operating conditions will have to be optimised depending on the required objectives: wall material composition, powder yield at different process scales, percentage of encapsulated aroma, and efficiency of encapsulation for specific release conditions.

References

Bhandari BR, Dumoulin ED, Richard HMJ, Noleau I, Lebert AM (1992) Flavor encapsulation by spray drying: application to citral and linalyl acetate. J Food Sci 57(1):217–221

Buffo RA, Probst K, Zehentbauer G, Luo Z, Reineccius GA (2002) Effects of agglomeration on the properties of spray-dried encapsulated flavours. Flavour Fragr J 17:292–299

Cho YH, Park J (2002) Characteristics of double-encapsulated flavor powder prepared by secondary fat coating process. Food Chem Toxicol 67:968–972

Clark J, Shen C (2004) Fast flavor release coating for confectionery. Patent WO/2004/077956

Finney J, Buffo R, Reineccius GA (2002) Effects of type atomization and processing temperatures on the physical properties and stability of spray-dried flavors. Food Eng Phys Prop 67:1108–1114

Fuchs M, Turchiuli C, Bohin M, Cuvelier ME, Ordonnaud C, Peyrat-Maillard MN, Dumoulin E (2006) Encapsulation of oil in powder using spray drying and fluidised bed agglomeration. J Food Eng 75:27–35

Gibbs BF, Kermasha S, Alli I, Mulligan CN (1999) Encapsulation in the food industry: a review. Int J Food Sci Nutr 50:213–224

Gouin S (2004) Microencapsulation: industrial appraisal of existing technologies and trends. Trends Food Sci Technol 15:330–347

Guignon B, Duquenoy A, Dumoulin E (2002) Fluid bed encapsulation of particles and practice. Drying Technol 20(2):419–447

Hetch JP, King CJ (2000) Influence of developing drop morphology on drying rates and retention of volatile substances. 1. Single-drop experiments. 2. Modeling. Ind Eng Chem Res 39:1756–1765 and 1766–1774

Jafari SM, Assadpoor E, He Y, Bhandari B (2008) Encapsulation efficiency of food flavours and oils during spray drying. Drying Technol 26:816–835

Jeon JK, Park CG, Lee MC, Kim YT (2008) Center-filled coated gum and a method of its preparation. Patent WO/2008/096945

Jimenez T, Turchiuli C, Dumoulin E (2006) Particles agglomeration in a conical fluidized bed in relation with air temperature profiles. Chem Eng Sci 61:5954–5961

King CJ (1995) Spray drying: retention of volatile compounds revisited. Drying Technol 13 (5–7):1221–1240

Krishnan S, Kshirsagar AC, Singhal RS (2005) The use of gum arabic and modified starch in the microencapsulation of a food flavoring agent. Carbohydr Polym 62:309–315

Lakkis JM (2007) Confectionery products as delivery systems for flavours, health and oral-care actives. In: Lakkis JM (ed) Encapsulation and controlled release technologies in food systems. Blackwell, Ames, Blackwell Pub, pp 171–200

Liu RT, Rushmore DF (1996) Process for making encapsulated sensory agents. US Patent 5,580,593

Madene A, Jacquot M, Scher J, Desobry S (2006) Flavour encapsulation and controlled release – a review. Int J Food Sci Technol 41(1):1–21

McNamee BF, O'Riordan ED, O'Sulivan M (2001) Effect of partial replacement of gum arabic with carbohydrates on its microencapsulation properties. J Agric Food Chem 49:3385–3388

Merrit CG, Wingerd WH, Keller DJ (1985) Encapsulated flavorant material, method for its preparation, and food and other composition incorporating same. US Patent 4,515,769

Pearl TT, Soper JC, Wampler DJ (1993) Heat-stable and fracturable spray-dried free-flowing flavour oil capsules, method of making and using in foods. Patent WO 93/19622

Reineccius GA (2004) The spray drying of food flavors. Drying Technol 22(6):1289–1324

Risch SJ, Reineccius GA (1995) Encapsulation and controlled release of food ingredients. In: Risch SJ, Reineccius GA (eds) ACS symposium series 590, American Chemical Society, Washington, DC, p 214

Rosenberg M, Kopelman IJ, Talmon Y (1990) Factors affecting retention in spray drying microencapsulation of volatile materials. J Agric Food Chem 38:1288–1294

Senoussi A, Dumoulin E, Berk Z (1995) Retention of diacetyl in milk during spray drying and storage. J Food Sci 60(5):894–897 and 905

Soottitantawat A, Bigeard H, Yoshii H, Furuta T, Ohgawara M, Linko P (2005) Influence of emulsion and powder size on the stability of encapsulated D-limonene by spray drying. Innov Food Sci Emerg Technol 6:107–114

Turchiuli C, Fuchs M, Bohin M, Cuvelier ME, Ordonnaud C, Peyrat-Maillard MN, Dumoulin E (2005) Oil encapsulation by spray drying and fluidised bed agglomeration. Innov Food Sci Emerg Technol 6:29–35

Vandamme T, Poncelet D, Subra-Paternault P (2007) Microencapsulation. Des sciences aux technologies. Ed. Tec & Doc, Lavoisier, p 355

Chapter 15
Advancements in Microbial Polysaccharide Research for Frozen Foods and Microencapsulation of Probiotics

Pavan Kumar Soma, Patrick D. Williams, BoKyung Moon, and Y. Martin Lo

15.1 Introduction

Nonstarch polysaccharides derive from several sources, including botanical, algal, microbial, and animal origins (Glicksman 1982; Williams and Phillips 2000; Hoefler 2004). Microbial polysaccharides are commonly used as thickeners (Mandala et al. 2004; Sikor et al. 2008), emulsifiers, and stabilizers (Garti and Leser 2001; Dickenson 2008; Mikkonen et al. 2009), as well as microencapsulating agents for flavor compounds and enzymes (Goud et al. 2005; Gouin 2004). In frozen foods, microbial polysaccharides help maintain or control textural stability such as ice recrystallization, minimize water distribution, and enhance viscosity (Freeland 2002; Goff 2006). A number of microbial polysaccharides, namely xanthan gum, gellan gum, and curdlan gum, when interacting with other biopolymers, such as locust bean gum, guar gum, microcrystalline cellulose (MCC), and carboxymethyl cellulose (CMC), can serve as texture modifiers that protect products from heat shock and deteriorations, function as bodying agents (Towle 1996; Marshall et al. 2003), help stabilize frozen foods (Sworn 2000), and reduce recrystallization (Goff 2006). For instance, when used as an emulsifier, stabilizer, or thickener, xanthan gum has the ability to retain its viscosity after being defrosted (Imeson 1997; Sworn 2000). Nevertheless, there are limited studies in the literature characterizing the textural effects of different gum combinations under freeze–thaw conditions.

On the other hand, an increasing number of studies have been directed toward microbial polysaccharides and their use for microencapsulation. Microencapsulation is the process by which a pure material or a mixture is coated or entrapped in

P.K. Soma • P.D. Williams • Y.M. Lo (✉)
Department of Nutrition and Food Science, University of Maryland, College Park, MD 20742, USA
e-mail: ymlo@umd.edu

B. Moon
Department of Food and Nutrition, Chung-Ang University, Anseoung 456-756, South Korea

S. Yanniotis et al. (eds.), *Advances in Food Process Engineering Research and Applications*, Food Engineering Series, DOI 10.1007/978-1-4614-7906-2_15,

another, which is called a wall material, membrane, carrier, or shell, to produce capsules in the micrometer to millimeter range known as microcapsules. The purpose of microencapsulation is to protect the functional core ingredient to be separated from the surrounding destructive environment until its release is desired (Anal and Stevens 2005). Natural examples of encapsulation include egg shells, plant seeds, and bacterial spores (Gibbs et al. 1999). Microencapsulation has been used for a variety of functional materials like cells, enzymes, and pharmaceutical drugs. Food ingredients like oleoresins, oxidation-sensitive vitamins, sweeteners, minerals, antioxidants, and proteins are often found encapsulated as well. The food industry mainly relies on microencapsulation technology for controlled-release applications, enhanced stability, flavor masking, protection against harsh conditions, and improved nutrition (Shahidi et al. 1993; Schrooyen et al. 2001). Release of the core material can be designed to be triggered by temperature, pH changes, osmotic shock, or a combination of factors. Wall material for capsule formation usually utilizes a combination of one or more sugars, proteins, natural and modified polysaccharides, lipids, and synthetic polymers. Different techniques used for encapsulation include spray drying, spray chilling or cooling, extrusion coating, fluidized bed coating, liposome entrapment, coacervation, inclusion complexation, and centrifugal extrusion (Gibbs et al. 1999). Chitosan, a natural biopolymer derived from chitin, has received significant research attention as an encapsulating agent due to its unique positive charges on side chains (Agullo et al. 2003; Shahidi 1999). Chitosan, with its ability to form polyionic complexes with negatively charged polysaccharides, has been used for microencapsulation of probiotics.

Probiotics are defined as live microorganisms (bacteria or yeasts) that, when ingested or locally applied in sufficient numbers, confer one or more specified demonstrated health benefits for the host (FAO/WHO 2001). Currently, the standard for any food sold with health claims from the addition of probiotics follows the FAO/WHO recommendation that it must contain per gram at least 10^6–10^7 cfu of viable probiotic bacteria. It is well recognized that the ability of probiotic microorganisms to survive and multiply in the host strongly influences their probiotic benefits. The bacteria need to reach the intestine in large numbers in order to provide health benefits. For instance, probiotic cells have to pass through the strong acidity in the stomach (pH 1–2) and the high concentration of bile for a long period of time (Dave and Shah 1997). Besides the adverse conditions in the gastrointestinal (GI) tract, food processing parameters such as dissolved oxygen content, temperature, and moisture content could also greatly impact the viability of probiotics. Therefore, in the last decade efforts have been devoted to developing microencapsulation techniques to improve the viability of probiotics during processing steps and in storage and against harsh conditions after ingestion. There is a pressing need for biodegradable natural polymers that are resistant to degradation in the upper GI tract and capable of releasing probiotic cells in the intestine.

The ideal polysaccharide for frozen food and microencapsulation applications needs to (1) have a gel or gel-like properties, either through junction zone formation or polyionic complexes; (2) have the ability to withstand harsh conditions of processing or the GI tract; and (3) retain the sensory qualities of the given food. The formation of junction zones or ionic bonds between two materials that produce the gel or gel-like properties give polysaccharides the ability to control ice recrystallization for frozen foods and provide the strength needed to encapsulate a material. Proper selection of polysaccharides is necessary because they differ in their ability to resist pH, temperature, electrolyte concentration, enzymatic degradation, and storage conditions (Williams and Phillips 2000). However, while these abilities are known in some polysaccharides, it is not uncommon to encounter certain limitations in order not to negatively impact the product's sensory quality. For example, polysaccharides must be added in an effective amount that provides cryoprotection and inhibition of water recrystallization during freezing and thawing conditions but that should not affect mouthfeel or result in significant syneresis in frozen foods. It is also desirable that, when used to protect and deliver probiotics, polysaccharide capsules should be as small as possible in order to be applicable in a variety of foods without being noticed. This review will provide an detailed overview of several nonstarch polysaccharides that are utilized in both frozen foods and microencapsulation of probiotics, while providing some insight into the reasons they are used in these particular sectors of the food industry. The key techniques capable of characterizing the conformation, structure, and textural function of these polysaccharides are also highlighted.

15.2 Nonstarch Polysaccharides

15.2.1 Seed-Derived Gums

15.2.1.1 Guar Gum

Native to India and Pakistan, guar gum is obtained from the guar plant, *Cyamposis tetragonolobus* of the family Leguminosae (Glicksman 1986; Wielinga et al. 2000). Currently, the plant is commercially grown annually in Texas, Oklahoma, Arizona, and the southern hemisphere in response to fluctuations in the availability of guar bean gum from foreign sources (Glicksman 1986; Wielinga et al. 2000). Made up of nearly all galactomannans, the nonionic structure of guar gum consists of a backbone with a (1 → 4)-linked β-D6 mannopyranosyl unit with a side chain unit consisting of (1 → 6)-linked α-D-galactopyranosyl at a ratio of 1.8:1, respectively (Fig. 15.1). The large amount of galactose substitutions prevents strong cohesion of the main backbone, and therefore extensive crystalline regions cannot be formed allowing hydration at and above room temperature (Wielinga et al. 2000). Hydration of guar gum can be correlated to inter- and intramolecular hydrogen bonding through the unsubstituted regions of the backbone (Goycoolea et al. 1995; Sandolo et al. 2008) and thus makes guar gum soluble in cold, highly agitated water while

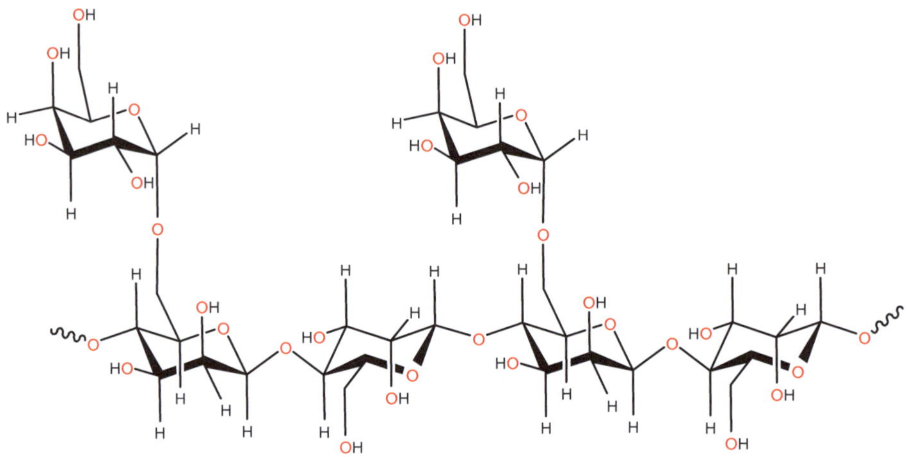

Fig. 15.1 Structure of guar gum

exhibiting pseudoplastic behavior. It is used as a formulation aid, stabilizer, firming agent, and thickener (Towle 1996) with a classification of generally recognized as safe (GRAS) by the U.S. Food and Drug Administration (FDA).

Wielinga et al. (2000) first reported that guar gum can help enhance the freeze–thaw stability of solutions. It was found that guar gum can be stable over two freeze–thaw cycles in aqueous, NaCl, $CaCl_2$, citric acid, acetic acid, and milk serum solutions at a concentration of 1.0 %. Another study found that when added to curdlan gum in the same amount, totaling 2.0 % (w/v), syneresis was undetectable over five freeze–thaw cycles (Williams et al. 2009). Galactomannans such as guar gum and locust bean gum are among the most frequently used ingredients for the stabilization of ice cream (Marshall et al. 2006). Among fresh and frozen vegetable puree, guar gum was found to be effective in reducing drip loss in potatoes, carrots, and turnips (Downey 2002). Moreover, when added to frozen dough, guar gum has shown improvements in bread texture, volume, and crumb structure (Ribotta et al. 2004).

The ability of guar gum to form a polymeric matrix that is resistant to degradation in the upper GI tract and susceptible to enzymatic action of colonic bacteria has led to extensive studies for its colon-specific drug-release properties by various researchers (Krishnaiah et al. 2001, 2002; Gliko-Kabir et al. 2000; Wong et al. 1997). These studies utilized varying compositions of guar gum (20–98 %) for the preparation of compression-based tablets. Tablets with more than 40 % guar gum resulted in limited release of drug even after 24 h dissolution in the presence of rat caecal contents; thus, 20–30 % was recommended for colon-specific release (Krishnaiah et al. 2001). Despite its ability to retain bioactive material in the harsh conditions of stomach gastric and intestinal fluids, very few studies have been reported on guar gum for intestinal delivery of probiotic bacteria. However, in a recent study, Ding and Shah (2009) reported microencapsulation of various

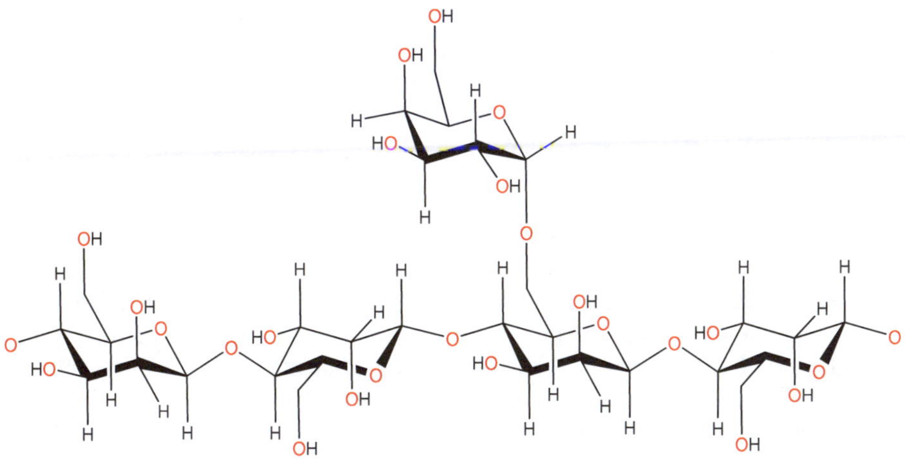

Fig. 15.2 Structure of locust bean gum

probiotic strains using guar gum. Encapsulation was achieved by an emulsion technique using $CaCl_2$ followed by 1 h of exposure to gastric conditions at pH 2 where viability decreased from 10^{10} to $<10^7$. The viability was decreased to <5 log cfu/ml after 8 h incubation in 3 % tauchloric acid. Permeability of encapsulated fluorescent dye was also least retained by guar gum compared to other gums, which could be due to the large pore size of guar gum, resulting in poor protection of probiotic bacteria.

15.2.1.2 Locust Bean Gum

Locust bean gum (LBG), also known as carob bean or St. John's gum, is obtained from the seeds of the carob tree, *Ceratonia siliqua*, grown in Mediterranean countries (Glicksman 1986; Wielinga et al. 2000). The galactomannan structure of LBG, similar to guar gum, consists of a backbone with (1 → 4)-linked β-D-mannopyranosyl units and a side chain consisting of a single (1 → 6)-linked α-D-galactopyranosyl unit with a ratio of 3.9:1 (Fig. 15.2). Unlike the structure of guar gum, LBG side chains are very unevenly substituted with sections of the backbone concentrated with substitutions and sections with no substitutions. Slightly soluble in water at room temperature, LBG must be heated to approximately 60–85 °C to achieve complete hydration. LBG solution exhibits pseudoplasticity (shear-thinning behavior), which shows reduced viscosity at increased shear rates, and can form a weak gel network with concentrations as low as 0.5 % (Dea et al. 1977). Classified as GRAS by the FDA, LBG is used for its stabilizing, thickening, and fat-replacing properties.

Unlike guar gum, there was little stability in a system containing LBG. For instance, LBG forms a gel after the first freeze–thaw cycle in the presence of NaCl, acetic acid, and milk serum, and syneresis occurs in both aqueous and $CaCl_2$ solutions (Wielinga et al. 2000). This could be attributed to the unevenly substituted sections of the backbone that create junction zones and thus a gel or gel-like network. It has been shown that freeze–thaw cycles and the rate of freezing or thawing affect its gel strength (Tanaka et al. 1998; Lozinsky et al. 2000, 2001; Zeira and Nussinovitch 2004). Therefore, LBG alone could not provide freeze–thaw stability over multiple cycles.

Several studies with LBG and xanthan gum have been performed because of their unique gelling behavior that is not observed with the other galactomannan, guar gum (Rocks 1971; Goycoolea et al. 1995; Wang et al. 2002b). Conformational studies via parameters such as intrinsic viscosity that reflects molecular occupancy in highly dilute solutions have been used to better understand this interaction. It was reported that the intrinsic viscosity exhibited lower values than weighted averages, suggesting that there is some flexibility in the structure of xanthan that may allow LBG to bond to the backbone (Wang et al. 2002b; Higiro et al. 2006). Even so, LBG is best known for its use in ice cream to prevent iciness during heat shock (Glicksman 1986; Towle 1996; Marshall et al. 2003) and many times combined with guar and carrageenan. However, few or no studies have been conducted to examine any potential synergistic effects provided by LBG and xanthan under freeze–thaw abuse. On the other hand, frozen dough containing LBG was found to enhance gluten quality and reduce proof times while increasing specific loaf volume with improved external and internal bread characteristics (Sharadanant and Khan 2003a, b; Mandala et al. 2008).

Since LBG forms very weak gels, it is not used for encapsulating probiotic bacteria by itself. However, strong gels due to a synergistic effect result when LBG is mixed with other gums such as k-carrageenan and xanthan. A recent study investigating the ability of an LBG matrix to protect probiotics reported that the viability of probiotic strains rapidly decreased from 10 log cfu/ml to 4.84 log CFU/ml within 2 h in simulated gastric juice (SGJ) at pH 2 and to <5 log cfu/ml after 8 h incubation in 3 % tauchloric acid, both less than the requirement of probiotic bacteria to confer their health benefits (Ding and Shah 2009).

15.2.2 Gums Derived from Seaweed Extracts

15.2.2.1 Alginate

The extraction and processing of brown algae from a host of species, including *Laminaria hyperborean*, *Macrocystis pyrifera*, *L. giditata*, *Ascophyllum nodosum*, *L. japonica*, *Eclonia maxima*, and others can produce the intermediate product alginic acid. After the process of neutralization with sodium carbonate or sodium hydroxide, alginic acid forms the more stable water-soluble product called sodium alginate (Onsoyen 1997; Draget 2000). The linear structure of alginate consists of either the homopolymeric blocks β-D-mannuric acid (M) and α-L-guluronic acid

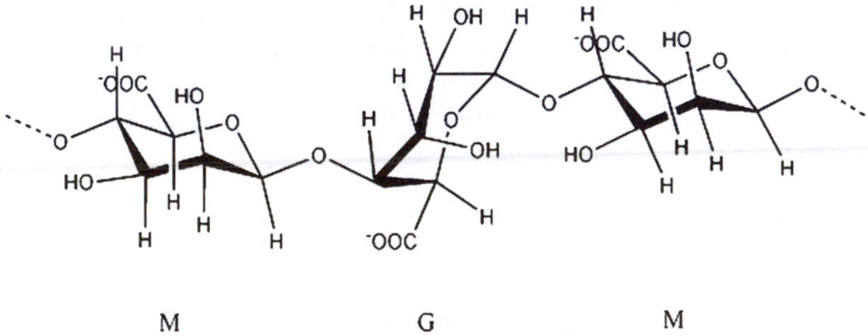

Fig. 15.3 Structure of alginate gum

(G) or the heteropolymeric blocks of alternating M and G linked by (1 → 4)-glycosidic linkages, making it a polyanionic hydrocolloid (Fig. 15.3). The amount of homo- and heteropolymeric blocks present in alginate is dependent upon the sources previously mentioned. Thermoirreversible gels can be formed by controlling the amount of calcium and acidity. On the other hand, low-acid solutions below pH 4.0 form thermoreversible gels when combined with high-methoxyl pectin. There are a variety of different viscosity grades of alginates due to the amount of G-blocks available for Ca^{2+} to form an "egg-box" structure at junction zones creating a strong gel. Due to their ability to form strong gels, they are used to control the shape of foods such as onion rings, pimiento and anchovy olive fillings, apple pieces for pie fillings, cocktail berries, meat chunks for pet food, shrimplike fish products, and fish patties (Onsoyen 1997).

Although temperature does not hinder gelation of sodium alginate, it affects the final gel properties; however, once it is a gel, it is heat and freeze–thaw stable (Onsoyen 1997; Draget 2000). Sodium alginate has been seen to improve frozen dough stability along with whey by increasing the specific loaf volume (Shon et al. 2009). Following each of seven freeze–thaw cycles, alginate increased dough development and water adsorption and reduced syneresis; however, it yielded different texture profiles with a firmer dough compared to control samples (Lee et al. 2008).

Alginate is the most common encapsulating agent used by researchers for probiotic bacteria mainly due to its ease of formation of capsules and release of probiotic bacteria using chelation in phosphate buffer. Alginate capsules are made using different techniques like extrusion and emulsification (Doleyres et al. 2004). Alginate capsules made with 1.8 % alginate solution provided protection to probiotic bacteria and improved viability up to 3 log cfu/ml for 2 h incubation at pH 1.2 gastric solution (Ross et al. 2008). Ding and Shah (2008) used an extra layer of poly-L-lysine with alginate microcapsules. This technique improved viability by 1 log cfu/ml compared to alginate capsules in gastric conditions; however, no improvement was observed in intestinal solution, suggesting an emulsion-breaking capacity of bile salts. Furthermore, poly-L-lysine-coated capsules retained double the amount of water-soluble fluorescent dye as compared to alginate capsules over a period of 6 weeks of storage. In another study, encapsulation in alginate capsules with diameters less than 100 μm did not improve survival of the acid-sensitive *Bifidobacteria* exposed to SGJ at pH 2. However, survival during refrigerated storage in milk was significantly improved for *Bifidobacterium longum* Bb-46 (Hansen et al. 2002).

15.2.2.2 Carrageenan

Three varieties of carrageenan, kappa (κ), lambda (λ), and iota (ι), can be extracted from red seaweeds. They are primarily extracted from the *Gigartina* species and *Chondrus crispus*, which produce kappa and lambda types, and *Eucheuma cottonii* and *spinosum* species, which produce kappa and iota types, respectively (Imeson 2000; Hoefler 2004). The general structure of carrageenan contains repeating galactose units and 3,6-anhydrogalactose, both with 15–40 % (w/w) ester sulfate content and a side chain consisting of alternating α-(1 → 3)- and β-(1 → 4)-glycosidic linkages. The three types of carrageenan exist not singly but as a combination of two types and are available with one predominating type or molecules containing structural components of more than one type. Each type of carrageenan has a unique set of characteristics, including gel strength, viscosity, temperature stability, synergism, and solubility. The solubility of each type of carrageenan depends on the number of sulfate groups, which increases water solubility, compared to anhydro bridges, which is hydrophobic. Having the least water solubility, κ-carrageenan has one sulfate group for every two galactose units and one anhydro bridge, ι-carrageenan has two sulfate groups for every two galactose units along with one anhydro bridge, and, with the highest solubility, λ-carrageenan has three sulfate groups for every two galactose units and no anhydro bridges (Imeson 2000; Hoefler 2004) (Fig. 15.4).

There are distinct differences between each type of carrageenan, but all types are soluble at high temperatures and stable above pH 4.5. Though the least soluble, κ-carrageenan provides the strongest, yet brittle, gel that allows for some syneresis. The ι-carrageenan type is more elastic and is freeze–thaw stable, allowing for no syneresis, and λ-carrageenan thickens without gelling. Moreover, due to syneresis, κ-carrageenan has poor freeze–thaw stability; however, with the right combination of κ- and ι-carrageenan, intermediate freeze–thaw stability can be acquired along with a range of gel textures without syneresis (Imeson 2000). In ice cream and other frozen milk-based products, a small amount of κ-carrageenan (0.01–0.3 %) is used to prevent phase separation of casein (Fox 1997; Marshall et al. 2003). κ-carrageenan alone or in combination with whey exhibited increased specific volume in bread after the dough was frozen during storage (Sharadanant and Khan 2003b; Shon et al. 2009).

Although carrageenan forms gels by itself, gel properties are influenced greatly by the addition of ions and other gums due to stabilization effects and synergistic effects, respectively. Potassium ions have been found to be most effective in increasing the gel strength and has been used for microencapsulation of probiotics.

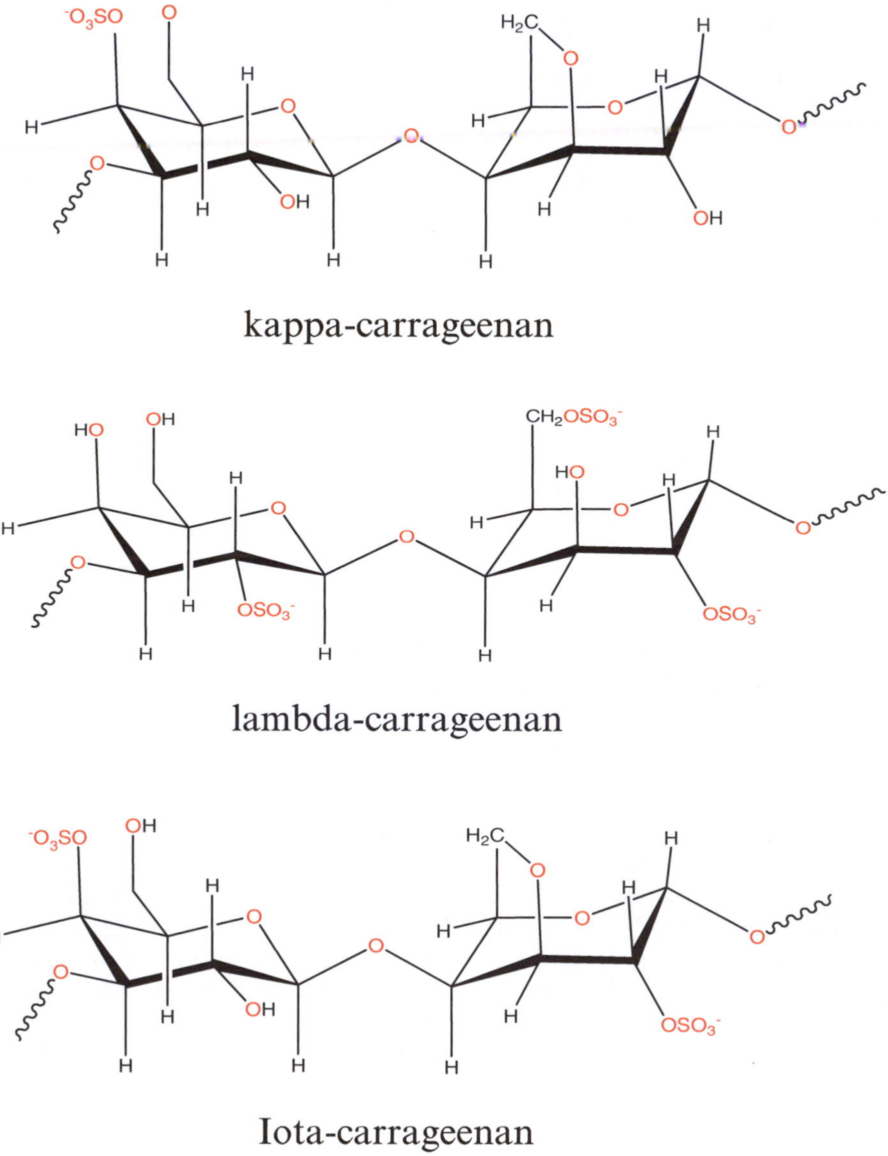

Fig. 15.4 Structures of kappa-, iota-, and lambda-carrageenan

In a recent study, the viability of *Lactobacillus acidophilus* cells were significantly improved when encapsulated with κ-carrageenan during fermentation of tomato juice and storage at 4 °C for 10 weeks as compared to free cells (Tsen et al. 2008). In another study, encapsulation with κ-carrageenan showed increased viability of *Bifidobacteria* in yogurt during storage up to 30 days compared to nonencapsulated

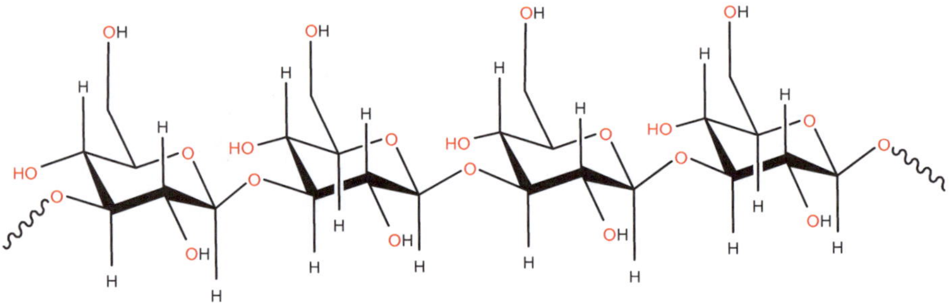

Fig. 15.5 Structure of curdlan

bacteria (Adhikari et al. 2000). The viability of ten different probiotic strains encapsulated in κ-carrageenan was significantly improved in simulated gastric and intestinal solutions (Ding and Shah 2009). κ-carrageenan, in combination with LBG, was able to maintain the viability of entrapped lactic acid bacteria in NaCl, glycerol solutions at 4 °C for at least 11 days before reaching 10^5 cfu/ml (Audet et al. 1991).

15.2.3 Microbial Exopolysaccharides

15.2.3.1 Curdlan

Alcaligenes faecalis var. *myxogenes*, now classified as *Agrobacterium* biovar. 1, produces the neutral, linear polysaccharide curdlan. Curdlan is the third microbial extracellular polysaccharide, following xanthan gum and gellan gum, to be approved for food uses in the USA (FDA 1996). Repeating units of (1 → 3)-β-glucan (Fig. 15.5) can form a low-set thermoreversible gel when heated between ~55 and 80 °C or a high-set, triple helix, and thermoirreversible gel when heated above ~80 °C and then cooled (Harada et al. 1994; Hirashima et al. 1997; Nakao 1997; McIntosh et al. 2005; Gagnon and Lafleur 2007). The gelation mechanism of the low- and high-set gels continues to be studied (Tada et al. 1999; Hatakeyama et al. 2006; Gagnon and Lafleur 2007). Curdlan lacks solubility in water, alcohols, and most organic solvents but shows solubility in alkaline solutions (Pederson 1979; McIntosh 2005). It is known to be able to hold moisture in processed meat and flour products. It is stable over a pH range of 2–12 and freeze–thaw conditions, and except as an aqueous solution there is syneresis (Nakao 1997).

Composed of a three-dimensional structure stabilized by cross links that connect junction zones between molecules (Harada et al. 1966; Saito et al. 1978; Marchessault and Deslandes 1978; Nakao et al. 1991; Jezequal 1998), curdlan gum was once the focus of extensive research efforts to expand its applications to improve the texture of various food products (Kanzawa et al. 1987; Nakao et al. 1991; Sanderson 1996) as well as the delivery of medicinal ingredients (Harada and Harada 1996; Na et al. 2000). To date, unfortunately, despite curdlan's ability to form viscous aqueous suspensions with shear-thinning flow behavior (Hirashima et al. 1997; Lopes da Silva et al. 1998; Funami et al. 1999), the applicability of curdlan in the US market remains scarce and limited, due mainly to its less profound viscosity than xanthan gum in solution and inferior gel-formation capacity compared with gellan (Sadar 2004).

Nevertheless, curdlan continues to be used as a texture modifier in Chinese and Japanese noodles and surimi-based products (Nakao et al. 1991; Nishinari 2000) as well as in processed meats such as pork, fried battered chicken, hamburger patties, and meatballs to yield juicier and softer products (Nakao 1997; Hsu and Chung 2000). Furthermore, curdlan is also capable of forming tasteless, odorless, and colorless hydrogel complexes with other polysaccharides (Lo and Ramsden 2000; Lee et al. 2002). This unique gelling mechanism could be used to increase retention or absorption of moisture and other ingredients (Lo et al. 2003) while withstanding the temperature extremes of freezing and retorting processes (Wielinga and Maehall 2000; McIntosh et al. 2005).

Though the gel strength of curdlan was consistent after one freeze–thaw cycle compared to carrageenan, aga-agar, and konjac, syneresis occurred (Nakao 1997). Xanthan in combination with curdlan has been demonstrated by our research group to reduce syneresis to a point where it is not detected and yields stable rheological and physical properties over freeze–thaw cycles when compared to curdlan combined with guar, locust bean, or k-carrageenan gum (Williams et al. 2009). The study measured viscosity, heat stability, modulus, gel strength, adhesiveness, and syneresis over five freeze–thaw cycles and found a stable system throughout each experiment, in addition to showing good water-holding capacity.

15.2.3.2 Xanthan Gum

Produced by the fermentation of *Xanthomonas campestris*, xanthan gum is an anionic microbial polysaccharide with a disaccharide backbone and a trisaccharide side chain. The backbone consists of two repeating $(1 \rightarrow 4)$-β-D-glucose units, the same as cellulose, and located on C3 of every other glucose; α-D-mannose, β-D-glucuronic acid, and β-D-mannose trisaccharide are attached (Fig. 15.6). Various terminal mannose residues are pyruvated depending on the *X. campestris* strain. Xanthan is soluble in water with the ability to hydrate in cold water, exhibiting pseudoplastic behavior. The viscosity of xanthan is stable over a range of pH values, salt concentrations, temperatures, and enzymatic breakdown (Whitcomb and Macosko 1978; Sworn 2000). Thickening, stabilizing, and emulsifying are some of the well-recognized characteristics of xanthan gum. As a thickening agent, low shear rates show stiff, aggregated, highly ordered molecules, and as shear is increased, the stiff molecules separate and align in the direction of the shear force. As a consequence, pseudoplasticity is exhibited. Because of its

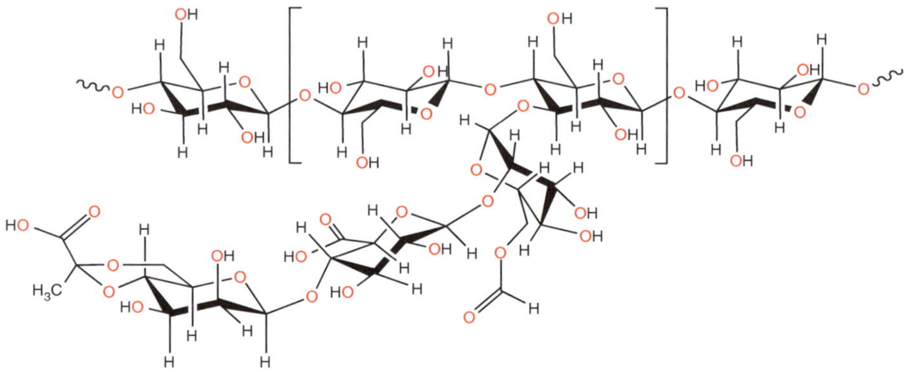

Fig. 15.6 Xanthan gum

resilience to pH and temperature fluctuations, xanthan gum has been extensively studied for its own properties and synergistic effects with other gums, starches, and food ingredients. The interactions between xanthan and galactomannans have been examined to identify their synergy, which alters viscous and gelation properties (Rocks 1971; Sworn 2000; Pai et al. 2002; Wang et al. 2002a, b; Richter et al. 2004; Higiro 2006). Synergy of xanthan has also been explored with xyloglucan (Kim et al. 2006) and glucomannan (Paradossi et al. 2002), and its positive effects with gum arabic and LBG in emulsions (Makri 2006) have been investigated, as has its effect on starch when combined with other gums (Chaisawang and Suphantharika 2006).

While 1.0 % xanthan solution produces gel-like consistency, as little as 0.1 % xanthan gum can increase viscosity and shows similar rheological properties. Giannouli and Morris (2003) suggested that xanthan can form stronger, more cohesive networks when frozen and thawed, similar to how Ca^{2+} enhances xanthan's weak gel network. Furthermore, the stability of frozen entrees and sauces was improved by xanthan gum with respect to syneresis and viscosity control over freeze–thaw cycles (Sworn 2000). The presence of xanthan gum in starch gels has significantly increased its freeze–thaw stability, according to Lo and Ramsden (2000), indicating an excellent compatibility with major food components. Xanthan gum has been shown to increase the specific volume of bread after 1 week of frozen dough storage and heating via a microwave oven (Mandala 2005), reducing crust deterioration when compared with LBG and guar (Mandala et al. 2008). Improved dough quality by the addition of xanthan could be attributed to the reduction of free water as indicated by a reduction in the fusion enthalpy (Matuda et al. 2008).

Xanthan gum was excellent in protecting probiotic bacteria from the acidic environment of gastric juice at pH 2 and 3 % taurocholic acid (Ding and Shah 2009). Xanthan has been used in combination with other polymers like gellan for immobilizing cells (Wenrong and Griffiths 2000), alginate for encapsulating urease

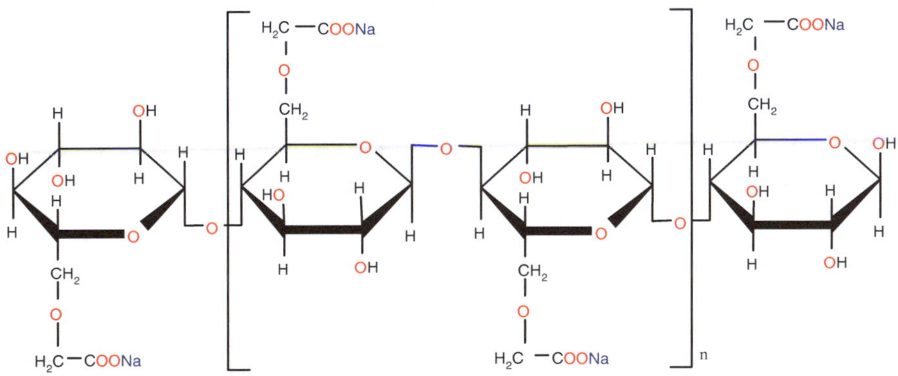

Fig. 15.7 Structure of carboxymethyl cellulose

enzyme (Elcin 1995), and chitosan for immobilizing xylanase (Dumitriu and Chornet 1997). Also, when combining xanthan with chitosan, our research group has shown the combination to be effective in protecting probiotic bacteria against gastric conditions at pH 2 followed by release in the presence of simulated intestinal conditions, making it an effective carrier for probiotic bacteria for colon-specific release (Soma and Lo 2009).

15.2.4 *Cellulose Derivatives*

15.2.4.1 Sodium Carboxymethyl Cellulose

Sodium carboxymethyl cellulose (CMC), or cellulose gum (Fig. 15.7), is produced by treating cellulose with sodium hydroxide, reacting it with sodium monochloroacetate, and, finally, washing it (Granz 1977). CMC is water soluble and has the ability to increase viscosity to 5 Pa s in a 1 % aqueous solution, which is considerably lower than that of xanthan gum. CMC solutions exhibit shear thinning (Murray 2000) and sometimes Newtonian behavior or shear thickening (Yaşar et al. 2007). It is currently used as a thickener or bodying agent for instant products, sauces, dressings, and soft drinks; however, high-viscosity forms of CMC can cause a "gummy" mouthfeel (Murray 2000). Similar to carrageenan and locust bean gum, CMC is used to stabilize frozen products like ice cream by inhibiting ice crystal formation (Murray 2000; Marshall 2003). Similar to LBG, CMC was found to reduce proof times and be more resistant to extension in previously frozen dough while improving the external and internal characteristics of bread, but it also increased a desired yellow color in the crust compared to control samples (Sharadanant and Khan 2003a, b).

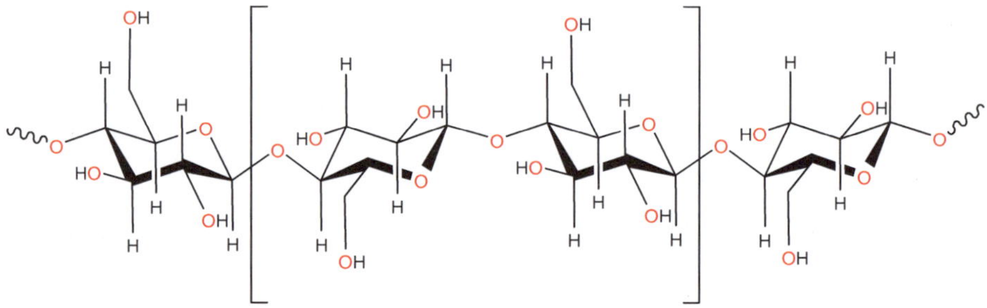

Fig. 15.8 Structure of cellulose

15.2.4.2 Microcrystalline Cellulose

Microcrystalline cellulose (MCC) is a linear structure consisting of anhydroglucose units linked by a (1 → 4) β-glycosidic bond. MCC is water insoluble and thus needs sufficient shear or a copolymer network to properly disperse in water (Iijima and Takeo 2000). Temperature has little to no effect on the functionality or apparent viscosity of networks and remains stable during high-temperature processing such as baking or retort. Not only can MCC be applied to high-temperature processes, but it can also be used as a texture modifier for low-fat desserts (Marshall et al. 2003) and prevents the growth of ice crystals in frozen foods during freeze–thaw cycles (Champion et al. 1982). Synergy is important to create and improve texture properties; however, this is more difficult to achieve with other materials or gums due to its highly linear structure. In many cases, coprocessing under specific conditions is required to form hydrogen bonds between MCC and other materials (Glicksman 1986) (Fig. 15.8).

15.2.5 Chitosan

Chitosan is the deactelyated form of chitin, the most abundant natural biopolymer after cellulose. Chitin is a copolymer of glucosamine and N-acetyl-d-glucosamine linked together by β(1,4) glycosidic bonds (Fig. 15.2). Chitin is the major structural component of the exoskeleton of invertebrates and the cell walls of fungi (Shahidi et al. 1999). Chitosan is a primary aliphatic amine with a pKa of 6.3 and can be protonated by selected acids. It is a biocompatible polymer and does not result in adverse reactions when in contact with human cells. It has seen many applications in the food industry, including the edible film industry, owing to the antifungal properties of chitosan (Ghaouth et al. 1992), water purification as a chelation ion exchange polymer (Jeuniaux 1986), and the clarification and deacidification of fruit juices (Soto-Perlata 1996). Chitosan, being a cationic polymer, has pharmaceutical applications, for example, as a nasal drug delivery agent due to its

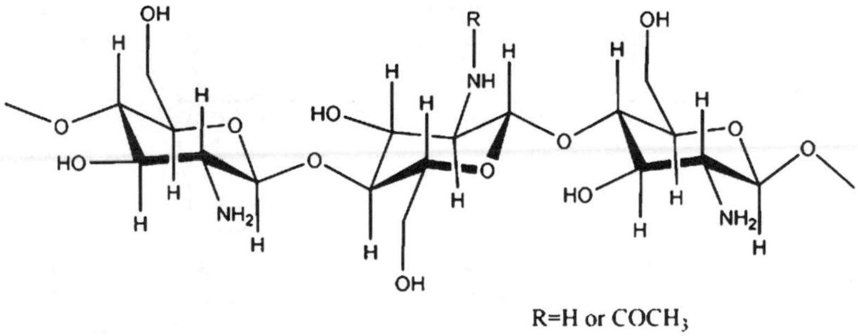

Fig. 15.9 Structure of chitosan

bioadhesive nature. It also interferes with the metabolic process of cholesterol (and other neutral lipids) by binding them with hydrophobic bonds (Kumar et al. 2004).

Chitosan received GRAS status by the US FDA in 1983 for use as an animal feed component. In 1992, the use of chitosan for purification of water was approved by the US Environment Protection Agency (EPA) up to a maximum concentration of 10 mg/l. Japan's health department approved the use of chitin and its derivatives as functional food ingredients (Shahidi and Abuzaytoun 2005) (Fig. 15.9).

As an edible film, chitosan has proven effective in coating frozen foods to reduce several negative effects of frozen storage. After 8 months of frozen storage, 1.0 % (w/v) chitosan significantly inhibited lipid oxidation compared to a control sample of salmon fillet (Sathivel et al. 2007). It has also been shown that as an edible coating chitosan reduces drip loss and maintains the textural quality of frozen strawberries after thawing (Han et al. 2004).

Chitosan-coated alginate beads provided a viability of at least 1 log cfu/ml greater than free probiotic bacteria in yogurts made with ultra-high-temperature treated and conventionally treated milk (Wunwisa et al. 2006). In another study, three different chitosans varying in molecular weight were used to coat alginate capsules. It was found that the viability of the probiotic bacteria *L. bulgaricus* KFRI 673 increased with increasing molecular weight of chitosan after sequential incubation in simulated gastric and intestinal fluids. However, high molecular weight chitosan-coated alginate capsules resulted in nonuniform-shaped capsules with partial collapse in the center compared to spherically shaped capsules with low molecular weight chitosan (Lee et al. 2004).

15.3 Characterization Tools

Several tools have been reported in the literature that are capable of characterizing different gums to aid in understanding any positive effects that could exist when more than one polymer is present in a system. Rheological properties, including viscosity and dynamic modulus, are often used to characterize the behavior of gum,

Table 15.1 Investigative methodology of hydrocolloidal systems

Investigative methodology	Test properties	Reference
Rheological	Flow characteristics, elasticity classification	Whitcomb (1978), Steffe (1992), Williams and Langdon (1996), Tada et al. (1999), Wang et al. (2002a, b), Giannouli et al. (2003), Ikeda et al. (2004), Jin et al. (2006), Kim et al. (2006)
Differential scanning calorimetry	Thermal analysis, gelation mechanisms explored	Annable et al. (1994), Williams and Langdon (1996), Ikeda et al. (2004), Hatakeyama et al. (2006), Jin et al. (2006), Kim et al. (2006)
Nuclear magnetic resonance	Molecular structure and structural changes	Vittadini et al. (2002), Richter et al. (2005), Kim et al. (2006)
Electron spin resonance	Molecular structure and structural changes	Shimada et al. (1993), Annable et al. (1994), Williams and Langdon (1996)
Dynamic light scattering	Particle size	Coviello and Burchard (1992), Rodd et al. (2001), Richter et al. (2004, 2005)
Atomic force microscopy	Surface morphology	Ikeda et al. (2004), Jin et al. (2006)
Scanning electron microscopy	Surface morphology	Kanzawa et al. (1989), Sanchez et al. (2000)

starch, and gum/starch combinations (Whitcomb and Macosko 1978; Williams and Langdon 1996; Wang et al. 2002a, b; Giannouli et al. 2003; Ikeda et al. 2004; Higiro et al. 2006; Kim et al. 2006). Along with rheological properties, certain physical properties such as gel strength and adhesiveness could also provide a broader view of the behavior and stability of the gum system.

To better describe the conformational and structural characteristics of nonstarch polysaccharides, methods such as magnetic resonance, light scattering, and microscopy could be useful (Table 15.1). Time correlation functions (TCFs) via dynamic light scattering (DLS) can provide clues to gelation behavior, as demonstrated by investigations with xanthan and LBG (Richter et al. 2004, 2005) or xanthan in the presence of metal ions (Rodd et al. 2001). Differential scanning calorimetry (DSC) can also provide gelation information through thermal analysis. DSC measures changes in the difference in heat flow rate to a sample and to a reference sample while they are subjected to a controlled temperature program. It measures a heat flow rate difference due to a temperature change, where a spike in heat flow provides the glass transition, crystallization, and melting temperatures.

Surface morphology can be investigated using scanning electron microscopy (SEM) to elucidate the network properties and microstructure of combination. It has been successfully employed for investigating skim milk with xanthan and locust bean gum (Sanchez 2000), tapioca starch modified by guar and xanthan gum (Chaisawang 2006), effects of xanthan, LBG, and guar on sucrose (Fernández et al. 2007), chitosan and methylcellulose (Pinotti et al. 2007), k-carrageenan and locust bean gum (Dunstan et al. 2001), calcium-induced k-carrageenan (MacArtain et al. 2003), and carrageenan/*O. ficus indica* (Medina-Torres et al. 2006). In several of these studies, a cryo-chamber was in place with a scanning electron microscope that froze the sample at around −160 °C, followed by sublimation; then the sample could be coated and examined through SEM. Morphology characteristics can also be revealed with atomic force microscopy (AFM), which uses a technique called single-molecule force spectroscopy. A fractal analysis can be performed based on the image produced by AFM; however, it requires proper adsorption of sample materials onto a supporting surface; thus, some affinity for a specific surface such as mica is needed before the samples can be examined.

To delve further into the nature and adjustments in the molecular structure of polysaccharides, nuclear magnetic resonance (NMR) has been used for molecular analysis in carbohydrates (Rao et al. 1998), including xanthan and xyloglucan (Kim et al. 2006), xanthan and locust bean gum (Richter et al. 2005), or fructose with locust bean gum (Martin et al. 1999). In these studies the spin-spin relaxation time, T_2, was monitored to identify changes in structure that could be attributed to the combining two substances using ^{1}H NMR and could also be related to bound water and water mobility (Hatakeyama and Hatakeyama 1992; Hofmann and Hatakeyama 1994; Ramakrishnan et al. 2004; Richter 2005). Similar to NMR, electron magnetic resonance can be used to measure structural adjustments using the free radicals inherent in the substance being investigated or the free radicals that can be attached to a substance (Takigami et al. 1993).

15.4 Conclusions

It is apparent that microbial polysaccharides play an important role in the texture, quality, and functionality of many foods. They are used in both frozen foods and microencapsulation due possibly to their gel or gel-like forming ability and capability of withstanding a wide range of processing conditions and surviving in the GI tract. Frozen foods utilize them for viscosity enhancement, just like in conventional foods, but they also serve the purpose of cryoprotection, helping to maintain the quality of frozen foods. There is a large body of studies on these polysaccharides along with many in combination with the others; however, few examine their stability over low-temperature abuse. These studies could provide information about the gums and gum combinations and their behavior as, has been done with xanthan and curdlan (Williams et al. 2009). The strength of the ionic complexation or junction zones and enzyme resistance of polysaccharides makes them ideal for carrying and delivering probiotics through the GI tract.

Though not perfectly meeting each criterion for utilization in frozen foods and the microencapsulation of probiotics, alginate, xanthan, guar, and chitosan are well suited for both instances. They form gel or gel-like materials and have the ability to withstand the harsh conditions of processing or the GI tract; however, few studies have examined their sensory qualities in frozen foods and foods containing probiotics. Guar gum could also be an excellent candidate, yet it lacks the ability to provide a firmer gel that could allow better protection of bacteria. Curdlan may be a good candidate for further microencapsulation experiments with its excellent freeze–thaw stability, gelling properties, and resistance to harsh conditions. Though both locust bean gum and carrageenan gums are used in frozen foods, they lack freeze–thaw stability and require further studies of ionic complexations or combining them with other materials to improve their applicability. Carrageenan can provide good protection for probiotics, but not LBG, thus making it the least suitable for both applications. CMC and MCC, like curdlan, have not been extensively studied for microencapsulation, most likely due to their inability to form gels.

Several tools can be used to help identify materials that are suitable for use in both frozen foods and microencapsulation. One such tool is rheology, particularly the modulus data that can show how elastic or gel-like a material is and give an idea of its consistency if the polysaccharide is subject to many types of abuse including temperature and pH. For frozen foods it will then be important to identify any syneresis. Also, SEM can show the surface morphology of frozen foods, which exhibits the network formed with the polysaccharide alone or in a food system. Furthermore, with NMR, water mobility can be examined when polysaccharides are involved in a system.

References

Adhikari K, Mustapha A, Grun IU, Fernando A (2000) Viability of microencapsulated bifidobacteria in set yogrut during refrigerated storage. J Dairy Sci 83:1946–1951

Agullo E, Rodriguez MS, Ramos V, Albertengo L (2003) Present and future role of chitin and chitosan in food. Macromol Biosci 3:521–530

Anal AK, Stevens WF (2005) Chitosan-alginate multilayer beads for controlled release of ampicillin. Int J Pharm 290:45–54

Annable P, Williams PA, Nishinari K (1994) Interaction in xanthan-glucomannan mixtures and the influence of electrolyte. Macromolecules 27:4204–4211

Audet P, Paquin C, Lacroix C (1991) Effect of medium and temperature of storage on viability of lactic acid bacteria immobilized in k-carrageenan-locust bean gim gel beads. Biotechnol Tech 5(4):307–312

Chaisawang M, Suphantharika M (2006) Pasting and rheological properties of native and anionic tapioca starches as modified by guar gum and xanthan gum. Food Hydrocoll 20:641–649

Coviello T, Burchard W (1992) Criteria for the point of gelation in reversibly gelling systems according to dynamic light-scattering and oscillatory rheology. Macromolecules 25:1011–1012

Dave RI, Shah NP (1997) Viability of yoghurt and probiotic bacteria in yoghurts made from commercial starter cultures. Int Dairy J 7:31–41

Dea ICM, Morris ER, Rees DA, Welsh EJ, Barnes HA, Price J (1977) Associations of like and unlike polysaccharides: mechanism and specificity in galactomannans, interacting bacterial polysaccharides, and related systems. Carbohydr Res 57:249–272

Dickenson E (2008) Hydrocolloids as emulsifiers and emulsion stabilizers. Food Hydrocoll 23:1473–1482

Ding WK, Shah NP (2008) An improved method of microencapsulation of probiotic bacteria for their stability in acidic and bile conditions during storage. J Food Sci 74(2):M53–M61

Ding WK, Shah NP (2009) Effect of various encapsulating materials on the stability of probiotic bacteria. J Food Sci 74(2):M100–M107

Doleyres Y, Fliss I, Lacroix C (2004) Increased stress tolerance of Bifidobacterium longum and lactococcus lactis produced during continuous mixed strain immobilized cell fermentation. J Appl Microbiol 97:527–539

Downey G (2002) Quality changes in frozen and thawed, cooked pureed vegetables containing hydrocolloids, gums and dairy powders. Int J Food Sci Technol 37:869–877

Draget KI (2000) In: Phillips GO, Williams PA (eds) Handbook of hydrocolloids. CRC Press LLC, Boca Raton, pp 379–397

Dumitriu S, Chornet E (1997) Immobilization of xylanase in chitosan-xanthan hydrogels. Biotechnol Prog 13:539–545

Dunstan DE, Chen Y, Liao M-L, Salvatore R, Boger DV, Prica M (2001) Structure Rheology of the κ-carrageenan/locust bean gum gels. Food Hydrocoll 15:475–484

Elcin YM (1995) Encapsulation of urease enzyme in xanthan-alginate spheres. Biomater 16:1157–1161

FAO/WHO Experts Report (2001) Health and nutritional properties of probiotics in food including powder milk with live lactic acid bacteria. United Nations

FDA (1996) 21 CFR 172: food additives permitted for direct addition to food for human consumption: Curdlan. Fed Regul 61:65941–65942

Fernández PP, Martino MN, Zaritzky NE, Guignon B, Sanz PD (2007) Effects of locust bean, xanthan and guar gums on the ice crystals of sucrose solution frozen at high pressure. Food Hydrocoll 21:507–515

Fox JE (1997) In: Imeson A (ed) Thickening and gelling agents for foods. Chapman & Hall, New York, pp 262–283

Freeland M (2002) Formulation tips on Hydrocolloids. Prepared Foods 171:69

Funami MF, Yada H, Nakao Y (1999) Rheological and thermal studies on gelling characteristics of curdlan. Food Hydrocoll 13:317–324

Gagnon MA, Lafleur M (2007) From curdlan powder to the triple helix gel structure: an attenuated total reflection-infrared study of the gelation process. Appl Spectrosc 61:374–378

Garti N, Leser NM (2001) Emulsification properties of hydrocolloids. Polym Adv Technol 12:123–135

Ghaouth EA, Arul J, Asselin A, Benhamou N (1992) Antifungal activity of chitosan on post harvest pathogens: induction of morphological and cytological alterations an Rhizopus Stolonifer. Mycol Res 96:769–779

Giannouli P, Morris ER (2003) Cryogelation of xanthan. Food Hydrocoll 17:495–501

Gibbs BF, Kermasha S, Alli I, Mulligan CN (1999) Encapsulation in the food industry: a review. Int J Food Sci Nutr 50:213–224

Glicksman M (1986) Food Hydrocolloids, vol 3. CRC Press, Inc., Boca Raton

Gliko-Kabir I, Yagen B, Baluom M, Rubinstein A (2000) Phosphated crosslinked guar for colon-specific drug delivery II. *In vitro* and *in vivo* evaluation in the rat. J Controll Release 63:129–134

Goff HD (2006) In: Williams PA, Phillips GO (eds) Gums and stabilisers for the food industry 13. Royal Society of Chemistry, Cambridge, pp 403–412

Goud K, Desai H, Park HJ (2005) Recent developments in microencapsulation of food ingredients. Drying Technol 23:1361–1394

Gouin S (2004) Microencapsulation: industrial appraisal of existing technologies and trends. Trends Food Sci Technol 15:330–347

Goycoolea FM, Morris ER, Gidley M (1995) Viscosity of galactomannans at alkaline and neutral pH: evidence of 'hyperentanglement' in solution. Carbohydr Polym 27:69–71

Granz AJ (1977) Cellulose hydrocolloids. In: Graham H (ed) Food colloids. The AVI Publishing Company, Inc., Westport, pp 382–417

Han C, Zhao Y, Leonard SW (2004) Traber. Edible coatings to improved storability and enhance nutritional value of fresh and frozen strawberries (*Fragaria x ananassa*) and raspberries (*Rubus ideaus*). Postharvest Biol Technol 33:67–78

Hansen LT, Allan-Wojtas PM, Jin YL, Paulson AT (2002) Survival of Ca-alginate microencapsulated bifidobacterium spp. in milk and simulated gastrointestinal conditions. Food Microbiol 19:35–45

Harada T, Harada A (1996) In: Dumitriu S (ed) Polysaccharides in medical applications. CRC Press, Boca Raton, pp 21–58

Harada T, Masada M, Fujimari K, Maeda I (1966) Production of a firm, resilient gel-forming polysaccharide by a mutant of *Alcaligenes faecalis* var. *myxogenes* 10C3. Agric Biol Chem 30:196–198

Harada T, Okuyama K, Konno A, Koreeda A, Harada A (1994) Effect of heating on formation of curdlan gels. Carbohydr Polym 24:101–106

Hatakeyama T, Hatakeyama H (1992) In: Glasser WG, Hatakeyama H (eds) Viscoelasticity and biomaterials. ACS symposium, p 329

Hatakeyama T, Ueda C, Hatakeyama H (2006) Structural change of water by gelation of curdlan suspension. J Therm Anal Calorim 85:661–668

Higiro J, Herald TJ, Alavi S (2006) Rheological study of xanthan and locust bean gum interaction in dilute solution. Food Res Int 39:165–175

Hirashima M, Takaya T, Nishinari K (1997) DSC and rheological studies on aqueous dispersions of curdlan. Thermochim Acta 306:109–114

Hoefler AC (2004) Hydrocolloids. Eagan Press, St. Paul, pp 7–25

Hofmann K, Hatakeyama H (1994) H-1-NMR relaxation studies and lineshape analysis of aqueous sodium carboxymethylcellulose. Polymers 35:2749–2758

Hsu SY, Chung HY (2000) Interactions of konjac, agar, curdlan gum, κ-carrageenan and reheating treatment in emulsified meatballs. J Food Eng 44:199–204

Iijima H, Takeo K (2000) In: Phillips GO, Williams PA (eds) Handbook of hydrocolloids. CRC Press LLC, Boca Raton, pp 331–346

Ikeda S, Nitta Y, Kim BS, Temsiripong T, Pongsawatmanit R, Nishinari K (2004) Single-phase mixed gels of xyloglucan and gellan. Food Hydrocoll 18:669–675

Imeson AP, Humphreys W (1997) In: Imeson AP (ed) Thickening and gelling agents for food. St Edmunsbury Press, Suffolk, pp 180–197

Imeson AP (2000) In: Phillips GO, Williams PA (eds) Handbook of hydrocolloids. CRC Press LLC, Boca Raton, pp 87–102

Jeuniaux C (1986) In: Muzzarelli RAA, Jeuniaux C, Gooday GW (eds) Chitin in nature and technology. Plenum Press, New York, pp 551–570

Jezequal V (1998) Curdlan: a new functional beta-glucan. Cereal Food World 43:361–364

Jin Y, Zhang HB, Yin YM, Nishinari K (2006) Comparison of curdlan and its carboxymethylated derivative by means of Rheology, DSC, and AFM. Carbohydr Res 341:90–99

Kanzawa Y, Harada T, Koreeda A, Harada A (1987) Curdlan gel formed by neutralizing its alkaline solution. Agric Biol Chem 51:1839–1843

Kanzawa Y, Koreeda A, Harada A, Harada T (1989) Electron microscopy of the gel-forming ability of polysaccharide food additives. Agric Biol Chem 53:979–986

Kim B, Takemasa M, Nishinari K (2006) Synergistic interaction of xyloglucan and xanthan investigated by rheology, differential scanning calorimetry, and NMR. Biomacromolecules 7:1223–1230

Krishnaiah YSR, Karthikeyan RS, Sankar VG, Satyanarayana V (2002) Three-layer guar gum matrix tablet formulations for oral controlled delivery of highly soluble trimetazidine dihydrochloride. J Control Release 81:45–56

Krishnaiah YSR, Raju PV, Kumar BD, Bhaskar P, Satyanarayana V (2001) Development of colon targeted drug delivery systems for mebendazole. J Control Release 77:87–95

Kumar RMNV, Muzzarelli RAA, Muzzarelli C, Sashiwa H, Domb AJ (2004) 200Chitosan chemistry and pharmaceutical perspectives. Chem Rev 104:6017–6084

Lee JS, Cha DS, Park HJ (2004) Survival of freeze dried Lactobacillus bulgaricus KFRI 673 in chitosan coated calcium alginate microparticles. J Agric Food Chem 52:7300–7305

Lee MH, Baek MH, Cha DS, Park HJ, Lim ST (2002) Freeze-thaw stabilization of sweet potato starch gel by polysaccharide gums. Food Hydrocoll 16:345–352

Lo CT, Ramsden L (2000) Effects of xanthan and galactomannan on the freeze/thaw properties of starch gels. Nahrung 44:211–214

Lo YM, Robbins KL, Argin-Soysal S, Sadar LN (2003) Viscoelastic effects on the diffusion properties of curdlan gels. J Food Sci 68:2057–2063

Lopes Da Silva JA, Rao MA, Fu JT (1998) In: Rao MA, Hartel RW (eds) Phase/State transitions in foods. Marcel Dekker, Inc., New York, pp 111–157

Lozinsky VI, Damshkaln LG, Brown R, Norton IT (2000) Study of cryostructuring of polymer systems. XIX. On the nature of intermolecular links in the cryogels of locust bean gum. Polym Int 49:1434–1443

Lozinsky VI, Plieva FM, Galaev IY, Mattiasson B (2001) The potential of polymeric cryogels in bioseparation. Bioseperation 10:163–188

MacArtain P, Jacquier JC, Dawson KA (2003) Physical characteristics of calcium induced kappa-carrageenan networks. Carbohydr Polym 53:395–400

Mandala IG, Sawas TP, Kostaropoulos AE (2004) Xanthan and locust bean gum influence on the rheology and structure of a white model-sauce. J Food Eng 64:335–342

Mandala IG (2005) Physical properties of fresh and frozen stored, microwaved-reheated breads, containing hydrocolloids. J Food Eng 66:291–300

Mandala I, Kapetanakou A, Kostaropoulos A (2008) Physical properties of breads containing hydrocolloids stored at low temperature: II–Effect of freezing. Food Hydrocoll 22:1443–1451

Marchessault RH, Deslandes Y (1978) Fine structure of (1 → 3)-β-D-glucans: curdlan and paramylon. Carbohydr Res 75:231–242

Marshall RT, Goff HD, Hartel RW (2003) Ice cream, 6th edn. Kluwer Academic/Plenum Publishers, New York

Martin DR, Ablett S, Darke A, Sutton RL, Sahagian M (1999) Diffusion of aqueous sugar solutions as affected by locust bean gum studied by NMR. J Food Sci 64:46–49

Matuda TG, Chevallier S, Filho PAP, LeBail A, Tadini CC (2008) Impact of guar and xanthan gums on proofing and calorimetric parameters of frozen bread dough. J Cereal Sci 48:741–746

McIntosh M, Stone BA, Stanisich VA (2005) Curdlan and other bacterial (1 → 3)-β-D-glucans. Appl Microbiol Biotechnol 68:163–173

Medina-Torres L, Brito-De La Fuente E, Gómez-Aldapa CA, Aragon-Piña A, Toro-Vazquez JF (2006) Structural characteristics of gels formed by mixtures of carrageenan and mucilage gum from *Opuntia ficus indica*. Carbohydr Polym 63:299–309

Mikkonen KS, Tenkanen M, Cooke P, Xu C, Rita H, Willfo S, Holmbom B, Hicks KB, Yadav MP (2009) Mannans as stabilizers of oil-in-water beverage emulsions. LWT- Food Sci Technol 42:849–855

Murray JCF (2000) In: Phillips GO, Williams PA (eds) Handbook of hydrocolloids. CRC Press LLC, Boca Raton, pp 219–229

Na K, Park KH, Kim SW, Bae YH (2000) Self-assembled hydrogel nanoparticles from curdlan derivatives: characterization, anti-cancer drug release and interaction with a hepatoma cell line (HepG2). J Control Release 69:225–236

Nakao Y, Konno A, Taguchi T, Tawada T, Kasai H, Toda J, Terasaki M (1991) Curdlan: properties and application to foods. J Food Sci 56:769–772

Nakao Y (1997) Properties and food applications of curdlan. Agro-Food-Industry Hi-Tech, Tekno Scienze Publisher, Italy, pp 12–15

Nishinari HZ (2000) In: Phillips GO, Williams PA (eds) Handbook of hydrocolloids. CRC Press LLC, Boca Raton, pp 269–286

Onsoyen E (1997) In: Imeson A (ed) Thickening and gelling agents for food, 2nd edn. Chapman & Hall, New York, pp 22–44

Pai V, Srinivasarao M, Khan SA (2002) Evolution of microstructure and rheology in mixed polysaccharide systems. Macromolecules 35:1699–1707

Paradossi G, Chiessi E, Barbiroli A, Fessas D (2002) Xanthan and glucomannan mixtures: synergistic interactions and gelation. Biomacromolecules 3:498–504

Pederson J (1979) In: Blanshard JMV, Mitchell JR (ed) Polysaccharides in food. Butterworths, London, pp 219–227

Pinotti A, Garcia MA, Martino MN, Zaritzky NE (2007) Study on microstructure and physical properties of composite films based on chitosan and methylcellulose. Food Hydrocoll 21:66–72

Ramakrishnan S, Gerardin C, Prud'homme RK (2004) Syneresis of carrageenan gels: NMR and rheology. Soft Mater 2:145–153

Rao VSR, Qasba PK, Balaji PV, Chandrasekaran R (1998) Conformation of carbohydrates. Overseas Publishers Association, Amsterdam, p 29

Ribotta PD, Pérez GT, León AE, Añón MC (2004) Effect of emulsifier and guar gum on micro structural, rheological and baking performance of frozen bread dough. Food Hydrocoll 18:305–313

Richter S, Boyko V, Matzker R, Schröter K (2004) A thermoreversible gelling system: mixtures of xanthan gum and locust-bean gum. Macromol Rapid Commun 25:1504–1509

Richter S, Brand T, Berger S (2005) Comparative monitoring of the gelation process of a thermoreversible gelling system made of xanthan gum and locust bean gum by dynamic light scattering and ^{1}H NMR Spectroscopy. Macromol Rapid Commun 26:548–553

Rocks JK (1971) Xanthan gum. Food Technol 25:476–483

Rodd AB, Dunstan DE, Boger DV, Schmidt J, Burchard W (2001) Heterodyne and nonergodic approach to dynamic light scattering of polymer gels: aqueous xanthan in the presence of metal ions (aluminum (III)). Macromolecules 34:3339–3352

Ross GM, Gusils C, Gonzalez SN (2008) Microencapsulation of probiotic strains for swine feeding. Biol Pharm Bull 31(11):2121–2125

Sadar LN (2004) Rheological and textural characteristics of copolymerized hydrocolloidal solutions containing curdlan gum. Thesis for the Masters Degree, University of Maryland, College Park

Saito H, Miyata E, Sasaki Y (1978) A ^{13}C nuclear magnetic resonance study of gel-forming $(1 \rightarrow 3)$-β-Dglucans: molecular-weight dependence of helical conformation and of the presence of junction zones for association of primary molecules. Macromolecules 11:1244–1251

Sanchez C, Zuniga-Lopez R, Schmitt C, Despond S, Hardy J (2000) Microstructure of acid-induced skim milk-locust bean gum-xanthan gels. Int Dairy J 10:199–212

Sanderson GR (1996) Gums and their use in food systems. Food Technol 50:81–84

Sandolo C, Matricardi P, Alhaique F, Coviello T (2008) Effect of temperature and cross-linking density on rheology of chemical cross-linked guar gum at the gel point. Food Hydrocoll 23:210–220

Sathivel S, Liu Q, Huang J, Prinyawiwatkul W (2007) The influence of chitosan glazing on quality of skinless pink salmon (*Oncorhynchus gorbuscha*) fillets during frozen storage. J Food Eng 83:366–373

Schrooyen PMM, Meer VDR, Kruif CGD (2001) Microencapsulation: its application in nutrition. Proc Nutr Soc, Cambridge University Press 60:475–479

Shahidi F, Abuzaytoun A (2005) Applications of chitin and chitosan and their oligomers: Taylor, S. L. Adv Food Nutr Res 49:114–128

Shahidi F, Arachchi JKV, Jeon YJ (1999) Food applications of chitin and chitosan. Trends Food Sci Technol 10:37–51

Sharadanant R, Khan K (2003a) Effect of hydrophilic gums on frozen dough. I. Dough quality. Cereal Chem 80:764–772

Sharadanant R, Khan K (2003b) Effect of hydrophilic gums on frozen dough. II. Bread characteristics. Cereal Chem 80:773–780

Shon J, Yun Y, Shin M, Chin KB, Eun JB (2009) Effects of milk protiens and gums on quality of bread made from frozen dough. J Sci Food Agric 89:1407–1415

Sikor M, Badrie N, Deisingh AK, Kowalski S (2008) Sauces and dressings: a review of properties and applications. Crit Rev Food Sci Nutr 48:50–77

Soma PK, Lo YM (2009) Characterization of the diffusional properties of polyelectrolyte complex gel formed by xanthan and chitosan. Institute of Food Technologists, Chicago, IL, Annual Meeting and Food Expo, Food Engineering Division

Sworn G (2000) In: Phillips GO, Williams PA (eds) Handbook of hydrocolloids. CRC Press LLC, Boca Raton, pp 103–115

Tada T, Matsumoto T, Masuda T (1999) Dynamic viscoelasticity and small-angle X-ray scattering studies on the gelation mechanism and network structure of curdlan gels. Carbohydr Polym 39:53–59

Takigami S, Shimada M, Williams PA, Phillips GO (1993) E.s.r. study of the conformation transition of spin-labeled xanthan gum in aqueous solution. Int J Biol Macromol 15:367–371

Tanaka R, Hatakeyama T, Hatakeyama H (1998) Formation of locust bean gum hydrogel by freezeing-thawing. Polym Int 45:118–126

Towle GA (1996) In: Phillips GO, Williams PA, Wedlock DJ (eds) Gums and stabilisers for the food industry 8. Oxford University Press, Inc., New York, pp 79–87

Tsen ZH, Lin YP, Haung HY, King VAE (2008) Studies on the fermentation of tomato juice by using k-carrageenan immobilized *Lactobacillus acidophilus*. J Food Process Preserv 32:178–189

Vittadini E, Dickinson LC, Chinachoti P (2002) NMR water mobility in xanthan and locust bean gum mixtures: possible explanation of microbial response. Carbohydr Polym 49:261–269

Wang F, Wang Y-J, Sun Z (2002a) Conformational role of xanthan in its interaction with locust bean gum. J Food Sci 67:2609–2614

Wang F, Wang Y-J, Sun Z (2002b) Conformational role of xanthan in its interaction with guar gum. J Food Sci 67:3289–3294

Wenrong S, Griffiths MW (2000) Survival of bifidobacteria in yogurt and simulated gastric juice following immobilization in gellan-xanthan beads. Int J Food Microbiol 61:17–25

Whitcomb PJ, Macosko CW (1978) Rheology of xanthan gum. J Rheol 22:493–505

Wielinga WC, Maehall AG (2000) In: Phillips GO, Williams PA (eds) Handbook of hydrocolloids. CRC Press LLC, Boca Raton, pp 137–154

Williams PA, Langdon MJ (1996) The influence of locust bean gum and dextran on the gelation of kappa-carrageenan. Biopolymers 38:655–664

Williams PA, Phillips GO (2000) Handbook of hydrocolloids. CRC Press LLC, Boca Raton, pp 1–19

Williams PD, Sadar LN, Lo YM (2009) Texture stability of hydrogel complex containing curdlan gum over multiple freeze-thaw cycles. J Food Process Preserv 33:126–139

Wong D, Larrabee S, Clifford K, Tremblay J, Friend DR (1997) USP dissolution apparatus III (Reciprocating Cylinder) for screening of guar-based colonic delivery formulations. J Control Release 47:173–179

Yaşar F, Toğrul H, Arslan N (2007) Flow properties of cellulose and carboxymethyl cellulose from orange peel. J Food Eng 81:187–199

Zeira A, Nussinovitch A (2004) Mechanical properties of weak locust bean gum (LBG) gels under controlled rapid freeze-thawing. J Texture Stud 34:561–573

Chapter 16
Food Allergens and Processing: A Review of Recent Results

Milan Houska, Ivana Setinova, and Petr Kucera

16.1 Introduction

Allergenicity is not a classic food property. It is based on interactions between a food component or ingredient and the consumer's specific immune system. Basic principle of food allergenicity is expressed in Fig. 16.1. In this paper we will deal with true allergy syndromes and ignore such problems as lactose and other intolerances.

Most true allergens (epitope sources) are protein molecules. These proteins are found in foods such as milk [whey proteins (e.g., beta-lactoglobulin and alpha-lactalbumin)], egg white (ovalbumin), birch pollen (Bet v1) and its homologs found in apples (Mal d 1), carrots (Dau c 1), celery (Api g 1), and peanuts (Ara h 1) as well as many other foods (Breiteneder and Mills 2005; Fuchs 2007). The basic structure of the typical allergenic protein of apples, Mal d1, is apparent from Fig. 16.2.

The most dangerous allergens, due to severe patient reactions, are members of the prolamin family containing lipid transfer proteins (LTPs), which are heat stable. The most common LTP is Mal d 3, which is found in apples (the highest prevalence of sensitized patients are in southern European countries such as Spain and Italy). Another and similar prolamin family member is the peanut allergen Ara h2 (Mills and Mackie 2008).

M. Houska (✉)
Food Research Institute Prague, Prague, Czech Republic
e-mail: milan.houska@vupp.cz

I. Setinova
Centre of Allergology Imumed, Ltd., Prague, Czech Republic

P. Kucera
Department of Allergy and Clinical Immunology, University Hospital Kralovske Vinohrady, Prague, Czech Republic

Department of Immunology, 3rd Faculty of Medicine, Charles University, Prague, Czech Republic

S. Yanniotis et al. (eds.), *Advances in Food Process Engineering Research and Applications*, Food Engineering Series, DOI 10.1007/978-1-4614-7906-2_16,

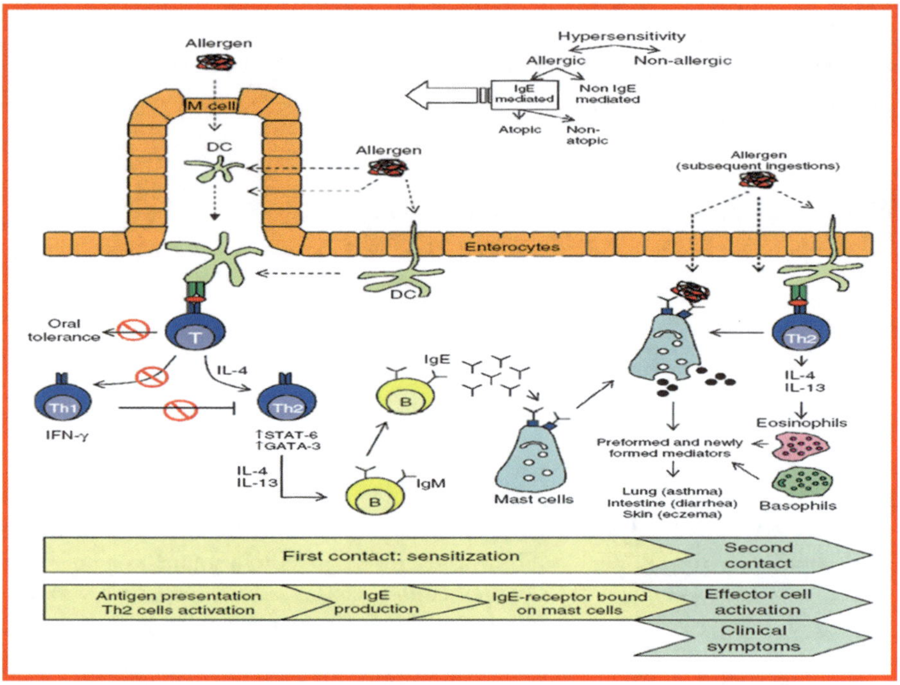

Fig. 16.1 Immune system and allergenicity development

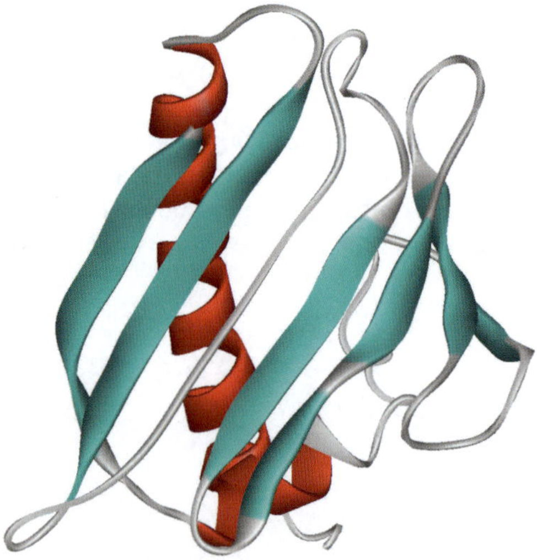

Fig. 16.2 Example of 3D structure of Mal d1 (main apple allergen), predicted secondary structure: alpha helices 21 %, beta sheets 39 %, 7 beta-strands, 3 alpha-helices, 6 loops (figure provided by courtessy of Dr. L. Smeller, Hungary)

The human immune system, as a "test tool," is not homogeneous: different reactions can be found using skin (skin-prick test), blood cells (basophile activation test), blood serum [quantification of specific IgE, demonstration of antigen-antibody reactions using SDS- polyacrylamide gel electrophoresis (PAGE) electrophoresis and western blot (WB)], or oral mucosa [double-blind placebo-control food challenge (DBPCFC)]. Therefore, a battery of tests and studies (e.g., protein structure, stability to pepsin digestion in vitro, protein content in the product, processing effects, specific IgE binding studies, or skin-prick test) must be done when testing the effect of processing in order to obtain conclusive results (Ladics 2008; Björksten et al. 2008; Host and Halken 2004).

A review by Mills and Mackie (2008) noted that many earlier papers dealing with the influence of processing effectiveness used only one method, mostly in vitro tests, and did not combine them with in vivo tests on well-defined patient groups and thus missed an opportunity for statistically significant evidence.

Several studies have reviewed the influence of processing on the allergenicity of treated foods (Fiocchi et al. 2004; Poms and Anklam 2004; Paschke and Besler 2002; Besler et al. 2001; Hefle 1999; Davis and Williams 1998); however, most did not mention the test methods used and did not assess the effects of novel processing methods like pulsed electric field (PEF), pulsed ultraviolet light irradiation (PUV), or high-pressure treatment (HPT). The main objective of this chapter is to provide an overview of recent achievements in food deallergization using different processing methods. The various allergenicity tests used in the studies are also reviewed.

16.2 Materials and Methods

A literature search was done using the Scopus database on the keywords "food allergenicity" and "process." The following processing methods were found during the search: heat processing, enzymatic hydrolysis and in vitro digestion, gamma irradiation, fermentation, refining (oils), microparticulation, genetic modification, PUV, polymerization of allergen proteins, HPT, and PEF treatment. Ultrafiltration was found only combined with a prior processing step involving protein coagulation. In the next section the results from processing experiments and the foods studied are discussed.

16.3 Results and Discussion

The processing methods, along with details of the foods or raw material processed, and the allergenicity tests used in the study are presented in Table 16.1. Figure 16.3 shows the percentage of reviewed papers using a particular allergenicity test in the study. Figure 16.4 is a histogram that plots the number of reviewed papers versus

Table 16.1 Overview of existing technologies tested for de-allergization of various foods and raw materials

Process	Food or food component/details of processing	Allergenicity test used (or analytical method applied): Electrophoresis and or SDS-PAGE	Immunodetection with patient sera (ELISA) or western blot (immunoblotting)	Quantitative gel permeation chromatography	Other methods	Skin prick test	Basophil activation test	DBPCFC	References
Heat processing									
Eggs									
Boiling	Egg white components: ovalbumin, ovomucoid, and conalbumin				Radioimmunoelectrophoresis (RIEP) and RAST				Hoffman (1983)
Addition of vinegar during cooking process	Egg, chicken, lentils		x			7 patients		1 patient	Armentia et al., (2010)
Legumes									
Boiling	Lentils	x	x						Ibáñez Sandín et al., (1999)
Heat treatment at 80 °C, 100 °C, and 120 °C for 30 min	Proteins of soybean 11S-, 7S-, and 2S-globulins				Radioallergosorbent test (RAST) and by RAST inhibition experiments				Shibasaki et al. (1980)
Fish and seafood									
Cooking	Ten fish species and raw and cooked protein extracts	x	x, x			x, 11 patients			Bernhisel-Broadbent et al., (1992a)
Canning	Cooked lyophilized fish extracts and canned tuna and salmon	x	x					x, 18 sensitized patients	Bernhisel-Broadbent et al., (1992b)
Smoking, salting/sugaring, canning, lye treating, and fermenting	Smoked, salted/sugar-cured, canned, lye-treated, and fermented cod, haddock, salmon,		x						Sletten et al., (2010)

	trout, tuna, mackerel, and herring and of hydrolysates based on salmon and whiting					
Heat sterilization	Mackerel (*Scomber scombrus*)		Ig E binding			Chopin et al., (2000)
Cooking	Razor shell			In vivo and in vitro studies		Martín-García et al., (2007)
Meat and meat products						
Autoclave treatment at 121 °C for 5, 10, and 30 min	Autoclaved market pork sausages digested using pepsin (30 min) and trypsin (5, 30, 60, 90, and 120 min)	x	x,x			Kim et al., (2009)
Steam-cooking, homogenization and freeze drying together with in vitro multienzymatic digestibility assay	Meat-based baby foods		x		x	Restani et al., (1997)
Peanuts and other nuts						
Roasting	Nonallergenic lectin was reacted with glucose or fructose at 50 °C for 28 days. Browning products from heat-treated peanuts were also examined		x,x			Chung and Champagne (1999)

(continued)

Table 16.1 (continued)

		Allergenicity test used (or analytical method applied)							
Process	Food or food component/details of processing	Electrophoresis and or SDS-PAGE	Immunodetection with patient sera (ELISA) or western blot (immunoblotting)	Quantitative gel permeation chromatography	Other methods	Skin prick test	Basophil activation test	DBPCFC	References
Roasting	Roasted and raw peanut extracts		x		Digestion by gastric secretion				Maleki et al., (2000)
Roasting	Peanut origin protein-bound end products or adducts such as advanced glycation end products (AGE), N-(carboxymethyl) lysine (CML), malondialdehyde (MDA), and 4-hydroxynonenal (HNE)		x						Chung and Champagne (2001)
Roasting, boiling, frying	Two varieties of peanut	x	x						Beyer et al., (2001)
Roasting	Main peanut allergen Ara h 2 purified from raw and roasted peanuts	x	x		Circular dichroism (CD) spectroscopy				Maleki et al. (2003)
Heat treatment	Recombinant peanut allergen Ara h2		x						Gruber et al., (2005)
Roasting and curing	Mature and immature roasted peanuts and peanuts cured at different temperatures (35–77 °C)		x		Advanced glycation end adducts (AGEs)				Chung et al., (2003)

Roasting and boiling	Protein extracts from raw, roasted, and boiled peanuts	x	x	Enzyme allergosorbent test (EAST)		Mondoulet et al., (2003)
Roasting and boiling in water	Peanut protein extracts and of major peanut allergens Ara h 1 and Ara h 2	x	x	Enzyme allergosorbent test (EAST) and EAST inhibition		Mondoulet et al., (2005)
Roasting, autoclaving, blanching, microwave heating	Nonpareil almonds	x	x,x			Venkatachalam et al., (2002)
Heat treatment	Four hazelnut varieties (Runde Romer, Levantiner, Neapler, Contorta)	x	x			Wigotzki et al. (2000)
Proteins generally						
Heat treatment	Proteins in foods					Davis et al., (2001)
Fruits and vegetables						
Heat treatment at 121 °C for 10 and 30 min, chemical lye peeling of fruits, and ultrafiltration of juice through membranes with suitable cutoff	Peach juice, nectar, jam, syrupy peach	x	x			Brenna et al., (2000)
Chemical peeling, thermal treatment, and syruping processes	Different varieties of cherry (*Prunus avium*)	x	x		x	Primavesi et al., (2006)
Heat treatment	Recombinant main cherry allergen Pru av 1		x			Gruber et al., (2004)

(continued)

Table 16.1 (continued)

		Allergenicity test used (or analytical method applied)							
Process	Food or food component/details of processing	Electrophoresis and or SDS-PAGE	Immunodetection with patient sera (ELISA) or western blot (immunoblotting)	Quantitative gel permeation chromatography	Other methods	Skin prick test	Basophil activation test	DBPCFC	References
Heat treatment	Extracts from six different potato strains, potato flakes prepared at different temperatures	x	x		Enzyme allergosorbent tests (EAST-inhibition)				Schubert et al., (2003)
Microwave heating	Extracts from native and microwaved (750 W, 30 min, 100 °C) celery root		x		Enzyme allergosorbent test (EAST)	x			Jankiewicz et al., (1996)
Cooking (ev. drying)	Extracts of raw, cooked celery and celery spice		x		Enzyme allergosorbent test (EAST)			x, 12 patients	Ballmer-Weber et al., (2002)
Cooking	Carrot		x			x			Gómez et al., (1993)
Cooking	Carrot		x			x			Quirce et al., (1997)
Pasta, grain and bakery products									
Drying at temperatures 20 °C, 60 °C, 85 °C, 110 °C and 180 °C and cooking in boiling water	Model pasta samples (durum wheat proteins)		x						De Zorzi et al., (2007)
Cooking, hypodeallergization (unspecified method)	Cooked and hypoallergenic rice	x	x		Mass spectrometry, Streptavidin ImmunoCAP system	x			Yum et al., (2006)

x, Ig E binding

Baking and in vitro digestion	Wheat dough and bread crumb and crust, before and after being in vitro digested					Simonato et al., (2001)
Boiling	Pastalike model samples of wheat flour mixed with egg white		x			Kato et al., (2001)
Heat treatment similar to food preparation	Protein allergens extracted from wheat, rye, barley, and oat flour	x	x			Varjonen et al., (1996)
Microwave heating	Commercial gliadins and wheat flour exposed to microwaves at power distribution of 70, 200, and 500 W for different time periods	x	x,x			Leszczynska et al. (2003a)
Milk and milk proteins						
Pasteurization, homogenization	Raw, pasteurized, and homogenized/pasteurized cow milk + hypoallergenic infant formula as a control			x	x 5 patients	Host and Samuelsscn (1988)
Boiling	Whole casein, gamma-globulin, serum albumin, beta-lactoglobulin, alfa-lactalbumin, and whole cow's milk			x 8 positive adults		Norgaard et al., (1996)
Enzymatic hydrolysis and in vitro digestion						

(continued)

Table 16.1 (continued)

Process	Food or food component/details of processing	Allergenicity test used (or analytical method applied) Electrophoresis and or SDS-PAGE	Immunodetection with patient sera (ELISA) or western blot (immunoblotting)	Quantitative gel permeation chromatography	Other methods	Skin prick test	Basophil activation test	DBPCFC	References
In vitro model of gastro-duodenal digestion	Major peanut allergen Ara h 1	x			T cell reactivity		x		Eiwegger et al., (2006)
Two-step in vitro hydrolysis by pepsin, followed by a trypsin/chymotrypsin (T/C) mixture performed in dialysis bags with molecular weight cutoffs (MWCO) of 1000 or 8000 Da	Peanut protein isolate (PPI) digested together with polysaccharides, i.e., gum arabic, low methylated pectin (LMP), and xylan	x	x						Mouécoucou et al., (2004)
Enzymatic decomposition, mechanical tissue disintegration, and heating during peeling, mash treatment, and pasteurization	Mango purees and nectars	x	x		Enzyme allergosorbent tests (EAST inhibition)				Dube et al., (2004)
In vitro digestion system modeling of the passage of food through stomach into duodenum	Two kiwifruit allergens, actinidin (Act d 1) and thaumatinlike protein (Act d 2)		x		Circular dichroism spectroscopy				Bublin et al., (2008)
			ELISA						

Enzymatic decomposition of cellulose- and protein-based allergens (cellulase at 50 °C for 1 h, actinase at 40 °C for 1 h applied)	Hypoallergenic wheat flour						Watanabe et al., (2000)
Bromelain treatment	Wheat flour		x				Tanabe et al., (1996)
Enzymatic hydrolysis	Cow's milk						Bousquet et al., (1998)
Proteinase treatment	Soybeans		x				Yamanishi et al., (1996)
Simulated gastric fluid (SGF) digestion assay	Purified known allergenic and nonallergenic proteins from leguminous crops	x	x		x		Misra et al., (2009)
Actinase treatment	Rice grains			Radioallergosorbent test (RAST)		x	Watanabe et al., (1990)
Gamma irradiation							
Gamma irradiation	Lectins			Hemagglutination assays			Vaz et al., (2011)
Gamma irradiation with 10 or 20 kGy	Hen egg albumin (ovalbumin, OVA) in white layer cakes containing egg white		x				Lee et al., (2005)
Gamma irradiation	Bovine alpha-casein and beta-lactoglobulin	x	x, x				Lee et al., (2001)
Gamma irradiation	Commercial gliadin powder and wheat flour were irradiated with doses between 2.2 and 12.8 kGy	x	x,x				Leszczynska et al. (2003b)

(continued)

Table 16.1 (continued)

		Allergenicity test used (or analytical method applied)							
Process	Food or food component/details of processing	Electrophoresis and or SDS-PAGE	Immunodetection with patient sera (ELISA) or western blot (immunoblotting)	Quantitative gel permeation chromatography	Other methods	Skin prick test	Basophil activation test	DBPCFC	References
Gamma irradiation	Shrimp heat-stable protein and fresh shrimp were irradiated	x	x,x						Byun et al., (2000)
Gamma irradiation	Ovalbumin and bovine serum albumin in solution (0.2 % in 0.01 M phosphate buffer, pH 7.4)	x	x						Kume and Matsuda (1995)
Gamma irradiation	Hen egg ovomucoid under basic pH, gamma irradiated at 10 kGy, heated at 100 °C for 15 min, or both		x						Lee et al., (2002)
Fermentation									
Fermentation	Alpha-lactalbumin and beta-lactoglobulin in whey from fermented milk		x			x			Jedrychowski and Wróblewska (1999)
Milk fermentation	Raw and pasteurized milk "naturally" fermented and industrially manufactured sour milk as well as "acidophilus milk"		x						Maier et al., (2006)

Fermentation	Soybean products (sprouts-Sp, tempeh-T, tofu-To, miso-M, mold hydrolyzed soy sauce-MHS, acid-hydrolyzed soy sauce-AHS, and hydrolyzed vegetable protein-HVP)			RAST inhibition	Herian et al., (1993)
Fermentation	Pea flour fermented with three lactic acid bacteria and two fungi		x		Barkholt et al., (1998)
Processing of soya to get soy flour, soy milk, and soy texturized protein	Three commercial products and two infant formulas were studied: soybean flour, soy milk, texturized soy proteins, two infant formulas, the first containing total proteins and the second containing a soy protein hydrolysate	x	x		Franck et al., (2002)
Fermentation (enzymatic treatment by microbial enzymes)	Fermented soybean seeds, fermented soybean paste, wheat flour dough, soaked rice from Thai fermented rice noodle		x		Phromraksa et al., (2008)
Fermentation in cabbage kimchi at temperatures 25 °C, 15 °C, and 5 °C	Raw shrimp (*Acetes japonicus*) and saeujeot (salted and fermented shrimp) in cabbage kimchi		x		Park et al. (2007)

(continued)

Table 16.1 (continued)

Process	Food or food component/details of processing	Allergenicity test used (or analytical method applied): Electrophoresis and or SDS-PAGE	Immunodetection with patient sera (ELISA) or western blot (immunoblotting)	Quantitative gel permeation chromatography	Other methods	Skin prick test	Basophil activation test	DBPCFC	References
Refining									
Refining	Peanut oil								Crevel et al., (2000)
Different steps of refining process: crude pressed state of oil, acidification, and neutralization, pregumming by centrifugation, washing, bleaching, gumming by filtration, and deodorization	Sunflower oil at different steps of refining process	x	x		Micro-Bradford assay				Zitouni et al., (2000)
Refining	Soy lecithin and refined and nonrefined soybean oils		x		EAST inhibition				Paschke et al., (2001)
Refining of lecithin extracted from soybean	Six commercial soy lecithins with extracts from raw and heat-treated soybeans	x	x		Allergosorbent test (EAST) and mediator release assay based on rat basophil leukemia cell line				Müller et al., (1998)
Microparticulation									
Microparticulation	Microparticulated egg white or cow's milk proteins used as fat substitute	x	x						Sampson and Cooke (1992)

Genetic modification						
Genetic modification	Transgenic soybeans with introduced methionine-rich 2S albumin from Brazil nut (Bertholletia excelsa)	x	x	RAST	x	Nordlee et al., (1996)
Genetic modification (peptide composition from individual amino acids)	Soybean protein			Serum IgE from soybean-allergic individuals was used to identify P34/Gly m Bd 30 K in the native and single amino acid substituted peptides with use of the SPOTS peptide synthesis technique to determine critical amino acids required for IgE binding		Helm et al., (2000)
Genetic modification	Transgenic rice	x	x	RNA blot analyses		Tada et al., (1996)
Genetic modification	Transgenic rice	x	x, x			Nakamura and Matsuda (1996)
Genetic modification	One peanut line, GT-C9, lacking several seed proteins, which were identified as Ara h 3 isoforms	1D + 2D PAGE	x	Protein extracts method		Guo et al., (2008)
Genetic modification (substitution of two amino acids in six peptides in third domain of ovomucoid)	Main hen egg white allergen ovomucoid (Gal d1)		x,x			Mine et al., (2003)

(continued)

Table 16.1 (continued)

Process	Food or food component/details of processing	Allergenicity test used (or analytical method applied)							References
		Electrophoresis and or SDS-PAGE	Immunodetection with patient sera (ELISA) or western blot (immunoblotting)	Quantitative gel permeation chromatography	Other methods	Skin prick test	Basophil activation test	DBPCFC	
Pulsed ultraviolet light treatment (PUV)									
Pulsed ultraviolet (PUV) light treatment	Peanut extracts and liquid peanut butter were PUV treated using a Xenon RS-3000C under: 3 pulses/s, 14.6 cm from central axis of lamp, 4 min (extract) or 3 min (liquid peanut butter); boiling treatment used as control	x	x						Chung et al., (2008)
Pulsed ultraviolet light treatment (PUV)	Soy extracts treated with PUV at various times (2, 4, and 6 min)	x	x						Yang et al., (2010)
Polymerization of allergen proteins									
Polymerization of allergens	Protein extracts from raw and roasted defatted peanut meals at pH 8 were incubated with and without peroxidase in presence of hydrogen peroxide at 37 °C for 60 min	x	x,x						Chung et al., (2004)

PolymeriZation of allergen with various fenolic acids oxidized by polyphenoloxidase	Recombinant Pru av 1	x	x				Gruber et al., (2004)
Polymerization of allergens with caffeic acid assisted with polyphenoloxidase	Peanut extracts treated with and without polyphenoloxidase/caffeic acid (pH 8, 37 °C for 1 h) and caffeic acid (pH 10.5, overnight)	x	x,x				Chung et al., (2005)
Polymerization of allergens with phytic acid	Extracts from raw and roasted peanuts were treated with and without phytic acid at various pH values	x	x				Chung and Champagne (2007)
Polymerization of allergens with phenolics	Peanut extracts and liquid peanut butter mixed with phenolics	x	x				Chung and Champagne (2009)
Oxidation of phenolics and polymerization	Juice from Golden Delicious apple	x	x	Open challenge test			Setinova et al., (2010)
Oxidation of phenolics and polymerization	Celery juice, celery juice:apple juice (1:5)	x	x		x	x	Novotna et al., (2011)
Oxidation with polyphenoloxidase at presence of catechin	Mal d1 extracted from peel and flesh of various sorts of apples		x				Garcia et al., (2007)
High pressure treatment (pulsed electric field)							
High-pressure treatment	Rice grains immersed in distilled water and pressurized at 100–400 MPa	x	x				Kato et al., (2000)

(continued)

Table 16.1 (continued)

Process	Food or food component/details of processing	Allergenicity test used (or analytical method applied)							References
		Electrophoresis and or SDS-PAGE	Immunodetection with patient sera (ELISA) or western blot (immunoblotting)	Quantitative gel permeation chromatography	Other methods	Skin prick test	Basophil activation test	DBPCFC	
High-pressure treatment	Apple and main apple allergen Mal d1 and hazelnut, carrot, cherry, peach, celery				Structure changes observation using circular dichroism (CD) and FTIR spectra method	x	x	x	Scheibenzuber (2003)
High-pressure treatment	Buffered solutions of rBet v1 and birch pollen extract with high pressure (450–550 MPa) for 10 min at temperature range 30–50 °C	x	x		Circular dichroism spectroscopy				Setinova et al., (2009a)
High-pressure treatment	Buffered rApi g1 treated at 500 MPa held for 10 and 20 mins at temperatures 30 °C, 40 °C, and 50 °C	x	x		Circular dichroism spectroscopy				Houska et al., (2009a)
High-pressure treatment	rDau c1 and carrot juice treated by 500 MPa for 10 min and different temperatures (30 °C, 40 °C, 50 °C) and pressures from 400 to 550 MPa for 3 and 10 min	x	x		Circular dichroism spectroscopy	x	x	x 19 patients	Heroldova et al., (2009)

High-pressure treatment	Apple extract (Golden Delicious) treated by various pressures up to 800 MPa + 10 months of storage	x			Competitive RAST inhibition tests				Fernandez et al., (2009)
High-pressure treatment	rMal d1 and rDau c1 were treated by 400, 450, 500, 550 MPa held for 3 and 10 min at 25 °C and at 500 MPa for 10 min at 30, 40, and 50 °C	x	x		Circular dichroism spectroscopy		x		Setinova et al., (2009b)
High-pressure treatment	rMal d1 and apple juice and homogenates prepared from Golden Delicious variety	X	x		Circular dichroism spectroscopy	x, 21 patients	x	x, 21 patients	Houska et al., (2009b)
High-pressure treatment	Alpha amylase inhibitor		x		x				Yamamoto et al. (2010a)
High-pressure treatment	Bovine gamma globulin, treated by 100–600 MPa, at temperature 5–7 °C, 5 min holding time on pressure		x, x		x		x		Yamamoto et al. (2010b)
Chymotrypsin and trypsin enzymatic hydrolysis under high pressure	β-lactoglobulin		x						Chicón et al., (2008)
Tryptic and chymotryptic enzymatic hydrolysis under high isostatic pressure 500 MPa	α-and β-casein, bovine serum albumin (BSA), β-lactoglobulin (β-Lg) and α-lactalbumin (α-La).			x	Streptavidin ImmunoCAP system				Beran et al., (2009)

(continued)

Table 16.1 (continued)

Process	Food or food component/details of processing	Allergenicity test used (or analytical method applied)							References
		Electrophoresis and or SDS-PAGE	Immunodetection with patient sera (ELISA) or western blot (immunoblotting)	Quantitative gel permeation chromatography	Other methods	Skin prick test	Basophil activation test	DBPCFC	
High-pressure/temperature treatment and pulsed electric field treatment	Native peanut Ara h 2, 6 and apple Mal d 3 and Mal d 1b		x	x	Circular dichroism spectroscopy				Johnson et al., (2010)
High-pressure treatment	rMal d 1dissolved in D_2O buffer, treatment at different pD				Fourier-transform infrared (FTIR) spectroscopy				Somkuti et al., (2010)
High-pressure-assisted sterilization	Mal d1, Mal d3, Api g1	x	x						Husband et al., (2011)

the number of allergenicity tests used in the research. It is apparent from Table 16.1 and Figs. 16.3 and 16.4 that only a limited number of studies had conclusions based on more than one or two allergenicity tests. The most frequent test found in the reviewed papers was the in vitro immunodetection test based on IgE reactions. In the next section we will describe the results in detail and provide opinions on the current state of the art.

16.3.1 Heat Processing

The most common process was the use of heat to transform raw material into consumable foods. Pasteurization, sterilization, boiling, and roasting were the most frequently tested methods. The effects of Maillard reaction products (from protein amino groups to saccharide reaction products), which are associated with heat processing, can be hazardous and have not been sufficiently studied (Davis et al. 2001).

16.3.1.1 Eggs

Hoffman (1983) found that egg whites, after boiling, still exhibited detectable allergenicity; however, cooking with various food extracts, such as vinegar, substantially decreased the size of wheals in prick-to-prick skin tests in sensitized individuals (Armentia et al. 2010).

16.3.1.2 Legumes

Boiled lentil extract was found to retain its allergenicity (Ibáñez Sandín et al. 1999), whereas soybean globulin lost part of its allergenicity when heated (Shibasaki et al. 1980).

16.3.1.3 Fish and Seafood

When fish was cooked, it caused denaturation and conglomeration of proteins, but some protein bands showing IgE binding (WB) persisted (Bernhisel-Broadbent et al. 1992a). Canning has been shown to reduce IgE binding (Bernhisel-Broadbent et al. 1992b). Sletten et al. (2010) recently published a study on the effect of smoking, salting/sugaring, canning, lye treatment, and fermentation processes and showed that allergenicity decreases were more dependent on the process than the species. Chemical processing of fish led to a loss in IgE-binding activity; however, some IgE binding was found to take place in response to the resulting modified or degraded peptides. Mackerel sterilization has been shown to lead to a

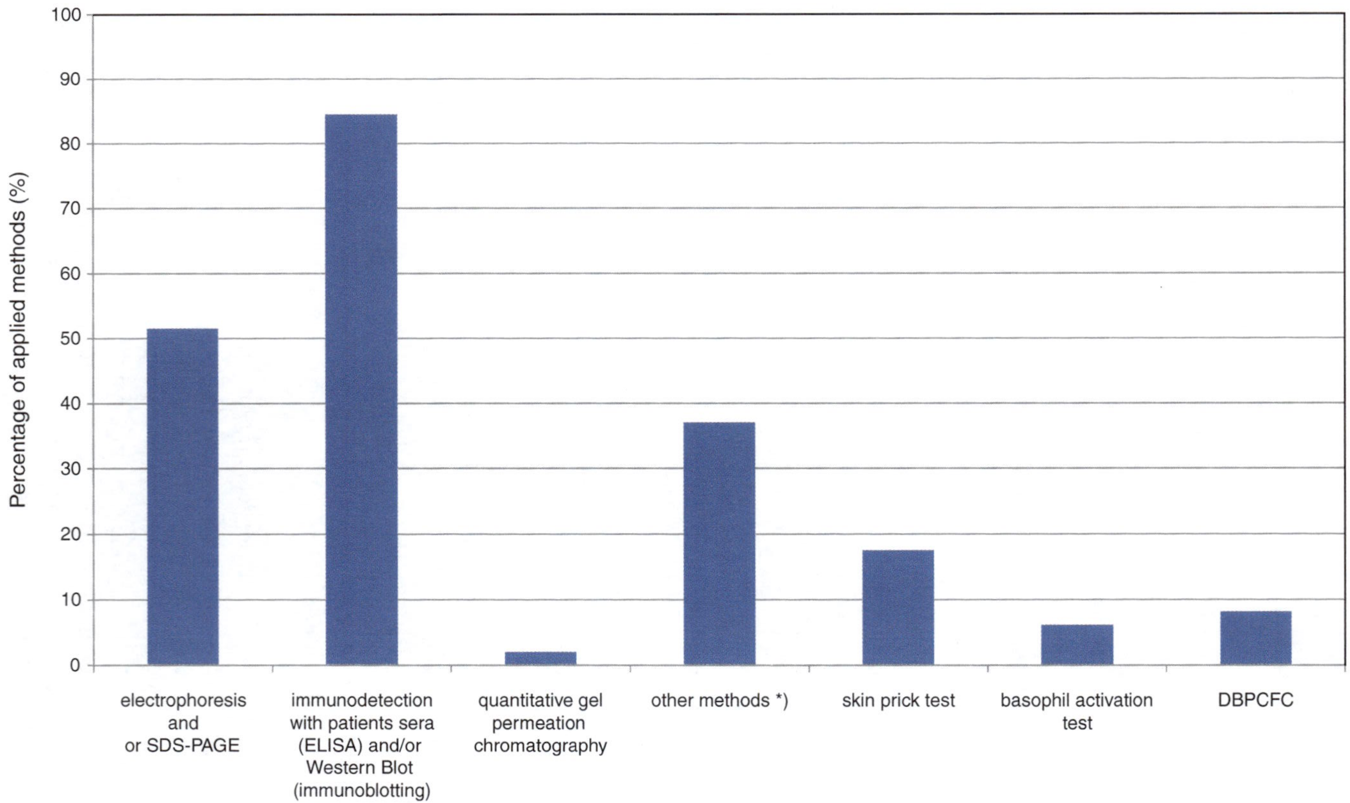

Fig. 16.3 Percentage of applied methods in reviewed papers (*) circular dichroism spectroscopy, streptavidin Immuno CAP system, competitive RAST inhibition test, open challenge test, protein extract method, RNA blot analyses, micro-Bradford assay, EAST inhibition, Hemagglutination assays, T cell reactivity

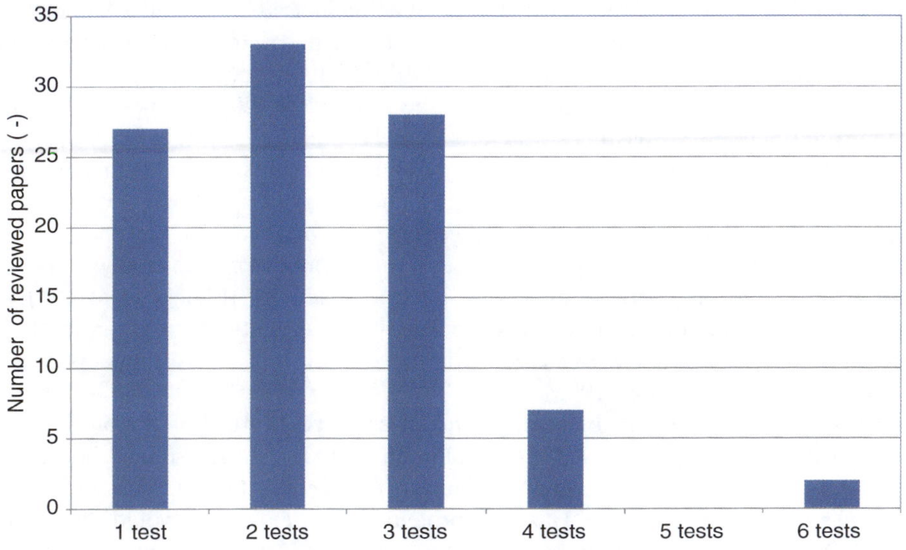

Fig. 16.4 Number of reviewed papers using a given number of tests

substantial reduction in allergenicity (Chopin et al. 2000). Items such as razor shell meat contain many high-molecular-weight and heat-stable proteins; therefore, it retains its allergenicity (Martín-García et al. 2007) even after processing.

16.3.1.4 Meat and Meat Products

The allergenicity of pork sausage digested using pepsin and trypsin was considerably decreased after autoclave treatment and was either stable or decreased after enzyme treatment. Regarding these foods, autoclave treatment represents a promising processing technology for the reduction in the allergenicity of a variety of related food products (Kim et al. 2009). Steam cooking, homogenization, or freeze drying together with in vitro multienzymatic digestion was shown to substantially decrease the allergenicity of meat-based baby foods (Restani et al. 1997).

16.3.1.5 Peanuts and Other Nuts

Initial studies on the roasting of peanuts showed that dry heat treatment produced Maillard reaction products, which had much higher IgE binding properties than untreated controls (Chung and Champagne 1999; Maleki et al. 2000). Peanut protein, protein-bound end products, or adducts, such as advanced glycation end products (AGE), N-(carboxymethyl)lysine (CML), malondialdehyde (MDA), and 4-hydroxynonenal (HNE), have been identified after roasting. A higher level of IgE binding correlated with higher levels of AGE adducts (Chung and Champagne 2001). Maturation and curing, in conjunction with peanut roasting, may be associated with allergenicity (Chung et al. 2003). Methods such as frying or boiling of peanuts, as practiced in China, appear to reduce the allergenicity of peanuts compared with dry roasting, which is widely practiced in the USA (Beyer et al. 2001). The decrease in allergenicity of boiled peanuts results mainly from a transfer of low-molecular-weight allergens to the water during cooking (Mondoulet et al. 2003, 2005). The main peanut allergen, Ara h 2, purified from raw and roasted peanuts, is homologous to and functions as a trypsin inhibitor. Roasting was found to cause a 3.6-fold increase in trypsin inhibitory activity (Maleki et al. 2003). Thermal treatment of recombinant Ara h 2 led to a substantial increase in the IgE binding reaction when reactive carbohydrates, or their breakdown products, were also present (Gruber et al. 2005).

Roasting, autoclaving, blanching, and microwave heating of almonds have demonstrated the antigenic stability of almond proteins [enzyme-linked immunosorbent assay (ELISA) and WB tests] compared to unprocessed nuts (Venkatachalam et al. 2002). Hazelnut protein, with a molecular weight less than 14 kDa, demonstrated high heat stability and was detected even after high-temperature (185 °C) processing (Wigotzki et al. 2000).

16.3.1.6 Fruits and Vegetables

Heat sterilization of peaches (121 °C for 10 and 30 min), chemical lye peeling of fruits, and ultrafiltration of juice through membranes with suitable cutoffs have been tested. Sterilization was not able to decrease the allergenicity of the Pru p 1 protein. Furthermore, the protein band was still present even after 60 min of reaction with two different acidic proteases. Chemical lye peeling of fruits and ultrafiltration of juice through membranes (with suitable cutoffs) was able to decrease the amount of the major allergen in peach juice (Brenna et al. 2000).

Chemical peeling, thermal treatment, and syruping processes were applied to different varieties of cherries. Chemical peeling successfully removed Pru av 3, a LTP responsible for allergic reactions in patients without pollinosis. The syruping process removed almost all allergenic proteins (Primavesi et al. 2006). Recombinant Pru av1 (r Pru av1) was thermally treated in the presence and absence of carbohydrates. Thermal treatment for 30 min had no effect on the IgE binding properties of the main cherry allergen, whereas reaction with glucose and ribose, as well as carbohydrate breakdown products, was able to decrease the binding capacity of the r Pru av1 protein (Gruber et al. 2004).

Different types of heat processing of potatoes only slightly lowered the allergenic activity (Schubert et al. 2003). Microwaved extracts of celery root demonstrated high heat lability for Api g 1, but profilin and carbohydrate epitopes appeared to be more resistant to heat compared to untreated controls (Jankiewicz et al. 1996). Extracts of raw, cooked celery and celery spice were studied by Ballmer-Weber et al. (2002). They found that the allergenicity of celery was preserved for four patients with a positive DBPCFC to celery even after the celery was cooked (76.07 min, 100 °C). Celery spice was found to be allergenic for patients with an allergy to raw celery. In another study, cooked carrots were found to be tolerated by sensitized patients (Gómez et al. 1993; Quirce et al. 1997).

16.3.1.7 Pasta, Grain, and Baked Products

Model pasta samples (durum wheat) were dried at temperatures of 20 °C, 60 °C, 85 °C, 110 °C, and 180 °C and then cooked in boiling water. The digestion process, together with previous heat treatments, was not able to completely inactivate the IgE-reactive peptides (De Zorzi et al. 2007).

Almost all rice proteins were excluded or weakened in the process of boiling; however, IgE binding activity remained, even in hypoallergenic rice (Yum et al. 2006). Wheat dough and bread crumbs and crust, before and after in vitro digestion, were tested for the presence of allergens (Simonato et al. 2001). During in vitro digestion, the IgE binding protein components of unheated dough disappeared; however, bread crumbs and crust-isolated proteins retained their IgE binding. The effects of baking must be taken into account when studying food allergies associated with wheat products. Pastalike model samples of wheat flour mixed with egg white were tested after boiling by Kato et al. (2001). Almost no antigenic activity of ovomucoid (coming from egg white) was detected in the extracts of heated samples. Protein allergens extracted from wheat, rye, barley, and oat flour were heat treated to mimic standard methods of food preparation; however, none of the processing conditions entirely abolished IgE-binding. Varjonen et al. (1996) found that heating of proteins reduced the allergenicity and, thus, the sensitivity during testing. Commercial gliadins and wheat flour were exposed to microwaves at power settings of 70, 200, and 500 W for different time periods (Leszczynska et al. 2003a). A significant increase in reactivity, over the untreated control sample, was observed for gliadins exposed to an energy dose of 40 kJ. Gliadins treated with higher energy doses showed a reactivity decrease during immune response testing.

16.3.1.8 Milk and Milk Proteins

Raw, pasteurized, and homogenized/pasteurized cow's milk and hypoallergenic infant formula, as a control, were tested by Host and Samuelsson (1988). This work provided evidence that heat treatment increased the ability of pasteurized and homogenized/pasteurized milk to evoke allergic reactions in patients allergic to milk. Whole casein, gamma-globulin, serum albumin, beta-lactoglobulin, alpha-lactalbumin, and whole cow's milk were tested for allergenicity after boiling by Norgaard et al. (1996). Boiling milk for 10 min, but not 2 min, eliminated the skin-prick-test response in subjects sensitized to heat labile serum albumin or beta-lactoglobulin, whereas subjects sensitized to the heat-stabile casein reacted equally to fresh and boiled cow's milk.

16.3.2 Enzymatic Hydrolysis and In Vitro Digestion

Enzymatic hydrolysis is currently the most successful method for the preparation of hypoallergenic and nonallergenic foods and is used mainly for baby food formulas (Von Berg 2007). The method is based on cutting proteins into peptides or even into individual amino acids with the aim of destroying the epitope(s) causing IgE binding in sensitized individuals. The food matrix effect and the number of different allergens complicate this type of processing.

16.3.2.1 Milk

A review by Bousquet et al. (1998) cited many papers on this topic and concluded that only an extensive enzymatic hydrolysate should be adopted as hypoallergenic milk for infant formula. Currently several commercial preparations are available with varying levels of protein hydrolysis (Fuchs 2007).

16.3.2.2 Peanut and Peanut Allergens

Gastroduodenal digestion fragments of Ara h 1 retain the T-cell stimulatory and IgE-binding and cross-linking properties found in the intact protein (Eiwegger et al. 2006). Peanut protein isolate digested together with polysaccharides (i.e., gum Arabic, low methylated pectin and xylan) was only partly hydrolyzed; this was probably due to nonspecific interactions between polysaccharides and peptides (Mouécoucou et al. 2004). In retentates, IgE binding was reduced by digestion and the presence of xylan. In dialysates, IgE binding was reduced by all of the polysaccharides. This work highlights an opportunity to control food allergenicity using properly formulated final products.

16.3.2.3 Fruit and Vegetables

Enzymatic decomposition, mechanical tissue disintegration, and heating during peeling, mash treatment, and pasteurization of mango product processes were tested by Dube et al. (2004) to determine their effects on allergenicity. No substantial decrease in allergenicity was observed in mango puree extracts and nectars. In addition, conventional processes did not completely eliminate the allergenicity of mango-pulp-containing products. Bublin et al. (2008) modeled digestion (passage from stomach to duodenum) using two kiwifruit allergens, actinidin (Act d 1) and thaumatin-like protein (Act d 2). Act d 1 was irreversibly destabilized in acidic solutions; however, heat-induced denaturation of Act d 2 (pH 2) was fully reversible. IgE binding to Act d 2, but not Act d 1, was detected in processed food products.

16.3.2.4 Wheat Flour

Wheat flour was enzymatically hydrolyzed by bromelain and was found to be effective in decomposing the epitope structure (Tanabe et al. 1996). Cellulase (50 °C, 1 h) and actinase (40 °C, 1 h) were used for enzymatic decomposition of cellulose- and protein-based allergens in wheat flour; the flour was then used to make flour batter through gelatinization of starch components. Results of ELISA tests showed negative allergenicity in most cases (Watanabe et al. 2000).

16.3.2.5 Legumes

Protease treatment of soybeans has been shown to markedly reduced antigenicity to monoclonal antibodies and allergenicity to sera from soybean-sensitive patients. These results evidenced that there is an opportunity for developing a hypoallergenic product (Yamanishi et al. 1996). Purified known allergenic and nonallergenic proteins from leguminous crops were studied using a simulated gastric fluid digestion assay (Misra et al. 2009). Most of the proteins that were stable in simulated gastric fluid were similar to characterized allergens based on their molecular weights (soybean, peanut, chickpea, and black gram).

16.3.2.6 Rice

Watanabe et al. (1990) treated rice grains with actinase. Allergenicity tests showed negative results, and the product, when clinically administered to seven patients with atopic dermatitis, failed to provoke an allergic reaction in six of the seven patients.

16.3.3 Gamma Irradiation

This process is extensively used for microbial decontamination of food additives and dry foods such as spices and dried herbs. Recently passed European legislation requires that any irradiated components must be declared on the packaging. As a result, currently, the method is practically unused; instead, steaming or dry-heat processing is used for microbial decontamination of these products. Regardless, some interesting results were obtained regarding the allergenicity of selected foods.

The principle of the effect can be demonstrated with ovalbumin and bovine serum albumin in solution (0.2 % in 0.01 M phosphate buffer, pH 7.4); the proteins were then irradiated with high doses (8 kGy). This process caused the formation of protein aggregates and degraded fragments with reactivity to specific antibodies. The main part of the conformation-dependent reactivity, i.e., the spatial antigenic structure (conformational epitope), was lost, but some antigenicity persisted (Kume and Matsuda 1995).

16.3.3.1 *Sebastiania jacobinensis* Bark Lectin

High doses of gamma irradiation (above 1 kGy) induced a significant loss of activity in *Sebastiania jacobinensis* bark lectin, caused by apparent changes in the hydrophobic surface. Gamma irradiation has also been shown to cause protein misfolding and aggregation (Vaz et al. 2011).

16.3.3.2 Milk Proteins

Bovine alpha-casein and beta-lactoglobulin, when irradiated, changed their allergenicity and antigenicity. The change was probably caused by agglomeration of proteins (Lee et al. 2001).

16.3.3.3 Egg Proteins

Cakes containing egg whites were gamma-irradiated with 10 or 20 kGy. After irradiation and processing, ovalbumin allergenicity was found to have decreased. Egg white, irradiated to reduce egg allergies, could be used to produce safer cakes (Lee et al. 2005). Hen egg ovomucoid, at a basic pH, was irradiated at 10 kGy or heated at 100 °C for 15 min. The combination of irradiation and heating was found to be very effective in reducing the amount of intact ovomucoid, regardless of pH conditions (Lee et al. 2002).

16.3.3.4 Gliadin

Commercial gliadin powder and wheat flour were irradiated with doses between 2.2 and 12.8 kGy. Surprisingly, irradiated gliadin increased its allergenicity. Gliadin extracted from irradiated wheat flour was found to exhibit higher immunoreactivity than pure gliadin irradiated with the same dose (Leszczynska et al. 2003b).

16.3.3.5 Shrimp

The heat-stable protein in shrimp was isolated and gamma-irradiated at 0, 1, 3, 5, 7, or 10 kGy in solution (1 mg/mL); fresh shrimp was also irradiated. The IgE binding rate was reduced with increasing doses of radiation. The main band disappeared, and traces induced from coagulation appeared at a higher-molecular-weight zone, as demonstrated using SDS-PAGE. The same results were also obtained for proteins extracted from irradiated shrimp (Byun et al. 2000).

16.3.4 Fermentation

This process is similar in many respects to enzymatic hydrolysis because microorganisms act via their enzymatic systems on food matrix components including proteins (Phromraksa et al. 2008). Fermentation is a complex process and was the reason for reviewing this topic separately.

16.3.4.1 Whey Proteins

Alpha-lactalbumin and beta-lactoglobulin, in whey from sterilized fermented milk, were tested for antigenicity and allergenicity. Whereas the antigenicity was substantially reduced, the allergenicity of the whey proteins only slightly decreased (Jędrychowski and Wróblewska 1999). The immunoreactivity of raw and pasteurized milk "naturally" fermented was only slightly reduced. Industrially manufactured sour milk as well as Acidophilus milk presented much lower levels of immunoreactivity (Maier et al. 2006).

16.3.4.2 Legumes

Soybean products such as sprouts, tempeh, tofu, miso, mold-hydrolyzed soy sauce, acid-hydrolyzed soy sauce, and hydrolyzed vegetable protein were tested for allergenicity using a competitive inhibition test. Results showed that fermentation could alter or destroy allergenic epitopes (Herian et al. 1993).

Pea flour, fermented with three lactic acid bacteria and the fungi *R. microspores*, had reduced residual antigenicity against antipea antibodies; however, reactions to antipea profilin and anti-Bet v 1 were still detectable after fermentation (Barkholt et al. 1998).

Soybean flour, soy milk, texturized soy proteins, and two infant formulas (the first containing total proteins and the second containing a soy protein hydrolysate) were tested. Immunoblotting showed a lack of allergenicity in the infant formulas, and the 30-kDa allergen (Gly m Bd 30) disappeared during production of texturized soy protein (Franck et al. 2002).

Gliadin was treated using concentrated crude enzyme from *Bacillus subtilis*. The *B. subtilis* isolated from fermented soybean foods served as the source of enzyme. The process reduced the allergenicity of gliadin by hydrolyzing the allergenic gliadin fragments detected using immunoblotting. Phromraksa et al. (2008) showed that *B. subtilis* could be used for the production of hypoallergenic wheat flour or milk food products.

16.3.4.3 Shrimp

Raw shrimp (*Acetes japonicus*) and saeujeot (salted and fermented shrimp) in cabbage kimchi were tested for allergenicity. The allergenicity of both raw shrimp and saeujeot, in kimchi, was shown to decrease during fermentation, but the decrease in allergenicity of saeujeot was greater than that observed of raw shrimp (Park et al. 2007).

16.3.5 Refining

This process involves mostly vegetable oils that contain traces of proteins derived from seeds of plants during oil pressing.

16.3.5.1 Peanut Oil

Refined peanut oil has been shown to be safe for the vast majority of people with peanut allergies, whereas unrefined oil can provoke reactions in some of the same individuals (Crevel et al. 2000).

16.3.5.2 Sunflower Oil

Sunflower oil was sampled in different stages of the refining process: crude pressed state of oil, acidification and neutralization, pregumming by centrifugation, washing, bleaching, gumming by filtration, and deodorization. The SDS-PAGE test identified five bands, from 67 to 145 kDa, with 67 kDa being the most visible band and representing the main allergen. The amount of this protein decreases as the refining process proceeds; unfortunately, traces were still present in the refined oil. Therefore, refined sunflower oil may represent a risk to people highly sensitized to sunflower seeds (Zitouni et al. 2000).

16.3.5.3 Soy Lecithin and Soybean Oils

No IgE-binding activity was found in refined soybean oils, although the nonrefined oils and soy lecithin showed residual IgE-binding activity. The lecithin extracts showed IgE-binding protein with a molecular mass of approximately 16 kDa (Paschke et al. 2001). Six commercial soy lecithins with extracts from raw and heat-treated soybeans were tested. Lecithins that contained residual proteins caused specific mediator release. These products may induce allergic symptoms. The results of Müller et al. (1998) showed that soybean lecithins were capable of introducing hidden allergens into processed foods and that the IgE binding potential corresponded to the total protein content.

16.3.6 Microparticulation

European legislation requires that novel foods and novel technologies be tested for adverse effects on the structure and composition of treated foods, including neoallergen generation, before being utilized in the food industry. The paper by Sampson and Cooke (1992) showed the results of such a test of microparticulation technology.

16.3.6.1 Fat Replacers

Microparticulated egg white or cow's milk proteins used as a fat substitute were tested for allergenicity. The same allergens as those found in the original products were identified in microparticulated products. No novel protein fractions were found in the processed test materials (Sampson and Cooke 1992).

16.3.7 Genetic Modification

This is not a process, in the food engineering sense, but we have included this methodology because of its great potential. It may influence the generation of allergenic proteins in the initial stages of plant breeding. Genetically modified organism (GMO) produce is broadly tested for the presence of neoallergens (Björksten et al. 2008). GMO breeding companies are therefore motivated to fund research on existing allergens in plants and compare them with newly designed plants with lower allergen content.

16.3.7.1 Legumes

Transgenic soybeans, with methionine-rich 2S albumin from the Brazil nut (*Bertholletia excelsa*), have been developed. The 2S albumin is probably a major Brazil-nut allergen, and the transgenic soybeans analyzed in this study contained this protein. It has been demonstrated that an allergen from a food known to be allergenic can be transferred to another food by genetic engineering (Nordlee et al. 1996).

Helm et al. (2000) genetically modified the composition of peptides. They substituted a single-site amino acid in the 5 immuno-dominant epitopes of Gly m Bd 30 K with alanine and caused an IgE-binding reduction or elimination of epitopes 6 and 16, as shown through serum testing of six soybean-sensitive patients.

16.3.7.2 Transgenic Rice

14–16 kDa allergen proteins and their transcripts of the seeds from several transgenic lines showed much lower concentrations than proteins isolated from parental wild-type rice. This reduction was stable in three generations of plants (Tada et al. 1996). Additionally, the 16-kDa allergen content of seeds from several transgenic rice plants was shown to be much lower than the 16-kDa allergen content found in parental wild-type rice (Nakamura and Matsuda 1996).

16.3.7.3 Peanut

Guo et al., (2008) identified one peanut line, GT-C9, lacking several seed proteins, which were identified as Ara h 3 isoforms by peptide sequencing, and called it iso-Ara h 3. A full-length sequence of iso-Ara h 3 (GenBank number DQ855115) was obtained. The deduced amino acid sequence of iso-Ara h 3 (ABI17154) had the first three of four IgE-binding epitopes of Ara h 3. Anti-Ara h 3 antibodies reacted with two groups of protein peptides. One exhibited a strong reaction and another had only a weak reaction. The peptide spots with a weak reaction to anti-Ara h 3 antibodies were subunits or iso-allergens of this potential peanut allergen iso-Ara h 3. A recent study suggested that Ara h 3 basic subunits may be more significant allergens than the acidic subunits.

16.3.7.4 Main Hen Egg White Allergen Ovomucoid (Gal d1)

Substitution of two amino acids in six peptides in the third domain of ovomucoid [the replacement of phenylalanine at position 37 (F37) with methionine] produced a significant loss of IgG and IgE binding as well as disruption of the α-helix structure. Substituting glycine at the 32nd position, together with F37, showed a synergistic effect in decreasing the antigenicity (Mine et al. 2003).

16.3.8 Pulsed Ultraviolet Light (PUV) Treatment

This method has great potential for use on an industrial scale due to its simplicity and effectiveness.

16.3.8.1 Peanut

Peanut extracts and liquid peanut butter were PUV treated with a xenon RS-3000C lamp, at a frequency of three flashes per second and positioned 14.6 cm from the central axis of the lamp. The treatment lasted 4 min (extract) or 3 min (liquid peanut butter). Boiling was used as a control. PUV treatment reduced the solubility of the 63-kDa peanut allergen, whereas 18- to 20-kDa allergen solubility was not affected. A reduced level of allergens in PUV-treated samples was caused by the insolubility of the aggregates. PUV light was effective in reducing IgE binding of peanut extracts and liquid peanut butter; however, the allergenicity reduction needs to be confirmed by clinical studies (Chung et al. 2008).

16.3.8.2 Soy

Soy extracts were treated with PUV at 2, 4, and 6 min. PUV treatment was shown to decrease the soy allergen (i.e., glycinin and beta-conglycinin) levels. PUV treatment also reduced the allergenic potency of soy extracts. The optimal PUV treatment time was found to be 4 min. This technology has the potential to be used for the production of less allergenic soybean beverages and products, but clinical studies are needed (Yang et al. 2010).

16.3.9 Polymerization of Allergen Proteins

This method is based on the oxidation of phenolics that naturally occur in a given food matrix or are intentionally added to a target food. The oxidation products polymerize with proteins and form complexes with a much larger molecular size than the original allergen. This makes the epitopes inaccessible for IgE binding, as discussed subsequently.

16.3.9.1 Peanut

Protein extracts from raw and roasted defatted peanut meal (pH 8) were incubated (37 °C, 60 min) with and without peroxidase in the presence of hydrogen peroxide. Peroxidase treatment had no effect on protein (isolated from raw peanuts) cross-linking. A decrease in the content of the major allergens, Ara h 1 and Ara h 2, in roasted peanuts after peroxidase treatment was observed, and the formation of polymers and reduction in IgE binding was observed (Chung et al. 2004).

Peanut extracts were treated with and without polyphenol oxidase (PPO)/caffeic acid (pH 8, 37 °C, 1 h) and caffeic acid (pH 10.5) overnight. Cross-links and a decrease in the levels of two major peanut allergens, Ara h 1 and Ara h 2, were observed. PPO/caffeic acid treatment reduced the allergenic properties of Ara h 1 and Ara h 2 by cross-linking and decreasing the levels of allergens (Chung et al. 2005).

Extracts from raw and roasted peanuts were treated with and without phytic acid at various pH values. Phytic acid formed complexes with the major peanut allergens. This made the Ara h 1 and Ara h 2 allergens inaccessible to IgE (complexes were insoluble under acidic and neutral conditions) that resulted in reduced allergenicity, and peanut butter slurry was shown to behave in a similar way (Chung and Champagne 2007).

Peanut extracts and liquid peanut butter, mixed with phenolics, led to irreversible precipitation of the major peanut allergens Ara h 1 and Ara h 2, and IgE binding was reduced approximately 10- to 16-fold (Chung and Champagne 2009).

16.3.9.2 Fruit and Vegetables, Juices

IgE-binding activity of recombinant cherry allergen, Pru av 1, was studied after reaction with oxidation products of various phenolic acids in the presence of polyphenol oxidase. Caffeic acid in combination with epicatechin was among the strongest inhibitor of rPru av1 protein IgE binding activity (Gruber et al. 2004).

Juice from Golden Delicious apples was treated via oxidation and followed by HPT. This process deallergized the apple juice main allergen Mal d 1. During 3 weeks of storage, the oxidized, pressure-treated apple juice did not recover its allergenicity, as determined by testing (WB test) on a group of people sensitized to Mal d 1 (Setinova et al. 2010). Details of WB test results and the final shape of deallergized apple juice are given in Figs. 16.5a and 16.6. The oxidation method is based on theory of Chung et al. (2005) (Fig. 16.7). All necessary components are frequently present in apple juice available on the market. Therefore, this method of oxidation can be applied here.

Celery juice and a celery/apple juice mixture (1:5) were oxidized. Oxidization failed to eliminate the allergenicity of pure celery juice. For the mixed juice, the WB test showed that oxidization by stirring for 120 min reduced the allergenicity of the mixture. However, the basophil activation test showed no reduction in the allergic response to the oxidized juice mixture. Skin-prick testing showed that the oxidized juice mixture stirred for 120 min exhibited a significantly lower reaction than the juice mixture stirred for 60 min or control juices. The pure celery and apple juices stabilized with ascorbic acid served as fully allergenic controls. Due to contradictory results in different tests, oxidized mixtures of juices cannot be declared safe (Novotná et al. 2011). This case shows how important it is to use a battery of tests for allergenicity testing.

The main apple allergen, Mal d1, in phosphate buffer, prepared from apple peel and flesh, was oxidized using endogenous and exogenous polyphenol oxidase and peroxidase with added catechin and studied using in vitro tests (Garcia et al. 2007). The strongest effect of the oxidation process was the effect of catechin and polyphenol oxidase on IgE binding of Mal d1.

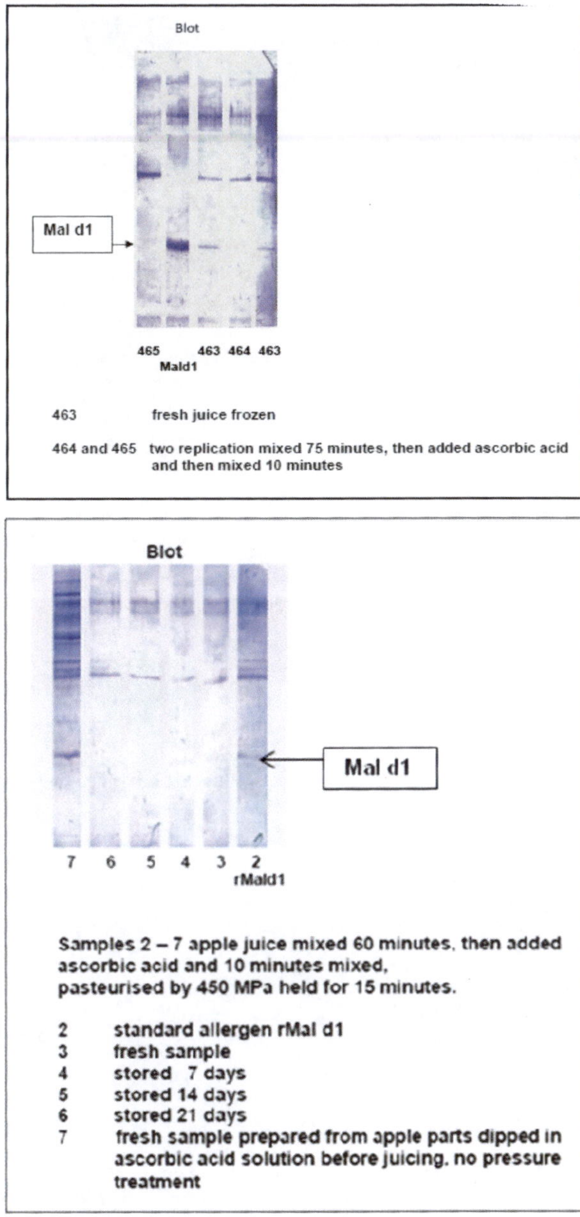

Fig. 16.5 Apple juice deallergization by oxidation – results regarding the juice prepared in industrial conditions, Open food challenge test done for 19 patients confirmed no reaction on 18 of them

16.3.10 High-Pressure Treatment and Pulsed Electric Field

Very likely there were great expectations for HPT relative to the allergenicity of fruit and vegetables based on results from the thesis of Scheibenzuber (2003). Rigorous tests on pure homolog allergens of birch, apple, carrot, and celery showed

Fig. 16.6 Deallerized apple pulp juice enriched by vitamin C and pasteurised by high pressure Tested during shelf life: Microbial stability proven (due to pasteurisation by high pressure) , Deallerisation stability proven , Vitamin C stability proven

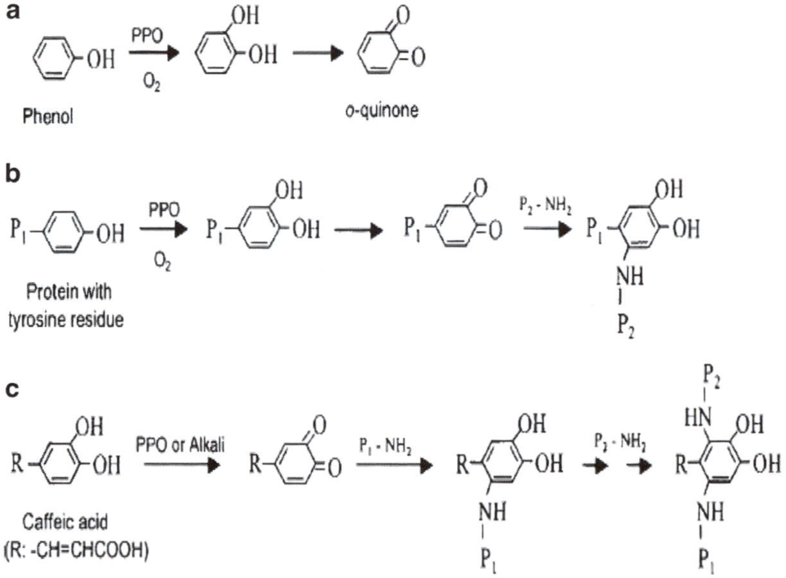

Fig. 16.7 The mechanism given by Chung et al (2005) requires presence of phenolic acids, PPO, low antioxidant content, oxygen (air) and time for reaction. These components are all present in apple flesh

that HPT itself is not a causative parameter of the observed changes in allergenicity. High-pressure conditions probably enhanced the reactions (oxidation, polymerization) that led to the observed decreased allergenicity of treated foods. These reactions were observed and described by Garcia et al. (2007).

Recent results presented in papers by Yamamoto et al. (2010a, b) showed that HPT could decrease allergenicity (IgE-based testing) of wheat alpha-amylase inhibitor and bovine gamma globulin.

Three earlier papers presented applications of high pressure as the tool for increasing the rate and depth of enzymatic hydrolysis reactions of the main food allergens in rice (Kato et al. 2000) and milk (Chicón et al. 2008; Beran et al. 2009).

High-pressure-assisted sterilization, which is an emerging technology, seems to be an effective tool for inactivation of the main apple allergens Mal d1 and Mal d3 and the celery allergen Api g1 (Husband et al. 2011).

16.3.10.1 Rice

Rice grains immersed in distilled water and pressurized at 100–400 MPa released a substantial portion of the proteins present in the grain. SDS-PAGE testing identified the major protein allergens as 16 kDa albumin, alpha-globulin, and 33-kDa globulin. Application of proteolytic enzymes led to complete removal of allergens from rice grains (Kato et al. 2000).

16.3.10.2 Alpha Amylase Inhibitor

Circular dichroism (CD) spectra, fluorescence, and ultraviolet (UV) spectroscopy were used to investigate the influence of HPT on the structure of the alpha amylase inhibitor. IgE-specific binding activities were also used to test the high-pressure influence on allergenicity of this protein. Pressure treatment over 300 MPa has been shown to change the tertiary protein structure. Pressure-treated protein exhibited decreased IgE-binding activities (dot blotting) (Yamamoto et al. 2010a). The authors of this paper expected that conformational changes in the tertiary structure would be responsible for lowering the allergenicity of pressure-treated protein.

16.3.10.3 Fruits and Vegetable, Juices, and Allergens

In his thesis, Scheibenzuber (2003) presented interesting results regarding the effect of HPT on allergenicity and structural changes in apple (main apple allergen Mad d 1), hazelnut, carrot, cherry, peach, and celery. Nineteen allergic patients tolerated high-pressure-treated apple (600 MPa, 5 min) without symptoms of oral allergy syndrome. Prick-to-prick tests showed a correlation between allergen inactivation and applied pressure levels. HPT caused obvious structural changes in allergens, as indicated by CD and Fourier-transform infrared (FTIR) spectra. Surprisingly, high-pressure and thermally treated Mal d 1 produced a higher degranulation ratio in the basophil test.

Buffered solutions of recombinant Bet v 1 (r Bet v 1) and birch pollen extract were treated using high pressure (450–550 MPa) for a period of 10 min in a temperature range of 30–50 °C. The greatest changes in the rBet v 1 structure were found in samples treated at 450 MPa at 30 °C. Samples treated at 500 and 550 MPa at 30 °C (10 min) showed no structural changes. Pressures of 450–550 MPa (10 min) at temperatures of 30 °C, 40 °C, and 50 °C did not change the allergenicity (WB test) of the rBet v1 protein or the birch pollen extract compared with untreated samples (Setinova et al. 2009a).

A buffered recombinant Api g 1 (rApi g 1) solution, which is the main celery allergen, was treated at 500 MPa for 10 and 20 min at temperatures of 30 °C, 40 °C, and 50 °C. The greatest changes in structure were found in samples treated at 500 MPa at 50 °C. Samples treated at 50 °C at 400, 450 and 500 MPa for 10 and 20 min demonstrated that protein structural changes were positively correlated with pressure. WB analysis showed that pressure treatments applied in the study did not change the allergenicity of the rApi g 1 protein compared with untreated samples (Houska et al. 2009a).

Recombinant main carrot allergen (rDau c 1) and carrot juice were treated at 500 MPa for 10 min and different temperatures (30 °C, 40 °C, and 50 °C) and pressures of 400–550 MPa for 3 and 10 min. Based on in vitro and in vivo tests, HPT had no effect on the allergenicity of rDau c 1 and carrot juice. Only structural changes were observed (Heroldova et al. 2009).

Extracts from Golden Delicious apples were treated at various pressures up to 800 MPa and stored for 10 months. The RAST inhibition tests, using sera of Spanish patients, fail to show any effects of HPT on apple antigenicity or IgE binding. The antigenicity of high-pressure-treated extract was preserved after 10 months of frozen storage. Immunoblots have confirmed that heat is more important than pressure in inducing irreversible changes in key epitopes (Fernandez et al. 2009).

The main apple and carrot allergens rMal d 1 and rDau c 1 were treated at 400, 450, 500, and 550 MPa for 3 and 10 min at 25 °C and at 500 MPa for 10 min at 30, 40, and 50 °C. There was no influence of high pressure on allergenicity of rMal d 1 and rDau c 1 proteins in vitro. A pressure of 500 MPa showed major spectral changes in protein rMal d 1 without the reduction of its in vitro reactivity (Setinova et al. 2009b).

rMal d 1 treated with high pressure (500 MPa, 10 min, 30 °C) showed the greatest changes in CD spectra compared with untreated samples. HPT in the range of the applied parameters was not capable of altering the allergenicity of rMal d1 in pure solutions in a given group of patients. Also, the allergenicity of apple juice and apple homogenates, prepared using this technique, were not substantially decreased (Houska et al. 2009b).

The combined effect of high pressure and temperature on the main apple allergens Mal d1 and Mal d3 and the main celery allergen Api g1, in their original matrices, was studied. The Mal d1 allergen was found to be labile, whereas Mal d3 and Api g1 exhibited relative stability during processing. Added pectin had a protective effect relative to the reduction of immunoreactivity, as shown using SDS-PAGE and WB tests. High pressure (700 MPa) combined with high temperature (up to 115 °C or 118 °C) showed promising results in reducing apple and celeriac allergenicity (Husband et al. 2011).

16.3.10.4 Bovine Milk Proteins

The following two papers can also be reviewed in the enzymatic methods section. High pressure is used as a reaction rate enhancer.

Chymotrypsin and trypsin enzymatic hydrolysis of beta-lactoglobulin was run under high pressure. Treatment of beta-lactoglobulin dissolved in buffer with chymotrypsin and trypsin under high pressure for relatively short times (20 min) was found to accelerate proteolysis by leading to a rapid removal of intact protein. However, the results did not support the notion that trypsin and chymotrypsin enzymatic hydrolysis, under high pressure, selectively addresses the allergenic regions of beta-lactoglobulin. Some residual IgE-binding properties were retained. However, it is possible to select the conditions that quickly produce hydrolysates with reduced potential allergenicity that could be used in hypoallergenic food production (Chicón et al. 2008).

Tryptic and chymotryptic enzymatic hydrolysis of alpha- and beta-casein, bovine serum albumin, beta-lactoglobulin, and alpha-lactalbumin under high isostatic pressure (500 MPa) has also been studied. Significant changes in peptide profiles and a progressive reduction in residual intact proteins was seen after high-pressure tryptic proteolysis of beta-lactoglobulin and bovine serum albumin and high-pressure chymotryptic proteolysis of beta-lactoglobulin, alpha-lactalbumin, and bovine serum albumin. A statistically significant decrease in the residual immunochemical reactivities of beta-lactoglobulin tryptic and alpha-lactalbumin chymotryptic hydrolysates, prepared under high pressure, compared to control samples hydrolyzed at atmospheric pressure, was also observed (Beran et al. 2009).

Bovine gamma globulin structural changes and IgE-specific binding activity have also been studied (Yamamoto et al. 2010b). Whereas the secondary structure of the protein was not changed by high pressure, the tertiary structure was changed and IgE-binding activity decreased.

16.3.10.5 Peanut and Apple Allergens

Native purified peanut Ara h 2, 6 and apple Mal d 3 and Mal d 1b allergens were treated using high pressure and PEF. No significant change in structure was observed (CD spectroscopy) after PEF treatment of any of the tested allergens. The structure of the Ara h 2, 6 and Mal d 3 did not change after HPT at 20 °C, and only minor changes were observed in the structure of Mal d 1b. Ara h 2, 6 was stable under HP at 80 °C, but the structures of Mal d 1b and Mal d 3 were changed during more severe treatment. An ELISA test of temperature-treated Mal d 3 showed that antibody reactivity correlated well with the loss of structure. HP and PEF processing exhibited little effect on the structure of purified allergen (Johnson et al. 2010).

Somkuti et al., (2010) used high-pressure FTIR spectroscopy to determine the pressure and temperature stability of recombinant Mal d 1. The protein unfolded at

55 °C when heated at normal atmospheric pressure. Pressure unfolding was measured under different pD conditions, and the effect of a sugar mixture, similar to that of apples, and the effect of ionic strength were also studied. In all cases it was shown that the allergen unfolded in a range of 150–250 MPa. Unfolding was irreversible and was followed by aggregation of the unfolded protein.

16.4 Conclusion

There are few successful processing methods for deallergization of foods on an industrial scale. Hypoallergenic or nonallergenic products currently on the market are mostly prepared by enzymatic hydrolysis (infant formula) or by intentional design from nonallergenic components (e.g., gluten-free bakery products, gluten-free beer).

Deallergization is a challenge for food engineers and those involved in developing novel food processing technologies. High-pressure or pulsed ultraviolet light treatment could be used in the near future to enhance enzymatic reactions or speed the oxidation and polymerization of phenolics and allergenic components that are naturally present in food or intentionally added to foods (e.g., apple juice, peanut butter).

Our review shows that study conclusions are seldom based on more than two allergenicity tests (Figs. 16.3 and 16.4). There are examples of where in vitro tests provided satisfactory results but in vivo tests yielded contradictory results. Great care should be taken in testing before food items are released onto the market and labeled as hypoallergenic products. The food industry should work to develop hypoallergenic foods using the methods described here; however, the new products will require thorough studies using the gold standard DBPCFC, skin-prick test, basophil test, and tests based on IgE immunodetection methods and must demonstrate lowered or diminished allergic reactions.

Additionally, genetic manipulation offers a great opportunity to breed plants with limited allergen content. However, this method could produce different proteins to which people would become sensitized after daily long-term consumption. Therefore, other methods should be pursued at the same time.

16.4.1 Insight into the Future

There are opportunities to design hypo- or nonallergenic foods from alternative components. A good example of this is gluten-free bread. Currently, hundreds of scientific papers and even more patent applications exist for gluten-free products. Tens of gluten replacers have been found, with maize being a good example (Brites et al. 2010).

An additional direction for allergy control is directed toward patients. One recent method was the vaccination of profilin-allergic patients (Westritschnig et al. 2007). A restructured recombinant vaccine has been developed for treatment using the strategy of tail-to-head reassembly of hypoallergenic allergen fragments within one molecule. This strategy is generally applicable for the production of allergy vaccines.

Another strategy that focused on sensitized humans was recently presented by Meyer-Pittroff et al. (2007). High-pressure-treated apple slices were consumed for several weeks by three patients who had a strong allergy to apples. Following immunotherapy, they were hyposensitized to apples.

The same principle was proposed by Tanabe (2008). Enzymatically modified wheat flour was used to prepare hypoallergenic cakes, which were consumed over a long period of time by patients. Surprisingly, more than half of them developed oral tolerance and eventually were able to eat normal wheat products.

A breakthrough method was presented recently by Lodinová-Zadniková et al. (2004). A probiotic strain of *Escherichia coli* was orally administered, after birth, to colonize the gut, change allergy markers, and affect the clinical manifestation of allergies in high-risk infants of allergic mothers. This method offers an opportunity to cut the closed circle of transfer of allergies from parents to children. The correlation between gut permeability and food allergies is mentioned in the recent paper by Perrier and Corthésy (2011). We can speculate that an early colonized gut, probably trained with *E. coli* strains, can strongly contribute to the regulatory effect of the immune system. Such a gut could better handle incoming protein antigens without accelerating the intestinal permeability by overshoot inflammatory reactions.

All these methods offer a chance to stem the steadily growing numbers of allergy sufferers and stabilize the concomitant treatment costs.

Acknowledgments This work was supported by Institutional Grant 00027022–02 of the Ministry of Agriculture of the Czech Republic.

References

Armentia A, Dueñas-Laita A, Pineda F, Herrero M, Martín B (2010) Vinegar decreases allergenic response in lentil and egg food allergy. Allergol Immunopathol 38:74–77

Ballmer-Weber BK, Hoffmann A, Wüthrich B, Lüttkopf D, Pompei C, Wangorsch A, Kästner M, Vieths S (2002) Influence of food processing on the allergenicity of celery: DBPCFC with celery spice and cooked celery in patients with celery allergy. Allergy Eur J Allergy Clin Immunol 57:228–235

Barkholt V, Jorgensen PB, Sorensen D, Bahrenscheer J, Haikara A, Lemola E, Laitila A, Frokiaer H (1998) Protein modification by fermentation: effect of fermentation on the potential allergenicity of pea. Allergy Eur J Allergy Clin Immunol 53:106–108

Beran M, Klubal R, Molik P, Strohalm J, Urban M, Klaudyova AA, Prajzlerova K (2009) Influence of high-hydrostatic pressure on tryptic and chymotryptic hydrolysis of milk proteins. High Press Res 29:23–27

Bernhisel-Broadbent J, Scanlon SM, Sampson HA (1992a) Fish hypersensitivity. I. In vitro and oral challenge results in fish-allergic patients. J Allergy Clin Immunol 89:730–737

Bernhisel-Broadbent J, Strause D, Sampson HA (1992b) Fish hypersensitivity. II: clinical relevance of altered fish allergenicity caused by various preparation methods. J Allergy Clin Immunol 90:622–629

Besler M, Steinhart H, Paschke A (2001) Stability of food allergens and allergenicity of processed foods. J Chromatogr B Biomed Sci Appl 756:207–228

Beyer K, Morrow E, Li X-M, Bardina L, Bannon GA, Burks AW, Sampson HA (2001) Effects of cooking methods on peanut allergenicity. J Allergy Clin Immunol 107:1077–1081

Björksten B, Crevel R, Hischenhuber C, Lovik M, Samuels F, Strobel S, Taylor SL, Wal JM, Ward R (2008) Criteria for identifying allergenic foods of public health importance. Regul Toxicol Pharmacol 51:42–52

Bousquet J, Björkstén B, Bruijnzeel-Koomen CAFM, Huggett A, Ortolani C, Warner JO, Smith M, Beukers M (1998) Scientific criteria and the selection of allergenic foods for product labelling. Allergy Eur J Allergy Clin Immunol 53:3–21

Breiteneder H, Mills ENC (2005) Molecular properties of food allergens. J Allergy Clin Immunol 115:4–24

Brenna O, Pompei C, Ortolani C, Pravettoni V, Farioli L, Pastorello EA (2000) Technological processes to decrease the allergenicity of peach juice and nectar. J Agric Food Chem 48:493–497

Brites C, Trigo MJ, Santos C, Collar C, Rosell CM (2010) Maize-based gluten-free bread: Influence of processing parameters on sensory and instrumental quality. Food Bioprocess Technol 3:707–715

Bublin M, Radauer C, Knulst A, Wagner S, Scheiner O, Mackie AR, Mills ENC, Breiteneder H (2008) Effects of gastrointestinal digestion and heating on the allergenicity of the kiwi allergens Act d 1, actinidin, and Act d 2, a thaumatin-like protein. Mol Nutr Food Res 52:1130–1139

Byun M-W, Kim J-H, Lee J-W, Park J-W, Hong C-S, Kang I-J (2000) Effects of gamma radiation on the conformational and antigenic properties of a heat-stable major allergen in brown shrimp. J Food Prot 63:940–944

Chicón R, Belloque J, Alonso E, Martín-Álvarez PJ, López-Fandiño R (2008) Hydrolysis under high hydrostatic pressure as a means to reduce the binding of β-lactoglobulin to immunoglobulin E from human sera. J Food Prot 71:1453–1459

Chopin C, Lardy N, Daniel A, Fleurence J (2000) Allergy to mackerel (Scomber scombrus): effect of sterilisation treatment. Sci Des Aliments 20:379–385

Chung SY, Champagne ET (1999) Allergenicity of Maillard reaction products from peanut proteins. J Agric Food Chem 47:5227–5231

Chung SY, Champagne ET (2001) Association of end-product adducts with increased IgE binding of roasted peanuts. J Agric Food Chem 49:3911–3916

Chung SY, Champagne ET (2007) Effects of phytic acid on peanut allergens and allergenic properties of extracts. J Agric Food Chem 55:9054–9058

Chung SY, Champagne ET (2009) Reducing the allergenic capacity of peanut extracts and liquid peanut butter by phenolic compounds. Food Chem 115:1345–1349

Chung SY, Butts CL, Maleki SJ, Champagne ET (2003) Linking peanut allergenicity to the processes of maturation, curing, and roasting. J Agric Food Chem 51:4273–4277

Chung SY, Maleki SJ, Champagne ET (2004) Allergenic properties of roasted peanut allergens may be reduced by peroxidase. J Agric Food Chem 52:4541–4545

Chung SY, Kato Y, Champagne ET (2005) Polyphenol oxidase/caffeic acid may reduce the allergenic properties of peanut allergens. J Sci Food Agric 85:2631–2637

Chung SY, Yang W, Krishnamurthy K (2008) Effects of pulsed UV-light on peanut allergens in extracts and liquid peanut butter. J Food Sci 73:C400–C404

Crevel RWR, Kerkhoff MAT, Koning MMG (2000) Allergenicity of refined vegetable oils. Food Chem Toxicol 38:385–393

Davis PJ, Williams SC (1998) Protein modification by thermal processing. Allergy Eur J Allergy Clin Immunol 53:102–105

Davis PJ, Smales CM, James DC (2001) How can thermal processing modify the antigenicity of proteins? Allergy 56(Suppl 67):56–60

De Zorzi M, Curioni A, Simonato B, Giannattasio M, Pasini G (2007) Effect of pasta drying temperature on gastrointestinal digestibility and allergenicity of durum wheat proteins. Food Chem 104:353–363

Dube M, Zunker K, Neidhart S, Carle R, Steinhart H, Paschke A (2004) Effect of technological processing on the allergenicity of mangoes (Mangifera indica L.). J Agric Food Chem 52:3938–3945

Eiwegger T, Rigby N, Mondoulet L, Bernard H, Krauth M-T, Boehm A, Dehlink E, Valent P, Wal JM, Mills ENC, Szépfalusi Z (2006) Gastro-duodenal digestion products of the major peanut allergen Ara h 1 retain an allergenic potential. Clin Exp Allergy 36:1281–1288

Fernandez A, Butz P, Tauscher B (2009) IgE binding capacity of apple allergens preserved after high pressure treatment. High Press Res 29:705–712

Fiocchi A, Bouygue GR, Sarratud T, Terracciano L, Martelli A, Restani P (2004) Clinical tolerance of processed foods annals of allergy. AsthmaImmunol 93(Suppl 5):S38–S46

Franck P, Moneret Vautrin DA, Dousset B, Kanny G, Nabet P, Guénard-Bilbaut L, Parisot L (2002) The allergenicity of soybean-based products is modified by food technologies. Int Arch Allergy Immunol 128:212–219

Fuchs M (2007) Allergy lurks in foods and drinks. Adela Publisher (In Czech), Plzeň, Czech Republic. ISBN 80-902532-2-9

Garcia A, Wichers JH, Wichers HJ (2007) Decrease of the IgE-binding by Mal d 1, the major apple allergen, by means of polyphenol oxidase and peroxidase treatments. Food Chem 103:94–100

Gómez M, Curiel G, Mendez J, Rodriguez M, Moneo I (1993) Hypersensitivity to carrot associated with specific IgE to grass and tree pollens. Allergy Eur J Allergy Clin Immunol 51:425–429

Gruber P, Vieths S, Wangorsch A, Nerkamp J, Hofmann T (2004) Maillard reaction and enzymatic browning affect the allergenicity of Pru av 1, the major allergen from cherry (Prunus avium). J Agric Food Chem 52:4002–4007

Gruber P, Becker W-M, Hofmann T (2005) Influence of the maillard reaction on the allergenicity of rAra h 2, a recombinant major allergen from peanut (Arachis hypogaea), its major epitopes, and peanut agglutinin. J Agric Food Chem 53:2289–2296

Guo B, Liang X, Chung SY, Holbrook CC, Maleki SJ (2008) Proteomic analysis of peanut seed storage proteins and genetic variation in a potential peanut allergen. Protein Pept Lett 15:567–577

Hefle SL (1999) Impact of processing on food allergens. Adv Exp Med Biol 459:107–119

Helm RM, Cockrell G, Connaughton C, West CM, Herman E, Sampson HA, Bannon GA, Burks AW (2000) Mutational analysis of the IgE-binding epitopes of P34/Gly m Bd 30K. J Allergy Clin Immunol 105:378–384

Herian AM, Taylor SL, Bush RK (1993) Allergenic reactivity of various soybean products as determined by RAST inhibition. J Food Sci 58:385–388

Heroldova M, Houska M, Vavrova H, Kucera P, Setinova I, Honzova S, Kminkova M, Strohalm J, Novotna P, Proskova A (2009) Influence of high-pressure treatment on allergenicity of rDau c1 and carrot juice demonstrated by in vitro and in vivo tests. High Press Res 29:695–704

Hoffman DR (1983) Immunochemical identification of the allergens in egg white. J Allergy Clin Immunol 71:481–486

Host A, Halken S (2004) Hypoallergenic formulas – when, to whom and how long: after more than 15 years we know the right indication! Allergy Eur J Allergy Clin Immunol 59:45–52

Host A, Samuelsson EG (1988) Allergic reactions to raw, pasteurized, and homogenized/ pasteurized cow milk: a comparison. Allergy 43:113–118

Houska M, Kminkova M, Strohalm J, Setinova I, Heroldova M, Novotna P, Honzova S, Vavrova H, Kucera P, Proskova A (2009a) Allergenicity of main celery allergen rApi g1 and high-pressure treatment. High Press Res 29:686–694

Houska M, Heroldova M, Vavrova H, Kucera P, Setinova I, Havranova M, Honzova S, Strohalm J, Kminkova M, Proskova A, Novotna P (2009b) Is high-pressure treatment able to modify the allergenicity of the main apple juice allergen, Mal d1? High Press Res 29:14–22

Husband FA, Aldick T, Van der Plancken I, Grauwet T, Hendrickx M, Skypala I, Mackie AR (2011) High-pressure treatment reduces the immunoreactivity of the major allergens in apple and celeriac. Mol Nutr Food Res (in press). doi:10.1002/mnfr.201000566

Ibáñez Sandín D, Martínez San Ireneo M, Maranõón Lizana F, Fernández-Caldas E, Alonso Lebrero E, Laso Borrego T (1999) Specific IgE determinations to crude and boiled lentil (Lens culinaris) extracts in lentil-sensitive children and controls. Allergy Eur J Allergy Clin Immunol 54:1209–1214

Jankiewicz A, Aulepp H, Baltes W, Bögl KW, Dehne LI, Zuberbier T, Vieths S (1996) Allergic sensitization to native and heated celery root in pollen-sensitive patients investigated by skin test and IgE binding. Int Arch Allergy Immunol 111:268–278

Jędrychowski L, Wróblewska B (1999) Reduction of the antigenicity of whey proteins by lactic acid fermentation. Food Agric Immunol 11:91–99

Johnson PE, Van Der Plancken I, Balasa A, Husband FA, Grauwet T, Hendrickx M, Knorr D, Mills ENC, Mackie AR (2010) High pressure, thermal and pulsed electric-field-induced structural changes in selected food allergens. Mol Nutr Food Res 54:1701–1710

Kato T, Katayama E, Matsubara S, Omi Y, Matsuda T (2000) Release of allergenic proteins from rice grains induced by high hydrostatic pressure. J Agric Food Chem 48:3124–3129

Kato Y, Oozawa E, Matsuda T (2001) Decrease in antigenic and allergenic potentials of ovomucoid by heating in the presence of wheat flour: dependence on wheat variety and intermolecular disulfide bridges. J Agric Food Chem 49:3661–3665

Kim S-J, Kim K-B-W-R, Song E-J, Lee S-Y, Yoon S-Y, Lee S-J, Lee C-J, Ann DH (2009) Effect of digestive enzymes on the allergenicity of autoclaved market pork sausages. Korean J Food Sci Anim Resour 29:238–244

Kume T, Matsuda T (1995) Changes in structural and antigenic properties of proteins by radiation. Radiat Phys Chem 46:225–231

Ladics GS (2008) Current codex guidelines for assessment of potential protein allergenicity. Food Chem Toxicol 46(Suppl 10):S20–S23

Lee J-W, Kim J-H, Yook H-S, Kang K-O, Lee S-Y, Hwang H-J, Byun M-W (2001) Effects of gamma radiation on the allergenic and antigenic properties of milk proteins. J Food Prot 64:272–276

Lee JW, Lee K-Y, Yook H-S, Lee S-Y, Kim H-Y, Jo C, Byun M-W (2002) Allergenicity of hen's egg ovomucoid gamma irradiated and heated under different pH conditions. J Food Prot 65:1196–1199

Lee J-W, Seo J-H, Kim J-H, Lee S-Y, Kim K-S, Byun M-W (2005) Changes of the antigenic and allergenic properties of a hen's egg albumin in a cake with gamma-irradiated egg white. Radiat Phys Chem 72:645–650

Leszczynska J, Łącka A, Szemraj J, Lukamowicz J, Zegota H (2003a) The effect of microwave treatment on the immunoreactivity of gliadin and wheat flour. Eur Food Res Technol 217:387–391

Leszczynska J, Łącka A, Szemraj J, Lukamowicz J, Zegota H (2003b) The influence of gamma irradiation on the immunoreactivity of gliadin and wheat flour. Eur Food Res Technol 217:143–147

Lodinová-Zadniková R, Prokesova L, Tlaskalova H, Kocourkova I, Zizka J, Stranak Z (2004) Influence of oral colonization with probiotic *E. coli* strain after birth on frequency of recurrent infections, allergy and development of some immunologic parameters. Long-term studies. Ceska Gynekol 69(Suppl 1):91–97

Maier I, Okun VM, Pittner F, Lindner W (2006) Changes in peptic digestibility of bovine β-lactoglobulin as a result of food processing studied by capillary electrophoresis and immunochemical methods. J Chromatogr B Analyt Technol Biomed Life Sci 841:160–167

Maleki SJ, Chung S-Y, Champagne ET, Raufman J-P (2000) The effects of roasting on the allergenic properties of peanut proteins. J Allergy Clin Immunol 106:763–768

Maleki SJ, Viquez O, Jacks T, Dodo H, Champagne ET, Chung SY, Landry SJ (2003) The major peanut allergen, Ara h 2, functions as a trypsin inhibitor, and roasting enhances this function. J Allergy Clin Immunol 112:190–195

Martín-García C, Carnés J, Blanco R, Martínez-Alonso JC, Callejo-Melgosa A, Frades A, Colino T (2007) Selective hypersensitivity to boiled razor shell. J Investig Allergol Clin Immunol 17:271–273

Meyer-Pittroff R, Behrendt H, Ring J (2007) Specific immuno-modulation and therapy by means of high pressure treated allergens. High Press Res 27:63–67

Mills ENC, Mackie AR (2008) The impact of processing on allergenicity of food. Curr Opin Allergy Clin Immunol 8:249–253

Mine Y, Sasaki E, Zhang JW (2003) Reduction of antigenicity and allergenicity of genetically modified egg white allergen, ovomucoid third domain. Biochem Biophys Res Commun 302:133–137

Misra A, Prasad R, Das M, Dwivedi PD (2009) Probing novel allergenic proteins of commonly consumed legumes novel allergenic legume. Immunopharmacol Immunotoxicol 31:186–194

Mondoulet L, Drumare MF, Ah-Leung S, Paty E, Scheinmann P, Wal JM, Bernard H (2003) Influence of thermal processing on the IgE binding capacity of peanut allergens. Revue Francaise d'Allergologie et d'Immunologie Clinique 43:486–491

Mondoulet L, Paty E, Drumare MF, Ah-Leung S, Scheinmann P, Willemot RM, Wal JM, Bernard H (2005) Influence of thermal processing on the allergenicity of peanut proteins. J Agric Food Chem 53:4547–4553

Mouécoucou J, Frémont S, Sanchez C, Villaume C, Méjean L (2004) In vitro allergenicity of peanut after hydrolysis in the presence of polysaccharides. Clin Exp Allergy 34:1429–1437

Müller U, Weber W, Hoffmann A, Franke S, Lange R, Vieths S (1998) Commercial soybean lecithins: a source of hidden allergens? Z Lebensm Unters Forsch 207:341–351

Nakamura R, Matsuda T (1996) Rice allergenic protein and molecular-genetic approach for hypoallergenic rice. Biosci Biotechnol Biochem 60:1215–1221

Nordlee JA, Taylor SL, Townsend JA, Thomas LA, Bush RK (1996) Identification of a Brazil-nut allergen in transgenic soybeans. N Engl J Med 334:688–692

Norgaard A, Bernadrd H, Wal JM, Peltre G, Skov IPS, Poulsen LK, Bindslev-Jensen C (1996) Allergenicity of individual cow milk proteins in DBPCFC-positive milk allergic adults. J Allergy Clin Immunol 97:237

Novotná P, Šetinová I, Heroldová M, Kmínková M, Průchová J, Strohalm J, Fiedlerová V, Winterová R, Kučera P, Houška M (2011) Deallergisation trials of pure celery juice and apple celery juice mixture by oxidation. Czech J Food Sci 29:190–200

Park J-G, Saeki H, Nakamura A, Kim K-B-W-R, Lee J-W, Byun M-W, Kim S-M, Lim S-M, Ahn D-H (2007) Allergenicity changes in raw shrimp (*Acetes japonicus*) and saeujeot (salted and fermented shrimp) in cabbage Kimchi due to fermentation conditions. Food Sci Biotechnol 16:1011–1017

Paschke A, Besler M (2002) Stability of bovine allergens during food processing. Ann Allergy Asthma Immunol 89(6 Suppl 1):16–20

Paschke A, Zunker K, Wigotzki M, Steinhart H (2001) Determination of the IgE-binding activity of soy lecithin and refined and non-refined soybean oils. J Chromatogr B Biomed Sci Appl 756:249–254

Perrier C, Corthésy B (2011) Gut permeability and food allergies. Clin Exp Allergy 41:20–28

Phromraksa P, Nagano H, Boonmars T, Kamboonruang C (2008) Identification of proteolytic bacteria from thai traditional fermented foods and their allergenic reducing potentials. J Food Sci 73:M189–M195

Poms RE, Anklam E (2004) Effects of chemical, physical, and technological processes on the nature of food allergens. J AOAC Int 87:1466–1474
Primavesi L, Brenna OV, Pompei C, Pravettoni V, Farioli L, Pastorello EA (2006) Influence of cultivar and processing on cherry (Prunus avium) allergenicity. J Agric Food Chem 54:9930–9935
Quirce S, Blanco R, Díez-Gómez ML, Cuevas M, Eiras P, Losada E (1997) Carrot-induced asthma: immunodetection of allergens. J Allergy Clin Immunol 99:718–719
Restani P, Fiocchi A, Restelli AR, Velonà T, Beretta B, Giovannini M, Galli CL (1997) Effect of technological treatments on digestibility and allergenicity of meat-based baby foods. J Am Coll Nutr 16:376–382
Sampson HA, Cooke S (1992) The antigenicity and allergenicity of microparticulated proteins: Simplesse®. Clin Exp Allergy 22:963–969
Scheibenzuber M (2003) Molekulare and klinische Auswirkungen einer Hochdruckbehandlung allergener Lebensmittel. Ph.D. thesis, TU Muenchen
Schubert S, Steinhart H, Paschke A (2003) The influence of different potato (Solanum tuberosum) strains and technological processing on allergenicity. Food Agric Immunol 15:41–53
Setinova I, Kminkova M, Strohalm J, Heroldova M, Novotna P, Honzova S, Vavrova H, Kucera P, Proskova A, Houska M (2009a) Allergenicity of main birch allergen rBet v1 and high-pressure treatment. High Press Res 29:680–685
Setinova I, Honzová S, Kváčová A, Trnková B, Heroldová M, Vávrová H, Kučera P, Houška M, Kmínková M, Gabrovská D, Strohalm J, Paulíčková I, Julínek O, Urbanová M (2009b) Influence of the high pressure on reduction of allergenicity of proteins rMal d1, rDau c1 and orientation quantification of their contents in apple Golden Delicious and in carrot. Alergie 11:102–114 (in Czech)
Setinova I, Kminkova M, Louckova K, Heroldova M, Vavrova H, Pruchova J, Strohalm J, Novotna P, Gresova P, Trnkova B, Kvacova A, Honzova S, Kucera P, Houska M (2010) Influence of storage and antioxidant on oxidative and polymerisation processes of apple juice and protein Mal d 1. 29th congress of the European academy of allergy and clinical immunology, London, 5–9 June 2010
Shibasaki M, Suzuki S, Tajima S, Nemoto H, Kuroume T (1980) Allergenicity of major component proteins of soybean. Int Arch Allergy Appl Immunol 61:441–448
Simonato B, Pasini G, Giannattasio M, Peruffo ADB, De Lazzari F, Curioni A (2001) Food allergy to wheat products: the effect of bread baking and in vitro digestion on wheat allergenic proteins. A study with bread dough, crumb, and crust. J Agric Food Chem 49:5668–5673
Sletten G, Van Do T, Lindvik H, Egaas E, Florvaag E (2010) Effects of industrial processing on the immunogenicity of commonly ingested fish species. Int Arch Allergy Immunol 151:223–236
Somkuti J, Houska M, Smeller L (2010) Pressure and temperature stability of the main apple allergen Mal d 1. Eur Biophys J 40:143–151
Tada Y, Nakase M, Adachi T, Nakamura R, Shimada H, Takahashi M, Fujimura T, Matsuda T (1996) Reduction of 14–16 kDa allergenic proteins in transgenic rice plants by antisense gene. FEBS Lett 391:341–345
Tanabe S (2008) Analysis of food allergen structures and development of foods for allergic patients. Biosci Biotechnol Biochem 72:649–659
Tanabe S, Arai S, Watanabe M (1996) Modification of wheat flour with bromelain and baking hypoallergenic bread with added ingredients. Biosci Biotechnol Biochem 60:1269–1272
Varjonen E, Björkstén F, Savolainen J (1996) Stability of cereal allergens. Clin Exp Allergy 26:436–443
Vaz AFM, Costa RMPB, Coelho LCBB, Oliva MLV, Santana LA, Melo AMMA, Correia MTS (2011) Gamma irradiation as an alternative treatment to abolish allergenicity of lectins in food. Food Chem 124:1289–1295
Venkatachalam M, Teuber SS, Roux KH, Sathe SK (2002) Effects of roasting, blanching, autoclaving, and microwave heating on antigenicity of almond (Prunus dulcis L.) proteins. J Agric Food Chem 50:3544–3548

Von Berg A (2007) The concept of hypoallergenicity for atopy prevention. Nestle Nutr Workshop Ser Pediatr Program 59:49–56

Watanabe M, Miyakawa J, Ikezawa Z, Suzuki Y, Hirao T, Yosihzawa T, Arai S (1990) Production of hypoallergenic rice by enzymatic decomposition of constituent proteins. J Food Sci 55:781–783

Watanabe M, Watanabe J, Sonoyama K, Tanabe S (2000) Novel method for producing hypoallergenic wheat flour by enzymatic fragmentation of the constituent allergens and its application to food processing. Biosci Biotechnol Biochem 64:2663–2667

Westritschnig K, Linhart B, Focke-Tejkl M, Pavkov T, Keller W, Ball T, Mari A, Hartl A, Stöcklinger A, Scheiblhofer S, Thalhamer J, Ferreira F, Vieths S, Vogel L, Böhm A, Valent P, Valenta R (2007) A hypoallergenic vaccine obtained by tail-to-head restructuring of timothy grass pollen profilin, Phl p 12, for the treatment of cross-sensitization to profilin. J Immunol 179:7624–7634

Wigotzki M, Steinhart H, Paschke A (2000) Influence of varieties, storage and heat treatment on IgE-binding proteins in hazelnuts (Corylus avellana). Food Agric Immunol 12:217–229

Yamamoto S, Takanohashi K, Hara T, Odani S, Suzuki A, Nishiumi T (2010a) Effects of a high-pressure treatment on the wheat alpha-amylase inhibitor and its relationship to elimination of allergenicity. J Phys Conf Ser 215, art. no. 012170

Yamamoto S, Mikami N, Matsuno M, Hara T, Odani S, Suzuki A, Nishiumi T (2010b) Effects of a high-pressure treatment on bovine gamma globulin and its reduction in allergenicity. Biosci Biotechnol Biochem 74:525–530

Yamanishi R, Tsuji H, Bando N, Yamada Y, Nadaoka Y, Huang T, Nishikawa K, Emoto S, Ogawa T (1996) Reduction of the allergenicity of soybean by treatment with proteases. J Nutr Sci Vitaminol 42:581–587

Yang WW, Chung SY, Ajayi O, Krishnamurthy K, Konan K, Goodrich-Schneider R (2010) Use of pulsed ultraviolet light to reduce the allergenic potency of soybean extracts. Int J Food Eng 6 (3), art. no.11

Yum H-Y, Lee KE, Choi SY, Yang HS, Sohn MH, Kim K-E, Lee S-I (2006) Wild rice, hypoallergenic rice – immunologic comparison. Allergy Asthma Proc 27:387–392

Zitouni N, Errahali Y, Metche M, Kanny G, Moneret-Vautrin DA, Nicolas JP, Fremont S (2000) Influence of refining steps on trace allergenic protein content in sunflower oil. J Allergy Clin Immunol 106:962–967

Part III
Novel Food Processes

Chapter 17
Emerging Technologies for Targeted Food Processing

D. Knorr, A. Froehling, H. Jaeger, K. Reineke, O. Schlueter, and K. Schoessler

17.1 Introduction

Taking advantage of the specific potentials and opportunities of new food processing technologies, including the understanding and control of the complex process-structure-function relationships, offers the possibility for a science-based development of tailor-made foods. In this chapter, we use high hydrostatic pressure (HP), pulsed electric fields (PEFs), ultrasound (US), and cold plasma (CP) to exemplify scalable and flexible food manufacturing techniques, discussing the state of the art regarding the research and application of these emerging technologies and demonstrating the potential of establishing new routes of process and product development by interfacing food science and food manufacturing.

Significant, science-based achievements have been made to better understand the basic principles underlying HP and PEF processing (Hendrickx and Knorr 2002; Raso and Heinz 2007).

The food and beverage industry offers manifold possibilities for the use of US. The basic phenomena, such as microstreaming and cavitation and the resulting hydrodynamic shear forces, make US an alternative technology for homogenization, dispersion, and emulsifying, as well as for the disintegration of tissue, to enhance mass transfer processes (Mason et al. 1996; Povey and Mason 1998).

A strong increase in plasma applications in medical device technology and therapeutic medicine is currently taking place, including applications such as plasma decontamination, and research is focusing on the interaction between

D. Knorr (✉) • H. Jaeger • K. Reineke • K. Schoessler
Department of Food Biotechnology and Food Process Engineering, Berlin University of Technology, Koenigin-Luise-Str. 22, D-14195 Berlin, Germany
e-mail: dietrich.knorr@tu-berlin.de

A. Froehling • O. Schlueter
Department of Horticultural Engineering, Leibniz Institute for Agricultural Engineering Potsdam, Max-Eyth-Allee 100, 14469 Potsdam, Germany

S. Yanniotis et al. (eds.), *Advances in Food Process Engineering Research and Applications*, Food Engineering Series, DOI 10.1007/978-1-4614-7906-2_17,

plasma and biological cells and tissue as well as on plasma diagnostics with regard to the understanding and control of the complex behavior of CP (Daeschlein et al. 2010; Weltmann et al. 2009). Similar research work is being undertaken in the field of food science to explore the potential for CP application in the food industry (Mastwijk and Nierop Groot 2010) which will be discussed below.

Understanding the impact and potential of such technologies on food systems at the cellular level will enable the design of tailor-made foods and to establish process-structure-function relationships. Based on this knowledge, completely new process designs, the incorporation of HP, PEF, US, and CP in traditional processes, and the generation of improved equipment design will be possible. Consequently, the use of such nonthermal processes for maintenance or even improvement of product quality via processing to meet the PAN (consumer preference, acceptance, and needs) concept of the European Technology Platform Food for Life (http://etp.ciaa.eu), and thus the reverse food engineering approach, will be a major innovative approach within the food industry.

17.2 High Pressure Processing

High pressure processing (HPP) is one of the leading nonthermal food processing technologies and often regarded as one of the major technological innovations in food preservation in recent decades. In recent years HPP has emerged as a viable commercial alternative for food pasteurization for high-quality food.

The pressure levels used range from several tens of megapascals in common homogenizers or supercritical fluid extractors to several hundred megapascals in ultra-high-pressure homogenizers or HP pasteurization units. Besides the inactivation of microorganisms to enhance the shelf life of the treated food, which is by far the most common HPP application, there are numerous other interesting applications like food structure engineering or HP biotechnology (Aertsen et al. 2009; Diels and Michiels 2006; Knorr et al. 2006; Sharma and Yadav 2008).

17.2.1 Process Description: High-Pressure Processing on an Industrial Scale

On the industrial scale, HPP is commonly used for the inactivation of vegetative microorganisms in packed meat (30 %) and vegetable (34 %) products as well as fruit juices and smoothies (13 %), often accompanied by a simultaneous inactivation of enzymes

The applied pressure levels tend from 200 to 350 MPa for seafood, to increase the shucking yield, up to 600 MPa for meat products, to increase the shelf life. The installations that are in commercial use have vessel volumes between 35 and 687 L.

Typical industrial HP units consist of a horizontal HP vessel and an external pressure-generating device. The simplest practical system of an intensifier is a single-acting, hydraulically driven pump (Rovere 2002).

For HP treatment, the packaged food is deposed in a carrier and automatically loaded into the HP vessel, and the vessel plugs are closed. The pressure-transmitting medium, usually water, is pumped into the vessel from one or both sides. When the desired maximum pressure is reached, the pumping is stopped, and in ideal cases no further energy input is needed to hold the pressure during dwell time. In contrast to thermal processes where temperature gradients occur, all molecules in HP vessels are subjected to the same amount of pressure at exactly the same time due to the isostatic principle of pressure transmission (Heinz et al. 2009; Rastogi et al. 2007).

Accompanied by an increase in pressure the temperature rises as well due to the thermodynamic effect of the adiabatic heat of compression. This temperature increase could be calculated as a function of thermophysical properties of the compressible product for simple systems by a rearrengemant of the Maxwell equations (Perry 1984; Reineke et al. 2008). This quasi-adiabatic heating or cooling occurs instantly. Hence, pressure-induced temperature changes are predictable and homogeneous throughout the product, assuming it is homogeneous in its composition. This ideal adiabatic process does not occur in practical applications, but the extent of the temperature increase could be estimated at 3–9 °C per 100 MPa depending on the treated food or food composition (Ting et al. 2002).

17.2.2 Research State of the Art

The main thermodynamic driving force behind reactions under pressure is the specific reaction volume. Hence, reactions, accompanied by a decrease in its reaction volume, are favored under pressure. Consequently, HPP does not affect covalent bonds at pressure levels below 2 GPa or the structure of small molecules, contrary to macromolecules or the more complex systems that will be discussed in this chapter.

17.2.2.1 Impact of HP on Biological Cells

To ensure food safety, tremendous research was done on the inactivation of biological cells and its underlying inactivation mechanisms. In general HPP is described as a nonthermal technology, but with regard to the adiabatic heat of compression, thermal effects could not be fully ruled out. Hence, the complex inactivation mechanism of biological cells and viruses is almost always connected to the process temperature (Smelt et al. 2001). Primarily, HP affects the properties of biological membranes over a broad temperature range (Winter and Jeworrek 2009). Although this lethal effect could be further accelerated at higher treatment

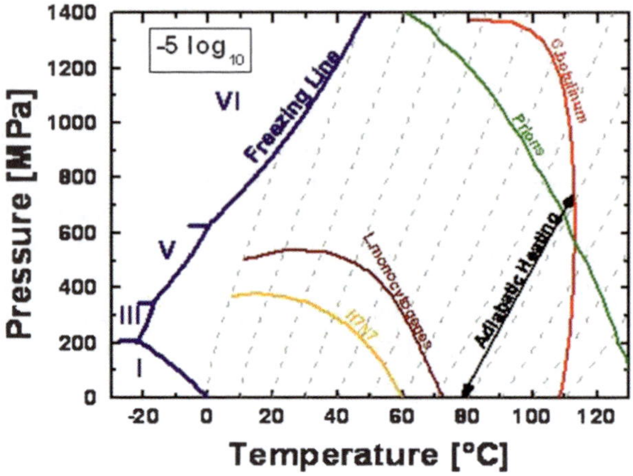

Fig. 17.1 Isorate lines for a 5 $\log_{10}$ reduction of microorganisms, prions, and viruses under various pressure and temperature conditions with adiabatic lines due to the compression (–) of water. H7N7, the surrogate for bird flu virus H5N1 (chicken meat slurry) (Isbarn et al. 2007), *Listeria monocytogenes* 75903 (in ham slurry), prions-PrP^{SC} (in raw meat), and *Clostridium botulinum* (in TRIS buffer pH 5.15) (Margosch et al. 2006). *Blue line*: freezing line between liquid water and different pressure-dependent ice modifications

temperatures, additional enzyme inactivation occurs (Ananta 2005; Ardia 2004) (Buckow and Heinz 2008).

However, pressure treatments do not necessarily inactivate biological cells. At low pressure levels increased resistance of vegetative cells was observed. Ananta and Knorr (2003) reported a higher thermotolerance of lactic acid bacteria after HP treatments up to 200 MPa. As a result of this phenomenon, pressure-induced stress response was found to offer promising processing options, such as pretreatment of lactic acid bacteria before spray or freeze drying for the purpose of starter culture production (Ananta 2005).

Higher pressures between 300 and 800 MPa and ambient treatment temperatures are suitable for the inactivation of several vegetative microorganisms in food products (Ananta et al. 2005; Hendrickx and Knorr 2002). The special shape of isorate lines at these conditions are shown in Fig. 17.1. A synergistic effect between pressure and temperature on the inactivation of vegetative cells is typical but was also observed for bacterial spores (Ardia 2004; Margosch et al. 2006), viruses (Isbarn et al. 2007), and proteins (Heinz and Kortschack 2002; Smeller 2002). By increasing the process temperature, it is possible to decrease the applied pressure, but unwanted reactions (e.g., based on residual enzyme activity) that would lead to quality losses must also be taken into account.

According to Smelt et al. (2001), the HP induced effects resulting in vegetative cell death can be summarized as follows:

Proteins and enzymes: HP induces unfolding of globular proteins. It is assumed that the combined, complete, or partial inactivation of numerous enzymes and metabolic pathways leads to an inability to proliferate and cell death, respectively (Bunthof 2002).

Membranes: besides the inactivation of enzymes, membrane damage is considered one of the key events related to microbial cell death. Membranes undergo phase transitions and solidify under pressure and perturbations are promoted (Schlueter 2004; Winter 1996). In addition, pressure leads to the detachment and inactivation of membrane proteins (Ulmer et al. 2002).

Ribosomes: the disintegration of ribosomes in their subunits is promoted by pressure and may be related to cell death (Niven et al. 1999).

pH: the maintenance of intracellular pH is crucial for the survival of cells. Some authors have related cell death predominantly to intracellular pH changes, which are related to inactivation of enzymes controlling acidity and membrane damage (Molina-Gutierrez et al. 2002).

Bacterial spores have a higher barotolerance than vegetative bacteria and survive pressures above 1,200 MPa at ambient temperatures (Ananta et al. 2001; Margosch et al. 2006). Early approaches to spore inactivation aimed at a pressure-induced germination at moderate pressure. However, combination processes with spore germination at pressures below 200 MPa and an additional moderate heat treatment could not guarantee sufficient inactivation since small populations of spores could not be germinated and remain in the dormant state (Heinz and Knorr 1996; Reineke et al. 2011).

Several papers indicate a synergism between pressure and temperature with regard to spore inactivation (Bull et al. 2009; Margosch et al. 2006; Mathys et al. 2009; Olivier et al. 2011; Rajan et al. 2006; Reineke et al. 2012). Consequently, the HP sterilization process is rapidly attracting industrial interest, especially due to the acceptance of pressure-assisted thermal processing (PATS) by the U.S. Food and Drug Administration in 2009, in which the adiabatic heat of compression is used to rapidly achieve 121.1 °C.

In contrast to the synergistic effect of pressure and temperature, several authors have stated that some spore strains are more resistant under certain pressure–temperature combinations (Margosch et al. 2006; Mathys et al. 2009; Wuytack et al. 1998), that their thermal resistance is nearly independent of the treatment pressure (proteolytic Typ B *Clostridium botulinum* TMW 2.357 for pressures below 1.2 GPa) (Margosch et al. 2006), or that their thermal/pressure resistance varies with the pH of the suspension medium or the food matrix being used, e.g., for *Bacillus coagulans* (Olivier et al. 2011; Vercammen et al. 2011).

The HP inactivation of bacterial spores is not yet fully understood and still of high relevance in today's HP sterilization research activities (Reineke et al. 2011a, b). A detailed discussion of HP-related spore inactivation mechanisms is given by Mathys (2008).

Furthermore, HPP below 0 °C enables a significant inactivation of vegetative microorganisms as well if the treated food is frozen under pressure. Here the dominant mechanism of microorganism inactivation is due to mechanical stress during the freezing process. Luscher (2008) suggests two different mechanisms that occur during high freezing rates and the occurrence of different ice modifications. Internal disintegration occurs during freezing at high freezing rates, for example, during high-pressure shift freezing (HPSF), when water in the intracellular room freezes. The increase in volume of the water during crystallization initiates high internal pressure and eventually causes lethal membrane rupture. However, the freezing rate must be high enough to allow crystallization of the intracellular water since in general high freezing rates are linked to the formation of small ice crystals, which reduces the lethal effect of the freezing process.

The second supposed inactivation mechanism is external disintegration, which takes place during the solid–solid phase change from ice III to ice I. The density decrease of about 19 % during recrystallization causes high mechanical stress to the bacterial cells. The increasing volume of the ice crystals potentially breaks the cell walls of bacterial cells that are incorporated into the ice matrix.

17.2.2.2 Impact of HP on Enzymes

HP is regarded as a mild process by which the primary structure of proteins is not affected at pressure below 2 GPa. However, it could have an impact on hydrophobic interactions as well as on intermolecular covalent bonds, such as disulfide bonds, which stabilize the quaternary and tertiary structure. Generally pressure-induced changes in proteins and enzymes between 100 and 300 MPa at ambient temperature are reversible and could additionally lead to an unfolding of protein chains and a dissociation of oligomeric proteins (Buckow 2006; Tauscher 1995). A further pressure increase above 400 MPa could lead to an irreversible unfolding and, consequently, to an inactivation of the enzymes, for example.

Due to conformational changes, the unfolding of an enzyme can alter its functionality and result in a decreased or increased biological activity and could even change its substrate specificity (Buckow and Heinz 2008; Ludikhuyze et al. 2002). The pressure stability of enzymes can vary significantly, ranging from pressure-sensitive enzymes, such as phosphohexoseisomerase from bovine milk ($p < 400$ MPa) (Rademacher and Hinrichs 2006), to extreme pressure-resistant enzymes like peroxidase from horseradish ($p > 700$ MPa) (Smeller and Fidy 2002). However, a categorization of enzymes as a result of their pressure stability is not appropriate since there is a structural variability among enzymes catalyzing the same reaction (Buckow and Heinz 2008). Even isoforms of an enzyme from the same origin can vary in their physical stability between several hundred megapascals (Buckow et al. 2005; Rodrigoa et al. 2006).

Additionally, the pressure–temperature resistance of enzymes shows a significant dependency on matrix conditions like, for example, the pH value (Riahi and Ramaswamy 2004; Zipp and Kauzmann 1973).

Interestingly, some enzymes show an enhanced thermostability under specific pressure–temperature conditions, which could be attributed to the often antagonistic effect of pressure and temperature (Heremans and Smeller 1998). Such stabilization of enzymes could occur when the volume difference between the folded and unfolded states of the protein are positive, which might be caused by the promoted formation of noncovalent bonds under the applied pressure.

17.2.2.3 Impact of HP on Food Constituents

Due to their relatively simple three-dimensional structure, low molecular weight molecules like peptides, lipids, vitamins, and saccharides are rarely affected by isostatic HP because of the very low compressibility of covalent bonds at pressures below 2 GPa (Cheftel and Culioli 1997; Oey et al. 2006; Van den Broeck et al. 1998). Conversely, HP can change the native structure of macromolecules, such as starch, similar to thermal-induced structural changes.

In particular, the impact of HP on gelatinization of starch could influence the processing properties of starch-based food systems, whereas pressure-induced gelatinization differs in comparison to the thermal-induced one (Stute et al. 1996). HP-induced gelatinization is characterized by a limited swelling of the starch granules, which is accompanied by a retention of its structure and a loss of birefringence under polarized light (Buckow et al. 2007). The extent of starch gelatinization is dependent on the applied pressure, the dwell time, and the treatment temperature (Rumpold and Knorr 2005), whereas the type of starch has the highest impact on its gelatinization behavior under pressure. Starches with a B-type crystalline structure such as potato, which is generally found in roots, are more resistant to pressure than A- or C-type starches like, for example, wheat starch (Stute et al. 1996).

Furthermore, HP could significantly affect the three-dimensional structure of proteins. Depending on the applied pressure level, this could lead to a reversible or irreversible unfolding of the proteins, whereas the primary structure itself would not be affected. In this regard, examples and mechanisms were discussed in the previous section.

17.2.3 Process-Structure Function Interactions

As mentioned in Sect. 3.2.3, the gelatinization of starches under pressure is significantly different from that induced by heat, and hence they offer unique functional properties, like, for example, the formation of weak gels, which could be used as a fat replacement in dietary foods (Sharma et al. 2008; Zhang et al. 2008).

Gels with unique properties are also formed by proteins, such as ß-lactoglobulin (Dumay et al. 1998). For such a system, the pressure-induced gel matrix yields small particles in highly packed ß-lactoglobulin, unlike the gels after heating, which exhibit finely stranded aggregates. Furthermore, Zeece et al. (2008) reported that a HP treatment from 400 to 800 MPa at 20 °C considerably improved the digestibility of ß-lactoglobulin. Thus, pressure-induced protein gels open up the possibility of generating new textures as they additionally retain their original flavor and color accompanied by a glossy appearance. Such gels can be applied for the manufacture of milk products to, for example, improve yogurt texture (Johnston et al. 1993) or increase cheese yield (López-Fandino et al. 1996).

A further possibility for the creation of a novel texture is to pressure treat mixtures of proteins and polysaccharides, which has been studied by Michel and Autio (2002).

17.2.4 Current Applications and Process Development

Actually, HP pasteurization of fresh products is the main application, with an annual production volume of more than 300,000 t/a (C. Tonelle, personal communication, NC Hiperbaric). Whereas most products are fresh-cut fruits and fruit purees (34 %), with the processing goal of an extended shelf life including reduced or no enzymatic browning, the pressure processing of meat. More than 30 % of the total vessel volume is used to process meat products like sliced ham, turkey or chicken cuts, and ready-to-eat-meals, primarily to inactivate *Listeria* and to increase the shelf life of the treated product.

17.2.5 Research Needs and Challenges

The continuous increase in HP research over the last several decades has already generated the bases for several commercially available HP-processed high-quality products. Besides "cold" pasteurization, isostatic HP can also be used to generate novel functional features, such as specific textures or health-promoting properties, to develop tailor-made foods. A very promising field for further research is the application of low and moderate HP to modulate microbial fermentations or enzymatic conversations. Under these conditions, biosyntheses pathways could be active, which could lead to the formation of product variants with novel functional properties (Aertsen et al. 2009). A better understanding of the pressure and temperature stability of proteins could enable the construction of HP-resistant enzymes. These findings may also be applicable to the behavior of nutrients, allergens, and food-spoiling viruses under defined matrix (Mathys and Knorr 2009) and treatment conditions.

A standardization of experimental protocols would increase the comparability of results, which could be used later on for modeling.

To successfully introduce HP thermal sterilization to the food industry, a pressure- and temperature-resistant indicator microorganism must be found and the microbial targets that lead to an inactivation of bacterial endospores understood.

It should also be noted that the physicochemical properties of food constituents such as water vary under high-pressure or high-temperature conditions, making process development and the understanding of mechanisms challenging (Mathys and Knorr 2009) and promoting or inhibiting desired and undesired chemical reactions in processed food.

To reduce processing costs and to investigate the temperature distribution during HP treatment, especially under sterilization conditions, modeling and simulation of the behavior of HP-treated biomaterials plus the temperature distributions in the HP vessel present a challenge (Delgado et al. 2008).

17.3 Pulsed Electric Fields

17.3.1 Process Description

When exposed to high electric field pulses, cell membranes develop pores that may be permanent or temporary, depending on the intensity and treatment conditions (Angersbach et al. 2000; Zimmermann et al. 1976; Zimmermann et al. 1974). Pore formation increases membrane permeability, which results in the loss of cell content or the intrusion of surrounding media (Phoon et al. 2008; Vorobiev and Lebovka 2008). Low-intensity treatment has the potential to induce stress reactions in plant cells, resulting in the promotion of a defense mechanism by increased production of secondary metabolites (Dörnenburg and Knorr 1993; Galindo et al. 2009; Gomez Galindo, et al. 2008). An irreversible perforation of the cell membrane reduces its barrier effect permanently and causes cell death, which can be applied to plant and animal raw material disintegration (Angersbach and Knorr 1998; Puértolas et al. 2010; Toepfl and Heinz 2007) and to the nonthermal inactivation of microorganisms (Lelieveld et al. 2007; Toepfl et al. 2007).

Pulsed electric field (PEF) processing consists of the application of very short electric pulses (1–100 μs) at electric field intensities in the range of 0.1–1 kV/cm (reversible permeabilization for stress induction in plant cells), 0.5–3 kV/cm (irreversible permeabilization of plant and animal tissue), and 15–40 kV/cm for the irreversible permeabilization of microbial cells. Depending on cell size and shape, the aforementioned field intensities lead to the formation of a critical transmembrane potential that is regarded as the precondition for membrane breakdown (Tsong 1996).

Since the mechanism of electroporation is mainly based on a mechanical electrocompressive force affecting the cell membrane, PEF technology is considered a nonthermal cell disintegration or preservation process. It provides an alternative to mechanical, thermal, and enzymatic cell disintegration of plant and animal raw materials, providing a short-duration (milliseconds), low-energy treatment, and an alternative to the traditional thermal pasteurization of liquid food products (Barbosa-Cánovas et al. 1999; Raso and Heinz 2006).

17.3.2 Research State of the Art

Effective inactivation for most spoilage and pathogenic microorganisms has been demonstrated, and colony count reductions, depending on treatment intensity, product properties, and type of microorganism in the range of 4–6 log-cycles, are comparable to traditional thermal pasteurization. Bacterial spores and viruses are not affected by PEF treatment (Lelieveld et al. 2007). Reports on the effects of PEFs on enzymes are limited, and different experimental setups and processing parameters make them difficult to compare (Van Loey et al. 2002). Thermal effects were also found to contribute to enzyme inactivation during PEF treatment (Jaeger et al. 2009; Jaeger et al. 2010). The first large-scale industrial applications were carried out for the disintegration of plant raw materials such as sugar beet and fruit mashes (Bluhm and Sack 2009). Industrial equipment is available up to capacities of a single system of 5,000 L/h for PEF processing of liquids for nonthermal pasteurization with total treatment costs of around US$0.6/kg and 50 t/h for cell disintegration applications with related total treatment costs of around 0.5 euro/t (DIL 2011).

17.3.2.1 Impact on Biological Cells

Until now there has been no clear evidence on the underlying mechanisms at the cellular level, but two main effects have been described as being triggered by the electric field: the ionic punch-through effect (Coster 1965) and the dielectric breakdown of the membrane (Zimmermann et al. 1974). The factors affecting microbial inactivation during PEF treatment are process factors, such as electric field intensity, pulse width and shape, and treatment time and temperature, microbial factors such as type, shape, size, concentration, and growth stage of the microorganism, and media factors such as pH, antimicrobials and ionic compounds, electrical conductivity, and ionic strength of the medium.

Membrane damage and inactivation of microorganisms due to PEF, first considered as an all-or-nothing event in some studies (Russel et al. 2000; Simpson et al. 1999; Wuytack et al. 2003; Yaqub et al. 2004), revealed a required differentiated approach even if the critical parameters for the electrical breakdown of cell membranes were exceeded. Membrane damage and sublethal injury is repairable under certain conditions, and the extent to which cells repair their injuries was found to depend on treatment intensity, the microorganism, and treatment medium pH (Garcia et al. 2005).

17.3.2.2 Impact on Enzymes

The evaluation of the effect of PEFs on enzymes is complex, and available reports on the mechanisms are limited; the various experimental setups and processing parameters make them difficult to compare (Schuten et al. 2004; Van Loey et al. 2001; Yang et al. 2004).

The observed effects of PEFs on enzymes by different research groups appear to depend, besides on the enzyme, on the characteristics of the PEF treatment system used and on the electric process parameters. PEF side effects, such as changes in pH at the electrode surface due to electrochemical reactions (Saulis et al. 2005) and the occurrence of temperature hot spots due to ohmic heating effects within a nonuniform electric field (Jaeger et al. 2009), may contribute to the observed overall enzyme inactivation during PEF processing. As the PEF effect on proteins, enzymes, and other food constituents remains small, possible applications could include the pasteurization of bioactive antimicrobial milk fractions, such as lactoperoxidase, lactoferrin, or immunoglobulins, as well as heat-sensitive vitamin solutions that are destroyed during thermal pasteurization.

Although enzymes do not contain membrane structures, which are the target of an inactivation based on electroporation, the possible impact of PEF side effects indicates that process modifications toward the inactivation of microorganisms and enzyme structures are also possible (Aguiló-Aguayo et al. 2010; Martín-Belloso and Elez-Martínez 2005). Improvements like treatment chamber design have offered the ability to selectively retain enzyme activities or to inactivate them via temperature effects caused by the electrodes and flow conditions in the treatment chamber.

A further discussion of PEF impact on proteins and enzymes is conducted in the following section.

17.3.2.3 Impact on Food Constituents

Few data are available regarding PEF effects on other food constituents, especially proteins (Barsotti et al. 2001). Perez and Pilosof (2004) reported a partial modification of the native structure of β-lactoglobulin when the concentrate is subjected to an electric field of 12.5 kV/cm. No effects of PEF treatment on the physicochemical properties of lactoferrin were found by Sui et al. (2010) for treatment intensities up to 35 kV/cm and a total specific energy input of 41 kJ/kg (treatment time 19 μs) with temperatures below 65 °C.

Fernandez-Diaz et al. (2000) studied the effects of PEFs on ovalbumin solutions (2 %; pH 7; 5 mS/cm) and dialyzed egg white (pH 9.2; 4–5 mS/cm) applying an electric field strength in the range of 27–33 kV/cm. Partial protein unfolding or enhanced SH ionization of ovalbumin was observed after PEF treatment and was found to increase with increases in the total specific energy input.

Recent investigations by Marco-Moles et al. (2009) focused on the PEF effect on proteins and lipids in liquid whole egg and the microstructure of these components studied by low-temperature scanning electron microscopy (Cryo-SEM). A partial denaturation and insolubilization of the protein was observed during conventional pasteurization and resulted in a thickening of the lipoprotein matrix as observed by Cryo-SEM. The microstructure of PEF-treated samples showed some discontinuities in the lipoprotein matrix.

Due to the application of PEFs, changes in the conformational state of proteins might cause changes in enzyme structure and activity (Bendicho et al. 2003; Tsong 1990). In general, the mechanisms involved in the inactivation of enzymes by PEFs are not fully understood (Ohshima et al. 2006). Possible mechanisms have been proposed (Castro et al. 2001; Perez and Pilosof 2004). If the duration of the pulse is long enough, then the effects of PEFs on proteins could entail polarization of the protein molecule, dissociation of noncovalently linked protein subunits involved in quaternary structures, changes in the protein conformation so that hydrophobic amino acid or sulfhydryl groups are exposed, attraction of polarized structures by electrostatic forces, and hydrophobic interactions or noncovalent bonds forming aggregates.

17.3.3 Process-Structure-Function Interactions

The effect of PEFs on food constituents such as proteins and carbohydrates and the resulting changes in the functional properties have been discussed in the section "Impact on Food Constituents." The following section will focus on the PEF effects on structured foods.

PEFs affect cell membranes and thus can be expected to influence the texture of products in which the structure is largely dependent on the integrity of cells.

Fundamental research on the modification of the textural properties of plant and animal raw materials represents the basis for further possible applications. Lebovka et al. (2004) studied the impact of PEFs on apple, carrot, and potato tissue. Stress deformation and relaxation tests were performed to analyze the changes in tissue texture. PEF treatment, in combination with a mild heat pretreatment, leads to the complete elimination of the textural strength of tissue. It was shown that by proper selection of PEF treatment conditions, it was possible to obtain a controlled degree of tissue softening.

Suitable methods such as impedance measurement (Angersbach et al. 1999) and acoustic impulse response (Grimi et al. 2010) were developed and are in use as efficient tools to quantify structural modifications.

Various concepts for the application of PEFs to improve mass transfer processes and to affect plant food material properties have been developed (Ade-Omowaye et al. 2003; Knorr et al. 2001, Lebovka et al. 2007; Puértolas et al. 2010).

However, studies on the effect of electropermeabilization on protein-based structures of solid foods such as fish and meat are limited.

Improved water binding during cooking of meat was found to occur after PEF pretreatment due to enhanced microdiffusion of brine and water binding agents.

Hydrocolloids will influence protein swelling and water binding activity, and their microdiffusion into meat tissue can be enhanced by PEF pretreatment (Toepfl 2006).

The impact of a PEF treatment on the microstructure and texture of salmon was investigated by Gudmundsson and Hafsteinsson (2001). Fish muscle was found to be more susceptible to gaping due to PEF treatment in comparison to chicken meat, probably due to the lower content of connective tissue (0.6 % in comparison to 2 % for chicken meat).

An understanding of the impact and potential of PEF technology on food systems at the cellular level will allow for the design of tailor-made foods and to establish process-structure-function relationships.

17.3.4 Current Applications and Process Development

PEF technology is on the verge of industrial application with various pilot-scale units available worldwide (Lelieveld et al. 2007; Raso and Heinz 2006).

Application of PEF technology as a short-duration (milliseconds), continuous operation will improve the sustainability of food processing or reduce energy requirements while maintaining or improving food quality and safety. Even if PEF treatment requires an additional input of electrical energy, it has beneficial effects on the total energy consumption of mass transfer processes, such as extraction, pressing, and drying. Processing times are reduced, utilization of production capacities is improved, and water and raw material consumption is diminished (Toepfl et al. 2006).

Application of PEFs, in combination with mild heat, seems to be a promising technique for a gentle, multihurdle preservation process.

17.3.4.1 Low-Intensity PEF Treatment

Controlled reversible permeabilization offers the potential for sublethal stress induction on biological cells triggering a metabolic response (Bonnafous et al. 1999; Gomez Galindo et al. 2008) and an increased production of secondary metabolites such as phenols or phytosterols, leading to increased antimicrobial and antioxidative effects (Dörnenburg and Knorr 1995). It also offers the potential for the "infusion" of precursors or other desired constituents into cells as well as the recovery of metabolites from cells while maintaining their viability and productivity (Tryfona and Bustard 2008). The irreversible rupture of plant membranes offers various applications to replace or support conventional thermal as well as enzymatic processes for cell disintegration (Vorobiev and Lebovka 2008). Irreversible permeabilization allows for the significant improvement of mass transfer, especially for drying, expression, concentration, and extraction, resulting in higher product yields, shorter processing times, and, consequently, reduced energy consumption (Toepfl et al. 2006). Solid–liquid extraction, pressing, or drying of food matrixes is strongly dependent on diffusion and mass transfer through cells and tissue and is characterized by the disintegration of the material. To achieve a high level of disintegration of solid tissue, thermal, mechanical, chemical, and enzymatic methods are commonly used. PEF technology could be applied as a new method for cell disintegration.

A comparison of juice yields obtained after PEF processing of apple and carrot mash is presented by Jaeger et al. (2012), with an emphasis on the integration of the technology into the whole juice winning process.

The extractability of intracellular pigments facilitated with PEF treatment has proven to be a very efficient process for the winning of these valuable components with beneficial antioxidant properties. PEF-induced cell permeabilization and the release of intracellular pigments (anthocyanins) from wine grapes were studied by Praporscic et al. (2007) and Puertolas et al. (2010). They showed a 30–40 % increase in total polyphenols and anthocyanin content in the juce obtained from Cabernet Sauvignon grapes.

PEF enhanced the permeability of potato tissue, which resulted in an improved mass transfer during dehydration, and therefore a shorter drying time was demonstrated by Lebovka et al. (2007). The process duration of the osmotic dehydration processes after PEF treatment of apples can be reduced by up to 40 % and the rehydration capacity could be increased substantially (Taiwo et al. 2002).

In the case of sugar beet processing, PEF treatment enables the sugar recovery process at lower extraction temperatures, which reduces the water binding properties of the treated couchettes, thereby doubling the dry matter content and significantly reducing energy costs for drying of the resulting pellets (Eshtiaghi and Knorr 2002; Lebovka et al. 2007).

17.3.4.2 High-Intensity PEF Treatment

Microbial inactivation of vegetative cells via PEF offers pasteurization with low energy input, selective inactivation of microorganisms depending on cell size or shape (Toepfl et al. 2007), and the retention of bioactive heat-sensitive food compounds while inactivating pathogenic microorganisms and increasing product shelf life and safety (Guerrero-Beltrán et al. 2010; Jaeger et al. 2009; Sui et al. 2010).

A number of studies have shown synergetic effects between PEF and heat treatment at nonlethal processing temperatures. The synergetic effect of temperature during PEF inactivation of microorganisms can be used to improve inactivation results or to reduce electrical energy costs (Craven et al. 2008; Riener et al. 2008). Energy savings derive from the lower PEF treatment intensity (treatment time and total specific energy input) required for a certain level of microbial inactivation at increased temperature and from the possibility to recover the electrical energy dissipated during PEF treatment in the form of thermal energy for preheating the incoming product. Another key aspect for the successful application of PEF pasteurization is its selective inactivation capability since the pore formation process and the required transmembrane potential depend on the size of the treated cell.

Larger cells like yeasts require a smaller intensity of the external electric field, and they are thus more sensitive to electropermeabilization. Hence, the inactivation of yeast in a product containing desirable probiotic bacteria can be realized without a loss of bacterial cell vitality.

The selectivity of PEF inactivation based on cell size and other factors that affect the electroporation mechanism are different from the susceptibility of microorganisms to thermal treatment. Thermotolerant microorganisms can be affected by PEFs, and inactivation of the more PEF-resistant species can be conducted by thermal effects; consequently, a combination of thermal and PEF treatment can improve inactivation effectiveness in addition to the synergistic temperature effect on PEF inactivation below thermal inactivation level.

17.3.5 Research Needs and Challenges

The application of PEFs to induce stress reactions in biological systems and a basic understanding of the underlying mechanisms at the cellular and metabolic levels will be the main focus of the research undertaken in the field of reversible permeabilization of plant cells. The first attempts at doing this, as described earlier, are already showing the potential for the modification and improvement of the production of valuable secondary metabolites.

The application of PEFs for the irreversible cell disintegration of plant and animal raw materials was limited by the availability of large-scale pulse modulators, but a forward-looking technical development was made in recent years to overcome production-scale limitations. To implement the cell disintegration processing step into existing processes, an integrative approach will be required that considers pre- and post-PEF processing unit operations, such as the mechanical disintegration of solid–liquid separation in the case of extraction of juice recovery in order to successfully translate the cell disintegration provided by PEF into improved process results such as higher juice yields.

For PEF-assisted pasteurization, the design and optimization of the PEF treatment chamber is the most challenging point with regard to different product properties, such as viscosity and electrical conductivity, and with regard to uniform treatment conditions in terms of electric field and temperature distribution. Many authors (Fiala et al. 2001; Gerlach et al. 2008; Jaeger et al. 2009; Lindgren et al. 2002; van den Bosch et al. 2002) have described the temperature distribution in a PEF treatment chamber and reported the occurrence of high local temperatures due to the inhomogeneous distribution of the electric field, limited flow velocity, and recirculation of the liquid. Numerical simulations using computational fluid dynamics have attracted increased interest for this purpose since experimental measurement of the related parameters is not possible in most cases due to the small dimensions of the treatment chamber as well as the interference of the measuring device with the product flow and electric field. Treatment homogeneity and the avoidance of overprocessing of the product, including the occurrence of local high temperatures, are key aspects in guaranteeing predictable cell disintegration and microbial inactivation while maintaining heat-sensitive food constituents.

In PEF systems working at higher electric field intensities, electrochemical reactions can occur at the electrode surface (Morren et al. 2003). Related undesired effects, such as a partial electrolysis of the solution, the corrosion of the electrode material, and the introduction of small particles of electrode material into the liquid, can be limited or avoided by a suitable selection of electrode materials and by adaptation of the electrical pulse shape and duration (Roodenburg et al. 2005; Saulis et al. 2005). Its consideration is a crucial prerequisite in the study of inactivation kinetics in order to exclude other simultaneously occurring side effects.

Protective effects existing in real food systems may limit the process effectiveness of PEFs compared to inactivation studies conducted in model solutions, and the occurrence of sublethally injured cells must be taken into account with regard to food safety aspects (Jaeger et al. 2009). A comprehensive statement concerning the food safety aspects of PEF treatment can be found in Knorr et al. (2008). Furthermore, in addition to the complexity of treatment media, the consideration of microbial growth state, adaptation to the treatment media, and the existence of inhomogeneous microbial populations with less sensitive subpopulations seem to be the most challenging aspects when transferring inactivation results to real products and industrial implementation.

17.4 Ultrasound

17.4.1 Process Description

Ultrasound (US) is defined as the energy emitted by sound waves with frequencies from 18 kHz up to the gigahertz range. Longitudinal sound waves can be transmitted to gases, fluids, or foodstuffs, causing cyclic compressions and rarefactions of the respective material. High-intensity, low-frequency (16–100 kHz) US can lead to cavitation and the creation, growth, and violent collapse of gas bubbles (Patist and Bates 2008). The bubble collapse is accompanied by high pressure and temperature peaks (up to 100 MPa and 5,000 K) as well as intense local shear (Clark 2008). Such high-power US treatments have the potential to improve a large range of key processes in food production.

In contrast, low intensities and high frequencies in the megahertz range lead to US treatments with acoustic streaming as the main mechanism (Patist and Bates 2008). Such low-energy US is used for nondestructive testing as well as for the stimulation of living cells.

17.4.2 Research State of the Art

17.4.2.1 Impact on Biological Cells

A bactericidal effect of US was first reported in the 1920s (Harvey and Loomis 1929). US-induced cell damage is primarily explained by cavitation phenomena such as shear disruption (microstreaming), localized heating, and free radical formation (Hughes and Nyborg 1962). The cell walls and membranes of biological cells can be damaged by surface rubbing, leading to fracture and leakage (Kinsloe et al. 1954) and separation of the cytoplasmic membrane (Alliger 1975).

US has an additive or even synergistic effect when combined with other lethal effects such as elevated temperatures (thermosonication – TS), pressure (manosonication – MS), or both (manothermosonication – MTS) (Knorr et al. 2004), which has been explained by the weakening of the cell wall, making it more susceptible to the effects of cavitation (Patist and Bates 2008). US pasteurization carried out at lowered temperatures could preserve the physicochemical properties of color and flavor (Patist and Bates 2008). However, the specific energy requirement of such combined treatments must be considered as the most important limitation for the application of US in food preservation processes today (Zenker et al. 2003).

Low-intensity US can improve fermentation processes due to increased mass transfer through cell walls and membranes and its influence on boundary layers (Sinisterra 1992). Increased fermentation rates could be observed because of the US-assisted removal of CO_2, which otherwise can inhibit fermentations (Matsuura et al. 1994).

17.4.2.2 Impact on Enzymes

Research on the inactivation of enzymes for food preservation revealed increased effects for TS, MS, and MTS compared to US alone. Nevertheless, the sensitivity can be very different from one enzyme to another. Among others, positive results have been reported for tomato pectic enzymes as well as for α-chymotrypsin and porcine lipase, whereas phospholipase A_2 was nearly insensitive to MTS and the sensitivity of trypsin was found to be temperature dependent (Vercet et al. 2001; Vercet et al. 2002).

Positive effects on enzyme activity can be achieved by the application of low-energy US. Substrates can be made available in large amounts, and the transport of substrate to immobilized enzymes is increased by microstreaming (Mason et al. 1996).

17.4.2.3 Impact on Food Constituents

Intense pressures, temperatures, and shear forces induced by US and cavitation can lead to the denaturation of proteins and the alteration of their structure and functional properties. Cavitation-induced tissue disruption results in the migration of proteins, minerals, and other components, and periodic oscillations lead to the softening of cell membranes (Jayasooriya et al. 2004), which is of importance for US-induced meat tenderization (Jayasooriya et al. 2007; Lyng et al. 1998; McClements 1995), where US is reported to release myofibrillar proteins and improve physical properties, such as water binding capacity, tenderness, and cohesiveness, and sensory properties (Jayasooriya et al. 2004; Roberts 1992).

On the other hand, US-induced cavitation may promote undesired changes depending on the treated product. In water-filled cavities, one of the primary reactions is the degradation of water molecules into H and OH radicals. Radical formation is fundamental for the explanation of US-induced oxidation reactions, adversely affecting fats, oils, vitamins, and color pigments (Roberts 1993). Critical results were obtained in the case of rabbiteye blueberries, with reduced product quality following US treatment (Stojanovic and Silva 2006). In contrast, a US treatment of blackberry juice resulted in only minor color changes and retention of 94 % of the anthocyanins, which led to a positive rating of the application of US for the preservation of fruit juices (Tiwari et al. 2009).

17.4.3 Process-Structure-Function Interactions

The impact of US on protein structures and their functional properties has already been discussed. In addition to meat tenderization, US can be applied to achieve certain textural characteristics of tailor-made foods. US-treated soy proteins showed significant textural changes in model food systems (Jambrak et al. 2009), such as changes in conductivity, increased solubility, and increased specific surface area, which is of importance for food texture and functionality.

In extrusion processes ultrasonic excitation can lead to a reduction in drag resistance, improved flow characteristics, and the formation of edible coatings due to a denaturation of proteins (Knorr et al. 2004; Mousavi et al. 2007).

Effects on viscosity are additionally observed for US treatments of macromolecular solutions due to degradation caused by hydrodynamic forces. Sonolytic degradation of aqueous carboxymethylcellulose polymers showed that the rate of degradation increased with increasing initial dynamic viscosity, higher molecular mass, and increased polymer concentration (Grönroos et al. 2008).

In crystallization processes cavitation can lead to crystal fracture, leading to overall smaller crystal size and improved crystal size distribution (Zheng and Sun 2006). This fracturing also affects fat globules or solid particles in liquids, leading to homogenization effects (Jafari et al. 2007; Villamiel and de Jong 2000).

When US is applied to plant-based tissues, cavitation damage, distortion of cells, and alteration of tissue structure appear, influencing the internal mass transfer characteristics of the product (Fernandes et al. 2009; Jambrak et al. 2007).

17.4.4 Current Applications and Process Development

In addition to US application to assist in food preservation processes using the synergistic effects of US, heat, and pressure on microorganisms and enzymes, US process development has covered a large range of food manufacturing where the structure and properties of the raw material are of importance.

Ultrasonic pulverization techniques are applied for the destruction of residual cell wall material and vegetable purees achieving significant modifications of textural and rheological properties, as by releasing pectin from cell walls, which contributes to the formation of continuous matrices (Bates et al. 2006; Mawson and Knoerzer 2007). In the case of the preparation of biomaterials for further processing by fermentation or enzyme digestion, US-assisted pulverization of cell matrices is used to facilitate the release of substrates or nutrients (Matsuura et al. 1994; Mawson and Knoerzer 2007; Wu et al. 2000).

Thermal processes can be improved by enhanced heat transfer, microstreaming at boundary layers, reduced fouling due to cavitation phenomena, and a faster formation of gas bubbles in evaporation processes. The positive effect on heat and mass transfer processes is used for the assistance of drying processes (García-Pérez et al. 2006; Schössler et al. 2012; Simal et al. 1998).

During extraction processes, cavitation can improve penetration with the solvent and disrupt cell walls when high intensities are applied (Li et al. 2004).

Cavitation, shear forces, and an influence on boundary layers provide the opportunity to improve emulsification and homogenization (Jafari et al. 2007; Villamiel and de Jong 2000).

A change in the US parameters can lead to the opposite effect, and emulsions can be split into their components (Pangu and Feke 2004). An acoustic radiation force holds particles in position in a stationary field, leading to coalescence (Masudo and Okada 2006).

US during freezing processes can promote ice nucleation and enhance heat and mass transfer processes (Zheng and Sun 2006). Crystal size distribution can be controlled, leading to reduced cell destruction, and cavitation effects can minimize fouling in surface freezers (Acton and Morris 1992).

US filtration systems are used as an add-on to existing vibratory screens, while the combination of US and membrane filtration is still in the early phase of development (Patist and Bates 2008).

Airborne US is used for the defoaming of carbonated beverages and fermentation systems (Gallego-Juárez 1998).

At lower treatment intensities US can be applied to the detection of foreign bodies in packaged and nonpackaged food (Knorr et al. 2004; Leemans and Destain

2009) and for nondestructive testing. Low-intensity US has been used to monitor the gelation process in milk and tofu (Dwyer et al. 2005; Ting et al. 2009) and to measure the mechanical properties of cheese products (Benedito et al. 2000).

17.4.5 Research Needs and Challenges

The presented overview of US research in the food industry underlines the versatility of US processes. One of the most important challenges for the industrial application of US is the definition of optimized parameters for all processes and products. Detailed knowledge about quality aspects, together with a thorough analysis of energy requirements, is the basis for the successful scale-up from laboratory tests to industrial scale.

17.5 Plasma Treatment

17.5.1 Process Description

Plasmas can be described as quasineutral particle systems in the form of gaseous or fluidlike mixtures of free electrons and ions, frequently also containing neutral particles (atoms, molecules), with a large mean kinetic energy of the electrons or all of the plasma components and a substantial influence of the charge carriers and their electromagnetic interactions on the system properties (Rutscher 2008).

According to Fridman et al. (2005), all varieties of plasma–chemical systems are traditionally divided into two major categories: thermal and nonthermal plasmas, both with specific advantages and disadvantages. Thermal plasmas [usually arcs or radiofrequency (RF) inductively coupled plasma discharges] are associated with Joule heating and thermal ionization and enable the delivery of high power (up to over 50 MW per unit) at high operating pressures. Besides other limitations, very high gas temperatures limit thermal plasmas' applicability to food systems.

In nonthermal plasmas, the electron temperature is much higher than the bulk gas temperature. While the electron temperature can reach several tens of thousands of degrees Kelvin, the gas temperature remains at temperature levels below 40 °C (Mastwijk and Nierop Groot 2010). Nonthermal plasmas may be produced by a variety of electrical discharges at different pressure levels. Working pressures below atmospheric conditions are mainly suitable for dried food materials or packaging materials since a vacuum will support liquid-to-gaseous phase changes in high-moisture food products. The most suitable systems for food processing are atmospheric-pressure plasma devices since mild conditions at low temperatures can be realized. Atmospheric-pressure plasma devices like corona discharges, dielectric barrier discharges (DBD), and plasma jets used for microbial inactivation were described in detail by Ehlbeck et al. (2011).

17.5.2 *Research State of the Art*

Recent research activities in food-related application of plasma have focused mainly on the inactivation of microbes, but little is known about the effect of plasma on food matrices. Since emitted reactive species react with bacteria, they may also affect food components such as water, lipids, proteins, and carbohydrates (Deng et al. 2007; Keener 2008).

17.5.2.1 Impact on Biological Cells

An overview of microbial inactivation in model systems using nonthermal plasma published during the last 3 years is given by Wan et al. (2009).

Although several reviews focus on the inactivation mechanisms of plasma (Boudam et al. 2006; Gaunt et al. 2006; Moisan et al. 2001; Moreau et al. 2008), they are not yet fully understood. In particular, the inactivation mechanisms of different plasma sources are difficult to compare because the effects depend on, for example, the plasma source, process parameters, and generated reactive species within the plasma due to the application of different process gases. Additionally, the inactivation mechanisms also depend on the type and the differentiation state of bacteria (e.g., spores, Gram-negative or Gram-positive bacteria, biofilm). Moisan et al. (2001) stated that three basic mechanisms are involved in plasma inactivation at low pressures: (1) UV irradiation of genetic material, (2) intrinsic photodesorption, and (3) etching. Although three-phase survivor curves were also reported in atmospheric-pressure plasma studies (Laroussi 2002), the explanation of the inactivation mechanism cannot be transferred to these studies. Most of the researches claim that UV plays a minor role in the inactivation of microorganisms at atmospheric pressure and the inactivation process is controlled by chemically reactive species. However, it was shown that in some cases UV photons can play a role in the inactivation process of microorganisms at atmospheric pressure (Boudam et al. 2006). Moreau et al. (2008) compared plasma inactivation effects to the effects of micro pulses. Similar to the effects of micro pulses, the cell membrane of microorganisms is perforated following plasma treatment. Besides the perforation of the cell membrane, the inactivation effect of plasma is induced by the bombardment of the cell membrane by radicals (OH17 or NO17). These radicals are absorbed onto the bacteria surface, and volatile components are formed and eliminated from the cells (etching). Xiong et al. (2011) evaluated the penetration depth of plasma in a biofilm. For a treatment with a plasma jet at atmospheric pressure a penetration depth of 15 μm was obtained and related to O and OH.

17.5.2.2 Impact on Enzymes

To date, information regarding the impact of plasma on enzymes has been given in only a small number of papers. Dudak et al. (2007) found a decrease in enzyme activity following RF glow discharge plasma treatment, with the highest decrease in activity within the first 10 min of treatment, including a fragmentation of proteins. A fragmentation of proteins was also found by Deng et al. (2007) after DBD treatment. In this process, atomic oxygen has been shown to play a dominant role in destruction and degradation reactions. Oxygen plasma generated by RF discharge led to a reduction in C-H and N-H bonds in casein protein and to a modification of the secondary protein structure (Hayashi et al. 2009).

17.5.2.3 Impact on Food Constituents

Although much work has already been performed on the effects of nonthermal plasma on microorganisms, information on plasma interactions with food components is scarce. This is mainly due to the fact that the application of plasma was for a long time limited to heat- and vacuum-resistant materials.

A time- and dose-dependent degradation has been observed for flavonoids, known for their high antioxidant activity protecting cells against the damaging effects of reactive oxygen species. The degradation rate strongly depended on the polyphenolic substitution pattern. While glycosidic flavonoids showed a rather inert behavior throughout plasma treatment, aglycosid derivates were quickly degraded (Grzegorzewski et al. 2010).

However, organoleptic analysis of plasma-treated nut samples likewise showed no relation between treatment and perceptual sensory character (Basaran et al. 2008). On a molecular level, SEM analysis of varius cabbage and lettuce species revealed that under certain conditions, plasma treatment may lead to changes in plant surface hydrophobic wax layers (Grzegorzewski et al. 2010). On the other hand, Ragni et al. (2010) observed no negative effects of plasma treatment on egg quality.

17.5.3 Process-Structure-Function Interactions

The complex process-structure-function interactions are not yet fully understood. Depending on the plasma source, process parameters, and process gases, the reactive species vary within the plasma, which makes reaction mechanisms difficult to compare.

Plasma treatment in air or using oxygen as feed gas led to a strong surface oxidation, resulting in the formation of carbonyl and carboxyl functions, as was demonstrated for several polymers (Morent et al. 2008). The modification of starch in a glow discharge argon plasma was manifested in a loss of OH groups, probably due to the crosslinking of α-D-glucose units (Zou et al. 2004).

17.5.4 Current Applications and Process Development

Plasma technologies in food processing are not yet established, but investigations using complex food raw materials have been performed. Some studies focus on the plasma-related decontamination of bacteria at the surface of several fruit and vegetable samples without evaluating the obtained product quality (Niemira and Sites 2008). For example, Critzer et al. (2007) showed the capability of a one atmosphere uniform glow discharge plasma to inactivate *E. coli* on mangos and *E. coli* O157:H7, *Salmonella*, and *L. monocytogenes* on apples, cantaloupe, and lettuce, respectively. Perni et al. (2008a, b) used a cold atmospheric plasma pen to inactivate *Saccharomyces cerevisiae, Pantoea agglomerans,* and *Gluconacetobacter liquefaciens* inoculated on pericaps of mango and melon or cut melon and mango pieces inoculated with *E. coli, S. cerevisiae, G. liquefaciens*, and *L. monocytogenes*.. A decontamination of the fruit pericaps was detected, whereas the efficiency on cut fruit surfaces was reduced.

17.5.5 Research Needs and Challenges

Approaches to studying the effects of plasmas in industrial plasma engineering often regard plasma as a black box with inputs and outputs. In addition, studies are focusing on plasma application to foods the desired output of a plasma-related process is mainly achieved by adjusting inputs until the desired result is obtained. In such approaches, no serious attempt is made to understand the plasma-physical, plasma-chemical, or plasma-biological processes occurring in this black box. Future research needs to involve more interdisciplinary studies to allow a better understanding of the complex interactions during plasma processing and, thus, the design of beneficial and controlled plasma applications for food processing, which might encompass microbial inactivation as well as the modification of functional food properties. Further, plasma processing must be considered as a surface treatment, and the impact of potential toxicological effects resulting from chemical reactions based on plasma-air-food surface interactions must be viewed in relation to the surface volume ratio of the particular product.

17.6 Conclusion

High HP treatment offers the promise to produce microbiologically safe, high-quality, tailor-made foods under gentle processing conditions. Process improvement can only be achieved by understanding and applying the different temperature, time, and pressure dependencies of desired and undesired reactions. Detailed studies regarding the inactivation kinetics and mechanisms of pathogens should be performed in the respective food matrix and must result in constant process parameters or, if that is not possible, in optimally controlled process parameters. Controlled and reproducible studies on the pressure effect of high pressure on nutrient biopolymers, toxins, and allergens are also needed.

The interactions between products and PEF processes and possible undesired changes during PEF treatment remain uncertain and require further investigation. For example, high-value products – such as enzyme or vitamin solutions or protein fractions isolated from milk, all of which are heat sensitive – are potential products for nonthermal pasteurization by PEFs. A combination of techniques that deliver effective preservation without the excessive use of any single conventional process parameter such as time or temperature will allow for the selective retention or inactivation of food constituents. The combination of PEFs with other stress factors like mild heat, antimicrobial compounds, pH, or organic acids, as well as their combination with other thermal or nonthermal decontamination techniques, will determine further development. The impact of PEFs on the structure of food matrices and on mass transfer within food matrices and the subsequent understanding of PEF-related process-structure relationships will allow the development of unique tailor-made foods.

US-assisted processing offers advantages for a large variety of food production processes. The unique characteristics of sound waves provide opportunities to treat products with specifically adapted parameters. For instance, US with a low maximum pressure amplitude will cause microstreaming in fluids, gently manipulating mass transfer, whereas high-pressure amplitudes cause cavitation associated with high pressure and temperature peaks, permitting changes in the cell structure as well as homogenization of disperse systems. However, the large range of attainable effects is one of the most important obstacles to the transfer of laboratory-scale results to the industrial scale. Equipment and parameters must be directly adapted to all products and objectives. Furthermore, undesired changes in product structure and quality must be known and minimized. A better understanding of the basic mechanisms of US treatments depending on the product and process parameters is necessary to allow general conclusions to be drawn and the design of new processes and applications to be simplified.

CP treatment at atmospheric pressures offers various opportunities in food processing, e.g., surface decontamination, modification of surface properties, and enhancement of mass transfer with respect to foods and food-related materials. Attempts to limit heat transfer to sensitive materials such as food products resulted in the development of new atmospheric plasma jets and will allow efficient in-line integration in production lines; however, further research is required in light of the lack of data regarding plasma-matrix interactions and to ensure the development of safe and tailor-made processes for food application. Further investigations focusing on, for example, the spatial composition of plasma, physicochemical reaction kinetics, and penetration depths, should be supported by validated mathematical models and simulation approaches.

Acknowledgments This work was supported by the German Federal Ministry of Economics and Technology (via AiF) and the FEI (Forschungskreis der Ernährungsindustrie e.V., Bonn), Project AiF 15610 N; by Bundesanstalt für Landwirtschaft und Ernährung (BLE), Project FriPlas, and Verbundprojekt Hochdrucktechnologie; by the German Federal Ministry of Education and Research, Project BioMed; by the Commission of the European Communities, Framework 6, Priority 5 "Food Quality and Safety," Integrated Project NovelQ FP6-CT-2006-015710, and Framework 7 Project Ultraveg.

References

Acton E, Morris GJ (1992) USA Patent No. W.O. 99/20420

Ade-Omowaye BIO, Taiwo KA, Eshtiaghi NM, Angersbach A, Knorr D (2003) Comparative evaluation of the effects of pulsed electric field and freezing on cell membrane permeabilisation and mass transfer during dehydration of red bell peppers. Innov Food Sci Emerg Technol 4:177–188

Aertsen A, Meersman F, Hendrickx M, Vogel R, Michiels CW (2009) Biotechnology under high pressure: applications and implications. Trends Biotechnol 27:434–441

Aguiló-Aguayo I, Soliva-Fortuny R, Martín-Belloso O (2010) Impact of high-intensity pulsed electric field variables affecting peroxidase and lipoxygenase activities of watermelon juice. LWT- Food Sci Technol 43:897–902

Alliger H (1975) Ultrasonic disruption. Am Lab 10:75–85

Ananta E (2005) Impact of enviromental factors on vitality and stability and high pressure pretreatment on stress tolerance of *Lactobacillus rhamnosus* GG (ATCC 53103) during spray drying. Technische Universitiät Berlin, Berlin

Ananta E, Knorr D (2003) Pressure-induced thermotolerance of *Lactobacillus rhamnosus* GG. Food Res Int 36:991–997

Ananta E, Heinz V, Schlüter O, Knorr D (2001) Kinetic studies on high-pressure inactivation of *Bacillus stearothermophilus* spores suspended in food matrices. Innov Food Sci Emerg Technol 2:261–272

Ananta E, Heinz V, Knorr D (2005) Assessment of high pressure induced damage on *Lactobacillus rhamnosus* GG by flow cytometry. Food Microbiol 21:567–577

Angersbach A, Knorr D (1998) Impact of high-intensity electric field pulses on plant membrane permeabilzation. Trends Food Sci Technol 9:185–191

Angersbach A, Heinz V, Knorr D (1999) Electrophysiological model of intact and processed plant tissues: cell disintegration criteria. Biotechnol Prog 15:753–762

Angersbach A, Heinz V, Knorr D (2000) Effects of pulsed electric fields on cell membranes in real food systems. Innov Food Sci Emerg Technol 1:135–149

Ardia A (2004) Process considerations on the application of high pressure treatment at elevated temperature levels for food preservation. Ph.D. thesis, Berlin University of Technology, Berlin, 94 pp

Barbosa-Cánovas GV, Góngora-Nieto MM, Pothakamury UR, Swanson BG (1999) Preservation of foods with pulsed electric fields. Academic, San Diego

Barsotti L, Dumay E, Mu TH, Fernandez Diaz MD, Cheftel JC (2001) Effects of high voltage electric pulses on protein-based food constituents and structures. Food Sci Technol 12:136–144

Basaran P, Basaran-Akgul N, Oksuz L (2008) Elimination of *Aspergillus parasiticus* from nut surface with low pressure cold plasma (LPCP) treatment. Food Microbiol 25:626–632

Bates DM, Bagnall WA, Bridges MW (2006) Patent No. US patent application 20060110503

Bendicho S, Barbosa-Cánovas GV, Martín O (2003) Reduction of protease activity in simulated milk ultrafiltrate by continuous flow high intensity pulsed electric field treatments. J Food Sci 68:952–957

Benedito J, Carcel JA, Sanjuan N, Mulet A (2000) Use of ultrasound to assess Cheddar cheese characteristics. Ultrasonics 38:727–730

Bluhm H, Sack M (2009) Industrial-scale treatment of biological tissues with pulsed electric fields. In: Vorobiev E, Lebovka N (eds) Electrotechnologies for extraction from food plants and biomaterials. Springer, New York, pp 237–269

Bonnafous P, Vernhes M-C, Teissié J, Gabriel B (1999) The generation of reactive-oxygen species associated with long-lasting pulse-induced electropermeabilisation of mammalian cells is based on a non-destructive alteration of the plasma membrane. Biochimica et Biophysica Acta (BBA) – Biomembranes 1461:123–134

Boudam MK, Moisan M, Saoudi B, Popovici C, Gherardi N, Massines F (2006) Bacterial spore inactivation by atmospheric-pressure plasmas in the presence or absence of UV photons as obtained with the same gas mixture. J Phys D: Appl Phys 39:3494–3507

Buckow R (2006) Pressure and temperature effects on the enzymatic conversion of biopolymers. Technsiche Universität Berlin, Berlin

Buckow R, Heinz V (2008) High pressure processing – a database of kinetic information. Chemie Ingenieur Technik 80:1081–1095

Buckow R, Heinz V, Knorr D (2005) Two fractional model for evaluating the activity of glucoamylase from *Aspergillus niger* under combined pressure and temperature conditions. Food Bioprod Process 83:220–228

Buckow R, Heinz V, Knorr D (2007) High pressure phase transition kinetics of maize starch. J Food Eng 81:469–475

Bull MK, Olivier SA, van Diepenbeek RJ, Kormelink F, Chapman B (2009) Synergistic inactivation of spores of proteolytic *Clostridium botulinum* strains by high pressure and heat is strain and product dependent. Appl Environ Microbiol 75:434–445

Bunthof CJ (2002) Flow cytometry, fluorescent probes, and flashing bacteria. Ph.D. thesis, Wageningen University, Wageningen, 160 pp

Castro AJ, Swanson BG, Barbosa-Cánovas GV, Zhang QH (2001) Pulsed electric field modification of milk alkaline phosphatase activity. In: Barbosa-Cánovas GV, Zhang QH (eds) Electric fields in food processing. Technomic, Lancaster, pp 65–82

Cheftel JC, Culioli J (1997) Effects of high pressure on meat: a review. Meat Sci 46:211–236

Clark JP (2008) An update on ultrasonics. Food Technol 62:75–77

Coster HGL (1965) A quantitive analysis of the voltage-current relationships of fixed charge membranes and the associated property of "punch-through". Biophys J 5:669–686

Craven HM, Swiergon P, Ng S, Midgely J, Versteeg C et al (2008) Evaluation of pulsed electric field and minimal heat treatments for inactivation of pseudomonads and enhancement of milk shelf-life. Innov Food Sci Emerg Technol 9:211–216, Food Innovation: Emerging Science, Technologies and Applications (FIESTA) Conference

Critzer FJ, Kelly-Wintenberg K, South SL, Golden DA (2007) Atmospheric plasma inactivation of foodborne pathogens on fresh produce surfaces. J Food Prot 70:2290–2296

Daeschlein G, Woedtke T, Kindel E, Brandenburg R, Weltmann K-D, Jünger M (2010) Antibacterial activity of an atmospheric pressure plasma Jet against relevant wound pathogens in vitro on a simulated wound environment. Plasma Processes Polym 7:224–230

Delgado A, Rauh C, Kowalczyk W, Baars A (2008) Review of modelling and simulation of high pressure treatment of materials of biological origin. Trends Food Sci Technol 19:329–336

Deng XT, Shi JJ, Chen HL, Kong MG (2007) Protein destruction by atmospheric pressure glow discharges. Appl Phys Lett 90

Diels AMJ, Michiels CW (2006) High-pressure homogenization as a Non-thermal technique for the inactivation of microorganisms. Crit Rev Microbiol 32:201–216

DIL (2011) Elcrack Generators and Technology. Presented at iFood2011, Osnabrueck

Dörnenburg H, Knorr D (1993) Cellular permeabilization of cultured plant cell tissues by high electric field pulses or ultra high pressure for recovery of secondary metabolites. Food Biotechnol 7:35–38

Dörnenburg H, Knorr D (1995) Strategies for the improvement of secondary metabolite production in plant cell cultures. Enzyme Microb Technol 17:674–684

Dudak FC, Kousal J, Seker UÖS, Boyaci IH, Choukourov A, Biederman H (2007) Influence of the plasma treatment on enzyme structure and activity. Presented at 28th ICPIG, Prague

Dumay EM, Kalichevsky MT, Cheftel JC (1998) Characteristics of pressure-induced gel of beta-lactoglobulin at various times after pressure release. Lebensm Wiss Technol 31:10–19

Dwyer C, Donnelly L, Buckin V (2005) Ultrasonic analysis of rennet-induced pre-gelation and gelation processes in milk. J Dairy Res 72:303–310

Ehlbeck J, Schnabel U, Polak M, Winter J, Woetke T et al (2011) Low temperature atmospheric pressure plasma sources for microbial decontamination. J Phys D: Appl Phys 44:013002

Eshtiaghi MN, Knorr D (2002) High electric field pulse treatment: potential for sugar beet processing. J Food Eng 52:265–272

Fernandes FAN, Gallão MI, Rodrigues S (2009) Effect of osmosis and ultrasound on pineapple cell tissue structure during dehydration. J Food Eng 90:186–190

Fernandez-Diaz MD, Barsotti L, Dumay E, Cheftel C (2000) Effects of pulsed electric fields on ovalbumin solutions and dialyzed egg white. J Agr Food Chem 48:2332–2339

Fiala A, Wouters PC, van den Bosch E, Creyghton YLM (2001) Coupled electrical-fluid model of pulsed electric field treatment in a model food system. Innov Food Sci Emerg Technol 2:229–238

Fridman A, Chirokov A, Gutsol A (2005) Non-thermal atmospheric pressure discharges. J Phys D: Appl Phys 38:R1–R24

Galindo F, Dejmek P, Lundgren K, Rasmusson A, Vicente A, Moritz T (2009) Metabolomic evaluation of pulsed electric field-induced stress on potato tissue. Planta 230:469–479

Gallego-Juárez JA (1998) Some applications of air-borne ultrasound to food processing. In: Povey MJW, Mason TJ (eds) Ultrasound in food processing. Thomson Science, London, pp 127–143

Garcia D, Gómez N, Manas P, Condon S, Raso J, Pagan R (2005) Occurence of sublethal injury after pulsed electric fields depending on the microorganism, the treatment medium pH and the intensity of the treatment investigated. J Appl Microbiol 99:94–104

García-Pérez JV, Cárcel JA, de la Fuente-Blanco S, Riera-Franco de Sarabia E (2006) Ultrasonic drying of foodstuff in a fluidized bed: parametric study. Ultrasonics proceedings of ultrasonics international (UI'05) and world congress on ultrasonics (WCU) 44:e539–e543

Gaunt LF, Beggs CB, Georghiou GE (2006) Bactericidal action of the reactive species produced by gas-discharge nonthermal plasma at atmospheric pressure: a review. IEEE Trans Plasma Sci 34:1257–1269

Gerlach D, Alleborn N, Baars A, Delgado A, Moritz J, Knorr D (2008) Numerical simulations of pulsed electric fields for food preservation: a review. Innov Food Sci Emerg Technol 9:408–417

Gomez Galindo F, Wadsö L, Vicente A, Dejmek P (2008) Exploring metabolic responses of potato tissue induced by electric pulses. Food Biophys 3:352–360

Grimi N, Mamouni F, Lebovka N, Vorobiev E, Vaxelaire J (2010) Acoustic impulse response in apple tissues treated by pulsed electric field. Biosyst Eng 105:266–272

Grönroos A, Pirkonen P, Kyllönen H (2008) Ultrasonic degradation of aqueous carboxymethylcellulose: effect of viscosity, molecular mass, and concentration. Ultrason Sonochem 15:644–648

Grzegorzewski F, Rohn S, Kroh LW, Geyer M, Schlüter O (2010a) Surface morphology and chemical composition of lamb's lettuce (Valerianella locusta) after exposure to a low-pressure oxygen plasma. Food Chem 122:1145–1152
Grzegorzewski F, Rohn S, Quade A, Schröder K, Ehlbeck J et al (2010) Reaction chemistry of 1,4-Benzopyrone derivates in non-equilibrium low-temperature plasmas. Plasma Processes Polym 9999:NA
Gudmundsson M, Hafsteinsson H (2001) Effect of electric field pulses on microstructure of muscle foods and roes. Food Sci Technol 12:122–128
Guerrero-Beltrán JÁ, Sepulveda DR, Góngora-Nieto MM, Swanson B, Barbosa-Cánovas GV (2010) Milk thermization by pulsed electric fields (PEF) and electrically induced heat. J Food Eng 100:56–60
Harvey EN, Loomis AL (1929) The destruction of luminous bacteria by high frequency sound waves. J Bacteriol 17:373–376
Hayashi N, Kawaguchi R, Liu H (2009) Treatment of protein using oxygen plasma produced by RF discharge. Presented at 14th international congress on plasma physics (ICPP2008), Journal of Plasma and Fusion Research SERIES
Heinz V, Knorr D (1996) High pressure inactivation kinetics of *Bacillus subtilis* cells by a three-state-model considering distribution resistance mechanisms. Food Biotechnol 10:149–161
Heinz V, Kortschack F (2002) Germany Patent No. WO 02/49460
Heinz V, Knoch A, Lickert T (2009) Product innovation by high pressure processing. New Food 2:43–44
Hendrickx M, Knorr D (2002) Ultra high pressure treatment of foods. Kluwer Academic/Plenum Publisher, New York
Heremans R, Smeller L (1998) Protein structure and dynamics at high pressure. Biochim Biophys Acta 1386:353–370
Hughes DE, Nyborg WL (1962) Cell disruption by ultrasound: streaming and other activity around sonically induced bubbles is a cause of damage to living cells. Science 138:108–114
Isbarn S, Buckow R, Himmelreich A, Lehmacher A, Heinz V (2007) Inactivation of avian influenza virus by heat and high hydrostatic pressure. J Food Prot 70:667–673
Jaeger H, Meneses N, Knorr D (2009a) Impact of PEF treatment inhomogeneity such as electric field distribution, flow characteristics and temperature effects on the inactivation of *E. coli* and milk alkaline phosphatase. Innov Food Sci Emerg Technol 10:470–480
Jaeger H, Meneses N, Knorr D (2009) Pulsed electric field preservation of heat sensitive products – food safety and qualtiy aspects. Presented at international conference on bio- and food electrotechnologies, Compiegne
Jaeger H, Schulz A, Karapetkov N, Knorr D (2009c) Protective effect of milk constituents and sublethal injuries limiting process effectiveness during PEF inactivation of Lb. rhamnosus. Int J Food Microbiol 134:154–161
Jaeger H, Meneses N, Moritz J, Knorr D (2010) Model for the differentiation of temperature and electric field effects during thermal assisted PEF processing. J Food Eng 100:109–118
Jaeger H, Schulz M, Lu P, Knorr D (2012) Adjustment of milling, mash electroporation and pressing for the development of a PEF assisted juice production in industrial scale. Innov Food Sci Emerg Technol 14:46–60
Jafari SM, He Y, Bhandari B (2007) Production of sub-micron emulsions by ultrasound and microfluidization techniques. J Food Eng 82:478–488
Jambrak AR, Mason TJ, Paniwnyk L, Lelas V (2007) Ultrasonic effect on pH, electric conductivity and tissue surface of button mushrooms, brussels sprouts and cauliflower. Czech J Food Sci 25:90–100
Jambrak AR, Lelas V, Mason TJ, Kresic G, Badanjak M (2009) Physical properties of ultrasound treated soy proteins. J Food Eng 93:386–393
Jayasooriya SD, Bhandari BR, Torley P, D'Arcy BR (2004) Effect of high power ultrasound waves on properties of meat: a review. Int J Food Prop 7:301–319

Jayasooriya SD, Torley PJ, D'Arcy BR, Bhandari BR (2007) Effect of high power ultrasound and ageing on the physical properties of bovine semitendinosus and longissimus muscles. Meat Sci 75:628–639

Johnston DE, Austin BA, Murphy RJ (1993) Properties of acid-set gels prepared from high-pressure treated skim milk. Milchwissenschaft 48:206–209

Keener KM (2008) Atmospheric non-equilibrium plasma. Encyclopedia Agric Food Biol Eng 1:1–5

Kinsloe H, Ackermann E, Reid JJ (1954) Exposure of microorganisms to measured sound fields. J Bacteriol 68:373–380

Knorr D, Angersbach A, Eshtiaghi M, Heinz V, Lee D-U (2001) Processing concepts based on high intensity electric field pulses. Trends Food Sci Technol 12:129–135

Knorr D, Zenker M, Heinz V, Lee D-U (2004) Applications and potential of ultrasonics in food processing. Trends Food Sci Technol 15:261–266

Knorr D, Heinz V, Buckow R (2006) High pressure application for food biopolymers. Biochim Biophys Acta 1764:619–631

Knorr D, Engel K-H, Vogel R, Kochte-Clemens B, Eisenbrand G (2008) Statement on the treatment of food using a pulsed electric field. Mol Nutr Food Res 52:1539–1542

Laroussi M (2002) Nonthermal decontamination of biological media by atmospheric-pressure plasmas: review, analysis and prospects. IEEE T Plasma Sci 30:1409–1415

Lebovka NI, Praporscic I, Vorobiev E (2004) Effect of moderate thermal and pulsed electric field treatments on textural properties of carrots, potatoes and apples. Innov Food Sci Emerg Technol 5:9–16

Lebovka NI, Shynkaryk MV, El-Belghiti K, Benjelloun H, Vorobiev E (2007a) Plasmolysis of sugarbeet: pulsed electric fields and thermal treatment. J Food Eng 80:639–644

Lebovka NI, Shynkaryk NV, Vorobiev E (2007b) Pulsed electric field enhanced drying of potato tissue. J Food Eng 78:606–613

Leemans V, Destain M-F (2009) Ultrasonic internal defect detection in cheese. J Food Eng 90:333–340

Lelieveld HLM, Notermans S, de Haan SWH (eds) (2007) Food preservation by pulsed electric fields. Woodhead Publishing, Abington

Li H, Pordesimo L, Weiss J (2004) High intensity ultrasound-assisted extraction of oil from soybeans. Food Res Int 37:731–738

Lindgren M, Aronsson K, Galt S, Ohlsson T (2002) Simulation of the temperature increase in pulsed electric field (PEF) continuous flow treatment chambers. Innov Food Sci Emerg Technol 3:233–245

López-Fandino R, Carrascosa AV, Olano A (1996) The effects of high pressure on whey protein denaturation and cheese making properties of raw milk. J Dairy Sci 79:929–936

Ludikhuyze L, Van Loey A, Indrawati I, Denys S, Hendrickx M (2002) Effects of high pressure on enzymes related to food quality. In: Hendrickx M, Knorr D (eds) Ultra high pressure treatments of food. Kluwer/Plenum Publisher, New York, pp 115–166

Luscher CM (2008) Effect of high pressure – low temperature phase transitions on model systems, foods and microorganisms. Ph.D. thesis, Berlin University of Technology, Berlin, 158 pp

Lyng JG, Allen P, McKenna B (1998) The effects of Pre- and post-rigor high-intensity ultrasound treatment on aspects of lamb tenderness. LWT- Food Sci Technol 31:334–338

Marco-Moles R, Perez-Munuera I, Quiles A, Hernando I (2009) Effect of pulsed electric fields on the main chemical components of liquid egg and stability at 4 °C. Czech J Food Sci 27:109–112

Margosch D, Ehrmann MA, Buckow R, Heinz V, Vogel RF, Gänzle MG (2006) High-pressure-mediated survival of *Clostridium botulinum* and *Bacillus amyloliquefaciens* endospores at high temperature. Appl Environ Microbiol 72(5):3476–3481

Martín-Belloso O, Elez-Martínez P (2005) Enzymatic inactivation by pulsed electric fields. In: Sun D-W (ed) Emerging technologies for food processing. Academic Press, London, pp 155–181

Mason TJ, Paniwnyk L, Lorimer JP (1996) The uses of ultrasound in food technology. Ultrasonics sonochemistry proceedings of the symposium on the chemical effects of ultrasound in the 1995 international chemical congress of pacific basin societies 3:S253–S260

Mastwijk HC, Nierop Groot MN (2010) Use of cold plasma in food processing. In: Heldman DR, Bridges A, Hoover DG, Wheeler MB (eds) Encyclopedia of biotechnology in agriculture and food. Taylor & Francis, New York

Masudo T, Okada T (2006) Particle separation with ultrasound radiation force. Curr Anal Chem 2:213–227

Mathys A (2008) Inactivation mechanisms of *Geobacillus* and *Bacillus* spores during high pressure thermal sterilization. Ph.D. thesis, Technische Universität Berlin, Berlin

Mathys A, Knorr D (2009) The properties of water in the pressure–temperature landscape. Food Biophys 4:77–82

Mathys A, Reineke K, Heinz V, Knorr D (2009) High pressure thermal sterilization – development and application of temperature controlled spore inactivation studies. High Pres Res 29:3–7

Matsuura K, Hirotsune M, Nunokawa Y, Satoh M, Honda K (1994) Acceleration of cell growth and ester formation by ultrasonic wave irradiation. J Ferment Bioeng 77:36–40

Mawson R, Knoerzer K (2007) A brief history of the application of ultrasonics in food processing. Presented at 19th international congress on acoustics, Madrid

McClements DJ (1995) Advances in the application of ultrasound in food analysis and processing. Trends Food Sci Technol 6:293–299

Michel M, Autio K (2002) Effects of high pressure on protein- and polysaccharide-based structures. In: Hendrickx MEG, Knorr D (eds) Ultra high pressure treatments of foods. Kluwer, New york, pp 189–214

Moisan M, Barbeau J, Moreau S, Pelletier J, Tabrizian M, Yahia LH (2001) Low-temperature sterilization using gas plasmas: a review of the experiments and an analysis of the inactivation mechanisms. Int J Pharm 226:1–21

Molina-Gutierrez A, Stippl V, Delgado A, Gaenzle MG, Vogel RF (2002) In situ determination of the intracellular pH of *Lactococcus lactis* and *Lactobacillus plantarum* during pressure treatment. Appl Environ Microbiol 68:4399–4406

Moreau M, Orange N, Feuilloley MGJ (2008) Non-thermal plasma technologies: new tools for bio-decontamination. Biotechnol Adv 26:610–617

Morent R, Geyter ND, Leys C, Gengembre L, Payen E (2008) Comparison between XPS- and FTIR-analysis of plasma-treated polypropylene film surfaces. Surf Interface Anal 40:597–600

Morren J, Roodenburg B, de Haan SWH (2003) Electrochemical reactions and electrode corrosion in pulsed electric field (PEF) treatment chambers. Innov Food Sci Emerg Technol 4:285–295

Mousavi SAAA, Feizi H, Madoliat R (2007) Investigations on the effects of ultrasonic vibrations in the extrusion process. J Mater Process Technol 187–188:657–661

Niemira BA, Sites J (2008) Cold plasma inactivates *Salmonella* stanley and *Escherichia coli* O157: H7 inoculated on golden delicious apples. J Food Prot 71:1357–1365

Niven GW, Miles CA, Mackey BM (1999) The effects of hydrostatic pressure on ribosome conformation in *Escherichia coli*: an in vivo study using differential scanning calorimetry. Microbiology 145:419–425

Oey I, Verlinde P, Hendrickx M, Van Loey A (2006) Temperature and pressure stability of l -ascorbic acid and/or [6s] 5-methyltetrahydrofolic acid: a kinetic study. Eur Food Res Technol 223:71–77

Ohshima T, Tamura T, Sato M (2006) Influence of electric field on various enzyme activities. J Electrostat 65:156–161

Olivier SA, Bull MK, Stone G, van Diepenbeek RJ, Kormelink F et al (2011) Strong and consistently synergistic inactivation of spores of spoilage-associated *Bacillus* and *Geobacillus* spp. By high pressure and heat compared with inactivation by heat alone. Appl Environ Microbiol 77:2317–2324

Pangu GD, Feke DL (2004) Acoustically aided separation of oil droplets from aqueous emulsions. Chem Eng Sci 59:3183–3193

Patist A, Bates D (2008) Ultrasonic innovations in the food industry: from the laboratory to commercial production. Innovative Food Science & Emerging Technologies 9:147–154, Food Innovation: Emerging Science, Technologies and Applications (FIESTA) Conference

Perez O, Pilosof AMR (2004) Pulsed electric field effects on the molecular structure and gelation of ß-lactoglobulin concentrate and egg white. Food Res Int 37:102–110

Perni S, Liu DW, Shama G, Kong MG (2008a) Cold atmospheric plasma decontamination of the pericarps of fruit. J Food Prot 71:302–308

Perni S, Shama G, Kong MG (2008b) Cold atmospheric plasma disinfection of cut fruit surfaces contaminated with migrating microorganisms. J Food Prot 71:1619–1625

Perry RH (1984) Perry's chemical engineers'handbook. McGraw- Hill Book Co, New York

Phoon PY, Galindo FG, Vicente A, Dejmek P (2008) Pulsed electric field in combination with vacuum impregnation with trehalose improves the freezing tolerance of spinach leaves. J Food Eng 88:144–148

Povey MJW, Mason TJ (1998) Ultrasound in food processing. Blackie Academic and Professional, London/Weinheim/New York/Tokyo/Melbourne/Madras

Praporscic I, Lebovka N, Vorobiev E, Mietton-Peuchot M (2007) Pulsed electric field enhanced expression and juice quality of white grapes. Sep Purif Technol 52:520–526

Puértolas E, López N, Condón S, Álvarez I, Raso J (2010) Potential applications of PEF to improve red wine quality. Trends Food Sci Technol 21:247–255

Rademacher B, Hinrichs J (2006) Effects of high pressure treatment on indigenous enzymes in bovine milk: reaction kinetics, inactivation and potential application. Int Dairy J 16:655–661

Ragni L, Berardinelli A, Vannini L, Montanari C, Sirri F et al (2010) Non-thermal atmospheric gas plasma device for surface decontamination of shell eggs. J Food Eng 100:125–132

Rajan S, Ahn J, Balasubramaniam VM, Yousef AE (2006) Combined pressure-thermal inactivation kinetics of *Bacillus amyloliquefaciens* spores in egg patty mince. J Food Prot 69:853–860

Raso J, Heinz V (eds) (2006) Pulsed electric fields technology for the food industry. Springer, Heidelberg

Raso J, Heinz V (2007) Pulsed electric fields technology for the food industry. Springer, New York

Rastogi NK, Raghavarao KSM, Balasubramaniam VM, Niranjan K, Knorr D (2007) Opportunities and challenges in high pressure processing of foods. Crit Rev Food Sci Nutr 47:1–44

Reineke K, Mathys A, Heinz V, Knorr D (2008) Temperature control for high pressure processes up to 1400 MPa. J Phys Conf 121:142012–142016

Reineke K, Doehner I, Schlumbach K, Baier D, Mathys A, Knorr D (2011a) The different pathways of spore germination and inactivation in dependence of pressure and temperature. Innov Food Sci Emerg Technol (in press)

Reineke K, Mathys A, Knorr D (2011b) The impact of high pressure and temperature on bacterial spores: inactivation mechanisms of *Bacillus subtilis* above 500 MPa. J Food Sci 76: M189–M197

Reineke K, Doehner I, Schlumbach K, Baier D, Mathys A, Knorr D (2012) The different pathways of spore germination and inactivation in dependence of pressure and temperature. Innov Food Sci Emerg Technol 13:31–41

Riahi E, Ramaswamy HS (2004) High pressure inactivation kinetics of amylase in apple juice. J Food Eng 64:151–160

Riener J, Noci F, Cronin DA, Morgan DJ, Lyng JG (2008) Combined effect of temperature and pulsed electric fields on apple juice peroxidase and polyphenoloxidase inactivation. Food Chem 109:402–407

Roberts RT (1992) High intensity ultrasonics. In: Johnston DE, Knight MK, Ledward DA (eds) The chemistry of muscle-based foods. Royal Society of Chemistry, Cambridge, pp 287–297

Roberts RT (1993) High intensity ultrasonics in food processing. Chem Ind 4:119–121

Rodrigoa D, Cortésb C, Clynenc C, Schoofsc L, Van Loeya A, Hendrickx M (2006) Thermal and high-pressure stability of purified polygalacturonase and pectinmethylesterase from four different tomato processing varieties. Food Res Int 39:440–448

Roodenburg B, Morren J, Berg HE, de Haan SWH (2005) Metal release in a stainless steel pulsed electric field (PEF) system. Part I. Effect of diferent pulse shapes; theory and experimental method. Innov Food Sci Emerg Technol (in press)

Rovere P (2002) Industrial-scale high pressure processing of foods. In: Hendrickx MEG, Knorr D (eds) Ultra high pressure treatments of foods. Kluwer Academic/Plenum Publishers, New York, pp 251–268

Rumpold BA, Knorr D (2005) Effect of salts and sugars on pressure-induced gelatinisation of wheat, tapioca, and potato starches. Starch – Stärke 57:370–377

Russel NJ, Colley M, Simpson RK, Trivett AJ, Evans RI (2000) Mechanism of action of pulsed high electric field (PHEF) on the membranes of food-poisoning bacteria is an 'all-or-nothing' effect. Int J Food Microbiol 55:133–136

Rutscher A (2008) Characteristics of low-temperature plasmas under nonthermal conditions – a short summary. In: Hippler R, Kersten H, Schmidt M, Schoenbach KH (eds) Low temperature plasmas – fundamentals, technologies, and techniques. Wiley-VCH Verlag GmbH & Co KGaA, Weinheim, pp 1–14

Saulis G, Lape R, Praneviciute R, Mickevicius D (2005) Changes of the solution pH due to exposure by high-voltage electric pulses. Bioelectrochemistry 67:101–108

Schlueter O (2004) Impact of high pressure – low temperature processes on cellular materials related to foods. Technische Universität Berlin, Berlin, p 172

Schössler K, Jäger H, Knorr D (2012) Effect of continuous and intermittent ultrasound on drying time and effective diffusivity during convective drying of apple and red bell pepper. J Food Eng 108:103–110

Schuten H, Gulfo-van Beusekom K, Pol I, Mastwijk H, Bartels P (2004) Enzymatic stability of PEF processed orange juice. Presented at safe consortium seminar: novel preservation technologies in relation to food safety. Brussels

Sharma A, Yadav BS (2008) Resistant starch: physiological roles and food applications. Food Rev Int 24:193–234

Sharma A, Yadav BS, Ritika BS (2008) Resistant starch: physiological roles and food applications. Food Rev Int 24:193–234

Simal S, Benedito J, Sánchez ES, Rosselló C (1998) Use of ultrasound to increase mass transport rates during osmotic dehydration. J Food Eng 36:323–336

Simpson RK, Whittington R, Earnshaw RG, Russell NJ (1999) Pulsed high electric field causes 'all or nothing' membrane damage in listeria monocytogenes and *Salmonella typhimurium*, but membrane H+-ATPase is not a primary target. Int J Food Microbiol 48:1–10

Sinisterra JV (1992) Application of ultrasound to biotechnology: an overview. Ultrasonics 30:180–185

Smeller L (2002) Pressure-temperature phase diagram of biomolecules. BBA – Biochimica et Biophysica Acta 1595:11–29

Smeller L, Fidy J (2002) The enzyme horseradish peroxidase is less compressible at higher pressures. Biophys J 82:426–436

Smelt JP, Hellemons JC, Patterson M (2001) Effects of high pressure on vegetative microorganisms. In: Hendrickx M, Knorr D (eds) Ultra high pressure treatments of foods. Kluwer, New York, pp 55–76

Stojanovic J, Silva JL (2006) Influence of osmoconcentration, continuous high-frequency ultrasound and dehydration on properties and microstructure of rabbiteye blueberries. Dry Technol Int J 24:165–171

Stute R, Klingler RW, Boguslawski S, Knorr D, Eshtiaghi MN (1996) Effects of high pressures treatment on starches. Starch – Starke 48:399–408

Sui Q, Roginski H, Williams RPW, Versteeg C, Wan J (2010) Effect of pulsed electric field and thermal treatment on the physicochemical properties of lactoferrin with different iron saturation levels. Int Dairy J (in press), Corrected Proof

Taiwo KA, Angersbach A, Knorr D (2002) Influence of high intensity electric field pulses and osmotic dehydration on the rehydration characteristics of apple slices at different temperatures. J Food Eng 52:185–192

Tauscher B (1995) Pasteurization of food by hydrostatic high pressure: chemical aspects. Lebensmittel-Untersuchung und Forschung 200

Ting E, Balasubramaniam VM, Raghubeer E (2002) Determining thermal effects in high pressure processing. J Food Technol 56:31–35

Ting C-H, Kuo F-J, Lien C-C, Sheng C-T (2009) Use of ultrasound for characterising the gelation process in heat induced CaSO4 · 2H2O tofu curd. J Food Eng 93:101–107

Tiwari BK, O'Donnell CP, Cullen PJ (2009) Effect of sonication on retention of anthocyanins in blackberry juice. J Food Eng 93:166–171

Toepfl S (2006) Pulsed electric fields (PEF) for permeabilization of cell membranes in food- and bioprocessing – applications, process and equipment design and cost analysis. Ph.D. thesis, University of Technology, Berlin. 180 pp

Toepfl S, Heinz V (2007) Application of pulsed electric fields to improve mass transfer in dry cured meat products. Fleischwirtschaft International 22:62–64

Toepfl S, Mathys A, Heinz V, Knorr D (2006) Review: potential of high hydrostatic pressure and pulsed electric fields for energy efficient and environmentally friendly food processing. Food Rev Int 22:405–423

Toepfl S, Heinz V, Knorr D (2007) High intensity pulsed electric fields applied for food preservation. Chem Eng Process 46:537–546

Tryfona T, Bustard MT (2008) Impact of pulsed electric fields on corynebacterium glutamicum cell membrane permeabilization. J Biosci Bioeng 105:375–382

Tsong TY (1990) Electrical modulation of membrane-proteins - enforced conformational oscillations and biological energy and signal transductions. Annu Rev Biophys Biophys Chem 19:83–106

Tsong TY (1996) Electrically stimulated membrane breakdown. In: Lynch PT, Davey MR (eds) Electrical manipulation of cells. Chapman & Hall, New York, pp 15–36

Ulmer HM, Herberhold H, Fahsel S, Gänzle MG, Winter R, Vogel RF (2002) Effects of pressure-induced membrane phase transitions on inactivation of HorA, an ATP-dependent multidrug resistance transporter, in *Lactobacillus plantarum*. Appl Environ Microbiol 68:1088–1095

van den Bosch HFM, Morshuis PHF, Smit JJ (2002) Temperature distribution in fluids treated by Pulsed Electric Fields. Presented at international conference on dielectric liquids, Graz

Van den Broeck I, Ludikhuyze L, Weemaes C, Van Loey A, Hendrickx M (1998) Kinetics for isobaric-isothermal degradation of l-ascorbic acid. J Agric Food Chem 46:2001–2006

Van Loey A, Verachtert B, Hendrickx M (2001) Effects of high electric field pulses on enzymes. Trends Food Sci Technol 12:94–102

Van Loey A, Verachtert B, Hendrickx M (2002) Effects of high electric field pulses on enzymes. Trends Food Sci Technol 12:94–102

Vercammen A, Vivijs B, Luriquin I, Michiels C (2011) Germination and inactivation of *Bacillus coagulans* and *Alicyclobacillus acidoterrestris* spores by high hydrostatic pressure treatment in buffer and tomato sauce. Int J Food Microbiol (in press)

Vercet A, Burgos J, Crelier S, Lopez-Buesa P (2001) Inactivation of proteases and lipases by ultrasound. Innov Food Sci Emerg Technol 2:139–150

Vercet A, Sánchez C, Burgos J, Montañés L, Lopez Buesa P (2002) The effects of manothermo-sonication on tomato pectic enzymes and tomato paste rheological properties. J Food Eng 53:273–278

Villamiel M, de Jong P (2000) Influence of high-intensity ultrasound and heat treatment in continuous flow on Fat, proteins, and native enzymes of milk. J Agric Food Chem 48:472–478

Vorobiev E, Lebovka N (eds) (2008) Electrotechnologies for extraction from plant foods and biomaterials. Springer, New York

Wan J, Coventry J, Swiergon P, Sanguansri P, Versteeg C (2009) Advances in innovative processing technologies for microbial inactivation and enhancement of food safety – pulsed electric field and low-temperature plasma. Trends Food Sci Technol 20:414–424
Weltmann K-D, Kindel E, Brandenburg R, Meyer C, Bussiahn R et al (2009) Atmospheric pressure plasma Jet for medical therapy: plasma parameters and risk estimation. Contrib Plasma Phys 49:631–640
Winter R (1996) High pressure effects on the structure and mesophase behaviour of supramolecular lipid aggregates and model membrane systems. Prog Biotechnol 13:21–28
Winter R, Jeworrek C (2009) Effect of pressure on membranes. Soft Matter 5:3157–3173
Wu H, Hulbert GJ, Mount JR (2000) Effects of ultrasound on milk homogenization and fermentation with yogurt starter. Innov Food Sci Emerg Technol 1:211–218
Wuytack EY, Boven S, Michiels CW (1998) Comparative study of pressure-induced germination of *Bacillus subtilis* spores at low and high pressures. Appl Environ Microbiol 64:3220–3224
Wuytack E, Phuong LDT, Aersten A, Reyns KMF, Marquenie D et al (2003) Comparison of sublethal injury induced in *Salmonella enterica* serovar Typhimurium by heat and by different nonthermal treatments. J Food Prot 66:31–37
Xiong Z, Du T, Lu X, Cao Y, Pan Y (2011) How deep can plasma penetrate into a biofilm? Appl Phys Lett 98:221503-1-03-3
Yang RJ, Li SQ, Zhang QH (2004) Effects of pulsed electric fields on the activity of enzymes in aqueous solution. J Food Sci 69:241–248
Yaqub S, Anderson JG, MacGregor SJ, Rowan NJ (2004) Use of a fluorescent viability stain to assess lethal and sublethal injury in food-borne bacteria exposed to high-intensity pulsed electric fields. Lett Appl Microbiol 39:246–251
Zeece M, Huppertz T, Kelly A (2008) Effect of high pressure treatment on in-vitro digestibility of beta-lactoglobulin. Innov Food Sci Emerg Technol 9:62–69
Zenker M, Heinz V, Knorr D (2003) Application of ultrasound-assisted thermal processing for preservation and quality retention of liquid foods. J Food Prot 66:1642–1649
Zhang G, Sofyan M, Hamaker BR (2008) Slowly digestible state of starch: mechanism of slow digestion property of gelatinized maize starch. J Agric Food Chem 56:4695–4702
Zheng L, Sun D-W (2006) Innovative applications of power ultrasound during food freezing processes–a review. Trends Food Sci Technol 17:16–23
Zimmermann U, Pilwat G, Riemann F (1974) Dielectric breakdown in cell membranes. Biophys J 14:881–899
Zimmermann U, Pilwat G, Beckers F, Riemann F (1976) Effects of external electrical fields on cell membranes. Bioelectrochem Bioenerg 3:58–83
Zipp A, Kauzmann W (1973) Pressure denaturation of metmyoglobin. Biochemistry 12:4217–4228
Zou J-J, Liu C-J, Eliasson B (2004) Modification of starch by glow discharge plasma. Carbohydr Polym 55:23–26

Chapter 18
Nonthermal Technologies to Extend the Shelf Life of Fresh-Cut Fruits and Vegetables

Iryna Smetanska, Dase Hunaefi, and Gustavo V. Barbosa-Cánovas

18.1 Introduction

Natural, healthy, and functional foods are becoming major features for markets in Europe and other developed countries. Consumers are increasingly seeking products that retain natural nutrition, flavor, texture, and color properties. The fresh-cut fruit and vegetable market has grown rapidly in recent decades. In 2003 the fresh-cut industry represented 50 % of fresh produce sales (Zaborowski 2003), and in 2007 retail sales accounted for $998.25 billion (Cook 2007) in the USA. Meanwhile, in Germany the turnover of fresh food industries was around €52 billion in 2000 (Entrup 2005), with justifiable expectations of continued increase due to the popularity of fresh food as a convenient and healthy food choice (Artés and Allende 2005; Rico et al. 2007).

The International Fresh-Cut Produce Association (IFPA) defines fresh-cut products as fruits or vegetables that have been trimmed or peeled, or cut, into a 100 % usable product that is bagged or prepackaged to offer consumers high nutrition, convenience, and flavor, while still maintaining freshness. The value of fresh-cut produce lies in the primary characteristics of freshness and convenience. Achieving food safety and retaining nutrition and sensory quality are also required, while providing extended shelf life (González-Aguilar et al. 2004).

I. Smetanska (✉)
Berlin University of Technology, Berlin, Germany

University of Applied Science, Weihenstephan-Triesdorf, Triesdorf, Germany
e-mail: iryna.smetanska@hswt.de

D. Hunaefi
Berlin University of Technology, Berlin, Germany

G.V. Barbosa-Cánovas
Washington State University, Pullman, USA

S. Yanniotis et al. (eds.), *Advances in Food Process Engineering Research and Applications*, Food Engineering Series, DOI 10.1007/978-1-4614-7906-2_18,

Interestingly, one of the most challenging problems in the commercialization of these products is usually their short shelf life and declining quality after processing (Du et al. 2009; Jacxsens et al. 1999). The shelf life of fresh-cut fruits and vegetables varies from 3 to 7 days and depends on the product; however, a longer shelf life is better to make safe distribution feasible. Shelf life is the time a product can retain its quality under standardized storage conditions before its attributes drop below the acceptance limit (Tijskens 2000). It is broadly believed that the processing of fruits and vegetables promotes faster physiological deterioration, biochemical changes, and microbial degradation of the product, even when minimal processing operations are applied (O'Beirne and Francis 2003). These changes, such as degradation of color, texture, and flavor, will reduce the shelf life of these products compared to unprocessed products (Watada et al. 1990).

Furthermore, providing information about the safety of these products is still an issue of heightened concern to consumers. Meanwhile, fresh-cut fruits and vegetables are highly perishable since a large proportion of those damaged have exposed surface areas without epidermis, i.e., the outer protective layer of the tissue (Sandhya 2010; Watada and Qi 1999). Therefore, the fresh-cut produce industry is challenged with potential outbreaks of illness that could be associated with microbial growth during the extended shelf life of these products (Alzamora and Guerrero 2003). This threat is due to the fact that these products provide a viable medium for the growth of microorganisms.

In response to these challenges, many efforts using recent science and technologies have been devoted to prolonging the shelf life of these products and to improving their quality and safety. Extended shelf life of foods has traditionally been linked with thermal processing. Heat treatment, however, leads to the destruction of freshness and to nutrient losses. As a consequence, it is imperative that new preservation methods be found that cause only minimal adverse changes in foods, with limited (or no) use of additives, while offering desirable benefits derived from an increased shelf life (Camelo et al. 2003). One approach that promises to increase the shelf life of fresh-cut products is to use nonthermal technologies, which, to some extent, achieve a freshlike quality and safe product and retain high nutritional value. The emergence of novel nonthermal technologies, which are often regarded as *invisible preservation*, allows the production of high-quality products with improvements in shelf life and freshlike characteristics. The present review summarizes recent contributions reported on minimal processing of plant food products using nonthermal technologies to maintain freshness and achieve longer shelf life.

18.2 Physiological Effects of Processing Operations on Fresh-Cut Plant Products

Fresh-cut fruits and vegetables are products partially prepared so that no additional preparation is necessary for their use. Fresh-cut products have also been referred to as *lightly processed*, *minimally processed*, *fresh processed*, *partially processed*, or *preprepared* products (Garcia and Barrett 2005). Generally, it is unavoidable that the overall quality is impaired during processing and storage. The increasing demand for convenient fresh-cut products has a detrimental effect on product quality. Physiologically, the operations of preparation damage the integrity of cells, promoting contact between enzymes and substrates, the entry of microorganisms, and creation of stress conditions (Fonseca et al. 1999). The consequences of wounding are (1) an increase in the respiration rate, (2) the production of ethylene, (3) oxidative browning, (4) water loss, and (5) degradation of membrane lipids (Brecht 1995, cited in Fonseca et al. 1999). Therefore, fresh-cut fruits and vegetables tend to be more perishable than the commodity from which they were prepared (Kader 2002).

Cutting is the main factor responsible for the deterioration of these products during storage, which, as a consequence, is more rapid than in unprocessed (whole) products (Degl'Innocenti et al. 2007). Damaged plant tissues exhibit an increased respiratory rate (Fonseca et al. 2002). Treatment applied after, or before, wounding can affect the respiration rate (Del Nobile et al. 2006). Therefore, cutting induces deteriorative changes associated with plant tissue senescence and a consequential decrease in shelf life relative to unprocessed produce.

It is also well established that a faster rate of enzymatic browning by cutting is considered to be one of the major limitation factors detrimental to many fresh-cut products, such as apples, bananas, potatoes, and lotus root, since browning is easily detected by consumers, with evident consequences for marketing (Degl'Innocenti et al. 2007; Luo and Barbosa-Cánovas 1996). Polyphenoloxidase (PPO, EC1.14.18.1) activity is known to be the main cause involved in browning (Walker and Ferrar 1998).

Another important enzyme influenced by cutting is lipoxidase, which catalyzes the peroxidation reaction, thereby causing the formation of numerous bad smelling aldehydes and ketones (Varoquaux and Wiley 1994). Furthermore, a faster rate of softening of tissue results from an increase in ethylene production during this fresh-cut processing due to the fact that ethylene contributes to the biosynthesis of enzymes in fruit maturation and is partially responsible for the physiological changes in sliced fruit.

These physiological changes may be accompanied by flavor loss, cut-surface discoloration, color loss, decay, increased rate of vitamin loss, rapid softening, shrinkage, and a shorter storage life. Increased water activity and mixing of intracellular and intercellular enzymes and substrates may also contribute to flavor and texture change/loss during and after processing (Sandhya 2010). By understanding the physiology-changing mechanisms of fresh-cut-product processing and controlling factors that have a detrimental effect on quality, good-quality fresh-cut products with sufficient shelf life can be obtained.

18.3 Nonthermal Technologies

The term *nonthermal processing* is often used to designate technologies that are effective at ambient or sublethal temperatures (Pereira and Vicente 2009) or when the main stress factor for microbial inactivation is not heat. An increase in consumer demand for fresher foods, which at the same time provide a high degree of safety, has driven the growing interest in nonthermal preservation techniques because of their capability to inactivate microorganisms and enzymes in foods (Mertens and Knorr 1992). High capital investment in nonthermal technologies is one of the main issues for commercial application; fortunately, it is usually believed that consumers are willing to pay more when better quality and added value are offered.

Fresh-cut fruits and vegetables have been found to be more prone to pathogen contamination (Farkas et al. 1999). Therefore, nonthermal technologies may play an important role in producing more convenient freshlike products and safe foods as an alternative technology to thermal processing. This review concentrates on physical treatment alternatives to thermal processing. In the following sections, the main emphasis will be on processes already being applied commercially or that are likely to become so in the foreseeable future.

18.4 High-Pressure Processing

Many studies have been conducted to evaluate the safety of high pressure food processing (Zook et al. 1999). It has been suggested that the major cause of damage in microbial inactivation subjected to high pressure is modification of the molecules in cell membranes, leading to permeability (Hoover 1993). Changes in cell morphology involve collapse of intercellular gas vacuoles, cell elongation, and cessation of microorganisms' movement; thus, high-pressure processing (HPP) damages microbial cell membranes (Barbosa-Cánovas and Rodriguez 2002). HPP is based on the principle of *Le Chatelier-Braun* and the state transition theory (Welti-Chanes et al. 2005). The following equations illustrate the work of HPP:

$$\Delta G = G^{(\mathrm{products})} - G^{(\mathrm{reactants})}, \tag{18.1}$$

where ΔG is free energy, and when ΔG is negative, the reaction is spontaneous.

$$\mathrm{d}G = -\mathrm{SdT} + \mathrm{VdP} + \sum \mu_i \mathrm{dn_i}, \tag{18.2}$$

where S is entropy, V is volume, P is pressure, T is temperature, μ is chemical potential, and n is the number of moles of a compound (Welti-Chanes et al. 2005). The preceding equation is the basic principal equation of Gibbs, while the next equation correlates the influence of pressure on G with volume:

$$V = \left[\frac{\partial G}{\partial P}\right]_{\mathrm{T,ni}}. \tag{18.3}$$

The increase in pressure leads to damage to the microbial cell membranes, which eventually causes the microbial cell to become inactive; however, this transition condition could avoid the destruction of the covalent bonds that retain natural flavor, taste, color, and nutrients (Hayashi 1995). Thus, the physical structure of most fresh-cut fruit and vegetable products remains unchanged after exposure to HPP.

It has been reported that some problems may arise associated with the use of HPP on certain plant material systems because HPP affects the integrity of porous products. The air confined in the food matrix is subjected to compression and expansion during pressurization and decompression, disrupting food plant tissues, thereby making this unit operation not entirely suitable for all plant products (Palou et al. 2000).

On the other hand, Torres and Velazquez (2005) showed that fresh-cut products represented one of many opportunities in which HPP has a clear competitive advantage. Nishin Oil Mills, a Japanese food manufacturer, has produced nontropical fruits using HPP at pressures of 50–200 MPa ("freezing" temperature, −18 °C) [Campden New Technology Bulletin No. 14; cited in Ohlsson and Bengtsson (2002)]. In addition, Ting and Marshall (2002) illustrated that HPP treatment of fresh-cut fruit produces a very high-quality product with extended shelf life that can also be produced without heating or chemical additives. Palou et al. (2002) added that HPP treatments could preserve the delicate sensory attributes of avocados, which are used in the preparation of guacamole, one of the first HPP-treated food plant products to be commercialized in the USA, demonstrating a reasonably safe and stable shelf life. Several companies (Avure: http://www.nchyperbaric.com and NC Hyperbaric: http://www.avure.com) that manufacture HPP equipment offer promising safety and extended shelf life for fresh-cut fruits and vegetables, such that these products will become more profitable.

The application of HPP on fresh-cut products can be seen in Table 18.1.

In terms of microbial inactivation using HPP treatment on fresh-cut plant products, Alemán et al. (1994) reported application of HPP on fresh-cut pineapple in combination at a pressure of only 270 MPa for 15 min, resulting in more than two decimal reductions at a temperature of 38 °C. Meanwhile, Alemán et al. (1998) reported that step pulsed pressure could reduce the amount of yeast and bacteria at a pressure of 270 MPa for a shorter time, 100 s, as compared to static pressure of 270 MPa for 15 min. Zong and Zhang (2009) emphasized that HPP treatment could effectively kill microorganisms and had a slight influence on the firmness of the product. Moreover, a few years ago, Ramaswamy et al. (2005) stated that spore inactivation was a primary challenge in HPP application; today, the challenge has been answered. Bermudez-Aguirre and Barbosa-Cánovas (2011) emphasized that by utilizing a smart combination of pressure, temperature, and time without exceeding quality parameters values it is possible to achieve partial inactivation of spores. Methods used to achieve full inactivation of spores using HPP have yet to

Table 18.1 Application of HPP to fresh-cut fruits and vegetables

Conditions	Results	Author
Fresh-cut pineapple		
200, 270, and 340 MPa; −4 °C, 21 °C, and 38 °C	340 MPa (15 min) for all temperatures tested: ~ 3 decimal reduction: 270 MPa at 38 °C for 15 min: > two decimal reduction: other combinations resulted in no significant reduction	Alemán et al. 1994
270 MPa for 9,000 s	Inoculated with $10^{4–5}$ CFU g^{-1} *Saccharomyces cerevisiae*, packed in heat-sealed polyethylene pouches, and subjected to static and step pulsed pressure; step pulsed pressure resulted in higher destruction of microorganisms tested	Alemán et al. 1998
Fresh-cut jujube fruit		
600 MPa for 10 min	Could inhibit polygalacturonase (PG) activity and decrease hydrolyzation of non-water-soluble pectin when stored at 4 °C for 9 days	Zong and An 2008
Fresh-cut kiwi		
0.1, 200, 400, and 600 MPa for 10 min	Could effectively kill microorganisms and inhibit PPO activity, but has little influence on firmness of slices and ascorbic acid content	Zong and Zhang 2009
Fresh-cut Chinese yam		
600 MPa for 10 min	Effectively prevented browning and microbial growth when stored at 4 °C for 9 days	An and Li 2007
Fresh-cut cubes of Granny Smith and Pink Lady apples		
600 MPa for 1–5 min at 22 °C	Vacuum packed in barrier bags with 50 % (v/v) pineapple juice; significantly reduced residual PPO activity	Perera et al. 2010
Packaged fresh-cut melon		
600 MPa for 10 min	Enhancement of β-carotene and minimal loss of sensory properties and health-promoting phytochemicals	Wolbang 2008

be developed. Therefore, there is a need to develop and standardize HPP process parameters for fresh-cut products with respect to microbial inactivation. This development is essential for the commercial success of this technology.

As shown in Table 18.1, HPP has several advantages as a treatment for fresh-cut fruits and vegetables. HPP allows in-package fresh-cut fruit and vegetable processes that reduce the microbial load and extend the product's shelf life. Wolbang et al. (2008) investigated the effect of HPP on the nutritional content of packaged cut melon. It was proven that HPP enhanced not only microbial inactivation, but also the level of health-promoting phytochemicals, with a minimal loss of sensorial qualities. Recent findings by Perera et al. (2010) using HPP on fresh-cut apple cubes in vacuum packages showed significantly reduced residual activity of polyphenoloxidase (PPO). They reported that a combined treatment of 5 min/HPP and 50 % pineapple juice inactivated approximately 40 % PPO in Granny Smith apples. This means that HPP treatment has the potential to prevent browning of fresh-cut fruits.

This result is in agreement with the finding by An and Li (2007), who found that HPP treatment employed at 600 MPa for 10 min on fresh-cut Chinese yam effectively prevented browning and growth of microorganisms when stored at 4 °C for 9 days. In addition, Zong and An (2008) reported that HPP treatment at 600 MPa for 10 min on fresh-cut jujube fruits successfully inhibited the activity of polygalacturonase (PG) and successfully prevented firmness of jujube pieces associated with a decrease in hydrolyzation of non-water-soluble pectin when stored at 4 °C for 9 days.

Enzymes are inactivated by HPP as a result of the conformational changes taking place in enzyme active sites (Min and Zhang 2005). Several enzymes that are partially responsible for oxidative degradation reactions in fresh-cut fruits and vegetables, such as PPO, lypoxygenase (LOX), pectimethylesterase (PME), have been inhibited by HPP treatment (Garcia-Viguera and Bridle 1999; Garzon and Wrolstad 2002). However, during HPP application, it should be pointed out that the residual potent activity or the activation of endogenous enzymes depends on the pressure, temperature, and time at which HPP is applied (Del Pozo-Insfran et al. 2007).

The critical process factors in HPP include pressure, time at pressure, time to achieve treatment pressure, decompression time, treatment temperature (including adiabatic heating), initial product temperature, vessel temperature distribution at pressure, product pH, product composition, product water activity, packaging material integrity, and concurrent processing aids (Hui and Nip 2004).

Although currently only a small number of commercial HPP-treated fresh-cut fruits and vegetables is available, surveys show that demand for these products is increasing, in line with the growth in the number of food companies adopting this HPP technology in Europe and North America. Lastly, HPP has proven to be applicable to fresh-cut fruit and vegetable products in order to extend product shelf life and to achieve product safety without altering the freshlike quality and nutrition of the products.

18.5 Pulsed Electric Fields

Pulsed electric fields (PEFs) have been used as an alternative to heat pasteurization for processing real food systems, including fruit and vegetable tissues (Knorr and Angersbach 1998). A number of experiments have been conducted to study the mechanism and safety of microbial inactivation under PEF treatment (Buchanan et al. 1998; Jin et al. 2001; Raso et al. 1998). Membrane permeability (i.e., damage to cell membranes) has been proposed as the causative factor behind microbial inactivation (Harrison et al. 2001). Another approach was reported by Harrison et al. (1997) in which damaged organelles and lack of ribosomes after PEF treatment are viewed as an alternative inactivation mechanism. However, research or application of this approach using fresh-cut fruits or vegetables as a medium for inactivation of the bacteria is very limited.

In terms of enzyme inactivation mechanisms, extensive studies of PEF have been conducted (Giner et al. 2001). Some studies have used fruits and vegetables as media for the inactivation of enzymes. Changes in enzyme conformational structure specifically have been reported by several authors (Ho et al. 1997; Vega-Mercado et al. 1995). Further, enzymes have different sensitivities to PEF treatments, and many factors influence PEF enzymatic inactivation (Yeom and Zhang 2001).

Some of the technical drawbacks in PEF application that have been further discussed by Barbosa-Cánovas et al. (1999) include residual activity of enzymes and breakdown of food. Textural changes with PEF treatment such as tissue softening in apples, potatoes, and carrots have been reported by Lebovka et al. (2004). The tissue-softening effect of PEF, based on cell membrane electropermeabilization and loss of turgor (Fincan and Dejmek 2003), can be utilized to reduce the energy required for cutting plant material as well as to enhance juice extraction. Therefore, PEF technology is mainly intended for the preservation of pumpable fluids (Hui and Nip 2004; Qin et al. 1996), predominantly fruit and vegetable juices (Min et al. 2003; Heinz et al. 2003). In fact, the adoption of PEF for commercial nonthermal pasteurization of fruit juices was first implemented by Genesis Juice, Springfield and Eugene, OR, USA (Clark 2006).

The results of PEF inactivation of microorganisms and enzymes have been encouraging, suggesting that this technology is an alternative nonthermal method for processing fruits and vegetables (Barbosa-Cánovas et al. 1997). However, for shelf-life extension of fresh-cut fruits and vegetables, PEF technology alone seems far from promising, as it is highly recommended that PEF be combined with other preservation methods.

18.6 Irradiation

Irradiation of food is not a new idea. It has been established as a safe and effective method of food processing and preservation after more than five decades of research and development (Korkmaz and Polat 2005). Irradiation literally means exposure to radiation (Grandison 2006). Food irradiation is a physical treatment in which food is exposed to ionizing radiation, i.e., radiation of sufficient energy to expel electrons from atoms and ionize molecules (Niemira and Deschenes 2005). Irradiation is a nonthermal technology or *cold process* because foods increase by only a few degrees from the radiation energy absorbed, even at the higher doses used for sterilization. Consequently, irradiation treatment causes minimal changes in appearance and provides good nutrition.

This technology has been used to eliminate foodborne pathogens and natural contaminants from fresh fruits and vegetables (Thayer and Rajkowski 1999). The mechanism of decontamination by irradiation is that it directly harms the genetic material of the living cell, leading to mutagenesis and eventually to cell death (Barkai-Golan 2001). Irradiation affects microorganisms such as bacteria, yeasts, and molds by causing lesions in the genetic material of cells, effectively preventing the cells from carrying out the biological processes necessary for continued cell existence (Rahman 1999). Irradiation can reduce the incidence of pathogenic microorganisms associated with fresh-cut products without changing the minimally processed character of the products (Farkas et al. 1999). Thus, irradiation is one of the potential effective technologies that can inactivate foodborne pathogens in fresh-cut fruits and vegetables.

The U.S. Food and Drug Administration (FDA) approved the use of irradiation on wheat and potatoes in 1963; on spices, pork, fruits, and vegetables in the 1980s; poultry in the early 1990s; and red meats in (beef, veal, lamb) (Majchrowicz 1999). In accordance with international regulations such as the *Codex Standards for Food Irradiation*, the ionizing radiation permissible for irradiating foods, including fruits and vegetables, is limited to (a) gamma rays from radioisotope Cobalt 60 or Cesium-137, (b) x-rays generated from machine sources operated at or below an energy level of 5 MeV, and (c) electrons generated from machine sources operated at or below an energy level of 10 MeV (Lacroix et al. 2003; Sharma 2004).

Table 18.2 shows that low-dose irradiation has been successfully applied for extending the shelf life of many fresh-cut products. Several investigators have proven that irradiation can be a suitable method for extending shelf life while ensuring product safety by inactivation of microorganisms on fresh-cut fruits and vegetables (Buchanan et al. 1998; Hagenmaier and Baker 1997).

Lu et al. (2005) reported that fresh-cut celery treated with low-dose gamma irradiation has more vitamin C, soluble solids, and total sugars, as well as better sensory quality, than that of nonirradiated celery. They summarized that low-dose gamma irradiation, 1 kGy, was successfully utilized to extend the shelf life and ensure the safety of fresh-cut celery. Irradiation has also been applied to fresh-cut carrots stored in microporous bags, resulting in limited respiration increase due to wounding and reduced ethylene production and increasing the shelf life of the product (Chervin et al. 1992).

Many other researchers have reported the successful application of irradiation as an effective method to extend the shelf life of fresh-cut products, as summarized in Table 18.2. Application of irradiation on fresh-cut products is very dependent on the dose of irradiation. At higher doses, a higher reduction in microbiology as well as higher rate of destruction of the quality of fresh-cut plant products have been observed. As shown in Table 18.2, irradiation doses should be kept as low as possible to achieve the safe quality of products and extend their shelf life without decreasing their organoleptic quality. Grandison (2006) stated that overdoses of irradiation generate textural problems resulting from radiation-induced depolymerization of cellulose, hemicellulose, starch, and pectin, leading to the softening of tissue. Other dose-dependent disorders include discoloration of skin, internal browning, and increased susceptibility to chilling injury, although nutrition losses are minimal. Some studies have also shown that ionizing radiation can cause the development of an off-odor (Spoto et al. 1997).

The main issue with this technology is the unfavorable image or public perception of irradiated food. Nevertheless, low-dose irradiation is considered to have high potential and is probably one of the most successful and versatile processes for

Table 18.2 Application of low-dose irradiation for extending shelf life of fresh-cut fruits and vegetables

Dosage	Results	Author
Fresh-cut lettuce		
1.0 kGy	Number of aerobic mesophyllic bacteria was reduced by 2.35 log; sensory quality was maintained best during storage for 8 days at 4 °C	Zhang et al. 2006
0.5 and 1.0 kGy	Similar firmness and vitamin C and antioxidant contents to that of controls after 14 and 21 days of storage	Fan et al. 2003
0.19 kGy	8 days after irradiation microbial population was much lower compared to nonirradiated sample	Hagenmaier and Baker 1997
0.15, 0.38, or 0.55 kGy	5 log reduction in *E. coli* counts and lack of adverse effects on sensory attributes, indicating that low-dose irradiation can improve the safety and shelf life of fresh-cut iceberg lettuce for retail sale or food service	Foley 2002
1.0 kGy	Eliminated bacterial contamination from lettuce sample without any sensorial quality defects	Kim 2006
1.0, 1.5 and 3.2 kGy	Potential fungicidal effect of low-dose irradiation (1.0 kGy) on packaged romaine lettuce hearts without altering overall quality	Han et al. 2004
0.3–1.2 kGy	1.15 and 0.51 kGy with subsequent refrigerated storage (4 °C) reduced inoculated-microbial populations by $>$ 5 and $>$ 2 log, respectively	Mintier and Foley 2006
Fresh-cut romaine and iceberg lettuce and endive		
0, 0.5, 1.0, and 2.0 kGy	Irradiation increased phenolic content and antioxidant capacity of both tissue types in all vegetables at days 4 and 8 at 7–8 °C	Fan 2005
Fresh-cut cantaloupe		
1.0, 1.5, and 3.1 kGy	Samples irradiated at dose levels 1.0–1.5 kGy had better quality attributes than nonirradiated samples	Castell-Perez et al. 2004
Up to 1.5 kGy	Irradiated samples resulted in a much higher reduction, 3 log TPC number; shelf life was extended by up to 11 days of storage	Boynton et al. 2005
0.5 and 1.0 kGy	Color and texture remained stable after 20 days of storage; low-dose irradiation may increase shelf life while maintaining high product quality	Boynton 2004
0.5 kGy	Low microbial load similar to that of cubes prepared from hot-water-treated fruit	Fan et al. 2006
0.7 and 1.4 kGy	Irradiation reduced total aerobic microbial counts with increasing doses	Palekar et al. 2004
Up to 10.0 kGy	D10 values of Poliovirus Type 1 and MS2 bacteriophage were 4.76 and 4.54 kGy, respectively; results suggest that E-beam irradiation can be used to destroy enteric viruses on cantaloupe surfaces	Pillai et al. 2006
1.0 and 3.0 kGy	1.0 kGy caused 2–3 log reductions in microbial counts; Brix, pH, and sensory attributes were also unaffected	Horrick 2002
Fresh-cut apple		
Up to 5.0 kGy	Threshold for maintaining firmness of sliced apples was 0.34 kGy	Gunes et al. 2001

(continued)

Table 18.2 (continued)

Dosage	Results	Author
0.5 and 1.0 kGy	Enhanced microbial food safety while maintaining quality of fresh-cut apple slices	Fan et al. 2005
1.6 kGy	Did not significantly affect color, soluble solid content, titratable acidity, or apple aroma intensity	Fan et al. 2005
1.2 kGy	Doses less than 1.2 kGy had no effect on rates of CO_2 production and O_2 consumption	Gunes et al. 2000
Fresh-cut celery		
1.0 kGy	Number of bacteria and fungi in fresh-cut celery was decreased by orders of 10^2 and 10^1	Lu et al. 2005
0.5 and 1.0 kGy	Sensory shelf life of 1.0 kGy treated celery was 29 days compared to 22 days for control (chlorinated)	Prakash et al. 2000
0.5, 1.0, 1.5, and 2.0 kGy	Dose of 1 kGy reduced bacteria and fungi contamination to acceptable levels without changing acceptability of samples during recommended shelf life (8 days at 5 °C)	López et al. 2005
Fresh-cut green onion		
0.5, 1.0, and 1.5 kGy	Irradiated samples maintained a relatively low decay percentage at 14 days of storage	Kim et al. 2005
1.0 kGy	Reduced aerobic plate counts, yeasts and molds, and psychrotrophic counts by 3–5 log CFU g^{-1} without negatively affecting quality (14 days of storage)	Butris et al. 2003
Fresh-cut tomato		
0.7 and 0.95 kGy	Reduction of 1.8 and 2.2 log10 CFU g^{-1} on tomatoes inoculated with *Salmonella* for 0.7 and 0.95 kGy	Schmidt 2004
Up to 0.9 kGy	D-value ranged from 0.26–0.39 kGy, indicating that a 5 log_{10} CFU g^{-1} reduction in *Salmonella* spp. in diced tomatoes would require a dose of 1.3–1.95 kGy	Prakash et al. 2007
Fresh-cut (diced) bell pepper and tomato		
Up to 3.7 kGy	Total pectin content in tomatoes and bell peppers remained relatively constant with irradiation during storage for 12 days at 4 °C	Costa et al. 2001
Fresh-cut Chinese cabbage		
up to 2.0 kGy	1–2 kGy ensured microbial safety of minimally processed Chinese cabbage without significant loss in quality for 3 weeks	Ahn et al. 2005
Fresh-cut broccoli head		
1.0, 2.0, and 3.0 kGy	Treatment up to 3 kGy maintains overall quality of fresh broccoli	Gomes et al. 2008
Fresh-cut mint (inoculated with *E. coli* O157:H7, *Salmonella*, and *MS2 bacteriophage*)		
Up to 2.0 kGy	Population decreased up to 5.8 log CFU g^{-1}, and 1.0–2.0 kGy appears promising as a method for improving microbiological quality of fresh mint without compromising visual and color attributes	Hsu et al. 2010
Fresh-cut carrot		
2.0 kGy	Twofold to fourfold increase in shelf life at refrigeration temperature	Kamat et al. 2005

(continued)

Table 18.2 (continued)

Dosage	Results	Author
2.0 kGy	Prevented losses of orange color and carotenes; sensory analysis demonstrated preferences for irradiated vegetables	Chervin and Boisseau 1994
2.0 kGy	Significant reduction in respiration by 50 % and ethylene production by 80 %	Chervin et al. 1992
0.5 kGy	Shredded carrots treated with irradiation had a mean microbial population of 1,300 CFU g^{-1} at expiration date (9 days after irradiation) compared to 87,000 CFU g^{-1} for nonirradiated, chlorinated controls	Hagenmaier and Baker 1998
Fresh-cut carrot, cucumber, lettuce, mixed vegetable salad, green bean, celery, mixed peas/diced carrots, and fruits (pear and apple)		
Up to 6.0 kGy	Large reduction in microbial counts, shelf-life extension, and elimination of foodborne pathogens without impairing quality, at following doses: 1.5 kGy for celery (14 days of storage); 2 kGy for lettuce, green beans, apples, pears (15 days of storage); 3 kGy for cucumbers, mixed vegetables salad, mixed peas with diced carrots (15 days of storage); and 4 kGy for carrots (21 days of storage)	Hammad et al. 2006
Fresh-cut fruits (pineapple, jackfruit, mixed fruits) and vegetables (onion and cucumber)		
0.0–3.0 kGy	Low-dose irradiation eliminated or reduced risk of foodborne pathogens and did not affect the color (lightness, redness and greenness) of minimally processed onion and cucumber	Faridah et al. 2006
Polyethylene packaged fresh-cut cucumber, tomato, carrot, cabbage, apple, melon, cauliflower, bitter gourd		
Up to 2.5 kGy	Maintained shelf life for 2 weeks at refrigerated temperature: 2 kGy for carrots, 1.5 kGy for cauliflower and 2.5 kGy for apples, cucumber and cabbage; 1 week shelf life at refrigerated temperature: 2.5 kGy for tomato, 2 kGy for bitter gourd, and 1 kGy for melon	Bibi et al. 2006
Fresh-cut coriander, lettuce, mint, parsley, turnip, watercress, melon, and watermelon inoculated with *E. coli* 0157:H7 and *L. innocua*		
Up to 1.0 kGy	No important differences were found overall in sensory and physicochemical properties after irradiation up to 1 kGy	Trigo et al. 2006
Fresh-cut (shredded) carrot, mixed salad samples (radicchio and butterhead lettuce), red leaf lettuce, green lettuce and soybean sprouts (inoculated with *L. monocytogenes, E. coli* 0157:H7, and *Salmonella*)		
1.0–1.5 kGy	Inactivation of pathogens, no effect on sensory attributes during storage	Basbayraktar et al. 2006
Fresh-cut ready-to-use vegetables: cucumber, blanched and seasoned spinach (inoculated with *Salmonella typhimurium*, *E. coli*, *Staphylococcus aureus*, and *L. ivanovii*)		
Up to 3.0 kGy	Low-dose irradiation (3 kGy or less) can improve microbial safety of ready-to-use vegetables	Young Lee et al. 2006
Fresh-cut cauliflower		
1.5 kGy	Treatment maintained quality of fresh-cut cauliflower	Nunes et al. 2009

(continued)

Table 18.2 (continued)

Dosage	Results	Author
Fresh-cut red beet		
1.0 or 2.0 kGy	Pigments seemed to overcome irradiation dose of 1 kGy	Latorre et al. 2010
Fresh-cut witloof chicory		
3.0 kGy	Irradiation-induced browning of cut chicory may be due to increased phenolic metabolism and reduced antioxidant capacities; increased membrane permeability may also allow substrate-enzyme contact after 5 days of storage at 10 °C	Hanotel et al. 1995

application in the commercialization of fresh-cut products; however, its application on food is conditional and regulated based on the appropriate dose levels. Other factors associated with this technology are the initial capital investment and operational costs

18.7 Ozone

Washing is a critical step in processing fresh-cut fruits and vegetables. Generally, chlorine is used to sanitize fruits and vegetables in fresh-cut processing. However, the use of chlorine is questionable because of the possible formation of carcinogenic chlorinated compounds in water. The use of ozone as an alternative sanitizer to chlorine is attracting more interest (Guzel-Seydim et al. 2004). There are a number of reviews about the utilization of ozone technology in food processing (Kim et al. 2003; Khader et al. 2001) and, in particular, fresh-cut processing (Strickland et al. 2010). Ozone, with its high oxidizing power, is considered to be a strong antimicrobial agent with high reactivity and thus is applicable in ensuring the microbial safety of raw produce (Khader et al. 2001). Due to its instability, ozone does not produce persistent disinfection residuals (Kim et al. 2003). The spontaneous autodecomposition of a nontoxic product in the absence of ozone residue is one of the advantages of using this nonthermal technology.

Ozone technology has long been used as a treatment for drinking water (Muthukumarappan et al. 2000) and was deemed Generally Recognized As Safe (GRAS) in July 1997 by an independent panel of scientists (Majchrowicz 1999). Fruits and vegetables have the advantage of having smooth surfaces with low ozone demand (Khader et al. 2001), making them potentially and particularly suitable for aqueous ozone treatments. Ozone-sanitized fresh produce has recently been introduced in markets in the USA (Artés and Allende 2005), and numerous researchers have also focused on ozone application to other fresh-cut products (Table 18.3).

Table 18.3 Ozone application studies on fresh-cut fruits and vegetables (O_{3A} = aqueous O_3, O_{3G} = gaseous O_3)

Conditions	Results	Author
Fresh-cut lettuce		
0.5–4.5 ppm O_{3A} for 0.5–3.5 min	2 ppm O_{3A} treatment for 2 min was optimal for O_3 disinfection of *L. monocytogenes* on green leaf lettuce, sensory quality during cold storage	Ölmez and Akbas 2009
1.11 and 1.15 mg L^{-1} O_{3A} for 2 min pre-washing	1 mg L^{-1} was too low to achieve germ reduction of trimmed lettuce heads, unlike that achieved with 200 mg L^{-1}chlorine for 2 min common in commercial practice	Baur et al. 2004
0–10 mg L^{-1} O_{3A} in cold water (2 °C) in ice—slurry system	3 mg L^{-1} O_{3A} was optimal, causing 3–4 log unit in reductions of microorganisms, similar to results achieved with 100 mg L^{-1} chlorinated water	Kim 2006
5 ppm O_{3A} for 5 min	Reduction of microbial count by 1.8 log units	Selma 2007
4 mg L^{-1} O_{3A}	Reduction of mesophilic and psychrotrophic bacteria by 1.7 and 1.5 $\log_{10}$ CFU g^{-1}, respectively	Akbas and Ölmez 2007
10, 20, and 10 mg L^{-1} O_{3A} at 4 °C and UV	O_{3A} could be an alternative sanitizer to chlorine for fresh-cut lettuce due to good retention of sensory quality and browning control with no detrimental reduction in antioxidant constituents	Beltran et al. 2005
Aqueous O_3 (3, 5, and 10 ppm)	Bacterial reduction of 1.4 log CFU g^{-1} up to 5 ppm aqueous O_3	Koseki and Isobe 2006
1, 3, and 5 ppm O_{3A} for 0.5, 1, 3, or 5 min	1.09 and 0.94 log reductions of *E. coli* O157:H7 and *L. monocytogenes*, respectively	Yuk et al. 2006
1 mg L^{-1} O_{3A} at 18–20 °C	Enzymatic activity decreased significantly	Rico et al. 2006
O_{3A} at 4 °C for 6 days	Industrial trial: O_3 generator was integrated into a commercial lettuce-washing facility; quality of lettuce during storage was unaffected; only limited destruction in microbial population	Hassenberg et al. 2007
Fresh-cut celery		
O_{3A} at different doses	0.18 ppm O_{3A} dipp resulted in highest reduction of microbial population	Zhang et al. 2005
Fresh-cut cilantro		
O_{3A} followed by AcEW or chlorine wash	Sequential wash is effective in reducing initial total aerobic plate count and maintaining relatively low microbial count; O_{3A} treatment effectively maintained typical aroma and overall quality of fresh-cut cilantro leaves	Wang et al. 2004
Fresh-cut cantaloupe		
10,000 ppm O_{3G} for 30 min	Treatment reduced total mesophilic and psychrotrophic bacteria and mold counts by 1.1, 1.3, and 1.5 log units, respectively; combination of hot water and O_{3G} was most	Selma et al. 2008

(continued)

Table 18.3 (continued)

Conditions	Results	Author
	effective in controlling microbial growth, achieving 3.8, 5.1, 2.2, and 2.3 log reductions for mesophilic and psychrotrophic bacteria, molds, and coliforms, respectively	
0.4 mg L^{-1} O_{3A} combined with peracetic acid	O_{3A} for 3 or 5 min had lower efficiency in reducing microbial count; nevertheless, the combination of peracetic acid with O_{3A} (3 min) was effective at reducing microbial count and maintaining antioxidant compounds such as vitamin C and DPPH	Silveira et al. 2010
10,000 ppm O_{3G} for 30 min under vacuum	Reduction and spoilage of fresh and fresh-cut melon	Selma et al. 2008
Fresh-cut onion, escarole, carrot, and spinach		
O_{3G} flow of 80 mg min^{-1}	O_{3G} flow was effective in reducing microbial counts of bacteria, molds, and yeasts; maximum total microbial reductions achieved 5.9 log CFU mL^{-1}	Selma et al. 2008
Fresh-cut broccoli		
1 ppm O_{3A}	50 min O_{3A} treatment reduced microorganisms better, compared to 10 min treatment, and did not influence quality parameters	Zhuang et al. 1996
O_{3A}, tap water, chlorinated water, and electrolyzed water	O_{3A} for 90 s was not very effective in reducing microbial population compared to chlorinated and electrolyzed water; however, samples washed with O_{3A} for 180 s showed the lowest number of total aerobic and coliform plate counts	Das and Kim 2010
Fresh whole and fresh-cut (sliced) tomato		
4 mu L L^{-1} O_{3A} applied cyclically for 30 min every 3 h	Bacteria: 1.1–1.2 log reduction Fungi: 0.5 log reduction O_3 caused no damage or off-flavor in sliced or whole tomatoes	Aguayo et al. 2006
Fresh-cut green asparagus		
1 mg L^{-1} O_{3A}	Enzyme activity inhibited	An et al. 2007
Fresh-cut baby leaf *Brassica* species (salad rocket, wild rocket, mizuna, and watercress) stored at 1 °C, 4 °C, 8 °C, and 12 °C		
10 mg L^{-1} O_{3A} and UV	O_{3A} combined with UV was the most efficient washing treatment in reducing total mesophyllic count; heat shock treatment was ineffective	Martínez-Sánchez et al. 2008
Fresh-cut green bean and bell pepper		
2 ppm O_{3A}	Peroxidase inactivation, color of vegetables unaffected	Elisabete and Joana 2006

As shown in Table 18.4, extending shelf life and preserving the freshlike quality of fresh-cut products can be achieved by ozone treatment, particularly in the form of aqueous ozone.

Ozone dissolved in water is a thermodynamically ideal condition, as expressed in the following equation:

$$Cs = \beta M \times P\gamma, \tag{18.4}$$

where Cs is the saturation concentration (kg O_3 m^{-3} water), β (absorption coefficient) is the volume of ozone dissolved per unit volume of water (at a given temperature) in the presence of equilibrating ozone at 1 atm pressure, M is the mass volume of ozone (kg m^{-3}), and $P\gamma$ is the partial pressure of ozone in the gas phase (Bablon et al. 1991 cited in Khader et al. 2001). In practical terms, dissolution of ozone in water can also be expressed in the solubility ratio (Sr) as follows:

$$\text{Sr} = \frac{\text{mg L}^{-1}\ \text{O}_3\ \text{in water}}{\text{mg L}^{-1}\ \text{in the gas phase}}. \tag{18.5}$$

An excellent assessment of microbial inactivation by aqueous ozone has been reviewed by Khader et al. (2001). According to Zambre et al. (2010), the longer shelf life of ozone-treated samples is mainly due to a reduction in surface microbial count. Ozone is considered to be an efficient decontamination agent for fresh products, demonstrating a broad-spectrum bactericidal attack on the bacterial cell wall and outer cell membrane (Bialka et al. 2008). The antimicrobial action of ozone is due to the strong oxidizing activity of either the molecular ozone itself or its decomposition products. As shown in Table 18.3, ozone was effective in eliminating inoculated microorganisms and extending shelf life, for example, fresh-cut lettuce inoculated by *Listeria monocytogenes* (Ölmez and Akbas 2009) and *Escherichia coli* O157:H7, *Salmonella enterica serovar Typhimurium*, *L. monocytogenes*, and *Staphylococcus aureus* (Kim et al. 2006). A low level of ozone (2–3 ppm) for short exposure times (2–5 min) was found to be an optimum processing condition for disinfection of fresh-cut lettuce in terms of reducing microbial load and maintaining sensory quality. Furthermore, Olmez and Akbas (2009) reported that ozone treatment was found to be more effective than chlorine and organic acid treatments in maintaining sensory quality.

Kim et al. (1999) concluded that ozone is more efficient at lower concentrations and shorter treatment times than more standard sanitizers, including chlorine. Singh et al. (2002) reported that gaseous treatments of ozone were more effective than aqueous ozone in reducing *E. coli* O157:H7 on lettuce. Similar results were also reported by Bialka and Demirci (2007) in which reductions of *E. coli* O157:H7 and *Salmonella* were achieved on blueberries treated with gaseous ozone, among other treatments, and by Akbas and Ozdemir (2008), who found that a decreasing number of *E. coli* and *Bacillus cereus* cells treated with gaseous ozone was achieved at low concentrations. Moreover, Klockow and Keener (2009) reported that ozone has the

Table 18.4 Extending shelf life of fresh-cut fruits and vegetables using UV light

Conditions	Results	Author
Fresh-cut apple		
5.6, 8.4, and 14.1 kJ m^{-2} for 10, 15, and 25 min	Breakage of cellular membranes in UV-treated samples; lower microbial count compared to untreated samples	Gómez et al. 2010
13.8 W m^{-2} for 5 min at 4 °C	Enzyme inactivation was limited to a thin surface layer without affecting interior of tissue	Manzocco et al. 2009
Fresh-cut Tommy Atkins mango		
250–280 nm UV for 10 min	Improvement in total antioxidant capacity of fresh-cut mango	Gonzalez-Aguilar et al. 2007
Fresh-cut baby spinach leaves		
Double-sided UV (2.4–24 kJ m^{-2})	At low doses, UV was effective in reducing initial psychrotrophic and *enterobacteria* counts and kept *L. monocytogenes* at low levels during storage without affecting sensory quality	Escalona et al. 2010
Fresh-cut watermelon		
4.1 kJ m^{-2}	In commercial trials, exposing packaged watermelon cubes to UV produced a $>$ 1 log reduction in microbial populations by end of product shelf life without affecting juice leakage, color, or overall visual quality	Fonseca and Rushing 2006
1.6, 2.8, 4.8, and 7.2 kJ m^{-2}	No significant effect on vitamin C content; catalase activity and total polyphenol content declined considerably throughout storage	Artés-Hernández et al. 2010
Fresh-cut tropical fruits: honey pineapple, banana *pisang mas*, and guava		
2.1 J m^{-2} for 10, 20, and 30 min	Increase in total phenol and flavonoids, decrease in vitamin C	Alothman et al. 2009
Fresh-cut cantaloupe		
UV (254 nm) for 15 and 60 min	Increase concentrations of terpenoids in cantaloupe tissue	Beaulieu 2007
Fresh-cut (tissue slices) zucchini squash		
UV (250–280 nm) for 1, 10, and 20 min	Delay of senescence and deterioration in tissues, reduction in microbial population, improved storage quality	Erkan et al. 2001
Fresh-cut spinach leaves		
4.54, 7.94, and 11.35 kJ m^{-2}	Low to moderate UV can be an effective alternative to sanitizing with chlorine for minimally processed spinach leaves and can preserve quality	Artés-Hernández et al. 2010
Fresh-cut pear		
Up to 87 kJ m^{-2}	Large log reduction rates were observed at doses between 0 and 15 kJ m^{-2}	Schenk et al. 2008
Fresh-cut tomato		
3.2–19.2 kJ m^{-2}	Reduction in development of microbial populations, increased phenolic content, and delayed degradation of vitamin C after 7 days of storage at 4–6 °C	Kim et al. 2008

(continued)

Table 18.4 (continued)

Conditions	Results	Author
Fresh-cut pineapple		
UV for 15 min, 1–2 cm^2 slices	UV Fotodyne Model 3–3000 Transilluminator (Fotodyne, Hartland, WI, USA); UV-induction of phytoalexin compounds in pineapple; potential use of UV to elicit natural defense mechanisms of fresh-cut fruits	Lamikanra and Richard 2004

capability of reducing *E. coli* O157:H7 in packaged spinach (3–5 log cfu per leaf) with low concentrations (1.6 and 4.3 ppm) for short exposure (5 min).

Indeed, ozone technology for shelf-life extension of fresh-cut salad mix products has been applied successfully since 2000 by Strickland Produce (Strickland et al. 2010). Significant improvements in product quality, plus savings in water and labor, were achieved by the company with the recovery of its investment in ozonation in less than 2 years. As with much of the newest information related to ozone application and regarding the relatively lower capital cost compared to irradiation, commercial use of ozone systems for fresh-cut fruits and vegetables, in particular for sanitizing systems, is applicable.

18.8 Ultraviolet Radiation

Excellent reviews about the benefits and adequacy of ultraviolet (UV) light processing in food applications have been published by several authors (Guerrero-Beltrán and Barbosa-Cánovas 2004). Book chapters on ultraviolet processing of food are also available (Lopez-Malo and Palou 2005). Bintsis et al. (2000) indicated that some companies use UV-treated water in the decontamination of fresh produce such as shredded lettuce. Utilization of UV light in the food industry has several advantages over other agents because it leaves no residue, does not affect moisture and temperature, and is economical. Another advantage is that no excessive protection for workers is necessary and there is no residual radioactivity, even at high levels of exposure (Bintsis et al. 2000).

Microbial inactivation with UV light is a well-known surface treatment. For that reason, the technology has been more applied on the surface of fresh-cut fruits and vegetables than in liquid plant products (Bermudez-Aguirre and Barbosa-Cánovas 2011). The microbial cell surface has been suggested as the primary target of UV treatment. Inactivation is caused by the cross-linking of thymine dimers of cell DNA, thus preventing repair and reproduction (Sizer and Balasubramanian 1999). The effectiveness of microbial disinfection with UV light in extending the shelf life of fresh-cut products depends on several factors, including product profile, wavelength, transmissivity of the product, geometric configuration of the reactor, and radiation path length (Hui and Nip 2004). High UV doses may promote damage to the treated tissue (Rico et al. 2007), and in certain plants UV can easily cause off-flavors (Ohlsson and Bengtsson 2002).

The decrease in intensity when monochromatic UV light is transmitted through the medium is described by the Lambert-Beer law:

$$I = Io.e^{ad}, \tag{18.6}$$

where I is the attenuated intensity, Io is the incident monochromatic UV intensity, and d is the depth reached by the UV light (Guerrero-Beltrán and Barbosa-Cánovas 2004).

Survival of inoculated microorganisms after UV treatment in fresh-cut samples is measured by the first-order kinetics model describing the relationship between survival microorganisms and doses, based on the following equation:

$$log\frac{N}{No} = -kD, \tag{18.7}$$

where No is the initial concentration of microorganisms, N is the concentration of microorganisms after UV treatment, k is a constant, and D is the dose (J m^{-2}) (Chang et al. 1985). Another mathematical model has been described as follows:

$$ln\frac{N}{No} = -klt, \tag{18.8}$$

where N is the number of survival microorganisms after UV radiation, No is the initial load of microorganisms, k is a constant that depends on the type of microorganisms and environmental conditions, I is the intensity (W m^{-2}), and t is the exposure time (s) (Stermer et al. 1987).

The application of UV radiation in extending the shelf life of fresh-cut vegetables has generated numerous studies (Table 18.4) describing the effects of UV radiation on fresh-cut products. Recent research by Artés-Hernández et al. (2010) concluded that UV radiation is a promising tool for maintaining the overall quality of fresh-cut watermelon. Additional findings were reported by Artés-Hernández et al. (2010), who found that low to moderate UV radiation could be used in preserving the quality of minimally processed spinach leaves instead of using chlorine as a sanitizer. Another successful finding has been reported by Lamikanra et al. (2005), who indicated that UV radiation could improve the shelf life and quality of fresh-cut cantaloupe melon. Application of low-dose UV light has been shown to be effective against common foodborne pathogens (Sastry et al. 2000). Additionally, Yaun et al. (2004) found that low-dose UV has a strong beneficial effect on the microbial safety of fresh-cut fruits and vegetables in compliance with Good Agricultural Practices and Good Manufacturing Practices.

Allende et al. (2006) reported that at low-dose two-sided UV radiation, 1.18, 2.37 and 7.11 kJ m^{-2} were effective in reducing the natural microflora of fresh-cut lettuce and therefore UV could improve the shelf life of minimally processed red oak leaf lettuce without compromising quality. Meanwhile, Vicente et al. (2005) suggested that UV treatments could be useful in reducing tissue decay and maintaining the quality of fresh-cut bell peppers. Furthermore, they explained that the incidence and severity of chilling injury could be reduced by short UV treatments. Kasim et al. (2008) found that UVC at a low dose may control the rate of decay in fresh-cut green onions and preserve fresh-like quality. Exposure to UV also may enhance the synthesis of bioactive compounds; for example, UVC irradiation appears to be a good technique to improve the total antioxidant capacity of fresh-cut mango. Similar results have been reported by Lamikanra and Richard (2004) who found that UV-treated fresh-cut pineapple induced production of phytoalexin compounds, demonstrating the potential use of UV radiation to elicit natural defense mechanisms in fresh-cut fruits, which could also enhance the shelf life of fresh-cut pineapple. Finally, these examples and a commercial trial by Fonseca and Rushing (2006) demonstrating shelf life extension of fresh-cut products show that appropriate UV radiation dosage has a positive advantage in providing more ideal fresh-cut products with fresh-like characteristics.

18.9 Ultrasound

Investigation of ultrasound has recently been introduced for preservation purposes in food processing due to its ability to inactivate microorganisms without the deleterious effects of heat on the flavor, color, and nutrition of food (Piyasena et al. 2003). Ultrasound consists of "elastics waves" with frequencies that are above the threshold of human hearing ($\cong$ 20 kHz) (Mulet et al. 2002).

The bacterial inactivation effect of ultrasound has long been observed [Harvey and Loomis 1929, cited in Feng and Yang (2006)]. The mechanical effects of power ultrasonics are mainly attributed to cavitation, whose forces have a dramatic effect on biological systems (Mason and Paniwnyk 2003). The inactivation effect of ultrasound has also been attributed to intracellular cavitation; these mechanical shocks can disrupt cellular structural and functional components to the point of cell lysis (Butz and Tauscher 2002).

As summarized in Table 18.5, ultrasound is applied to fresh-cut products to reduce the initial microbial load and inhibit endogenous enzymes of minimally processed fruits and vegetables. Seymour et al. (2002) investigated the effectiveness of power ultrasound (25–70 kHz) on the microbial decontamination of minimally processed fruits and vegetables for iceberg lettuce, whole cucumber, cut baton carrot, capsicum pepper, white cabbage, spring onion, strawberry, curly leaf parsley, and mint and other herbs. They reported that ultrasound treatment in the range 25–70 kHz does not completely eliminate the risk of pathogens on fresh-cut products. Zhou et al. (2009) concluded that ultrasonication enhanced the reduction of *E. coli* cells on spinach in all treatments (0.7–1.1 log) over that of washing with a sanitizer alone.

Table 18.5 Ultrasonic application on fresh-cut fruits and vegetables

Condition	Results	Author
Fresh-cut apple		
40 kHz with 1 % ascorbic acid	In combination with ascorbic acid had positive effect on maintaining quality	Jang et al. 2009
Fresh whole spinach leaves		
21.2 kHz	Significant reduction of *E. coli* cells on spinach in all treatments by 0.7–1.1 log compared to washing with a sanitizer alone	Zhou et al. 2009
Fresh-cut apple and lettuce (inoculated with *Salmonella* and *E. coli* O157:H7)		
Combined with chlorine decontamination	Treatment time can be increased to bring temperature to sublethal point without damaging the food, increasing decontamination efficacy	Huang et al., 2006
Fresh-cut Royal Gala apples and romaine lettuce (inoculated with *E. coli* O157:H7 and *Salmonella*)		
120 and 170 kHz	Combination with chlorine dioxide increased decontamination	Xu 2005
Fresh-cut fruits and vegetables (iceberg lettuce, whole cucumber, cut baton carrot, capsicum pepper, white cabbage, spring onion, strawberry, curly leaf parsley, mint and other herbs) inoculated with *Salmonella typhimurium*, *L. monocytogenes*, and *E. coli*		
25, 32–40, and 62–70 kHz for 30 min	Frequency of ultrasound treatment had no significant effect on decontamination efficiency	Seymour et al. 2002

Recently, the effects of ultrasound treatment on fruit decay and the physiological quality of strawberries were described by Cao et al. (2010), who reported that a 40 kHz ultrasound treatment resulted in a significant decrease in decay incidence and number of microorganisms. Further, the fresh quality of strawberries was maintained in terms of firmness, total soluble solids, vitamin C, and titratable acidity. The researchers concluded that ultrasound treatment possessed the potential ability to extend the shelf life and maintain the quality of strawberries. In addition, the use of ultrasound in combination with ascorbic acid on fresh-cut apples inhibited 98 % of PPO enzymes (Jang et al. 2009). It was shown that ultrasound technology could inhibit microorganisms and enzymes, prolonging the shelf life and maintaining the quality of fresh-cut apples. However, other studies have shown relatively low inactivation of microorganisms and enzymes (Sala et al. 1999); therefore, it is widely accepted that ultrasound alone is not an effective method for inactivating bacteria in food (Piyasena et al. 2003). In combination with acid (i.e., ascorbic acid) or mild heat, pressure, and other nonthermal processes, power ultrasound has a synergistic effect on inactivation and may be relatively more effective for pathogen removal and inactivation. Also, these combination treatments may help to prolong shelf life and retain the quality of fresh produce, such as retention of vitamins and other heat-sensitive ingredients.

18.10 Acidic Electrolyzed Water

Acidic electrolyzed water (AcEW) is a new concept technology that was developed in Japan and has been classified as so-called functional water. Some scientists use the alternative term electrolyzed oxidizing water (EOW) (Al-Haq et al. 2005), while at present the term neutral electrolyzed water (NEW) has gained more interest. This technology offers the potential ability to inactivate microorganisms with minimal effects on nutritional content. For this reason the application of electrolyzed water on foods such as fresh-cut products in particular is currently a subject of major interest for food decontamination as an alternative to chlorine decontamination. To date, electrolyzed water has been applied on fresh-cut fruits and vegetables, mainly as a washing and disinfection agent to reduce the level of microorganism contamination to ensure food safety and extend shelf life. Washing and disinfection are major critical steps in the processing chain for fresh-cut fruits and vegetables.

EOW production requires only water and salt (sodium chloride). EOW has the following advantages over other traditional cleaning agents: effective disinfection, practicality, relatively low expense, and environmental friendliness. The main advantage of EOW is its safety. EOW, which is also a strong acid element, is different from hydrochloric acid or sulfuric acid in that it is not corrosive to the skin, mucous membranes, or organic material. Electrolyzed water has been tested and used as a disinfectant in the food industry and other applications (Huang et al. 2008).

The utilization of this technology on fresh-cut fruits and vegetables by microbial decontamination is summarized in Table 18.6.

Unfortunately, research on the impact of this technology on enzyme inactivation is scarce. Most of the described studies were carried out in laboratory settings only for microbial disinfection. Koseki et al. (2001) suggested the use of electrolyzed water for controlling microorganisms for plant products as an alternative to chlorine sanitizer. Moreover, Bari et al. (2003) concluded that acidic electrolyzed water could be useful in controlling pathogenic microorganisms. Very interesting results were obtained by Achiwa et al. (2004) in the bacterial inactivation of fresh-cut salads treated with acidic electrolyzed water. AcEW has a considerable effect on higher initial bacterial counts in salad products (without packaging), but when applied to takeout sealed packaged products, this treatment was not very effective, although reduction in the number of bacteria was noted. However, a recent finding by Issa-Zacharia et al. (2010) indicated that AcEW can potentially reduce the amount of *E. coli* and *S. aureus* cell contamination within a short period of time, which makes it a potential replacement for NaOCl solutions commonly used in the food industry. It appears that this novel disinfection system could represent an alternative to chlorine (sodium hypochlorite) disinfection, which is currently the most widespread method used by fresh-cut industries. Thus, EOW shows promise

Table 18.6 Application of electrolyzed water (EW) and acidic electrolyzed water (AcEW) in extending shelf life of fresh-cut fruits and vegetables

Conditions	Results	Author
Fresh-cut apple and cantaloupe (dip-inoculated with *E. coli O157:H7*)		
AcEW (pH 2.7), 68–70 mg L^{-1} free chlorine, 8 min	1 log CFU cm^{-2} reduction	Wang et al. 2008
Fresh-cut cabbage		
AcEW (pH 2.5-2.7), 20–40 mg kg^{-1} available chlorine	Reduced *E. coli* by approx. 1 log CFU g^{-1}	Achiwa et al. 2004
AcEW (pH 6.1), 20 mg L^{-1} available chlorine	Reduction of 1.5 log CFU g^{-1} for total aerobic bacteria and 1.3 log CFU g^{-1} for molds and yeasts	Koide et al. 2009
AcEW (pH 2.5–2.7), 20–40 mg kg^{-1} available chlorine	Effective treatment without producing trihalomethanes or causing abnormality in odor or taste, even when used on foods supplied directly to consumers	Achiwa et al. 2003
Fresh-cut leaves of romaine or iceberg lettuce (inoculated with *E. coli O157:H7*)		
AcEW (pH 2.6), 50 ppm free available chlorine	Reductions of less than 1 log CFU g^{-1}	Keskinen et al. 2009
Fresh-cut lettuce and fresh whole spinach leaves (inoculated with *E. coli O157:H7*, *Salmonella Typhimurium*, and *L. monocytogenes*)		
AcEW, alkaline EW water, and deionized water	Alkaline EW 3 min treatment reduced 3 tested pathogens	Park et al. 2008
Fresh-cut lettuce (inoculated with *L. monocytogenes*)		
AcEW-ice (20, 50, 100, and 200 ppm available chlorine)	AcEW-ice, prepared by freezing at −40 °C with 30, 70, 150, and 240 ppm chlorine gas, generated 70–240 ppm of Cl_2, reducing *L. monocytogenes* and *E. coli O157:H7* by 1.5 log CFU g^{-1}	Koseki et al. 2004
Fresh-cut lettuce leaves (inoculated with *L. monocytogenes* and *E. coli*)		
AcEW (45 ppm residual chlorine) compared to chlorinated water	Difference between bactericidal activity of AcEW and chlorinated water was not significant	Park et al. 2001
Fresh-cut cabbage and lettuce		
AcEW, 150 ppm NaOCl, tap water	Loss in quality of AcEW-treated cabbage and lettuce was equivalent to NaOCl and tap water treatment	Koseki and Itoh 2001
Fresh-cut lettuce and cabbage		
AcEW	Decrease at 1 °C, 5 °C, and 10 °C storage decreased microbial populations	Koseki and Itoh 2001
Fresh-cut cilantro		
AcEW, 24 % NaOCl, deionized water	NaOCl and deionized water were pumped into generator chamber; final concentration of sodium chloride solution passing through electrodes in AcEW generator chamber was 0.2 %; AcEW alone resulted in moderate control of aerobic bacterial growth during storage	Wang et al. 2004

(continued)

Table 18.6 (continued)

Conditions	Results	Author
Fresh-cut vegetables: carrot, spinach, bell pepper, Japanese radish, potato, and cucumber		
EW (pH 6.8), 20 ppm available chlorine	Effective as a disinfectant for tested fresh-cut vegetables without causing discoloration	Izumi 1999
Green onion and tomato (inoculated with *E. coli O157:H7, Salmonella typhimurium*, and *L. monocytogenes*)		
AcEW treatment 5, 10, 15, and 20 ml L^{-1} for 15 s, 30 s, 1 min, 3 min and 5 min	Reduced levels of cells to below detection limit (0.7 log CFU g^{-1}) within 3 min	Park et al. 2009
Fresh-cut salad products (cabbage, lettuce, carrot, red cabbage, paprika, onion, *wakame* seaweed, cucumber, red leaf lettuce, tomato, asparagus, corn, green pepper, onion, potato, tomato with tuna)		
AcEW (pH 2.5–2.7), available chlorine 20–40 mg kg^{-1}	Group A: tested commercial salads from supermarket – no significant bactericidal effect Group B: tested salads from restaurants – strong bactericidal effects	Achiwa et al. 2004
Fresh-cut mixed leafy greens (spinach and lettuce) (inoculated with *E. coli* O157:H7, *Salmonella*, and *L. monocytogenes*)		
AcEW (pH 2.1), 30–35 ppm free chlorine	Oxygen reduction potential 1,100 mV, 2.1–2.8 log CFU g^{-1} reduction	Stopforth et al. 2008
Fresh-cut salad: carrot, endive, iceberg lettuce, 'Four seasons' salad, corn salad		
NEW (pH 8.6), 50 ppm free chlorine	Reduction in amount of free chlorine used to disinfect fresh-cut product, with same microbial reduction effectiveness as sodium hypochlorite	Abadias et al. 2008
Fresh-cut spinach, lettuce (inoculated with *E. coli, Salmonella typhimurium, Staphylococcus aureus, L. monocytogenes*, and *Enterococcus faecalis*)		
NEW (pH 6.3), 1 % NaCl, 20, 50, 100, and 120 ppm total residual chlorine for 10 min	Resulted in 100 % inactivation of all 5 organisms (6.7 log_{10} CFU mL^{-1})	Guentzel et al. 2008

as an environmentally friendly broad-spectrum microbial decontamination agent. However, in terms of extending the shelf life of fresh-cut products, it is likely that further research will be needed due to the low power of microbial inactivation and scarce reportage on enzyme inactivation.

18.11 An Emerging Nonthermal Technology: Cold Plasma

Cold plasma is considered a relatively new decontamination technology in the field of food processing. Plasma is often referred to as the fourth state of matter after solid, liquid, and gas and is actually a mix of ionized gas molecules and free electrons. Cold plasma is electrically energized matter in a gaseous state, and it

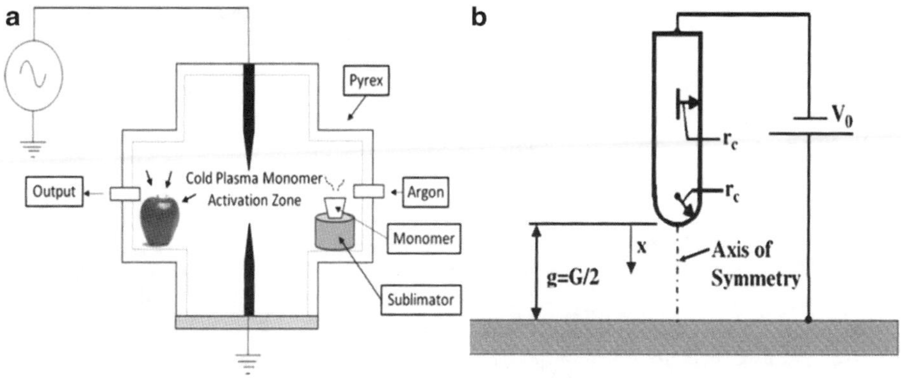

Fig. 18.1 (**a**) Schematic illustration of an experiment utilizing cold plasma to treat red delicious apples. (**b**) Simplified model used to approximate the on-axis electric field in Fig. 18.1. (**a**) Parameter r_c is the radius of curvature of the needle tip as described by Fernandez-Gutierrez et al. (Song et al. 2009)

can be generated by electrical discharge (Song et al. 2009). When a gas passes through plasma, the gas becomes excited, ionized, and full of dissociated electrons, leading to the formation of active species such as atomic oxygen, ozone, and free radicals (e.g., hydroxyl, superoxide, and nitrogen oxides). These reactive species have been shown to have antimicrobial activity (Critzer et al. 2007; Chunqi et al. 2009) by oxidation, although the exact mechanism of microbial cell inactivation is not fully understood (Fernandez-Gutierrez et al. 2010).

Formerly, stable glow discharge plasma could only be generated under vacuum or with gases such as helium and argon. However, researchers have now developed methods for electrically generating nonthermal gas plasma at ambient conditions, resulting in reduced costs and, more practically, offering a potentially new process for ensuring microbial safety and extending the shelf life of food products. Moreover, the relatively low discharged plasmas make their use suitable for heat-sensitive products (Kim et al. 2011).

For the calculation of the electric field intensity E of a needle, as illustrated in Fig. 18.1, the following equation may be used:

$$Ex(x) = E(x) = \frac{-2gVo}{ln\frac{(4g)}{r_c}\left[g(2x + r_c) - x^2\right]}. \tag{18.9}$$

To estimate the number of free electrons (as a function of distance from the negative needle point) associated with an avalanche initiated at the negative needle tip in a nonattaching (electropositive) gas such as argon, this can be written as

$$N(x) = N(0)exp\left[\int_0^x \alpha(E(x))dx\right], \tag{18.10}$$

Table 18.7 Studies on microbial inactivation of fruit and vegetables using cold plasma

Conditions	Results	Author
Mango and melon (cut fruit surface inoculated *Escherichia coli, Saccharomyces cerevisiae, Pantoea agglomerans,* and *Gluconacetobacter liquefaciens*)		
8 kV, 30 kHz	3 log reduction	Perni et al. 2008
Golden Delicious apple (inoculated *Salmonella Stanley* and *E. coli 0157:H7*)		
15 kV, 60 Hz	3.4–3.6 log reduction	Kim et al. 2011
Golden Delicious apple (inoculated with *Listeria innocua*)		
10 kV, 115–150 mA	0.39–1.1 log reduction	Niemira et al. 2005, cited in Niemira and Sites 2008
Seeds: tomato, wheat, bean, chick pea, soybean, barley, oat, rye, lentil, corn (inoculated with *Aspergillus paraciticus* 798. and *Penicillum* MS 1982)		
20 kV, 1 kHz	3 log reduction	Selcuk et al. 2008
Almond (inoculated with *E. coli*)		
30 kV, 2 kHz	5 log reduction	Deng et al. 2007
Peanut and pistachio (inoculated with *A. parasiticus* and aflatoxins)		
20 kV, 1 kHz with total applied power approximately 300 W	5 log reduction	Basaran et al. 2008
Apple, cantaloupe, and lettuce (inoculated with *E. coli* O157:H7, *Salmonellan* and *Listeria monocytogenes*)		
9 kV, 6 kHz	5 min treatment resulted in > 5 log reduction	Critzer et al. 2007
Lettuce (inoculated with *E. coli* ATCC 11775)		
5.7 kV RMS in Argon, 0,5;3.0 and 5.0 min	1.2 log reduction	Bermudez-Aguirre et al. 2011

where $N(x)$ and $N(0)$ represent the numbers of free electrons at locations x and $x = 0$, respectively, and α is Townsend's first ionization coefficient given by

$$\frac{\alpha}{p} = Aexp\left(-\frac{C}{E/p}\right) \tag{18.11}$$

with $A = 1{,}200\ \mathrm{m}^{-1} \cdot \mathrm{torr}^{-1}$ and $C = 20{,}000\ \mathrm{V} \cdot \mathrm{m}^{-1} \cdot \mathrm{torr}^{-1}$, where p is the pressure in torr, and the E/p range for A and C is $C/2 < (E/p) < 3C$ (Fernandez-Gutierrez et al. 2010).

The term cold plasma is a relative term that, in the context of food processing, refers to the capability to generate and apply the plasma nonthermally (Niemira and Sites 2008). This technology is experiencing a heightened interest and research focus by both plasma science researchers and food scientists. Comprehensive review articles and excellent book chapters about the utilization of cold plasma in food processing are available (Misra et al. 2011; Knorr et al. 2011); unfortunately, information about the application of cold plasma on fruit and vegetable products is largely lacking. Table 18.7 summarizes several studies on the utilization of cold plasma with fruit and vegetable products; however, for fresh-cut products, the literature remains scarce.

The advantage of cold plasma in food application is that it allows for microbial inactivation at a low temperature by generating free radicals in the plasma environment, provides uniform treatment, and diffuses into complicated structures with a relatively short processing time. In addition, plasma does not require chemicals, and hence does not leave toxic residues, and it is harmless for operators (Selcuk et al. 2008).

Based on the studies summarized in Table 18.7, cold plasma has a promising future application in the extension of the shelf life of fresh-cut fruits and vegetables. This extended shelf life is attributed to the microbial disinfection provided by this technology at low temperatures on food surfaces by the generation of free radicals that inactivate microorganisms, resulting in food products that are free of any residues and retain the freshlike characteristics of fresh-cut products.

18.12 Final Remarks

Current techniques used by the fresh-cut industry have improved the overall quality of products; however, extending shelf life while guaranteeing safety is still an issue of great concern. Development of novel approaches to assuring the quality and safety of fresh-cut produce has led to the utilization of nonthermal technologies. The benefits of using these technologies lie in the minimal adverse effects on product quality since these decontamination processes do not produce heat. The issue of shelf-life extension of fresh-cut produce using nonthermal technology cannot be separated from and goes hand in hand with safety. The longer shelf life of fresh-cut products in this case is achieved by reducing and minimizing microbial contamination.

Research focus on the utilization of nonthermal emerging technologies has mainly been on extending shelf life and ensuring the safety of products, which has led to several promising nonthermal technologies for later advanced commercial application, though some are commercially available today. The potential of commercial nonthermal technologies in prolonging fresh-cut-produce shelf life is summarized in Table 18.8.

To sum up, most studies have reported that nonthermal processing is a promising means of extending the shelf life, enhancing the safety, and maintaining the freshlike characteristics of fresh-cut fruits and vegetables. On the other hand, due to the extensive current studies on microbial inactivation using nonthermal technologies, as noted in Table 18.8, most exist at the laboratory level with application to certain bacteria and several fresh-cut products. There is relatively limited explanation on the scale-up of these technologies, and available comprehensive data on microbial decontamination using nonthermal processes in the fresh-cut industry are scarce. Therefore, to make commercial adoption easier for the

Table 18.8 Application of shelf-life extension of fresh-cut products using nonthermal technologies

Type of nonthermal technology	Research on fresh-cut fruits and vegetables	Commercial potential
High-pressure processing	Proven to extend shelf life without compromising freshlike quality; adequate microbial decontamination at appropriate levels of pressure for most fresh-cut fruits and vegetables	++
Pulsed electric fields	Several drawbacks noted; this technology is suitable for liquid foods and can assist in drying fresh-cut foods; for all practical purposes as of today, this technology is not appropriate for fresh-cut products	–
Irradiation	Low-dose irradiation has proven effective in extending shelf life and quality	+++
Ozone	Has proven to be an effective decontamination agent as compared to chlorine; environmentally friendly disinfectant	++
UV radiation	Low-dose UV radiation has proven effective for microbial decontamination in combination with other preservation methods	+
Ultrasound	Has proven to have a microbicidal effect in combination with other preservation methods	+
Electrolyzed water	New system and a decontamination agent; could be a replacement for sodium hypochlorite; environmental friendly disinfectant	+
Cold plasma	Microbial inactivation by generating free radicals at low temperature, resulting in food products free of all residue	+

fresh-cut fruit and vegetable industry, future work focusing on the extensive scale-up and guidelines describing potential nonthermal processes for commercial use to extend the shelf life of fresh-cut products will definitely be required.

References

Abadias M, Usall J, Oliveira M, Alegre I, Viñas I (2008) Efficacy of neutral electrolyzed water (NEW) for reducing microbial contamination on minimally-processed vegetables. Int J Food Microbiol 123(1–2):151–158

Achiwa N, Katayose M, Abe K (2003) Efficacy of electrolyzed acidic water for disinfection and quality maintenance of fresh-cut cabbage. Food Preserv Science 29(6):341–346

Achiwa N, Katayose M, Kusakari S, Abe K (2004a) Effects of storage time and temperature after inoculation on bacterial reduction of fresh-cut cabbage treated with electrolyzed water. Food Preserv Science 30(2):69–75

Achiwa N, Katayose M, Yoshida K, Kusakari S, Abe K (2004b) Viable bacterial counts on fresh-cut salads and bactericidal effect of electrolyzed acidic water. Food Preserv Science 30(4):185–190

Aguayo E, Escalona VH, Artés F (2006) Effect of cyclic exposure to ozone gas on physicochemical, sensorial and microbial quality of whole and sliced tomatoes. Postharvest Biol Technol 39(2):169–177

Ahn H-J, Kim J-H, Kim J-K, Kim D-H, Yook H-S, Byun M-W (2005) Combined effects of irradiation and modified atmosphere packaging on minimally processed Chinese cabbage (*Brassica rapa* L.). Food Chem 89(4):589–597

Akbas MY, Ölmez H (2007) Effectiveness of organic acid, ozonated water and chlorine dippings on microbial reduction and storage quality of fresh-cut iceberg lettuce. J Sci Food Agric 87(14):2609–2616

Akbas MY, Ozdemir M (2008) Application of gaseous ozone to control populations of *Escherichia coli*, *Bacillus cereus* and *Bacillus cereus* spores in dried figs. Food Microbiol 25(2):386–391

Alemán G, Farkas DF, Torres JA, Wilhelmsen E, McIntyre S (1994) Ultra-high pressure pasteurization of fresh cut pineapple. J Food Prot 57:931–934

Alemán GD, Ting EY, Farkas DF, Mordre SC, Hawes ACO, Torres JA (1998) Comparison of static and step-pulsed ultra-high pressure on the microbial stability of fresh cut pineapple. J Sci Food Agric 76(3):383–388

Al-Haq MI, Sugiyama J, Isobe S (2005) Applications of electrolyzed water in agriculture and food industries. Food Sci Technol Res 11(2):135–150

Allende A, McEvoy JL, Luo Y, Artes F, Wang CY (2006) Effectiveness of two-sided UV-C treatments in inhibiting natural microflora and extending the shelf-life of minimally processed *'Red Oak Leaf'* lettuce. Food Microbiol 23(3):241–249

Alothman M, Bhat R, Karim AA (2009) UV radiation-induced changes of antioxidant capacity of fresh-cut tropical fruits. Innov Food Sci Emerg Technol 10(4):512–516

Alzamora SM, Guerrero S (2003) Plant antimicrobials combined with conventional preservatives for fruit products. In: Roller S (ed) Natural antimicrobials for the minimal processing of foods. CRC Press LLC, Boca Raton, pp 235–249

An GJ, Li B (2007) Effect of ultra-high pressure on the quality and safety of fresh-cut Chinese yam. Orient Acad Forum, Marrickville, pp 941–945

An J, Zhang M, Lu Q (2007) Changes in some quality indexes in fresh-cut green asparagus pretreated with aqueous ozone and subsequent modified atmosphere packaging. J Food Eng 78(1):340–344

Artés F, Allende A (2005) Minimal fresh processing of vegetables, fruits and juices. In: Da-Wen S (ed) Emerging technologies for food processing. Academic Press, London, pp 677–716

Artés-Hernández F, Robles PA, Gómez PA, Tomás-Callejas A, Artés F (2010) Low UV-C illumination for keeping overall quality of fresh-cut watermelon. Postharvest Biol Technol 55(2):114–120

Barbosa-Cánovas GV, Rodriguez JJ (2002) Update on nonthermal food processing technologies: pulsed electric field, high hydrostatic pressure, irradiation and ultrasound. Food Aust 54(11):8

Barbosa-Cánovas GV, Qin B-L, Swanson BG (1997) The study of critical variables in the treatment of foods by pulsed electric fields. Chapman and Hall/ITP International Thomson Publishing, New York

Barbosa-Canovas GV, Gongora-Nieto MM, Pothakamury UR, Swanson BG (1999) Preservation of foods with pulsed electric fields. Academic, Sandiego

Bari ML, Sabina Y, Isobe S, Uemura T, Isshiki K (2003) Effectiveness of electrolyzed acidic water in killing *Escherichia coli* O157:H7, *Salmonella enteritidis*, and *Listeria monocytogenes* on the surfaces of tomatoes. J Food Prot 66:542–548

Barkai-Golan R (2001) Postharvest diseases of fruits and vegetables development and control. Elsevier, Amsterdam

Basaran P, Basaran-Akgul N, Oksuz L (2008) Elimination of *Aspergillus parasiticus* from nut surface with low pressure cold plasma (LPCP) treatment. Food Microbiol 25(4):626–632

Basbayraktar V, Halkman H, Yucel P, Cetinkaya N (2006) Use of irradiation to improve the safety and quality of minimally processed fruits and vegetables. In: Proceedings of a final research coordination meeting Islamabad 22–30 July 2005: Use of irradiation to ensure the hygienic quality of fresh, pre-cut fruits and vegetables and other minimally processed food of plant origin, the Joint FAO/IAEA programme of nuclear techniques in food and agriculture, Food & Environmental Protection Section International Atomic Energy Agency, Vienna

Baur S, Klaiber R, Hammes WP, Carle R (2004) Sensory and microbiological quality of shredded, packaged iceberg lettuce as affected by pre-washing procedures with chlorinated and ozonated water. Innov Food Sci Emerg Technol 5(1):45–55

Beaulieu JC (2007) Effect of UV irradiation on cut cantaloupe: terpenoids and esters. J Food Sci 72(4):S272–S281

Beltran D, Selma MV, Marin A, Gil MI (2005) Ozonated water extends the shelf life of fresh-cut lettuce. J Agric Food Chem 53(14):5654–5663

Bermudez-Aguirre D, Barbosa-Cánovas GV (2011) Recent advances in emerging nonthermal technologies. In: Aguilera JM (ed) Food engineering interfaces. Springer, New York

Bermudez-Aguirre D, Wemlinger E, Barbosa-Canovas G, Pedrow P, Garcia-Perez M (2011) Inactivation of *Eschericia coli* ATCC in fresh produce using atmospheric pressure cold plasma. Bulletin of the American Physical Society 53rd annual meeting of the APS division of plasma physics, 56 (16)

Bialka KL, Demirci A (2007) Utilization of gaseous ozone for the decontamination of *Escherichia coli* O157:H7 and *Salmonella* on raspberries and strawberries. J Food Prot 70:1093–1098

Bialka KL, Demirci A, Puri VM (2008) Modeling the inactivation of *Escherichia coli* O157:H7 and *Salmonella enterica* on raspberries and strawberries resulting from exposure to ozone or pulsed UV-light. J Food Eng 85(3):444–449

Bibi N, Khattak MK, Badshah A, Chaudry MA (2006) Radiation treatment of minimally processed fruits and vegetables for ensuring hygienic quality. In: Proceedings of a final research coordination meeting Islamabad 22–30 July 2005: Use of irradiation to ensure the hygienic quality of fresh, pre-cut fruits and vegetables and other minimally processed food of plant origin, the Joint FAO/IAEA programme of nuclear techniques in food and agriculture, Food & Environmental Protection Section International Atomic Energy Agency, Vienna

Bintsis T, Litopoulou-Tzanetaki E, Robinson RK (2000) Existing and potential applications of ultraviolet light in the food industry – a critical review. J Sci Food Agric 80(6):637–645

Boynton BB (2004) Determination the effects of modified atmosphere packaging and irradiatiom om semsory characteristics. microbiology, texture and colour of fresh-cut cantaloupe using modeling for package design. University of Florida, USA, Florida

Boynton BB, Welt BA, Sims CA, Balaban MO, Brecht JK, Marshall MR (2005) Effects of low-dose electron beam irradiation on respiration, microbiology, color, and texture of fresh-cut cantaloupe. Hort Technol 15(4):802–807

Buchanan RL, Edelson SG, Snipes K, Boyd G (1998) Inactivation of *Escherichia coli* O157:H7 in apple juice by irradiation. Appl Environ Microbiol 64(11):4533–4535

Butris S, Foley DM, Caporaso F, Prakash A (2003) Effect of low dose gamma and electron beam irradiation on the quality characteristics of fresh cut green onions. In: 2003 IFT annual meeting, Chicago

Butz P, Tauscher B (2002) Emerging technologies: chemical aspects. Food Res Int 35(2–3):279–284

Camelo AL, Gomez P, Artez-Calero F (2003) Use of a* and b* colour parameters to assess the effect of some growth regulators on carotenoid biosynthesis during postharvest tomato ripening. Acta Hortic 599:305–308

Cantwell M (1998) Introduction to fresh-cut products: maintaining quality and safety. Postharvest Horticulture, Series No. 10, (Section 1), University of California

Cao S, Hu Z, Pang B, Wang H, Xie H, Wu F (2010) Effect of ultrasound treatment on fruit decay and quality maintenance in strawberry after harvest. Food Control 21(4):529–532

Castell-Perez E, Moreno M, Rodriguez O, Moreira RG (2004) Electron beam irradiation treatment of cantaloupes: effect on product quality. Food Sci Technol Int 10(6):383–390

Chang JC, Ossoff SF, Lobe DC, Dorfman MH, Dumais CM, Qualls RG, Johnson JD (1985) UV inactivation of pathogenic and indicator microorganisms. Appl Environ Microbiol 49(6):1361–1365

Chervin C, Boisseau P (1994) Quality maintenance of "Ready-to-eat" shredded carrots by gamma irradiation. J Food Sci 59(2):359–361

Chervin C, Triantaphylides C, Libert MF, Siadous R, Boisseau P (1992) Reduction of wound-induced respiration and ethylene production in carrot root tissues by gamma irradiation. Postharvest Biol Technol 2(1):7–17

Chunqi J, Meng-Tse C, Schaudinn C, Gorur A, Vernier PT, Costerton JW, Jaramillo DE, Sedghizadeh PP, Gundersen MA (2009) Pulsed atmospheric-pressure cold plasma for endodontic disinfection. Plasma Sci IEEE Trans 2009 37(7):1190–1195

Clark J (2006) Pulsed electric field processing. Food Technol 60(1):66–67

Cook R (2007) Trends in the marketing of fresh produce and fresh-cut products. In November 2007 ed.; Agricultural Marketing Resource Center, Agricultural Issues Center University of California

Costa SD, Prakash A, Munimbazi C (2001) The effect of low dose gamma irradiation on pectic substances and texture of diced bell peppers and tomatoes. In: 2001 IFT annual meeting, New Orleans

Critzer FJ, Kelly-Wintenberg K, South SL, Golden DA (2007a) Atmospheric plasma inactivation of foodborne pathogens on fresh produce surfaces. J Food Prot 70(10):2290–2296

Critzer F, Kelly-Wintenberg K, South S, Golden D (2007b) Atmospheric plasma inactivation of foodborne pathogens on fresh produce surfaces. J Food Protect 70(10):2290

Das BK, Kim JG (2010) Microbial quality and safety of fresh-cut broccoli with different sanitizers and contact times. J Microbiol Biotechnol 20(2):363–369

Degl'Innocenti E, Pardossi A, Tognoni F, Guidi L (2007) Physiological basis of sensitivity to enzymatic browning in 'lettuce', 'escarole' and 'rocket salad' when stored as fresh-cut products. Food Chem 104(1):209–215

Del Nobile MA, Baiano A, Benedetto A, Massignan L (2006) Respiration rate of minimally processed lettuce as affected by packaging. J Food Eng 74(1):60–69

Del Pozo-Insfran D, Del Follo-Martinez A, Talcott ST, Brenes CH (2007) Stability of copigmented anthocyanins and ascorbic acid in muscadine grape juice processed by high hydrostatic pressure. J Food Sci 72(4):S247–S253

Deng S, Ruan R, Mok CK, Huang G, Lin X, Chen P (2007) Inactivation of *Escherichia coli* on almonds using nonthermal plasma. J Food Sci 72(2):M62–M66

Du J, Fu Y, Wang N (2009) Effects of aqueous chlorine dioxide treatment on browning of fresh-cut lotus root. LWT- Food Sci Technol 42(2):654–659

Elisabete MC, Joana F (2006) Effect of ozone on the quality of fresh-cut beans (*Phaseolus vulgaris* L.) and bell peppers (*Capsicum annuum* L.). 7º Encontro de Quimica dos Alimentos, Viseu, pp 1–6

Entrup ML (2005) Advanced planning in fresh food industries: integrating shelf life into production planning. Physica-Verlag/Springer, Heidelberg

Erkan M, Wang CY, Krizek DT (2001) UV-C irradiation reduces microbial populations and deterioration in Cucurbita pepo fruit tissue. Environ Exp Bot 45:1–9

Escalona VH, Aguayo E, Martínez-Hernández GB, Artés F (2010) UV-C doses to reduce pathogen and spoilage bacterial growth in vitro and in baby spinach. Postharvest Biol Technol 56(3):223–231

Fan X (2005) Antioxidant capacity of fresh-cut vegetables exposed to ionizing radiation. J Sci Food Agric 85:995–1000

Fan X, Toivonen PMA, Rajkowski KT, Sokorai KJB (2003) Warm water treatment in combination with modified atmosphere packaging reduces undesirable effects of irradiation on the quality of fresh-cut iceberg lettuce. J Agric Food Chem 51(5):1231–1236

Fan X, Niemera BA, Mattheis JE, Zhuang H, Olson DW (2005) Quality of fresh-cut apple slices as affected by low-dose ionizing radiation and calcium ascorbate treatment. J Food Sci 70(2): S143–S148

Fan X, Annous BA, Sokorai KJB, Burke A, Mattheis JP (2006) Combination of hot-water surface pasteurization of whole fruit and low-dose gamma irradiation of fresh-cut cantaloupe. J Food Prot 69:912–919

Faridah MS, Ilida MN, Asiah AS, Mahmud M (2006) Effect of gamma radiation on the safety and quality of selected minimally processed fruits and vegetables. In: Proceedings of a final research coordination meeting Islamabad 22–30 July 2005: Use of irradiation to ensure the hygienic quality of fresh, pre-cut fruits and vegetables and other minimally processed food of plant origin, the Joint FAO/IAEA Programme of Nuclear Techniques in Food and Agriculture, Food & Environmental Protection Section International Atomic Energy Agency, Vienna

Farkas J, Meszaros L, Mohacsi-Farkas C, Saray T, Andrassy E (1999) The role of ionizing radiation in minimal processing of precut vegetables with particular reference to the control of *Listeria monocytogenes*. CRC Press LLC, Boca Raton

Feng H, Yang W (2006) Power ultrasound. In: Hui YH (ed) Handbook of food science, technology, and engineering. Taylor/Francis Group LLC, Boca Raton

Fernandez-Gutierrez SA, Pedrow PD, Pitts MJ, Powers J (2010) Cold atmospheric-pressure plasmas applied to active packaging of apples. Plasma Sci IEEE Trans 38(4):957–965

Fincan M, Dejmek P (2003) Effect of osmotic pretreatment and pulsed electric field on the viscoelastic properties of potato tissue. J Food Eng 59(2–3):169–175

Foley DM, Dufour A, Rodriguez L, Caporaso F, Prakash A (2002) Reduction of *Escherichia coli* 0157:H7 in shredded iceberg lettuce by chlorination and gamma irradiation. Radiat Phys Chem 63(3–6):391–396

Fonseca JM, Rushing JW (2006) Effect of ultraviolet-C light on quality and microbial population of fresh-cut watermelon. Postharvest Biol Technol 40(3):256–261

Fonseca SC, Oliveira FAR, Brecht JK, Chau KV (1999) Development of perforation-mediated modified atmosphere packaging for fresh-cut vegetables. In: Oliveira FAR et al (eds) Processing foods: quality optimazation and process assessment. CRC Press LLC, Boca Raton

Fonseca SC, Oliveira FAR, Brecht JK (2002) Modelling respiration rate of fresh fruits and vegetables for modified atmosphere packages: a review. JFood Eng 52(2):99–119

Garcia E, Barrett DM (2005) Fresh-cuts fruits. In: Barrett DM et al (eds) Processing fruits: science and technology, 2nd edn. CRC Press LLC, Boca Raton

Garcia-Viguera C, Bridle P (1999) Influence of structure on colour stability of anthocyanins and flavylium salts with ascorbic acid. Food Chem 64(1):21–26

Garzon GA, Wrolstad RE (2002) Comparison of the stability of pelargonidin-based anthocyanins in strawberry juice and concentrate. J Food Sci 67(4):1288–1299

Giner J, Gimeno V, Barbosa-Canovas GV, Martin O (2001) Effects of pulsed electric field processing on apple and pear polyphenoloxidases. Food Sci Technol Int 7(4):339–345

Gomes C, Da Silva P, Chimbombi E, Kim J, Castell-Perez E, Moreira RG (2008) Electron-beam irradiation of fresh broccoli heads (*Brassica oleracea* L. italica). LWT- Food Sci Technol 41(10):1828–1833

Gómez PL, Alzamora SM, Castro MA, Salvatori DM (2010) Effect of ultraviolet-C light dose on quality of cut-apple: microorganism, color and compression behavior. J Food Eng 98(1):60–70

González-Aguilar GA, Ayala-Zavala JF, Ruiz-Cruz S, Acedo-Félix E, Díaz-Cinco ME (2004) Effect of temperature and modified atmosphere packaging on overall quality of fresh-cut bell peppers. Lebensm Wiss Technol 37(8):817–826

Gonzalez-Aguilar GA, Villegas-Ochoa MA, Martinez-Tellez MA, Gardea AA, Ayala-Zavala JF (2007) Improving antioxidant capacity of fresh-cut mangoes treated with UV-C. J Food Sci 72(3):S197–S202

Grandison AS (2006) Irradiation. In: Brennan J (ed) Food processing handbook. Wiley-VCH Verlag Gmbh &Co. KGaA, Weinheim

Guentzel JL, Liang Lam K, Callan MA, Emmons SA, Dunham VL (2008) Reduction of bacteria on spinach, lettuce, and surfaces in food service areas using neutral electrolyzed oxidizing water. Food Microbiol 25(1):36–41

Guerrero-Beltrán JA, Barbosa-Cánovas GV (2004) Review: advantages and limitations on processing foods by UV Light. J Food Sci Tech Int 10(3):137–147

Gunes G, Watkins CB, Hotchkiss JH (2000) Effects of irradiation on respiration and ethylene production of apple slices. J Sci Food Agric 80(8):1169–1175

Gunes G, Hotchkiss JH, Watkins CB (2001) Effects of gamma irradiation on the texture of minimally processed apple slices. J Food Sci 66(1):63–67

Guzel-Seydim ZB, Greene AK, Seydim AC (2004) Use of ozone in the food industry. Lebensm Wiss Technol 37(4):453–460

Hagenmaier RD, Baker RA (1997) Low-dose irradiation of cut iceberg lettuce in modified atmosphere packaging. J Agric Food Chem 45(8):2864–2868

Hagenmaier RD, Baker RA (1998) Microbial population of shredded carrot in modified atmosphere packaging as related to irradiation treatment. J Food Sci 63(1):162–164

Hammad AA, Elnour SAA, Salah A (2006) Use of irradiation to ensure hygienic quality of minimally processed vegetables and fruits. In: Proceedings of a final research coordination meeting Islamabad, 22–30 July 2005: Use of irradiation to ensure the hygienic quality of fresh, pre-cut fruits and vegetables and other minimally processed food of plant origin, the Joint FAO/IAEA Programme of Nuclear Techniques in Food and Agriculture, Food & Environmental Protection Section International Atomic Energy Agency, Vienna

Han J, Gomes-Feitosa CL, Castell-Perez E, Moreira RG, Silva PF (2004) Quality of packaged romaine lettuce hearts exposed to low-dose electron beam irradiation. Lebensm Wiss Technol 37(7):705–715

Hanotel L, Fleuriet A, Boisseau P (1995) Biochemical changes involved in browning of gamma-irradiated cut witloof chicory. Postharvest Biol Technol 5(3):199–210

Harrison SL, Barbosa-Cánovas GV, Swanson BG (1997) *Saccharomyces cerevisiae* structural changes induced by pulsed electric field treatment. Lebensm Wiss Technol 30(3):236–240

Harrison SL, Barbosa-Cánovas GV, Swanson BG (2001) Pulsed electric field and high hydrostatic pressure induced leakage of cellular material from *Saccharomyces cerevisae*. In: Barbosa-Canovas G, Zhang QH (eds) Pulsed electric fields in food processing. Technomic Publishing Company, Inc, Lancester

Hassenberg K, Idler C, Molloy E, Geyer M, Plöchl M, Barnes J (2007) Use of ozone in a lettuce-washing process: an industrial trial. J Sci Food Agric 87(5):914–919

Hayashi R (1995) Advances in high pressure food processing technology in Japan. In: Gaonkar A (ed) Food processing recent development. Elsevier, Amsterdam

Heinz V, Toepfl S, Knorr D (2003) Impact of temperature on lethality and energy efficiency of apple juice pasteurization by pulsed electric fields treatment. Innov Food Sci Emerg Technol 4(2):167–175

Ho SY, Mittal GS, Cross JD (1997) Effects of high field electric pulses on the activity of selected enzymes. J Food Eng 31(1):69–84

Hoover DG (1993) Pressure effects on biological systems. Food Technol 47:150–155

Horrick JM, Foley DM, Caporaso F, Prakash A (2002) Effects of low-dose gamma irradiation on the shelf-life and quality characteristics of fresh cut cantaloupe. In: 2002 IFT annual meeting and food expo, Anaheim

Hsu W-Y, Simonne A, Jitareerat P Jr, Marshall MR (2010) Low-dose irradiation improves microbial quality and shelf life of fresh mint (*Mentha piperita L.*) without compromising visual quality. J Food Sci 75(4):M222–M230

Huang T-S, Xu C, Walker K, West P, Zhang S, Weese J (2006) Decontamination efficacy of combined chlorine dioxide with ultrasonication on apples and lettuce. J Food Sci 71(4): M134–M139

Huang Y-R, Hung Y-C, Hsu S-Y, Huang Y-W, Hwang D-F (2008) Application of electrolyzed water in the food industry. Food Control 19(4):329–345

Hui YH, Nip W (2004) New technology, vegetable processing and microbial inactivation. In: Hui YH et al (eds) Handbook of vegetable preservation and processing. Marcel Dekker, Inc, New York

Issa-Zacharia A, Kamitani Y, Morita K, Iwasaki K (2010) Sanitization potency of slightly acidic electrolyzed water against pure cultures of *Escherichia coli* and Staphylococcus aureus, in comparison with that of other food sanitizers. Food Control 21(5):740–745

Izumi H (1999) Electrolyzed water as a disinfectant for fresh-cut vegetables. J Food Sci 64(2):536–539

Jacxsens L, Devlieghere F, Debevere J (1999) Validation of a systematic approach to design equilibrium modified atmosphere packages for fresh-cut produce. Lebensm Wiss Technol 32(7):425–432

Jang JH, Kim ST, Moon KD (2009) Inhibitory effects of ultrasound in combination with ascorbic acid on browning and polyphenol oxidase activity of fresh-cut apples. Food Sci Biotechnol 18(6):1417–1422

Jin ZT, Tuhela L, Zhang H, Sastry SK (2001) Inactivation of Bacillus subtilis spores using high voltage pulsed electric fields. In: Barbosa-Canovas G, Zhang QH (eds) Pulsed electric fields in food processing. Technomic Publishing Company, Inc, Lancester

Kader AA (2002) Quality parameters of fresh-cut fruits and vegetables products. In: Lamikanra O (ed) Fresh-cut fruits and vegetables: science, technology and safety. CRC Press LLC, Boca Raton

Kamat AS, Ghadge N, Ramamurthy MS, Alur MD (2005) Effect of low-dose irradiation on shelf life and microbiological safety of sliced carrot. J Sci Food Agric 85:2213–2219

Kasim MU, Kasim R, Erkal S (2008) UV-C treatments on fresh-cut green onions enhanced antioxidant activity, maintained green color and controlled "telescoping". Int J Food Agric Environ 6(3–4):63–67

Keskinen LA, Burke A, Annous BA (2009) Efficacy of chlorine, acidic electrolyzed water and aqueous chlorine dioxide solutions to decontaminate *Escherichia coli* O157:H7 from lettuce leaves. Int J Food Microbiol 132(2–3):134–140

Khader MA, Yousef AE, Kim JG (2001) Microbiological aspects of ozone applications in food: a review. J Food Sci 66(9):1242–1252

Kim J-G, Yousef AE, Dave S (1999) Application of ozone for enhancing the microbiological safety and quality of foods: a review. J Food Prot 62:1071–1087

Kim J-G, Yousef AE, Khadre MA (2003) Ozone and its current and future application in the food industry. Adv Food Nutr Res 45:167–218

Kim HJ, Feng H, Toshkov SA, Fan X, USDA A (2005) Effect of sequential treatment of warm water dip and low-dose gamma irradiation on the quality of fresh-cut green onions. J Food Sci 70(3):M179

Kim BS, Kwon JY, Kwon KH, Cha HS, Jeong JW, Kim GH (2006) Antimicrobial effect of cold ozonated water washing on fresh-cut lettuce. In: Batt PJ (ed) Proceedings of the 1st international symposium on improving the performance of supply chains in the transitional economies, pp 235–242

Kim J-H, Lee J-W, Kim J-H, Seo J-H, Han S-B, Chung H-J, Byun M-W (2006b) Effect of gamma irradiation on Listeria ivanovii inoculated to iceberg lettuce stored at cold temperature. Food Control 17(5):397–401

Kim HJ, Fonseca JM, Kubota C, Kroggel M, Choi JH (2008) Quality of fresh-cut tomatoes as affected by salt content in irrigation water and post-processing ultraviolet-C treatment. J Sci Food Agric 88(11):1969–1974

Kim B, Yun H, Jung S, Jung Y, Jung H, Choe W, Jo C (2011) Effect of atmospheric pressure plasma on inactivation of pathogens inoculated onto bacon using two different gas compositions. Food Microbiol 28:9–13

Knorr D, Angersbach A (1998) Impact of high-intensity electric field pulses on plant membrane permeabilization. Trends Food Sci Technol 9(5):185–191

Knorr D, Froehling A, Jaeger H, Reineke K, Schlueter O, Schoessler K (2011) Emerging technologies in food processing. Annu Rev Food Sci Technol 2:203–235

Koide S, Takeda J-i, Shi J, Shono H, Atungulu GG (2009) Disinfection efficacy of slightly acidic electrolyzed water on fresh cut cabbage. Food Control 20(3):294–297

Korkmaz M, Polat M (2005) Irradiation of fresh fruit and vegetables. CRC Press LLC and Woodhead Publishing Limited, Boca Raton

Koseki S, Isobe S (2006) Effect of ozonated water treatment on microbial control and on browning of iceberg lettuce (Lactuca sativa L.). J Food Prot 69:154–160

Koseki S, Itoh K (2001a) The effect of acidic electrolyzed water on the quality of cut vegetables. J Jpn Soc Food Sci Technol 48(5):365–639

Koseki S, Itoh K (2001b) Prediction of microbial growths in fresh-cut vegetables treated with acidic electrolyzed water during storage under various temperature conditions. J Food Prot 64 (12):1935–1942

Koseki S, Yoshida K, Isobe S, Itoh K (2001) Decontamination of lettuce using acidic electrolyzed water. J Food Prot 64:652–658

Koseki S, Isobe S, Itoh K (2004) Efficacy of acidic electrolyzed water ice for pathogen control on lettuce. J Food Prot 67(11):2544–2549

Lacroix M, Marcotte M, Ramaswamy H (2003) Irradiation of fruits, vegetables, nuts and spices. Marcel Dekker, Inc, New York

Lamikanra O, Richard OA (2004) Storage and ultraviolet-induced tissue stress effects on fresh-cut pineapple. J Sci Food Agric 84:1812–1816

Lamikanra O, Richard OA, Parker A (2002) Ultraviolet induced stress response in fresh cut cantaloupe. Phytochemistry 60(1):27–32

Lamikanra O, Kueneman D, Ukuku D, Bett-Garber KL (2005) Effect of processing under ultraviolet light on the shelf life of fresh-cut cantaloupe melon. J Food Sci 70(9):C534–C539

Latorre ME, Narvaiz P, Rojas AM, Gerschenson LN (2010) Effects of gamma irradiation on bio-chemical and physico-chemical parameters of fresh-cut red beet (Beta vulgaris L. var. conditiva) root. J Food Eng 98(2):178–191

Lebovka NI, Praporscic I, Vorobiev E (2004) Effect of moderate thermal and pulsed electric field treatments on textural properties of carrots, potatoes and apples. Innov Food Sci Emerg Technol 5(1):9–16

López L, Avendaño S, Romero J, Garrido S, Espinoza J, Vargas M (2005) Effect of gamma irradiation on the microbiological quality of minimally processed vegetables. Arch Latinoam Nutr 55(3):287–292

Lopez-Malo A, Palou E (2005) Ultraviolet light and food preservation. In: Barbosa-Cánovas GV et al (eds) Novel food processing technologies. CRC PRESS/Marcel Dekker, New York

Lu Z, Yu Z, Gao X, Lu F, Zhang L (2005) Preservation effects of gamma irradiation on fresh-cut celery. J Food Eng 67(3):347–351

Luo Y, Barbosa-Cánovas GV (1996) Preservation of apple slices using ascorbic acid and 4-hexylresorcinol/Preservación de rodajas de manzana con ácido ascorbico y 4-hexilresorcinol. Food Sci Technol Int 2(5):315–321

Majchrowicz A (1999) Innovative technologies could improve food safety. Food Rev Int 22(2):16–20

Manzocco L, Dri A, Quarta B (2009) Inactivation of pectic lyases by light exposure in model systems and fresh-cut apple. Innov Food Sci Emerg Technol 10(4):500–505

Martínez-Sánchez A, Allende A, Cortes-Galera Y, Gil MI (2008) Respiration rate response of four baby leaf Brassica species to cutting at harvest and fresh-cut washing. Postharvest Biol Technol 47(3):382–388

Mason TJ, Paniwnyk L (2003) Ultrasound as a preservation technology. In: Zeuthen P, Bogh-Sorensen L (eds) Food preservation techniques. Woodhead Publishing Limited/CRC Press LLC, Cambridge, England

Mertens B, Knorr D (1992) Developments of nonthermal processes for food preservation. Food Technol 46(5):124–133

Min S, Zhang Q (2005) Packaging for nonthermal food processing. In: Han JH (ed) Innovation in food packaging. Elsevier Academic Press, San Diego

Mintier AM, Foley DM (2006) Electron beam and gamma irradiation effectively reduce *Listeria monocytogenes* populations on chopped romaine lettuce. J Food Prot 69:570–574

Misra NN, Tiwari BK, Raghavarao KSMS, Cullen PJ (2011) Nonthermal plasma inactivation of food-borne pathogens. Food Eng Rev 3:159–170

Mulet A, Carcel JA, Benedito J, Sanjuan N (2002) Application of low-intensity ultrasonics in the dairy industry. In: Welti-Chanes J et al (eds) Enginnering and food for 21st century. CRC Press LLC, Boca Raton

Muthukumarappan K, Halaweish F, Naidu AS (2000) Ozone. In: Naidu AS (ed) Natural food antimicrobial systems. CRC Press LLC, Boca Raton, pp 783–800

Niemira BA, Deschenes L (2005) Ionising radiation processing of fruits and fruit products. CRC Press LLC, Boca Raton

Niemira BA, Sites J (2008) Cold plasma inactivates *Salmonella stanley* and *Escherichia coli* O157:H7 inoculated on golden delicious apples. J Food Prot 71:1357–1365

Nunes TCF, Rogovschi VD, Santillo AG, Michels R, Pitombo RNM, Villavicencio ALC H (2009) ^{60}Co irradiation effect on collor minimally processed cauliflower (Brassica spp). In: 2009 international nuclear Atlantic conference – INAC 2009

O'Beirne D, Francis GA (2003) Analysing pathogen survival in modified atmosphere packaged (MAP) produce. In: Ahvenainen R (ed) Novel food packaging techniques. Woodhead Publishing, London

Ohlsson T, Bengtsson N (2002) Minimal processing of foods with nonthermal methods. In: Ohlsson T, Bengtsson N (eds) Minimal processing technologies in the food industry. Woodhead Publishing Limited and CRC Press, Boca Raton

Ölmez H, Akbas MY (2009) Optimization of ozone treatment of fresh-cut green leaf lettuce. J Food Eng 90(4):487–494

Palekar MP, Cabrera-Diaz E, Kalbasi-Ashtari A, Maxim JE, Miller RK, Cisneros-Zevallos L, Castillo A (2004) Effect of electron beam irradiation on the bacterial load and sensorial quality of sliced cantaloupe. J Food Sci 69(9):M267–M273

Palou E, Lopez-Malo A, Barbosa-Cánovas GV, Welti-Chanes J (2000) High hydrostatic pressure and minimal processing. In: Alzamora M et al (eds) Minimally processed fruits and vegetables. Aspen Publisher, Inc, Mayland

Palou E, Lopez-Malo A, Welti-Chanes J (2002) Innovative fruit preservation methods using high pressure. In: Welti-Chanes J et al (eds) Engineering and food for the 21st century. CRC Press, Boca Raton

Park C-M, Hung Y-C, Doyle MP, Ezeike GOI, Kim C (2001) Pathogen reduction and quality of lettuce treated with electrolyzed oxidizing and acidified chlorinated water. J Food Sci 66(9):1368–1372

Park EJ, Alexander E, Taylor GA, Costa R, Kang DH (2008) Effect of electrolyzed water for reduction of foodborne pathogens on lettuce and spinach. J Food Sci 73(6):M268–M272

Park EJ, Alexander E, Taylor GA, Costa R, Kang DH (2009) The decontaminative effects of acidic electrolyzed water for *Escherichia coli* O157:H7, *Salmonella typhimurium*, and *Listeria monocytogenes* on green onions and tomatoes with differing organic demands. Food Microbiol 26(4):386–390

Pereira RN, Vicente AA (2009) Environmental impact of novel thermal and nonthermal technologies in food processing. Food Res Int (in press), Corrected Proof

Perera N, Gamage TV, Wakeling L, Gamlath GGS, Versteeg C (2010) Colour and texture of apples high pressure processed in pineapple juice. Innov Food Sci Emerg Technol 11(1):39–46

Perni S, Liu D, Shama G, Kong MG (2008a) Cold atmospheric plasma decontamination of the pericarps of fruit. J Food Prot 71:302–308

Perni S, Shama G, Kong MG (2008b) Cold atmospheric plasma disinfection of cut fruit surfaces contaminated with migrating microorganisms. J Food Prot 71:1619–1625

Pillai S, Garcia N, Maxim J (2006) Electron beam inactivation of enteric viruses on cantaloupe surfaces. In: Proceedings of a final research coordination meeting Islamabad, 22–30 July 2005: Use of irradiation to ensure the hygienic quality of fresh, pre-cut fruits and vegetables and other minimally processed food of plant origin, the Joint FAO/IAEA Programme of Nuclear Techniques in Food and Agriculture, Food & Environmental Protection Section International Atomic Energy Agency, Vienna

Piyasena P, Mohareb E, McKellar RC (2003) Inactivation of microbes using ultrasound: a review. Int J Food Microbiol 87(3):207–216

Prakash A, Inthajak P, Huibregtse H, Caporaso F, Foley DM (2000) Effects of low-dose gamma irradiation and conventional treatments on shelf life and quality characteristics of diced celery. J Food Sci 65(6):1070–1075

Prakash A, Johnson N, Foley D (2007) Irradiation D values of *Salmonella* spp. in diced tomatoes dipped in 1 % calcium chloride. Foodborne Pathog Dis 4(1):84–88

Qin B-L, Pothakamury UR, Barbosa-Cánovas GV, Swanson BG, Peleg M (1996) Nonthermal pasteurization of liquid foods using high-intensity pulsed electric fields. Crit Rev Food Sci Nutr 36(6):603–627

Rahman MS (1999) Irradiation preservation of foods. In: Rahman MS et al (eds) Handbook of food preservation. Marcel Dekker, Inc, New York

Ramaswamy H, Chen C, Marcotte M (2005) Novel processing technologies for food preservation. In: Barrett DM et al (eds) Processing fruits: science and technology, 2nd edn. CRC Press LLC, Boca Raton

Raso J, Calderon ML, Gongora M, Barbosa-Cánovas GV, Swanson BG (1998) Inactivation of *Zygosaccharomyces bailii* in fruit juices by heat, high hydrostatic pressure and pulsed electric fields. J Food Sci 63(6):1042–1044

Rico D, Martín-Diana AB, Frías JM, Henehan GTM, Barry-Ryan C (2006) Effect of ozone and calcium lactate treatments on browning and texture properties of fresh-cut lettuce. J Sci Food Agric 86(13):2179–2188

Rico D, Martín-Diana AB, Barat JM, Barry-Ryan C (2007) Extending and measuring the quality of fresh-cut fruit and vegetables: a review. Trends Food Sci Technol 18(7):373–386

Sala FJ, Burgos J, Condon S, Lopez P, Raso J (1999) Effect of heat and ultrasound on microorganisms and enzymes. In: Gould G (ed) New methods of food preservation. Aspen Publishers, Inc, Gaithersburg

Sandhya M (2010) Modified atmosphere packaging of fresh produce: current status and future needs. LWT- Food Sci Technol 43(3):381–392

Sastry SK, Datta AK, Worobo RW (2000) Ultraviolet light. J Food Sci 65(12):90–92

Schenk M, Guerrero S, Alzamora S (2008) Response of some microorganisms to ultraviolet treatment on fresh-cut pear. Food Bioprocess Technol 1(4):384–392

Schmidt HM (2004) Improving the microbiological quality and safety of fresh-cut tomatoes by low dose electron beam irradiation. Texas A&M University, College Station

Selcuk M, Oksuz L, Basaran P (2008) Decontamination of grains and legumes infected with *Aspergillus* spp. and *Penicillum* spp. by cold plasma treatment. Bioresour Technol 99(11):5104–5109

Selma MV, Beltrán D, Allende A, Chacón-Vera E, Gil MI (2007) Elimination by ozone of Shigella sonnei in shredded lettuce and water. Food Microbiol 24(5):492–499

Selma MV, Allende A, López-Gálvez F, Conesa MA, Gil MI (2008a) Disinfection potential of ozone, ultraviolet-C and their combination in wash water for the fresh-cut vegetable industry. Food Microbiol 25(6):809–814

Selma MV, Ibáñez AM, Allende A, Cantwell M, Suslow T (2008b) Effect of gaseous ozone and hot water on microbial and sensory quality of cantaloupe and potential transference of *Escherichia coli* O157:H7 during cutting. Food Microbiol 25(1):162–168

Seymour IJ, Burfoot D, Smith RL, Cox LA, Lockwood A (2002) Ultrasound decontamination of minimally processed fruits and vegetables. Int J Food Sci Technol 37:547–557

Sharma A (2004) Post-harvest processing of fruits and vegetables by ionizing radiation. In: Dris R, Jain S (eds) Production and quality assessment of food crops: vol.4, postharvest treatment and technology. Kluwer, Dordrecht

Silveira AC, Aguayo E, Artés F (2010) Emerging sanitizers and clean room packaging for improving the microbial quality of fresh-cut 'Galia' melon. Food Control 21(6):863–871

Singh N, Singh RK, Bhunia AK, Stroshine RL (2002) Efficacy of chlorine dioxide, ozone, and thyme essential oil or a sequential washing in killing *Escherichia coli* O157:H7 on lettuce and baby carrots. Lebensm Wiss Technol 35:720–729

Sizer CE, Balasubramanian VM (1999) New intervention process for minimally processed juices. Food Technol 53(10):64–67

Song HP, Kim B, Choe JH, Jung S, Moon SY, Choe W, Jo C (2009) Evaluation of atmospheric pressure plasma to improve the safety of sliced cheese and ham inoculated by 3-strain cocktail *Listeria monocytogenes*. Food Microbiol 26:432–436

Spoto MHF, Domarco RE, Walder JMM, Scarminio IS, Bruns RE (1997) Sensory evaluation of orange juice concentrate as affected by irradiation and storage. J Food Process Preserv 21(3):179–191

Stermer RA, Lasater-Smith M, Brasington CF (1987) Ultraviolet radiation – an effective bactericide for fresh meat. J Food Prot 50(2):108–111

Stopforth JD, Mai T, Kottapalli B, Samadpour M (2008) Effect of acidified sodium chlorite, chlorine, and acidic electrolyzed water on *Escherichia coli* O157:H7, *Salmonella*, and *Listeria monocytogenes* inoculated onto leafy greens. J Food Prot 71:625–628

Strickland W, Sopher CD, Rice RG, Battles GT (2010) Six years of ozone processing of fresh cut salad mixes. Ozone Sci Eng 32:66–70

Thayer DW, Rajkowski KT (1999) Developments in irradiation of fresh fruits and vegetables. Food Technol 53:62–65

Tijskens LMM (2000) Acceptability. In: Shewfelt LR, Bruckner B (eds) Fruit and vegetable quality: an integrated view. CRC Press, New York, pp 125–143

Ting EY, Marshall RG (2002) Production issues related to UHP Food. In: Welti-Chanes J et al (eds) Engineering and food for 21st century. CRC Press LLC, Boca Raton

Torres JA, Velazquez G (2005) Commercial opportunities and research challenges in the high pressure processing of foods. J Food Eng 67(1–2):95–112

Trigo M, Ferreira A, Sapata M, Sousa M, Curado T, Andrada L, Erreira E, Botelho M, Veloso M (2006) Improving quality and safety of minimally processed fruits and vegetables by gamma irradiation. In: Proceedings of a final research coordination meeting Islamabad, 22–30 July 2005: use of irradiation to ensure the hygienic quality of fresh, pre-cut fruits and vegetables and other minimally processed food of plant origin, the Joint FAO/IAEA Programme of Nuclear Techniques in Food and Agriculture, Food & Environmental Protection Section International Atomic Energy Agency, Vienna

Varoquaux P, Wiley R (1994) Biological and biochemical changes in minimally processed refrigerated fruits and vegetables. In: Wiley RC (ed) Minimally processed refrigerated fruits and vegetables. Chapman and Hall, New York, pp 226–268

Vega-Mercado H, Powers JR, Barbosa-Cánovas GV, Swanson BG (1995) Plasmin inactivation with pulsed electric fields. J Food Sci 60(5):1143–1146

Vicente AR, Pineda C, Lemoine L, Civello PM, Martinez GA, Chaves AR (2005) UV-C treatments reduce decay, retain quality and alleviate chilling injury in pepper. Postharvest Biol Technol 35(1):69–78

Walker JRL, Ferrar PH (1998) Diphenol oxidases, enzyme-catalysed browning and plant disease resistance. Biotechnol Genet Eng Rev 15:457–498

Wang H, Feng H, Luo Y (2004) Microbial reduction and storage quality of fresh-cut cilantro washed with acidic electrolyzed water and aqueous ozone. Food Res Int 37(10):949–956

Wang HUA, Feng HAO, Luo Y (2006) Dual-phasic inactivation of *Escherichia coli* 157:H7 with peroxyacetic acid, acidic electrolyzed water and chlorine on cantaloupes and fresh-cut apples. J Food Saf 26:335–347

Watada AE, Qi L (1999) Quality of fresh-cut produce. Postharvest Biol Technol 15:201–205
Watada AE, Abe K, Yamauchi N (1990) Physiological activities of partially processed fruits and vegetables. Food Technol 44(116–118):120–122
Watada AE, Ko NP, Minott DA (1996) Factors affecting quality of fresh-cut horticultural products. Postharvest Biol Technol 9(2):115–125
Welti-Chanes J, Lopez-Malo A, Palou E, Bermudez D, Guerrero-Beltran JA, Barbosa-Cánovas GV et al (2005) Fundamentals and applications of high pressure processing of foods. In: Barbosa-Cánovas GV (ed) Novel food processing technologies. CRC PRESS/Marcel Dekker, New York
Wolbang CM, Fitos JL, Treeby MT (2008) The effect of high pressure processing on nutritional value and quality attributes of *Cucumis melo* L. Innov Food Sci Emerg Technol 9(2):196–200
Xu C (2005) Decontamination of *Escherichia coli* O157:H7 and *Salmonella* in lettuce, chicken, and apples by chlorine dioxide and ultrasound. Auburn University, Auburn
Yaun BR, Sumner SS, Eifert JD, Marcy JE (2004) Inhibition of pathogens on fresh produce by ultraviolet energy. Int J Food Microbiol 90(1):1–8
Yeom HW, Zhang QH (2001) Enzymatic inactivation by pulsed electric fields: a review. In: Barbosa-Cánovas GV, Zhang QH (eds) Pulsed electric fields in food processing. Technomic Publishing Company, Inc, Lancester
Yeom HW, Zhang QH, Dunne CP (1999) Inactivation of papain by pulsed electric fields in a continuous system. Food Chem 67(1):53–59
Young Lee N, Jo C, Hwa Shin D, Geun Kim W, Woo Byun M (2006) Effect of gamma-irradiation on pathogens inoculated into ready-to-use vegetables. Food Microbiol 23(7):649–656
Yuk H-G, Yoo M-Y, Yoon J-W, Moon K-D, Marshall DL, Oh D-H (2006) Effect of combined ozone and organic acid treatment for control of *Escherichia coli* O157:H7 and *Listeria monocytogenes* on lettuce. J Food Sci 71(3):M83–M87
Zaborowski D (2003) Fresh-cut produce industry Ontario marketing opportunities. The Grower 53(10):7
Zambre SS, Venkatesh KV, Shah NG (2010) Tomato redness for assessing ozone treatment to extend the shelf life. J Food Eng 96(3):463–468
Zhang L, Lu Z, Yu Z, Gao X (2005) Preservation of fresh-cut celery by treatment of ozonated water. Food Control 16(3):279–283
Zhang L, Lu Z, Wang H (2006) Effect of gamma irradiation on microbial growth and sensory quality of fresh-cut lettuce. Int J Food Microbiol 106(3):348–351
Zhou B, Feng H, Luo YG (2009) Ultrasound enhanced sanitizer efficacy in reduction of *Escherichia coli* O157:H7 population on spinach leaves. J Food Sci 74(6):M308–M313
Zhuang H, Lewis LS, Michelangeli C, Hildebrand DF, Payne FA, Bastin S, Barth MM, Int Inst R (1996) Ozone water treatments for preserving quality of packaged, fresh-cut broccoli under refrigeration. In: New developments in refrigeration for food safety and quality, pp 267–276
Zong W, An GJ (2008) Effect of ultra high pressure on softening of fresh cut jujube. ISHS Acta Horticulturae 840, I International Jujube symposium
Zong W, Zhang HH (2009) Effect of ultra high pressure on quality characteristic of fresh cut kiwi fruit. Acad Serv Group Ltd, London, pp 7–10
Zook CD, Parish ME, Braddock RJ, Balaban MO (1999) High pressure inactivation kinetics of *Saccharomyces cerevisiae* ascospores in orange and apple juices. J Food Sci 64(3):533–535

Chapter 19
Enhancing Extraction from Solid Foods and Biosuspensions by Electrical Pulsed Energy (Pulsed Electric Field, Ohmic Heating, and High-Voltage Electrical Discharge)

Eugène Vorobiev and Nikolai Lebovka

19.1 Introduction

Recently, the use of electrical pulsed energy (EPE) in the treatment of food and agricultural products has become very popular (Vorobiev and Lebovka 2008, 2010; Barbosa-Cánovas and Cano 2004). Disruption of nanometer-sized biological membranes by EPE application usually results in a critical acceleration of mass transfer. Acceleration of mass transfer in foods by traditional methods (e.g., ultrasonic, pressure, thermal, chemical) is commonly accompanied by many undesirable factors, such as degradation, necessity of using unsafe organic solvents, enzymes, detergents, temperature elevation, or unreasonably high energy consumption. The moderate energy requirements and potential for inactivation of food pathogenic and spoilage microorganisms demonstrate the supplementary attractiveness of EPE-assisted methods (Barbosa-Cánovas and Cano 2004). The modern practical realizations of EPE-assisted processing include pulsed electric field, pulsed ohmic heating, and high-voltage electrical discharge techniques.

Laboratory experiments have already demonstrated many promising examples of the EPE-assisted extraction of juices, sugars, colorants, polyphenolic substances, and oils. EPE applications to various vegetable and fruit tissues, e.g., potato, sugar beet and sugar cane, red beet, carrot, apple, grape, chicory, oil, and fat-rich plants, have been analyzed. EPE treatment also offers unique possibilities for the extraction

E. Vorobiev (✉)
Département de Génie Chimique, Université de Technologie de Compiègne,
Centre de Recherche de Royallieu, Compiègne, France
e-mail: eugene.vorobiev@utc.fr

N. Lebovka
F. D. Ovcharenko Institute of Biocolloidal Chemistry, National Academy
of Sciences of Ukraine, Kyiv, Ukraine
e-mail: lebovka@gmail.com

S. Yanniotis et al. (eds.), *Advances in Food Process Engineering Research and Applications*, Food Engineering Series, DOI 10.1007/978-1-4614-7906-2_19,

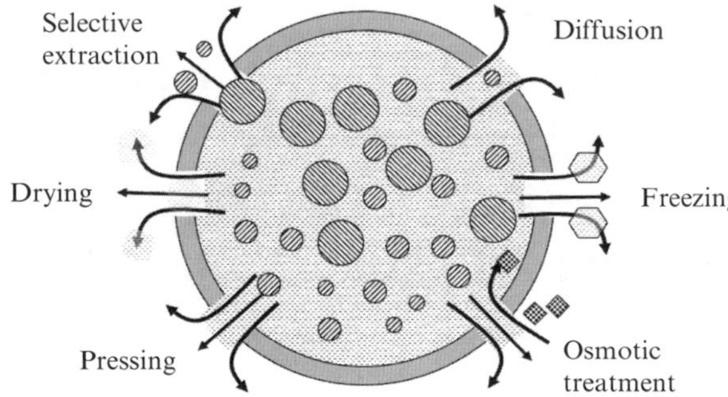

Fig. 19.1 Application of PEF and HVED treatments to plants or microorganisms for assistance with different processes

of valuable proteins, cytoplasmic enzymes, and polysaccharides from cells in suspensions (e.g., *Saccharomyces cerevisiae*, *Escherichia coli*) (Pakhomov et al. 2010).

EPE technology is not simple in application and has a long history. The most important developments were the discoveries of bioelectricity by Luigi Galvani in 1791 and electroporation in the 1960s and 1970s. Many efforts have been aimed at industrial implementations of ohmic heating (OH) using alternative (AC) or direct (DC) currents, pulsed electric fields (PEF), and high-voltage electrical discharge (HVED). They started in the early twentieth century with electric current application for the killing of microbes and canning and, later on, to treat prunes, apples, grapes, and sugar beets. Very important steps were made in the 1960s and 1970s, when the first applications of PEF and industrial AC setups were implemented and the membrane electroporation concept was theoretically worked out. Starting in the early 1990s, many new practical PEF and HVED techniques were tested, and their usability for microbial killing, food preservation, and acceleration of different processes was demonstrated (Fig. 19.1.). In the same period, new types of higher-voltage PEF generator, new treatment chamber designs, and new pilot schemes were developed (e.g., Vorobiev and Lebovka 2008; Barbosa-Cánovas and Cano 2004).

This contribution reviews the modern state of the art, the existing fundamental knowledge of the mechanism of EPE-induced effects in biomaterials, the impact of EPE on functional food ingredients, recent experimental efforts in the field, practical EPE applications and their examples in different food materials, and prospects of industrial applications of EPE-assisted extraction techniques.

19.2 EPE-Induced Effects in Foods

19.2.1 Electroporation

The impact of EPE on biomaterials may reflect loss of barrier functions by membranes. This phenomenon is called electroporation or electropermeabilization (Pakhomov et al. 2010). The degree of electroporation depends on the transmembrane potential, u_{m}. Electroporation requires some threshold value of the potential difference across a membrane u_m, typically 0.5–1.5 V. The value of u_m is directly proportional to the cell radius, R. Thus, larger cells get damaged before smaller ones.

In practice, the electropermeabilization degree also depends on the properties of the material [e.g., cell size, electrical conductivity of cells before (σ_i) and after treatment (σ_{d},)], and details of the pulse protocol. Significant damage to plant tissues can be observed at $E = 500$–1,000 V/cm and treatment time within 10^{-4} to 10^{-1} s (Vorobiev and Lebovka 2008); for microbial killing larger fields ($E = 20$–50 kV/cm) and smaller treatment times (10^{-5}–10^{-4} s) are required (Barbosa-Cánovas and Cano 2004). In plant tissues, the observed threshold value E is noticeably higher for materials with a small electrical conductivity contrast, $k = \sigma_d/\sigma_i$ (Shynkaryk 2007). High-intensity PEF treatment causes irreversible damage to the cell membrane; however, electroporation may be reversible and membrane integrity may be recovered within several seconds after pulse termination at low values of E. Long-term changes in conductivity following PEF treatment application can also be related to osmotic flow and moisture redistribution inside the sample.

19.2.2 Supplementary Effects

Supplementary effects can be observed in the application of HVED treatment. HVED treatment (at $E = 40$–60 kV/cm, 2–5 μs pulses) can produce a breakdown of liquid and fragmentation of particles. The mechanisms of HVED treatment include electrical breakdown of water, as well as propagation of streamers, shock waves, cavitation of bubbles, and many other secondary phenomena. The electroporation mechanism of HVED treatment is important for food particles; however, the cell structure can also be damaged mechanically owing to the shock waves.

19.2.3 Techniques for Characterization of EPE-Induced Effects

Experimental determination of the degree of damage Z, defined as the volume fraction of the damaged cells, is not an easy task. Various techniques were recently developed for estimation of EPE impact on plant tissues and colloidal biosuspensions (Vorobiev and Lebovka 2008; Pakhomov et al. 2010).

Microscopic observation is the most direct way to determine the degree of damage. It has been demonstrated that PEF does not noticeably affect the structure of cell walls. This important conclusion was supported by scanning electron microscope (SEM) images, where a similarity in the structure of the cell walls,

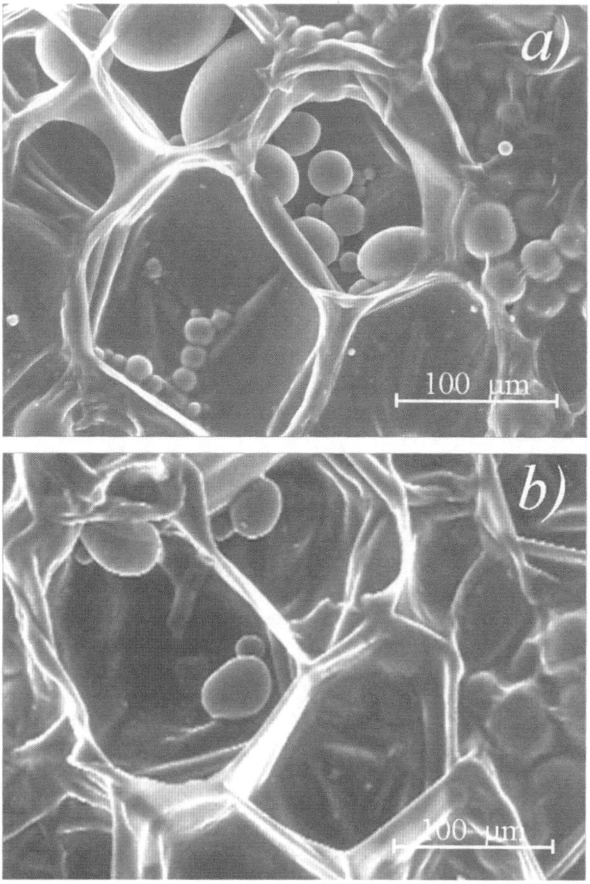

Fig. 19.2 SEM micrographs of untreated (**a**) and PEF pretreated (**b**) potato tissues (Reproduced from Ben Ammar et al. (2010) with permission from Springer – Verlag)

area, and morphology of starch granules in untreated and PEF-treated potato tissues was observed (Ben Ammar et al. 2010).

Figure 19.2 compares the SEM micrographs of untreated (Fig. 19.2a) and PEF pretreated (Fig. 19.2b) potato, which appeared to be rather similar. The cells were polyhedrally shaped; the starch was in the form of oval granules approximately 10–100 μm in size, their surface was smooth and showed no damage or pores, and no visible changes were observed in the structure of the tissue subjected to PEF pretreatment. Therefore, the results presented here provide visual evidence for the absence of PEF-induced cell wall disruption that might have reduced the cell area or changed the shape and size of the starch granules.

However, application of microscopic methods is not simple due to difficulties related to preparing the samples and the sensibility of the method to pH and the conductivity of the solution used in mounting the epidermis.

Different electrical-conductivity-based methods have also been proposed (e.g., Vorobiev and Lebovka 2008). The electrical conductivity of a material increases with the growth of its degree of damage; thus, electrical conductivity measurements can be easily applied for continuous monitoring of the damage degree. However, electrical conductivity is sensible to, for example, the spatial redistribution of air and moisture content and membrane resealing. Diffusion- and texture-based techniques may reflect electrically induced changes in a complex form, while definitive relations between these parameters and cell damage degree have not yet been established. Moreover, these methods are indirect and invasive (Vorobiev and Lebovka 2008).

Successful application of the acoustic technique for the characterization of PEF-treated tissues was recently reported (Grimi 2009). Note that the advantages of the acoustic technique for the characterization of PEF-induced effects may be important when fruits and vegetables are processed as whole unpeeled samples. It was shown that the acoustically estimated disintegration index Z is better for damage degree characterization than the conductivity disintegration index Z (Grimi 2009).

19.2.4 Damage Kinetics

The characteristic damage time τ is defined as the time necessary to inflict half-damage on a material (i.e., $Z = 0.5$) (Bazhal 2001); this measure is useful for a crude characterization of the damage kinetics. The τ versus E dependencies were measured for different vegetables and fruits (Grimi 2009; Bazhal 2001; Ben Ammar et al. 2011).

The examples of Z versus the time of PEF application t_{PEF} (Fig. 19.3a) and characteristic damage time τ versus the electric field strength E (Fig. 19.3b) for different vegetable and fruit samples are presented in Fig. 19.3 (Grimi 2009). For example, it was shown that onion and orange had stronger resistance to PEF treatment and required a longer treatment time or higher electric field strength as compared with tomatoes and apples. The observed effects reflected the specific structure of the cellular materials, the differences in the size of their cells, and the relative electrical conductivities of the products and the aqueous media.

19.2.5 Synergy of Simultaneous Electrical and Thermal Treatments

An obvious synergy of simultaneous electrical and thermal treatments of food products was observed in PEF treatment and OH experiments (Praporscic 2005; Shynkaryk 2007). This synergy was most evident for electroprocessing at a moderate electric field strength ($E < 100$ V/cm) under ambient conditions.

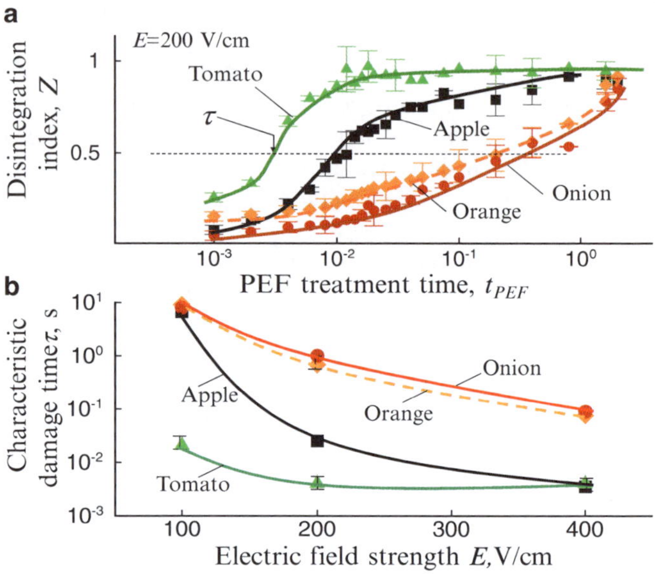

Fig. 19.3 Disintegration index Z versus time of PEF application t_{PEF} (**a**) and characteristic damage time τ versus electric field strength E (**b**) for different vegetable and fruit samples (These data were estimated from an acoustic technique for the characterization of PEF-treated tissues (Grimi 2009)

A rather complex kinetics with an intermediate saturation step (when disintegration index Z reaches a plateau, $Z = Z_s$) was often observed for long PEF treatment by a moderate electric field ($E < 300$ Vcm^{-1}) and at a moderate temperature ($T < 50$ °C). A steplike behavior of $Z(t)$ was also observed for inhomogeneous tissues, such as in red beetroots, which possibly reflected the presence of a wide distribution of cell survivability related to the distributions in cell geometries and sizes. The synergy of simultaneous PEF and thermal treatments with increases in temperature T or electric field strength E (or both) was also demonstrated by the presence of a drastic drop in the characteristic damage time by many orders of magnitude (Shynkaryk 2007). Moreover, the electroporation activation energy W of tissues was a decreasing function of the electric field strength E as a result of electrothermal synergy (Loginova et al. 2010).

19.2.6 Power Consumption

Power consumption Q is the most important measure for estimation of the industrial attractiveness of any electrotechnology, and the values of Q were already reported for different EPE experiments. The experimentally estimated power consumptions Q for PEF-treated tissues were found to be rather low and typically lying within 1–15 kJ/kg. For example, they were 6.4–16.2 kJ/kg ($E = 0.35$–3.0 kV/cm) for potato, 0.4–6.7 kJ/kg ($E = 2$–10 kV/cm) for grape skins, 2.5 kJ/kg (7 kV/cm) for red beetroot, 3.9 kJ/kg (7 kV/cm) for sugar beet, and 10 kJ/kg (400–600 V/cm) for chicory root [these data were collected from different sources; for references see, e.g., Vorobiev and Lebovka (2008) and Toepfl (2006)].

Thus, from the standpoint of power consumption, the PEF method is practically ideal for the production of damaged plant tissue compared with other methods of plant tissue damage like mechanical methods (20–40 kJ/kg), enzymatic methods (60–100 kJ/kg), and heating or freezing/thawing ($>$ 100 kJ/kg). However, bacterial inactivation and food preservation require high electric field strength values ($E = 15$–40 kV/cm), and this naturally results in a noticeably higher specific power consumption, of the order of 40–1,000 kJ/kg. Such power consumption is typical also for HVED treatment. OH treatment at a moderate electric field ($\approx$ 20–80 V/cm) requires a high power consumption that is comparable with heating or freezing/thawing (20–40 kJ/kg).

19.3 EPE-Assisted Extraction Applications

19.3.1 PEF-Assisted Extraction

19.3.1.1 Vegetable and Fruit Tissues

PEF treatment has also been applied to various vegetable and fruit tissues. In raw food plants, the valuable compounds are initially enclosed in cells, which should be damaged for facilitation of intracellular matter recovery.

The most investigated materials are potato and apple. They were used as model systems in many electrically assisted experiments for testing electroporation effects in plant tissue (Grimi 2009; Bazhal 2001; Ben Ammar et al. 2011; Praporscic 2005; Shynkaryk 2007; Praporscic 2005; Toepfl 2006; Chalermchat et al. 2010). Potato was used for studies of reversible electroporation, transient viscoelastic behavior, various stress-induced effects, metabolic responses, and electrostimulated effects (e.g., Vorobiev and Lebovka 2008; Grimi 2009; Chalermchat et al. 2010). The effects of PEF treatment on the textural and compressive properties of potato have been studied in detail (Grimi 2009). It has also been shown that application of only PEF treatment is not sufficiently effective for the complete elimination of textural strength; however, mild thermal pretreatment at 45–55 °C increased the PEF efficiency (Praporscic 2005). Potato was also used to investigate the temperature and PEF protocol effects on characteristic damage time, dehydration, freezing, and drying (Ben Ammar et al. 2010; Bazhal 2001; Praporscic 2005; Shynkaryk 2007). Attempts have been made to use an extraction-oriented practical application of PEF. For example, applications of PEF to facilitate starch extraction from potato and enhance the extractability of an anthocyanin-rich pigment were reported (Toepfl 2006).

Models of dielectric breakage were developed and tested using experimental data obtained for PEF-treated apple tissue (Bazhal 2001). The effects of apple tissue anisotropy and orientation with respect to the applied electric field on electropermeabilization were reported (Chalermchat et al. 2010). It was shown that elongated cells (taken from the inner region of the apple parenchyma) responded to the electric field in a different manner, while no field orientation dependence was observed for round cells (taken from the outer region of the parenchyma). The textural relaxation data support higher apple damage efficiency at longer pulse durations. The effect of the synergy of PEF and thermal treatments on the textural properties of apple tissue and apple juice expression was demonstrated (Praporscic 2005). It was shown that mild thermal treatment makes it possible to increase the damage efficiency of PEF treatment, and apple tissue preheated at 50 °C and treated by PEF at $E \approx 500$ V/cm exhibits a noticeable enhancement of juice extraction by pressing.

Different aspects of PEF-assisted pressing and aqueous extraction from the sugar beets were recently studied (Bouzrara 2001; El-Belghiti 2005; Loginova et al. 2011a). The efficiency of so-called cold pressing of PEF-treated sugar beet cossettes was demonstrated (Bouzrara 2001). Diffusion experiments with PEF-treated cossettes prepared from sugar beet by industrial knives were recently conducted with the electric field strength E varied between 100 and 600 V/cm and a total time of PEF treatment of 50 ms (Loginova et al. 2011a). The sugar beet pulp could be well exhausted by a cold or mild thermal extraction of cossettes treated by PEF, and the pulp obtained by cold extraction of PEF-treated cossettes had a noticeably higher (more than 30 %) dryness than the pulp obtained by a conventional hot water extraction technique. The estimated energy surplus for cold extraction with temperature reduction from 70 to 30 °C (i.e., by $\Delta T = 40$ °C) was ≈ 46.7 kWh/t and was noticeably higher than the power consumption required for PEF treatment (≈ 5.4 kWh/t). Power consumption can even be reduced by the further optimization of the PEF parameters and minimization of the liquid-to-solid ratio during PEF treatment (Loginova et al. 2011a).

PEF treatment has demonstrated the possibility of accelerated extraction of thermally sensible colorants, for example, betalains of the red beet (*Beta vulgaris* L.) (Loginova et al. 2011b). This is important, because 1 h of thermal extraction at 80 °C resulted in a nearly complete damage to betalains. PEF treatment seems to be very promising and opens up encouraging possibilities for very efficient "cold" diffusion with a high level of colorant extraction and low level of colorant degradation (5–10 %) (Loginova et al. 2011b).

Many studies have been devoted to the investigation of the effect of PEF treatment on juice expression from carrots, apples, and grapes (Grimi 2009; Bazhal 2001; Praporscic 2005; El-Belghiti 2005; Turk 2010; Grimi et al. 2011). PEF treatment noticeably improved the juice yield and soluble matter content in the juice. It was demonstrated for carrot and apple that juice characteristics and yield are directly related to the size of slices (Praporscic 2005). For example, the yield of juice from 2 × 3.5 × 55 mm Golden Delicious apple slices increased after PEF treatment ($E = 400$ V/cm, $t_{PEF} = 0.1$ s) by 28 %, while the juice yield increase by only 5 % when the apple slices were finer (1 × 1.9 × 55 mm) (Grimi 2009; Grimi et al. 2011). Juice yield *Y* increased significantly after PEF treatment of large apple mash (Y = 71.4 %) at $E = 450$ V/cm as compared with the control sample with small apple mash (45.6 %) (Turk 2010). PEF treatment did not alter the acid-sweet balance and pH; however, a decrease in the yield of native polyphenols was observed following PEF treatment (control: 9.6 %; treated: 5.9 % for small mash). The PEF pretreatment was accompanied by a noticeable improvement in apple juice clarity, an increase in the total soluble matter and polyphenol content, and intensification of the antioxidant capacities of the juice (Grimi et al. 2011).

A principal possibility of PEF-assisted selective extraction of water-soluble components (soluble sugars) and production of a sugar-free concentrate rich in vitamins and carotenoids was demonstrated for carrot (Grimi 2009). PEF-induced electroporation of wine grapes was shown to be a promising alternative process leading to prudent extraction of colorants and valuable constituents (Praporscic 2005; Puértolas et al. 2010). The PEF treatment ($E = 750$ V/cm, $t_{PEF} = 0.3$ s) resulted in an increase of the final juice yield up to 73–78 % as compared to 49–54 % for the untreated grapes (muscadelle, sauvignon, and semillon) (Praporscic 2005). PEF treatment enhanced the compression kinetics and extraction of polyphenols from Chardonnay grapes (Grimi 2009). PEF treatment application holds promise for a reduction in maceration time during the vinification and production of wines with better characteristics. Some aspects and potential applications of PEF technology in the winemaking industry were recently reviewed in Puértolas et al. (2010). PEF treatment was also applied to other tissues, such as red bell pepper and paprika, fennel, chicory, alfalfa, and red cabbage (e.g., Vorobiev and Lebovka 2008; Loginova et al. 2010). These studies demonstrated an improvement in juice yield and extraction of dry matter, carotenoids, vitamins, sucrose, proteins, and inulin.

19.3.1.2 Biosuspensions

Disruption of some microorganisms is a very important step in the industrial extraction of valuable proteins, cytoplasmic enzymes, and polysaccharides, which are present in cells. It is possible to produce the valuable proteins in cultures of recombinant host cells (*S. cerevisiae*, *E. coli*) containing foreign genes, and important medical materials can be synthesized in such cells. However, extraction of intracellular proteins is not an easy task and requires application of special techniques for cell disruption and further purification of proteins. Nowadays, there exist many examples of PEF application for killing and disrupting microorganisms (Barbosa-Cánovas and Cano 2004). Electroporation-assisted extraction from cells is expected to be highly selective with respect to low and high molecular intracellular components and promising for the recovery of homogeneous and heterogeneous intracellular proteins having wide biotechnological applications (Pakhomov et al. 2010). However, the efficiency of PEF-assisted extraction in its applications to biosuspensions may depend on many factors.

It was recently demonstrated that leakage of cytoplasmic ions during PEF application to suspensions of *S. cerevisiae* and *E. coli* influences the ionic concentration of the medium and its electrical conductivity (El Zakhem 2007).

The PEF treatment of *S. cerevisiae* at 5 kV/cm made it possible to attain a high level of the conductivity disintegration index, $Z \approx 1$, with more intensive release of peptides and proteins than nucleic bases (El Zakhem 2007). However, in PEF-treated suspensions of wine yeast cells (*S. cerevisiae bayanus*, strain DV10), a relatively small release of proteins was observed even at a high level of $Z \approx 0.8$ (Shynkaryk et al. 2009). Moreover, it was demonstrated that a high level of membrane disintegration ($Z > 0.8$) in these yeasts required rather strong PEF treatment (at $E = 10$ kV/cm using 2.10^5 pulses of 100 μs). Thus, the efficiency of the PEF-assisted extraction was dependent on the yeast strain. This may reflect the presence of hard cell walls in addition to cell membranes that can restrict extraction of intracellular compounds. More effective extraction of the high molecular weight content (e.g., proteins) from electrically resistant strains requires more powerful mechanical disintegration of cell walls, which is provided by HVED and high-pressure homogenization (HPH) techniques. In principle, HPH permitted better extraction than HVED (Shynkaryk et al. 2009; Loginov et al. 2009; Liu et al. 2010).

However, a synergistic enhancement of protein release from yeasts can be attained using combined disruption techniques. It was shown that a combined HVED + HPH technique made it possible to reach a high level of protein extraction from wine yeast cells (*S. cerevisiae bayanus*, strain DV10) at smaller pressure or with a smaller number of passes through the homogenizer (Shynkaryk et al. 2009). At high cell concentrations, the PEF disruption efficiency was affected by the formation of large aggregates (El Zakhem 2007). It was demonstrated in *S. cerevisiae* suspensions that intact cells have a negative charge as compared with the positive charge of damaged cells. Thus, PEF treatment can provoke an electrostatic attraction between intact and damaged cells and the formation of large aggregates.

Experimental studies of PEF treatment of flowing concentrated aqueous suspensions of *E. coli* (1 wt%) at E = 5–7.5 kV/cm and medium temperatures within a range of 30–50 °C were carried out (El Zakhem 2007). A noticeable disruption of cells was observed at a PEF treatment time t_{PEF} of 0–0.2 s and a thermal treatment time t_T of 0–7,000 s. It was shown that disruption of *E. coli* was accompanied by a cell size decrease and release of intracellular components. An absorbance analysis of supernatant solutions evidenced the leakage of nucleic acids. Thermal treatment alone at $T = 30$–50 °C was ineffective for disruption of *E. coli* cells and required a long treatment time ($t_T >> 1$ h). There was an evident synergism between the simultaneously applied electric and thermal treatments. After 1 h of PEF and thermal treatment the electrical conductivity disintegration index Z reached ≈ 0.22; ≈ 0.83, and ≈ 0.99 at temperatures of $T = 30$, 40, and 50 °C, respectively. It was also shown that surfactant additives (Triton X-100) additionally improved disruption of cells in *E. coli* suspensions (El Zakhem 2007). The disruption efficiency of cells in *E. coli* suspensions was also noticeably improved by the addition of organic (citric, malic, and lactic) acids in small concentrations ($\leq$ 0.5 g/L) and an 8 log cycle reduction was reached using 0.375 g/L of lactic acid. It was assumed that lactic acid might act as a very effective potentiator of PEF effects in membranes of *E. coli* cells.

Recently, the high selectivity of extraction of ionic substances, proteins, and nucleic acids from the wine *S. cerevisiae* (bayanus) yeasts by PEF and HVED techniques was demonstrated (Liu et al. 2013). For example, electrical pulsed treatments at $E = 40$ kV/cm and $N = 500$ allowed the extraction of $\approx$ 80 % and $\approx$ 70 % of ionic substances, $\approx$ 4 % and $\approx$ 1 % of proteins, and $\approx$ 30 % and $\approx$ 16 % of nucleic acids for HVED and PEF modes, respectively.

19.3.2 HVED-Assisted Extraction

The HVED method is based on the treatment of liquid food placed in a chamber between a needle electrode and a plated grounded electrode using short pulses (typically 40–60 kV/cm, 2–5 μs). HVED treatment was used to accelerate the extraction of solutes from soybeans, potato, tea leaves, peat, fennel, and other food materials (Vorobiev and Lebovka 2008).

However, the most promise is held by the possibility of HVED application for enhancing the extraction of oil from oilseeds (Grémy-Gros et al. 2008). In these experiments, linseeds were crushed and pressed by hydraulic press, and then the resulting press cake was dispersed in demineralized water at ambient temperatures. Then the mixture was treated by HVED and centrifuged. The process was optimized by varying the number of pulses, pH, water/press-cake ratio, and temperature.

HVED treatment permitted the extraction of 74 % of oil after 1,640 pulses and the enhancement of mucilage extraction from the entire linseed (Grémy-Gros et al. 2008). However, oil-protein-water emulsions, obtained by HVED treatment, require posterior separation. The membrane-assisted technique of separation was recently proposed for emulsions created by HVED (Grémy-Gros et al. 2008).

Recent studies have shown that HVED treatment is promising also for the extraction of polyphenols from grape pomace (Boussetta et al. 2009, 2011). Grape pomace (composed of stems, seeds, and skins) has a high content of polyphenols. These polyphenols have attracted considerable interest because they exhibit antibacterial, antiviral, and antioxidant properties and can prevent cardiovascular diseases. HVED was applied to samples of pomace mixed with distilled water (the liquid-to-pomace ratio was 3) at temperatures between 20 and 60 °C using 80 successive discharges (40 kV, pulse repetition rate 0.5Hz). It was shown that both temperature and HVED treatment improved the extraction of polyphenols. It was also demonstrated that HVED treatment allowed a noticeable increase in the final yields of solutes, a decrease in the time of diffusion, and an enhancement of the yield of polyphenols not just from fresh grape pomace but also from sulfured and frozen pomaces.

19.4 Conclusion

EPE may be useful in various extraction applications related to the treatment of food and agricultural products. For example, pulsed electric fields (PEFs) made it possible to preserve the integrity of cell walls, color, flavor, vitamin C, and important nutrients in food.

Recent laboratory experiments have already provided many promising examples of the PEF-assisted extraction of juices, sugars, colors, polyphenolic substances, and oils from solid foods (e.g., sugar beets, apples, grapes). PEF treatment also offer unique possibilities for the extraction of valuable proteins, cytoplasmic enzymes, and polysaccharides from cells in suspensions and opens new prospects for the production of valuable proteins in cultures of recombinant host cells (*S. cerevisiae*, *E. coli*).

Accounting for the observed synergy of simultaneous electrical and thermal treatments, pulsed OH can also be applied to accelerate extraction from foods using both moderate electric fields (below 500 V/cm) and moderate temperatures (below 60 °C). In some applications, HVED treatment may be useful. HVED treatment makes it possible to overcome some restrictions related to the PEF treatment of oil-saturated products (e.g., oilseeds), and it is very promising for the production of polyphenols from grape pomace.

The first important steps in the practical implementation of PEF-assisted technologies at the pilot plant scale have already been taken. Future technologies will make it possible to overcome the difficulties associated with mechanical, thermal, or chemical pretreatments presently used in food processing and provide new opportunities for innovative technologies with the potential for the production of foods with excellent sensory and nutritional qualities.

Acknowledgments The authors thank Dr. N. S. Pivovarova for her help with the manuscript preparation.

References

Barbosa-Cánovas GV, Cano MP (2004) Novel food processing technologies. Marcel Dekker, New York

Bazhal M (2001) Etude du mécanisme d'électropermeabilisation des tissus végétaux. Application à l'extraction du jus des pommes (Study of electropermeabilisation mechanisms of vegetable tissue. Application for extraction of potato juice). Ph.D. dissertation, University of the Technology of Compiegne, Compiègne

Ben Ammar J, Lanoisellé J-L, Lebovka NI, Van Hecke E, Vorobiev E (2010) Effect of a pulsed electric field and osmotic treatment on freezing of potato tissue. Food Biophys 5(3):247–254

Ben Ammar J, Lanoisellé J-L, Lebovka NI, Van Hecke E, Vorobiev E (2011) Impact of a pulsed electric field on damage of plant tissues: effects of cell size and tissue electrical conductivity. J Food Sci 76:E90–E97

Boussetta N, Lanoisellé J-L, Bedel-Cloutour C, Vorobiev E (2009) Extraction of polyphenols from grape pomace by high voltage electrical discharges: effect of sulphur dioxide, freezing process and temperature. J Food Eng 95(1):192–198

Boussetta N, Reess T, Vorobiev EI, Lanoisellé J-L (2011) Pulsed electrical discharges: principles and application to extraction of biocompounds. In: Lebovka N, Vorobiev E, Shemat F (eds) Enhancing extraction processes in the food industry. CRC Press/Taylor & Francis LLC, Boca Raton, pp 145–172

Bouzrara H (2001) Amélioration du pressage de produits végétaux par Champ Electrique Pulsé. Cas de la betterave à sucre. Ph.D. dissertation, University of the Technology of Compiegne, Compiègne

Chalermchat Y, Malangone L, Dejmek P (2010) Electropermeabilization of apple tissue: effect of cell size, cell size distribution and cell orientation. Biosyst Eng 105(3):357–366

El Zakhem H (2007) Inactivation de suspensions microbiennes de *Saccharomyces Cerevisiae* et d' *Escherichia coli* par champs électriques pulsés modérés. Ph.D. dissertation, University of the Technology of Compiegne, Compiègne

El-Belghiti K (2005) Effets d'un champ électrique pulsé sur le transfert de matière et sur les caractéris-tiques végétales. Ph.D. dissertation, University of the Technology of Compiegne, Compiègne

Grémy-Gros C, Lanoisellé J-L, Vorobiev E (2008) Application of high-voltage electrical discharges for the aqueous extraction from oilseeds and other plants. In: Vorobiev E, Lebovka N (eds) Electrotechnologies for extraction from food plants and biomaterials. Springer, New York, pp 217–236

Grimi N (2009) Vers l'intensification du pressage industriel des agroressources par champs électriques pulsés: étude multi-échelles (Towards the intensification of industrial pressing of agricultural resources by pulsed electric fiels: multiscale study). Ph.D. dissertation, University of the Technology of Compiegne, Compiegne

Grimi N, Boussetta N, Mamouni F, Lebovka N, Vorobiev E, Vaxelaire J (2011) Impact of apple processing modes on extracted juice quality: pressing assisted by pulsed electric fields. J Food Eng 103(1):52–61

Liu D, Savoire R, Vorobiev E, Lanoisellé J-L (2010) Effect of disruption methods on the dead-end microfiltration behavior of yeast suspension. Sep Sci Technol 45(8):1042–1050

Liu D, Lebovka N, Vorobiev E (2013) Impact of electrical pulsed treatment on selective extraction of intracellular compounds from *Saccharomyces cerevisiae* yeasts. Food Bioprocess Technol 6:576–584 (springerlink:10.1007/s11947-011-0703-7)

Loginov M, Lebovka N, Larue O, Shynkaryk M, Nonus M, Lanoisellé J-L, Vorobiev E (2009) Effect of high voltage electrical discharges on filtration properties of *Saccharomyces cerevisiae* yeast suspensions. J Membr Sci 346(2):288–295

Loginova KV, Shynkaryk MV, Lebovka NI, Vorobiev E (2010) Acceleration of soluble matter extraction from chicory with pulsed electric fields. J Food Eng 96:374–379

Loginova KV, Vorobiev E, Bals O, Lebovka NI (2011a) Pilot study of countercurrent cold and mild heat extraction of sugar from sugar beets, assisted by pulsed electric fields. J Food Eng 102(4):340–347

Loginova KV, Lebovka NI, Vorobiev E (2011b) Pulsed electric field assisted aqueous extraction of colorants from red beet. J Food Eng 106:127–133

Pakhomov AG, Miklavcic D, Markov MS (eds) (2010) Advanced electroporation techniques in biology and medicine. CRC Press, Boca Raton

Praporscic I (2005) Influence du traitement combiné par champ électrique pulsé et chauffage modéré sur les propriétés physiques et sur le comportement au pressage de produits végétaux. Ph.D. dissertation, University of the Technology of Compiegne, Compiègne

Puértolas E, López N, Condón S, Álvarez I, Raso J (2010) Potential applications of PEF to improve red wine quality. Trends Food Sci Technol 21(5):247–255

Shynkaryk MV (2007) Influence de la perméabilisation membranaire par champ électrique sur la performance de séchage des végétaux. Ph.D. dissertation, University of the Technology of Compiegne, Compiègne

Shynkaryk MV, Lebovka NI, Lanoisellé J-L, Nonus M, Bedel-Clotour C, Vorobiev EI (2009) Electrically-assisted extraction of bio-products using high pressure disruption of yeast cells (*Saccharomyces cerevisiae*). J Food Eng 92:189–195

Toepfl S (2006) Pulsed Electric Fields (PEF) for permeabilization of cell membranes in food- and bioprocessing – applications, process and equipment design and cost analysis. Ph.D. dissertation, Institut für Lebensmitteltechnologie und Lebensmittelchemie, Berlin

Turk M (2010) Vers une amélioration du procédé industriel d'extraction des fractions solubles de pommes à l'aide de technologies électriques. Ph.D. dissertation, Université de Technologie de Compiègne, France, Compiegne

Vorobiev EI, Lebovka NI (eds) (2008) Electrotechnologies for extraction from food plants and biomaterials. Springer, New York

Vorobiev E, Lebovka N (2010) Enhanced extraction from solid foods and biosuspensions by pulsed electrical energy. Food Eng Rev 2(2):95–108

Chapter 20
Food Structure Engineering for Nutrition, Health and Wellness

Stefan F.M. Kaufmann and Stefan Palzer

20.1 Introduction

Global consumers are increasingly demanding healthier and more nutritious food products that address specific health needs and are sustainably produced according to high ethical standards. As for many consumers, healthy eating is linked to natural products, and a growing demand for these products has been observed, particularly in Europe and North America. To improve the quality of lives of consumers in developing countries, highly nutritious products are needed that are relevant and that can be offered at an affordable price. It goes without saying that pleasure should – of course – never be compromised. Food, unlike supplements and pharmaceutical products, is consumed not only for nourishment or specific health purposes but also for the enjoyment of its taste, flavor, texture, and color.

Depending on the product category and the targeted consumer health concerns, certain nutritional, health, and wellness aspects can be linked to different product attributes like low energy density, high content of whole grains/fiber and proteins, or the addition of micronutrients and bioactive compounds. However, the corresponding products often show sensorial defects or a limited physical or chemical stability compared to the "standard" products. Some of these issues can be resolved by an appropriate design of the matrix structure of the targeted food product. However, it must always be ensured that the composition and processing of foods are carefully balanced to ensure the optimal nutritional properties (Turgeon and Rioux 2011).

Parallel to the structural design, the future oral breakdown of foods must be taken into account. Texture results from a dynamic process in which textural

S.F.M. Kaufmann (✉)
Nestlé Research Center, Nestec Ltd., Vers-chez-les-Blanc,
P.O. Box 44, 1000 Lausanne 26, Switzerland
e-mail: stefan.kaufmann@rdls.nestle.com

S. Palzer
Nestlé Product Technology Centre for Confectionery, Nestec York Ltd.,
P.O. Box 204, Haxby Road, York YO91 1XY, UK

S. Yanniotis et al. (eds.), *Advances in Food Process Engineering Research and Applications*, Food Engineering Series, DOI 10.1007/978-1-4614-7906-2_20,

attributes are continuously analyzed by the oral sensory system during mastication. It can be shown that for each food a so-called texture pathway can be built based on the dynamics of texture perception during food consumption (Lenfant et al. 2009).

Finally, the digestion behavior of food structures can be modulated, and thus different physiological responses can be triggered.

20.2 Designing Complex Food Structures with Tailored Properties

20.2.1 Energy Density/Salt Reduction

To address severe global public health concerns like obesity (overnutrition), the reduction of the volumetric energy density of a targeted food product can be achieved by either increasing its air and water content or by decreasing its sugar and fat content.

Air or other gases dispersed in the form of small bubbles act as structural elements in many liquid, semiliquid, and solid foods including ready-to-drink beverages, mousses and other chilled or frozen desserts, chocolate, and extruded breakfast cereals. Gelatin gels, for example, may be aerated by ultrasound. This dispersed air provides an additional phase within gel-type foods and potentially accommodates new textural and functional demands (Zúñiga et al. 2011).

An increasing amount of water can be bound in a food matrix through emulsion or gel structuring. Water-in-oil-in-water (W/O/W) double emulsions show much potential in the formulation of fat-reduced products such as mayonnaise or dressings. Food biopolymers like proteins and polysaccharides have been successfully incorporated into the internal and external aqueous phases of this type of double emulsion to improve the stability and yield of model systems (Dickinson 2011).

Such disperse systems need to be effectively stabilized by surface active molecules like biopolymers (e.g., proteins) or by solid particles situated at the interface. Today, relatively little research is directed at answering the question about what happens when both surface active polymers and particles are present together (Murray et al. 2011).

To compensate for sensorial changes, a particular micro- and macrostructural design is required. The structure of emulsions and foams needs to be adapted in such a way that the final rheological properties (mainly responsible for the mouth-feel) and the release characteristics of flavors match the original product.

When reducing the sugar content of products, their perceived sweetness can be replaced by natural noncaloric sweeteners. For instance, steviol glycoside extracts (i.e., from the leaves of *Stevia rebaudiana*) are approximately 350 times sweeter than sugar and have become more and more available as a natural and healthy sugar alternative. The physicochemical impact of sugar in products, however, needs to be

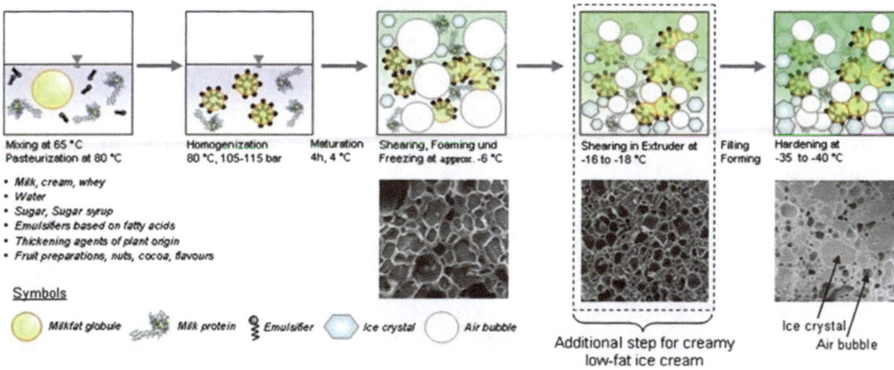

Fig. 20.1 Generation of microstructure for fat-reduced ice cream (Reprinted from Palzer (2009a) with permission from Elsevier)

replaced by a combination of bulking agents to compensate for the loss in matrix structure (or simply volume). This complex task involves a detailed understanding of each bulking agent's contribution to the structural buildup and the related impact on certain sensorial properties.

Low-temperature, frozen low-fat ice cream is an example of an energy-density-reduced food in which the combination of a significant change in the recipe (decreased fat content, adjusted stabilization) and an additional processing step (twin screw extrusion) leads to a texture that is perceived to be creamier compared to an ice cream produced conventionally (Fig. 20.1). The additional high shear forces (i.e., exceeding those in conventional scraped-surface heat exchangers by approximately three orders of magnitude) applied to the complex multiphase system of ice cream result in a reduction of air bubble and ice crystal sizes while increasing the functionality of the destabilized fat droplets (Eisner et al. 2005). The enhanced fat destabilization, which is increasingly dependent on the type of emulsifier used, promotes the fat structuring and thus, ultimately, improves the overall stability and meltdown behavior (Bolliger et al. 2000).

That these structural changes have a significant impact on quality characteristics can be proven by the measurement of the rheological properties of the product. Sensorial studies have demonstrated, for example, a close correlation between loss moduli G'' measured in an oscillatory thermorheometry test and typical quality parameters like scoop ability and creaminess (Wildmoser et al. 2004).

Another potential way to provide fat-reduced textures is through the use of microparticulated proteins or hydrocolloid structures that mimic full-fat food properties.

Additional ideas about engineering approaches to fight obesity are summarized in Norton et al. (2007).

Similar to sugar and fat, elevated salt intake is a major concern for public health authorities, and its reduction has become a major initiative within the packaged food industry in recent years. Salt release and perception is very much influenced by changes in the food matrix. A recent study provided evidence for the important role of structure and texture on salt mobility and perception and the interaction between fat and salt in model dairy products (Sagalowicz and Leser 2009). In addition to these concerns, it must not be forgotten that a reduced salt content can also heavily influence food preservation and, thus, food safety, which is beyond the scope of this review.

20.2.2 Stabilization and Bioavailability of Bioactive Compounds/Encapsulation of Aroma Compounds

Enriching food products with vitamins, minerals, or bioactive plant compounds requires the homogeneous dispersion or dissolution of the active substances in their respective food matrices. Furthermore, their active constituents must be stabilized as they are often susceptible to oxidation or other degradative reactions. Generally, bioactives need to be protected against oxygen and light and be physically separated from potential reaction partners such as iron. Further, uncontrolled release or reduced bioavailability must be prevented.

Several encapsulation systems based on protective carbohydrate structures for the dissolution and stabilization of such substances in solid matrices are available and generally well mastered (Palzer 2009a). A far more challenging task is the targeted delivery of active ingredients in a fluid aqueous phase (Sagalowicz and Leser 2009). A major issue is the different solubility characteristics exhibited by the bioactive molecules in different matrices. Additionally, encapsulation techniques for hydrophilic components differ considerably from systems targeting the delivery of lipophilic components. Figure 20.2 shows some encapsulation/delivery systems for hydrophilic and lipophilic bioactive molecules and micronutrients developed by the food industry and suppliers.

The food industry could benefit from the pharmaceutical industry in regards to the selection of appropriate delivery systems and strategies for the inclusion of bioactives in food products. The molecular dispersion of a drug, just as a bioactive compound used in food, is a prerequisite for its absorption across biological membranes. After eating (i.e., oral administration, in pharmaceutical terms), this dictates that the active substance must first dissolve within the gastrointestinal tract before partitioning into and then across the enterocyte (Porter et al. 2007). Colloidal structures that may even be formed during the digestive process (as in the case of lipids) have the potential to prevent precipitation and enhance absorption. Enhancing solubilization in the intestinal milieu is one way to enhance the bioavailability of bioactives with poor water solubility.

Encapsulation techniques can also be used to prevent the loss of aroma compounds that can affect flavor intensity and, thus, positively affect food quality and perception. Therefore, the physicochemical characteristics of volatile

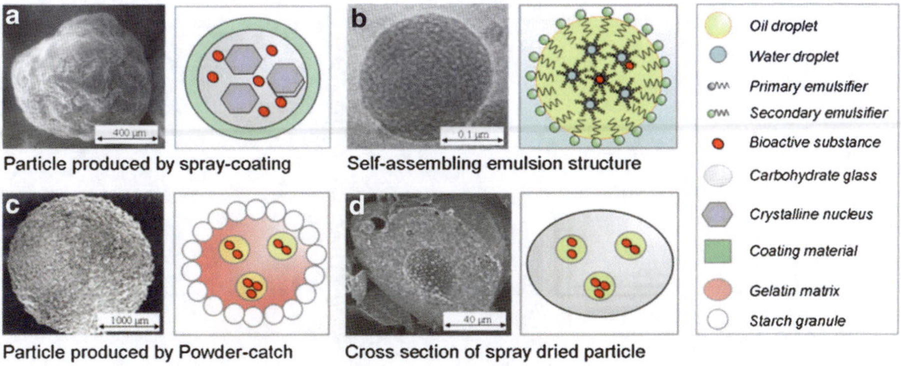

Fig. 20.2 Various encapsulation/delivery systems used for bioactives and micronutrients (including (**a**) spray-coated particle: courtesy Givaudan AG Dübendorf/CH; (**b**) self-assembling structured emulsion; (**c**) powder-catch particle: courtesy Hoffmann La Roche AG Basel/CH; (**d**) spray-dried particle containing encapsulated lipophilic bioactives) (Reprinted from Palzer (2009a) with permission from Elsevier)

compounds and their interaction with the food matrix need to be considered. As an example, the encapsulation of bioactive compounds with edible films not only allows for the control of flavor loss but is also a technique for controlled flavor release of aroma compounds with time (Marcuzzo et al. 2010). This controlled-release effect can be used to either maintain a certain quality of the food product over shelf life and consumption or to provide a special flavor experience for the consumer (e.g., a flavor burst upon chewing food).

20.2.3 Fortification and Encapsulation

Today roughly one-third of the world's population suffers from micronutrient deficiencies leading to increased mortality, impairment, and underperformance.

More than 1.5 billion people suffer from iron deficiency, which leads to impaired children, physical weakness, still birth, and increased maternal and child mortality (De Benoist et al. 2008b). Vitamin A deficiency has been shown to be present in190 million preschool children, who then suffer from blindness and immune impairment (WHO 2009). More than two billion people suffer from iodine deficiency, which can cause severe mental impairment (De Benoist et al. 2008a). Today 20 % of the world's population suffers from zinc deficiency. Inadequate dietary zinc intake might lead to delayed growth in children, immune impairment, and increased child mortality (Hotz and Brown 2004). Accordingly, undersupply of micronutrients is a huge global nutritional problem.

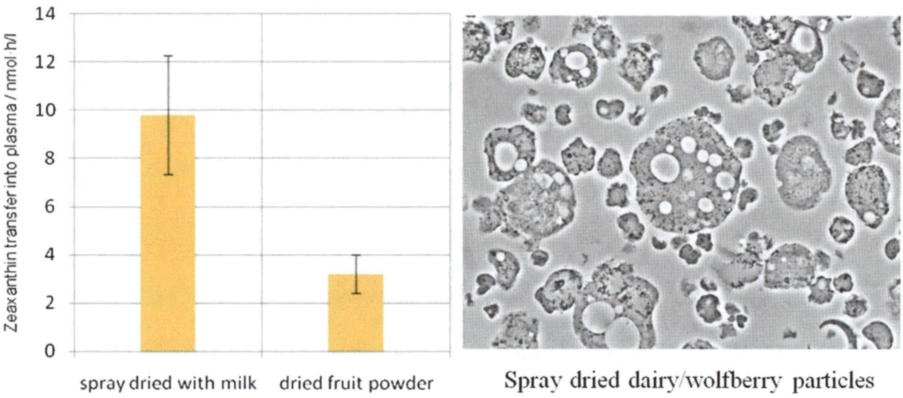

Fig. 20.3 Bioavailability of zeaxanthine in cospray-dried dairy/wolfberry extract compared to native fruit powder (*left image*; according to Benzie et al. (2006)) and spray-dried extract particles (*right*)

The following technological solutions may be applied to tackle this problem:

1. Ensuring a higher bioavailability of nutrients naturally present in raw materials;
2. Reducing nutrient degradation during processing and shelf life through improved processing technology and encapsulation technologies;
3. Designing host food matrices providing improved bioavailability and stability of micronutrients;
4. Enriching the vitamin content of plant raw materials by plant selection and breeding technologies;
5. Fortification of food by the addition of vitamins and mineral compounds.

The bioavailability and stability of micronutrients depends strongly on their chemical form and the chosen particle size (Hilty et al. 2010). The choice of optimal compound can be difficult because often the most bioavailable forms are in the meantime also the most sensitive or reactive compounds. Vitamins or fruit and vegetable extracts can also be cospray-dried with dairy components to enhance the bioavailability of the active component. This has been demonstrated for zeaxanthine derived from wolfberries, a vitamin-rich Asian fruit. Extracting the fruit with milk or solutions of dairy ingredients and spray drying of the obtained extract delivers roughly three times more bioavailability of zeaxanthine compared to whole fresh or dried fruits (Fig. 20.3) (Benzie et al. 2006).

Vitamins are often oxygen- and photosensitive, and their stability can be improved significantly through encapsulation. Introducing efficient oxygen or light barriers reduces degradation during manufacturing and shelf-life. Metal compounds such as iron compounds tend to react with components of the food matrix in which they are embedded. This often causes an alteration of the organoleptic properties of the food matrix and significant color deviations might be observed (Hilty et al. 2010). Thus vitamins and mineral compounds are usually encapsulated before they are added to liquid and solid food matrices.

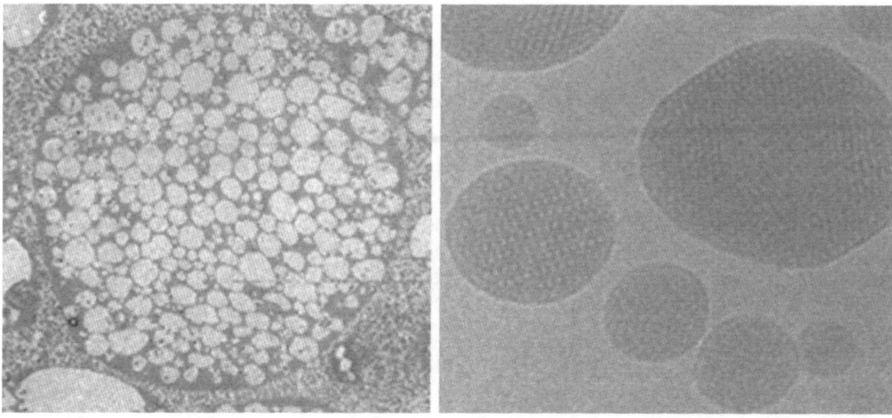

Fig. 20.4 Multiple-emulsion (*left*) and internal self-assembly emulsion structure (*right*, according to Sagalowicz et al. (2006))

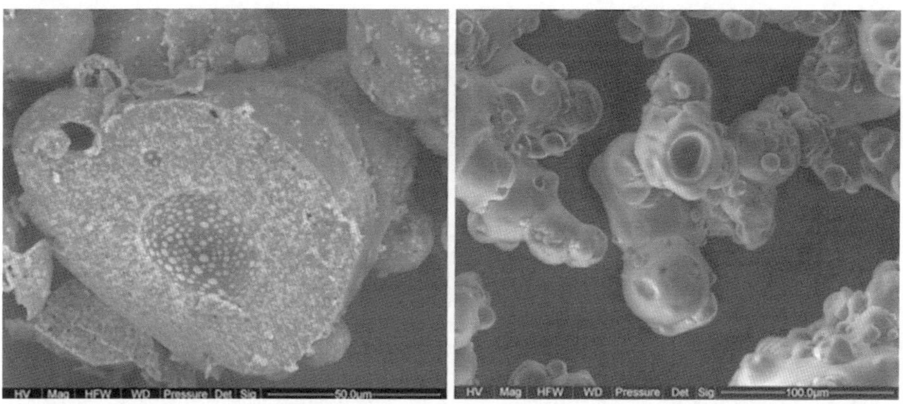

Fig. 20.5 Amorphous spray-dried particle with oil inclusions (*left*) and amorphous spray-dried particles containing microorganisms (*right*)

Sagalowicz et al. (2006) discussed structured emulsions as delivery systems for liquid foods. Emulsions, multiple emulsions, micelles, lamellar phases, cubic phases, and hexagonal mesophases can be used for encapsulation in liquid matrices.

Figure 20.4 shows images of a multiple-emulsion system.

Ribeiro et al. showed that the choice of emulsion system has a significant effect on the bioavailability of different carotenoids (2006).

For delivery in solid foods micronutrients are often encapsulated in amorphous particles (Ubbink and Kruger 2006; Palzer et al. 2006; Palzer 2009b). Figure 20.5 shows encapsulation systems suitable for solid foods.

Such amorphous carrier particles can be produced by spray drying, spray cooling, agglomeration, granulation, vacuum drying, melt extrusion, or fluid-bed coating. Often lipophilic active molecules are dissolved into carrier lipids. The lipid solution is then emulsified into a concentrated amorphous carbohydrate solution. The obtained oil-in-water emulsion is then dried. Hydrophilic micronutrients can be dissolved directly in the carbohydrate concentrate.

Fast dehydration generates glassy amorphous structures in which the micronutrients are dispersed in the form of single molecules or lipid droplets containing the active substance. Such glassy matrices show aging during storage and will eventually crystallize (Descamps et al. 2009). Amorphous matrices can be very effective for stabilizing oxygen-sensitive bioactives because their permeability can be reduced to a minimum by combining mono-, di-, oligo-, and polymers to a blend with a very low free specific volume (Kilburn et al. 2005). Adding an increasing amount of substances with low molecular weight to oligo- or polymers reduces the permeability of the molecular matrix but also increases its moisture sensitivity. Water also affects the density and mechanical properties of water-soluble molecular matrices. Increasing the water content in amorphous matrices first increases the density because the water molecules occupy the free volume between the larger matrix molecules. Further increasing the moisture content will increase the molecular mobility of the matrix-forming molecules, which eventually decreases the density of the encapsulation matrix again. Accordingly, oxygen and water diffusion through water-soluble encapsulation matrices depends on the moisture content. Water exhibits an autocatalytic diffusion process. The higher the water content, the faster is the diffusion of water molecules. Thus, sometimes amorphous carrier particles are coated with 5–20 % of solidifying fat or wax in fluid-bed granulators.

20.2.4 Pro- and Prebiotics in Food

The human body contains between 1 and 3 kg of bacteria mass. Roughly 1,000 bacteria are found in 1 ml stomach volume, and in the small intestine up to 10^6 bacteria are found per milliliter. Since bacteria cells are rather small, their number even exceeds the number of human cells in the body. As has been demonstrated in a number of scientific studies, this bacteria population has a significant impact on human health (Rolfe 2000). The bacteria inside the human gastrointestinal system ferment food and overgrow pathogens (e.g., *Helicobacter pylori*, EHEC bacteria), and their metabolites might stimulate the immune systems (Benyacoub et al. 2003) of humans and pets. Furthermore, it has been suggested that probiotics positively affect allergies, increase the immune response, facilitate digestion, and reduce symptoms of diarrhea. They also supposedly affect obesity, enhance nutrient delivery, and contribute to skin health and beauty.

The microbial population inside the human gastrointestinal system can be influenced through the diet. Accordingly, food is enriched with living probiotic bacteria strains. Examples of probiotic bacteria strains that are added to food are *Staphylococcus xylosus*, *Lactobacillus johnsonii*, *Streptococcus thermophilus*, *Bifidobacterium longum*, *L. rhamnosus*, and *L. para-casei* (Fig. 20.6).

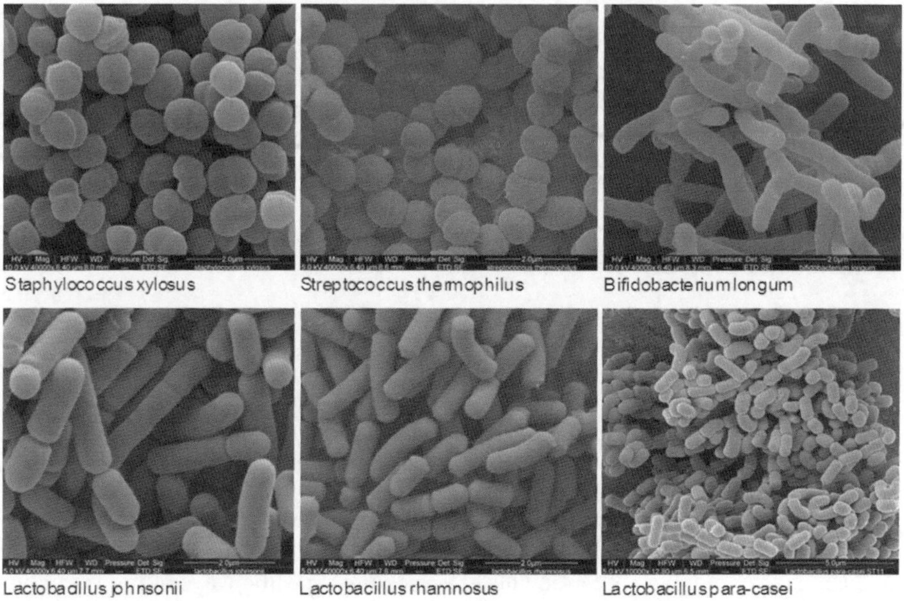

Fig. 20.6 Probiotic bacteria strains applied in human and pet food

Alternatively, carbohydrates that favor the development of probiotic bacteria species are added to functional food products. Such carbohydrates (e.g., oligosaccharides and lactulose) are called prebiotics. Such prebiotics can be either simply added to the food or even generated in situ by enzymatic conversion of food components (Schuster-Wolff-Buehring et al. 2009).

In the case of chilled dairy, the probiotic bacteria can be added directly to the liquid dairy product. In the case of dehydrated foods, probiotic bacteria are spray or freeze dried before being added to the food powder or are dried together with the food after being dispersed in the liquid food concentrate. It is desirable to keep the bacteria cells alive during drying and storage of the food product. To ensure a maximum of surviving bacteria cells, they are frequently encapsulated before being added to products. The aim is to reduce the contact of the living cells with air, which would accelerate metabolism and reduce their life span during storage. Alternatively, living cells can be stabilized by the addition of protecting substances prior to drying (Santivarangkna et al. 2008). These substances (typically carbohydrates, amino acids, or peptides) retain, for example, the cell membrane of the bacteria intact during drying and subsequent storage. Rupture of cell membranes would result in an inactivation of the concerned cells. Furthermore, industry continually searches for more robust strains and tries to identify by means of genomics and metabolomics the reasons for the improved stability of certain strains. Once the metabolic mechanisms providing increasing stability of cultures have been identified, they can be instrumental in increasing stability through strain selection and propagation or adjusting cultivation conditions. The complete genomic maps of

B. longum and *L. Johnsonii* have been decoded (e.g., Schell et al. 2002). These maps are useful for strain identification (required, for example, for quality control) or identification of metabolic characteristics responsible for cell stability.

Today a large number of foods are enriched with probiotic bacteria. The segment of probiotic products is comprised of chilled dairy products (milk shots, yogurt), ice cream, chocolate, pet food, beverages, and spreads.

In symbiotic products, prebiotic carbohydrates are applied together with probiotic bacteria strains in the same food matrix. There are very few symbiotic food products on the market.

20.2.5 Improved Nutritional Profile/Nutritional Application Addressing Special Health Care Needs

Improving the general nutritional profile of food can be achieved by increasing the overall content of (plant) proteins, dietary fiber, or whole grains; however, the inclusion of these types of ingredients often requires specific ways of food structuring and, consequently, the adjustment of process conditions during food production.

The consumption of whole grain cereals, fruits, and vegetables, for example, represents the most common and natural source of dietary fiber. Unfortunately, whole-grain-derived fiber is the most difficult to incorporate into products without compromising on the final texture and taste of the product (Redgwell 2010). A typical example is the negative effect of the incorporation of wheat bran in extruded cereal products. The addition of bran significantly changes the physicochemical properties of the starch like the glass transition temperature, melting temperature, and sorption isotherm. As a consequence of these changes, expansion properties are considerably reduced (Robin et al. 2011). A study focusing on the effect of whole grain flours from various origins (wheat, rye, triticale, barley, tritordeum) on the production of cakes showed similar results and found significant correlations between water absorption and specific volume, symmetry, and firmness (Gómez et al. 2010).

Food structures composed of natural and, thus, less purified raw materials are often physically and chemically less stable. Nevertheless, their stability can also be improved by means of structuring. Protective layers can be applied (Shih et al. 2011), natural stabilizing components can be added to the recipes, and the supramolecular structure of single ingredients can be changed through physical processing.

Increasing the protein content of foods is normally accompanied by a rise in viscosity. Conversely, the controlled unfolding and aggregation of protein macromolecules can be used for targeted structuring that can lead to a reduction of this negative effect. Depending on the pH, temperature, and the presence of ions, the macromolecular proteins can adopt different shapes and structures such as fibrils, spherical microgels, microparticles, and fractal aggregates of gels (Schmitt et al. 2007). Another example of how mixtures of whey protein microgels and soluble aggregates can be used as building blocks to control rheology and structure can be found in Donato et al. (2011). Nicolai et al. (2011) reviewed the use of β-lactoglobulin and whey protein isolate aggregates in cold-set gels, foams, and emulsions as key structuring elements.

In most cases, the quality of food is defined by sensorial characteristics and consumer-driven preference. Then again, there are also nutritional applications addressing specific health care needs. Patients suffering from neurogenic dysphagia, the medical term meaning difficulty in swallowing, are known to be at high risk of respiratory and nutritional complications. Increasing the bolus viscosity greatly improves the swallowing function (Clavé et al. 2006). Quality of life can be tremendously improved if the flow behavior of normal foods can be modified by adding certain viscosity-modifying supplements. This requires a profound understanding of the swallowing process (e.g., controlling mechanisms, determination criteria of bolus swallowing) as well as a rheological understanding of individual ingredients and their relation to the "food processing" in the mouth (Chen and Lolivret 2011; Fischer and Windhab 2011). Health care products that are normally very nutrient dense or exhibit a particular rheological behavior during manufacture or consumption can only be designed by targeted modification of their supramolecular structure.

20.2.6 Modulated Digestion

Finally, food microstructures can be designed in such a way that their modulated digestion behavior triggers different physiological responses – a research area that is attracting increasing attention within the food industry as in vitro models become more sophisticated.

Proteins are reportedly the most satiating of all macronutrients, but not all proteins are alike. Some are digested and absorbed rapidly, while others may impact metabolism and glucose control. Acheson et al. (2011) compared, for example, the effects of various protein sources (whey, casein, soy) on energy metabolism, satiety, and glucose control in humans. It was shown that protein-rich meals promote greater energy expenditure than carbohydrate-rich meals of equal caloric content. These findings confirm the assumption that increased protein content in the diet can promote weight control. Colloidal interactions (like those with proteolytic enzymes and physiological surfactants in the gastrointestinal tract) play a key role in protein digestion and need to be well understood to produce food with optimum nutrition for the consumer (Mackie and Macierzanka 2010).

The nutritional quality of starch strongly depends on the processing and the physicochemical state of the starch. Slowly digestible starch (SDS), such as native maize starch, offers the advantage of a slow increase in postprandial blood glucose levels and sustained blood glucose levels over time when compared to rapidly digestible starch with its fast increase, peak, and subsequent decline (Zhang and

Hamaker 2009). This more constant supply of energy during digestion might influence satiety and physical and mental performance and can have implications for diabetes management (Lehmann and Robin 2007).

The ability to regulate lipid uptake is seen as beneficial in several areas like improved nutrition, especially for the young and the elderly, or the reduction of disease risk. Emulsion systems with increased structural complexity are seen as a mechanism by which lipid uptake may be controlled (Golding and Wooster 2010; van Aken 2010). As already mentioned in Porter et al. (2007), lipid digestion and the resulting colloidal structures can be designed to elicit particular physiological properties, such as the targeted delivery of bioactives.

20.3 Conclusions

Most food products are structurally complex. It can be concluded that during the development of products addressing specific nutritional, health, and wellness needs, the structure of the food matrix and, consequently, product processing need to be adapted. A detailed understanding of the time-dependent transient changes in all of the structural aspects of food matrices from raw material harvesting to product processing to the point of breakdown during shelf life, consumption, and final digestion needs to be developed. Furthermore, at each point in time within this cycle there exists a broad range of length scales ranging from the molecular and supramolecular to the micro- and macrostructural level determining the mechanical, physical, and chemical properties of food structures. Only in this manner can a tailor-made buildup and controlled breakdown of food products be achieved and the specific nutritional, physical, and sensorial properties engineered.

Examples given in this review showed that a targeted structuring of processed food can partially compensate for the elimination of fat and sugar from foods (i.e., in turn addressing overnutrition) or enable a higher content of ingredients with an improved nutritional profile (e.g., whole grains/dietary fibers, or high-nutrient-dense foods addressing moderate undernutrition). Furthermore, structuring of the food matrix is an important tool to improve the stability of healthier formulations and the bioavailability of bioactive compounds. Finally, food microstructures can be designed in such a way that their modulated digestion behavior triggers different physiological responses.

Acknowledgments Parts of this chapter were published in Procedia Food Science (Kaufmann and Palzer 2011).

The authors would like to thank Martine Rouvet (Nestlé Research Center) for preparing the micrographs.

References

Acheson KJ, Blondel-Lubrano A, Oguey-Araymon S, Beaumont M, Emady-Azar S, Ammon-Zufferey C et al (2011) Protein choices targeting thermogenesis and metabolism. Am J Clin Nutr 93:525–534

Benyacoub J, Czarnecki-Maulden G, Cavadini C, Sauthier T, Anderson R, Schiffrin E, von der Weid T (2003) Supplementation of food with Ent. Faecium (SF68) stimulates immune functions in dogs. J Nutr 133:1158–1162

Benzie I, Chung W, Wang J, Richelle M, Bucheli P (2006) Enhanced bioavailability of zeaxanthine in a milk-based formulation of Wolfberry (Gou Qi Zi; *Fructus barbarum* L.). Br J Nutr 96(1):154–160

Bolliger S, Kornbrust B, Goff HD, Tharp BW, Windhab EJ (2000) Influence of emulsifiers on ice cream produced by conventional freezing and low-temperature extrusion processing. Int Dairy J 10(7):497–504

Chen J, Lolivret L (2011) The determining role of bolus rheology in triggering a swallowing. Food Hydrocoll 25(3):325–332

Clavé P, de Kraa M, Arreola V, Girvent M, Farré R, Palomera E et al (2006) The effect of bolus viscosity on swallowing function in neurogenic dysphagia. Aliment Pharmacol Ther 24:1385–1394

De Benoist B, McLean E, Andersson M, Rogers L (2008a) Iodine deficiency in 2007: global progress since 2003. In: WHO documents on Vitamin and Mineral Nutrition Information System (VMNIS). The United Nations University, Tokyo, Japan

De Benoist B, McLean E, Egli I, Cogswell M (2008b) Worldwide prevalence of anaemia 1993–2005. WHO global database on anaemia. In: WHO documents on Vitamin and Mineral Nutrition Information System (VMNIS). The United Nations University, Tokyo, Japan

Descamps N, Palzer S, Zürcher U (2009) The amorphous state of spray dried maltodextrin: sub-sub Tg enthalpy relaxation and impact of temperature and water annealing. Carbohydr Res 344:85–90

Dickinson E (2011) Double emulsions stabilized by food biopolymers. Food Biophys 6(1):1–11

Donato L, Kolodziejcyk E, Rouvet M (2011) Mixtures of whey protein microgels and soluble aggregates as building blocks to control rheology and structure of acid induced cold-set gels. Food Hydrocoll 25(4):734–742

Eisner MD, Wildmoser H, Windhab EJ (2005) Air cell microstructuring in a high viscous ice cream matrix. Colloids Surf A Physicochem Eng Asp 263(1–3):390–399

Fischer P, Windhab EJ (2011) Rheology of food materials. Curr Opin Colloid Interface Sci 16 (1):36–40

Golding M, Wooster TJ (2010) The influence of emulsion structure and stability on lipid digestion. Curr Opin Colloid Interface Sci 15(1–2):90–101

Gómez M, Manchón L, Oliete B, Ruiz E, Caballero PA (2010) Adequacy of wholegrain non-wheat flours for layer cake elaboration. LWT- Food Sci Technol 43(3):507–513

Hilty F, Arnold M, Hilbe M, Teleki A, Knijnenburg J, Ehrensperger F, Hurrell R, Pratsinis S, Langhas W, Zimmermann M (2010) Iron from nanocompounds containing iron and zinc is highly bioavailable in rats without tissue accumulation. Nat Nanotechnol 5:374–380

Hotz C, Brown K (2004) Assessment of the risk of zinc deficiency in populations and options for its control. Food Nutr Bull 25:91–204

Kaufmann SFM, Palzer S (2011) Food structure engineering for nutrition, health and wellness. Procedia Food Sci 1:1479–1486

Kilburn D, Claude J, Schweizer T, Alam A, Ubbink J (2005) Carbohydrate polymers in amorphous state: an integrated thermodynamic and nanostructural investigation. Biomacromolecules 6:864–879

Lehmann U, Robin F (2007) Slowly digestible starch – its structure and health implications: a review. Trends Food Sci Technol 18(7):346–355

Lenfant F, Loret C, Pineau N, Hartmann C, Martin N (2009) Perception of oral food breakdown. The concept of sensory trajectory. Appetite 52(3):659–667

Mackie A, Macierzanka A (2010) Colloidal aspects of protein digestion. Curr Opin Colloid Interface Sci 15(1–2):102–108

Marcuzzo E, Sensidoni A, Debeaufort F, Voilley A (2010) Encapsulation of aroma compounds in biopolymeric emulsion based edible films to control flavour release. Carbohydr Polym 80 (3):984–988

Murray BS, Durga K, Yusoff A, Stoyanov SD (2011) Stabilization of foams and emulsions by mixtures of surface active food-grade particles and proteins. Food Hydrocoll 25(4):627–638

Nicolai T, Britten M, Schmitt C (2011) β-Lactoglobulin and WPI aggregates: formation, structure and applications. Food Hydrocoll 25(8):1945–1962

Norton I, Moore S, Fryer P (2007) Understanding food structuring and breakdown: engineering approaches to obesity. Obes Rev 8:83–88

Palzer S (2009a) Food structures for nutrition, health and wellness. Trends Food Sci Technol 20 (5):194–200

Palzer S (2009b) Bioactive ingredients – maximising stability, solubility and bioavailability. biotechnology. Transkript Spec "Weisse Biotechnologie" 6(15):66–67

Palzer S, Rekhif N, Rabe S (2006) Food microencapsulation. Keynote lecture 4th industrial trade fair in microencapsulation. Strasbourg

Porter CJH, Trevaskis NL, Charman WN (2007) Lipids and lipid-based formulations: optimizing the oral delivery of lipophilic drugs. Nat Rev Drug Discov 6(3):231–248

Redgwell RJ (2010) Dietary fibre in food fabrication: a changing landscape for consumer and industry. Food Sci Technol 24(4):18–20

Ribeiro H, Guerrero J, Briviba K, Rechkemmer G, Schuchmann H, Schubert H (2006) Cellular uptake of carotenoid-loaded oil-in-water emulsions in column carcinoma cells in vitro. J Agric Food Chem 54:9366–9369

Robin F, Théoduloz C, Gianfrancesco A, Pineau N, Schuchmann HP, Palzer S (2011) Starch transformation in bran-enriched extruded wheat flour. Carbohydr Polym 85(1):65–74

Rolfe R (2000) The role of probiotic cultures in the control of gastrointestinal health. J Nutr 130:396S–402S

Sagalowicz L, Leser ME (2009) Delivery systems for liquid food products. Curr Opin Colloid Interface Sci 15(1–2):61–72

Sagalowicz L, Leser M, Watzke H, Michel M (2006) Monoglyceride self assembly structures as delivery vehicles. Trends Food Sci Technol 17:204–214

Santivarangkna C, Higl B, Foerst P (2008) Protection mechanism of sugars during different stages of preparation process of dried lactic acid starter cultures. Food Microbiol 25:429–441

Schell M et al (2002) The genome sequence of *Bifidobacterium longum* reflects its adaptation to the human gastrointestinal tract. Proc Natl Acad Sci USA 99(22):14422–14427

Schmitt C, Bovay C, Rouvet M, Shojaei-Rami S, Kolodziejczyk E (2007) Whey protein soluble aggregates from heating with NaCl: physicochemical, interfacial, and foaming properties. Langmuir 23(8):4155–4166

Schuster-Wolff-Buehring R, Jaindl K, Fischer L, Hinrichs J (2009) A new way to refine lactose: enzymatic synthesis of prebiotic lactulose. Eur Dairy Mag 21:25–27

Shih FF, Daigle KW, Champagne ET (2011) Effect of rice wax on water vapour permeability and sorption properties of edible pullulan films. Food Chem 127(1):118–121

Turgeon SL, Rioux L-E (2011) Food matrix impact on macronutrients nutritional properties. Food Hydrocoll 25(8):1915–1924

Ubbink J, Kruger J (2006) Physical approaches for the delivery of active ingredients in foods. Trends Food Sci Technol 17:244–254

van Aken GA (2010) Relating food emulsion structure and composition to the way it is processed in the gastrointestinal tract and physiological responses: what are the opportunities? Food Biophys 5(4):258–283

WHO (2009) Global prevalence of vitamin A deficiency in populations at risk 1995–2005. WHO global database on Vitamin A deficiency. In: WHO documents on Vitamin and Mineral Nutrition Information System (VMNIS). World Health Organization, Geneva, Switzerland

Wildmoser H, Scheiwiller J, Windhab EJ (2004) Impact of disperse microstructure on rheology and quality aspects of ice cream. Lebensm Wiss Technol 37(8):881–891

Zhang G, Hamaker BR (2009) Slowly digestible starch: concept, mechanism, and proposed extended glycemic index. Crit Rev Food Sci Nutr 49(10):852–867

Zúñiga RN, Kulozik U, Aguilera JM (2011) Ultrasonic generation of aerated gelatin gels stabilized by whey protein β-lactoglobulin. Food Hydrocoll 25(5):958–967

Chapter 21
Transfer of Water and Volatiles at Interfaces: Application to Complex Food Systems

Andrée Voilley, Sonia Lequin, Alicia Hambleton, David Chassagne, Thomas Karbowiak, and Frédéric Debeaufort

21.1 Introduction

The behaviour of small molecules such as water or aroma compounds at interfaces in complex food systems (with or without edible barriers) plays a key role in the quality and stability of food products (Voilley and Souchon 2006; Cayot et al. 2008).

This paper focuses on the thermodynamics (e.g. phase equilibrium, partition coefficients) and kinetics (e.g. diffusivity, permeability, transfer coefficients) phenomena occurring at interfaces with or without changes in the physical state at the interfaces. Each case will be discussed and illustrated by examples involving small volatile compounds at food or packaging material interfaces. The physicochemical and structural characteristics of both small molecules and matrices are taken into

A. Voilley (✉) • T. Karbowiak
EA EMMA, Université de Bourgogne, 1 esplanade Erasme, 21000 Dijon, France

Agrosup Dijon, 26 Boulevard Docteur Petitjean, 21079 Dijon, France
e-mail: a.voilley@agrosupdijon.fr; andree.voilley@u-bourgogne.fr; thomas.karbowiak@u-bourgogne.fr

S. Lequin • D. Chassagne
EA EMMA, Université de Bourgogne, 1 esplanade Erasme, 21000 Dijon, France

Institut Universitaire de la Vigne et du Vin "Jules Guyot", Université de Bourgogne, 21078 Dijon, France
e-mail: sonia.lequin@u-bourgogne.fr

A. Hambleton
EA EMMA, Université de Bourgogne, 1 esplanade Erasme, 21000 Dijon, France

F. Debeaufort
EA EMMA, Université de Bourgogne, 1 esplanade Erasme, 21000 Dijon, France

IUT Génie Biologique, Université de Bourgogne, 7 Bd Docteur Petitjean, 21078 Dijon, France
e-mail: frederic.debeaufort@u-bourgogne.fr

S. Yanniotis et al. (eds.), *Advances in Food Process Engineering Research and Applications*, Food Engineering Series, DOI 10.1007/978-1-4614-7906-2_21,

account and related to the properties of the interfaces. The objectives are to present some recent research related to the role of surface properties of different materials in mass transfers.

21.2 Thermodynamics of Sorption

21.2.1 Aroma Partitioning Between Food and Air

The release of aroma from a food product strongly influences the perception and then the quality. This release depends on mass transfers at the solid–vapour or liquid–vapour interface and on the hydration level of the food product.

The nature of the matrix and the volatility of the aroma compounds affect the partition coefficient at equilibrium, whereas the interface has a significant effect on the kinetics of the release.

A multiphasic model of fruit puree/yoghurt prepared as a bilayer that consisted of a pectin and a dairy gel were developed in order to study the partition of strawberry flavour compounds between the two phases (Nongonierma et al. 2008). The parameters that modified the partitioning of ethyl esters between the pectin and dairy gel (i.e. fat content of the dairy gel and flavour concentration) also affected the perception of the fruity odour attribute; the storage temperature had no effect on the sensory perception. An increase in the initial flavour concentration in the pectin gel increased the fruity odour intensity of both gels. Conversely, the presence of fat in the dairy gel decreased the odour intensity of both gels by reducing the flavour compound volatility in the dairy gel and decreasing the flavour concentration in the pectin gel. The addition of sucrose or glucose syrup to the low-fat dairy gel slightly increased the retention of the flavour compounds; this was attributed to physicochemical interactions with pectin and sucrose. In the presence of fat (5 wt%), the flavour compounds were solubilised in the fat and their release was not affected by the composition of the dispersing medium.

Beta-lactoglobulin is a protein present in dairy products which can control aroma partitioning. To understand the interactions between the aroma compounds and this protein, we studied the hydration of the protein and the effect of water activity on the sorption of the volatiles (Seuvre et al. 2007). It was shown that water is necessary in the formation of aroma–protein interactions. The presence of β-lactoglobulin in aqueous solution increased the retention of the four aroma compounds considered. They modify the flux at the air–aqueous solution interface. These results could be applied to food technology such as in emulsions, which could be stabilized with proteins, and particularly in dairy products, which are generally flavoured.

Then, to better focus on the phenomenon at the interface between two non-miscible liquids as in emulsions, the Harvey et al. (1995) method was set up to measure the mass transfers at the interface; it will be developed later.

21.2.2 *Study of Cork Interactions with Wine Active Compounds*

Based on studies (Lequin et al. 2009, 2010) related to the general problem of white wine oxidation, the first step was primarily to interest in the reactivity of cork stoppers toward wine active molecules at thermodynamic equilibrium.

Sulphur dioxide (SO_2) is the exogenous antioxidant most commonly added to wine to limit oxidation. However, its possible interactions with cork, in both liquid and gas phases, have never been studied. Therefore, a complete thermodynamic study of gaseous SO_2 sorption on cork was performed. The study aimed at (1) answering the hypothesis that the decrease in SO_2 content observed during the storage of wine in a bottle may be due to sorption on cork and (2) gaining a better understanding of the interactions involved.

In addition, wine is mainly composed of water (approximately 88 % v/v). So it seemed essential to study also the sorption of SO_2 on hydrated cork. Some results concerning the sorption of these two compounds (SO_2 and H_2O) are given in the following sections.

A thermodynamic study of sorption of small active compounds (water and sulphur dioxide) on cork stoppers used for wine bottling was performed by thermogravimetry (McBain and Bakr 1926). Additional information on the energy of the sorption process was obtained by determining the sorption heats by differential scanning calorimetry coupled with manometry (Weber et al. 2008).

21.2.2.1 Sorption of Single Compounds

Sorption isotherms of H_2O and SO_2 on cork powder at 298 K are shown in Fig. 21.1. The curves display a type II shape of the International Union of Pure and Applied Chemistry (IUPAC) classification, which is typical of sorption on non-porous or macroporous solids.

The maximum molar quantity is observed for H_2O (around 4 mmol.g^{-1}), whereas it is less for SO_2 (around 0.3 mmol.g^{-1}). During the desorption process (results not shown), complete desorption is observed for water. It reveals a physisorption process for this molecule. In contrast, desorption remains incomplete for SO_2. In this case, it can be explained by a chemisorption phenomenon. Moreover, a hysteresis is observed. This can be attributed to a structural change in the cork. A more detailed study of the sorption–desorption process of SO_2 and water was recently published (Lequin et al. 2009, 2010).

To better characterize the physisorption or chemisorption phenomenon, a thermal study was carried out by calorimetry.

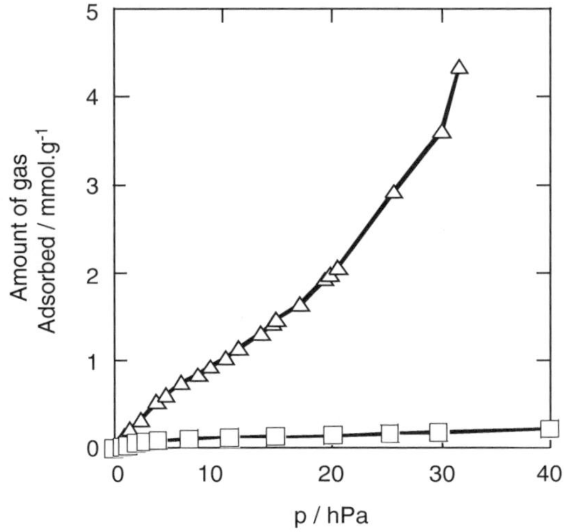

Fig. 21.1 Sorption isotherms of H_2O (*triangles*) and SO_2 (*squares*), on cork powder at 298 K

21.2.2.2 Sorption Enthalpies of Single Compounds

The sorption enthalpies of sulphur dioxide and water as a function of loading are displayed in Fig. 21.2. Concerning the sorption enthalpy of sulphur dioxide, the heat of sorption at low filling is high, around 100 kJ·mol^{-1}. This points to a chemisorption of the first sorbed molecules on the cork surface. Then the sorption heat sharply decreases as the filling increases. Above 0.2 mmol.g^{-1} of sorbed SO_2, it finally tends to the enthalpy of liquefaction of SO_2, which is equal to 22.9 kJ·mol^{-1}, as expected for a physisorption process (Lequin et al. 2009).

For water, when loading tends to zero, the sorption enthalpy is around 65 kJ·mol^{-1}. This indicates that cork–water interactions are rather strong compared to water–water interactions occurring in liquid bulk. However, this interaction energy is close to that usually involved in physisorption.

21.2.2.3 Sorption Isotherm of SO_2 on Hydrated Cork

A sorption isotherm of SO_2 on hydrated cork is displayed in Fig. 21.3. The isotherm is significantly modified when SO_2 is sorbed on hydrated cork. The sorption capacity is divided by a factor of 3. At first glance, this seems surprising since sulphur dioxide is highly soluble in water. Therefore, we suspect that the decrease in the sorption capacity of SO_2 in the presence of water is the result of a competitive sorption between water and sulphur dioxide. After presorption of water, many sites would not be able to interact directly with SO_2 since they are already occupied by water molecules.

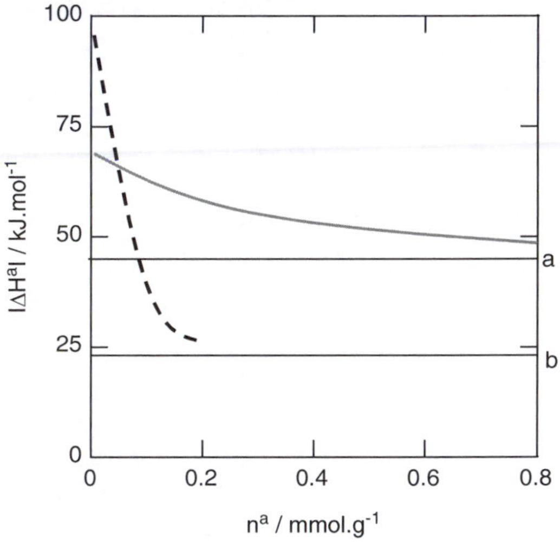

Fig. 21.2 Sorption enthalpy (absolute value) of H_2O (———) and SO_2 (— —) on cork powder at 298 K. *a*: liquefaction enthalpy of $H_2O = 44$ kJ·mol^{-1} *b*: liquefaction enthalpy of $SO_2 = 22.9$ kJ·mol^{-1}

Fig. 21.3 Sorption isotherms of SO_2 at 298 K. *Dots*: on dry cork. *Squares*: on hydrated cork (containing 5 wt% water content) (Adapted from Lequin et al. 2009)

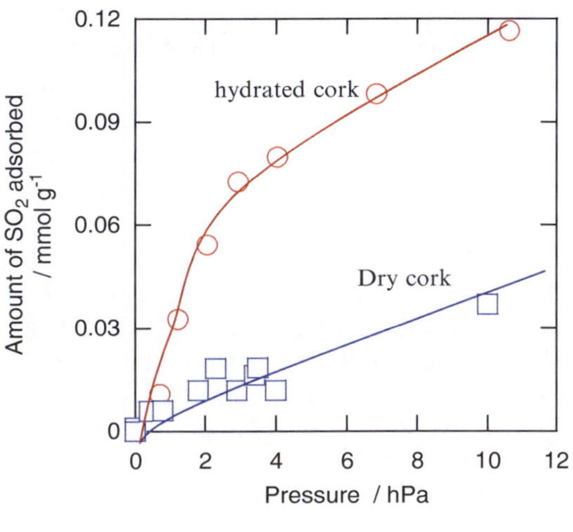

21.2.2.4 Sorption Enthalpy of SO_2 on Hydrated Cork

The sorption enthalpy of SO_2 on hydrated cork powder, measured in a larger pressure range than for dry cork (up to 60 hPa), is displayed in Fig. 21.4.

Note that the shape of the calorimetric curve is very different in the presence of water that has already been sorbed. At loading close to zero, the sorption heat is much lower than that observed on dry cork (60 versus 100 kJ·mol^{-1}). Then the sorption heat increases with the loading to reach a maximal value of around

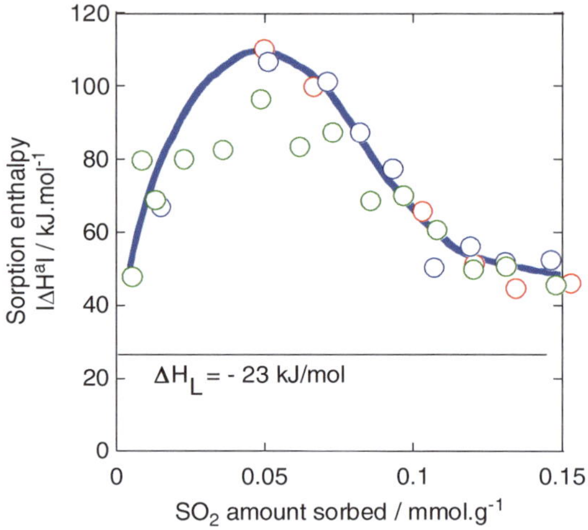

Fig. 21.4 Sorption enthalpy of gaseous SO_2 by partially hydrated cork powder (5 wt% water content) at 298 K as a function of the loading. Also indicated is the enthalpy of liquefaction of SO_2 (Adapted from Lequin et al. 2009)

110 kJ·mol^{-1} before decreasing sharply to approximately 50 kJ·mol^{-1}. It is worth noting that the maximal value of the sorption heat is observed at a filling of 0.35 wt %, which is exactly the value of the amount of SO_2 chemisorbed on the surface, as observed by thermogravimetry. Thus, it is tempting to attribute the ascending part of the calorimetric curve to the chemisorption of SO_2 on sites previously occupied by water molecules. The heat involved would correspond on the one hand to the exothermic chemisorption of SO_2 and on the other hand to the endothermic desorption of H_2O. This would explain why the resulting sorption heats are lower than those measured on dry cork. It is difficult to explain why the sorption heat increases in this range of loading. It could be due to sorbate–sorbate interactions, especially between water and sulphur dioxide, which certainly form hydrogen bonds. The descending part of the calorimetric curve would correspond to the physisorption of SO_2 on sites probably free of H_2O once the chemisorption sites are completely saturated. However, this remains a hypothesis.

In conclusion, this work shows that cork is not an inert material with respect to wine active molecules. Water is sorbed according to a physisorption process. Sulphur dioxide sorption follows a chemisorption and a physisorption process.

21.3 Mass Transfers Between Phases

21.3.1 Liquid/Liquid: Migration Phenomenon in Emulsions

The migration phenomenon in emulsion has to do with the liquid–liquid interface. The transfer of solutes between the dispersed phase and the continuous phase in emulsions is driven by the partition coefficient between oil and the aqueous phase and by the structure of the globule or droplet surface. An experimental design, the

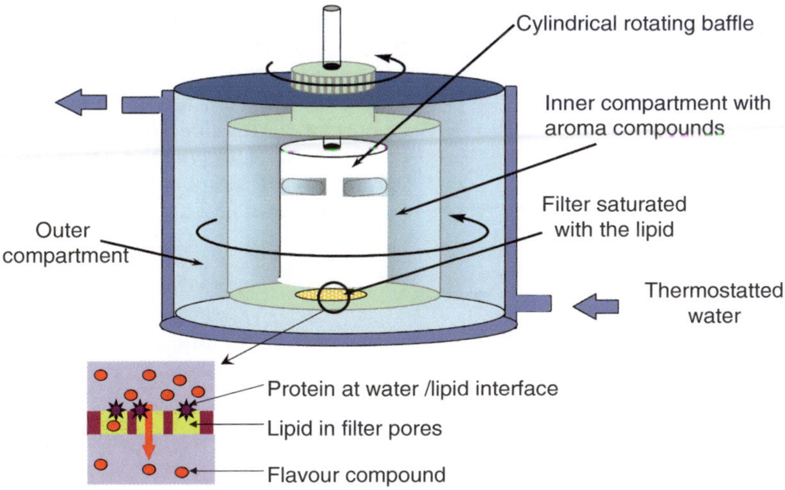

Fig. 21.5 Liquid-liquid mass transfer through a lipid layer – proteins at water/lipid interface: diagram of the rotative diffusion cell

so-called rotating diffusion cell, was developed by Harvey et al. (1995) and based on the Levich equations. It permits one to determine the resistances of mass transfer at the interfaces between two liquids such as in emulsions. The rotating diffusion cell (Fig. 21.5) makes it possible to determine the overall mass transfer coefficient, the mass transfer resistance of the interfaces, and the diffusion coefficient through the stagnant layers at the surface of the droplets, according to the equations in Fig. 21.6. When proteins are dispersed in the aqueous continuous phase or when the water content increases, the migration of aroma compounds varies according to their physicochemical properties such as, for example, solubility and hydrophobicity. Table 21.1 displays the effect of protein content on diffusivity in the aqueous phase, in oil, and at the interface. The interfacial resistance for benzaldehyde increases significantly only if β-lactoglobulin is added to 3 % (w/w) in the emulsion (Rogacheva et al. 1999).

21.3.2 *Liquid (or Vapour)–Solid: Mass Transfer Through Packaging Films*

The food industry has a strong interest in composite foods that present texture and composition contrasts such as stuffed biscuits or sandwiches. In these products, moisture transfers are often responsible for texture, colour and taste modifications. It is therefore important to control mass transfers at the liquid (or gel)–solid interface. This can be achieved through the use of edible barriers (films or coatings). Recent works show that spectroscopic techniques allow measuring the moisture

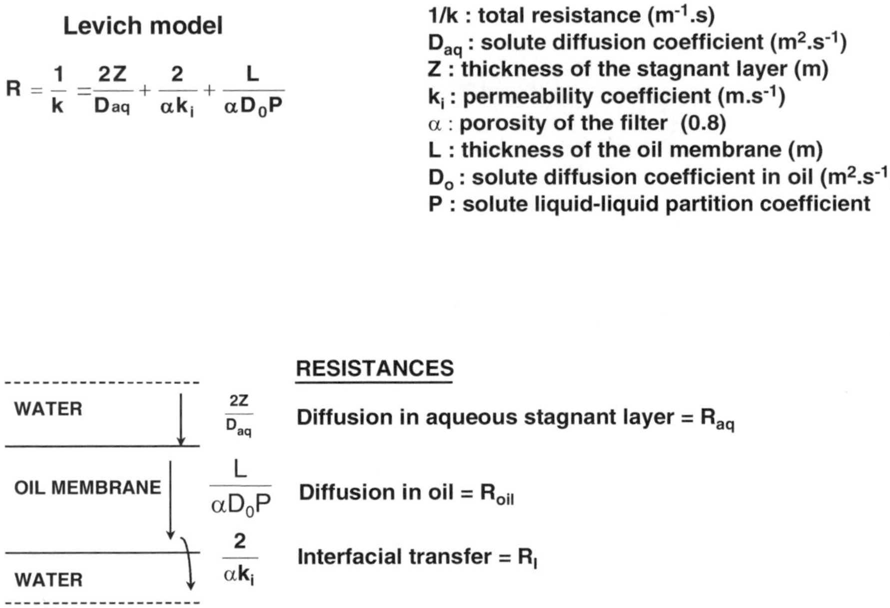

Fig. 21.6 Levich model for determining the diffusivity in the different phases and at interface

Table 21.1 Resistance to the mass transfer of some aroma compounds (Levich model) in water-lipids systems (Rogacheva et al. 1999)

Aroma compound	β-lactoglobulin (%, w/w)	R_{aq} (%)	R_{oil} (%)	R_I (%)
Benzaldehyde	0	89.2	5.9	4.9
Benzaldehyde	3	78.3	4.8	16.9
2-nonanone	0	55.7	0.3	44.0
2-nonanone (pH 3)	3	38.9	0.6	60.5
2-nonanone (pH 9)	3	64.1	0.9	35.0

diffusivity at the micrometric scale, which is very close to the interface between the different compartments of composite foods (Karbowiak et al. 2011). Figure 21.7 shows the principle of the apparent diffusivity measurement applied to films. This Fourier transform infra-red (FTIR) spectroscopic technique was used to study the apparent diffusivity of water in thin layers in relation to the surface properties. Figure 21.8 shows the importance of the scale of measurement and penetration depth of an infra-red signal for the analysis of results. From this technique, Karbowiak et al. (2011) demonstrated the importance of the film surface properties on the mass transfer. However, the interface structure can change over the course of the analysis, which can cause a wrong estimation of the diffusivity.

Moreover, the state of water affects moisture transfers through edible films in association with interfacial modifications. Hambleton et al. (2011) confirmed that the Schroeder paradox occurs in alginate films due to swelling and partial

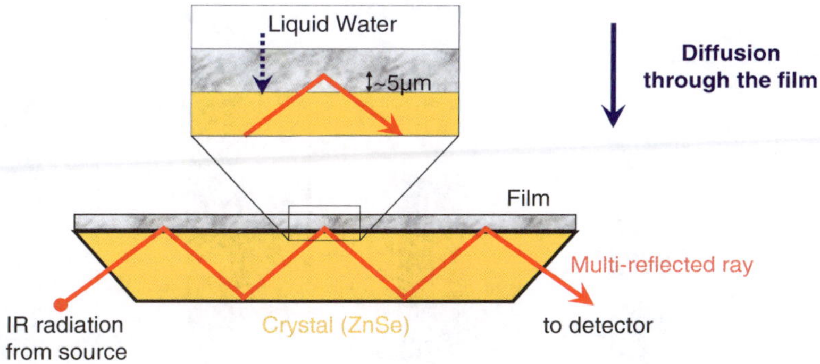

Fig. 21.7 Attenuated Total Reflectance (*ATR*) sampling unit for measuring apparent diffusivity in thin layer membranes

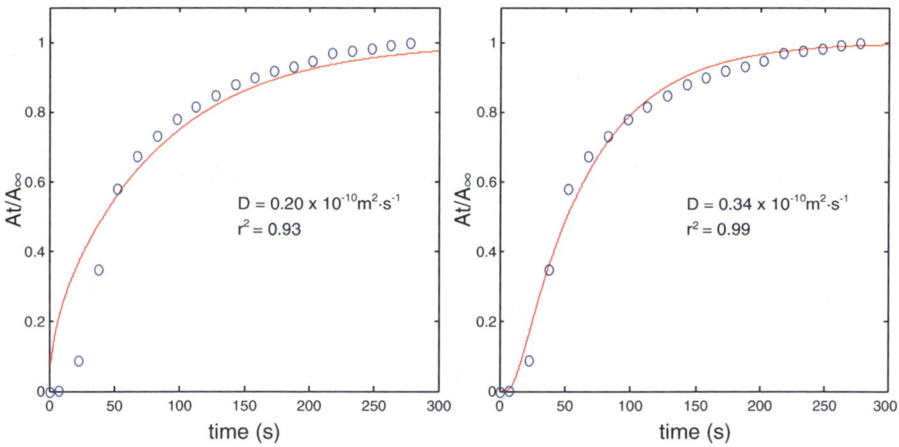

Fig. 21.8 Kinetics of the absorbance ratio obtained for a iota-carrageenan film: experimental data (○) and modeling (—) assuming a Fickian transfer, from analytical solution *(left graph)*, and by taking into account the penetration depth of the infrared radiation *(right graph)* (Adapted from Karbowiak et al. 2011)

solubilisation, as displayed in Fig. 21.9. Schroeder stated that the physical state (vapour or liquid) of the diffusing molecule at the ingress surface of a membrane or film strongly affects its transfer rate for the same chemical potential differential between the two sides of the film. He often observed a strong increase in the diffusivity when a liquid contact occurred.

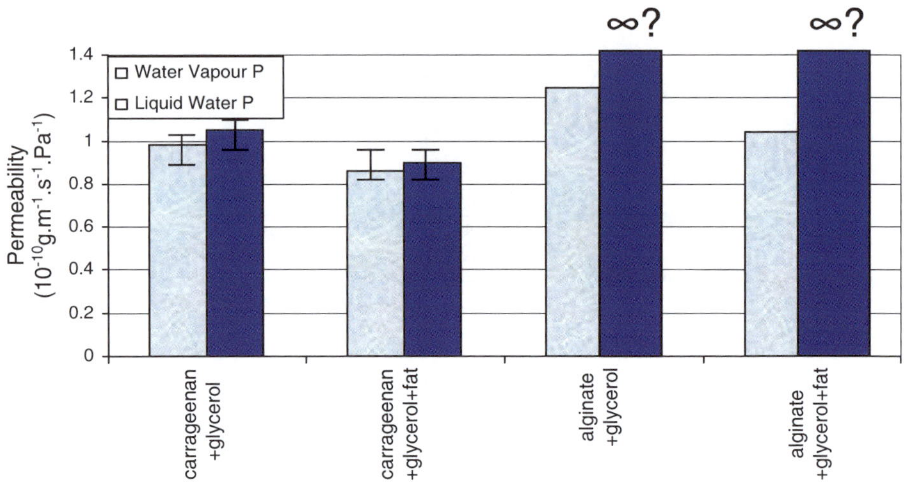

Fig. 21.9 Liquid and vapour water contact permeability in carrageenan and alginate films with glycerol and/or fat (Hambleton et al. 2011)

21.4 Application to Encapsulation of Aroma Compounds

Retention of six aroma compounds was studied following the dehydration of ternary mixtures of aroma water and beta-cyclodextrin (Goubet et al. 2001). The maximal retention of a mole of aroma per mole of beta-cyclodextrin was observed for most of the aroma compounds, whereas retention of benzyl alcohol could be twice as high. When aroma compounds were blended, it was noted that when volatile compounds compete for the same binding sites on beta-cyclodextrin, ethyl hexanoate, 2-methylbutyric acid, and benzyl alcohol are, respectively, better retained than ethyl propionate, hexanoic acid, and hexanol. The preferential retention that was observed with esters can be simply explained by the differences in their physicochemical properties (hydrophobicity), but for the acids and alcohols a study at the molecular scale was necessary. The better retention of 2-methylbutyric acid can be explained by differences in the nature of the interactions between the acids and their carrier. The selectivity of retention observed for the alcohol could be due to a difference in the location of the guest or a difference in the number of aroma molecules that could be bound per polysaccharide molecule. These results bring about changes in the aroma profile perception when encapsulated flavours are incorporated into food because of the different mechanisms involved for their encapsulation.

Encapsulation of aroma compounds (or other active compounds) with edible films could be used with the aim of better controlling their release and preventing their loss and degradation (Marcuzzo et al. 2010). A biopolymer network and the presence of fat make it possible to control the release kinetics of various aroma compounds. The hydrophobicity, the solubility, or the partition coefficient cannot explain the kinetics, which seems to depend mainly on structural characteristics. This study made it possible to investigate the possibility of using edible films as active packaging. In the same way, Gharsallaoui et al. (2012) also showed that pea proteins improve aroma retention in emulsions during spray drying.

21.5 Conclusion

All of these mass transfer phenomena in the various interface cases are capable of modifying the sensory quality of food. They can be coupled together or combined with chemical reactions in some complex systems. An integrated approach is often necessary to assess mass transfers in food systems by taking into account kinetic and thermodynamic parameters and the structural and interfacial characteristics of the matrices.

Acknowledgments We dedicate this paper to the memory of our colleague and friend David Chassagne.

References

Cayot N, Dury-Brun C, Karbowiak T, Savary G, Voilley A (2008) Measurement of transport phenomena of volatile compounds: a review. Food Res Int 41(4):349–362

Gharsallaoui A, Roudaut G, Beney L, Chambin O, Voilley A, Saurel R (2012) Properties of spray-dried food flavours microencapsulated with two-layered membranes: roles of interfacial interactions and water. Food Chem 132:1713–1720

Goubet I, Dahout C, Sémon E, Guichard E, Le Quéré J-L, Voilley A (2001) Competitive binding of aroma compounds by beta-cyclodextrin. J Agric Food Chem 49(12):5916–5922

Hambleton A, Perpiñan-Saiz N, Fabra MJ, Voilley A, Debeaufort F (2011) The Schroeder paradox or how the state of water affects the moisture transfers through edible films. Food Chem 132:1671–1678

Harvey B, Druaux C, Voilley A (1995) Effect of protein on the retention and transfer of aroma compound at the liquid water interface. In: Dickinson E, Lorient D (eds) Food macromolecule and colloids. The Royal Society of Chemistry, London, pp 154–163

Karbowiak T, Ferret E, Debeaufort F, Voilley A, Cayot P (2011) Investigation of water transfer across thin layer biopolymer films by infrared spectroscopy. J Membr Sci 370(1–2):82–90

Lequin S, Karbowiak T, Brachais L, Chassagne D, Bellat JP (2009) Adsorption equilibria of sulphur dioxide on cork. Am J Enol Viticult 60(2):138–144

Lequin S, Chassagne D, Karbowiak T, Rg G, Brachais L, Bellat J-P (2010) Adsorption equilibria of water vapor on cork. J Agric Food Chem 58(6):3438–3445

Marcuzzo E, Sensidoni A, Debeaufort F, Voilley A (2010) Encapsulation of aroma compounds in biopolymeric emulsion based edible films to control flavour release. Carbohydr Polym 80 (3):984–988

McBain JW, Bakr AM (1926) A new sorption balance. J Am Chem Soc 48:690–695

Nongonierma AB, Colas B, Springett M, Le Quéré J-L, Voilley A (2008) Influence of flavour transfer between different gel phases on perceived aroma. Food Chem 100(1):297–305

Rogacheva S, Espinoza-Diaz MA, Voilley A (1999) Transfer of aroma compounds in water-lipid systems: binding tendency of b-lactoglobulin. J Agric Food Chem 47:259–263

Seuvre AM, Espinosa Diaz MA, Voilley A (2007) Retention of aroma compounds by β-lactoglobulin in different conditions. Food Chem 77(4):421–429

Voilley A, Souchon I (2006) Flavour retention and release from the food matrix: an overview. In: Etiévan P, Voilley A (eds) Flavour in food. CRC Press, Boca Raton, pp 117–132

Weber G, Benoit F, Bellat J-P, Paulin C, Mougin P, Thomas M (2008) Selective adsorption of ethyl mercaptan on NaX zeolite. Micropor Mesopor Mat 109(1–3):184–192

Part IV
Modeling and Control of Food Processes

Chapter 22
Modeling Food Process, Quality and Safety: Frameworks and Challenges

Ashim Datta and Ashish Dhall

22.1 Introduction: Overview of Modeling Frameworks

Food materials, processes, and equipment cover a very large set of scenarios. Modeling of food processes has also been approached in a number of ways, from completely empirical to completely physics-based approaches. The question is whether we can develop a few general frameworks for modeling food processes that can effectively handle a large number of practical situations and be easily implemented (preferably in a commercial software) for widespread use of computer-aided food process engineering. In other words, what is a methodical and the most effective approach to modeling a process?

Over the years, the author's research group has modeled temperature and moisture transport in microwave heating, baking, deep frying, puffing, and meat cooking (Halder et al. 2011b). These processes cover a range of physics in addition to simple conduction and diffusion processes, for example, bread baking pressure generation inside the dough due to both evaporation and gas from fermentation, and large volume changes due to the ensuing pressures. Meat cooking involves significant moisture loss, which results in large volume changes and transport of water, while the driving force for transport of water is swelling pressure (instead of capillary pressure). In deep frying, strong evaporation inside the food matrix leads to pressure-driven flow, while oil pickup is an additional transport process of interest. Likewise, microwave heating entails enhanced pressure-driven transport of moisture from the interior to the surface. It appears that the same general framework of multiphase transport in deformable/swellable porous media with rapid evaporation is at work in this spectrum of processes. A survey of most of the existing models in the food literature shows that of the fundamental physics-based approaches, this framework is broad, flexible in accommodating many

A. Datta (✉) • A. Dhall
Cornell University, Ithaca, NY, USA
e-mail: akd1@cornell.edu

S. Yanniotis et al. (eds.), *Advances in Food Process Engineering Research and Applications*, Food Engineering Series, DOI 10.1007/978-1-4614-7906-2_22,

different processes, and easier to understand and implement in commercial software. By looking at additional processes, such as transport plus deformation plus reaction kinetics, an investigation was undertaken into how this same framework could include other processes and thus indeed be a general framework. Also investigated were the auxiliary relationships (e.g., evaporation) that would be needed for such frameworks. Finally, the frameworks were extended to quality and safety prediction.

22.2 Frameworks

The frameworks that can be used to model an arbitrary food process are summarized here. As shown in Fig. 22.1, first, all formulations are divided into three major categories – continuum formulation, porous media formulation, and deformable porous media formulation. Each of these may be called a framework. The typical continuum formulation (Framework F1) will not be discussed here, nor will the large pore formulation (Framework F2), which has been discussed elsewhere (Datta 2007). Here, we elaborate more on the small pore formulation and, especially, how to handle deformation. Thus, Fig. 22.2 shows a schematic of various possibilities for dealing with mass transport, heat transport, and deformation (Frameworks F3 and F4).

22.2.1 Framework F1: Single-phase and Multiphase Continuum Equations

The most common set of transport equations is for single-phase transport equations, as shown in any standard textbook on transport processes. Examples of processes that can be modeled using single-phase and multiphase continuum equations are shown in Table 22.1.

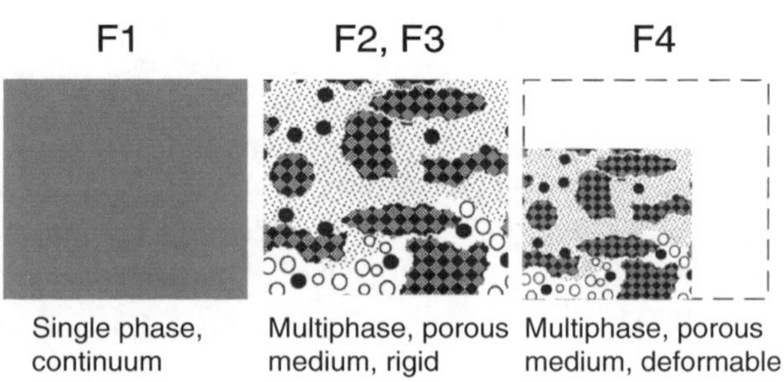

Fig. 22.1 Schematic showing three major classes of problems that are referred to as three formulations: *F1*, *F2*, and *F3* and *F4*

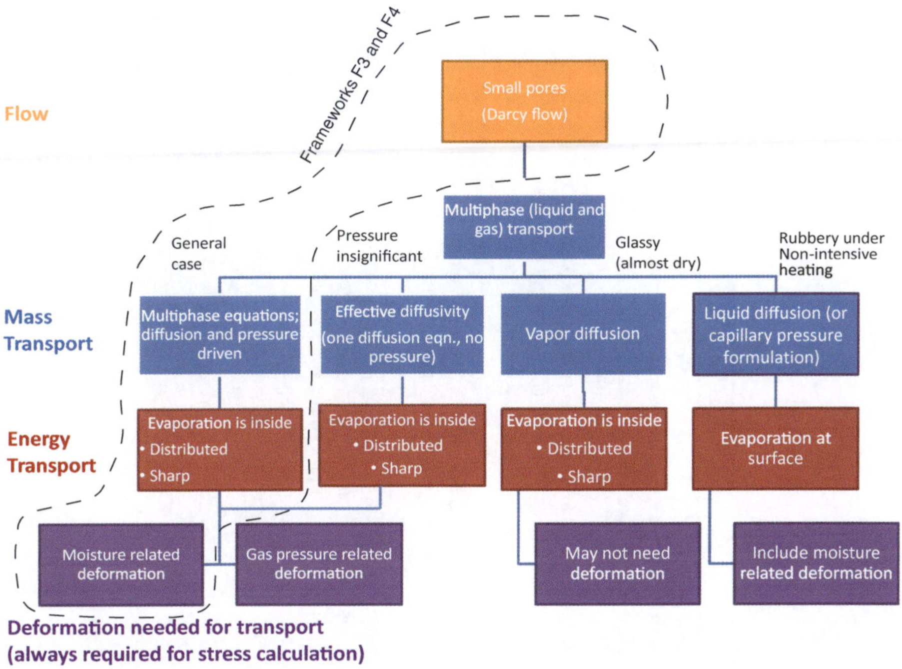

Fig. 22.2 Various formulations for food process modeling and their interrelations. For isothermal problems, the energy equation is ignored. While a deformation is always required for stress calculations, it may or may not be needed for transport

22.2.2 *Framework F2: Large Pores: Multiphase Transport with Navier–Stokes Equivalent of Darcy Equation*

An example of a problem formulation in large pores is the cooling of stacked bulk produce such as potatoes, chicory roots, and pears. The Navier-Stokes (N–S) analog of the Darcy equation, together with species transport and energy equations, is used in these studies. It is discussed elsewhere (Datta 2007) in more detail.

22.2.3 *Summary Framework F3: Multiphase, Rigid Porous Media Continuum Equations*

An enormous range of food processes can be viewed as involving the transport of heat and mass through porous media. Examples include drying, frying, microwave heating, meat roasting, puffing, and rehydration of solids. Most solid food materials can be treated as hygroscopic and capillary-porous. A porous media formulation homogenizes the real porous material and treats it as a continuum where the

Table 22.1 Frameworks for food process modeling discussed here

Framework description		Examples of food processes	Governing equations to use
F1	Single and multiphase	Simple heating of solids at lower temperatures, canning of solids, liquids, and mixtures	Momentum (N–S equation for fluid flow; species equations for each component; energy equation)
F2	Multiphase, porous media (large pores)	Cooling of stacked produce such as potatoes/strawberries in a cold room	N–S analog of Darcy equation; species equations; energy equation
F3	Multiphase, porous media (small pores), capillary pressure or gas pressure, rigid	Microwave heating, drying, frying, and other heating situations with significant moisture loss but where dimension changes can be ignored	Darcy equation replacing momentum equation; species equations; energy equation
F3.1	Effective diffusivity of combined liquid water and vapor for rigid or rubbery material	Ideally only for less intensive heating	Effective diffusivity for a combined (water + vapor) equation; energy equation
F3.2	Liquid diffusivity for rigid material or rubbery state	Less intensive heating or intensive heating of wet (rubbery) material where evaporation is only at surface (meat cooking)	Can be swelling pressure, capillary pressure, or their combination (no gas pressure); species equation for liquid phase; energy equation
F3.3	Vapor diffusivity for rigid material or glassy state	Very little liquid moisture, e.g., last stages of drying	No liquid transport; species equations for combined (primarily liquid) phase with vapor diffusivity; energy equation
F4	Multiphase, porous media (small pores), with deformation	Bread baking, puffing, rehydration	Darcy equation replacing N–S equation; species equations; energy equation; solid mechanics equations for small or large deformation providing solid velocity

pore-scale information is no longer available. Since this framework can cover a large range of situations and be simplified to various levels, the general set of equations is presented first, followed by various adaptations, which are the simplified versions.

In the most general version (for rigid porous media), individual phases of water, vapor, air, and energy are tracked using their conservation equations:

$$Water \ \frac{\partial c_w}{\partial t} + \nabla \cdot (\vec{n}_w) = -\dot{I}, \tag{22.1}$$

$$Vapor \ \frac{\partial c_v}{\partial t} + \nabla \cdot (\vec{n}_v) = \dot{I}, \tag{22.2}$$

$$Air \ \frac{\partial c_a}{\partial t} + \nabla \cdot (\vec{n}_a) = 0, \tag{22.3}$$

$$Energy \ (\rho c_p)_{eff} \frac{\partial T}{\partial t} + \left(c_{pv}\vec{n}_v + c_{pa}\vec{n}_a + c_{pw}\vec{n}_w\right) \cdot \nabla T = \nabla(k_{eff} \nabla T) - \lambda \dot{I} + \dot{Q}. \tag{22.4}$$

Here

$$\vec{n}_w = -\rho_w (k_w/\mu_w) \nabla p_g - D_{w,c_w} \nabla c_w - D_{w,T} \nabla T \tag{22.5}$$

is the flux of water due to gas pressure, the moisture dependence of capillarity, and the temperature dependence of capillarity, respectively. Similarly, flux of vapor is given by

$$\vec{n}_v = -\rho_v (k_g/\mu_g) \nabla p_g - (c^2/\rho_g) M_v M_a D_{bin} p_v / p_g^2 \nabla p_g - D_{v,c_w} \nabla c_w - D_{v,T} \nabla T \tag{22.6}$$

due to gradients in gas pressure and vapor pressure (decomposed into three separate effects representing the last three terms – binary diffusion, driven by liquid concentration and by temperature) (Dhall and Datta 2011). There is no distinct flow equation because Darcy's law used in water, vapor, and air transport equations is a replacement for the fluid flow or momentum equation. To complete the system, we need an additional equation. This additional information provides the rate of evaporation and can be formulated in one of two ways (Halder et al. 2011b):

$$p_v = p_{v,eq}(M, T), \tag{22.7}$$

$$\dot{I} = K(\rho_{v,eq} - \rho_v) S_g \varphi. \tag{22.8}$$

Equation 22.7 is the equilibrium relation for material relating the vapor pressure to moisture and temperature. Equation 22.8 is a nonequilibrium formulation that approaches Eq. 22.7 for large values of K. A number of food processes (e.g., drying, baking, frying, microwave heating) have been modeled by several researchers using mostly Eq. 22.7, but some have used Eq. 22.8.

22.2.3.1 Framework F3.1 Small Pores: Effective Diffusivity of Combined "Moisture" Phase

The liquid water and water vapor phases in the previous section can be combined into an effective "moisture" phase with the following equations that perhaps have been the most common formulations used in food:

$$\frac{\partial c_w}{\partial t} = \nabla \cdot (D_{c_w} \nabla c_w). \tag{22.9}$$

Here the pressure gradient inside the food is ignored and the transport of liquid or vapor due to the temperature gradient is also ignored. The diffusivity, D_{c_w}, is due to liquid and vapor diffusion from the gradient in the water content, $D_{c_w} = D_{w,c_w} + D_{v,c_w}$, which includes the capillary-pressure dependence on the moisture content in D_{w,c_w} and vapor-pressure dependence on the moisture content in D_{v,c_w}. Note that the rate of evaporation, $\dot{I}$, is not required if temperature is not a concern. If temperatures are required, Eq. 22.4 for energy will have convective terms contributed by the water and vapor.

22.2.3.2 Framework F3.2 Small Pores: Transport of Liquid Phase Only

Evaporation is considered only from the surface and ignores evaporation inside the domain. This can happen for intensive surface heating of a very wet material. The energy equation will have a convective term due to water only. Transport of water will be given by

$$\frac{\partial c_w}{\partial t} = \nabla \cdot (D_{w,c_w} \nabla c_w). \tag{22.10}$$

The liquid diffusivity, D_{w,c_w}, includes parameters for Darcy flow and those relating pressure to concentration. For meat cooking (Dhall and Datta 2011), for example, the liquid diffusivity includes parameters related to Darcy flow and those relating swelling pressure to water concentration (van der Sman 2007a). Additional temperature-related terms can also be present on the right-hand side (Dhall and Datta 2011).

22.2.3.3 Framework F3.3 Small Pores: Transport of Vapor Phase Only

This is a special case when very little liquid moisture is present, as near the end of a drying process. Transport is dominated by that of vapor:

$$\frac{\partial c_w}{\partial t} = \nabla \cdot (D_{v,c_w} \nabla c_w). \tag{22.11}$$

Since little liquid is present, the transport of liquid can be ignored, which then leads to the equation for $\dot{I} = -\partial c_w / \partial t$, which can be substituted into the energy equation.

22.2.3.4 Framework F4: Multiphase, Deformable Porous Media Continuum Equations

A deforming (shrinking/swelling) porous medium is essentially handled by treating all the fluxes discussed earlier for a rigid porous medium as being relative to the solid skeleton and combining this with a velocity of the solid skeleton that comes from the deformation obtained from a solid mechanical analysis. Since the solid has a finite velocity, $\vec{v}_{s,G}$, the mass flux of a species, i, with respect to a stationary observer, $\vec{n}_{i,G}$, (used in Eqs. 22.14–22.17) can be written as the sum of the flux with respect to solid and that due to the movement of the solid with respect to the stationary observer:

$$\vec{n}_{i,G} = \vec{n}_{i,s} + c_i \vec{v}_{s,G}. \tag{22.12}$$

The movement of the solid, in turn, is obtained from a stress–strain analysis. If $\bar{\sigma}'$ is the effective stress on the solid skeleton, then for an unsaturated material where the capillary pressure is dominant, and the swelling pressure and gravity can be ignored, $\bar{\sigma}'$ can be written in terms of fluid pressures as

$$\nabla . \bar{\sigma}' = \nabla p_g - \nabla (S_w p_c), \tag{22.13}$$

where the first term on the right-hand side is the gas pressure gradient (due to the evaporation of water or to gas release, as with carbon dioxide in baking) and p_c in the second term on the right-hand side is the capillary pressure that would be a function of the temperature and moisture content of the food material. In this equation, S_w is the water saturation. Kelvin's law can be used to estimate p_c from water activity. In situations where a swelling pressure is present, as in meat cooking, the right-hand side of the preceding equation has different terms, and an example of such details can be found in (Dhall and Datta 2011; van der Sman 2007a). A general form of such pressures is discussed in Datta (2012). Food material can be treated as elastic or viscoelastic, and the corresponding stress–strain relationship can be used with the appropriate solid momentum equation.

The transport equations in Framework F3 are generalized for this situation, using Eq. 22.12, as

$$\frac{\partial c_w}{\partial t} + \nabla.\left(\vec{n}_{w,G}\right) = -\dot{I} \tag{22.14}$$

$$\frac{\partial\left(c_g \omega_v\right)}{\partial t} + \nabla.\left(\vec{n}_{v,G}\right) = \dot{I} \tag{22.15}$$

$$\frac{\partial c_g}{\partial t} + \nabla.\left(\vec{n}_{g,G}\right) = \dot{I} \tag{22.16}$$

$$\left(\rho_{eff} c_{p,eff}\right)\frac{\partial T}{\partial t} + \sum_{i=w,v,g}\left(\vec{n}_{i,G}.\nabla\left(c_{p,i}T\right)\right) = \nabla.\left(k_{eff}\nabla T\right) - \lambda\dot{I} + Q \tag{22.17}$$

Here the air equation is replaced by a gas (vapor plus air) equation (Eq. 22.16), which is an equivalent form.

22.3 Advantages of Modeling Framework (F3 or F4)

Frameworks F3 and F4 described earlier are physics-based and mechanistic and use standard conservation laws and poromechanics formulations. The transport model is a departure from the effective diffusivity-based models often seen in the food process modeling literature. There are very few empirical factors in the preceding formulations. Food material aspects, such as a glassy or rubbery state, are included by having material properties that vary with the state of the material. A significant aspect of the formulations is that since they are based on standard forms of the equations, they are readily implementable in standard commercial software. Such an ability to be implemented should enable the food process modeling community to significantly reduce the time and effort needed in building a process model, bringing the full utility of modeling as a tool to design and optimization.

22.4 Challenges

Several challenges must be faced in the large-scale use of the modeling framework discussed earlier; these challenges are not insurmountable, but they would demand continuous improvement in the coming years. Such challenges can involve (1) obtaining the required properties, (2) obtaining appropriate driving forces,

(3) obtaining the appropriate boundary conditions, (4) meeting computational challenges, (5) making sure all relevant phenomena are adequately included, and (6) validating the models.

One of the bottlenecks in the large-scale use of the preceding framework is the lack of material properties. Since food material changes in temperature and composition during processing, the material must possess properties as a function of temperature and composition. Such exhaustive data will never be available from direct experimentation; predictive models are needed for properties, as compiled in Gulati and Datta (2012). Such prediction can come from thermodynamics, a soft-matter approach, effective medium theory, and homogenization, in addition to empirical correlations obtained directly from experiment. Significant work is being done with respect to a soft-matter approach, which has been successful in modeling moisture sorption and phase transitions in mixtures of water, polysaccharides, and proteins (van der Sman 2012; van der Sman and Meinders 2011) and computing the water-holding capacity of meat proteins (van der Sman 2007a,b, 2012). Another promising approach is to obtain properties through homogenization. This is also a multiscale approach (Nicolai et al. 2012) in which a detailed microstructure is obtained and the microscale problem is solved using a framework similar to that described here. The results are then homogenized to obtain macroscale properties. As the equipment and related software to obtain a complete three-dimensional microstructure and its analysis become readily available, this approach to obtaining macroscale properties is quite likely to become an important one.

The liquid pressure, p_w, in Eq. 22.13 is the driving force for liquid transport – it is part of the closure relations needed to solve the governing transport equations. In general, liquid pressure is composed of gas pressure (from evaporation or gas generation, as in bread baking), capillary pressure (in unsaturated materials), swelling pressure (arising from osmotic, elastic, and ionic components), and gravity. This is discussed in detail in Datta (2012). Difficulties arise in deciding which of these pressures are significant in a given situation and, for some of the pressures, in obtaining their numerical values.

Obtaining boundary conditions on a porous media surface can be a challenge, at least in some situations. For example, in frying, the surface heat and mass transfer coefficients vary with time in a complex manner that is difficult to determine even experimentally. In some situations, as in microwave heating of a food (porous medium) with surrounding air, the heat transfer coefficient was obtained by formulating the porous medium and its surrounding as a conjugate problem (Halder and Datta 2012).

There can be a number of numerical challenges. When one of the phases disappears (e.g., a pore is completely saturated with liquid so the gas phase becomes zero), a numerical workaround is needed. For example, the disappearing phase can be set to not drop below some very small amount (so the overall computation is not significantly affected). In some software, implementation of Darcy's law is straightforward, whereas in other software, Navier–Stokes equations need to be solved for each fluid phase, leading to a significant increase in computing time.

Although the framework is general enough and includes transport, solid mechanics, and their coupling, it is not clear whether the phenomenon of Case II diffusion can be captured in the framework. Based on the competing theories used to explain Case II diffusion (Bargmann et al. 2011), which are not unlike the framework presented here, whether Case II diffusion is included in the framework is far from obvious. Of course, a very important question in this context has to do with the product-process combination for which Case II diffusion is relevant (and must be included), which has also not been adequately answered. Planning is under way to compare the results from the framework presented here with those from a theory (Pawan 2011) that includes Case II diffusion more explicitly.

Detailed porous media models generate information on many different physical variables, including, for example, temperature, moisture, pressure, porosity development, and evaporation zones. Both the spatial and temporal variations of these quantities are available. While experimental verifications are possible, with reasonable effort, for temperatures at one physical location (using a thermocouple), spatial variation on an outside surface (using an infrared camera), loss of total moisture with time (from simple weight measurements), total change in volume with time (using a volume displacement method), and other predictions from the model can be quite difficult to verify. For example, spatial temperature and moisture profiles with time are possible (Rakesh et al. 2010) using magnetic resonance imaging (MRI), but this is hardly a routine process and the experimental work can be more time and resource consuming than the model. It would be very difficult to verify, for example, the spatial variation of the evaporation rate. Thus, comprehensive validation of these models will be difficult for years to come.

22.5 Framework for Quality Modeling

Once the process modeling framework is developed, quality and safety models can be related to the process models in a somewhat straightforward way as long as the relationship between quality parameters and temperature/moisture/composition or their histories are available. Although there are missing areas, such relationships for the kinetics of quality changes are available in a number of important areas (van Boekel 2009). A framework for the eventual prediction of quality from temperature and moisture is shown in Fig. 22.3a. We consider an effective quality attribute (e.g., texture, color) as a composite of a local (i.e., at a spatial location) quality attribute, obtained either by following relevant chemical reactions and transport or by following measured local properties. The local quality attribute, in turn, is predicted from the local temperature and composition available from the process models. The process and quality prediction models are thus coupled, closing the gap between process prediction and quality prediction for complex food processes. An application of this framework to the modeling of texture in the deep frying of potatoes is illustrated in Fig. 22.3b (Thussu and Datta 2012). Here the temperature and moisture variations from the process

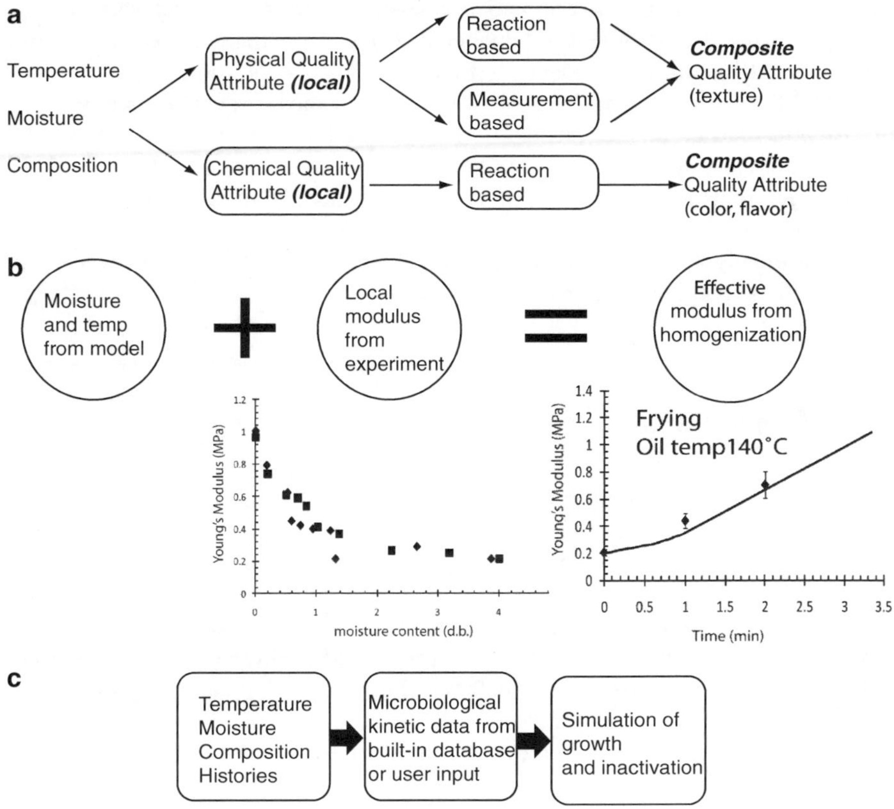

Fig. 22.3 (**a**) Quality framework. (**b**) Application of quality framework to texture prediction. (**c**) Safety framework

model are combined with the relationship of Young's modulus to the moisture obtained from thin sections of potatoes dried to various moisture contents to obtain a profile of Young's modulus throughout the cross section of the fried potatoes. Using this varying Young's modulus, a small deformation mechanical analysis is performed to obtain the effective Young's modulus of the entire cross section of the fried potato that correlates with texture.

22.6 Framework for Safety Modeling

In a very analogous way, the safety prediction framework will couple microbiological (or chemical) kinetics to temperature, moisture, and composition information obtained from the process model. The main issues in building such a framework include the selection of predictive models, associations of various food types with pathogens (as determined from outbreak histories), and variability in data from different experiments (Halder et al. 2010). More than 1,000 data sets from the published literature were analyzed and grouped by microorganism and food type. The final grouping of data consisted of the 8 most prevalent pathogens for 14 different food groups, covering all of the foods (more than 7,000) listed in the USDA National Nutrient Database. The primary advantage in obtaining group-specific kinetic data is that it confers the ability to extend microbiological growth and death simulation to a large array of product and process possibilities while still being reasonably accurate. This integration has been included in a software to be made publicly available (Halder et al. 2011a).

22.7 Conclusions

A deformable porous media framework can effectively model a large number of complex food processes. Such a generalized framework that is also easily implementable in existing software can go a long way toward enabling computer-aided food process engineering. Simplifications are possible in this framework, which resembles commonly used formulations but, starting from the most general equations, makes it easier to see the assumptions that are involved in the simplified formulations. A synthesis is provided for frameworks for modeling the largest collection of food processes. A framework for process models can be included in an overall diffusion-reaction framework to predict quality and safety.

Acknowledgements This project was partially supported by National Research Initiative Grant 2008-35503-18657 from the U.S. Department of Agriculture Cooperative State Research, Education, and Extension Service Competitive Grants program.

References

Bargmann S, McBride AT, Steinmann P (2011) Models of solvent penetration in glassy polymers with an emphasis on case II diffusion. A comparative review. Appl Mech Rev 64(1). doi:10.1115/1.4003955

Datta AK (2007) Porous media approaches to studying simultaneous heat and mass transfer in food processes. I: problem formulations. J Food Eng 80:80–95

Dhall A, Datta AK (2011) Transport in deformable food materials: a poromechanics approach. Chem Eng Sci 66(24):6482–6497

Datta AK, van der Sman R, Gulati T, Warning A (2012) Soft matter approaches as enablers for food macroscale simulation. Faraday Discuss 158:435–459

Gulati T, Datta AK (2013) Enabling computer-aided food process engineering: property estimation equations for transport phenomena-based models. J Food Eng 116:483–504

Halder A, Datta AK (2012) Surface heat and mass transfer coefficients for multiphase porous media transport models with rapid evaporation. Food Bioprod Process 90(c3):475–490

Halder A, Black DG, Davidson PM, Datta AK (2010) Development of associations and kinetic models for microbiological data to be used in comprehensive food safety prediction software. J Food Sci 75(6):R107–R120

Halder A, Dhall A, Datta AK, Black DG, Davidson PM, Li J, Zivanovic S (2011a) User-friendly general-purpose predictive software package for food safety. J Food Eng 104:173–185

Halder A, Dhall A, Datta AK (2011b) Modeling transport in porous media with phase change: applications to food processing. J Heat Transf Trans Am Soc Mech Eng 133(3):031010-1–031010-13

Ho QT, Carmeliet J, Datta AK, Defraeye T, Delele MA, Herremans E, Opara L, Ramon H, Tijskens E, van der Sman R, Van Liedekerke P, Verboven P, Nicolai BM (2013) Multiscale modeling in food engineering. J Food Eng 114(3):279–291

Nicolai BM, Datta AK, Defraeye T, Delele MA, Ho QT, Opara L, Ramon H, Tijskens E, van der Sman R, Liedekerke PV, Verboven P (2012) Multiscale modeling in food engineering. J Food Eng (To be submitted)

Pawan ST (2011) Hybrid mixture theory based moisture transport and stress development in corn kernels during drying: coupled fluid transport and stress equations. J Food Eng 105(4):663–670

Rakesh V, Seo Y, Datta AK, McCarthy KL, McCarthy MJ (2010) Heat transfer during microwave combination heating: computational modeling and MRI experiments. Am Inst Chem Eng J 56 (9):2468–2478

Thussu S, Datta AK (2012) Texture prediction during deep frying: a mechanistic approach. J Food Eng 108:111–121

van Boekel MAJS (2009) Kinetic modeling of reactions in foods. CRC/Francis & Taylor, Boca Raton, FL

van der Sman RGM (2007a) Soft condensed matter perspective on moisture transport in cooking meat. AIChE J 53(11):2986–2995

van der Sman RGM (2007b) Moisture transport during cooking of meat: an analysis based on Flory-Rehner theory. Meat Sci 76:730–738

van der Sman RGM (2012) Thermodynamics of meat proteins. Food Hydrocoll 27:529–535

van der Sman RGM, Meinders MBJ (2011) Prediction of the state diagram of starch water mixtures using the Flory–Huggins free volume theory. Soft Matter 7(2):429–442

Chapter 23
Mathematical Modeling of Transport Phenomena for Simulation and Optimization of Food Processing Operations

Ferruh Erdoğdu

23.1 Introduction

Modeling is defined as a new solution to "old" questions that has been made possible by required mathematical background and fundamental changes used to approach the problems (Wedzicha and Roberts 2006). Simulation of a food processing operation, on the other hand, can be described as imitation of the given process via use of mathematical models as a quick way of studying and evaluating this process for possible design and optimization purposes. It refers to applying computational mathematical models to predict physical events or the behavior of a system (Erdoğdu 2011), and various approaches are possible with their own challanges, interest, and difficulties (Trystram 2012). The objective is to replicate the transport phenomena governing a system to predict what will happen in processing conditions to further improve the process (Erdoğdu 2011).

Mathematical models, as a fundamental tool of the simulations, can be observation (*depending upon the availability of experimental data*) or physics based. Observational models are limited by their predictive capabilities, and hence it is necessary to define the transport phenomena governing the process by formulating the momentum, energy, and continuity equations with the initial-boundary conditions and thermophysical properties of a system (Fig. 23.1). For this reason, physics-based models are accepted as stand-ins for a mathematical model in a more descriptive way. Datta (2008) states the advantages of physics-based models as providing insight into a process that might be observable through experimentation, conferring predictive capabilities to answer any what-if scenarios, and providing improved process automation and control capabilities with and optimization of the given process. As indicated by Datta (2008), particularly physics-based modeling is an important tool to food product, process, and equipment designers that reduces

---

F. Erdoğdu (✉)
Department of Food Engineering, University of Mersin, Mersin, Turkey
e-mail: ferruherdogdu@yahoo.com

S. Yanniotis et al. (eds.), *Advances in Food Process Engineering Research and Applications*, Food Engineering Series, DOI 10.1007/978-1-4614-7906-2_23,

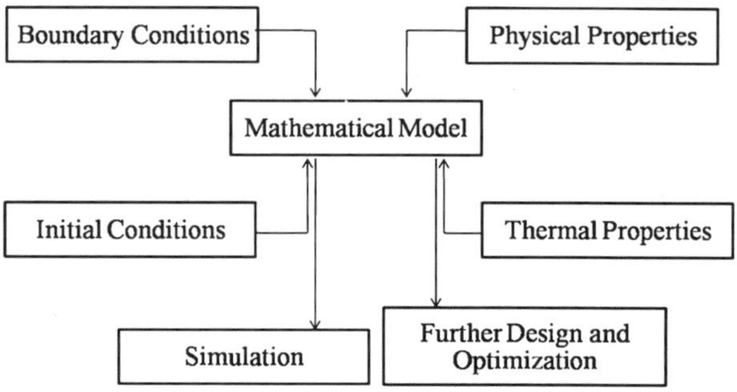

Fig. 23.1 Aspects of a mathematical model

the time and expenses involved in experimentation. To benefit from the given advantages and to overcome the possible variations in the input parameters, frequent computations are required when applying the mathematical models in simulations in a certain systematic way. Therefore, with the availability of high-power personal computers, mathematical modeling has emerged as a practical tool providing a convenient means for process evaluation, design, and optimization (Datta and Halder 2008).

23.2 Mathematical Modeling of Food Processing Operations

As reported by Datta (2002), the mathematical modeling of food processing operations has been widely applied in the food engineering literature starting with Teixeira et al. (1969), where the optimization of nutrient retention in the thermal processing of conduction-heated foods was carried out using a finite difference numerical methodology. Since that time, mathematical models have become significant for design and optimization purposes.

The mathematical modeling of food processing operations is based upon fundamental physical mechanisms governing a given process and has the benefit of providing a basic understanding of the process (Teixeira et al. 1969). In addition, quality- and safety-related issues should be implemented with a process model (Datta 2008) with a consideration of the required kinetics of the changes occurring in the system. With given fundamentals, processes can be described with a set of differential equations. Modeling then involves solving these equations for heat transfer to determine temperature distribution (*which can then be directly used in, for example, evaluating microbial destruction or the formation of a food safety issue*), for mass transfer to determine variations in concentration of certain components (*e.g., moisture content distribution*), and momentum transfer to determine the variations in velocity and momentum distributions of a fluid (*this directly affects temperature or mass concentration in the given process*). In addition, reaction kinetics might also play a significant role in simulation studies if the destruction of a certain nutrient, variation in textural or sensory attributes, or the emergence of a food safety problem as a result of applied processing technology (e.g., formation of acrylamide in french fries as a result of the frying process) is the objective to determine. Therefore, as indicated by Datta and Halder (2008), food process modeling is recognized as an interdisciplinary approach that involves combining engineering methodologies with chemistry, reaction kinetics, and predictive microbiology.

The essential aspects of a mathematical model are as follows (Singh and Vijayan 1998):

- Stating the problem with its rationale and defining any physical, chemical, or biological changes;
- Developing a mathematical basis of the process with appropriate assumptions;
- Converting the model into a numerical scheme if necessary;
- Conducting experiments or using literature data for the required properties of the product; and
- Validating the model.

Using the models, experimental methodologies can be carried out virtually in an economical and time-saving way, which brings the advantage of being able to obtain results for various processing conditions. The following equations, obtained by applying the conservation of mass and energy and Fourier's laws on a differential volume, are given for a complete heat transfer model:

- Continuity equation:

$$\frac{\partial U_j}{\partial x_j} = 0, \tag{23.1}$$

- Conservation of momentum:

$$\frac{\partial U_i}{\partial t} + \frac{\partial U_i U_j}{\partial x_j} = -\frac{1}{\rho} \cdot \frac{\partial P}{\partial x_i} + \frac{\partial}{\partial x_j}\left[\nu \cdot \left(\frac{\partial U_i}{\partial x_j} + \frac{\partial U_j}{\partial x_i} - \left\langle u_i' u_j' \right\rangle\right)\right], \tag{23.2}$$

- And conservation of energy:

$$\rho \cdot c_v \cdot \frac{\partial T}{\partial t} + \rho \cdot U_j \cdot c_p \cdot \frac{\partial T}{\partial x_j} = \frac{\partial}{\partial x_j}\left[k \cdot \frac{\partial T}{\partial x_i} - \rho \cdot c_p \cdot \left\langle u_j' T' \right\rangle\right], \tag{23.3}$$

where U is the average velocity (m/s), u' is the turbulent component of velocity (m/s), $\left\langle u_i' u_j' \right\rangle$ is the average value of the fluctuating component of velocity, T is the average temperature (K), T' is the fluctuating component of temperature (K), P is the pressure (Pa), ρ is the density (kg/m^3), ν is the kinematic viscosity of the fluid

(m^2/s), and c_p and c_v are the heat capacity ($J/kg{-}K$) at constant pressure and volume, respectively, and

$$\begin{aligned} \frac{\partial}{\partial x_i} &= \frac{\partial}{\partial x}, \frac{\partial}{\partial y}, \frac{\partial}{\partial z}, \\ \frac{\partial}{\partial x_j} &= \frac{\partial}{\partial x} + \frac{\partial}{\partial y} + \frac{\partial}{\partial z}, \\ U_j \cdot \frac{\partial}{\partial x_j} &= U_x \cdot \frac{\partial}{\partial x} + U_y \cdot \frac{\partial}{\partial y} + U_z \cdot \frac{\partial}{\partial z}. \end{aligned} \tag{23.4}$$

The viscous dissipation term on the right-hand side of the energy equation can be neglected since it is significant for high-speed flows where its magnitude is comparable to that of the conduction term. When internal heat generation is present, as in the case of microwave, radio-frequency, and ohmic heating processes, an additional term, Q (W/m^3), in the energy equation is required representing the heat-source situation. For modeling purposes in these cases, the heat generation term might be related to the electric field (E) using Maxwell's equations. Microwave and radio-frequency heating of foods is also related to the conversion of electromagnetic energy to heat within a food product. In ohmic heating processes, however, an electric current is passed through the product where it serves as electrical resistance, and internal heat is generated as a function of the applied electrical field or voltage distribution, which can be obtained using Maxwell's equations or combining Ohm's law and the continuity equation for the electric current.

In internal or external turbulent flow conditions, the velocity magnitude fluctuates with time, and these fluctuations are known as the turbulence, where the velocity term can be divided into average and turbulent components. The decomposition of the flow field into average and turbulent (fluctuating) components isolates the effects of fluctuations on the average flow. However, the addition of the turbulence results in additional terms, known as Reynolds stresses, leading to a closure problem. This obviously increases the number of unknowns to solve. For this purpose, a mathematical path for the calculation of turbulence quantities must be provided. Special turbulence models exist for resolving this issue, e.g., $\kappa - \varepsilon$, $\kappa - \omega$, Reynolds stress models, and many others. Olsson et al. (2004) gives detailed information on comparisons of these models with the experimental data available in the literature. A general approach to solving the given equations with the given terms and applied initial and boundary conditions is to use numerical solutions or a computational fluid dynamics (CFD) methodology.

In addition to the heat-mass-momentum-transfer-related food processing modeling issues, a promising field, due to improvements in imaging and magnetic-electronic beams as observational and analytical methods, concerns the structural changes in food products as a result of processing conditions (Trystram 2012). Bread baking, as demonstrated by Mondal and Datta (2008), illustrates the challenges in the mathematical modeling of structural changes occurring in a baking process. A general approach to solving the given set of equations with the applied initial and boundary conditions describing a process would be, again, to use numerical solutions or a CFD methodology via an available software program.

23.2.1 Numerical Solutions

Numerical solutions produce discrete mathematical approximations of time, spatial variations, and boundary conditions defined by partial differential equations. In these solutions, the governing partial differential equations are transformed into a series of difference equations and solved by applying different mathematical procedures. Since these solutions require a discrete number of points (a so-called grid) as a consequence of discretization, the result becomes an approximation rather than an exact solution (Nicolai et al. 2001). The error in this approximation can be reduced by increasing the number of discretized points in both space and time at the expense of computation time (Nicolai et al. 2001).

A frequently used finite difference method is the standard explicit method (*forward difference*). In its application to heat transfer problems, temperature changes over a time interval from m to $m + 1$ are due to the rate of heat flow at a time step of m. In this approach, the time step used must be small enough to prevent the violation of the second law of thermodynamics and stability problems in the solution. The standard implicit method (*backward difference*), however, is based on the rate of heat flow at a time step of $m + 1$, eliminating the possibility of the violation of the second law of thermodynamics. The standard implicit method requires the simultaneous solution of a set of equations, increasing the number of calculations, especially for two- and three-dimensional problems, to determine the temperature distribution in a given computational domain. Palazoğlu and Erdoğdu (2009) report detailed information on finite difference numerical solutions focusing on the standard explicit and implicit methods with examples.

The first step in numerical modeling is the discretization of the physical, or so-called computational, domain where discretization is carried out by division of the domain into differential elements. The number of these differential elements is based on the required accuracy of the solution since the increased number of differential elements is at the expense of computational time and plays a significant role in selecting the required time step, especially in the explicit solution methodology (Palazoğlu and Erdoğdu 2009).

While finite difference methods have been widely used, especially for regular shaped geometries, finite element and finite volume methods performed better for irregular geometries. The main difference between these two methods is the discretization method over the physical domain represented by a mesh. Finite volume methods were reported to combine the flexibility of a finite element method with the execution speed of a finite difference method (Nicolai et al. 2001; Palazoğlu and Erdoğdu 2009), and they have greater flexibility in discretizing complex geometries compared to the finite element methodology (Martins et al. 2009). Numerical modeling with finite element and finite volume methodologies was more complex compared to finite difference methods. Therefore, standard commercial software has been utilized mostly for numerical simulations using finite difference and finite volume methods (Martins et al. 2009).

Stability, convergence, and consistency are important issues in numerical solutions. A solution is said to be unstable if errors (round-off, truncation, and discretization errors) introduced at one time step grow unboundedly in the following steps; conversely, the solution is stable if these errors die out. Consistency is the property whereby truncation errors approach zero as the spatial and time increments approach zero. A finite difference representation is said to be consistent if truncation errors tend to approach zero with an increasing number of differential elements. Consistency and stability are necessary conditions for convergence, and a numerical method is convergent if the solution approaches the exact solution as both the time and space increments are reduced (Palazoğlu and Erdoğdu 2009).

23.2.2 Computational Fluid Dynamics Solutions

Besides finite difference, finite element, and finite volume methods, the CFD methodology is another powerful numerical tool that is widely used to simulate many processes in the food industry. The availability of CFD software programs relieves researchers of the difficulty of writing code to carry out numerical solutions (Datta et al. 2002). CFD is a general term used to describe numerical solutions including multiphase materials. In CFD calculations, continuity, momentum (Navier–Stokes equations), and energy equations are numerically solved for velocity, temperature, and pressure variation in a given system along with other required parameters using the finite element or finite volume methodologies.

In solving these equations with a CFD program, a general approach is to assume the flow to be incompressible where the density remains constant versus isothermal pressure changes (Datta 2008). For example, water density at 20.265 MPa differs by less than 1 % compared to a value of 1 atm. Additionally, the Boussinesq approach is used to describe the buoyancy effect as a result of changes in density in natural convection heat transfer problems. In this approach, fluid density is assumed to be the only parameter varying as a function of temperature.

The computational procedure of most CFD programs is based on using a variety of simulation approaches, including finite difference, finite element, and finite volume methods. Finite volume methods combine the flexibility of a finite element method with the execution speed of a finite difference method (Nicolai et al. 2001; Palazoğlu and Erdoğdu 2009). Datta (2008) discussed the interest in developing food operation simulation tools using the various CFD programs and highlighted their advantages, and Trystram (2011) also acknowledged the use of CFD programs for various comprehensive cases. A general solution procedure in a CFD methodology starts by composing the computational domain and continues with the preparation of a grid (mesh) structure. This is a significant part of a CFD study since nonphysical solutions might be obtained with the use of an improperly prepared mesh (grid) domain. After preparing the computational domain and mesh structure, it remains to identify the initial and boundary conditions with the thermal and physical properties and to apply a suitable solver depending ton the physics of the given process.

23.2.3 Exact (Analytical) Solutions

In all numerical calculations, the result is an approximation rather than an exact solution, and the governing fundamental physics of food processing operations have been well understood, and given governing differential equations cannot be solved analytically except in some simplified cases (Nicolai et al. 2001):

- The process involves a solid object conforming to a regular geometry (i.e., slab, cylinder, or sphere) during the entire process (with no significant changes in the physical geometry);
- The governing heat transfer mechanism is conduction;
- The initial temperature distribution is uniform;
- The surrounding medium temperature with the required parameters of the given boundary conditions is constant; and
- The thermal and physical properties are constant and isotropic.

Obtaining analytical (exact) solutions involving a nonzero volumetric heat generation term or with a nonuniform initial temperature distribution and variable thermal properties is also possible, but the results might be mathematically complex depending on the nature of the problem. Therefore, for conditions that do not conform to the foregoing simplified cases, it would be better to apply numerical solutions or CFD methodologies. In cases with a volumetric heat generation term (as in the case of ohmic, microwave, and radio-frequency heating), numerical solutions or CFD methodologies are also preferred since modeling for these cases also involves additional complexities due to the problem nature (i.e., temperature dependent and nonhomogeneous thermal and physical properties).

Given certain assumptions, analytical solutions can be obtained using different solution techniques, such as separation of variables, Green functions, and Laplace transform. Separation of variables has been widely used in obtaining analytical solutions, and this method is based on expanding a function in terms of Fourier series. The method is applied when the governing equation and differential equations representing boundary and initial conditions are homogeneous and linear, and a dependent variable, temperature, is assumed to be the product of independent variables (time and location). In cases of nonhomogeneous conditions, superposition techniques are applied to subdivide the problem into simpler problems (Çengel 2007; Özışık 1993), where the orthogonal coordinate system should first be chosen to coincide with boundary surfaces. For example, a cartesian coordinate system is used for rectangular bodies, a spherical coordinate system is used for spherical shapes, and a cylindrical coordinate system is used for cylindrical shapes.

23.2.4 Boundary Conditions

To solve the governing equations, the boundary conditions for the given process with initial conditions must be described. The boundary conditions usually encountered in heat transfer problems are the prescribed surface temperature, prescribed heat flux, and convection boundary condition. The last one, also called a boundary condition of the third kind, tends to be the prescribed heat flux for an insulation or symmetry condition, and for higher heat transfer coefficient values it leads to the prescribed surface temperature boundary condition. In the case of an additional heat source, e.g., radiant heat, this boundary condition can also be modified to include the heat source in the boundary condition. In the case of convection heat transfer, momentum equations with a continuity equation, in addition to the energy equation, enter the picture, and instead of analytical solutions, numerical solutions or CFD methodologies are preferred, and a no-slip condition at the walls might be applied. The no-slip boundary condition states that the velocity of a fluid in contact with a solid wall is equal to the wall velocity. As the name implies, there is no slip between the wall and the fluid itself. In other words, fluid particles adjacent to the wall adhere there and move with the same velocity. In conditions where the wall is motionless, fluid adjacent to the wall has zero velocity. In addition, the constant velocity profiles at the inlets and pressure outlets are defined for the solution. Inlets might also involve turbulence intensities depending on the nature of the flow. The thermal processing of food products (baking, frying, roasting, and grilling) must sometimes be coupled with moisture loss, and this effect might be required in solution methodologies via the use of an additional term in the boundary condition. In some cases, it is also possible to apply this effect in the energy equation itself. Different versions of this approach can be found in the literature for different simultaneous heat and mass transfer problems. In addition to all these complexities in food process modeling problems, the thermal and physical properties of the given system are also required.

23.2.5 Thermal and Physical Properties

Thermal and physical properties for mathematical modeling purposes can be found easily in the literature or obtained experimentally to obtain a more precise model. Excellent reviews on these properties for experimental and predictive methodologies are available. Thermal conductivity (k), specific heat (c_p), and density (ρ) are required parameters in modeling conduction heat transfer processes. In freezing and thawing processes, enthalpy is used since specific heat becomes infinitely high at the initial freezing or thawing point. For modeling convection, the viscosity of a fluid product is needed to solve the momentum equations.

Additional properties required for modeling infrared, ohmic, microwave, and radio-frequency heating processes are the radiative properties of food products (emissivity, spectral transmittance, and reflectance), dielectric properties, and electrical conductivity. In microwave and radio-frequency heating, the dielectric properties of food products, the dielectric constant and relative dielectric loss, are taken into consideration.

The heat transfer coefficient, in addition to the thermal and physical properties of food products, is another parameter required for the effect of the convective boundary condition. It depends on the thermophysical properties of the medium, the shape, dimensions, surface temperature, and surface roughness of the food product, and the velocity and turbulence of the fluid flow.

A volumetric or dimensional decrease, called the shrinkage problem, is also encountered in the thermal processing of foods, mostly due to changes in the protein structure, which lead to moisture loss. In some cases, it would be important to include shrinkage in models since changes in physical dimensions can have a large effect on the results. In some cases, it would be important to include shrinkage in models since the changes in physical dimensions can be quite effective in the results. Erdoğdu et al. (1998) evaluated changes in the physical dimensions of shrimp during thermal processing and included these data in their simulations based on a finite difference numerical model where grid reconstruction was required at each time step to map the temperature distribution in the reduced volume.

As explained, accurate modeling of a food processing operation is a complex task and requires a number of properties with some reasonable assumptions. Sensitivity analysis is sometimes needed to make a judgment regarding the acceptability of an applied assumption (Wang and Sun 2003). Experimental validation of the mathematical models is also required to be sure of the importance of the computational results.

23.3 Experimental Validation

Numerical models and CFD solutions have the ability to solve complex problems, but their results should be validated with experimental data. This could include mesh convergence, checking for mistakes, checking whether the results could be explained by commonsense physics, and comparing with the experimental data (Datta 2008). As reported by Jousse (2008), simulation studies tend to "oversell" by claiming predictability, despite the fact that the underlying models with their assumptions have not been fully validated.

Experimental validation is a difficult task, and it might also be time consuming. In some cases, analytical solutions can also be used to validate the results of numerical solutions when the models are run to simulate simpler problems. In the case of CFD solutions, Datta and Halder (2008) suggest checking the convergence of the solution, using common sense about the physics of the process, checking the results of a simpler problem (in this case, the results of an analytical solution

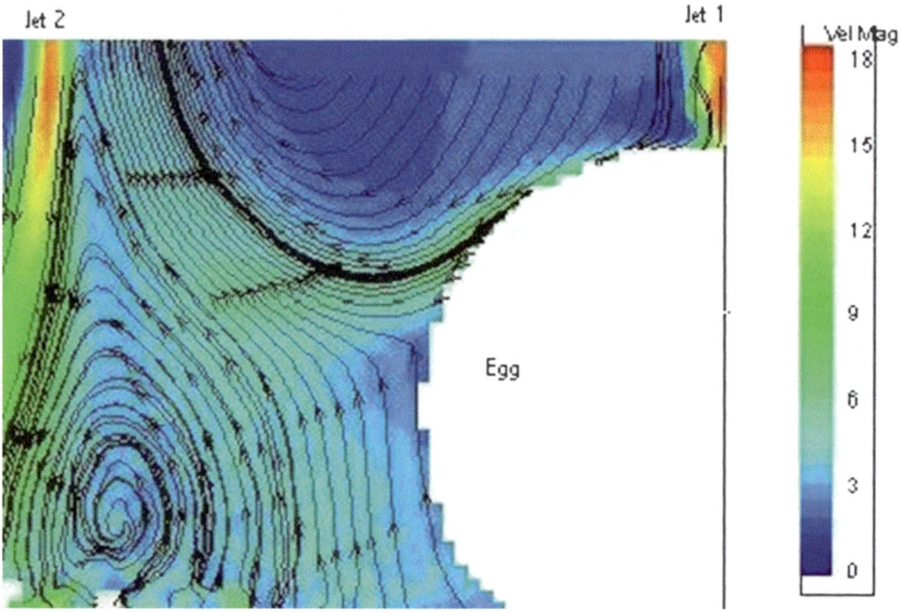

Fig. 23.2 Flow field obtained around egg with particle image velocimetry (velocity magnitude scale is in meters per second) (Reprinted from Journal of Food Engineering, Vol. 79, Erdoğdu et al. (2007), Air-impingement cooling of boiled eggs: analysis of flow visualization and heat transfer, pp. 920–928, © copyright 2007, with permission from Elsevier Ltd.)

can also be applied), and comparing the results with the experimental data. For conduction heat transfer, model validation seems to be rather easy with the suggested solutions since using analytical solutions or temperature measurements at a certain location of the food product is a simple task. However, for convection heating processes, more sophisticated technologies should be applied. For example, understanding and measuring flow field is significant for validating and improving upon the developed mathematical models. The accurate identification of the velocity field yields a correct description of its consequences (Datta 2008). In this case, particle image displacement velocimetry (PIDV), or, as it is now most often called, particle image velocimetry (PIV), is used to record the displacement of small particles embedded in a region of a fluid. A flow field image obtained using a PIV system around eggs cooled under an impinged air jet is shown in Fig. 23.2. The experimental data for the flow field, obtained in this study by Erdoğdu et al. (2007), was then used to validate the air impingement cooling model and to determine the heat flux distribution around the cooling eggs to demonstrate the significance of the variation in the flow field for the heat transfer.

Magnetic resonance imaging (MRI) is also a useful technique when temperature changes are difficult to measure. Temperature measurement by using MRI methodology is based on that the selected nuclei with a magnetic moment to be excited with a radio frequency in a magnetic field. MRI can also provide a direct measurement of flow field maps. Another method of determining the flow field maps is the use of a positron emission particle tracking technique.

Measuring the surface temperature in food processing studies is also not easy. For this purpose, liquid crystal termography methods can be applied. A liquid crystal is a chemical compound that changes its color with temperature and is available in a temperature range of −30 °C to 115 °C. Infrared cameras can also be recommended for surface temperature measurement.

The use of time-temperature indicators is another way to validate models. With a successful approach, for example using biochemical indicators (microorganisms or enzymes), the thermal history of a processed food can be evaluated without measuring temperature changes. Besides experimental validation, sensitivity analysis gives an idea of the acceptability of an assumption applied in the models (Wang and Sun 2003).

23.4 A Mathematical Modeling Example: Thermal Processing of Canned Foods

The thermal processing of canned foods is one widely used method of food preservation, and it consists of heating foods in a container in retorts under pressure for a certain amount of time. The process time is determined based on accomplishing a certain degree of sterilization. In the meantime, the degradation of nutrients occurs with changes in the texture and sensory properties of products. Thus, the thermal processing of foods has two opposing effects, both of which are time and temperature dependent. The first is characterized by a desired reduction, while the second one is characterized by an undesired reduction. To achieve optimal processing in terms of quality and safety, the mathematical modeling of heat transfer is required to determine the temperature changes of a product using the obtained sterilization value and resulting reductions of a nutrient concentration.

If the desired function of the model were just temperature changes, it would be possible to use an analytical solution for a canned solid food product (on the assumption that it was heated with conduction and the can's material had negligible effects on the heat transfer). However, in the given case, using a numerical finite difference model would be much more efficient to determine the temperature distribution and resulting nutrient retention. To this end, the required energy equation for cylindrical coordinates can be easily discretized using a finite difference scheme applying a convective boundary condition and uniform initial temperature distribution. The can material (due to its low thickness and higher thermal conductivity) might be assumed to have negligible effects on the heat transfer, or its thermophysical properties might also be included in the numerical model.

The thermophysical properties of canned food products can be obtained from the literature using the required thermal kinetic values for microbial sterility and nutrient degradation, or they can be experimentally determined. After obtaining the temperature distribution, center (the slowest heating point) temperature change of the product can be used to determine the sterilization value of the coldest point using the following equation:

$$F_0 = \int_0^t 10^{\frac{T_c(t)-T_R}{z}} \cdot dt, \tag{23.5}$$

where $T_c(t)$ is the temperature change at the coldest point, T_R is the reference temperature (generally 121.1 °C), z is the z-value of the given microorganism, and F_0 is the sterilization value. Retention of any nutrient in the overall volume of the product after processing is given by

$$C = \frac{1}{V}\int_0^V \left[10^{-\frac{1}{D_R}\int_0^t 10^{\frac{T(v,t)-T_R}{z}} \cdot dt}\right] \cdot dV, \tag{23.6}$$

where C is the retention of the nutrient, V is the volume of the product, T_R is the reference temperature, z is the z-value of the given nutrient, D_R is the reference thermal death time (time required to destroy 90 % of the nutrient at a given reference temperature), and $T(v,t)$ is the temperature distribution over time through the whole volume of the product. All these calculations should be carried out with a view toward developing code for the solution of finite difference equations and integrations or applying a CFD methodology.

It might be assumed that food products like canned tuna, thick syrups, purees, and concentrates were heated by conduction. In this case, the required processing time is determined by an analytical or numerical solution to the heat conduction equation as explained earlier. In the case of liquid products, however, natural convection heating should be modeled using the continuity, energy, and momentum equations with the appropriate boundary conditions. In this case, a CFD methodology is the best and easiest approach because the continuity, energy, and momentum equations must be solved simultaneously. For convection-heated foods, the slowest heating region, rather than the coldest point, should be considered due to the effect of the evolved natural convection currents. Figure 23.3 shows the temperature distribution in a vertical can for conduction-heated (Fig. 23.3a) and convection-heated liquid food product (Fig. 23.3b) in the vertical cross section of a can.

Fig. 23.3 Temperature distribution in a vertical can for conduction-heated (**a**) and convection-heated liquid food product (**b**) in vertical cross section of a can (left-hand side is the center line where the axisymmetric boundary conditions are applied) (Reprinted from the Journal of Food Process Engineering, Vol. 33, Erdoğdu et al. (2010), Experimental comparison of natural convection and conduction heat transfer, pp. 85–100, © copyright 2010, with permission from Wiley)

23.5 Optimization

Optimization is the choice of the best alternative from a specified set of alternatives. A formal description of any optimization problem has three parts: sets of variables to control and specify the alternatives, sets of requirements (constraints) to achieve and satisfy and a certain measure of performance, and an objective function to compare the alternatives satisfying the requirements (Erdoğdu 2009). Deciding on the constraints is a significant part of the optimization problem because this decision is dependent on the process characteristics. For example, Saguy (1988) reported a discussion on the constraints on quality optimization in aseptic processing.

Finding an optimal solution for a food processing problem by selecting the best condition, satisfying the given requirements from a large set of alternatives, is a difficult process, and the presence of a mathematical model in an optimization routine is needed to accomplish this task. Powerful mathematical-model-based optimization procedures use rigorous time-dependent models instead of applying a simple trial-and-error technique or an exhaustive statistical model (Banga et al. 2008).

Since food processing operations might involve several coupled time-dependent transport phenomena, the optimization of such a process requires a physics-based mathematical model for use in a systematic search technique while applying explicit and implicit constraints.

As indicated by Banga et al. (2008), optimization is more than simply finding an improvement over an existing condition; it means finding the best solution. Hence, as occasionally reported in the food engineering literature, simple trial and error or exhaustive searches are not proper optimization methods because of their fundamental limitations as, for example, in the case of a response surface methodology (Banga et al. 2008). Model-based optimization for food processes was first applied by Teixeira et al. (1969) and subsequently by Harper and Wanninge (1970) and Saguy (1982), and it is still quite common to find statistical-empirical equation-based models being used as optimization models in various food processing operations. For food process optimization, the first step is to develop a physics-based model with a good representation of the process validated under various situations.

The major challenge in developing an optimization model for a food processing operation lies in the possible presence of multiple suboptimal (due to nonconvexity) and equivalent (nonidentifiability) solutions (Banga et al. 2008). The use of global optimization methodologies might help solve the nonconvexity problem, but non-identifiability is one complex problem to deal with. A physics-based optimization model might be a powerful tool for food process design and optimization. In addition, the economic impact on efficiency, product quality, and safety, among all variables and constraints, should be taken into consideration to establish a corrective control action for achieving the desired process performance (Rogers 1985).

23.6 Conclusion

Knowledge of mathematical modeling techniques provides significant information for simulation and further design and optimization purposes. For this reason, mathematical modeling approaches to food processing simulation were summarized by focusing on heat transfer and fluid dynamics. As explained, numerical methods have their own complexities and advantages over analytical (exact) solutions, while CFD methodologies introduce new and innovative approaches into the picture. In conclusion, mathematical models for simulation and optimization studies can be accepted as a perfect way of creating alternative plans and addressing what-if questions for a given food processing operation without laborious and time-consuming experiments.

References

Banga JR, Balsa-Canto E, Alonso AA (2008) Quality and safety models and optimization as part of computer-integrated manufacturing. Compr Rev Food Sci Food Saf 7:168–174
Çengel YA (2007) Heat transfer: a practical approach. McGraw Hill Inc., New York
Datta AK (2008) Status of physics-based models in the design of food products, processes and equipment. Compr Rev Food Sci Food Saf 7:121–129
Datta AK, Halder A (2008) Status of food process modeling and where do we go from here (synthesis of the outcome from brainstorming). Compr Rev Food Sci Food Saf 7:117–120
Datta AK (2002) Simulation-based design of food products and processes. In: Welti-Chanes J (ed) Engineering and food for the 21st century. CRC Press, Boca Raton
Erdoğdu F (2009) Optimization: an introduction. In: Erdoğdu F (ed) Optimization in food engineering. CRE Press – Taylor & Francis Group, Boca Raton
Erdoğdu F (2011) Mathematical models for simulation of food processing operations. In: Heldman DR, Morau CI (eds) Encyclopedia of agricultural, food and biological engineering, 2nd edn. CRC Press – Taylor & Francis Group, Boca Raton
Erdoğdu F, Balaban MO, Chau KV (1998) Modeling of heat conduction in elliptical cross-section II. Adaptation to thermal processing of shrimp. J Food Eng 38:241–258
Erdoğdu F, Ferrua M, Singh SK, Singh RP (2007) Air impingement cooling of boiled eggs: analysis of flow visualization and heat transfer. J Food Eng 79:920–928
Erdogdu F, Uyar R, Palazoglu TK (2010) Experimental comparison of natural convection and conduction heat transfer. J Food Process Eng 33:85–100
Harper JM, Wanninge LA (1970) Process modelling and optimization. 3. Optimization. Food Technol 24(5):590
Jousse F (2008) Modelling to improve the efficiency of product and process development. Compr Rev Food Sci Food Saf 7:175–181
Martins RC, Lopes VV, Vicente AA, Teixeira JA (2009) Numerical solutions-finite element and finite volume methods. In: Erdoğdu F (ed) Optimization in food engineering. CRC Press, Boca Raton, Chap. 4
Mondal A, Datta AK (2008) Bread baking: a review. J Food Eng 86:465–474
Nicolai BM, Verboven P, Scheerlinck N (2001) Simulation-based design of food products and processes. In: Tijskens LMM (ed) Food process modeling. Woodhead Publishing Ltd, Cambridge, UK
Olsson EEM, Ahrne LM, Tragardh AC (2004) Heat transfer from a slot air jet impinging on a circular cylinder. J Food Eng 63:393–404
Özışık MN (1993) Heat conduction. Wiley, New York
Palazoğlu TK, Erdoğdu F (2009) Numerical solutions-finite difference methods. In: Erdoğdu F (ed) Optimization in food engineering. CRC Press, Boca Raton, Chap. 3
Rogers JA (1985) Optimizing process and economic aims. Chem Eng 9(23):95–99
Saguy I (1982) Optimization theory, techniques and their implementation in the food industry-introduction. Food Technol 36(7):87
Saguy I (1988) Constraints to quality optimization in aseptic processing. J Food Sci 53 (306–307):310
Singh RP, Vijayan J (1998) Predictive modeling in food process design. Food Sci Technol Int 4:303–310
Teixeira AA, Dixon JR, Zahradnik JW, Zinsmeister GE (1969) Computer optimization of nutrient retention in the thermal processing of conduction-heated foods. Food Technol 23(6):137–142
Trystram (2012) Modelling of food and food processes. J Food Eng 110:269–277
Wang L, Sun D-W (2003) Recent developments in numerical modelling of heating and cooling processes in the food industry – a review. Trends Food Sci Technol 14:408–423
Wedzicha M, Roberts C (2006) Modeling: a new solution to old problems in the food industry. Food Manuf Effic 1:1–17

Chapter 24
Food Preservation Process Design

Dennis R. Heldman

24.1 Introduction

Food preservation processes have evolved over a significant period of time. Early forms of preservation were achieved using solar radiation for food dehydration and the use of reduced temperatures to extend the storage life of a food. A more defined quantitative approach to designing the preservation process has evolved with the use of elevated temperatures to ensure microbiological safety and control product spoilage. The design of preservation processes has become quantitative as more information about the response of microbial populations to elevated temperatures has become available. In addition, quantitative process design has been enhanced through application of thermal energy transfer models and availability of reliable physical properties data for foods.

The overall objective of this chapter is to present and discuss the current status of preservation processes for foods. The specific objectives are as follows:

1. To present the key expressions and models for the description of microbial populations during preservation processes,
2. To present and discuss the transport phenomenon that occurs within food products during preservation processes,
3. To present and discuss the process design models needed to ensure target reductions in microbial populations and achieve maximum retention of product quality attributes.

The contents of this paper will focus on the quantitative evaluation of preservation processes for food products. The process design concepts to be presented will build on the long and successful history of thermal process design but will extend the analysis to combination processes and to nonthermal technologies, such as

D.R. Heldman (✉)
The Ohio State University, 43215 Columbus, OH, USA
e-mail: drheldman@earthlink.net

S. Yanniotis et al. (eds.), *Advances in Food Process Engineering Research and Applications*, Food Engineering Series, DOI 10.1007/978-1-4614-7906-2_24,

ultra-high-pressure and (UHP) pulsed electric fields (PEFs). In addition, the analysis will include concepts needed to estimate the impact of a process on food components, including nutrients and other product quality attributes. Finally, the opportunities for optimization of preservation processes to achieve process efficiency and product quality retention will be explored.

24.2 Process Design Models

24.2.1 Kinetic Models

The models available for application to preservation processes include kinetic models that describe changes in microbial populations and product quality attributes. During any preservation process, most components of the food are changed in some manner. The quantitative changes can be described by reaction kinetic models. These models describe the impact of the process on the microbial populations (both pathogens and spoilage) in the product as well as impact on product quality attributes.

24.2.1.1 Microbial Kinetics

The first-order model to describe the reduction in a microbial population as a function of time during a preservation process is

$$N = N_o \exp(-kt), \tag{24.1}$$

where N is the population at any time (t), N_o is the initial population, and k is the first-order rate constant. Preservation processes based on the use of thermal processes have used the following form of the first-order model:

$$N/N_o = 10^{-t/D}, \tag{24.2}$$

where D is the decimal reduction time.

There exists an inverse relationship between the decimal reduction time (D) and the first-order rate constant (k). More recently, models to account for the influence of non-log-linear microbial survivor curves on process design have been proposed. One of the models proposed for non-log-linear microbial survivor curves is (Van Boekel 1996)

$$\log N_o - \log\ N = [\ t/D']^n, \tag{24.3}$$

where D' is a time constant similar to the decimal reduction time (D) and n is a coefficient to account for deviations from a log-linear relationship.

A typical model used to describe the influence of temperature on the rate constant is the Arrhenius equation:

$$k = k^{o} \exp[-E_A/R\mathbf{T}], \tag{24.4}$$

where k^{o} is a preexponential factor, R is the gas constant, **T** is the absolute temperature, and E_A is the activation energy.

Pressure has emerged as a potential preservation technology, and the intensity of pressure will impact the rates of inactivation for microbial populations. The expression proposed for this relationship is

$$k = k^{o} \exp[-\Delta V\ P/R\mathbf{T}], \tag{24.5}$$

where P is pressure, R is the gas constant, k is the rate constant, **T** is the absolute temperature, and ΔV is the activation volume (m^3/mol). When the magnitude of V is negative, the reaction rate increases with increasing pressure.

The traditional approach to describing the influence of temperature on the rate of reduction in a microbial population defines a thermal resistance coefficient (z). The relationship between the decimal reduction time (D) and temperature is expressed by the following equation:

$$\mathrm{Log}[\ D_1/D_2] = (T_2 - T_1)/z, \tag{24.6}$$

where D_1 is the decimal reduction time at T_1, D_2 is the decimal reduction time at T_2, and z is the thermal resistance coefficient (z). A comparison of Eqs. 24.5 and 24.6, reveals a relationship between the thermal resistance coefficient (z) and the activation energy constant (E_A):

$$E_A = 2.303\ R\mathbf{T}^2/z, \tag{24.7}$$

where **T** is the absolute temperature. Although the range of temperature for Eq. 24.7 may not be specified, it must be limited to the range of temperatures used to generate the thermal resistance coefficient (z).

24.2.1.2 Quality Attributes

In general, changes in food quality attributes during a preservation process can be described by kinetic models similar to those presented for microbial populations. The applications of models to describe changes in quality attributes are associated with concerns about the retention of food quality attributes during preservation processes. The magnitudes of most kinetic parameters for quality attributes indicate that the use of higher-temperature processes for shorter time periods will improve

Table 24.1 Typical kinetic parameters for retention of heat-sensitive vitamins in foods

Preservation process	Vitamin	Rate constant	Agent intensity	Agent coefficients	Substrate
Thermal	Ascorbic acid	0.00900/min	132.2 °C	164.4 kJ/mol	Canned peas
Thermal	Ascorbic acid	0.0967/min	150 °C	117.6 kJ/mol	Orange juice
Thermal	Thiamine	0.002511/min	98 °C	113.4 kJ/mol	Meatloaf
Thermal	Thiamine	0.0435/min	138 °C	97.1 kJ/mol	Peas
Thermal	5-Methyl	0.249/min	70 °C	33.1 kJ/mol	Apple juice
Pressure	Ascorbic acid	0.005744/min	850 MPa	74.6 kJ/mol[a]	Tomato juice
Pressure	Ascorbic acid	0.010289/min	850 MPa	84.1 kJ/mol[a]	Orange juice

[a]Agent coefficient represents impact of temperature on rate constant from Heldman (2011)

the retention of quality attributes while still achieving the goals of the preservation process.

Villota and Hawkes (2007) presented an excellent review of the reaction kinetics associated with changes in the quality attributes in foods. The changes in vitamin concentration during preservation processes, storage, and distribution have been described in terms of first-order kinetics, and the kinetic parameters include first-order rate constants and activation energy constants. The magnitudes of typical kinetic parameters for vitamin retention are presented in Table 24.1. It should be noted that limited amounts of data are available for preservation processes other than thermal processes.

Similar kinetic parameters for the retention of pigment intensity during preservation processes, storage, and distribution have been assembled. The kinetic parameters for the retention of quality attributes in foods should be viewed as representative of the retention of most product quality attributes.

24.2.1.3 Physical Transport Models

The intensity of the preservation processes are evaluated by measurement of one or more physical parameters within the product. Many of these parameters can be predicted by physical transport models and are a function of product composition. Ultimately, the models can be used to predict the intensity of the process parameter at any location within the product structure.

The design of a preservation process depends on the physical properties of the product and the associated physical phenomenon. For a thermal process, the application depends on the transport of thermal energy within the product structure as required to increase the product temperature, as well as the decline in temperature during cooling of the product. The application of an UHP process requires the application of the pressure to the product structure and the uniform distribution of pressure throughout the product structure. In addition, the thermal energy associated with an increase in pressure will result in a temperature increase, and the resulting temperature distribution within the product must be analyzed.

The applications of transport phenomenon models for the prediction of temperature distribution histories within a food product structure depend on access to reliable physical property data. The relationships published by Choi and Okos (1986) are used to predict reliable physical properties of foods based on product composition. During unsteady-state heat transfer, Pflug et al. (1965) proposed the following solution to the appropriate partial differential equations to be used to predict temperature a a function of time and location with the product.

$$\log(T_a - T) = -\frac{t}{f_h} + \log[j_c(T_a - T_i)], \tag{24.8}$$

where the magnitudes of the two parameters f_h and j_c can be evaluated experimentally or predicted from physical the properties of the product and the boundary conditions. These parameters can be predicted from relationships with the Biot number:

$$N_{Bi} = \frac{hd_c}{k}, \tag{24.9}$$

where d_c is a characteristic dimension and k is the thermal conductivity of the product.

The magnitude of Biot numbers ranges from less than 1.0 to over 40, depending on the limiting conditions for heat transfer. The characteristic dimension (d_c) applies to all geometries and is the shortest distance from the geometric center to the surface. The use of these models to predict temperature distributions within food products was demonstrated in Heldman (2011).

24.2.2 Preservation Models

24.2.2.1 Preservation Process Design

The models for the design of a preservation process involve the integration of the appropriate kinetic model with the appropriate physical transport model. Although there have been limited attempts to demonstrate a quantitative process design for the prediction of changes in food quality attributes, the approach can be used in process optimization.

The process for the preservation of a food product depends on the target reduction in the microbial population and the physical phenomenon causing the reduction in the microbial population. Models to describe the preservation process require the integration of a model to predict reductions in the microbial population, with a model to predict the intensity of the physical phenomenon within the product mass. The output from the integration is the process time. The impact of the process on the quality attributes of the food product can be evaluated using the same integration steps.

24.2.2.2 Applications of Models

The preservation process time is the time required to ensure that the microbial population is reduced to a target level. Although a straightforward application of the survivor curve equation would define a time, the definition of the appropriate process time must account for the location within the product mass and the variability in the intensity of the physical phenomenon within the product structure. Most of the quantitative guidance for the development of the process time for preservation processes has been derived from the design of thermal processes (commercial sterilization) for shelf-stable foods. The process time is important for the operator of the process and the specific process conditions associated with ensuring that the target microbial population is reduced to the target level by the thermal process.

In general, the process time for a thermal process is established using Eqs. 24.1, 24.4, and 24.8. Equation 24.1 can be used to estimate the reduction in the microbial population for a given time increment when the rate constant is given. As indicated by Eq. 24.4, the rate constant for any time will be a function of the temperature during the time increment. Equation 24.8 can be used to predict the temperature at a specific location within the product at the specified time increment. Using these three equations during a stepwise integration of the temperature at the geometric center of the product during a thermal process, the reduction in the microbial population at that location can be predicted. The process time becomes the time required to reduce the microbial population to the target level.

The published literature for thermal processes provides detailed descriptions of procedures for integrating kinetic models with models for transport phenomena. These same procedures can be adapted for alternative preservation technologies. Most often, these processes are based on first-order survivor curves. If the survivor curve cannot be described by a first-order model, then an appropriate alternative model is selected.

The same approach can be used to predict the changes in a quality attribute of the food product during the preservation process. A model equivalent to Eq. 24.1 is used to predict the retention of a quality attribute as a function of time and location within the product structure. After integration over the dimensions of the product structure, the mass average retention of the quality attribute as a function of time is established. Ultimately, the quality attribute retention at the process time based on the target reduction in the microbial population is obtained.

Specific expressions for the prediction of the temperature distribution history within the product during thermal processes depend on the geometry of the product (or package/container). For liquid foods, the expressions are unique for the heat exchanger used for the process. For nonthermal preservation processes, appropriate alternative expressions would be selected to describe the distribution of agent intensity within the product structure as a function of time.

24.2.2.3 Optimization

Although the primary purpose of a preservation process is to ensure food product safety or acceptable levels of product spoilage, the impact of the process on product quality attributes has been receiving increased attention. Due to the different mechanisms associated with various processes, the impacts on product quality will vary, but these changes should be considered and minimized during the design of the preservation process. In general, a more intense preservation process is expected to be more detrimental to the quality attributes of the food product. This is most evident with thermal processes, when high temperatures accelerate the loss of temperature-sensitive food components. All processes must be carefully evaluated using the kinetic parameters of the target microbial populations as well as similar parameters of the temperature-sensitive product quality attributes. Due to the differences in the magnitudes of the kinetic parameters, opportunities for optimization of these processes become evident. Process optimization refers to defining the process needed to ensure product safety or an acceptable spoilage rate while ensuring the maximum retention of a product quality attribute. This approach to process optimization can be illustrated using the traditional preservation process design for a thermal process.

When a thermal process is applied to a product in a container or package, the temperature distribution within the product structure may be significant. In these situations, a process time is established by the time needed to ensure that the microbial population at the slowest heating location is reduced to an established target level. All other locations within the product structure are exposed to higher temperatures or longer times, and more intense processes than that required for the target reduction in the microbial population. It follows that the quality retention at other locations within the product structure will be less than at the slowest heating location. In general, quality retention will be greater within the internal regions of the product, as compared to those portions of the product located near the container surface in direct contact with the heating medium. The overall quality retention is predicted by the integration over the entire volume of the product structure or container.

The concept of optimization of a thermal process for conduction-heating food products was illustrated by Teixeira et al. (1969). The results of this investigation demonstrated that a defined temperature–time process could be identified for the maximum retention of a quality attribute. The optimum process was established by comparing a series of temperature–time processes. Each combination of heating medium temperature and process time provided the same reduction in a microbial population at the geometric center of the container. The results of the analysis indicated that the maximum retention of thiamine occurred at a specified heating medium temperature for a given process time. The thiamine retention is based on mass or volume average within the product container and considers significant reductions in thiamine retention in the outer regions of the container (near the heating medium) as compared to the higher retention in the regions at a greater

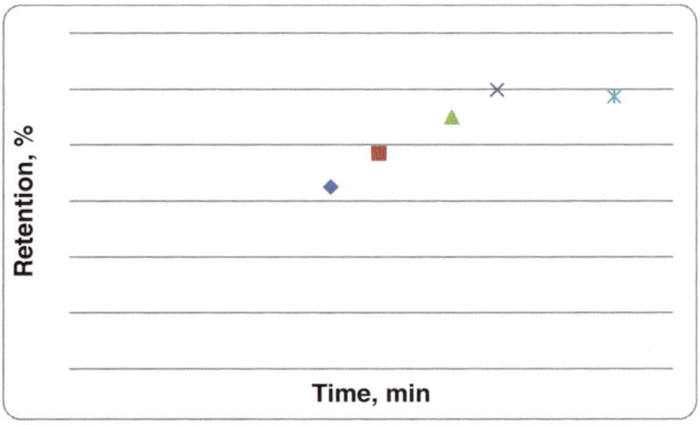

Fig. 24.1 Optimization of a thermal process for maximum quality retention [From Heldman (2011)]

distance from the surface. Since a greater proportion of the product mass is located in the outer regions of the container, the influence of quality retention in these locations on overall retention is more important and will influence the magnitudes of retention for the optimum process.

As illustrated in Fig. 24.1, the maximum quality retention (49.9 %) occurs when the heating medium temperature is 120 °C and the process time is 113.4 min. When the process time is decreased to 101.4 min (heating medium temperature = 125 °C), the quality retention decreases to 45.1 %. When the process time is increased to 144.5 min (heating medium temperature = 115 °C), the quality retention decreases to 48.7 %. The distribution of retentions with process time and heating medium temperature will be different for different container dimensions and shapes and for different microbial populations and kinetics of product quality attributes.

24.3 Conclusions

The design of most preservation processes involves the reduction in the microbial population within the food product. The reduction in the microbial population can be expressed in terms of kinetic parameters, and the magnitude of the kinetic parameters is a function of the preservation agent intensity. The design must consider the distribution of preservation intensity within the product structure and ensure that the target reduction in the microbial population is achieved at all locations. Appropriate analysis of the transport phenomenon for the preservation agent is used to evaluate the distribution of the preservation agent. The preservation process affects the quality of the product, and the changes in quality can be described by kinetic parameters for one or more of the quality attributes. An optimization analysis based on the kinetic parameters can lead to the identification of the process showing the maximum retention of product quality while meeting the expectations of the microbial safety of the product.

References

Choi Y, Okos MR (1986) Effects of temperature and composition on thermal properties of food. In: Le Maguer M, Jelen P (eds) Food engineering and process applications, Vol. 1, Transport phenomenon. Elsevier Applied Science, London, pp 93–101

Heldman DR (2011) Food preservation process design. Elsevier, San Diego

Pflug IJ, Blaisdell JL, Kopelman IJ (1965) Developing temperature-time curves for objects that can be approximated by a sphere, infinite plate or infinite cylinder. ASHRAE Trans 71 (1):238–248

Teixeira AA, Dixon JR, Zahradnik JW, Zinsmeinster GE (1969) Computer optimization of nutrient retention in thermally processing of conduction-heating foods. Food Technol 23:137

Van Boekel MAJS (1996) Statistical aspects of kinetic modeling for food science problems. J Food Sci 61(3):477–489

Villota R, Hawkes JG (2007) Reaction kinetics in food systems. In: Heldman DR, Lund DB (eds) Handbook of food engineering, 2nd edn. CRC Press, Taylor & Francis Group, Boca Raton

Chapter 25
Advanced Sensors, Quality Attributes, and Modeling in Food Process Control

Michael J. McCarthy and Kathryn L. McCarthy

25.1 Introduction

Food production and processing has evolved into a global system that is complex, immense, and rapidly changing. Producers and processors require methods to trace raw materials, determine purity or adulteration, validate safety, optimize utilization of raw materials, monitor product quality, and incorporate process sensors and mathematical models that allow modifying processing conditions in real time. Considerable advances have been made in developing sensor technology specific to measuring food-ingredient and product properties. Research efforts in the last 20 years in the area have focused on developing nondestructive sensors for application in raw material testing, process control, and final product quality evaluation. A wide range of technologies exist that can be used to measure food quality parameters including ultrasound, Raman spectroscopy, near infrared spectroscopy (NIR), mid-infrared spectroscopy, hyperspectral imaging, and magnetic resonance spectroscopy (Irudayaraj and Reh 2008). While there is a considerable range of literature and research-based demonstrations of the utility of these sensors, few industrial applications have been implemented.

Viscosity is a physical property affecting final product quality for a wide range of fluids in the food industry. Viscosity measurements are critical to process control of unit operations and final end use (Steffe 1996). Current inline viscometers typically measure viscosity only at one or a few shear rates (Cullen et al. 2000). For instance, conventional tube viscometry yields only one viscosity data point from a single measurement. Control of the process and product quality would be improved significantly if viscosity over a wide range of shear rates could be measured for non-Newtonian fluids. Magnetic resonance imaging (MRI) can be used as a viscometer, based on analysis of a measured velocity profile of fluid

M.J. McCarthy (✉) • K.L. McCarthy
Department of Food Science and Technology, University of California, Davis, CA, USA
e-mail: mjmccarthy@ucdavis.edu

S. Yanniotis et al. (eds.), *Advances in Food Process Engineering Research and Applications*, Food Engineering Series, DOI 10.1007/978-1-4614-7906-2_25,

flowing in a tube coupled with a simultaneous measurement of the pressure drop driving the flow (Maneval et al. 1996; McCarthy 1994; Choi et al. 2002). This type of measurement is well suited for rheological characterization of non-Newtonian fluids.

In this chapter we will describe the development of an MRI-based rheological measurement that is suitable for inline, online, or at-line implementation. This device is especially useful for measuring the rheological properties of non-Newtonian fluid foods. The strengths of the device include the direct measurement of apparent yield stress and apparent slip velocity. The technique usually provides a rapid noninvasively measured rheogram covering three decades of shear rates.

Nuclear magnetic resonance (NMR) and MRI are spectroscopic techniques that are based on the interaction between nuclear magnetic moments and applied external magnetic fields. NMR and MRI can be used to measure composition, structure, molecular mobility, molecular diffusion, and bulk material motion. In an MRI experiment, a sample is placed in a magnetic field within a radio-frequency probe, energy is added in the radio-frequency range, and the response of the material to that energy is recorded in terms of its attenuation, frequency, and phase. NMR measurements are averages over the entire volume of the sample. MRI is a modification of the NMR experiment; the NMR signal becomes a function of the spatial position within the sample. The display of MRI data yields an image. MRI data can be made sensitive to a variety of variables including position, displacement, diffusion, velocity, density, relaxation times, or combinations of these. The MRI signal intensity, S, in the velocity-encoded images described in this chapter is given by

$$S(k_x, q_z; T) = \int \left[\rho(x) exp(i2\pi k_x x) \times \int P(\Delta z, x; T) exp(i2\pi q_z \Delta z) d\Delta z\right] dx \quad (25.1)$$

and is due to protons in the fluid, primarily from water. The expression for signal intensity (Eq. 25.1) has a density component, a position component, and a displacement component. The variable k_x is the reciprocal space vector with units of 1/m and given by a product of the magnetogyric ratio (γ), the phase encoding gradient duration, and the applied phase encoding gradient. The variable q_z is the reciprocal space vector of displacement with units of 1/m and is given by the product of γ, duration of the displacement encoding gradient, and the applied displacement gradient vector. A two-dimensional Fourier transform with respect to q_z and k_x produces a map of $P(\Delta z, x; T)\rho(x)$ with respect to displacement Δz and position x. The product $P(\Delta z, x; T)\rho(x)$ is the conditional probability density that a nucleus at x will displace Δz within the pulse sequence time interval T, which is referred to as the flow time. The position-dependent density of spins attributed to each displacement is given by $\rho(x)$. The measurement of fluid velocity is accomplished through this displacement component (Δz) by measuring the distance a volume of fluid has moved in a specific time (T). When the motion is steady, the velocity is calculated from the ratio of the distance to the elapsed time. Details of the technique and applications are given by Callaghan (1991) and McCarthy (1994).

The combination of MRI velocity measurements with a pressure drop measurement permits a relationship between shear rate and shear stress to be developed and thus yields a rheological measurement. The applications of MRI measurements to the measurement of fluid rheological properties will be described in the following sections by several case studies.

25.2 Case Studies

A number of studies have been performed in the lab to assess the flow behavior of fluids relevant to food and personal care products. These studies have been performed using variations of the flow loop illustrated in Fig. 25.1. For all studies, a small tank serves as a fluid reservoir that feeds a positive displacement pump. Depending on the circumstances, the fluid may be agitated in the tank to ensure homogeneity. If the fluid is evaluated at temperatures other than room temperature (20 °C) or if viscous heating is anticipated, the fluid is pumped through a coil heat exchanger to maintain a constant and known temperature. The fluid then flows at a constant and known flow rate into a section of nonmagnetic pipe that is centered in an MRI magnet/spectrometer. Studies since 2008 have been performed using an Aspect Imaging 1 Tesla MRI spectrometer and industrially compatible permanent magnet (Aspect Imaging, Shoham, Israel). The magnet is designed to be compatible with process environments and has essentially zero external magnetic field. In other words, the magnetic field at the surface of the magnet is on the order of a few Gauss. The system has 30 G/cm peak gradient strength. Typically a solenoid radio-frequency coil imparts and receives radio-frequency energy. Velocity profiles are obtained noninvasively using a velocity-encoded pulsed gradient spin echo sequence (PGSE). Data can be acquired as fast as one image every 5 s. Typical measurement times are on the order of 1 min. In addition to the MR velocity image, an independent pressure drop is obtained over the straight length of pipe positioned in the magnet. A typical pipe diameter is 20 mm, though smaller and larger diameters have also been used. The pipe diameter is limited by the magnet construction. For this 1 Tesla Aspect unit, the maximum diameter (OD) is 59 mm. Alternate designs of the magnet can accommodate larger pipe diameters.

Usually the fluid is recirculated, as shown in Fig. 25.1. However, a number of studies have also been performed with fluids that have been single pass, which is important when evaluating shear-sensitive fluids. Image and data analyses are performed using MATLAB (MathWorks, Natick, MA, USA). Graphical user interface (GUI) programs (Choi et al. 2005) have been developed to analyze data and to communicate results. The data are reported in a form similar to that acquired from a standard research-grade laboratory rheometer (shear stress vs. shear rate).

Three case studies follow. The first case study focuses on the evaluation of a Carbopol solution. This is a model fluid for foods and for personal care products and

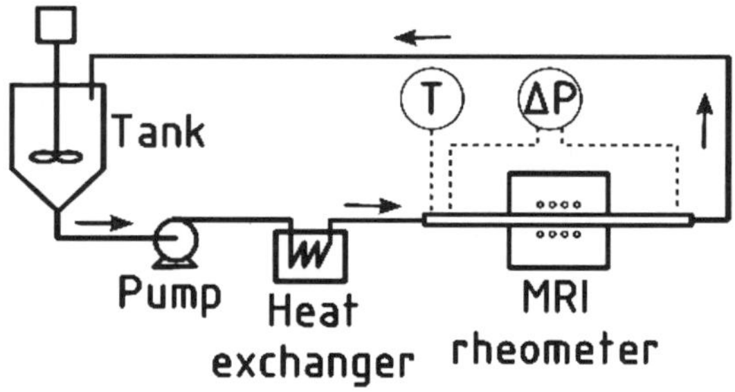

Fig. 25.1 Flow loop for MRI rheometer studies

will demonstrate the power of the technique and the format of data acquisition and analysis. The rheological characterization is real time and illustrates that data acquisition and processing can be adapted to a factory setting. The second case study is yogurt. The intent of this example is to illustrate the shear-sensitive nature of this fluid and the means to quantify the time-dependent flow behavior. The third case study extends the application of the MRI process viscometer to illustrate the link between the measurement of a fundamental fluid property (that is, apparent viscosity) and quality assessment of a final product.

25.2.1 Carbopol Solution

25.2.1.1 Carbopol Solution: Materials and Methods

The MRI process viscometry requires that a well-defined flow field be established. To evaluate shear viscosity in tube (or capillary) flow, an incompressible fluid must undergo steady pressure-driven flow in the laminar regime. The conservation of linear momentum, which equates pressure forces to viscous forces, provides the relationship between the shear stress, σ, and radial position, r:

$$\sigma(r) = \frac{-(\Delta P)}{2L} r, \tag{25.2}$$

where ΔP is the pressure drop over the tube length L. In tomography-based methods, the shear rate is obtained at the same radial position using the velocity profile obtained from a flow image. The expression for the shear rate in tube flow is

$$\dot{\gamma}(r) = \left| \frac{dv(r)}{dr} \right|, \tag{25.3}$$

where v is the axial velocity. Using Eqs. 25.2 and 25.3, the apparent viscosity η is determined by the ratio of shear stress to shear rate:

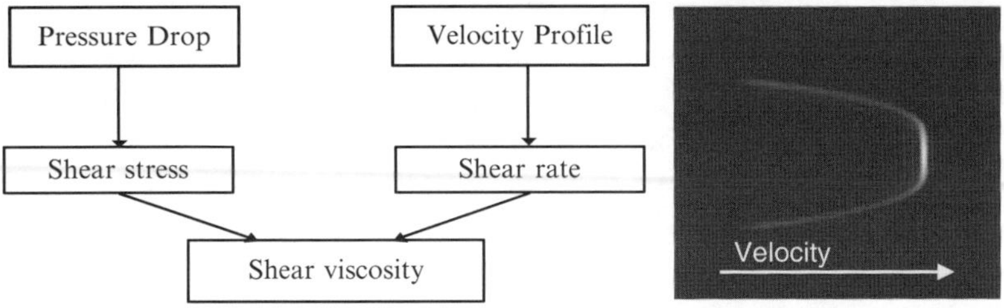

Fig. 25.2 Data processing procedure and a sample velocity profile image

$$\eta(r) = \frac{\sigma(r)}{\dot{\gamma}(r)}. \tag{25.4}$$

GUI programs were developed in the lab to analyze data and display results. This automation of data analysis provided rapid and consistent evaluation of multiple data sets. A schematic of the data processing steps and a sample velocity image are shown in Fig. 25.2. Major steps in the data processing procedure include calculating the shear stress as a function of radial position in the pipe, processing the velocity profile image to obtain a velocity profile, calculating the shear rate as a function of radial position from the velocity profile, and generating the rheogram by plotting the shear stress against the shear rate (Arola et al. 1997). Calculating the shear stress is straightforward (Eq. 25.2). Extracting the velocity profile from the image and calculating the shear rate data presents several challenges. The image data need to have sufficient signal-to-noise and sufficient velocity resolution to achieve a desired range of shear rates. The maximum shear rate is determined in the same manner as a conventional capillary viscometer (that being the shear rate at the wall). The minimum shear rate depends upon the velocity resolution (Arola et al. 1998). After the appropriate velocity resolution is set and the image acquired, a velocity profile is extracted. Shear rates are calculated as a function of radial position by taking the derivative of the velocity profile. At each radial position the shear rate is matched to its shear stress to create a rheogram. As with most rheological measurements, the shear stress/shear rate data are modeled with a specific constitutive expression to determine rheological parameters, e.g., Herschel–Bulkley (H–B) parameters, or, alternatively, the shear viscosity versus shear rate is plotted.

The test fluid in this first case study is a solution that contained a Carbopol polymer (Carbopol 940 T, The Lubrizol Corporation). This polymer family of products is based on crosslinked acrylic acid chemistry and is widely used in personal care and household products as rheological modifiers. A 0.2 % w/w Carbopol solution was prepared by slowly sifting a weighted amount of polymer into deionized water in a stirred tank. The Carbopol solution was neutralized with a 50 % NaOH solution to pH 7. The neutralization process is important because of the strong dependence of the flow behavior of Carbopol solutions on pH. The neutralization allows the solution to achieve its maximum viscosity as the polymer swells. The rheological properties of the test solution were measured offline using a TA

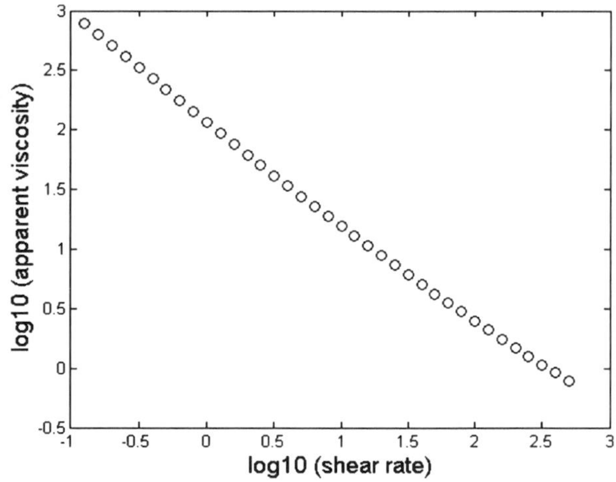

Fig. 25.3 Log-log plot of shear rate and apparent viscosity obtained by offline measurement for 0.2 % Carbopol solution

Instruments AR-G2 rheometer (New Castle, DE, USA) with a standard Couette geometry (14 mm diam. × 42 mm height) at 25 °C. A steady-state shear rate ramp from 0.1 and 500 1/s was performed in logarithmic mode with 10 points/decade and 5 % tolerance. Results are given in Fig. 25.3 as a log-log plot of shear rate and apparent viscosity. At low shear rate values, the apparent viscosity is on the order of 1000 Pa s; at high shear rate values, the apparent viscosity is on the order of 1 Pa s. The data were well described by a H–B model for shear stress,

$$\sigma = \sigma_o + K\dot{\gamma}^n, \tag{25.5}$$

with a consistency index K of 27.7 Pa s^n, flow behavior index n of 0.384, and yield stress σ_o of 87.0 Pa over the shear rate range of 0.1–500 1/s ($R^2 > 0.999$).

The MRI was performed as described in Sect. 25.2 using a 19.05 mm ID tube; the pressure drop was acquired over a 1.21 m length of straight pipe. Specific imaging parameters for the Carbopol solution were as follows: field of view (FOV) 30 mm, velocity sweep width (VSW) 40 cm/s, and repetition time (TR) 50 ms.

25.2.1.2 Carbopol Solution: Results and Discussion

Representative MRI images for 0.2 % Carbopol solutions for no-flow and flow conditions are given in Fig. 25.4a, b, respectively. The MRI no-flow image (Fig. 25.4a) is acquired prior to starting the pump. The signal intensity depicts the axial velocity as a function of radial position. The vertical bright region in the center of the no-flow image spans the width of the pipe, which indicates the position of the pipe walls. There is no signal intensity external to the pipe wall. The horizontal positioning is indicative of the axial velocity. The bright region position in the center of the image under a no-flow condition is positioned at a voxel number of 128 ($v = 0$). Signal intensities positioned to the right are positive velocities;

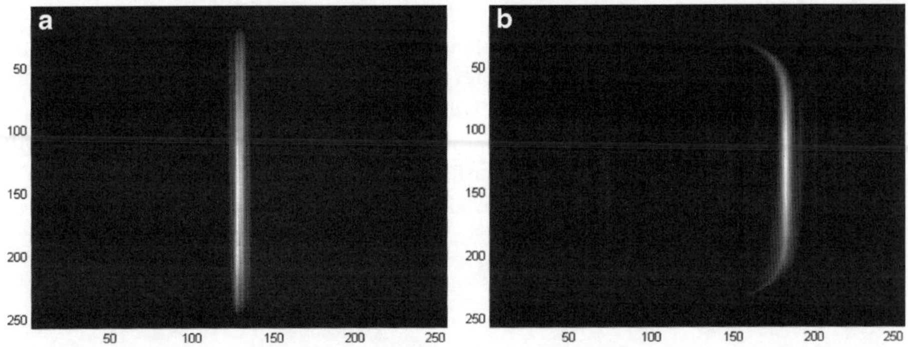

Fig. 25.4 Representative MRvelocity images for 0.2 % Carbopol solutions for (**a**) no-flow and (**b**) flow conditions

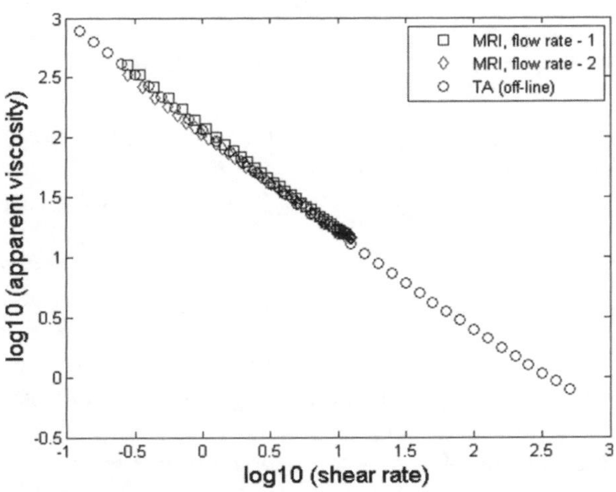

Fig. 25.5 Log-log plot of shear rate and apparent viscosity with MR data superimposed on offline data for 0.2 % Carbopol solution

signal intensities to the left of center are negative velocities. The magnetic resonance flow image (Fig. 25.4b) is typical of the velocity profile of a non-Newtonian fluid. These fluids exhibit a characteristic blunted profile, with the maximum velocity at centerline and no-slip conditions at the tube wall.

Typical MRI data are given in Fig. 25.5. Data collected under two similar, but not identical, flow conditions are plotted with the offline measurements. The apparent viscosity curves for the 2 MR data sets and the AR-G2 rheometer data are virtually indistinguishable over the shear rate range of 0.2–15 1/s. Since the MRI-based measurement is performed in a pipe, the range of shear rates obtained is that normally encountered in pipe flow (e.g., two to four orders of magnitude). The upper shear rate is limited by the maximum at the tube wall. The lower shear rate is limited by the velocity resolution of the flow image since no data are acquired on shear rates below the minimum calculated using the velocity resolution. The range of shear rates for an MRI-based measurement can be adjusted by changing the data acquisition parameters to decrease the value of the velocity resolution and extend the lower end of the shear rate range. Thus the range of shear rates accessible with MRI can match those needed for process control and quality assurance.

25.2.2 Yogurt

25.2.2.1 Yogurt: Materials and Methods

Yogurt, as the second case study, will illustrate the capability of MRI to capture the shear-sensitive nature of this fluid and to quantify the time-dependent flow behavior. In general, three kinds of yogurt products are commercially available: (1) firm yogurt, (2) stirred yogurt, and (3) drinking yogurt. The flow behavior of the three types of product differ and is unique due to the complexity of the three-dimensional gel structure.

Measuring the rheological behavior of gels in general presents difficulties due to (1) poor reproducibility, (2) sensitivity to gel preparation, (3) sensitivity to shear history, (4) limited range of linear viscoelastic response, and (5) wall slip (Larson 1999). The first four of these difficulties are due to the nonequilibrium structure of flocculated gels. When the association between particles requires a long relaxation time, the gel structures are unable to rearrange within the experimental time scale. Hence the gel structure and rheological behavior show dependence on shear history.

Characterizing the flow behavior of yogurt is necessary for design equipment and for the control of the yogurt manufacturing process. For example, rheological parameters are relevant to the postfermentation cooling process performed in a continuous heat exchanger. The product is cooled from a fermentation temperature between 40 °C and 42 °C to 15–20 °C, at which point fruit or other ingredients may be added (Varnam and Sutherland 1994).

The rheological properties of yogurt have been extensively studied with conventional rotational viscometers (Shoemaker et al. 1992; Benzech and Maingonnat 1994; Fangary et al. 1999). According to these studies, yogurt shows shear thinning behavior with yield stress and is best fit with the H–B model. In addition, researchers have recognized the thixotropic behavior of yogurt and the importance of controlling the state of the yogurt prior to and during testing (Suwonsichon and Peleg 1999).

In a 2002 study, Yoon and McCarthy used MRI velocimetry techniques to characterize the flow behavior of yogurt. The test fluid was commercial whole milk yogurt (Mountain High plain yogurt, Meadow Gold Dairies, Englewood, CO, USA). The composition of the yogurt was 3.5 % total fat, 7.9 % carbohydrate, and 4.8 % protein. The flow behavior of the yogurt was evaluated during pipe flow under isothermal conditions, at both 25 °C and 35 °C. The MRI velocity profiles

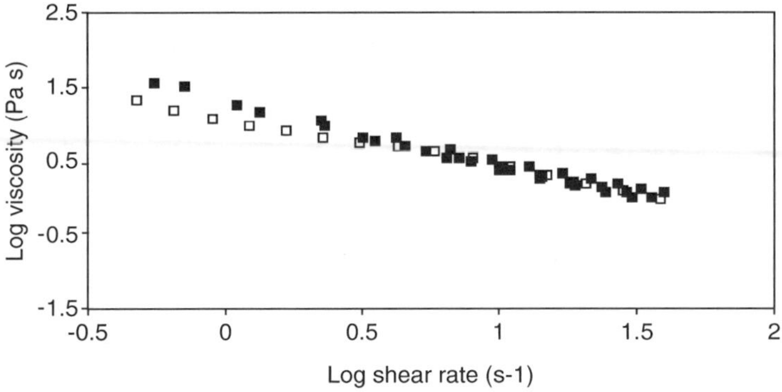

Fig. 25.6 Log-log plot of shear rate and apparent viscosity with MR data superimposed on offline data for whole milk yogurt (From Yoon and McCarthy (2002))

were obtained using a SMIS NMR spectrometer (Surrey Medical Imaging Systems, Surrey, UK) connected to a 0.1 Tesla electromagnet. As with the Aspect spectrometer (described in Sect. 25.2), velocity profiles were obtained noninvasively using a PGSE sequence. The pressure measurements required for the shear stress values were assessed by a different means than described earlier in Sect. 25.2 (Yoon and McCarthy 2002). Due to the single-pass noncirculating experimental setup, the flow behavior of the yogurt was not assessed for time dependency.

25.2.2.2 Yogurt: Results and Discussion

As with studies by other researchers, the yogurt was characterized as a H–B fluid. In addition, wall slip velocity was evaluated directly by assessing the fluid velocity at the liquid–solid interface of the pipe within the spatial resolution of the MRI image. In general, wall slip occurs when a thin layer of fluid having a viscosity lower than the bulk of the fluid forms at a liquid–solid surface (Steffe 1996). Wall slip, not considered a material property, is a physical phenomenon associated with the flow of weak gel systems or suspensions. In general, it is difficult and time consuming to evaluate apparent slip velocity by conventional (e.g., rotational) rheological measurements. However, not suspecting or correcting for slip velocity can result in significant underestimation of rheological parameters.

In the study by Yoon and McCarthy (2002), the slip velocity was a linear function of wall stress over the range of wall stress evaluated. This functionality was consistent with slip correction given in Steffe (1996) for tube viscometry. Apparent viscosity was corrected accordingly.

Figure 25.6 illustrates representative results at 25 °C from that study (Yoon and McCarthy 2002). The apparent viscosity from a conventional rotational viscometer (Bohlin CVO, Bohlin Instruments, Gloucestershire, UK) is plotted with the MRI-based information. The apparent viscosity values at 25 °C from both

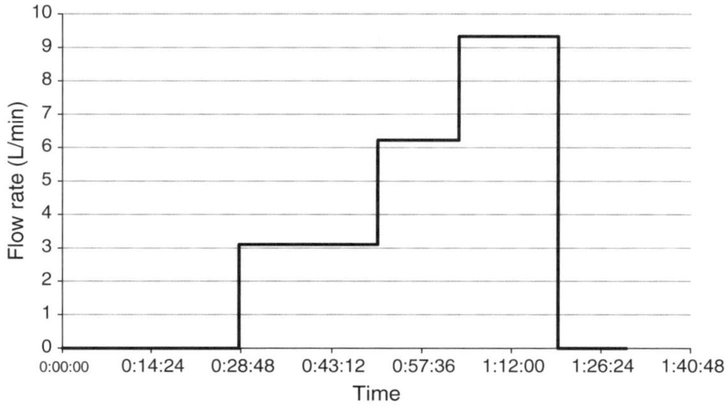

Fig. 25.7 Time interval at each pump setting to evaluate thixotropic behavior of yogurt

measurements were similar, though rotation viscometer values (open symbols) were somewhat lower at lower shear rates than the MRI-based measurements (closed symbols). This 2002 study provided validation of the MRI process viscometer for rheologically complex foods.

More recent work has evaluated the same test material, whole milk yogurt (Mountain High plain yogurt). The test was performed in a flow loop with an agitated tank that fed a progressive cavity pump. The yogurt was pumped through a heat exchanger consisting of a copper coil immersed in a thermostatic bath set at 20 °C and into a 19.0 mm ID Polymethyl methacrylate (PMMA) acrylic pipe. The pipe was centered in the imaging section of the Aspect MRI spectrometer. A differential pressure was acquired over a 2.04 m straight length of pipe. The pump flow rate was initially kept constant while flow images and pressure drops were obtained at intervals of 1 min. The flow rate was then increased. This occurred three times for three sets of multiple MRI velocity images and pressure drops. In contrast to the work by Yoon and McCarthy (2002), these trials were performed to observe the effect of flow rate and flow history on rheology.

The three pump settings (3, 6, and 9) corresponded to 3.3, 6.3, and 9.2 L/min, which yielded average velocities of 0.19, 0.37, and 0.54 m/s respectively. The time interval at each pump setting was approximately 10 min and is shown in Fig. 25.7. MRI images were acquired for each of the pump settings consecutively. Representative images are illustrated in Fig. 25.8. Similar to the set of images for the Carbopol solution (Fig. 25.4), the MRI image in Fig. 25.8a is a no-flow image, an image acquired prior to starting the pump. The vertical bright region in the center spans the width of the pipe, clearly indicating the position of the pipe walls as well as the centerline of the image at $v = 0$. High signal intensities to the right are positive velocities; high signal intensities to the left of the centerline are negative velocities. The MRI velocity image in Fig. 25.8b was acquired at a pump setting of 3. Characteristic of non-Newtonian behavior, the profile is blunted, rather than the parabolic shape distinctive of a Newtonian fluid.

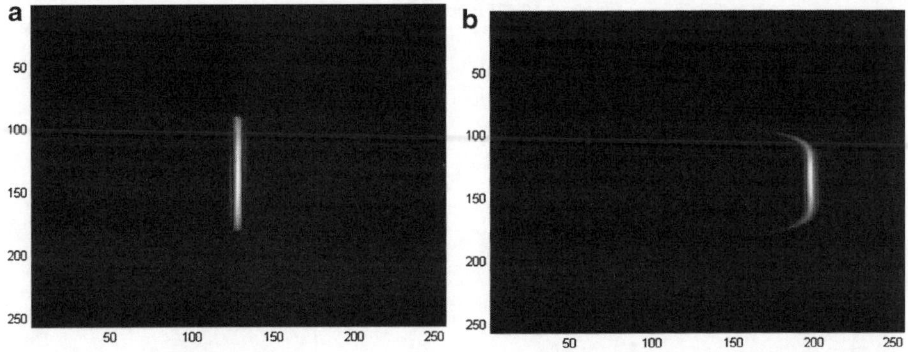

Fig. 25.8 Representative MR velocity images for yogurt for (**a**) no-flow and (**b**) flow conditions

Fig. 25.9 Pressure drop values over time acquired for yogurt at pump setting 3

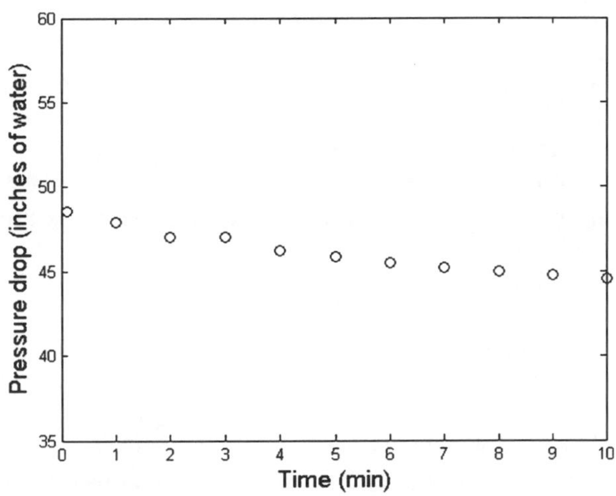

In contrast to time-independent fluids, pressure drop values acquired as a function of time over the straight length of pipe decreased as the gel structure was continually sheared. Figure 25.9 illustrates the pressure drop at pump setting 3 over the 10 shown in Fig. 25.7. Over the 10 min of data acquisition, the pressure drop decreased from 48 in. H_2O to roughly 44 in. of H_2O, an 8 % decrease that was reflected in the shear stress values. The MRI images did not noticeably reflect changes in the velocity profile. At a given time and pump setting, the flow behavior was best modeled by the power law expression

$$\sigma = K\dot{\gamma}^n \tag{25.6}$$

with a consistency index K and flow behavior index n. The log-log plots of apparent viscosity (Eq. 25.4) versus shear rate for the three pump settings is given in

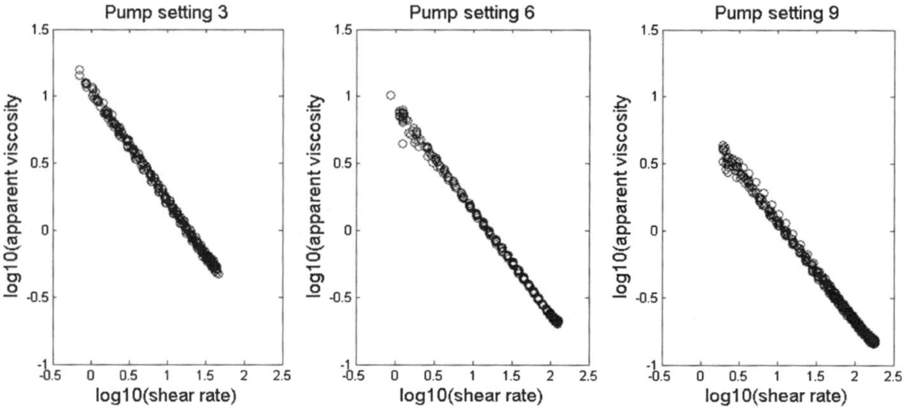

Fig. 25.10 Log-log plot of shear rate and apparent viscosity of yogurt obtained using MRI rheometer for three flow rates (pump settings)

Fig. 25.10. For each of the pump settings, the data are given at MRI acquisitions times of 0, 3, 7, and 10 min relative to the change in pump setting. There is a noticeable decrease in apparent viscosity as the pump setting increases from 3 to 9. At pump setting 3, the apparent viscosity is similar to the values in Fig. 25.6 from the Yoon and McCarthy (2002) study. These values reflect consistency in the yogurt product under relatively low shearing conditions. The higher pump settings increasingly break down the gel structure of the yogurt and apparent viscosity values decrease.

At first glance, the apparent viscosity curves in each subplot fell essentially on top of one another. A closer inspection of each subplot (Fig. 25.10) shows a slight downward shift in the curve from $t = 0$ to $t = 10$ min that documents the thixotropic behavior of the yogurt. Mullineux and Simmons (2007) represented the thixotropic behavior of yogurt with a model that incorporated temperature and time, as well as shear rate and shear stress. For isothermal conditions, it can be simplified and expressed in terms of apparent viscosity as

$$\eta = at^{-m}\dot{\gamma}^{n-1}, \tag{25.7}$$

where the product at^{-m} is the consistency index and n is the flow behavior index. The parameter a is a constant under given shearing conditions (e.g., pump setting) and the thixotropic behavior is described by the exponent m. The value of m is positive, but not necessarily an integer. Therefore, the apparent viscosity decreases with both time and shear rate. Figure 25.11 illustrates the change in viscosity as a function of pump setting in the three subplots and as functions of shear rate and time in each of the surface plots. The axes are the same for each subplot to clearly indicate the decrease in apparent viscosity from15 Pa s at pump setting 3 to well under 5 Pa s at pump setting 9. In addition, the surface plots clearly show the change in shear rate range as the flow rate changes. Table 25.1 gives the curve fitting results

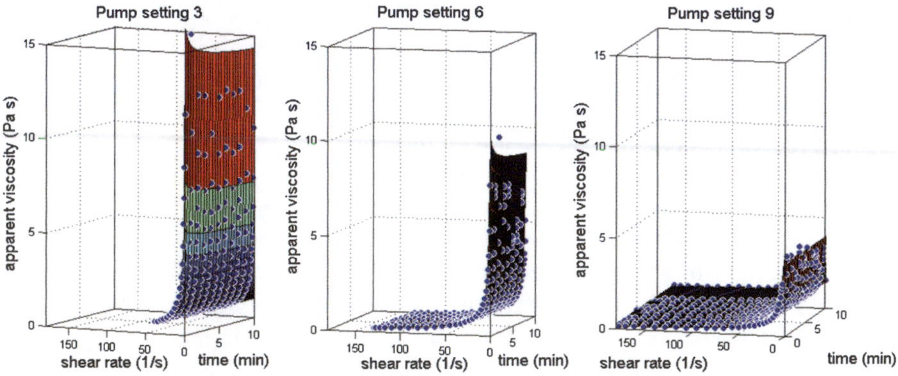

Fig. 25.11 Surface plots for yogurt at three flow rates (pump settings)

Table 25.1 Curve fitting results for model parameters to describe yogurt thixotropy

Pump setting	a	m	n	Shear rate range (1/s)	Adj. R^2
3	11.5	0.0276	0.180	0.70–45	0.999
6	8.57	0.0381	0.233	1.1–130	0.988
9	6.41	0.0145	0.276	2.0–180	0.986

that describe the thixotropic behavior (Eq. 25.7) as well as the adjusted coefficient of determination. Values of a and n decreased, primarily reflecting the shear thinning nature of yogurt. There is a general decrease in m, which reflects greater time-dependent behavior. The trend should, however, be interpreted cautiously because the shearing occurred as the pump settings were continuously increased (Fig. 25.7) while the yogurt was recirculated.

Incorporating this type of viscometric measurement inline would eliminate the potential of different shear rate histories on grab samples compared to the final product. Actual changes in product rheology could then be readily determined. Additionally, if the product is sufficiently sheared, the velocity profile becomes nonsymmetric (data not shown), indicating a material with nonuniform properties.

25.2.3 Tomato Concentrate

Ultimately, measurements of properties acquired by sensors must be linked to some quality specification of the product. Translating the sensor measurement to a specific quality feature ranges from being straightforward (e.g., speed of sound to density of a liquid) to challenging (e.g., rheological properties of suspensions to quality assurance measurements). It is important to note that quality assessment is a human construct based on food properties and characteristics. When a quality grade is assigned to a food product, this grade implies a certain degree of excellence and suitability for specific uses. Additionally, a quality grade is normally specified in contracts and used in marketing. Often quality grades and standards are constructed from tests that mimic product function. The Bostwick consistometer test for tomato concentrate consistency is a good example of a quality assurance test used to characterize product quality and is used in business contracts.

Several potential steps in tomato-based concentrate production and the subsequent formulation of tomato concentrate products could benefit both economically and in terms of improved product quality by incorporating process viscometry. This case study establishes a relationship between tomato concentrate consistency (e.g., Bostwick value) and the consistency of a product, ketchup, made from that concentrate. This type of relationship would assist in maintaining a product of uniform grade.

25.2.3.1 Tomato Concentrate: Materials and Methods

Many tomato products are formulated using tomato concentrates that have been stored in bins. The concentration and storage process allows maximum throughput during harvest and then reprocessing throughout the year. These storage bins contain material of different mass fraction solids and different Bostwick values. The Bostwick value is obtained from a standard test. The test is a batch process in which tomato concentrate at 12 °Brix (a measure of soluble solids) is loaded into a compartment at the end of a channel (Fig. 25.12a). A gate is released and the food product flows down the channel length (Fig. 25.12b). The length of flow of the product in a specified period of time (e.g., 30 s) is recorded and referred to as the Bostwick value. This instrument provides a single parameter that is based on the extent of fluid flow that characterizes the material consistency. Processors often need to mix different bins of concentrate. Knowledge of the relationship between raw product, intermediate product, and final product is vital.

In a study sponsored by ConAgra Foods (Helm, CA, USA), 50 kg of two commercial tomato concentrates differing in Bostwick values were provided. The two concentrates were blended at different mass fractions and diluted to 12 °Brix. The test fluids were pumped through the MRI flow loop. The flow loop system consisted of a 19.05 mm ID plexiglass tube with a 1.0 m straight section upstream of the imaging plane to ensure sufficient distance for fully developed flow. The pressure drop across the distance L of 2.08 m was measured at two pressure taps on both sides of the magnet bore using a differential pressure transducer. The fluid temperature was maintained at 60 $\pm$ 2 °C using a Haake circulator (DC10, Thermo Fisher Scientific, Waltham, MA, USA) in an open 4 L jacketed reservoir with foam insulation on all piping.

Similar to the yogurt work described in Sect. 25.2.2, the MRI was performed using the Aspect Imaging spectrometer/permanent magnet system. Two-dimensional proton images of tomato concentrates were obtained using a velocity-encoded PGSE sequence. The specific imaging parameters for the tomato concentrates were as follows: FOV 50 mm, VSW 95 cm/s, and TR 130 ms.

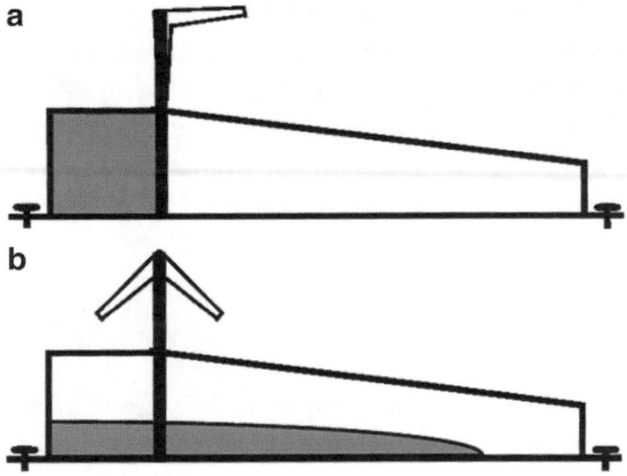

Fig. 25.12 Bostwick consistometer (**a**) before gate is opened and (**b**) after gate is opened

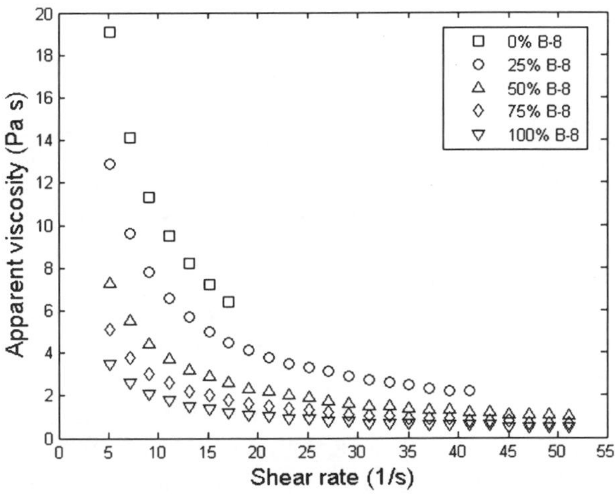

Fig. 25.13 Shear viscosity of tomato concentrate blends of concentrates with Bostwick values of 1 and 8 cm. Percentage of Bostwick value of 8 cm concentrate (B-8) is given in legend

25.2.3.2 Tomato Concentrate: Results and Discussion

Figure 25.13 illustrates shear viscosity versus shear rate for five different tomato concentrate blends. The data were obtained by data acquisition and analyses previously described. The blends were different percentages of the concentrate characterized by a Bostwick measurement of 8 cm (B-8) and the concentrate characterized by a Bostwick measurement of 1 cm (B-1). These data sets obtained for different blends of the two commercial concentrates clearly document that 12 ºBrix blends can have extremely different flow behaviors.

Previous work from the laboratory established the relationship between the Bostwick consistometer measurement and the physical properties of the fluid. Specifically, the flow in a Bostwick consistometer was treated as a gravity current

(McCarthy and Seymour 1993). Gravity currents occur when a fluid flows primarily horizontally due to gravitational forces. The gravitational force acting on fluid in a Bostwick consistometer is opposed by viscous forces, limiting the distance the fluid front travels during the measurement time. Generalizing the analysis developed for Newtonian fluids to non-Newtonian fluids yielded a relationship between L' and the apparent viscosity (McCarthy and Seymour 1994):

$$L' = c\left(\frac{\eta}{\rho}\right)^{-1/5}, \tag{25.8}$$

where L' is the total length of the gravity current and corresponds to the Bostwick measurement plus 5 cm (adding the length of the sample compartment), ρ is the fluid density, and c is a constant. The apparent viscosity, η, is evaluated at a characteristic shear rate,

$$\dot{\gamma} = \frac{(L')^2}{qt}, \tag{25.9}$$

where q is the volume/width of the sample compartment and t is the measurement time of 30 s. A linear relationship between L' and $(\eta/\rho)^{-1/5}$ has been established for a number of Newtonian and homogeneous power law fluids and extended to tomato concentrates in the range of 5 to 24 °Brix (Milczarek and McCarthy 2006) and a blend of 12 °Brix tomato concentrates (McCarthy et al. 2008).

To extend the work to final product quality, ketchup at three levels of natural tomato soluble solids (NTSS) were prepared from the five tomato concentrate blends shown in Fig. 25.13. These five 12 °Brix blends had Bostwick values in a range of 2–4 cm and were expected to yield good-quality ketchup, as characterized by ketchup Bostwick measurements between 3.0 and 7.0 cm. Although all ketchup formulations (6, 7, and 8 NTSS) met the USDA requirement of 33 % or more total solids, several had Bostwick measurements outside the 3.0–7.0 range specified by USDA (1992) for tomato ketchup.

To address the ability to predict final ketchup quality from intermediate product quality, the inline viscosity measurements were evaluated at the shear rate characteristic of flow in the consistometer (Eq. 25.9). The ketchup Bostwick measurement was then correlated to the ratio of $(\eta/\rho)^{-1/5}$ of the 12 °Brix tomato concentrate, yielding coefficients of determination of 0.97, 0.97, and 0.91 for NTSS levels of 6 %, 7 %, and 7.8 %, respectively (McCarthy and McCarthy 2009). The 50 % B-8 blend yielded acceptable ketchup Bostwick measurements in the 3.0–7.0 cm range for all three NTSSs. The 25 % B-8 blend yielded acceptable ketchup Bostwick values between 6 % and 7 % NTSS. Additionally, the 75 % B-8 blend produced acceptable ketchup in a range of 7–8 % NTSS. The 0 % B-8 (100 % B-1) blend produced an acceptable ketchup only at 6 % NTSS.

In summary, ketchup was used as a model to evaluate the feasibility of developing a correlation for tomato-based sauces. This data demonstrate that the inline viscometer approach can be extended to the remanufacture process to predict final ketchup grade. A rapid inline measurement of apparent viscosity permits the manufacturer to minimize the use of tomato concentrate or optimize the use of concentrates with different flow behaviors.

25.3 Current Challenges

The current challenge is to accomplish the engineering to incorporate the MRI-based sensor in a process line. A nonmetallic section of pipe that is compatible with food processing requirements (e.g., clean-in-place) is in the design phase. Also, some other minor changes will need to be made to incorporate the device into a process line or into a slipstream off the main production line.

25.4 Additional Measurement Opportunities

In addition to the measurement of rheological properties, NMR/MRI may be used to measure and quantify physical and chemical properties directly and indirectly, providing multiple quality assessments using a single sensor (McCarthy 1994). Direct measurements of composition are often possible using simple one-dimensional spectra or the initial signal intensity in the time domain. Indirect measurements of composition may often be achieved through correlations with relaxation times. These decay times for the NMR signal may be strong functions of moisture content, textural properties, or physical phases. Diffusion coefficient measurements have a strong dependence on food material structure such as particle size and other structural features in colloidal-based fluid foods. MRI measurements of component spatial distribution give insight into the rates of transient processes that impact final product quality and process efficiency. Collectively NMR and MRI measurement protocols permit a wide range of quality measurements in food systems.

During the development of the MRI viscometer for the tomato processing industry, the focus was to evaluate the rheological properties of tomato concentrates. However, it became apparent that there were considerable opportunities for evaluating processing tomato quality. For instance, bruising is a common defect in Roma variety processing tomatoes. Bruising arises during harvesting, transport, and conveying into the processing facility. These processing tomatoes are used to make concentrate, diced product, and a whole-peeled pack product. If a tomato is bruised and used for diced or whole-peeled product, the production efficiency decreases. There is a variety of potential methods to monitor for bruising, including hyperspectral imaging, MRI, and NIR. The current method for evaluating the extent of bruising is visual analysis, and this is done on a subjective basis since no scale exists to train the evaluator. Hence one of the initial steps in developing a sensor for this defect is to form a quantitative method for bruise volume and extent of cellular damage (Milczarek et al. 2009). Once bruise volume and extent of damage are quantified, the efficiency of using MRI or other techniques can be assessed. MRI is sensitive to a wide range of defects in fresh processing tomatoes; these include insect damage, viral infection, and mold damage. Most studies concentrate on a correlation for one quality parameter. The next challenge is to develop correlations to multiple quality factors from a single (or multiple) sensor measurement.

25.5 Conclusion

The issues described by the examples of developing sensors for the processing tomato industry are very similar to issues that must be faced by other food processors. In addition, the issues of traceability, purity, adulteration, and microbial safety are critical to the food industry. A more universal approach similar to that adopted in other processing industries should be considered. The pharmaceutical, chemical, and oil industries have adopted approaches of Quality by Design (QbD) and process analytical technology (PAT). QbD is a systematic approach to product development that includes design, formulation, and manufacturing process. The goal of QbD is to ensure a predefined product quality. PAT is a procedure to design, analyze, and control manufacturing by the measurement of critical process parameters that affect critical quality attributes. PAT involves the development and implementation of sensors and novel data analysis routines. Both QbD and PAT approaches have been extremely effective at improving production efficiency and final product quality, reducing product variability, and generally decreasing the cost of production. The food industry would benefit from utilizing approaches like QbD and PAT to enhance safety and quality and improve cost efficiency.

References

Arola DF, Barrall GA, Powell RL, McCarthy KL, McCarthy MJ (1997) Use of nuclear magnetic resonance imaging as a viscometer for process monitoring. Chem Eng Sci 52(13):2049–2057

Arola DF, Powell RL, Barrall GF, McCarthy MJ (1998) A simplified method for accuracy estimation of nuclear magnetic resonant imaging. Rev Sci Instrum 69(9):3300–3307

Benzech T, Maingonnat JF (1994) Characterization of the rhdological properties of yoghurt – a review. J Food Eng 21:447–472

Callaghan PT (1991) Principles of nuclear magnetic resonance microscopy. Clarendon, Oxford

Choi YJ, McCarthy KL, McCarthy MJ (2002) Tomographic techniques for measuring fluid flow properties. J Food Sci 67(7):2718–2724

Choi YJ, McCarthy KL, McCarthy MJ (2005) A MATLAB graphical user interface program for tomographic viscometer data processing. Comput Electron Agric 47(1):59–67

Cullen PJ, Duffy AP, O'Donnell CP, O'Callaghan DJ (2000) Process viscometry for the food industry. Trends Food Sc Technol 11:451–457

Fangary YS, Barigou M, Seville JPK (1999) Simulation of yoghurt flow prediction of its end-of-process properties using rheological measurements. Trans Inst Chem Eng 77:33–39, Part C

Irudayaraj J, Reh C (eds) (2008) Nondestructive testing of food quality. Blackwell Publishing, Ames

Larson RG (1999) The structure and rheology of complex fluids. Oxford University Press, New York

Maneval JE, McCarthy KL, McCarthy MJ, Powell RL (1996) Nuclear magnetic resonance imaging rheometer. US Patent No. 5,532,593

McCarthy MJ (1994) Magnetic resonance imaging in foods. Chapman & Hall, New York

McCarthy KL, McCarthy MJ (2009) Relationship between in-line viscosity and Bostwick measurement during Ketchup production. J Food Sci 74(6):E291–E297

McCarthy KL, Seymour JD (1993) A fundamental approach for the relationship between the Bostwick measurement and Newtonian fluid viscosity. J Texture Stud 24:1–10

McCarthy KL, Seymour JD (1994) Gravity current analysis of the Bostwick consistometer for power laws foods. J Texture Stud 25:207–220

McCarthy KL, Sacher RF, Garvey TC (2008) Relationship between rheological behavior and Bostwick measurement during manufacture of ketchup. J Texture Stud 39:480–495

Milczarek RR, McCarthy KL (2006) Relationship between the Bostwick measurement and fluid properties. J Texture Stud 37(6):640–654

Milczarek RR, Salveit ME, Garvey TC, McCarthy MJ (2009) Assessment of tomato pericarp mechanical damage using multivariate analysis of magnetic resonance images. Postharvest Biol Technol 52:189–195

Mullineux G, Simmons MJH (2007) Effects of processing on shear rate of yoghurt. J Food Eng 79:850–857

Shoemaker CF, Nantz J, Bonnas S, Noble AC (1992) Rheological characterization of dairy products. Food Technol 46:98–104

Steffe JF (1996) Rheological methods in food process engineering, 2nd edn. Freeman Press, East Lansing

Suwonsichon T, Peleg M (1999) Rheological characterization of almost intact and stirred yogurt by imperfect squeezing flow viscometry. J Sci Food Agric 79:911–921

United States Department of Agriculture (USDA), Agricultural Marketing Service (1992) United States standards for grades of tomato ketchup. United States Department of Agriculture (USDA), Agricultural Marketing Service, Washington, DC

Varnam AH, Sutherland JP (1994) Milk and milk products: technology, chemistry, and microbiology. Chapman & Hall, London

Yoon WB, McCarthy KL (2002) Rheology of yogurt during pipe flow as characterized by magnetic resonance imaging. J Texture Stud 33:431–444

Part V
Modeling and Control of Food Safety and Quality

Chapter 26
Predictive Modeling of Textural Quality of Almonds During Commercial Storage and Distribution

Li Z. Taitano and R. Paul Singh

26.1 Introduction

Water plays an important role in determining the quality of food products during storage, transport, and distribution. The migration of water between the surrounding medium and the food can significantly affect the textural properties of the product. Due to their hygroscopic nature, when foods are exposed to a moist environment during storage, they adsorb water and lose brittleness. It is necessary to understand water migration in food products and its influence on textural properties. The relationship of equilibrium between water activity and the moisture content of foodstuffs can be described by the moisture sorption isotherm, which depends on many factors, such as temperature, relative humidity, and the composition of food materials. Sorption isotherm data have been fitted into numerous mathematical models such as Brunauer–Emmett–Teller (BET), Guggenheim–Anderson–de Boer (GAB), Oswin, Halsey, and Henderson, with two or three parameters to explain the behavior of food products under the effects of water activity (a_w), moisture content of samples, and ambient temperature (Wani et al. 2006). The GAB sorption equation has been widely used to describe water sorption isotherms of food over a wide range of water activity, and monolayer moisture-content-related coefficients can be obtained from the GAB model.

Additional thermodynamic properties can be calculated from the moisture sorption isotherm, such as the isosteric heat of sorption (q_{st}), free energy (ΔG), enthalpy (ΔH), and entropy (ΔS). The isosteric heat of sorption (q_{st}) indicates the moisture-dependent binding strength of water adsorbed by solid particles. The Clausius–Clapeyron equation can be used to determine the isosteric heat of sorption (q_{st}) and estimate

L.Z. Taitano • R.P. Singh (✉)
Department of Biological and Agricultural Engineering, University of California, Davis, USA
e-mail: rpsingh@ucdavis.edu

S. Yanniotis et al. (eds.), *Advances in Food Process Engineering Research and Applications*, Food Engineering Series, DOI 10.1007/978-1-4614-7906-2_26,

the energy requirements of a process. The forces of attraction or repulsion of water molecules to food materials can be associated with the differential entropy (ΔS), which is calculated from the Gibbs–Helmholtz equation (Goneli et al. 2010). The sorbent's affinity for water can be described by the free energy change (ΔG), a criterion to determine whether water sorption is a spontaneous process (Avramidis and Vancouver 1992; Goula et al. 2008). The compensation theory proposes a linear relationship existing between enthalpy (ΔH) and entropy (ΔS), which is frequently used to investigate the water adsorption phenomena in food materials (Sharma et al. 2009).

The mechanism of moisture adsorption is a mass transfer phenomenon that is influenced by many factors including temperature, moisture content of the food material, and its geometry. Water movement is caused by a force due to the water potential gradient between the food and its surrounding environment. Many studies have been conducted on the rate and amount of mass transfer. The diffusion of water adsorption can be obtained by Fick's second law (Crank 1975). In an unsteady-state diffusion, a sample shape can be approximated as an infinite slab, infinite cylinder, or sphere to determine the effective diffusion coefficient (D_{eff}). Ruiz-Beviá and others (1999) calculated the effective diffusion coefficient value (D_{eff}) in almonds by assuming the geometry of a parallelepiped object using Fick's second law. Other empirical models, such as the Peleg model (Gowen et al. 2007) and the exponential models (Kashaniejad et al. 2007), are also used to study the kinetics of water adsorption.

The mechanical changes in almonds during storage and transport can be attributed to the effects of relative humidity (a_{w} = relative humidity/100) and temperature. Food texture can be determined using analytical or sensory methods, and therefore the textural property is often used as an attribute to predict the food quality. Some mathematical models based on Fermi's distribution function have been developed to describe the relationship between water activity and the mechanical properties of food materials (Peleg 1993), such as raw and roasted coffee beans (Pittia et al. 2007), legumes, nuts (Borges and Peleg 1997), and brittle cellular cereal foods (Harris and Peleg 1996). In addition, temperature can also be included in the Fermi model to be associated with moisture content or water activity (Pittia and Sacchetti 2008).

The goal of this study was to develop a predictive model of the textural properties of almonds as influenced by the environmental temperature and relative humidity in constant or dynamic storage conditions. Moisture sorption isotherms and the textural properties of equilibrated almonds were obtained within a temperature range of 7–50 °C and a_{w} of 0.11–0.98. In addition, the weight changes in almonds under six a_{w} levels (from 0.41 to 0.98) and four temperatures (7 °C, 25 °C, 35 °C, and 50 °C) were measured during storage. Based on the moisture adsorption isotherm, the diffusion coefficient of water in almonds, and Fermi's distribution function, a mathematical model was developed to predict the moisture content and mechanical properties of almonds during storage. The model was then tested using selected storage experiments with varying temperature and humidity.

26.2 Materials and Methods

26.2.1 Material Preparation

Pasteurized, unpasteurized, and blanched almonds of Nonpareil and Monterey variety were supplied by the Almond Board of California (Modesto, CA, USA). Samples were dried in desiccators over anhydrous calcium sulfate at room temperature (22 °C) for 15 days prior to the experiment. Nine laboratory grade salts, including LiCl, $KC_2H_3O_2$, $MgCl_2$, K_2CO_3, $Mg(NO_3)_2$, $NaNO_2$, NaCl, KCl, and K_2SO_4, were obtained from Fisher Scientific (Hanover Park, IL, USA), which were used to maintain water activities ranging from 0.11 to 0.98. The water activities of these saturated salt solutions at different temperatures were obtained from the earlier research of Greenspan (1977), as shown in Table 26.1.

26.2.2 Experimental Procedures

26.2.2.1 Water Sorption Isotherm

The static gravimetric method was used to obtain the equilibrium moisture content (EMC) of almonds at 7 °C, 25 °C, 35 °C, and 50 °C. Plastic jars with 500 mL capacity were used as sample containers, and the saturated salt solution was placed under the sample in each jar. An approximately 20 g sample was placed in each jar, and the weights were periodically measured in duplicate. When the difference of three consecutive weight measurements was less than 0.001 g, the sample was assumed to have reached equilibrium conditions. Crystalline thymol was used to avoid the growth of mold when the relative humidity was over 70 % (Wolf et al. 1985; Menkov 2000).

Table 26.1 Water activities (a_w) of saturated salt solutions at 7 °C, 25 °C, 35 °C and 50 °C (Greenspan 1977)

	Temperature (°C)			
Salt	7	25	35	50
LiCl	0.113	0.113	0.113	0.111
CH_3COOK	0.234	0.225	0.215	0.182
$MgCl_2$	0.335	0.328	0.321	0.305
K_2CO_3	0.431	0.432	0.431	0.409
$Mg(NO_3)_2$	0.588	0.529	0.499	0.454
$NaNO_2$	0.660	0.654	0.627	0.576
NaCl	0.757	0.753	0.749	0.744
KCl	0.867	0.843	0.83	0.812
K_2SO_4	0.982	0.973	0.967	0.958

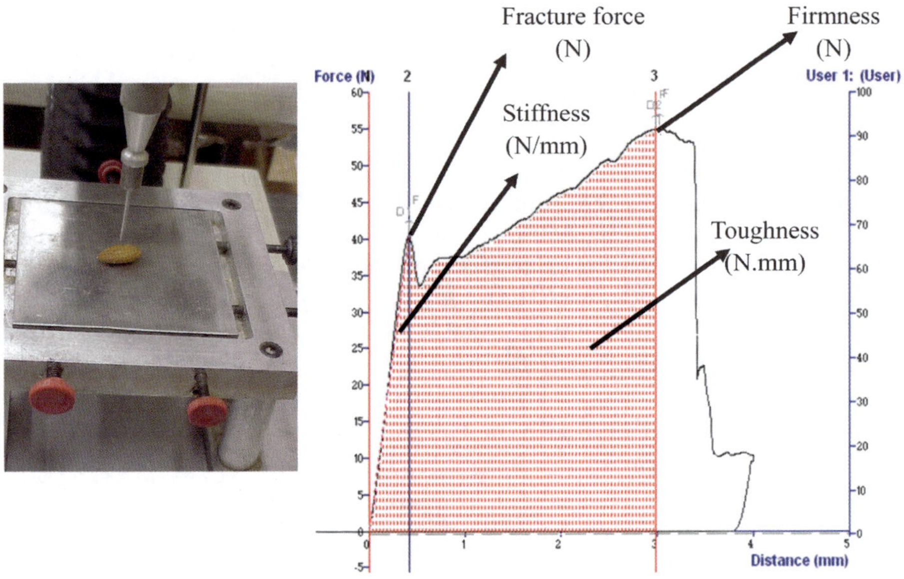

Fig. 26.1 Determination of various mechanical parameters obtained from penetration tests

26.2.2.2 Kinetics of Water Migration

In the kinetic study, whole kernel samples were stored under four temperatures (7 °C, 25 °C, 35 °C and 50 °C) and six relative humidities (43 %, 53 %, 65 %, 75 %, 85 %, and 98 %). Almonds were weighed at predetermined time intervals. Weight measurements were taken every day at 25 °C, 35 °C, and 50 °C. At 7 °C, measurements were done every 7 days for the first 3 weeks and then every 4 days for the remaining storage period. All experiments were conducted in duplicate. The moisture content was measured in triplicate by drying approximately 2 g of ground almonds to a constant weight at 95–100 °C under pressure 100 mmHg (13.3 KPa) for 5 h in a vacuum oven (AOAC 1990).

26.2.2.3 Mechanical Testing

The textural properties of almonds were determined using a TA.XT2 Texture Analyzer (TA.XT2, Texture Technologies, Scarsdale, NY/Stable Micro Systems, Godalming, Surrey, UK) equipped with a load cell of 50 kg. The penetration test was used to determine the mechanical properties of almonds with ten replicates. A 2 mm diameter cylinder probe was used to penetrate the sample for a distance of 4 mm with a test speed of 1.0 mm/s and 400 data points per second. The mechanical properties of equilibrated almonds were obtained from the force–displacement curve, as shown in Fig. 26.1: (1) fracture force (N): force at the first peak;

(2) firmness (N): maximum force obtained from the curve; (3) toughness (N · mm): area under force-deformation curve, which can be considered as the penetration work, and an area of 3 mm deformation was used as the index of the toughness (Kong and Singh 2009); (4) stiffness (N/mm): the relative stiffness determined from the gradient of the first peak.

26.2.2.4 Data Analysis

26.2.2.4.1 Modeling Sorption Isotherms

The moisture adsorption data of almonds were fitted into the GAB model, defined as follows (García-pérez, et al. 2008):

$$M = \frac{mCKa_w}{(1 - Ka_w)(1 - Ka_w + CKa_w)}, \tag{26.1}$$

where

M is the equilibrium moisture content (kg H_2O/kg solid);
m is the monolayer moisture content (kg H_2O/kg solid);
a_w is the water activity of almonds;
C and K are constants related to the sorption heat of the first layer and multilayer, respectively.

An Arrhenius-type equation was used to correlate the coefficient (C, K, and m) in the GAB model with temperature equations as follows (Maroulis et al. 1988):

$$m = m_0 \exp(q_m/RT), \tag{26.2}$$

$$C = C_0 \exp[(H_m - H_n)/RT], \tag{26.3}$$

$$K = K_0 \exp[(\lambda - H_n)/RT)], \tag{26.4}$$

where

q_m, H_m, and H_n are the coefficients related to the heat of adsorption as a function of temperature; m_O, C_O, and K_O are the adjusted constants for temperature; λ is the heat of condensation of pure water (J/mol); T is the absolute temperature in K; R is the universal gas constant, 8.314 J/(mol · K).

26.2.2.4.2 Diffusion Coefficient (D_{eff}) of Water in Almonds

Fick's second law of diffusion is commonly used to calculate the D_{eff} of materials in an unsteady state (Rastogi and Raghavarao 1997; Despond et al. 2001):

$$\frac{\partial m}{\partial t} = D_{eff}\frac{\partial^2 m}{\partial x^2}, \tag{26.5}$$

where

m is the moisture content on a dry basis (kg H_2O/kg dry solid),
t is the storage time (h),
x is the distance from the surface (m),
D_{eff} is the effective diffusion coefficient (m^2/h) of water.

For an infinite slab, it was assumed that the samples had a uniform initial moisture distribution, the external resistance to mass transfer was negligible, and volume did not change during water adsorption.

The initial condition was

$$m = m_i(\mathrm{t} = 0, -\mathrm{L}/2 < \mathrm{x} < +\mathrm{L}/2). \tag{26.6}$$

The boundary condition was

$$m = m_{\mathrm{e}}(\mathrm{t} > 0,\ \mathrm{x} = \mathrm{L}/2). \tag{26.7}$$

The analytical solution to Eq. 26.5 for boundary conditions in one dimension of a semi-infinite slab can be defined as follows (Lomauro et al. 1985):

$$m = m_{\mathrm{e}} + (m_0 - m_{\mathrm{e}})\frac{8}{\pi^2}\sum_{n=0}^{\infty}(2n+1)^{-2}\exp\frac{-(2n+1)^2\pi^2 D_{eff}t}{4L^2}, \tag{26.8}$$

where

m is the moisture content (kg H_2O/kg dry solid) at time t,
m_0 is the initial moisture content (kg H_2O/kg dry solid),
m_{e} is the equilibrium moisture content (kg H_2O/kg dry solid),
L is the thickness of the slab (m).

In this study, the water diffusion in the almonds was assumed to behave isotropically, and D_{eff} was determined by using the equation of a parallelepiped as follows (Ruiz-Beviá et al. 1999):

$$\begin{aligned}\frac{m - m_{\mathrm{e}}}{m_0 - m_{\mathrm{e}}} = \frac{8^3}{\pi^6} \times \Bigg[&\sum_{n=0}^{\infty}(2n+1)^{-2}\exp\frac{-D_{eff}(2n+1)^2\pi^2 t}{4a^2}\\ &\times\sum_{n=0}^{\infty}(2n+1)^{-2}\exp\frac{-D_{eff}(2n+1)^2\pi^2 t}{4b^2}\\ &\times\sum_{n=0}^{\infty}(2n+1)^{-2}\exp\frac{-D_{eff}(2n+1)^2\pi^2 t}{4c^2}\Bigg],\end{aligned} \tag{26.9}$$

where

a, b, and c equal the half distance of the dimension (length, width, height) of almonds (m);
m is the moisture content (dry basis, kg H_2O/kg dry solid) of almonds at time t (h);
m_0 is the initial moisture content of almond (dry basis, kg H_2O/kg dry solid);
m_e is the equilibrium moisture content of almonds (dry basis, kg H_2O/kg dry solid).

The temperature dependency of D_{eff} with activation energy was determined using an Arrhenius-type equation, (Eq. 26.10), as follows:

$$\ln D = \ln D_0 - \frac{E_a}{RT}, \tag{26.10}$$

where D_0 is a pre-exponential factor, which can be calculated from the intercept; E_a is the activation energy (J/mol), which can be obtained from the slope; R is the universal gas constant [8.314 J/(mol · K)]; and T is the absolute temperature (K).

26.2.2.4.3 Mechanical Model

The relationship between mechanical parameters and water activity was described by the Fermi distribution function (Peleg 1993):

$$Y_{(a_w,T)} = \frac{Y_0}{1 + e^{\frac{a_w - a_{wc}}{b_{(T)}}}}, \quad b_{(T)} = b_0 \exp(-k_1 T), \tag{26.11}$$

where Y ($a_{w,\,T}$) is the mechanical parameter, a function of water activity (a_w) and temperature (T);
Y_0 is the magnitude in a dry state;
a_{wc} is the critical water activity, where a drastic change in texture happens;
$b_{(T)}$ is a constant as a function of temperature;
k_1 is a dimensionless constant.

Fermi's distribution function has been found to successfully describe the relationships between a_w (moisture content or temperature) and mechanical parameters including stiffness, toughness, breaking force (Borges and Peleg 1997), compressive modulus (Chang, et al. 2000; Pittia et al. 2007), and sensory crunchiness (Peleg 1993; Peleg 1994a, b, c).

26.2.2.4.4 Verification of Predictive Model

A computer-aided model was developed to predict textural changes in almonds during storage (Fig. 26.2). The input information of the predictive model included almond variety, ambient temperature, water activity, initial moisture content (dry basis, kg H_2O/kg dry solid), and storage time. By applying both storage conditions

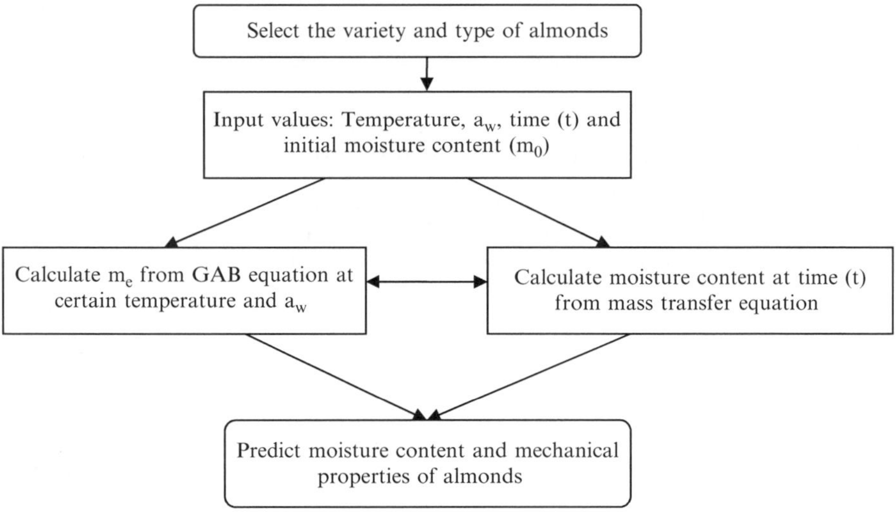

Fig. 26.2 Flow diagram for predictive model of almond

(temperature and relative humidity) and the initial moisture content of the almonds, the moisture content and textural properties of almonds could be calculated from the model. The experimental textural properties under constant and fluctuating ambient conditions were determined and compared with the predicted values to validate the predictive model.

The mean relative percentage of deviation modulus E (%) was calculated to evaluate the goodness of fit of the model, which was defined as follows (Chen and Morey 1989; Peng et al. 2007; Kaya and Kahyaolgu 2007):

$$E(\%) = \frac{100}{n} \sum_{i=1}^{n} \frac{\left|Y_{\exp} - Y_{pri}\right|}{Y_{\exp}}, \tag{26.12}$$

where

$Y_{\exp}$ is the experimental value of various parameters;
Y_{pri} is the predicted value from the predictive model;
n is the number of observations.

The smaller the E (%) value, the better the fit of the model is.

The percentage of deviation (%) was also used to compare the experimental result with the predicted value as the following expression:

$$\begin{aligned} Deviation(\%) &= \frac{100}{n} \times \sum_{i=1}^{n} \frac{Y_{pri} - Y_{\exp}}{Y_{\exp}} \quad \text{for one condition} \\ &= \frac{Y_{pri} - Y_{\exp}}{Y_{\exp}} \times 100 \quad \text{for each data set} \end{aligned} \tag{26.13}$$

where

Y_{exp} is the experimental value of various parameters;
Y_{pri} the predicted value from the predictive model.

Negative deviations indicate that the predicted value is less than the experimental value.

26.3 Results and Discussion

26.3.1 Moisture Adsorption of Almonds

In previous study, the moisture adsorption of almonds was investigated under a temperature range of 7–50 °C. The moisture adsorption curves of Nonpareil pasteurized and unpasteurized almonds showed a sigmoid-shaped isotherm (Fig. 26.3). The equilibrium moisture content (EMC) of almonds linearly increases with an a_w range of 0.1–0.5; thereafter, EMCs begin to increase exponentially. With an increase in temperature, EMC decreases at a constant a_w. High temperature accelerates the movement of water molecules; therefore, water easily breaks away from the water-binding sites at higher temperatures (Samapundo et al. 2007), and almonds become less hygroscopic. Similar phenomena have been reported for other foods (Lahsasni et al. 2003). The solid and dashed lines in Fig. 26.3 correspond to the predicted moisture content from the GAB model, which shows a high goodness of fit to the experimental data ($R^2 > 0.986$). The coefficients of the GAB model were solved by nonlinear regression, which can be used in the predictive model to calculate the moisture content at any level of water activity.

The estimated monolayer moisture content (*m*) values decrease with increasing temperature, within a range of 0.020–0.029 for Nonpareil unpasteurized almonds; the monolayer moisture content of other variety of almonds (Monterey and Carmel) are all in a similar range of 0.022–0.035. Water molecules are bonded with carboxyl groups or amino groups through ionic bonds at the monolayer moisture content; therefore, water molecules are unavailable as a reactant, and any associated reaction is almost negligible. The constant C decreases with temperature in a much wider range of 10.35–110.00; the constant K is less than 1, which is similar to the behavior of other food products (Chen and Morey 1989; Peng et al. 2007). These results are also comparable to the moisture adsorption studies of other almond varieties (Shands et al. 2010; Pahlevanzadeh and Yazdani 2005). An Arrhenius-type equation was used to correlate the constants of *C*, *K*, and *m* in the GAB equation with temperature; therefore, the moisture content can be calculated at any temperature within the studied range.

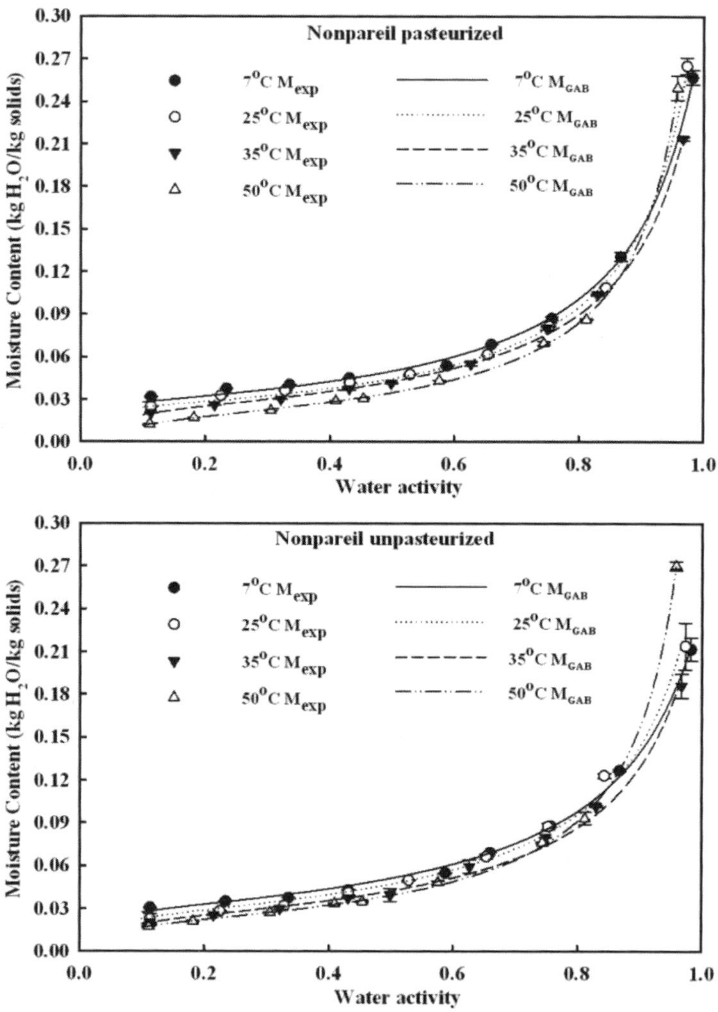

Fig. 26.3 Experimental data of EMC for Nonpareil pasteurized and unpasteurized almonds at different temperatures; *lines* correspond to GAB model

26.3.2 Diffusion of Water in Almonds

The moisture content of almonds increases rapidly at the initial stage of the water uptake process, as shown in Fig. 26.4. When the moisture content of almonds tends to be at a plateau stage, the sample is equilibrated with the environmental moisture, and the weight of the sample almost remains constant. With an increase in temperature, it takes less time to reach equilibrium conditions. At the initial moisture

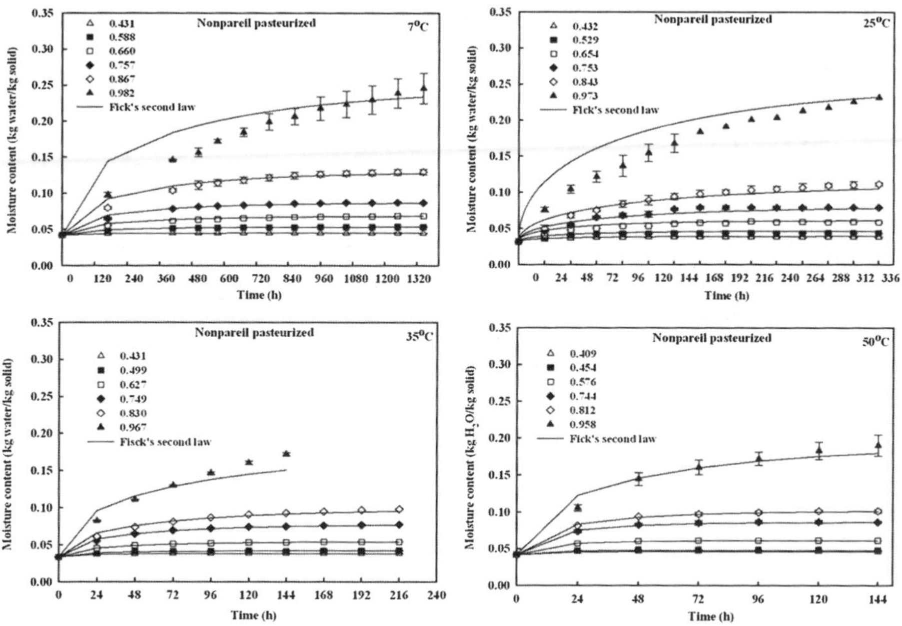

Fig. 26.4 Water adsorption curves of Nonpareil pasteurized almonds at different water activities and temperatures; *lines* correspond to predicted moisture contents from Fick's second law

adsorption point, plenty of free capillaries on the surface of almonds are available; the more water molecules are adsorbed on the surface, the less the water capacity is. Therefore, the rate of moisture adsorption decreases as the moisture content of the almonds increases. Until all the free capillaries are filled with water, the sample is saturated with moisture and reaches equilibrium (Hsu et al. 1983; Rhim 2003; Alakali et al. 2009).

The almond is assumed to be a parallelepiped shape, and water migration through the kernel is assumed to be in the liquid phase. The effective diffusion coefficient (D_{eff}) was calculated from Eq. 26.9 based on Fick's second law of diffusion. It was found that D_{eff} increased with temperature under a constant equilibrium relative humidity. The higher the temperature, the more active the water molecules (Van den Berg and Bruin 1981; Lahsasni et al. 2003). Therefore, the D_{eff} values of water in almonds at 50 °C are much higher than those of the other samples at a lower temperature. Under a constant temperature, D_{eff} decreased with an increase in relative humidity, which agreed with the observed phenomenon that longer time was needed for the sample to reach equilibrium at higher relative humidity.

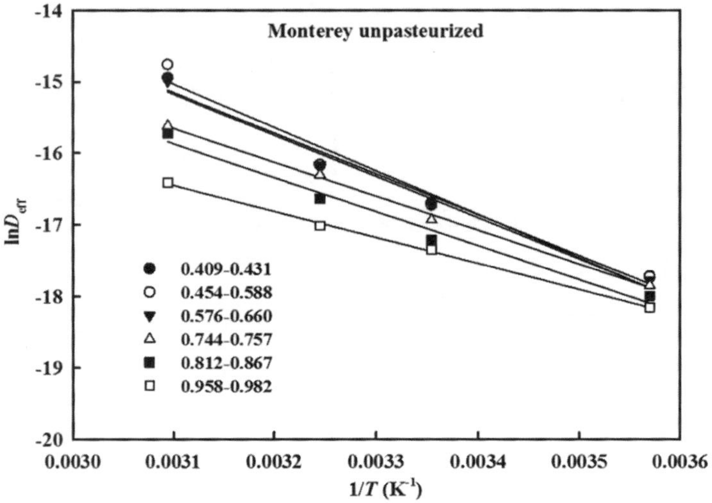

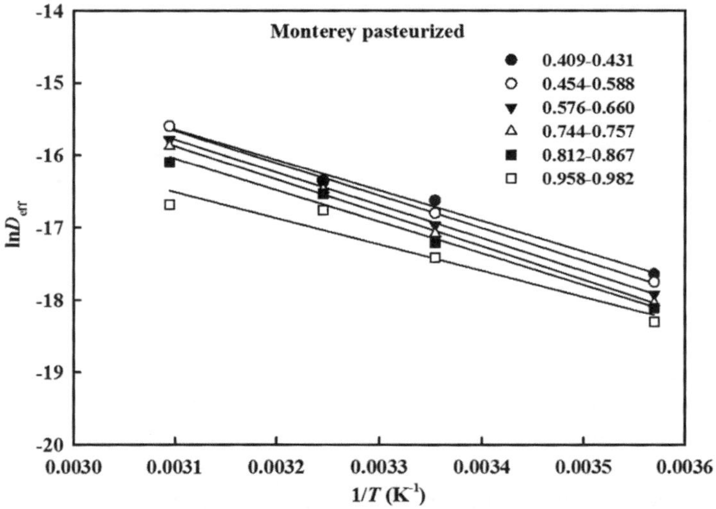

Fig. 26.5 Arrhenious plot of D_{eff} for Monterey pasteurized and unpasteurized almonds

A linear relationship was found between D_{eff} values and temperature using an Arrhenius-type equation (Eq. 26.10). Figure 26.5 shows the Arrhenius plots of the diffusion coefficient (D_{eff}) for the Monterey variety with an R^2 range of 0.930–0.999, indicating the existence of the temperature dependence of D_{eff}. The activation energy (E_{a}) for the selected almond varieties ranged from 30.14 to 52.19 KJ/mol and varied with a_{w}. The relationship between E_a and D_0 can be described as a linear

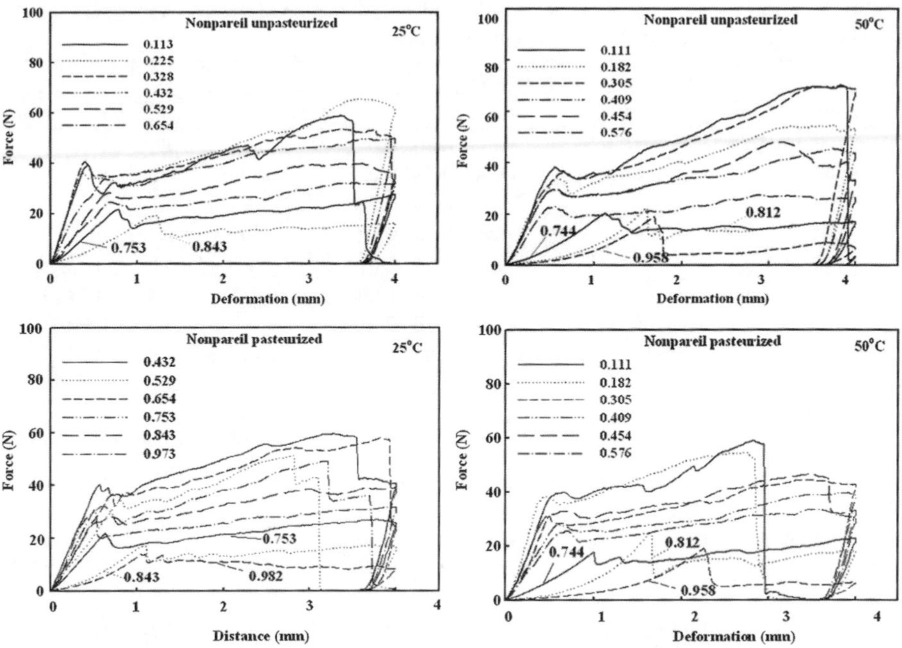

Fig. 26.6 Representative force-distance curves obtained from penetration test of Nonpareil pasteurized and unpasteurized almonds stored at 25 °C and 50 °C

expression (Eq. 26.10), known as the kinetic compensation effect (Özilgen 1993; Rhim 2002). Since the compensation theory was proved to be valid for water adsorption of almonds, the activation energy (E_a) can be used in the predictive model to determine D_o, and then the diffusion coefficient of water in almonds can be calculated at any temperature within the range of this study.

26.3.3 Textural Changes in Almonds During Storage

The effect of moisture on the textural properties of almonds can be monitored through the mechanical parameters derived from the force-displacement curves. The typical force-distance relationship curves of the penetration test for equilibrated almonds stored at 25 and 50 °C are shown in Fig. 26.6. For clarity, only one of the replicates for each a_w level was selected in the textural profiles. There is a decrease in the force after fracture occurring at the first peak, and this point is defined as the bio-yield point (Poulsen 1978; Gupta and Das 2000). After the bio-yield point, the force continued to increase with a lower slope until the penetration distance reached the fixed distance of 4 mm at water activity above 0.65.

Table 26.2 Regression parameters of Fermi's model as a function of a_w and temperature applied to mechanical parameters of almonds obtained from penetration tests

Almonds	Mechanical parameter	Regression parameter						
		Y_0	a_{wc}	b	k_1	SEE	R^2	$E\%$
Nonpareil pasteurized	Firmness	58.35	0.70	0.058	−0.005	7.10	0.75	15.27
	Fracture force	44.30	0.78	0.008	−0.013	4.38	0.71	12.71
	Toughness	102.15	0.70	0.080	−0.002	15.24	0.78	22.57
	Stiffness	68.96	0.65	0.005	−0.013	10.41	0.67	26.87
Nonpareil unpasteurized	Firmness	60.14	0.70	0.019	−0.008	7.75	0.75	14.74
	Fracture force	44.85	0.74	0.004	−0.015	4.46	0.74	12.37
	Toughness	115.49	0.70	0.010	−0.009	15.05	0.80	17.30
	Stiffness	68.79	0.65	0.017	−0.008	11.04	0.74	22.41
Monterey pasteurized	Firmness	71.80	0.62	0.041	−0.006	10.91	0.71	18.54
	Fracture force	42.34	0.76	0.013	−0.011	4.79	0.69	13.69
	Toughness	120.53	0.67	0.014	−0.008	17.706	0.77	17.99
	Stiffness	63.59	0.66	0.052	−0.004	10.413	0.72	23.55
Monterey unpasteurized	Firmness	74.58	0.61	0.027	−0.007	10.323	0.76	17.19
	Fracture force	44.82	0.69	0.008	−0.013	4.427	0.75	14.13
	Toughness	132.31	0.63	0.024	−0.007	16.366	0.83	16.79
	Stiffness	64.92	0.64	0.007	−0.011	9.835	0.76	23.41
Monterey blanched	Firmness	56.02	0.66	0.175	−0.001	7.188	0.74	15.32
	Fracture force	32.81	0.75	0.000571	−0.01944	4.266	0.71	13.83
	Toughness	89.12	0.75	0.070	−0.002	16.497	0.63	19.38
	Stiffness	64.04	0.65	0.004	−0.012	13.219	0.64	31.79

The deformation at the bio-yield point increased with water activity, indicating that almonds adsorbed water and began to lose their brittleness. The shorter the deformation at the bio-yield point, the crisper the almond is. After the water activity reached 0.75, force decreased dramatically after the bio-yield point, indicating a woody texture in almonds after moisture adsorption (Kong and Singh 2009).

Water acts as plasticizer only at a high hydration level with a_w above a specific value (0.65–0.75). The mechanical properties (firmness, fracture force, stiffness and toughness) of almonds had no significant differences until the a_w is above 0.32. The deformation significantly increased after a_w reached 0.75, and other mechanical properties dramatically decreased, showing the plasticization effect of moisture in almonds. The fracture force and stiffness of unpasteurized almonds were larger than those of pasteurized almonds because of the effects of heating during pasteurization. The change in textural properties was influenced not only by moisture but also temperature. Higher temperature can accelerate the plasticization effect of water in almonds. In Fig. 26.6, the deformation of the yield point under 50 °C is larger than that under 25 °C as a_w is over 0.7, and the stiffness is also much lower, showing the almonds lost their brittleness and turned into a rubbery state.

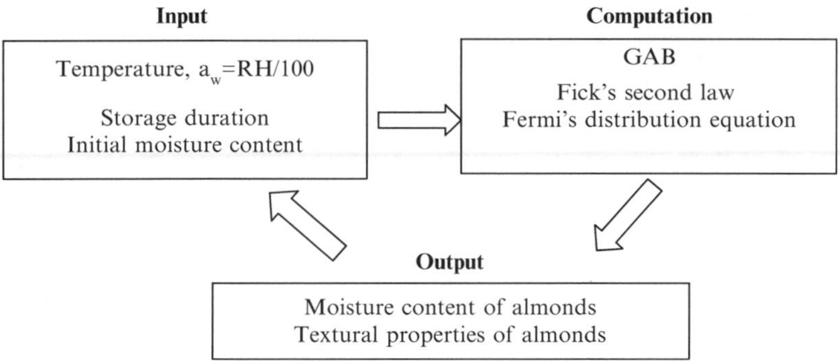

Fig. 26.7 Determination of moisture content and textural properties of almonds in predictive model

Fermi's distribution function (Eq. 26.11) was used to fit the mechanical parameters of almonds as the effects of a_w and temperature with R^2 ranged from 0.63 to 0.83, as shown in Table 26.2. Fracture force had lower (Pahlevanzadeh et al 2005), $SEE = \sqrt{\frac{\sum_{i=1}^{n}(m_i - \hat{m}_i)^2}{df}}$ and $E\%$ values than other mechanical parameters. The critical water activity (a_{wc}) obtained by Fermi's model suggested that a drastic textural change occurred at the a_w range of 0.61–0.78, which was in good agreement with the observed a_w range. Based on the experimental data from the moisture adsorption isotherms, the kinetic study of moisture uptake and textural properties of equilibrated almonds, a computer-aided model was developed to predict the moisture contents and textural properties of almonds during storage and distribution.

26.3.4 Verification of the Predictive Model

In this predictive model, the computation process consists of three equations, the moisture adsorption GAB equation, Fick's second law of diffusion, and Fermi's distribution function, as shown in Fig. 26.7. The initial conditions included the initial moisture content of almonds, storage temperature, relative humidity (a_w = relative humidity/100), and storage duration.

The moisture content values can be calculated from Fick's second law using the initial moisture content, equilibrium moisture content, and storage duration. The experimental moisture contents of Nonpareil pasteurized almonds stored at 7 °C were compared with the predicted moisture contents under different water activities (Fig. 26.8). The predicted moisture contents are very similar to the experimental

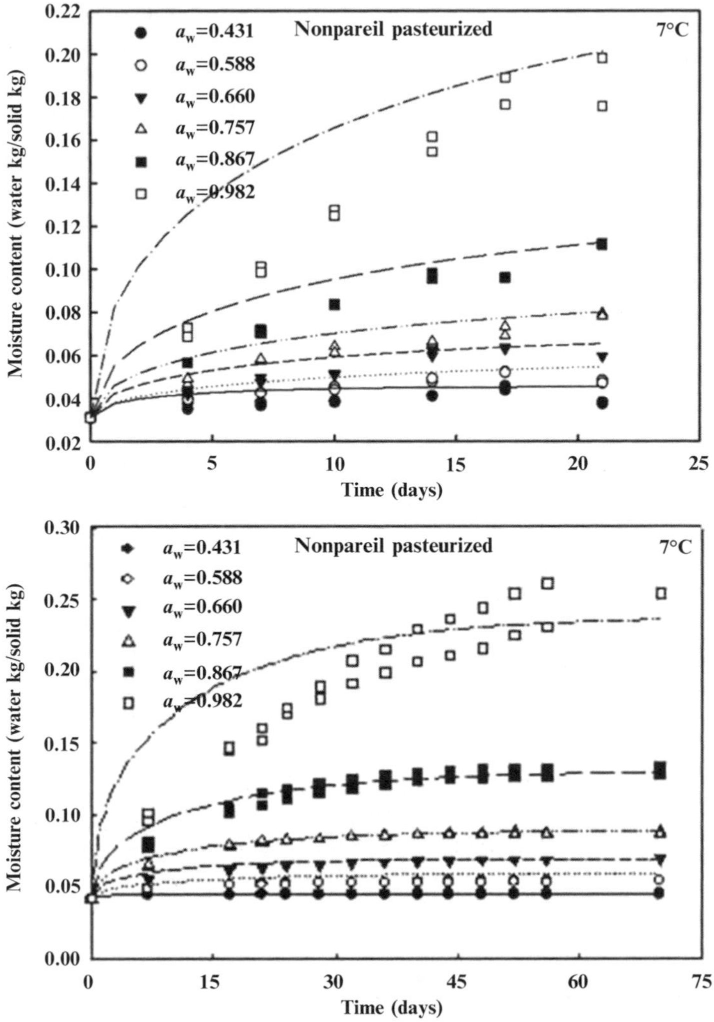

Fig. 26.8 Comparison of experimental moisture content with predicted values of nonpareil pasteurized almonds at different water activity levels during storage

values except that at a_w =0.982. The predicted mechanical properties of almonds were also compared with the experimental values in the same conditions. In Table 26.3, the predicted moisture content has a percentage deviation of −7.68 to −12.1 % in a range of water activity from 0.431 to 0.867 in Table 26.3. The minus sign indicates that the predicted value is larger than the experimental data. There is

Table 26.3 Percentage deviation (%) of predicted variables for nonpareil pasteurized almonds stored at 7 °C for 21 days

D% \ a_w	0.431	0.588	0.66	0.757	0.867	0.982
Moisture content (kg H_2O / kg solids)	-10.55	-7.68	-8.55	-9.5	-12.1	-24.8
Firmness (N)	9.29	7.11	8.35	13.61	17.8	20.00
Fracture force (N)	5.85	3.23	3.17	4.75	0.6	4.34
Toughness (N.mm)	3.38	5.65	9.05	16.09	26.1	34.67
Stiffness (N/mm)	3.80	1.44	-0.09	5.76	10.0	12.42

a larger variation between the experimental moisture content and predicted moisture content at $a_w = 0.982$, with a percentage deviation of -24.8 %.

The predicted fracture force and stiffness had good agreement with the experimental data with less deviation compared to the other two parameters (firmness and toughness) a range of water activities of 0.431–0.982. The percentage deviation of firmness and toughness are larger when the water activity is above the critical value of 0.75, at which the structure of almonds begins to convert into a rubbery state, and they lose their brittleness. Figure 26.9 shows the predicted fracture force with the experimental fracture force values of Nonpareil pasteurized almonds under different water activity during storage at 7 °C. The variation between the predicted fracture force and the experimental value is large at either a lower water activity of 0.431, when almonds are more fragile, or a higher water activity of 0.982, when the textural property of almonds turns into a rubbery state.

To investigate the effect of temperature on the predicted performance of the model, the predicted moisture content and mechanical properties were compared with the experimental data at two temperatures (25 °C and 50 °C), as shown in Figs. 26.10 and 26.11, respectively. A higher temperature can shorten the time for the sample to reach the equilibrium condition under constant water activity, which agrees with the observation in the previous study that the diffusion coefficient (D_{eff}) increases with temperature under constant water activity. The mechanical properties of almonds stored at a constant water activity showed little variation under the two temperatures of 25 and 50 °C. Therefore, the moisture adsorption and textural properties of almonds are more dependent on water

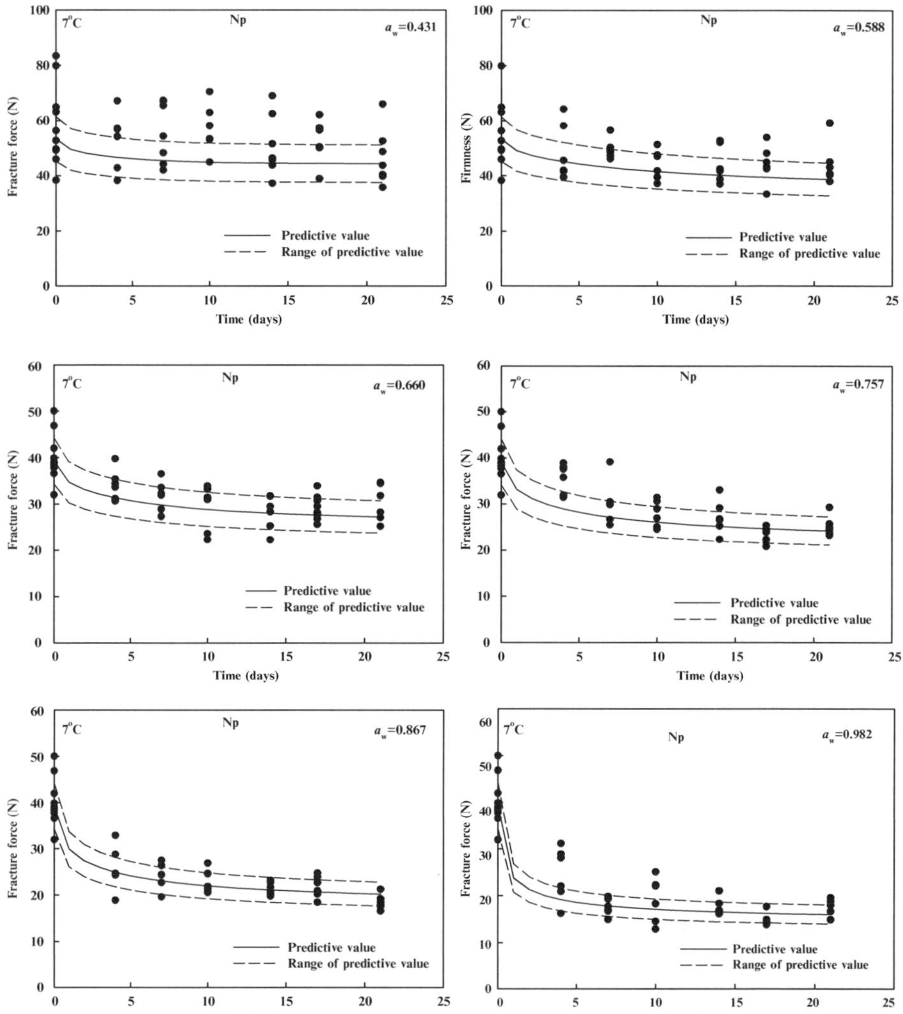

Fig. 26.9 Comparison of experimental fracture force with predicted values of Nonpareil pasteurized (NP) almonds at different water activity levels during storage

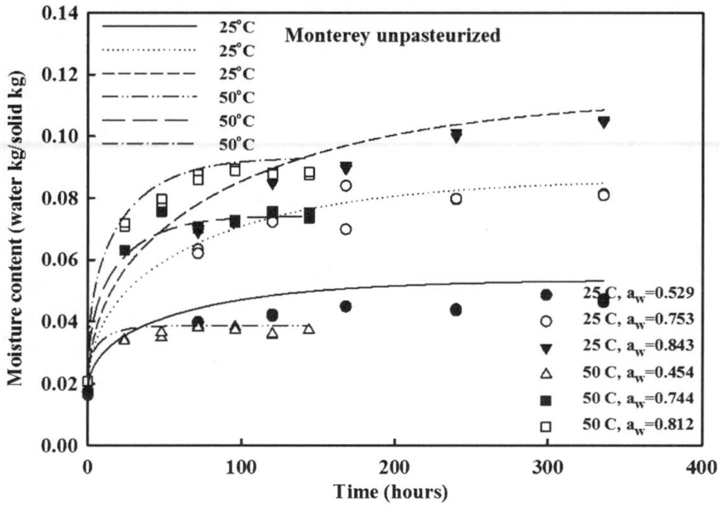

Fig. 26.10 Comparison of experimental moisture content with predicted values of Monterey unpasteurized almonds at 25 °C and 50 °C

activity than temperature, but temperature can accelerate the moisture adsorption of almonds.

Furthermore, Monterey unpasteurized almonds were stored under fluctuating ambient conditions. Two trials were conducted, and there were four stages for each trial listed in Table 26.4. The moisture contents and textural qualities of almonds during storage were measured and compared with the predicted values generated from the model. Figure 26.12 shows the experimental moisture content of almonds stored for 42 days at different relative humidities and temperatures in each trial. The lines correspond to predicted values calculated from the mathematical model, which has good agreement with the experimental data for both moisture adsorption and desorption of almonds. Although there are some variations between the predicted moisture content and experimental data, the mechanical profile is close to the experimental value even under fluctuating conditions. The mean relative percentage deviation moduli E (%) were less than 10 % in the second trial, indicating the goodness of fit of the predictive model (Table 26.5) (Lomauro et al. 1985; Corrêa et al. 2007).

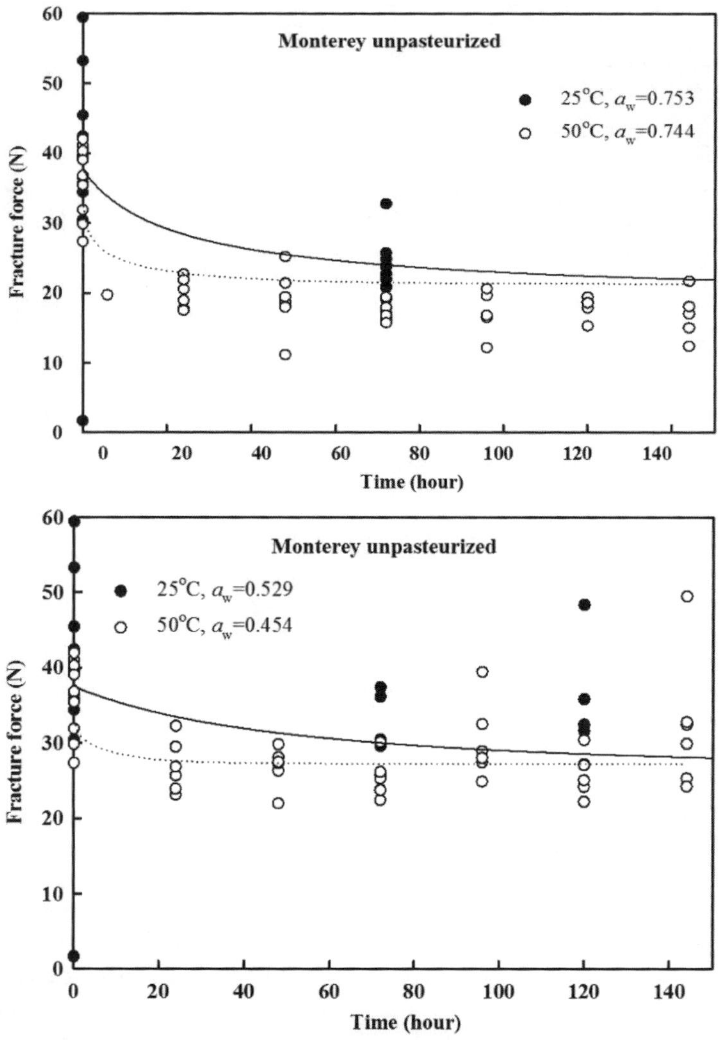

Fig. 26.11 Comparison of experimental fracture force with predicted values of Monterey unpasteurized almonds at 25 °C and 50 °C

Table 26.4 Storage conditions for each stage for each trial

Parameter	Stage I	Stage II	Stage III	Stage IV
(i) First trial				
Temperature (°C)	25	45	30	15
Water activity	0.654	0.745	0.836	0.559
Storage time (days)	15	6	6	15
(ii) Second trial				
Temperature (°C)	25	45	30	15
Water activity	0.654	0.817	0.514	0.741
Storage time (days)	15	6	6	15

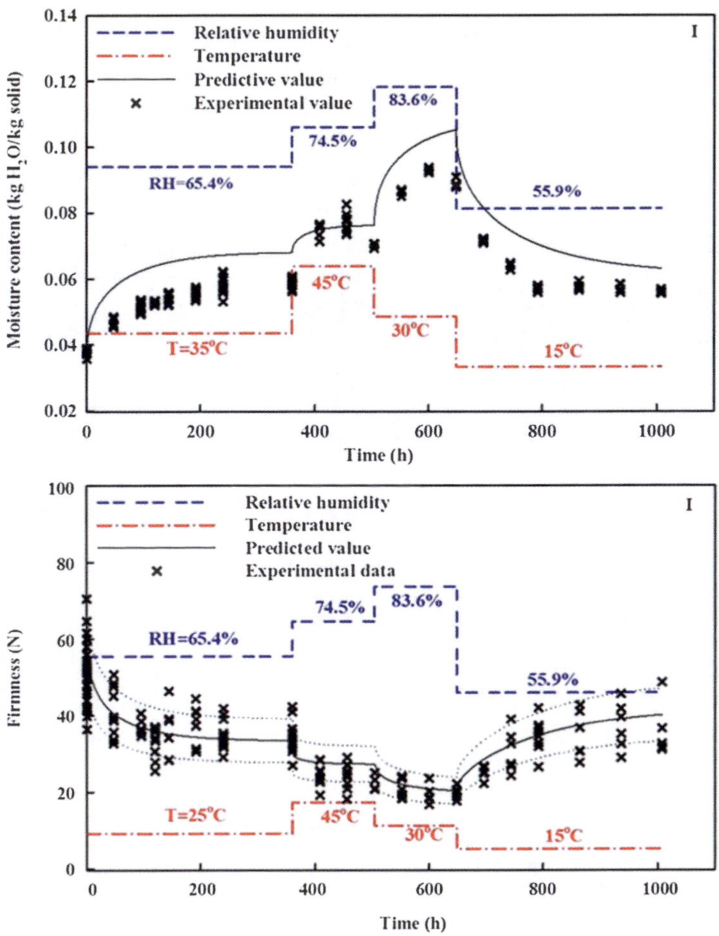

Fig. 26.12 Comparison of experimental data of moisture content and textural properties of Monterey unpasteurized almonds with predicted data

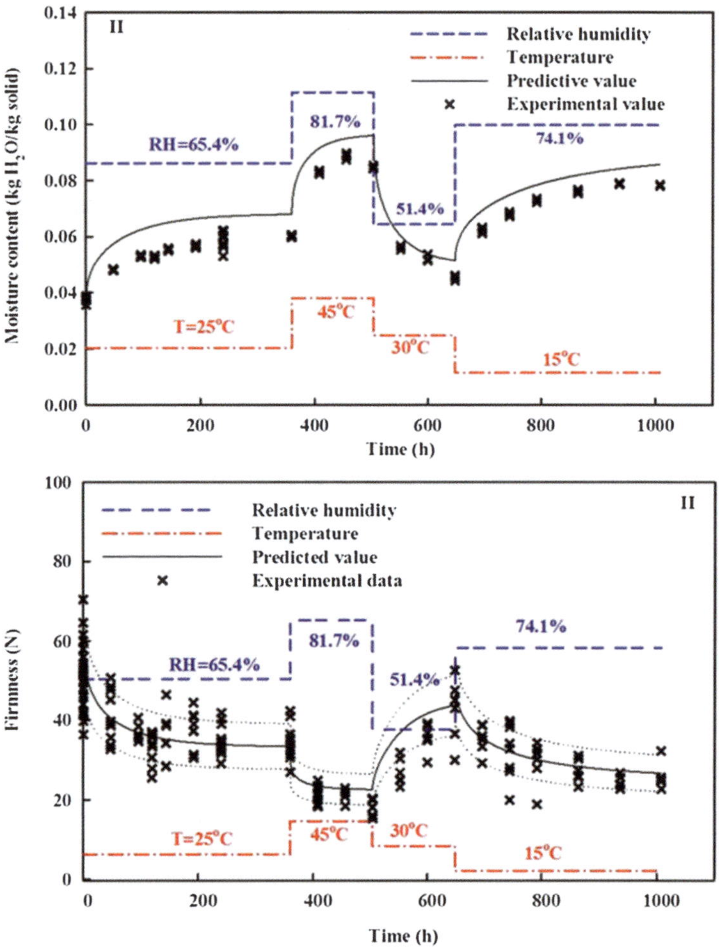

Fig. 26.12 (continued)

Table 26.5 Mean relative percentage deviation modulus *E* (%) for moisture content and mechanical properties of almonds in first and second trials

E%	First trial	Second trial
Moisture content	13.91	4.82
Firmness	24.18	6.18
Fracture force	16.86	6.42
Toughness	24.35	6.18
Stiffness	28.29	8.34

26.4 Conclusion

A computer-aided mathematical model was developed to estimate the changes in both moisture content and textural properties of almonds during storage and distribution. This predictive computer-aided model, incorporating typical storage and distribution conditions, showed good agreement with experimentally determined moisture content and textural properties of almonds. Both the moisture content and mechanical properties of almonds highly depend on the water activity level, and temperature can accelerate the quality change in almonds. Therefore, the predicted values are more sensitive to water activity than to temperature. The simulation of the moisture content and textural properties of almonds can be helpful to practitioners engaged in storing or distributing almonds in their search for protocols that minimize the quality deterioration of almonds during storage and distribution.

Acknowledgements The authors would like to thank the Almond Board of California for providing financial support and almonds for this research.

References

Alakali JS, Ariahu CC, Kucha EI (2009) Kinetics of moisture uptake of osmo-foam-mat dried mango powders and application of sorption isotherms to shelf-life prediction. Am J Food Technol 3:119–125
AOAC (1990) Official methods of analysis, 15th edn. Association of Official Analytical Chemists, Washington, DC
Avramidis S, Vancouver BC (1992) Enthalpy-entropy compensation and thermodynamic considerations in sorption phenomena. Wood Sci Technol 26:329–333
Borges A, Peleg M (1997) Effect of water activity on the mechanical properties of selected legumes and nuts. J Sci Food Agric 75:463–471
Corrêa PC, Goneli ALD, Jaren C, Ribeiro DM, Resende O (2007) Sorption isotherms and isosteric heat of peanutpods, kernels and hulls. Sci Tech Int 13:231–238
Chang YP, Cheah PB, Seow CC (2000) Variations in flexural and compressive fracture behaviour of brittle cellular food (dried bread) in response to moisture sorption. J Texture Stud 31:525–540
Chen CC, Morey RV (1989) Comparison of four EMC/ERH equations. Trans Am Soc Agric Eng 1989(32):983–990
Crank J (1975) The mathematics of Diffusion. 2nd Ed. Oxford University Press, Oxford, UK
Despond S, Espuche E, Domard A (2001) Water sorption and permeation in chitosan films: relation between gas permeability and relative humidity. J Polym Sci, Part B: Polym Phys 39:3114–3127
García-pérez JV, CárceL JA, Clemente G, Mulet A (2008) Water sorption isotherms for lemon peel at different temperature and isosteric heats. LWT- Food Sci Technol 41:18–25
Goneli ALD, Corrêa PC, Horta DOGH, Ferreira GC, Mendes BF (2010) Water sorption isotherms and thermodynamic properties of pearl millet grain. Int J Food Sci Technol 45(4):828–838
Goula AM, Karapantsios TD, Achilias DS, Adamopoulos KG (2008) Water sorption isotherms and glass transition temperature of spray dried tomato pulp. J Food Eng 85(1):73–83

Gowen A, Abu-Ghannam N, Frias J, Oliveira J (2007) Influence of pre-blanching on the water absorption kinetics of soybeans. J Food Eng 78:965–971

Greenspan L (1977) Humidity fixed points of binary saturated aqueous solutions. J Res Nat Bur Stand A Phys Chem 81A:89–96

Gupta RK, Das SK (2000) Fracture resistance of sunflower seed and kernel to compressive loading. J Food Eng 46(1):1–8

Harris M, Peleg M (1996) Patterns of textural changes in brittle cellular cereal foods caused by moisture sorption. Cereal Chem 73(2):225–231

Hsu KH, Kim CJ, Wilson LA (1983) Factors affecting water uptake of soybeans during soaking. Cereal Chem 60:208–211

Kashaniejad M, Maghsoudlou Y, Rafiee S, Khomeiri M (2007) Study of hydration kinetics and density changes of rice (Tarom Mahali) during hydrothermal processing. J Food Eng 79:1383–1390

Kaya S, Kahyaolgu T (2007) Moisture sorption and thermodynamic properties of safflower petals and tarragon. J Food Eng 78:413–421

Kong FB, Singh RP (2009) Digestion of raw and roasted almonds in simulated gastric environment. Food Biophys 4:365–377

Lahsasni S, Kouhila M, Mahrouz M, Fliyou M (2003) Moisture adsorption–desorption isotherms of prickly pear cladode at different temperatures. Energ Convers Manage 44:923–936

Lomauro CJ, Bakshi AS, Labuza TP (1985) Evaluation of food moisture sorption isotherm equation. Part II: Milk, coffee, tea, nuts oilseeds and starchy foods. Lebensm.-Wiss.u.-Technol 18:118–124

Maroulis ZB, Tsami E, Marinos-kouris D (1988) Application of the GAB model to the moisture sorption isotherms for dried fruits. J Food Eng 7:63–78

Menkov ND (2000) Moisture sorption isotherms of chickpea seeds at several temperatures. J Food Eng 45:189–194

Özilgen M (1993) Enthalpy-entropy and frequency factor-activation energy compensation relations for diffusion in starch and potato tissue. Starch-Stärke 45:48–51

Pahlevanzadeh H, Yazdani M (2005) Moisture adsorption isotherms and isosteric energy for almond. J Food Process Eng 28:331–345

Peleg M (1993) Mapping the stiffness-temperature-moisture relationship of solid biomaterials at and around their glass transitions. Rheologica Acta 32(6):575–580

Peleg M (1994a) A mathematical model for crunchiness/crispness loss in breakfast cereals. J Texture Stud 25:403–410

Peleg M (1994b) A model of mechanical changes in biomaterials at and around their glass transition. Biotechnol Prog 10:385–388

Peleg M (1994c) Mathematical characterizations and graphical presentation of the stiffness–temperature–moisture relationship of gliadin. Biotechnol Prog 10:652–654

Peng G, Chen XG, Wu WF, Jiang XJ (2007) Modeling of water sorption isotherm for corn starch. J Food Eng 80:562–567

Pittia P, Sacchetti G (2008) Antiplasticization effect of water in amorphous foods. A review. Food Chem 106(4):1417–1427

Pittia P, Nicoli MC, Sacchetti G (2007) Effect of moisture and water activity on textural properties of raw and roasted coffee beans. J Texture Stud 38(1):116–134

Poulsen GR (1978) Fracture resistance of soybeans to compressive loading. Transactions of the ASAE (Am Soc Agric Eng) 21:1210–1216

Rastogi NK, Raghavarao KSMS (1997) Water and solute diffusion coefficients of carrot as a function of temperature and concentration during osmotic dehydration. J Food Eng 34:429–440

Rhim JW (2002) Kinetic compensation relations for texture changes in sweet potatoes during heating. Food Sci Biotechnol 11:29–33

Rhim JW (2003) Hydration kinetics of soybeans. Food Sci Biotechnol 12:303–306

Ruiz-Beviá F, Fernández-Sempere J, Gómez-Siurana A, Torregrosa-Fuerte E (1999) Determination of sorption and diffusion properties of peeled almond nuts. J Food Eng 41(3):209–214

Samapundo S, Devlieghere F, Meulenaer BD, Atukwase A, Lamboni Y, Debevere JM (2007) Sorption isotherms and isosteric heats of sorption of whole yellow dent corn. J Food Eng 79:168–175

Shands J, Lam C, Labuza TP (2010) Moisture sorption properties of commercially processed almonds. In Institute of food technologies annual conference on food engineering division-product development posters; 146.06

Sharma P, Singh RRB, Singh AK, Patel AA, Patil GR (2009) Sorption isotherms and thermodynamics of water sorption of ready-to-use Basundi mix. LWT- Food Sci Technol 42:441–445

Van den Berg C, Bruin S (1981) Water activity and its estimation in food systems: theoretical aspects. In: Rockland LB, Stewart GE (eds) Water activity: influence on food quality. Academic, New York, pp 1–61

Wani AA, Sogi DS, Shivhare US, Ahmed I, Kaur D (2006) Moisture adsorption isotherms of watermelon seed and kernels. Drying Technol 24:99–104

Wolf W, Spiess WEL, Jung G (1985) Standardization of isotherm measurements (cost-project 90 and 90 bis). In: Stimatos D, Multon JL (eds) Properties of water in foods in relation to quality and stability. Martinus Nijhoff Publishers, Dordrecht, pp 661–679

Chapter 27
Developing Next-Generation Predictive Models: Systems Biology Approach

D. Vercammen, E. Van Derlinden, F. Logist, and J.F. Van Impe

27.1 Introduction

In predictive microbiology, an important area of food microbiology, the focus is on the mathematical description and prediction of the evolution (growth, survival, and inactivation) of pathogenic and spoilage microorganisms in food products. Since the 1980s, there has been a remarkable increase of interest in predictive microbiology (McMeekin et al. 2002). Implementation of these predictive models contributes to the improved control of food safety and spoilage, for example, by quantifying the effect of storage and distribution on microbial proliferation via the hazard analysis and critical control point (HACCP) system. Recently, predictive microbiology has been accepted as a tool to define the safety of (certain) food products in Europe. Predictive models are also being applied in software packages (e.g., ComBase [UK-USA], Sym'Previus [France], and the Pathogen Modeling Program (PMP) [USA]) that are useful in both academic and industrial environments. In addition, predictive models can be an essential tool for risk control in the optimization of food engineering processes.

The first publications in the domain of predictive microbiology mainly focused on the growth and inactivation dynamics of species exposed to (a constant value of) a single environmental condition. This behavior is described by a primary predictive model, which quantifies the microbial cell count as a function of time. In the classic approach, bacterial growth is modeled autonomously by a logistic-type equation. These models do not include mechanistic knowledge to describe the transitions between the different growth phases. The stationary phase, for example, is described by including the asymptotic value, i.e., the maximum cell number, as a

D. Vercammen • E. Van Derlinden • F. Logist • J.F. Van Impe (✉)
BioTeC -Chemical and Biochemical Process Technology and Control,
Department of Chemical Engineering, Katholieke Universiteit Leuven,
W. de Croylaan 46, B-3001 Heverlee, Belgium
e-mail: jan.vanimpe@cit.kuleuven.be

S. Yanniotis et al. (eds.), *Advances in Food Process Engineering Research and Applications*, Food Engineering Series, DOI 10.1007/978-1-4614-7906-2_27,

parameter. Primary models developed in the 1990s are still widely used but are mainly empirical (e.g., Baranyi and Roberts 1994; Buchanan et al. 1997; Zwietering et al. 1990; Geeraerd et al. 2000).

In a second step, secondary models are developed that describe the influence of changing environmental conditions on primary models, i.e., on their parameters. Most currently used secondary models can be subdivided into four classes: (1) -square-root models (e.g., Ratkowsky et al. 1982, 1983; Ross et al. 2003); (2) cardinal parameter models (e.g., Rosso et al. 1993, 1995; Sautour et al. 2001); (3) neural networks (e.g., Geeraerd et al. 1998; Panagou et al. 2007); and (4) response surface models (e.g., Baranyi et al. 1996; Geeraerd et al. 2004). According to Geeraerd et al. (2004), secondary models fall into two groups. The first group consists of models that include (some) biologically or graphically interpretable parameters and can be extended to more environmental factors via a multiplicative approach. Furthermore, they are parsimonious and have a high fitting quality. The second group of secondary models, i.e., neural networks and response surface models, do not presume a priori knowledge of the underlying relationship. These models are characterized by a high flexibility.

Most existing primary and secondary models enable an accurate description of microbial dynamics under (nonstressing) dynamic conditions for liquid systems. However, in the last decade, it has been widely recognized that these models fail when applied to real food products and under more realistic, more stressing conditions. The aforementioned models consider rather simple liquid systems, mainly controlled by temperature, pH, water activity, acids, and preservatives. However, more complex elements, like background flora, microbial competition, stress and stress adaptation, and the physicochemical properties of the food structure, are rarely taken into account. This is known as the *completeness error* and is considered (one of) the largest source(s) of error in predictive microbiology (McMeekin and Ross 2002).

Based on this generally accepted analysis, a quest for more mechanistically based predictive models has started (McMeekin et al. 2008; Brul et al. 2008). Examples can be found for different aspects of microbial dynamics. (1) Van Impe et al. (2005) introduced a novel class of macroscopic predictive microbial growth models that do take the (micro)biological phenomena governing the microbial growth process into account. Their work focused on the transition of the exponential growth phase to the stationary phase, which can be induced through toxic product accumulation or substrate exhaustion. This modeling approach was also employed to describe the more complex case study of coculture inhibition of *Listeria innocua* mediated by lactic acid production of *Lactococcus lactis* (Poschet et al. 2005). (2) To unravel the initial lag-phase dynamics, focus is on the individual cell dynamics and how they relate to the overall population behavior (e.g., McKellar 2001; Baranyi 2002; Baranyi et al. 2009). Baranyi and Pin (2001) constructed a mathematical relation between the stochastic individual cell dynamics and a deterministic model describing the population. Individual-based modeling techniques have been developed to gain insights into single-cell and population lag-phase dynamics (Standaert et al. 2007; Prats et al. 2008). (3) When microbial populations are exposed to (severe) stress (e.g., heat and acid), sigmoid growth curve patterns are often disturbed and typical primary and secondary models and modeling approaches are no longer valid. These unexpected growth curves can be attributed to microbial population heterogeneity induced by environmental conditions. Although population heterogeneity is a widely recognized phenomenon of particular interest for, for example, food safety and quality, few attempts have been made to include population heterogeneity in the modeling of microbial kinetics (e.g., Nikolaou and Tam 2005; Van Derlinden et al. 2009, 2010). (4) In structured (solid) foods, microbial growth can strongly depend on the position in the food, and the general assumption of homogeneity cannot be accepted, i.e., space must be considered as an independent variable. Dens and Van Impe (2001) presented a model that takes into account the variability of microbial growth with respect to space. The presented model describes two phenomena: the local evolution of biomass and the transfer of biomass through the medium.

(Unexpected) cell dynamics observed in heterogeneous environments or under stress conditions due to heterogeneous populations and stress, adaptation phenomena cannot be explained using the macroscopic approach generally applied in predictive microbiology. In the future, predictive microbiology must take the modeling one step further by including more micro- or mesoscopic information to understand these cell dynamics. The applicability and reliability of existing models under more realistic conditions will definitively be improved by looking inside the black box and unraveling the underlying mechanisms (Brul and Westerhoff 2007). Incorporating intracellular information in predictive models, following a top-down systems biology approach, will result in more widely applicable mechanistic models. The intrinsic complexity of biochemical processes, which consist of extensive reaction networks with numerous metabolites, is not reflected by simple macroscopic-level models, which are classically used in predictive microbiology. However, while more knowledge about the underlying mechanisms of biochemical processes becomes available, new opportunities arise, for instance by using (microscopic-level) metabolic network models to build next-generation predictive models. Relevant (extracellular) process conditions and key metabolic reactions/pathways can be identified, which is valuable information in the development of predictive models for more complex and realistic situations. Exploitation of metabolic flux analysis (MFA) as a technique to develop accurate mathematical models in the field of predictive microbiology is a largely unexplored domain that is presented in this chapter.

27.2 Methodology

Fundamental microbial research is in general conducted at three levels, i.e., the macroscopic, the mesoscopic, and the microscopic levels. At the *macroscopic level*, the overall population characteristics and behavior are studied. Macroscopic predictive models describe the growth and inactivation dynamics of populations.

As was stated in the introduction, macroscopic-level models are able to accurately predict population dynamics under nonstressing conditions in liquid food model systems. For process control, monitoring, and optimization purposes, macroscopic models are preferred as they have a rather simple structure, i.e., a limited number of model components and parameters. The *mesoscopic level* studies small populations, part of the population like subpopulations or colonies in structured environments. Due to environmental or population heterogeneity, differences in the microbial response are observed and all cells – or their dynamics – can no longer be assumed to be identical. Examples of more mesoscopic models can be found in McKellar (1997) and Skandamis et al. (2002). To completely unravel the mechanisms underlying the specific microbial response to, for example, stressing environments or environmental gradients, information is collected at the *microscopic level*, i.e., cellular or even intracellular level. In recent decades, for instance, much research has focused on the intracellular stress response, for example, the heat shock response of *E. coli* (Arsene et al. 2000; Yuk and Marshall 2003; Chung et al. 2006).

The metabolic network-based modeling approach presented in this work will ultimately link microscopic-level information, enclosed in a metabolic network, with macroscopic-level models that, in the end, are mechanistically inspired but still rather simple to use in practice.

27.2.1 Modeling at the Macroscopic Level

At the macroscopic level, the most important variables defining the microbial dynamics are the cell number N_{macro} (CFU/mL), the limiting substrate concentration $C_{S,macro}$ (mol/mL), and the metabolic product concentration $C_{P,macro}$ (mol/mL), with μ (1/h), σ, and π (mol/CFU h) the specific rates for cell growth, substrate consumption, and product formation, respectively. The effect of intrinsic or extrinsic conditions [henceforth denoted by (·) in equations] like, for example, temperature, water activity, and background flora, on microbial behavior is included in the specific rates using specific kinetic models like the cardinal temperature model with inflection (Rosso et al. 1993) or Monod- or Haldane-type relations:

$$\frac{dN_{macro}}{dt} = \mu(\cdot) \cdot N_{macro}, \tag{27.1}$$

$$\frac{dC_{S,macro}}{dt} = -\sigma(\cdot) \cdot N_{macro}, \tag{27.2}$$

$$\frac{dC_{P,macro}}{dt} = \pi(\cdot) \cdot N_{macro}. \tag{27.3}$$

The generic relation between the macroscopic reaction rates is described as follows:

$$\sigma(\cdot) = \frac{\mu(\cdot)}{Y_{X/S}} + \frac{\pi(\cdot)}{Y_{P/S}} + m, \tag{27.4}$$

with *YX/S* and *YP/S* the yield coefficients of biomass on the substrate and product on the substrate, respectively, and with *m* the maintenance factor. By concatenating all macroscopic concentrations (metabolites as well as cell number) into one macroscopic concentration vector C_{macro}, a concise vector representation of the macroscopic level model results:

$$\frac{dC_{macro}}{dt} = R_{macro} \cdot N_{macro}, \tag{27.5}$$

where *dt* with R_{macro} is the vector of specific rates for the various elements of C_{macro}. Note that the macroscopic concentration vector C_{macro} can contain as many concentration variables as needed, for example, multiple substrate/product concentrations. This matrix formulation of the model equations will be used throughout this chapter.

27.2.2 Basics of Metabolic Networks

A *metabolic network* is a graphical representation of (a subset of) the metabolic reactions that take place inside a cell. These reactions comprise intracellular reactions between intracellular metabolites and transport reactions between the intracellular and extracellular space. The *m* metabolites, both intracellular and extracellular, are represented by the nodes in the network. The *n* reactions are represented as links between these metabolites. All information about the stoichiometry of these reactions can be represented by a *stoichiometric matrix* S. The elements of this matrix are the stoichiometric coefficients of the reactions, i.e., the element at row *i* and column *j* is the stoichiometric coefficient of metabolite *i* in reaction *j*. This relationship between metabolic network and stoichiometric matrix is exemplified for a toy network in Fig. 27.1. The stoichiometric matrix can be subdivided into S_{int} and S_{ext}, respectively the rows of S representing intracellular metabolites and extracellular metabolites (including biomass). Furthermore, the reaction rates of all reactions are summarized in the flux vector v (mol/CFU h). Multiplying the stoichiometric matrix S by the reaction fluxes v renders the series of linear relations describing the metabolic network.

Metabolic networks exist in a wide range of sizes. Genome-scale metabolic networks are derived from the genome of an organism, from the set of enzymes that can be produced by the organism. They contain all possible reactions that can occur inside the cell. Because of this, they are very extensive and their stoichiometric matrices are very large. As an example, iAF1260 (Feist et al. 2007), a genome-scale metabolic network for *E. coli* K-12 MG1655, contains 2,077 reactions between

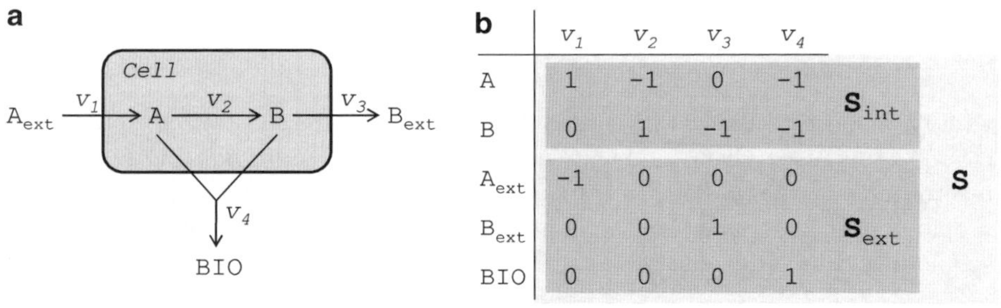

Fig. 27.1 Relationship between metabolic network and stoichiometric matrix for a toy network. (**a**) Toy network including two intracellular metabolites, three extracellular metabolites (including biomass), and four reactions. (**b**) Stoichiometric matrix S for network in (**a**), with subdivision into S*int*, the rows of S representing intracellular metabolites, and S*ext*, the rows of S representing extracellular metabolites. Note that the metabolite standing for biomass is always considered an extracellular metabolite

1,039 metabolites. In most applications, only a small part of this network is studied, so the genome-scale network is simplified to a smaller and computationally less demanding size. Mahadevan et al. (2002), for example, use a medium-scale network consisting of 85 reactions between 54 metabolites. For *Salmonella*, knowledge of the underlying network is less abundant, and only a limited number of networks, mainly genome-scale, have been published (e.g., AbuOun et al. 2009). At this moment, intracellular information for Listeria is incomplete. However, research to unravel its metabolic/genome network is ongoing and, based on the available information, (preliminary) networks have been constructed that can be used to study its intracellular dynamics (e.g., Schauer et al. 2010).

27.2.3 Modeling at the Microscopic Level

At the microscopic level, a distinction is made between intracellular and extracellular dynamics. The intracellular dynamics can be characterized by the intracellular metabolite concentrations C_{int} (mol/CFU), the intracellular reaction fluxes, i.e., the cell-specific reaction rates v, and the intracellular stoichiometry matrix S_{int}:

$$\frac{dC_{\text{int}}}{dt} = S_{\text{int}} \cdot v - \mu \cdot C_{\text{int}}. \tag{27.6}$$

In this equation, the last term is a dilution term. This term originates from the conversion of the system in absolute quantities (mol) to relative quantities or concentrations per cell unit (mol/CFU). Because these intracellular concentrations are expressed per cell unit, the growth of the cells (i.e., the increasing of the number of cells) introduces a dilution of these intracellular metabolites and, hence, a decrease in their concentrations.

Extracellular dynamics are characterized by the extracellular metabolite concentrations C_{ext}. This is a vector comprising concentrations of all extracellular metabolites, substrates as well as products, and possibly metabolites that can both be synthesized and taken up, for example, amino acids. It also contains the cell number N, which is seen as an extracellular metabolite:

$$\frac{dC_{ext}}{dt} = S_{ext} \cdot v \cdot N. \qquad (27.7)$$

The coupling between the two systems (Eqs. 27.6 and 27.7) is made at the level of the fluxes. To be able to simulate this system of ordinary differential equations, all fluxes must be related to extrinsic and intrinsic factors via kinetic expressions:

$$v = f(C_{ext}, C_{int}, T, pH, \ldots; \boldsymbol{\Phi}), \qquad (27.8)$$

with $\boldsymbol{\Phi}$ the set of kinetic parameters. Even for a small network, it is very demanding to identify these functions and their parameters, both experimentally and computationally. Typical small-scale networks describe around 50 fluxes. This means that 50 kinetic expressions must be identified, along with the kinetic parameters, which are typically a multiple of the number of kinetic expressions. Techniques of model reduction are thus essential to be able to successfully use these models in a simulation, prediction, and control environment.

27.2.4 Linking the Microscopic and Macroscopic Levels

The variables at the microscopic level are only valid locally, i.e., the extracellular metabolite concentrations and cell number $Cext$ are only valid in a small environment around the cell. To link this microscopic level with the macroscopic level, some measure must be taken at the mesoscopic level, i.e., the level between cell and population. One possibility is to integrate the microscopic level into a description of the mesoscopic level, for example, via individual-based modeling, linking the microscopic and macroscopic descriptions through the mesoscopic level. A second possibility is to short-cut the mesoscopic level. In this case, two assumptions/simplifications must be made. (1) By assuming that all cells are equal to one average cell, it is possible to limit the microscopic level to this one average cell. The behavior of this average cell is then modeled by means of the microscopic representation. In this case, the macroscopic cell concentration N_{macro} is equal to the microsopic cell concentration N. All cells are assumed to express the same flux pattern. (2) A second assumption is perfect mixing of a liquid medium. The microscopic metabolite concentrations will then be equal in the complete medium and turn into macroscopic variables. If these assumptions are made, the final system of model equations looks like this:

$$\frac{dC_{macro}}{dt} = S_{ext} \cdot v \cdot N_{macro}, \tag{27.9}$$

$$\frac{dC_{int}}{dt} = S_{int} \cdot v \cdot -\mu \cdot C_{int}, \tag{27.10}$$

$$v = f(C_{macro}, C_{int}, T, pH, \ldots; \Phi). \tag{27.11}$$

The unknown values of the model parameters are typically estimated by requiring that the model predictions be as close as possible to the experimental data. As a result, a *dynamic optimization* problem must be solved due to the presence of the dynamic model. Typically this leads to a nonlinear program (NLP). Depending on whether or not the states are discretized, two classes of methods exist: sequential and simultaneous approaches.

27.2.4.1 Sequential Approaches: Single Shooting

In single shooting (Schittkowski 2002) the states are not discretized and only values for the parameters must be determined. For each parameterization the differential equations are solved using a standard integration algorithm, and the objective function is evaluated. The parameter values are then updated employing a standard optimization algorithm. Consequently, the solution of the differential equations and the minimization are decoupled and proceed sequentially within one iteration step of the optimizer. This leads to small-scale NLPs, which can be solved using deterministic [e.g., sequential quadratic programming (SQP) (Schittkowski 2002)] or stochastic [e.g., evolutionary algorithms (Moles et al. 2003)] optimization routines. The most important advantage of the sequential approach is its straightforward implementation, resulting in rather small-scale NLPs. However, the sequential approach may be slow, especially in the presence of path constraints on the states, since these constraints cannot be enforced directly.

27.2.4.2 Simultaneous Approaches: Multiple Shooting and Collocation

Simultaneous approaches also discretize the states. Consequently, the simulation and optimization are performed simultaneously in the space of both the parameter values and states, yielding a large-scale NLP that requires tailored numerical methods. As a result, only deterministic gradient-based approaches can be exploited.

Two different approaches exist: multiple shooting (Bock 1983; Lohmann et al. 1992) and collocation (Tjoa and Biegler 1991; Betts and Huffman 2003). In the former case, the total integration range is split into a finite number of intervals on which integration of the states is continuous. The value of the control and the initial value of the states in each interval are chosen by the NLP solver in each iteration while trying to ensure the continuity of the states between the different intervals. In the latter case, also the states are fully discretized based on (orthogonal) polynomials. Hence, the minimal objective value must be found while satisfying the discretized differential equations. Clearly, multiple shooting and collocation allow a more direct enforcement of the state constraints. However, the size of the NLPs significantly increases, requiring tailored optimization algorithms that exploit the problem's structure and sparsity. Obviously, the largest but also sparsest NLPs are encountered with collocation. In view of parameter estimation, simultaneous approaches form a natural choice because the measurements of the states can be used directly for initializing the discretized problem (Balsa-Canto et al. 2008). Applications to biochemical engineering can be found in, for example, Moles et al. (2003), Rodriguez-Fernandez et al. (2006), and citetBalsaCanto10.

27.2.5 *Metabolic Network Model Reduction*

Metabolic networks are a blueprint of all the reactions that occur inside a microbial cell during the biochemical process. As a result of the complexity of the *microscopic-level* models, problems arise mainly due to insufficient calculation power and a shortage of intracellular experimental data. The latter limitation in particular is the major challenge when solving a metabolic network. For application of metabolic network models in the development of predictive models, these models must first be simplified.

27.2.5.1 Mathematical Model Reduction of Metabolic Network

As a first step, it is important that the metabolic network being used in the model be as complex as necessary and as simple as possible and that it be reduced in such a way that all redundancy is removed (i.e., S_{int} should be of full rank). Only a set of carefully selected key reactions (e.g., branch points), vital to describe the studied dynamics, is retained to build a structured model within the limits of acceptable complexity. Optimally, model complexity is reduced exploiting data-driven methods, such as *balanced truncation* (Hardin and van Schuppen 2006; Liebermeister et al. 2005), *intrinsic low-dimensional manifolds* (Zobeley et al. 2005), and *singular perturbation* (Zagaris et al. 2004), that can identify the excess reactions in the network. In addition, sensitivity analysis, elementary flux mode determination, convex analysis theory (e.g., Provost et al. 2006), and similar methods can be addressed to identify the final network.

27.2.5.2 Assuming Pseudo-Steady-State

A first simplification is obtained by acknowledging the fact that intracellular dynamics are much faster than macroscopic dynamics. This implies that, with respect to the macroscopic dynamics, a pseudo-steady state can be assumed at the intracellular level. If the dilution term in Eq. 27.10 is also neglected with respect to the fluxes affecting the same metabolite, the microscopic level mass balances reduce to the so-called *general equation* [see, e.g., Llaneras and Pico (2008) for a review]:

$$S_{\text{int}} \cdot v = 0. \tag{27.12}$$

In most metabolic networks, the number of intracellular metabolites m is smaller than the number of fluxes n. In that case, Eq. 27.12 describes an underdetermined system of linear equations, as the number of equations (one for each of the m metabolites) is smaller than the number of unknowns (the n fluxes). Such a system does not have one unique solution, but rather a subspace of possible flux solutions, which is called the *flux cone*. The dimension of this subspace is (if $Sint$ is of full rank) $n - m$. This means that the fluxes can be rewritten as a linear function of $n - m$ so-called *free fluxes* u:

$$v = M \cdot u, \tag{27.13}$$

with M ($n \times n - m$) a suitable basis for the null space of S_{int}. This linear relationship can be incorporated into the macroscopic part of the two-level model (Eq. 27.9), resulting in the following reduced model:

$$\frac{dC_{macro}}{dt} = S_{ext} \cdot M \cdot u \cdot N_{macro}, \tag{27.14}$$

$$u = f(C_{macro}, T, pH, \ldots; \Phi). \tag{27.15}$$

Note that this system now fully describes the macroscopic part by means of the metabolic network but no longer describes the intracellular dynamics as these are assumed to be at steady state. The number of kinetic expressions is reduced by m (from n to $n - m$). For most metabolic networks, this is a significant reduction, for example, for a small-scale metabolic network of 50 metabolites and 70 fluxes, the number of kinetic expressions is reduced from 70 to 20.

27.2.5.3 Assuming Optimal Cell Behavior

Further model reduction is obtained by assuming that the cell has evolved to achieve optimal behavior (Feist and Palsson 2010). Mathematically, this quest for optimal behavior is represented as an optimization problem via a technique called flux balance analysis (FBA). Within the flux cone (which is the feasible set of the optimization problem), a flux solution is defined by optimizing a mathematical objective function (e.g., Burgard and Maranas 2002; Holzhütter 2004; Schuetz et al. 2007). Commonly implemented objective functions are (1) maximization of the cell concentration (Boyle and Morgan 2009), (2) maximization of the ATP flux (Ramakrishna et al. 2001), (3) product maximization (e.g., Varma et al. 1993), and (4) maximization of the growth rate (e.g., Schilling et al. 2002; Covert et al. 2001). If the FBA problem is implemented in the microscopic description of the two-level model (Eq. 27.10), the model has the following form:

$$\frac{dC_{macro}}{dt} = S_{ext} \cdot M \cdot u \cdot N_{macro}, \tag{27.16}$$

with u the solution of

$$\underset{u}{\text{maximize}}\ J(\mathbf{u}) \tag{27.17}$$

with J(u) the objective function that the cell tries to maximize. Further biochemical knowledge is easily implemented in this model by incorporating extra constraints in the optimization problem, for example, irreversibility constraints and capacity constraints on the fluxes (Gianchandani et al. 2010):

$$u_{min} \leq u \leq u_{max}. \tag{27.18}$$

Additional constraints are introduced by defining one or more fluxes explicitly by kinetic relations, such as Michaelis–Menten, or by defining specific relations between certain fluxes (Haag et al. 2005).

Depending on the form of the objective function J(u), this optimization problem is a linear program (LP), a quadratic program (QP), or an NLP. Basic objective functions like maximization of growth rate are linear in the fluxes, resulting in an LP that can be efficiently solved. LPs and QPs are specific subclasses of convex optimization problems (Boyd and Vandenberghe 2004). Excellent algorithms exist to solve large-scale problems for these classes. In addition, it is guaranteed that the optimum found is the global optimum. For NLP this guarantee is no longer present as multiple and local minima can exist. A typical way to solve these kinds of problems involves SQP.

The FBA modeling approach has two drawbacks. First, after incorporating all available biochemical knowledge, it is possible that the solution of the FBA problem still results in multiple optima. If this occurs, extra experiments must be performed to identify extra constraints to implement in the optimization problem. The second and most important drawback is the fact that in some cases, the optimal solution does not correspond to the actual flux distribution. In this case, the assumption of optimal behavior is not justified, and the FBA framework cannot be used.

From the foregoing discussion it is clear that the validity of the assumptions made during the construction of the metabolic network may jeopardize the optimal solution. Therefore, specific attention must be paid to all hypotheses. This can be done by quantifying the effect of a microscopic-level hypothesis or optimization criterion on the optimal result. The construction of a more generic objective function, i.e., an objective function that can reflect more than one specific objective depending on the values of the degrees of freedom, will yield a more realistic flux distribution as a function of time and environmental conditions. For instance, in the case of non-growth-associated production, it is expected that these degrees of freedom depend on time or, even better, cell state, thus reflecting the macroscopic shift from growth phase to production phase. Degrees of freedom can be added to the FBA model structure by parameterizing the objective function and the constraints:

$$\frac{d\mathrm{C}_{macro}}{dt} = \mathrm{S}_{ext} \cdot \mathrm{M} \cdot \mathrm{u} \cdot N_{macro}, \tag{27.19}$$

with u the solution of

$$\begin{aligned} &\max_{\mathrm{u}} J(\mathbf{u}, \Phi_J) \\ &\text{s.t. } \mathrm{G}(\mathbf{u}, \Phi_G) \le 0, \end{aligned} \tag{27.20}$$

with J the parameterized objective function, depending on the free fluxes u and the vector of objective function parameters Φ_J, and with G the parameterized constraints, with Φ_G the vector of constraint parameters. To identify these parameters, again a dynamic parameter estimation problem must be solved. This time, however, the dynamic model itself contains an optimization problem. The parameter estimation problem then results in a *bilevel optimization problem*. These can be solved in a sequential or simultaneous manner. In the former approach, the inner optimization problem is solved for each time step in the simulation required during the outer optimization (e.g., Hanly and Henson 2011). In the latter approach, the problem is reformulated into a single-level dynamic optimization problem by replacing the inner problem with its first-order optimality conditions (e.g., Joy and Kremling 2010). The former approach is more straightforward than the latter, but it is also more time consuming. The latter gives rise to mathematical programs with complementarity conditions for which reformulation strategies are required (Baumrucker et al. 2008).

27.3 Case Study: Mechanisms Approach to (the Transitions Between) Microbial Growth Phases

Techniques of MFA can be exploited to describe the different phases of a bacterial growth curve. When microorganisms are grown under constant environmental conditions, the growth curve, i.e., the (natural) logarithm of cell number versus time, will typically have a sigmoidal shape, as presented in Fig. 27.2. After an adaptation or lag phase, the bacteria start the exponential phase, i.e., the log-linear part of the growth curve. After some time, the growth rate decreases to zero, so that the asymptotic maximum cell density is reached. This is the stationary phase.

27.3.1 Phase 1: Exponential Growth Phase

As mentioned earlier, FBA is performed under the assumption of pseudo-steady state. During the exponential growth phase, cells are assumed to be in a condition of balanced growth. This is a biological condition during which the internal composition of the cells is in balance with the medium composition, i.e., the intracellular metabolic reaction network is essentially operating at steady-state conditions. As a consequence, the pseudo-steady-state assumption is very likely to hold for the

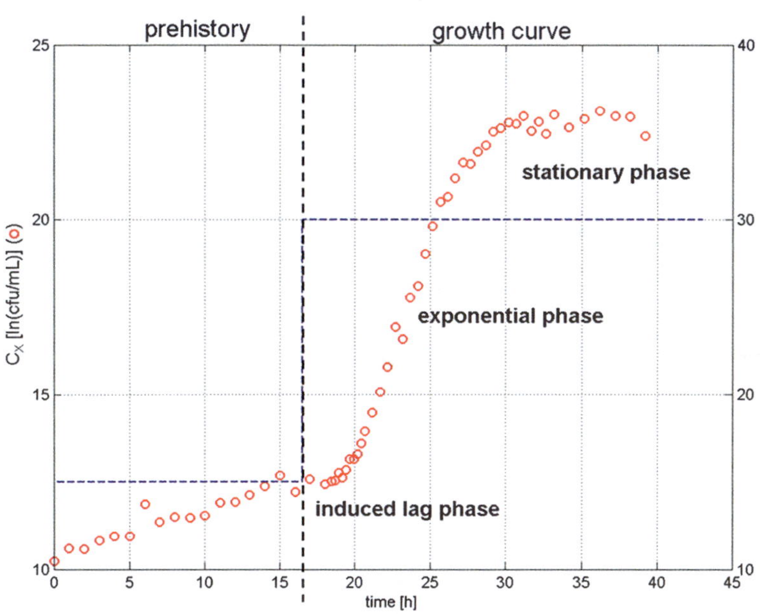

Fig. 27.2 Microbial growth curve for *E. coli* K-12 MG1655 in batch mode with sudden shift in temperature. *Blue profile*: bioreactor temperature; *red dots*: experimental measurements of cell count

exponential growth phase. Starting from a valid metabolic network, most likely the exponential growth phase can be described accurately by selecting the correct objective function.

Several FBA objective functions for describing the exponential growth phase have been reported in the literature. Different studies show that simulations based on the objective of maximizing the biomass growth rate are in good agreement with experimental data [e.g., Edwards et al. (2001) show that the error between simulations and experimental data was, in their specific case, around 10 %]. Pramanik and Keasling (1997) used a biomass composition that was dependent on growth rate. In other studies, other objective functions are found to be better descriptors of experimental data. Schuetz et al. (2007) performed an extensive analysis of 11 objective functions. They determined that in unlimited conditions, maximization of ATP yield per flux unit describes experimental data well, whereas under nutrient scarcity maximization of overall ATP or biomass yields performed better. In general, the analysis of objective functions depends on the situation: organism, growth mode (batch or continuous), and environmental conditions. It is for all applications a good practice to compare simulations with experimental data (Feist and Palsson 2010).

27.3.2 Phase 2: Stationary Phase

At a certain moment in the exponential phase, microbial growth is slowed down and finally stops due to (1) a shortage in nutrients or (2) the accumulation of toxic metabolites. At this point, cells go into the stationary phase. The results obtained for the exponential growth phase can be the starting point to calculate the flux distribution of the next phase, i.e., the stationary phase. In a first stage, the same objective function can be used as for the exponential phase to calculate an optimal flux distribution for the stationary phase. However, most likely, the objective function must be adapted to enable transition to the stationary phase. It is generally known that cells in the stationary phase undergo a morphological change that increases their chance of survival under stressful conditions (e.g., Ou et al. 2004; Nyström 2004; Zinser and Kolter 2004; Rees et al. 1995; Pichereau et al. 2000). Compared to exponential-phase cells, stationary-phase cells tend to be smaller and more spherical, whereby their metabolic system is reformed to enhance multiple stress resistance, acquired by the production of protecting proteins through the expression of specific genes. To yield a flux distribution that is able to describe the reformed metabolic activities in the stationary phase, an adaptation of the objective function constructed in phase 1 is required. Published results on the influence of environmental factors on the transition from exponential growth to the stationary phase (e.g., Kwiatkowska et al. 2008; Vadasz and Vadasz 2007; Theys et al. 2009) can be a good starting point for the selection of an accurate objective function.

27.3.3 Phase 3: Lag Phase

Subjecting microbial populations to environmental changes can induce an adaptation or lag phase. Lag phenomena are of major interest for predictive microbiology since the shelf life of food products can be extended by inducing longer lag phases. When microorganisms are transferred from one environment to another, many factors, such as dilution, nutritional shifts, or a shift in pH or temperature, can impose stress and trigger an adaptation phase. The lag phase is a very complex phenomenon because it is determined by the previous and the current environment. Characterization of the underlying mechanisms requires sufficient knowledge of both the prehistory and the new environment. A good starting point to study the flux distribution accounting for the lag phase is to use the large collection of experimental data during which a controlled lag phase was induced (Fig. 27.2) (e.g., Swinnen et al. 2005; Mellefont and Ross 2003). In those experiments, a lag phase was induced by changing only one of the environmental factors, i.e., a sudden shift in temperature. An important advantage of this approach is the availability of measurements of both the prehistory before the temperature shift and the following phase, i.e., the exponential growth phase.

Results obtained based on this metabolic modeling approach will yield more insight into the mechanisms behind microbial growth dynamics. Therefore, this approach forms an excellent starting point to develop accurate predictive models for more complex systems, such as cocultures and structured environments, based on the same top-down systems biology approach presented in this proposal.

27.4 Conclusion

The applicability and reliability of existing predictive models under more realistic conditions will definitely be improved by looking inside the black box and unraveling the underlying mechanisms. MFA is an excellent tool to gain in-depth insight (i.e., at the intracellular level) on the impact of process conditions on (the fluxes in) the cell metabolism and growth dynamics. Following a top-down systems biology approach, the link between the intracellular fluxes and the extracellular measurements is established by techniques of MFA.

Once this systems biology approach is successfully validated, it will be extended to develop next-generation predictive models for more complex systems, such as cocultures and structured environments.

While the potential is indeed large, a major shift in research objectives and intensified joint research within the entire predictive microbiology community is needed to fully understand the power and limitations (and thus take full advantage) of the systems biology approach: this approach is obviously not new per se, but additional challenges (at both the experimental and modeling levels) are faced when applying this approach to (non-steady-state) bacterial communities in foods.

Acknowledgments This research was supported in part by Project PFV/10/002 (Center of Excellence OPTEC Optimization in Engineering) and OT/10/035 of the KU Leuven Research Fund, Project KP/09/005 (SCORES4CHEM) of the KU Leuven Industrial Research Fund, Project FWOG-08-00360 of the Research Foundation-Flanders, and the Belgian Program on Interuniversity Poles of Attraction, initiated by the Belgian Federal Science Policy Office. J.F. Van Impe holds the chair in Safety Engineering sponsored by the Belgian Chemistry and Life Sciences Federation Essenscia. D. Vercammen was supported by a doctoral grant of the Agency for Innovation through Science and Technology (IWT).

References

AbuOun M, Suthers P, Jones G, Carter B, Saunders M, Maranas C, Woodward M, Anjum M (2009) Genome scale reconstruction of a *Salmonella* metabolic model. J Biol Chem 284(43):29,480–29488

Arsene F, Tomoyasu T, Bukaua B (2000) The heat shock response of *Escherichia coli*. Int J Food Microbiol 55:3–9

Balsa-Canto E, Peifer M, Banga J, Timmer J, Fleck C (2008) Hybrid optimization method with general switching strategy for parameter estimation. BMC Syst Biol 2:26–34

Baranyi J (2002) Stochastic modelling of bacterial lag phase. Int J Food Microbiol 73:203–206

Baranyi J, Pin C (2001) A parallel study on modelling bacterial growth and survival curves. J Theor Biol 210:327–336

Baranyi J, Roberts TA (1994) A dynamic approach to predicting bacterial growth in food. Int J Food Microbiol 23:277–294

Baranyi J, Ross T, McMeekin TA, Roberts TA (1996) Effects of parametrization on the performance of emperical models used in 'predictive microbiology'. Food Microbiol 13:83–91

Baranyi J, George S, Kutalik Z (2009) Parameter estimation for the distribution of single cell lag times. J Theor Biol 259:24–30

Baumrucker B, Renfro J, Biegler L (2008) MPEC problem formulations and solution strategies with chemical engineering applications. Comput Chem Eng 32(12):2903–2913

Betts J, Huffman W (2003) Large scale parameter estimation using sparse nonlinear programming methods. SIAM J Optimiz 14:223–244

Bock H (1983) Recent advances in parameter identification techniques for ODE. In: Deuflhard P, Hairer E (eds) Numerical treatment of inverse problems in differential and integral equations. Birkhäuser, Boston, pp 95–121

Boyd S, Vandenberghe L (2004) Convex optimization. University Press, Cambridge

Boyle NR, Morgan JA (2009) Flux balance analysis of primary metabolism in *Chlamydomonas reinhardtii*. BMC Syst Biol 3:4

Brul S, Westerhoff H (2007) Systems biology and food science. In: Brul S, van Gerwen S, Zwietering M (eds) Modelling microorganisms in food. Woodhead, Cambridge, pp 250–288

Brul S, Mensonides FIC, Hellingwerf KJ, Teixeira de Mattos MJ (2008) Microbial systems biology: new frontiers open to predictive microbiology. Int J Food Microbiol 128(1):16–21

Buchanan RL, Whiting RC, Damert WC (1997) When is simple good enough: a comparison of the Gompertz, Baranyi, and three-phase linear models for fitting bacterial growth curves. Food Microbiol 14:313–326

Burgard A, Maranas C (2002) Optimization-based framework for inferring and testing hypothesized metabolic objective functions. Biotechnol Bioeng 82:670–677

Chung H, Bang W, Drake M (2006) Stress response of *Escherichia coli*. Compr Rev Food Sci Food Saf 5(3):52–64

Covert M, Schilling C, Palsson B (2001) Regulation of gene expression in flux balance models of metabolism. J Theor Biol 213:309–325

Dens E, Van Impe J (2001) On the need for another type of predictive models in structured foods. Int J Food Microbiol 64:247–260

Edwards JS, Ibarra RU, Palsson BO (2001) In silico predictions of *Escherichia coli* metabolic capabilities are consistent with experimental data. Nat Biotechnol 19(2):125–130

Feist AM, Palsson BO (2010) The biomass objective function. Curr Opin Microbiol 13(3):344–349

Feist AM, Henry CS, Reed JL, Krummenacker M, Joyce AR, Karp PD, Broadbelt LJ, Hatzimanikatis V, Palsson BO (2007) A genome-scale metabolic reconstruction for *Escherichia coli* K-12 MG1655 that accounts for 1260 ORFs and thermodynamic information. Mol Syst Biol 3:121

Geeraerd A, Herremans C, Cenens C, Van Impe J (1998) Application of artificial neural networks as a non-linear modular modeling technique to describe bacterial growth in chilled food products. Int J Food Microbiol 44:49–68

Geeraerd AH, Herremans CH, Van Impe JF (2000) Structural model requirements to describe microbial inactivation during a mild heat treatment. Int J Food Microbiol 59:185–209

Geeraerd AH, Valdramidis VP, Devlieghere F, Bernaert H, Debevere J, Van Impe JF (2004) Development of a novel approach for secondary modelling in predictive microbiology: incorporation of microbiological knowledge in black box polynomial modelling. Int J Food Microbiol 91:229–244

Gianchandani E, Chavali A, Papin J (2010) The application of flux balance analysis in systems biology. Wiley Interdiscip Rev Syst Biol Med 2(3):372–382

Haag J, Vande Wouwer A, Bogaerts P (2005) Dynamic modeling of complex biological systems: a link between metabolic and macroscopic description. Math Biosci 193:25–49

Hanly T, Henson M (2011) Dynamic flux balance modeling of microbial co-cultures for efficient batch fermentation of glucose and xylose mixtures. Biotechnol Bioeng 108(2):376–385

Hardin H, van Schuppen J (2006) System reduction of nonlinear positive systems by linearization and truncation. Posit Syst 341:431–438

Holzhutter HG (2004) The principle of flux minimization and its application to estimate stationary fluxes in metabolic networks. Eur J Biochem 271:2905–2922

Joy J, Kremling A (2010) Study of the growth of *Escherichia coli* on mixed substrates using dynamic flux balance analysis. In: Proceedings of the 11th IFAC Symposium on computer applications in biotechnology, Leuven, 7–10 July 2010

Kwiatkowska J, Matuszewska E, Kuczynska-Wisnik D, Laskowska E (2008) Aggregation of *Escherichia coli* proteins during stationary phase depends on glucose and oxygen availability. Res Microbiol 159:651–657

Liebermeister W, Bauer U, Klipp E (2005) Biochemical network models simplified by balanced truncation. FEBS J 272:4034–4043

Llaneras F, Pico J (2008) Stoichiometric modelling of cell metabolism. J Biosci Bioeng 105:1–11

Lohmann T, Bock H, Schlöder J (1992) Numerical methods for parameter estimation and optimal experiment design in chemical reaction systems. Ind Eng Chem Res 31:54–57

Mahadevan R, Edwards J, Doyle F (2002) Dynamic flux balance analysis of diauxic growth in *Escherichia coli*. Biophys J 83(3):1331–1340

McKellar RC (1997) A heterogeneous population model for the analysis of bacterial growth kinetics. Int J Food Microbiol 36:179–186

McKellar R (2001) Development of a dynamic continuous-discrete-continuous model describing the lag phase of individual bacterial cells. J Appl Microbiol 90:407–413

McMeekin TA, Ross T (2002) Predictive microbiology: providing a knowledge-based framework for change management. Int J Food Microbiol 78:133–153

McMeekin TA, Olley J, Ratkowsky DA, Ross T (2002) Predictive microbiology: towards the interface and beyond. Int J Food Microbiol 73:395–407

McMeekin TA, Bowman J, McQuestin O, Mellefont L, Ross T, Tamplin M (2008) The future of predictive microbiology: strategic research, innovative applications and great expectations. Int J Food Microbiol 128:2–9

Mellefont LA, Ross T (2003) The effect of abrupt shifts in temperature on the lag phase duration of *Escherichia coli* and *Klebsiella oxytoca*. Int J Food Microbiol 83:295–305

Moles C, Mendes P, Banga J (2003) Parameter estimation in biochemical pathways: a comparison of global optimization methods. Genome Res 13:2467–2474

Nikolaou M, Tam VH (2005) A new modeling approach to the effect of antimicrobial agents on heterogeneous microbial populations. J Math Biol 52:154–182

Nyström T (2004) Stationary-phase physiology. Annu Rev Microbiol 58:161–181

Ou J, Wang L, Ding X, Du J, Zhang Y, Chen H, Xu A (2004) Stationary phase protein overproduction is a fundamental capability of *Escherichia coli*. Biochem Biophys Res Commun 314:174–180

Panagou EZ, Tassou CC, Saravanos EKA, Nychas GJE (2007) Application of neural networks to simulate the growth profile of lactic acid bacteria in green olive fermentation. J Food Prot 70:1909–1916

Pichereau V, Hartke A, Auffray Y (2000) Starvation and osmotic stress induced multiresistances influence of extracellular compounds. Int J Food Microbiol 55:19–25

Poschet F, Vereecken KM, Geeraerd AH, Nicola¨ı BM, Van Impe JF (2005) Analysis of a novel class of predictive microbial growth models and application to coculture growth. Int J Food Microbiol 100:107–124

Pramanik J, Keasling JD (1997) Stoichiometric model of *Escherichia coli* metabolism: incorporation of growth-rate dependent biomass composition and mechanistic energy requirements. Biotechnol Bioeng 56(4):398–421

Prats C, Giro A, Ferrer J, Lopez D, Vives-Rego J (2008) Analysis and IbM simulation of the stages in bacterial lag phase: basis for an updated definition. J Theor Biol 252:56–68

Provost A, Bastin G, Agathos S, Schneider YJ (2006) Metabolic design of macroscopic bioreaction models: application to Chinese hamster ovary cells. Bioprocess Biosyst Eng 29:349–366

Ramakrishna R, Edwards JS, Mcculluch A, Palsson BO (2001) Flux-balance analysis of mitochondrial energy metabolism: consequences of systemic stoichiometric constraints. Am J Physiol – Regulatory, Integrative Comp Physiol 280:R695–R704

Ratkowsky DA, Olley J, McMeekin TA, Ball A (1982) Relationship between temperature and growth rate of bacterial cultures. J Bacteriol 149:1–5

Ratkowsky DA, Lowry RK, McMeekin TA, Stokes AN, Chandler RE (1983) Model for bacterial culture growth rate throughout the entire biokinetic temperature range. J Bacteriol 154:1222–1226

Rees CED, Dodd CER, Gibson PT, Booth IR, Stewart GSAB (1995) The significance of bacteria in stationary phase to food microbiology. Int J Food Microbiol 28:263–275

Rodriguez-Fernandez M, Mendes P, Banga J (2006) A hybrid approach for efficient and robust parameter estimation in biochemical pathways. Biosystems 83(2–3):248–265

Ross T, Ratkowsky DA, Mellefont LA, McMeekin TA (2003) Modelling the effects of temperature, water activity, pH and lactic acid concentration on the growth rate of *Escherichia coli*. Int J Food Microbiol 82:33–43

Rosso L, Lobry JR, Flandrois JP (1993) An unexpected correlation between cardinal temperatures of microbial growth highlighted by a new model. J Theor Biol 162:447–463

Rosso L, Lobry JR, Bajard S, Flandrois JP (1995) Convenient model to describe the combined effects of temperature and pH on microbial growth. Appl Environ Microbiol 61:610–616

Sautour M, Dantigny P, Divies C, Bensoussan M (2001) A temperature-type model for describing the relationship between fungal growth and water activity. Int J Food Microbiol 67:63–69

Schauer K, Geginat G, Liang C, Goebel W, Dandekar T, Fuchs T (2010) Deciphering the intracellular metabolism of *Listeria monocytogenes* by mutant screening and modelling. BMC Genomics 11:573

Schilling C, Covert M, Famili I, Church G, Edwards J, Palsson B (2002) Genome-scale metabolic model of *Helicobacter pylori* 26695. J Bacteriol 184:4582–4593

Schittkowski K (2002) Numerical data fitting in dynamical systems: a practical introduction with applications and software, Applied optimization. Kluwer, Dordrecht

Schuetz R, Kuepfer L, Sauer U (2007) Systematic evaluation of objective functions for predicting intracellular fluxes in *Escherichia coli*. Mol Syst Biol 3:119

Skandamis PN, Davies KW, McClure PJ, Koutsoumanis K, Tassou C (2002) A vitalistic approach for non-thermal inactivation of pathogens in traditional Greek salads. Food Microbiol 19:405–421

Standaert AR, Francois K, Devlieghere F, Debevere J, Van Impe JF, Geeraerd AH (2007) Modeling individual cell lag time distributions for *Listeria monocytogenes*. Risk Anal 27:241–254

Swinnen IAM, Bernaerts K, Gysemans K, Van Impe JF (2005) Quantifying microbial lag phenomena due to a sudden rise in temperature: a systematic macroscopic study. Int J Food Microbiol 100:85–96

Theys TE, Geeraerd AH, Devlieghere F, Van Impe JF (2009) Extracting information on the evolution of living-and dead-cell fractions of *Salmonella Typhimurium* colonies in gelatin gels based on microscopic images and plate-count data. Lett Appl Microbiol 49:39–45

Tjoa I, Biegler L (1991) Simultaneous solution and optimization strategies for parameter estimation of differential-algebraic equation systems. Ind Eng Chem Res 30:376–385

Vadasz P, Vadasz AS (2007) Biological implications from an autonomous version of Baranyi and Roberts growth model. Int J Food Microbiol 114:357–365

Van Derlinden E, Bernaerts K, Van Impe JF (2009) Unraveling *E. coli* dynamics close to the maximum growth temperature through heterogeneous modeling. Lett Appl Microbiol 49(6):659–665

Van Derlinden E, Bernaerts K, Van Impe J (2010) Quantifying the heterogeneous heat response of *E. coli* under dynamic temperatures. J Appl Microbiol 108(4):1123–1135

Van Impe JF, Poschet F, Geeraerd AH, Vereecken KM (2005) Towards a novel class of predictive microbial growth models. Int J Food Microbiol 100:97–105

Varma A, Boesch B, Palsson B (1993) Biochemical production capabilities of *Escherichia coli*. Biotechnol Bioeng 42:59–73

Yuk HG, Marshall DL (2003) Heat adaptation alters *Escherichia coli* O157:H7 membrane lipid composition and verotoxin production. Appl Environ Microbiol 69(9):5115–5119

Zagaris A, Kaper H, Kaper T (2004) Analysis of the computational singular perturbation reduction method for chemical kinetics. J Nonlinear Sci 14:59–91

Zinser ER, Kolter R (2004) *Escherichia coli* evolution during stationary phase. Res Microbiol 155:328–336

Zobeley J, Lebiedz D, Kammerer J, Ishmurzin A, Kummer U (2005) A new time dependent complexity reduction method for biochemical systems. Trans Comput Syst Biol vol LNBI 3880:90–110

Zwietering M, Jongenburger I, Rombouts F, van't Riet K (1990) Modeling of the bacterial growth curve. Appl Environ Microbiol 56:1875–1881

Chapter 28
Dynamic Approach to Assessing Food Quality and Safety Characteristics: The Case of Processed Foods

Teresa R.S. Brandão, Maria M. Gil, Fátima A. Miller, Elsa M. Gonçalves, and Cristina L.M. Silva

28.1 Introduction

Foods often undergo processing, which has three major aims: to produce safe foods while providing products with the highest quality attributes; to transform food into forms that are more convenient or more appealing; and to extend shelf-life. Temperature plays an important role in food processing: high temperatures are key factors for microbial death or inactivation (safety point of view), whereas low temperatures are imperative for long-term food preservation, preventing microbial growth and retarding reactions of quality alterations (from a joint perspective of safety and quality).

Alterations of microbial loads and degradation of quality attributes of a food occur throughout thermal processes. Kinetic studies are those devoted to the rates of such alterations. Such studies are often carried out under constant temperature conditions. However, food processes, such as thermally based ones (e.g. pasteurization or drying) or frozen storage, occur in time-varying temperature conditions. The almost non-existence of isothermal processes in the food industry may compromise the application of kinetic results, obtained under constant temperature conditions, for dynamic situations. The assessment of time-varying

---

T.R.S. Brandão • F.A. Miller • C.L.M. Silva (✉)
CBQF – Centro de Biotecnologia e Química Fina, Laboratório Associado, Escola Superior de Biotecnologia, Universidade Católica Portuguesa, Rua Dr. António Bernardino de Almeida, 4200-072 Porto, Portugal
e-mail: clsilva@porto.ucp.pt

M.M. Gil
Instituto Politécnico de Leiria, Campus 4, Santuário N.ªSra. dos Remédios, Apartado 126, 2524 – 909 Peniche, Portugal

E.M. Gonçalves
Unidade de Investigação de Tecnologia Alimentar, Instituto Nacional de Recursos Biológicos, Estrada do Paço do Lumiar 22, 1649-038 Lisboa, Portugal

S. Yanniotis et al. (eds.), *Advances in Food Process Engineering Research and Applications*, Food Engineering Series, DOI 10.1007/978-1-4614-7906-2_28,

temperature conditions in microbial inactivation is therefore important in the design of safe food processes. When realistic food processing conditions occur, the temperature history used in the process defines the microbial behaviour (Gil et al. 2006). Bacteria submitted to a non-isothermal heating process (i.e. in which temperature rises throughout the process time till a target value is reached) are more heat resistant than bacteria treated at constant target temperatures (Miller et al. 2011).

Aiming at long-term food preservation, freezing and frozen storage are key processes. However, post-processing temperature conditions and temperature fluctuations throughout storage and distribution greatly affect the rates of quality degradation and shelf-life of frozen foods, particularly important in the case of vegetables. Comprehensive studies on the effect of frozen storage conditions on the quality of vegetables (e.g. green beans, tomatoes, potatoes, strawberries and broccoli) have been carried out using measurements of particular compounds/characteristics, such as ascorbic acid degradation, transformation of pigments, and alterations in sensory characteristics like colour and texture (Aparicio-Cuesta et al. 1989; Lisiewska and Kimiecik 2000; Gormley et al. 2002). Modelling the degradation kinetics of quality losses during isothermal frozen storage are decisively important because the estimation of kinetic parameters will make it possible to predict the final quality of products, and, consequently, improvements may be attained (Martins and Silva 2003; Gonçalves et al. 2009). However, the extensive storage times for quality change assessment in frozen conditions are limiting. The use of accelerated life testing methodology overcomes these difficulties since experiments are performed under higher than usual stress conditions to induce rapid alterations (Martins et al. 2005).

The use of mathematical models that describe/predict changes in processed food characteristics with accuracy and precision in realistic dynamic conditions is an important tool for developing new products and control systems. The greatest modelling effort has been devoted to data obtained under constant (or static) environmental conditions. From a realistic point of view, this is somehow restrictive since the majority of thermal processes occur under time-varying environmental conditions, and kinetic parameters obtained under such circumstances may differ from those estimated in static conditions, which compromises safety control and quality prediction. Researchers are conscious of these difficulties and are trying to extend and structure models to deal with predictions in time-variant domains since they may constitute a useful tool for companies, universities and regulatory agencies.

In this chapter, mathematical models that include time-varying temperature conditions (i.e. dynamic approach) will be presented for two relevant situations in the domain of processed foods: the case of microbial thermal inactivation and the case of food quality alterations under frozen storage.

28.2 The Models

28.2.1 Food Safety: The Case of Microbial Thermal Inactivation

The bacterial spoilage of foods and the survival of pathogens are of major importance to the food process industries because they directly affect consumer health and safety. As the critical boundary for contamination is the exposed surface, heat treatments at the food surface can be an effective means of controlling pathogens. This makes effective pasteurization systems critical to the production of safe products. Consequently, thermal decontamination must be designed to provide an adequate margin of safety against food-borne pathogens. However, it is difficult to determine the exact amount of microbial inactivation when these treatments are applied. The use of accurate microbial inactivation models, which describe the behaviour of microorganisms in foods, would be of considerable help to the food industry in the development of surface pasteurization systems. This would in turn lead to safer foodstuffs with improved quality and shelf-life.

The development of accurate and precise models, able to predict the behaviour of pathogens or spoiling microorganism populations under stress factors, such as high temperature, particular ranges of pH and water activity, is a discipline that is gaining considerable importance in the food processing domain. It is commonly referred to as *predictive food microbiology* and is a field of study that combines elements of microbiology, mathematics and statistics to develop models that describe and predict the growth or decline of microbes under specified environmental conditions. From the ability to describe growth and inactivation behaviours emerges the capability to predict them under various environmental conditions where no experimental data exist.

Microbial modelling kinetics began in the 1920s with thermal death time calculations: the Bigelow model, with the well-known D and z parameters, was successfully used to ensure that canned foods were free from the risk of food poisoning by *Clostridium botulinum* spores. With the recent development of software applications in the field of predictive microbiology, microbial modelling has become an area of increasing interest because a model can now be easily used by food technologists and microbiologists, and predictions are literally at their fingertips. Nevertheless, the modelling of microbial kinetics that lead to reliable predictions of the safety and shelf-life of foods is not commonly considered by microbiologists.

Whereas growth kinetic models are helpful for estimating the time required for pathogens to reach dangerous levels under specific conditions, inactivation kinetic models allow prediction of pathogen survival under stressing environmental factors (such as high temperatures, low pH and reduced water activity values). Inactivation models may help to determine the extent to which existing thermal food processes could be modified in order to improve shelf-life and quality while maintaining safe standards.

First-order kinetic models have been extensively used to describe a log-linear microbial variation with time. However, the growth/survival curves of most microbial cells do not show such a tendency (Van Boekel 2002), and a sigmoidal behaviour is often observed. These tendencies can be characterized by three main features: (1) a lag time (or shoulder) prior to a (2) maximum growth/inactivation rate period, and a (3) tail (or residual population). The majority of predictive approaches are based on experimental data obtained in broth and in isothermal conditions. However, in food processes, these conditions may not be realistic. The kinetic parameters estimated under time-varying temperature conditions may differ from those predicted at constant temperatures. Using the later ones, in situations where the temperature varies with time, may affect the predictive ability of the model. This can be particularly important when the safety of a product is the final goal. Van Impe et al. (1992, 1995), Nicolaï and Van Impe (1996), and Geeraerd et al. (2000) were pioneers in the way they approached the modelling of microbial growth or inactivation under dynamically changing temperature conditions.

Among the mathematical models used to describe microbial inactivation, two mathematical expressions deserve to be mentioned: Baranyi and Gompertz models. Such models include parameters in their definition that are directly related to the sigmoidal characteristics of microbial inactivation. The versatility of the Gompertz model in describing kinetics with different shapes, varying from a log-linear tendency to a complete sigmoidal shape, makes it attractive for predictive purposes, both under static and dynamic temperature conditions.

Baranyi model (Xiong et al. 1999):

$$\begin{cases} \dfrac{dN}{dt} = -k_{max}(t)\ L(t)\ N\,N_{res} \quad (N > 0; t \geq 0), \\ \\ N(0) = N_0 \quad (N_0 > 0; t = 0). \end{cases} \tag{28.1}$$

Gompertz model (Gil et al. 2011):

$$\begin{cases} \dfrac{d\left(log\left(\frac{N}{N_0}\right)\right)}{dt} = -k_{max}(t)\ e\ exp\left(-\dfrac{k_{max}(t)\ e}{log\left(\frac{N_{res}}{N_0}\right)}(L(t) - t\ ) + 1\right) \\ \qquad exp\left(-exp\left(-\dfrac{k_{max}(t)\ e}{log\left(\frac{N_{res}}{N_0}\right)}(L(t) - t) + 1\right)\right), \\ N(0) = N_0 \quad (N_0 > 0; t = 0). \end{cases} \tag{28.2}$$

In the preceding expressions, N is the microbial cell density at a particular process time, t. The indexes 0 and res indicate the initial and residual (or tail) microbial cell densities, respectively; e is the Euler number [$e = exp(1)$]; L is the initial shoulder; and k_{max} is the maximum inactivation rate. These parameters are temperature dependent.

Several models can be used to describe the temperature dependence of the shoulder parameter, but the following equation is often applied (Ratkowsky et al. 1982):

$$L(t) = c\,(T(t) - d)^2, \tag{28.3}$$

where c and d are parameters.

The dependence of k_{max} on temperature is often expressed by the well-known Arrhenius equation, using a finite reference temperature (T_{ref}) to reduce the parameters' collinearity (Haralampu et al. 1985):

$$k_{max}(t) = k_{ref}\,exp\left(-\frac{E_a}{R}\left(\frac{1}{T(t)} - \frac{1}{T_{ref}}\right)\right), \tag{28.4}$$

where k_{ref} is the inactivation rate at a finite reference temperature, E_a the activation energy, and R the ideal gas constant.

For non-isothermal conditions, temperature varies with time. If the temperature history $T(t)$ is known, then the model that is valid for non-isothermal conditions can be obtained by including Eqs. 28.3 and 28.4 in model 28.1 or 28.2.

28.2.2 Food Quality: The Case of Frozen Stored Vegetables

Vegetables are perishable foods subject to extremely rapid deterioration. Moreover, aroma, texture, taste and flavour are determinant in consumer choices. The assurance of product quality is essential after harvesting and during storage. Lowering the temperature of the products is a key procedure since under such conditions the reactions of quality degradation are retarded and microbial growth is prevented. Freezing is an efficient process in food preservation, which decreases deterioration of food products by turning water into ice, making it unavailable for bacterial survival/growth. The process consists of a quick temperature decrease of the food to its freezing point in a very short time. Inevitably, the freezing process and the posterior frozen storage have negative impacts on food quality. The main physical changes associated with frozen storage are moisture migration and ice recrystallization (Van Buggenhout et al. 2006). Both phenomena are related to the stability of products' frozen water, which affects the vegetable texture, loss of nutrients and weight if thawing drip loss takes place (Pukszta and Palich 2007). Frozen storage is an integral and important part of a frozen food distribution chain. In general, frozen foods are industrially stored at temperatures around −25 °C and are distributed near −18 °C, which is the legally allowed limit in most countries. However, at this temperature, water in vegetable tissue is not completely frozen, and only when the glass transition temperature is achieved are vegetables completely inert. Temperatures of the frozen distribution chain are frequently higher than glass transition temperatures, and the occurrence of undesirable time-varying temperature conditions is also common throughout storage, transportation and distribution (Canet 1989). Consequently, significant quality losses may occur by chemical or physical means, affecting the shelf-life of frozen products.

Models that account for time-varying temperature conditions are therefore important for the quality assessment of frozen foods. The kinetics of quality changes may be described by a zero- or first-order kinetic model (Gonçalves et al. 2011a) or by a fractional conversion model (Martins et al. 2005; Gonçalves et al. 2011b), depending on the quality indicator considered. The dependence of the reaction rate on temperature is commonly described by an Arrhenius equation (Eq. 28.4; Gonçalves et al. 2011a, b). The kinetic models are therefore of the following forms:

***Zero order*:**

$$\begin{cases} \dfrac{dC}{dt} = -\,k_{ref}exp\left[-\dfrac{E_a}{R}\left(\dfrac{1}{T(t)} - \dfrac{1}{T_{ref}}\right)\right], \\ C(0) = C_0 \;\; (C_0 > 0;\ t = 0); \end{cases} \tag{28.5}$$

***First order*:**

$$\begin{cases} \dfrac{dC}{dt} = -\,k_{ref}exp\left[-\dfrac{E_a}{R}\left(\dfrac{1}{T(t)} - \dfrac{1}{T_{ref}}\right)\right]C, \\ C(0) = C_0 \;\; (C_0 > 0;\ t = 0); \end{cases} \tag{28.6}$$

***Fractional conversion model*:**

$$\begin{cases} \dfrac{d\left(C - C_{eq}\right)}{dt} = -\,k_{ref}exp\left[-\dfrac{E_a}{R}\left(\dfrac{1}{T(t)} - \dfrac{1}{T_{ref}}\right)\right]\left(C - C_{eq}\right), \\ C(0) = C_0 \qquad (C_0 > 0; \quad t = 0). \end{cases} \tag{28.7}$$

Here C denotes a given quality characteristic (the indexes 0 and eq are initial and equilibrium values respectively). Under time-varying temperature conditions, and if $T(t)$ is known, the previous equations allow for predictions of food quality alterations.

28.3 Case Studies

To assess the dynamic model outline and for a clear understanding of the approaches, data related to both situations were considered. Combinations of food/characteristics were used as case studies.

28.3.1 *Thermal Inactivation of* Listeria

Safety was evaluated using *Listeria innocua*, a non-pathogenic surrogate of *L. monocytogenes*, both in liquid medium (TSBYE) and on a food surface artificially inoculated (parsley; *Petroselinum crispum*). Heat treatments were carried out in thermostatic baths with temperature programmers. Temperature varied linearly from 20 °C to 65 °C for 25 min (heating rate of 1.8 °C/min) and was held constant at 65 °C for the last 10 min of the process. Data were obtained experimentally by Miller et al. (2011).

The Gompertz model was used to data fit, merging Eqs. 28.2, 28.3 and 28.4. By the separation of variables and integration, the model transforms into

$$\begin{aligned} log\left(\frac{N}{N_0}\right) = \int_0^t \Bigg[& -k_{ref}exp\left(-\frac{E_a}{R}\left(\frac{1}{T(t')}-\frac{1}{T_{ref}}\right)\right) \\ & e\ exp\left(-\frac{k_{ref}exp\left(-\frac{E_a}{R}\left(\frac{1}{T(t')}-\frac{1}{T_{ref}}\right)\right)e}{\log\left(\frac{N_{res}}{N_0}\right)}\left(c(T(t')-d)^2-t'\right)+1\right) \cdot \\ & exp\left(-exp\left(-\frac{k_{ref}exp\left(-\frac{E_a}{R}\left(\frac{1}{T(t')}-\frac{1}{T_{ref}}\right)\right)e}{log\left(\frac{N_{res}}{N_0}\right)}\left(c(T(t')-d)^2-t'\right)+1\right)\right)\Bigg] dt'. \end{aligned} \quad (28.8)$$

The parameters of the model for non-isothermal conditions [k_{ref}, E_a, c, d and log (N_{res}/N_0)], were estimated by non-linear regression analysis. The reference temperature in the Arrhenius equation was considered equal to 58.5 °C. All regression analysis procedures and calculations were performed in programs specially written in FORTRAN 77 language (Fortran 5.1, Microsoft, 1990) using the Simplex method for minimization of the sum of squares of the residuals between experimental data and data predicted by the model (Nelder and Mead 1965).

The results of the thermal inactivation of *L. innocua* in parsley and in a liquid heating medium (TSBYE) under time-varying temperature conditions are shown in Fig. 28.1. An initial lag followed by a maximum inactivation rate period was observed, with these characteristics being described by model parameters. Besides the sigmoidal tendency observed in both cases, inactivation behaviours were

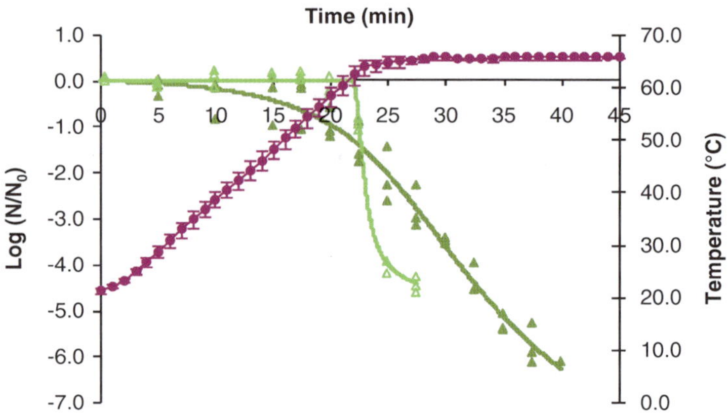

Fig. 28.1 Thermal inactivation data of *L. innocua*, in liquid medium (△) and in parsley (▲), using a heating rate of 1.8 °C/min. *Solid lines*: model fits; *bars*: standard deviation of temperature history (●)

Table 28.1 Estimated k_{ref}, Ea, c, d and tail parameter (± halved confidence intervals at 95 %) of *L. innocua* under non-isothermal treatments

Medium	$k_{58.5°C}$ (min)	Ea (J mol^{-1})	c (min)	d (K)	log(N_{res}/N_0)
TSBYE	2.11	84.6	1.02	338.8	−49.3
	$\pm 5.39 \times 10^6$	$\pm 3.38 \times 10^8$	$\pm 7.67 \times 10^4$	$\pm 4.47 \times 10^6$	$\pm 3.2 \times 10^7$
Parsley	1.33	52.4	5.00×10^{-3}	274.3	−7.4
	$\pm 5.77 \times 10^{-1}$	$\pm 5.45 \times 10^1$	$\pm 4.65 \times 10^{-2}$	$\pm 2.80 \times 10^2$	$\pm 1.1 \times 10^1$

considerably different. In experiments carried out in a liquid medium, inactivation started after a shoulder period of 22 min when the temperature reached 61 °C. When *Listeria* was artificially inoculated in parsley, the initial shoulder was reduced to 14 min, corresponding to a temperature of 45 °C. After these periods, the rate of inactivation was higher in the TSBYE medium than in the food.

Model parameters are presented in Table 28.1. The adequacy of model fits was tested by residual analysis (i.e. residuals were normally distributed with means equal to zero and constant variance; residuals were random). The coefficient of determination was above 0.97 in both cases. The quality of regression analysis was therefore assessed. However, parameter estimates lack precision since considerably high confidence intervals were obtained. This observation is not unexpected when a non-isothermal methodology is used in kinetic studies since the heating rate directly affects parameter precision (Brandão and Oliveira 1997). This occurrence may be circumvented by the use of a convenient experimental design, selecting experimental data points and heating rates based on statistical theories (Brandão et al. 2001).

Broth-based experiments highlight the importance of studying the influence of dynamic conditions for the thermal resistance of microorganisms since the heating-up phases can contribute to an increase in cell thermotolerance. Results obtained in parsley demonstrated that the product greatly affects bacterial thermal resistance;

although the heat resistance of *Listeria* increased (compared to the liquid medium), the inactivation began earlier.

Overall it can be stated that the model assumed has the ability to deal with time-varying temperature conditions, which is a key value in predicting the microbial loads of foods that undergo a thermal process.

28.3.2 Quality Degradation Under Frozen Conditions

Carrots (*Daucus carota* L. cv "Nantes") were selected for the assessment of quality alterations throughout frozen storage. Texture and drip loss were the characteristics selected.

The carrots were peeled, washed and sliced into 5 ± 1 mm length, using a cut machine. All samples were blanched in optimized conditions (85 °C for 6 min) before being frozen at −40 °C (average value), until the temperature of the centre of the cylindrical samples reached −35 °C. Frozen samples were packed into polyethylene bags and sealed.

After freezing, samples were stored under non-isothermal frozen conditions. This was considered an accelerated life test using a step-stress methodology (i.e. consecutive increases in temperature, by step levels; Gonçalves et al. 2011a, b). Samples were stored in sequence at −30 °C (for 21 days), −20 °C (for 16 days), −10 °C (for 11 days) and −5 °C (for 9 days), totalling 57 days. Samples were removed at heuristic time intervals (i.e. sampling tended to be equally spaced in the time scale) and analysed in terms of texture and drip loss.

Texture measurements were performed using a TA.HDi Texture Analyser (Stable Micro Systems, Godalming, UK). At least 19 measurements were performed per sample.

Frozen samples were laid over a weighed absorbent paper and allowed to thaw at 4 °C. Drip loss (DL) was evaluated in carrots by periodically weighing the absorbent paper until a constant value was reached:

$$DL\ (\%) = \frac{(W_t - W_0)}{W_s}\ 100, \tag{28.9}$$

where W_0 is the weight of the dry absorbent paper (g), W_t the weight of the wet absorbent paper at time t (g) and W_S the weight of the frozen sample (g). Measurements were made in triplicate per sample.

Experimental data were fitted to models previously presented according to their kinetic pattern. All regression analysis procedures and calculations were performed in programs specially written in the FORTRAN 77 language (Fortran 5.1, Microsoft, 1990). The simplex algorithm was used to minimise the sum of the squares of the residuals (Nelder and Mead 1965). The reference temperature assumed was −15 °C.

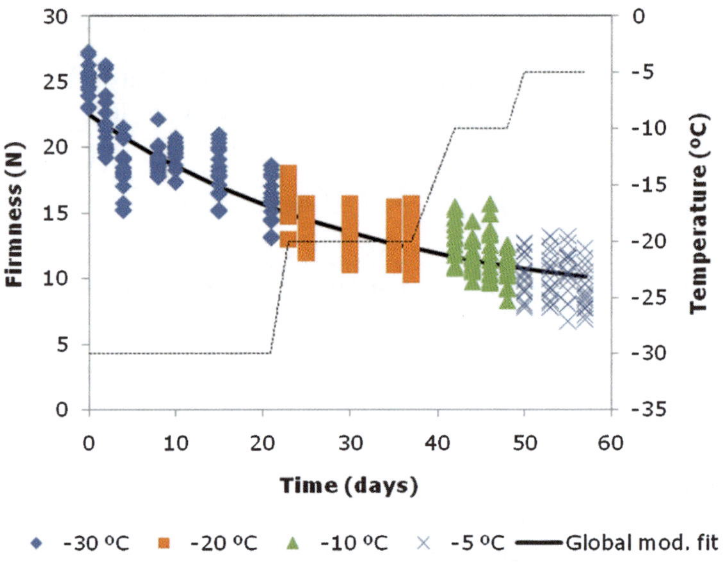

Fig. 28.2 Variation in carrot firmness throughout frozen storage under non-isothermal conditions (*dotted line*: temperature history)

28.3.2.1 Texture

Texture is a quality attribute greatly affected by freezing and frozen storage. Figure 28.2 shows the degradation of carrot firmness throughout frozen storage under the non-isothermal conditions imposed. The fractional conversion model (Eq. 28.7) was adequate in data fitting, which was used in the following form, after separation of variables and integration:

$$C = C_{eq} + \left(C_0 - C_{eq}\right) exp\left[-k_{ref}\int_0^t exp\left[-\frac{E_a}{R}\left(\frac{1}{T(t')} - \frac{1}{T_{ref}}\right)\right]dt'\right]. \qquad (28.10)$$

Here C denotes firmness.

The estimates of model parameters are in Table 28.2. Parameters lack precision, which was previously justified by the influence of the heating rates (Brandão and Oliveira 1997).

The equilibrium value for firmness (C_{eq}) was 7.51 N, which means that at the end of the storage period (57 days), a carrot lost approximately 67 % of its original texture ($C_0 = 22.54$ N).

The adequacy of the model was assessed by the randomness and normality of the residuals and by the coefficient of determination, R^2, which was 0.86. In this case study, the adequacy of the fractional conversion model to describe firmness changes of carrots stored under time-varying temperature conditions was proven.

Table 28.2 Kinetic parameters (± halved confidence intervals at 95 %) of quality changes due to non-isothermal frozen storage

	Kinetic parameter			
Quality factor	C_0	C_{eq}	$k_{-15\,°C} \times 10^3$ (days^{-1})	E_a (kJ mol^{-1})
Texture	22.54 ± 97.69	7.51 ± 3.53	31.10 ± 655.43	1.32×10^{-6a}
Drip loss	11.26 ± 0.24	–	42.66 ± 4.73	42.31 ± 6.89

[a]Meaningless large confidence interval; this lack of precision is expected if non-isothermal conditions are imposed

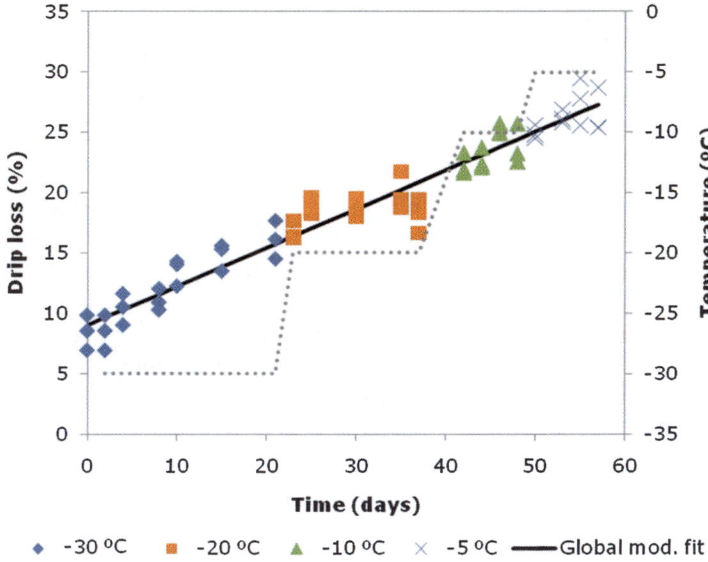

Fig. 28.3 Variation in carrot drip loss throughout frozen storage under non-isothermal conditions (*dotted line*: temperature history)

28.3.2.2 Drip Loss

Frozen storage destroys cell integrity; therefore, during thawing, membranes become very permeable and the diffusion of vacuolar content (such as salts, sugars and pigments) may occur. The quantity of exudates expelled depends on the nature of the frozen product, the freezing and thawing rates and the physiological state (International Institute of Refrigeration 1986).

The drip loss of carrots under non-isothermal frozen storage conditions is shown in Fig. 28.3. At the end of the storage period (57 days), drip loss increased by 215 % (in relation to initial values). This quality parameter appears to be greatly affected by temperature changes. Drip loss was well described by a zero-order kinetic model (Eq. 28.5) by separating the variables and integrating:

$$C = C_0 - \left[-k_{ref} \int_0^t exp\left[-\frac{E_a}{R} \left(\frac{1}{T(t)} - \frac{1}{T_{ref}} \right) \right] dt' \right], \tag{28.11}$$

where C is drip loss.

Estimated parameters are included in Table 28.2 and can be used for predictive purposes of drip loss alterations of carrots throughout time-varying frozen storage.

The normality and randomness of residuals were verified, and the coefficient of determination was satisfactorily high ($R^2 = 0.95$), which corroborate the adequacy of the model applied.

28.4 Concluding Remarks

Models that include the time effect on kinetic behaviour were satisfactorily developed. They were efficient in describing the thermal inactivation of microorganisms and quality alterations in products stored under frozen conditions. These approaches will be profitable in safety and quality predictions of food processes that occur under realistic dynamic temperature conditions.

Accelerated life tests applied were a satisfactory methodology for studying the kinetics of quality changes during frozen storage.

References

Aparicio-Cuesta P, Rivas-Gonzalo JC, Santos-Buelga C, Garcia Moreno C (1989) Quality of frozen green beans (*Phaseolus vulgaris*) subjected to different storage conditions. J Sci Food Agric 48(2):249–252

Brandão TRS, Oliveira FAR (1997) The influence of the temperature increase rate on the accuracy of diffusion parameters estimated under non-isothermal conditions. Int J Food Sci Technol 32:63–72

Brandão TRS, Oliveira FAR, Cunha LM (2001) Design of experiments for improving the precision in the estimation of diffusion parameters under isothermal and non-isothermal conditions. Int J Food Sci Technol 36:291–301

Canet W (1989) Quality and stability of frozen vegetables. In: Thorne S (ed) Developments in food preservation. Elsevier Science Publishing Inc., New York

Geeraerd AH, Herremans CH, Van Impe JF (2000) Structural model requirements to describe microbial inactivation during a mild heat treatment. Int J Food Microbiol 59:185

Gil MM, Brandão TRS, Silva CLM (2006) A modified Gompertz model to predict microbial inactivation under time-varying temperature conditions. J Food Eng 76(1):89–94

Gil MM, Miller FA, Brandão TRS, Silva CLM (2011) On the use of the Gompertz model to predict microbial thermal inactivation under isothermal and non-isothermal conditions. Food Eng Rev 3:17–25

Gonçalves EM, Cruz RMS, Brandão TRS, Silva CLM (2009) Biochemical and colour changes of watercress (*Nasturtium officinale R. Br.*) during freezing and frozen storage. J Food Eng 93 (1):32–39

Gonçalves EM, Abreu M, Brandão TRS, Silva CLM (2011a) Degradation kinetics of colour, total vitamin C and drip loss in frozen broccoli (*Brassica oleracea* L. ssp. *Italica*) during storage at isothermal and nonisothermal conditions. Int J Refrigeration 34:2136–2144

Gonçalves EM, Pinheiro J, Abreu M, Brandão TRS, Silva CLM (2011b) Kinetics of quality changes of pumpkin (*Curcurbita maxima* L.) stored under isothermal and non-isothermal frozen conditions. J Food Eng 106:40–47

Gormley R, Walshe T, Hussey K, Butler F (2002) The effect of fluctuating vs. constant frozen storage temperature regimes on some quality parameters of selected food products. Lebensm Wiss Technol 35(2):190–200

Haralampu SG, Saguy I, Karel M (1985) Estimation of Arrhenius model parameters using 3 least-squares methods. J Food Process Pres 9(3):129–143

International Institute of Refrigeration (IIR) (1986) Recommendations for the processing and handling of frozen foods. International Institute of Refrigeration, Paris

Lisiewska Z, Kimiecik W (2000) Effect of storage period and temperature on the chemical composition and organoleptic quality of frozen tomato cubes. Food Chem 70(2):167–173

Martins RC, Silva CLM (2003) Kinetics of frozen stored green bean (*Phaseolus vulgaris, L.*) Quality changes: texture, vitamin C, reducing sugars, and starch. J Food Sci 68(7):2232–2237

Martins RC, Lopes IC, Silva CLM (2005) Accelerated life testing of frozen green beans (*Phaseolus vulgaris, L.*) quality loss kinetics: colour and starch. J Food Eng 67(3):339–346

Miller FA, Ramos B, Gil MM, Brandão TRS, Teixeira P, Silva CLM (2011) Heat inactivation of *Listeria innocua* in broth and food products under non-isothermal conditions. Food Control 22:20–26

Nelder JA, Mead R (1965) A simplex-method for function minimization. Comput J 7(4):308–313

Nicolaï BM, Van Impe JF (1996) Predictive food microbiology: a probabilistic approach. Math Comput Simulat 42(2–3):287–292

Pukszta T, Palich P (2007) The effect of freezing conditions of strawberry storage on the level of thawing drip loss. Acta Agrophysica 9:203–208

Ratkowsky DA, Olley J, McMeekin TA, Ball A (1982) Relationship between temperature and growth-rate of bacterial cultures. J Bacteriol 149(1):1–5

Van Boekel MAJS (2002) On the use of the Weibull model to describe thermal inactivation of microbial vegetative cells. Int J Food Microbiol 74:139

Van Buggenhout S, Messagie I, Maes V, Duvetter T, Van Loey A, Hendrickx M (2006) Minimizing texture loss of frozen strawberries: effect of infusion with pectinmethylesterase and calcium combined with different freezing conditions and effect of subsequent storage/thawing conditions. Eur Food Res Technol 223:395–404

Van Impe JF, Nicolaï BM, Martens T, De Baerdemaeker J, Vandewalle J (1992) Dynamic mathematical model to predict microbial growth and inactivation during food processing. Appl Environ Microbiol 58(9):2901–2909

Van Impe JF, Nicolaï BM, Schellekens M, Martens T, Baerdemaeker J (1995) Predictive microbiology in a dynamic environment – a system theory approach. Int J Food Microbiol 25 (3):227–249

Xiong R, Xie G, Edmondson AS, Linton RH, Sheard MA (1999) Comparison of the Baranyi model with the Gompertz equation for modelling thermal inactivation of *Listeria monocytogenes* Scott A. Food Microbiol 16(3):269–279

Chapter 29
Hyperspectral Imaging Technology: A Nondestructive Tool for Food Quality and Safety Evaluation and Inspection

Di Wu and Da-Wen Sun

29.1 Introduction

Nowadays, consumer expectation for food products with high quality and safety are increasing. As a result, there is a need to determine the quality of food products accurately, rapidly, and objectively, which constitutes a central challenge for the food processing industry. In particular, online monitoring at critical processing points is important for food quality control. Human visual operation is still the most commonly used method in existing food quality inspection systems to evaluate the quality of food products. However, it is subjective, time consuming, laborious, tedious, and inconsistent. Other traditional methods of food quality inspection, including mass spectrometry (MS) and high-performance liquid chromatography (HPLC), are mainly analytical methods. Although these methods have accurate evaluation capabilities, they are destructive, expensive, time consuming, and inefficient. Therefore, accurate, reliable, and noninvasive systems are critical and necessary for the automatic quality evaluation and classification of food products.

Recently, optical sensing technologies have been investigated as potential tools for the aforementioned purposes. In particular, hyperspectral imaging, also called imaging spectroscopy or imaging spectrometry, has been widely studied and developed, resulting in many successful applications as a smart and promising analytical tool in food quality control. Compared to traditional methods, the hyperspectral imaging technique has the advantages of minimal sample preparation, noninvasive measurement, rapid data acquisition, and the capability to visualize the spatial distribution of numerous chemical compositions and physical features simultaneously. Although it faces many challenges, hyperspectral imaging is still expected to become one of the most preferred analytical tools in inspecting

D. Wu • D.-W. Sun (✉)
School of Biosystems Engineering, Food Refrigeration and Computerised Food Technology (FRCFT), University College Dublin, National University of Ireland, Agriculture & Food Science Centre, Belfield, Dublin 4, Ireland
e-mail: dawen.sun@ucd.ie; http://www.ucd.ie/refrig; http://www.ucd.ie/sun

S. Yanniotis et al. (eds.), *Advances in Food Process Engineering Research and Applications*, Food Engineering Series, DOI 10.1007/978-1-4614-7906-2_29,

the quality of foods and their authentication and is expected to be dominant in the future.

29.2 Fundamentals of Hyperspectral Imaging Technology

The use of hyperspectral imaging requires a good understanding of the basic principles of this tool. Therefore, some basic knowledge is presented in what follows.

29.2.1 Relationship Between Spectroscopy, Imaging, and Hyperspectral Imaging

In the past few decades, considerable efforts have been devoted to developing new sensing techniques for the online inspection of agricultural and food products to guarantee their quality and safety. Spectroscopy is a widely recognized technique for analyzing raw materials, product quality control, and process monitoring (Cen and He 2007). However, although spectroscopy can be used to determine the main composition of food products rapidly and noninvasively, it cannot detect compositional gradients with high spatial resolution. On the other hand, spatial information about food products can be obtained using a conventional imaging system or, more specifically, computer vision (Du and Sun 2006). However, a conventional imaging system cannot detect chemical properties or characteristics of food products, such as moisture, fat, and protein content. Therefore, hyperspectral imaging was developed; it can simultaneously obtain spectral and spatial information in one system, both of which are critical in determining the quality of agricultural and food products. Thus, using hyperspectral imaging one can determine the qualities of several food products, in addition to being to visualize content distribution within the same sample. As a result, the hyperspectral imaging technique provides unprecedented detection capabilities that are unavailable with either spectroscopy or imaging alone. If conventional imaging provides the answer to the question where and conventional spectroscopy provides the answer to the question what, hyperspectral imaging has can answer the question where is what.

29.2.2 Principles of Hyperspectral Imaging

A good understanding of the principles of hyperspectral imaging is necessary if one wishes to use such a powerful technique for food quality inspection. Some basic knowledge about hyperspectral imaging is provided in this section.

29.2.2.1 Classes of Spectral Imaging

A spectral image can be defined as a stack of images of the same object at different spectral wavelength bands. A spectral imaging field usually has three main classes: multispectral, hyperspectral, and ultraspectral imaging. Specifically, ultraspectral imaging usually refers to spectral imaging systems with a very fine spectral resolution. However, such systems commonly have a low spatial resolution of only a few pixels. Hyperspectral imaging has spectral bands over a contiguous wavelength range. Therefore, every pixel in a hyperspectral image has its own spectrum. Multispectral imaging systems differ from hyperspectral imaging systems in the number of spectral bands and spectral resolution. Usually there are fewer than ten spectral bands in the case of multispectral imaging systems, while hyperspectral images have hundreds of contiguous and regularly spaced bands.

29.2.2.2 Hyperspectral Cube

Several synonymous expressions are normally used to describe the form of the three-dimensional hyperspectral image, including hyperspectral cube, hypercube, spectral cube, spectral volume, datacube, or data volume. For example, a hyperspectral cube of a piece of meat was extracted from the corresponding hyperspectral image, as illustrated in Fig. 29.1. The cube consists of a series of contiguous subimages, one behind the other, at different wavelengths. Each subimage provides the spatial distribution and spectral intensity of the imaged specimen at a certain wavelength. Any spatial image within the spectral range of the system can be picked up from the hyperspectral cube at a certain wavelength within the wavelength sensitivity. Therefore, a hyperspectral image described as I (x, y, λ) can be viewed either as a separate spatial image I (x, y) at each individual wavelength (λ) or as a spectrum I (λ) at each individual pixel (x, y). Because a pixel-level spatial distribution is available with spectral responses covering all wavelengths in a hyperspectral image, a flexible selection of the regions of interest (ROIs) can be achieved on a target object. For example, if two different pixels extracted from two different compositional locations in a hypercube show different spectral signatures or fingerprints, then such a difference in the spectral signatures between the fat pixel and the lean meat pixel of the meat image is noticeably distinct, as shown in Fig. 29.1, without any further manipulation or preprocessing analysis of these spectra. On the other hand, the full spatial features of the imaged specimen could be viewed as a grayscale image at any single wavelength, as illustrated in Fig. 29.1. Although images at two adjacent wavelengths are very similar in a hyperspectral image, discrete information could be mined by analyzing images at a combination of featured wavelengths.

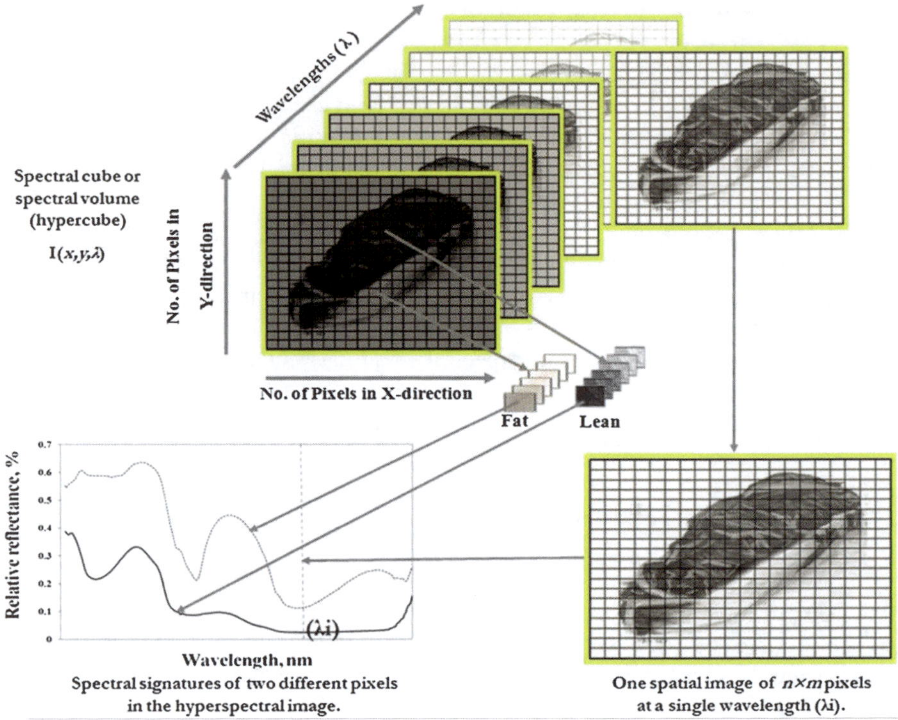

Fig. 29.1 Schematic diagram of hyperspectral image (hyperspectral cube) for a piece of meat. The cube consists of a series of contiguous subimages one behind the other at different wavelengths; meanwhile, every pixel in the cube has its own spectrum over the whole wavelength range

29.2.2.3 Generation of Hyperspectral Images

There are four approaches to acquiring three-dimensional hyperspectral image cubes [I (x, y, λ)] based on the methods of point scanning, line scanning, area scanning, and the single shot method. The point scanning method (also known as the whiskbroom method) scans a single pixel to acquire the spectrum of this pixel (Fig. 29.2). A hyperspectral image acquired using whiskbroom scanning is generated by moving either the detector or the sample in the direction of either the x- or y-dimension and is stored in band-interleaved-by-pixel (BIP) format. The second approach, illustrated in the top right corner of Fig. 29.2b, is called the line scanning method (also known as the pushbroom method), where a two-dimensional image (y, λ) with the whole spectral dimension (λ) and one spatial dimension (y) is acquired at a time. After the line is scanned along the direction of the x-dimension, a complete hyperspectral cube can be obtained and stored in band-interleaved-by-line (BIL) format. Unlike the whiskbroom and pushbroom spatial scanning methods, area or plane scanning (also known as band sequential method or

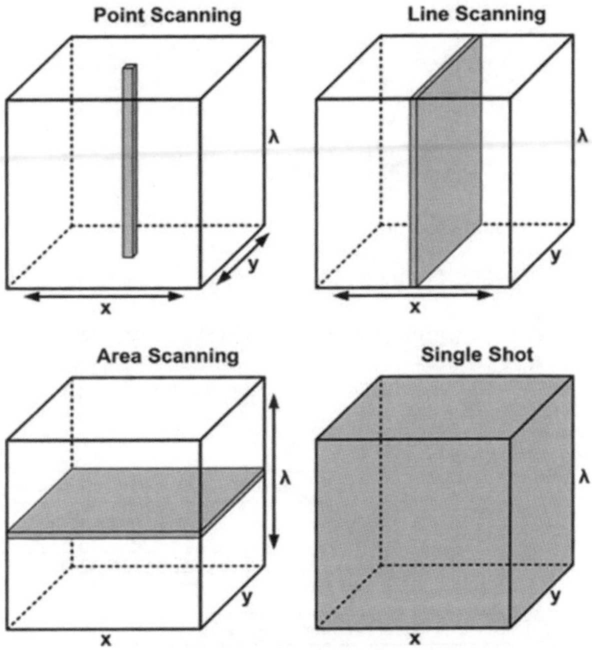

Fig. 29.2 Four approaches to acquiring three-dimensional hyperspectral image cubes [I (x, y, λ)]. *Arrows*: scanning directions; *grey areas*: data acquired each time

wavelength scanning) is a spectral scanning method, as illustrated in the lower left corner of Fig. 29.2c. To build up a hyperspectral cube, a two-dimensional monochrome image (x, y) is obtained at a single band each time. The hyperspectral cube is obtained by repeating the process at one wavelength after another to generate a stack of single-band images and stored in band sequential (BSQ) format. The single shot method has the ability to obtain both spatial and spectral information simultaneously. Although it is still in its infancy, the single shot method is very attractive if one needs to obtain hyperspectral images rapidly because of its ability to obtain both spatial and spectral information simultaneously.

29.2.2.4 Image Acquisition Modes

There are three common image acquisition modes for hyperspectral imaging: reflectance, transmittance, and interactance (Fig. 29.3). The reflectance mode obtains high relative information through the reflected light from a sample without any contact with the tested object, as illustrated in Fig. 29.3. A transmittance model captures transmitted light through the sample, which carries more valuable but weak information (Schaare and Fraser 2000). As illustrated in Fig. 29.3, a transmittance image is acquired with the light source positioned opposite to the detector and the sample in between. The interactance model is a combination of reflectance and transmittance (ElMasry et al. 2008), where both the light source and the detector are located on the same side and parallel to each other (Fig. 29.3).

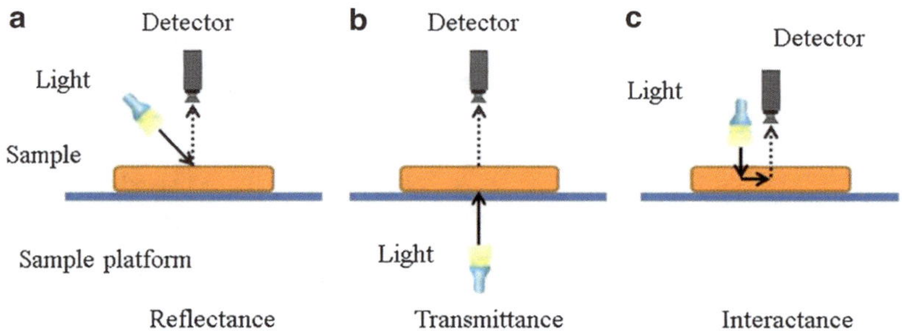

Fig. 29.3 Schematic of image acquisition modes; detector comprised of *lens*, *spectrograph*, and *camera*

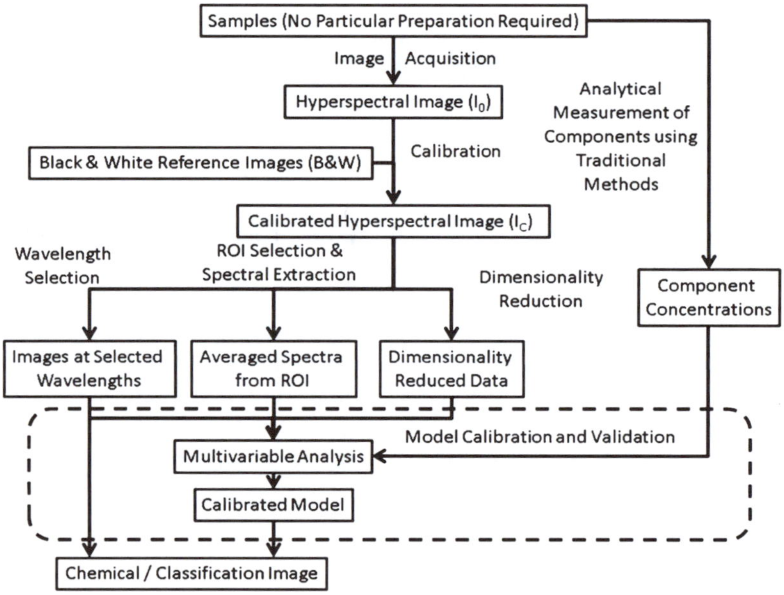

Fig. 29.4 Flowchart of a series of typical steps for analyzing hyperspectral image data

29.2.3 Hyperspectral Image Processing Methods

The hyperspectral imaging technique cannot work alone without the help of chemometric algorithms for data analysis. Figure 29.4 outlines typical steps taken to analyze the hyperspectral image data. The Environment for Visualizing Images (ENVI) software (Research Systems, Boulder, CO, USA), MATLAB (MathWorks, Natick, MA, USA), and Unscrambler (CAMO PROCESS AS, Oslo, Norway) are commonly used tools for hyperspectral image processing.

29.2.3.1 Correction of Hyperspectral Images

Hyperspectral imaging collects spectral data containing the detector signal intensity, not actual reflectance values. Therefore, the acquired hyperspectral images should be corrected to a reflectance or absorbance image based on a reflectance calibration (Burger and Geladi 2005). To obtain reflectance hyperspectral images, the original hyperspectral images should be calibrated with black-and-white reference images using the following equation:

$$I_C = \frac{I_0 - B}{W - B} \times 100, \tag{29.1}$$

where I_C is the corrected hyperspectral image in a unit of relative reflectance (%), I_0 is the original hyperspectral image, B is a dark image (approximately 0 % reflectance), and W is the white reference image (approximately 99.9 % reflectance). As a result, the raw hyperspectral image (I_0) with intensity values is corrected to reflectance values (ranging from 0 % to 100 %, unitless) with the aid of black-and-white reference images.

29.2.3.2 Image Enhancement and Spectral Preprocessing

Image enhancement is a necessary process for many hyperspectral image applications to relatively improve the visibility of certain specified image characteristics due to the noise inherent in hyperspectral images and nonideal hyperspectral imaging instruments. The commonly used image enhancement algorithms, such as noise filtering, edge and contrast enhancement, magnifying, pseudocoloring, and sharpening, are usually classified into two categories: spatial domain methods and frequency domain methods. Spatial domain methods operate on an objective region, for example, the histogram equalization method and local neighborhood operations based on convolution. Frequency domain methods transform images into frequency domains, such as discrete Fourier transforms (FTs) and wavelet transforms (WTs).

On the other hand, spectral preprocessing algorithms are commonly used to correct effects from random noise, length variation of light path, and light scattering resulting from instrumental effects or a sample's physical variability (Rinnan et al. 2009). The most widely used preprocessing algorithms are mainly classified into two categories, spectral derivatives and scatter-correction methods (Rinnan et al. 2009), such as Savitzky-Golay smoothing, first and second derivatives, standard normal variate, multiplicative scatter correction, FT, WT, and orthogonal signal correction.

29.2.3.3 Image Segmentation

Image segmentation is intended to isolate ROIs in the form of masks for further feature extraction and modeling (ElMasry et al. 2009). There are mainly two methods to perform image segmentation. The first is to segment an image into objects that have similar features according to predefined principles, such as thresholding, which includes global thresholding (Liu et al. 2007; Qin et al. 2009) and adaptive thresholding (Liu et al. 2002), and morphological processing, which includes erosion, dilation, open, closed, and watershed algorithms (Beucher and Meyer 1993). The second method is to segment an image based on sudden changes in intensity, such as edge-based segmentation, which includes gradient-based methods and Laplacian-based methods. Moreover, segmentation and classification can be integrated into a single process because of their closer relationship with hyperspectral image analysis, which is considered spectral image segmentation (ElMasry et al. 2011a; Kamruzzaman et al. 2011; Barbin et al. 2012).

29.2.3.4 Feature Extraction

Following image segmentation, ROIs can be extracted, which contain areas of intensity distributions. Because a hyperspectral image contains a huge amount of data, characteristic features, such as intensity-based, texture-based, and morphologically based features, should be extracted to further establish prediction models. The mean is widely used for obtaining intensity information (Park et al. 2006b; ElMasry et al. 2007; Qiao et al. 2007a, b), and it is calculated by averaging the intensity of pixels within the ROIs at each wavelength. Besides the mean, the standard deviation, skew, energy, and entropy are also representative first-order measures, while texture is a typical second-order measure based on joint distribution functions. The goal of texture is to represent the spatial arrangement of the pixel gray levels within the ROIs (IEEE 1990). The gray level cooccurrence matrix (GLCM) is the most commonly used texture attribute in hyperspectral imaging (ElMasry et al. 2007; Qiao et al. 2007a; Qin et al. 2009). There are several widely used statistical and probabilistic features derived from the GLCM (Haralick and Shapiro 1992), such as contrast (also called variance), inverse differential moment (IDM, also called homogeneity), angular second moment (ASM), entropy, and correlation. The two-dimensional Gabor filter is another effective tool for image texture extraction and analysis; it has the capability to achieve certain optimal joint localization properties in the spatial frequency domain and in the spatial domain (Daugman 1980, 1985).

29.2.3.5 Multivariate Analysis

Multivariate data analysis approaches are implemented to analyze the multiple variables of extracted features simultaneously and to predict and visualize the hidden quality information with the images of the tested samples based on the established calibrated regression/classification models (Bremner and Brodersen 2001). Generally, multivariate methods are classified into qualitative classification and quantitative regression.

Multivariate Classification. Multivariate classification (also called pattern recognition) is classified into two categories, unsupervised and supervised. The realization of unsupervised classifications does not require a priori knowledge about the classes information of the data, which is achieved according to the intrinsic characteristics of samples. Typical unsupervised multivariate classification algorithms for the analysis of hyperspectral data include principal component analysis (PCA) (Wu et al. 2008e; Yang et al. 2009), K-means clustering, and independent component analysis (Wu et al. 2007).

Unlike unsupervised pattern recognition, the goal of supervised pattern recognition is to group new samples into a classification model of predefined known classes according to their pattern of measurements. Typical unsupervised multivariate classification algorithms include linear discrimination analysis (LDA) (Park et al. 2007), partial least-squares discriminant analysis (DA) (Ariana and Lu 2008; Wu et al. 2008a), artificial neural networks (ANNs) (Nagata et al. 2006; Wu et al. 2008b; ElMasry et al. 2009), support vector machines (SVMs) (Fletcher and Kong 2003; Wu et al. 2007), and k-nearest neighbor (Kavdir and Guyer 2008).

Multivariate Regression. The function of multivariate regression in the application of spectral analysis is to establish a relationship between the spectrum response of the tested sample and the desired physical, chemical, or biological attributes with explanatory or predictive purposes. Multivariate regression generally has two categories, linear and nonlinear. The most widely used multivariate linear regression methods in quantitative analysis are multiple linear regression (MLR) (Wu et al. 2010, 2011a), principal component regression, and partial least-squares regression (Gerlach et al. 1979; Wu et al. 2011b).

Due to the effects of the physical properties of tested samples or instrumental effects, the relationship between the target features and spectra may have the nonlinear characteristics multivariate regression, and the solutions of such analysis get better through the use of nonlinear regression techniques such as ANNs (Fathi et al. 2009) and SVMs (Cogdill and Dardenne 2004). In particular, back propagation artificial neural network (BP-ANN) is a multilayer feedforward neural network that is frequently used due to its high nonlinear mapping ability, and least-squares support vector machine (LS-SVM) is an optimized version of the standard SVM, which can solve multivariate calibration problems in a relatively fast way (Wu et al. 2008d).

29.2.3.6 Optimal Wavelength Selection

The selection of optimal wavelengths is an important process of analyzing hyperspectral images because the acquisition of data of hyperspectral images is relatively slow, and the extraordinary data of hyperspectral images have a great degree of dimensionality with redundancy among contiguous variables (wavelengths). Besides the knowledge-based manual selection approach, there are several wavelength-variable selection algorithms for choosing optimum wavelengths, including some classic approaches such as correlation coefficients (Qiao et al. 2007b), loading and regression coefficients (Wu et al. 2008c), analysis of spectral differences (Liu et al. 2007), spectrum derivatives (Chao et al. 2007), stepwise regression (Cluff et al. 2008), DA (Nagata et al. 2006), sophisticated methods such as successive projections algorithm and uninformative variable elimination (Wu et al. 2011a, c), elaborate search-based strategies such as simulated annealing (Chen et al. 2009), ANNs (ElMasry et al. 2009), and genetic algorithms, and interval-based algorithms such as interval partial least squares (iPLS), windows PLS, and iterative PLS (Wu et al. 2010). Details of wavelength selection algorithms for spectral analysis can be found elsewhere (Zou et al. 2010).

29.2.3.7 Model Evaluation

For the analysis of hyperspectral image data, the multivariate data must be trained to build a calibration model, which is then evaluated for its performance. The performance of a model is usually evaluated in terms of standard error of calibration (SEC), root-mean-square error of calibration (RMSEC) and coefficients of determination of calibration (r_C^2) in the calibration process, root-mean-square error of cross-validation (RMSECV) and coefficients of determination of validation (r_V^2) in the validation process, and standard error of prediction (SEP), root-mean-square error of prediction (RMSEP), coefficients of determination of prediction (r_P^2), and residual predictive deviation (RPD). Generally, a good model should have higher values of r_C^2, r_V^2, r_P^2, and RPD, lower values of SEC, SEP, RMSEC, and RMSEP, and a small difference between SEC and SEP.

29.2.3.8 Visualization of Chemical Images

Recently, the concentration gradients of certain components of food products have been required for detailed food analysis. Hyperspectral imaging is a principal methodology to visualize the distribution of chemical components of food materials and has the advantages that it allows the investigator to know and understand the heterogeneity of food products. A typical way to obtain the quantitative distribution of the desired attributes of tested samples is to calculate the chemical contents of each pixel by applying one or more chemometric tools with the spectrum of the corresponding pixel. A problem for this method is that it is practically impossible to measure the precise concentration of compositions for every pixel within a sample. Therefore, reference values for each pixel might not be provided. The solution to this problem is to establish regression models on the basis of the average spectrum of all pixels within the object region where the corresponding reference value can be obtained. Following the calibration and validation processes, the established models can be applied to analyze the spectrum of each pixel within the object region for the determination of the composition content of each pixel, which can be used for further generation of chemical maps.

29.2.4 Hyperspectral Imaging Instruments

Instrumentation is basic for any reliable measurement. It is crucial to select the instruments and design their setup and calibrations for the performance of hyperspectral imaging systems. Some instruments are specifically designed for hyperspectral imaging. This section focuses primarily on instrumentation for a hyperspectral imaging technique.

29.2.4.1 Configuration of a Hyperspectral Imaging System

The essential components used for constructing a hyperspectral imaging system mainly include light sources, wavelength dispersion devices, and area detectors.

Light Sources. Light is an information carrier within image-based inspection systems. It is important to choose light sources and design the lighting apparatus to obtain reliable, high-quality hyperspectral images. Halogen lamps are commonly used as bright light sources. Typically a lamp filament made of tungsten wire is placed in a quartz glass bulb filled with halogen gas, which is called a quartz–tungsten–halogen lamp. As a solid-state sources, a light emitting diode (LED) emits light when a semiconductor is electrified. In recent years, it has advanced rapidly due to its many advantages, such as small size, fast response, long lifetime, low heat generation, low power consumption, robustness, and insensitivity to vibration. Lasers generate light by means of stimulated emission, which generally occurs inside a resonant optical cavity filled with a gain medium, such as a crystal, dye solution, gas, or semiconductor. Lasers have unique features like perfect directionality, highly concentrated energy, and real monochromatic emission. Moreover, because of their ability to produce narrowband pulsed or continuous light, LEDs are now used as excitation sources, although light generated from lasers have higher intensities and narrower bandwidths than that from LEDs. A tunable source sets the wavelength dispersion device in the illumination light path instead of the imaging light path. Therefore, it can do the area scanning to obtain both spatial and spectral information from the tested objects. Nowadays, the development of tunable light sources is still in an early stage.

The feasibility of implementing this new light source in the field of food quality inspection needs to be explored.

Wavelength Dispersion Devices. Wavelength dispersion devices are considered the heart of hyperspectral imaging systems. The function of wavelength dispersion devices is to disperse broadband light into different wavelengths. Typical examples include filter wheels, imaging spectrographs, acousto-optic tunable filters (AOTF), and liquid crystal tunable filters (LCTFs). A filter wheel is the most basic implementation, which uses a set of discrete bandpass filters to obtain spectral images at different wavelengths. However, there are limitations of filter wheels for hyperspectral imaging applications, such as the narrow spectral range and low resolution, mechanical vibration from moving parts, slow wavelength switching, and nonmatching images due to filter movement. An imaging spectrograph is a typical wavelength dispersion device that has the capability of dispersing incident broadband light into narrowband light at different wavelengths instantaneously for different spatial areas from the surface of a target. There are two main forms of imaging spectrograph, reflection gratings (i.e., gratings laid on a reflective surface) and transmission gratings (i.e., gratings laid on a transparent surface). An AOTF is a solid-state apparatus using an electronically tunable bandpass filter with light–sound interactions in a crystal. A LCTF is a solid-state device that is operated using electronically controlled liquid crystal cells inserted between two parallel polarizers. Unlike the fixed interference filters, electronically tunable filters like AOTFs and LCTFs can be flexibly controlled using a computer and do not have the problems of speed limitation, mechanical vibration, and image misregistration, which are constraints of rotating filter wheels.

Area Detectors. Detectors are used to quantify the intensity of acquired light by converting incident photons into electrons, which generates electrical signals. There are two major types of solid-state area detectors, charge-coupled devices (CCDs) and complementary metal-oxide-semiconductor (CMOS) cameras. A CCD sensor consists of mass photodiodes (called pixels) that are made of light-sensitive materials such as silicon (Si) or indium gallium arsenide (InGaAs). Each photodiode is an individual spot detector that converts radiation energy into an electrical signal that is proportional to the total light exposure. CCD area detectors are usually used in hyperspectral imaging systems for the acquisition of image data. A CMOS image sensor is the other main type of solid-state area detector that is considered to have the potential to compete with CCDs. Both a photodetector and a readout amplifier in each pixel are included with the CMOS image sensor, which is the main difference between these two types of detectors (Litwiller 2005). Besides fast signal transfer, there are many other advantages of CMOS image sensors, such as small size, random addressing, low cost, single power supply, and low power consumption, which make them competitive in the consumer electronics market.

Example of a Hyperspectral Imaging System. Line-scan systems (also called the pushbroom approach) are very common in a wide variety of food inspection

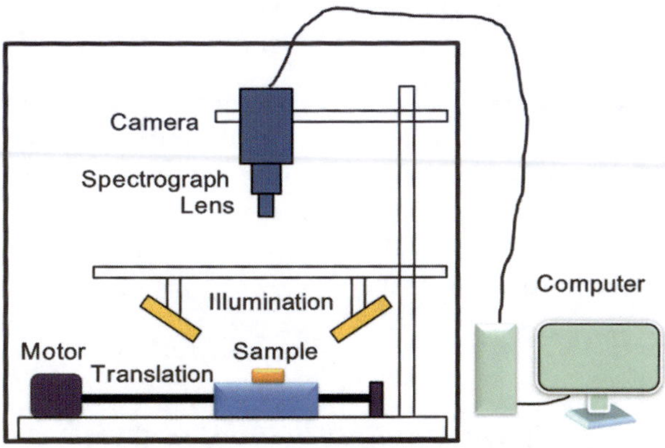

Fig. 29.5 Schematic diagram of a line-scan hyperspectral imaging system

approaches due to their consistency in online applications and are used as an example to illustrate a hyperspectral imaging system (Fig. 29.5). The system shown in Fig. 29.5 consists mainly of a high-performance, back-illuminated CCD or CMOS camera, an imaging spectrograph covering a certain spectral range, a specially assembled light source, a translation belt operated by a stepper motor, and a computer equipped with image acquisition software. In Fig. 29.5, the illumination and measurements are performed on the same side of the sample, where the illumination is focused on an area adjacent and parallel to the detector field of view. The detector performs the linear array scanning along the *y*-axis direction, as the samples move on the *x*-axis, thereby obtaining a hyperspectral image.

29.2.4.2 Calibration of a Hyperspectral Imaging System

To ensure the reliability of the acquired hyperspectral imaging data, an accurate calibration is an essential step to guarantee a consistent understanding of food products for a hyperspectral imaging system. The calibration process includes the standardization of the spectral axis of the hyperspectral image, the validation of the acceptability and reliability of the extracted spectral data, a determination of whether the hyperspectral imaging system is in running condition, the evaluation of the accuracy and reproducibility of the acquired data under different operating conditions, and a diagnosis of instrumental errors if necessary. However, all commercially available hyperspectral imaging systems are calibrated by their manufacturers, and usually there is no need to perform the calibration process unless the system is shaken or the system is assembled by the researchers themselves.

29.3 Applications

Nowadays, the hyperspectral imaging technique has attracted heightened interest as a nondestructive and fast inspection method for a wide range of food-quality-related applications, which include the nondestructive evaluation of food products and inspection of food for safety and quality. With its current success in these applications, it is evident that hyperspectral imaging can automate a variety of routine inspection tasks.

29.3.1 Meat Quality Evaluation Using Hyperspectral Imaging

The assessment of meat quality attributes has always been important for the meat industry because consumers are always demanding food products with superior quality at an affordable price. Recently, hyperspectral imaging has been widely studied and developed for the rapid and reliable evaluation of meat quality, resulting in many successful applications.

29.3.1.1 Beef

The prediction of tenderness of beef is a considerable challenge because it is a property of a cooked product, while the determination of this property is based on a fresh steak. Naganathan et al. (2008a, b) predicted tenderness as determined by shear force of 14-day aged, cooked beef from the hyperspectral images of fresh beef-ribeye steaks (*longissimus dorsi*) (LD) using two pushbroom hyperspectral imaging systems in a wavelength range of 400–1,000 nm and in a spectral range of 900–1,700 nm, respectively. In their first work (Naganathan et al. 2008b), a result of 96.4 % accuracy was obtained for the three tenderness categories (tender, intermediate, and tough). In their subsequent work (Naganathan et al. 2008a), the established model obtained an overall correct rate of 77.0 %.

Waterholding capacity (WHC) is an important attribute of beef. ElMasry et al. (2011b) evaluated the feasibility of using near-infrared (NIR) hyperspectral imaging (910–1,700 nm) to predict WHC in fresh beef, which was determined by a drip loss method. On the basis of a PLS model, good prediction results were obtained with a coefficient of determination (r_{CV}^2) of 0.89 and standard error estimated by cross validation (SECV) of 0.26 %. In addition, six important wavelengths of 940, 997, 1,144, 1,214, 1,342, and 1,443 nm were chosen according to regression coefficients of the PLS model, and a similar prediction was obtained based on these selected wavelengths (r_{CV}^2 of 0.89 and SECV of 0.28 %), which show the potential to predict drip loss nondestructively.

29.3.1.2 Pork

The high quality of pork is closely related to the factors that can affect the palatability and acceptability to consumers for both fresh and processed products and the processing of fat and lean tissues. The utilization of hyperspectral imaging was first tested for rapid and noninvasive determination of different quality grades (red, firm, normal (RFN), pale, soft, exudative (PSE), pale, firm, normal (PFN), red, soft, exudative (RSE)) of pork in 2007 (Qiao et al. 2007a). The main spectrum was extracted from the ROI for each sample following image calibration. The smoothing of a ten-point mean filter followed by second derivatives was executed as the spectral preprocessing for each spectrum to correct multiplicative scatter and correct the baseline. PCA was then performed for the reduction of the dimensionality of the extracted spectral data. Different numbers of principal components obtained from the PCA process were used to establish the classification model using cluster analysis and ANN, respectively. The classification results obtained using cluster analysis were between 75 and 80 %, and those obtained using ANN models were 69 and 85 % with first five and then ten principal components s, respectively. Recently, Barbin et al. (2012) utilized a method called dancing pixels to obtain a ROI comprising only the loin eye area. Following spectral extraction, six significant wavelengths were identified from second-derivative spectra, which were 960, 1,074, 1,124, 1,147, 1,207 and 1,341 nm. An overall accuracy of 96 % was then obtained by PCA with these particular wavelengths.

29.3.1.3 Fish

In aquiculture fields, the term quality is defined as the esthetic appearance and freshness or degree of spoilage that the product has undergone (Huidobro et al. 2001). As the first attempts, Wold et al. (2006) investigated the potential of using a high-speed transflectance spectral imaging system for noninvasively assessing the moisture concentration of dried salted coalfish (*bacalao*). With the same system, ElMasry and Wold (2008) assessed the water and fat distribution in six species of fish filets (Atlantichalibut, catfish, cod, mackerel, herring, and saithe). This industrial online implementation acquired hyperspectral images in a VIS (visible) and NIR range of 460–1,040 nm with a spectral resolution of approximately 20 nm at the speed of 10,000 spectra per second. The hyperspectral image of each entire fish was required by a pushbroom scanning device line by line. The distributions of fat and water contents with different concentration gradients in the fish filets were visualized by building chemical images in accordance with hyperspectral image data. Although determining the fat and water distribution in a filet is very difficult using naked eye, the spatial visualizations of fat and water were realized by a hyperspectral imaging system. Moreover, it is possible to see the concentration distribution of fat and water within the same filet.

29.3.1.4 Chicken and Poultry

Due to increasing consumption of chicken and the limitations of manual inspection, the development of a rapid, stable, noncontact, and accurate inspection system is required for the chicken and poultry industry. The USDA's Agricultural Research Service is the pioneer research group in utilizing hyperspectral and multispectral imaging techniques for the determination of different contaminants on poultry carcasses. Its specific works include the calibration of hyperspectral imaging systems, determination of spectral fingerprints of various contaminants in the VIS and NIR regions, development of algorithms for image processing and fecal detection, and exploitation of online inspection systems (Park et al. 2002, 2006a).

Besides the inspection of poultry carcasses, hyperspectral imaging systems have also been studied for the fast determination and characterization of turkey ham quality. ElMasry et al. (2011a) developed a hyperspectral imaging system in the NIR region (900–1,700 nm) to determine the quality of cooked turkey hams with various ingredient and processing parameters. The main ROI was obtained according to the final "ham mask" resulting from segmentation step. The average spectral data were then extracted from only the ham parts by ignoring the fat covering layer. The results showed that it was feasible to distinguish different quality turkey hams using hyperspectral imaging on the basis of chemical composition at a few spectral wavelengths. In addition, visualization of ham slice components (Fig. 29.6) was achieved using the construction of indexed pseudocolor images with a color map of only a limited number of colors instead of using the full color range. The real color images are shown in the left column of Fig. 29.6a. Note that it is difficult to distinguish the classes and characteristics of ham slices by conventional RGB imaging or via human visual inspection. The spectral image of the same ham slices at a single wavelength of 1,215 nm is shown in the middle column of Fig. 29.6b. The final visualized image of the same ham slices according to the PCA score image routine is shown in the right column of Fig. 29.6c, where pixels with similar spectral features appear in the same colors. The analyzed ham slices have different appearances due to the different processing regimes as well as different moisture and brine contents. The results proved the feasibility of using NIR hyperspectral imaging as an objective and noninvasive tool for the classification of cooked turkey ham slices.

29.3.1.5 Lamb

Kamruzzaman et al. (2011) evaluated the feasibility of using a NIR hyperspectral imaging system for the discrimination of three types of lamb muscle. A pushbroom hyperspectral imaging system with a spectral range of 900–1,700 nm was used for the image acquisition of samples from *semitendinosus* (ST), LD, and *psoas major* (PM) of the *Charollais* breed. Muscles were cut into slices of 1 in. thickness by a scalpel and cutting machine. There were 105 lamb samples including PM (35), ST

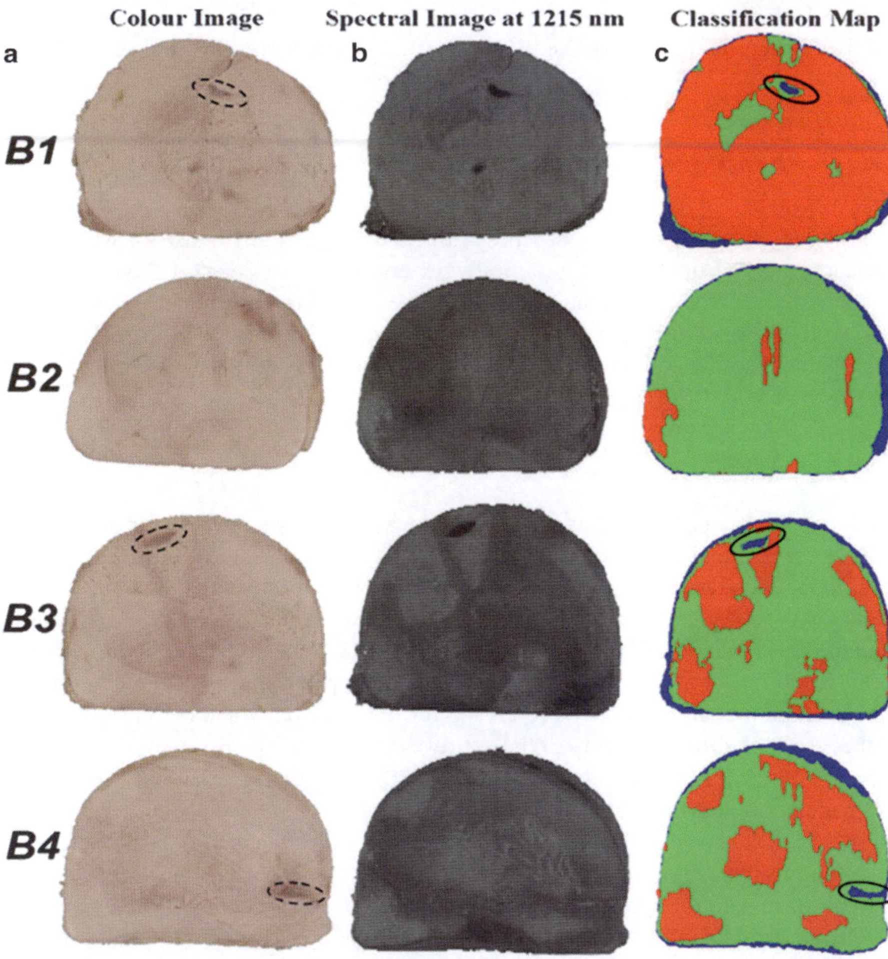

Fig. 29.6 Visualization of components within ham slices: (**a**) real color images, (**b**) spectral images at wavelength of 1,215 nm, and (**c**) visualized classification map of ham blocks B1, B2, B3, and B4. *Fat* areas that originated from turkey skin during ham processing are marked with *dashed circles* in the color images and are then colored *blue* in the final visualized classification map

(35), and LD (35) used for further research. PCA was performed to reduce the high dimensionality, select optimal wavelengths, and visualize chemical images. Six optimal wavelengths were selected, which can be used to further design a multi-spectral imaging system. An overall accuracy of 100 % was obtained based on a LDA calibration model with these six optimum wavelengths. A high overall accuracy in a pixelwise scale was also obtained for visualizing the classification results using an image processing algorithm. The results of this study show the great potential of using NIR hyperspectral imaging together with chemometrics as an efficient and reliable tool for the discrimination of lamb muscles.

29.3.2 Fruit and Vegetables

Quality is also important for the fruit and vegetable industry, where the measurements of quality factors are always crucial for the market. Hyperspectral imaging has proven its great potential in determining the quality of fruit and vegetable physical and chemical features, including the presence of bruises, firmness, soluble solids content, dry matter, sugar contents, and pH.

29.3.2.1 Apples

From harvest to market, apples are liable to be affected by several kinds of damage, such as mechanical damage and invasion of bacteria, molds, or diseases. However, traditional detection processes generally operated by experienced human inspectors cannot easily detect certain kinds of damage by visual examination. ElMasry et al. (2008) investigated the potential of a VIS/NIR hyperspectral imaging system (400–1,000 nm) for early detection ($<$12 h) of apple bruises. PLS and stepwise discrimination analyses (SDA) were used for the essential wavelength selection to reduce the high dimensionality of the entire spectral data. The two methods selected the same three optimal wavelengths of 750, 820, and 960 nm, all in the NIR region. To evaluate the feasibility of bruise determination using the selected wavelengths, the PCA scores were calculated on the hyperspectral images at only the effective wavelengths instead of the entire range of spectra. Hyperspectral imaging attained a high performance for apples presenting both recent (1 h) and old ($>$3 days) bruises.

Firmness and soluble solid content (SSC) are the most important quality factors of apples. Noh and Lu (2007) studied the feasibility of using hyperspectral imaging (500–1,040 nm) to predict the firmness and SSC of Golden Delicious apples in fluorescence and reflectance modes. The results showed that the integration of two kinds of modes obtained better prediction models than either reflectance or fluorescence. Zhao et al. (2009) successfully applied a hyperspectral imaging (685–900 nm) system to predict the sugar content of apples. The optimal spectral range of 704–805 nm was obtained and its PLS model produced a result with an r of 0.91.

29.3.2.2 Citrus and Oranges

Early disease infection detection is especially important for the citrus and orange industry. Currently, the detection of decay caused by fungi is done by trained operators who visually inspect the fruits as they pass. Gomez-Sanchis et al. (2008) applied a hyperspectral imaging technique as an efficient and reliable tool to detect mandarins, which were infected with *Penicillium (p) digitatum* spores. Hyperspectral images in a wavelength range of 460–1,020 nm with a spectral resolution of 10 nm were obtained as a series of 57 monochromatic images as soon as the rot appeared. After comparing the results using different numbers of bands, the maximum success rate was approximately 92 % on the basis of LDA with 57 bands, while the success rate increased to 95 % using only 20 bands with classification and regression trees (CART). This work shows the potential of using a hyperspectral sorting machine to replace the manual inspection of rotten fruits.

Besides the identification of fungal diseases, a hyperspectral imaging technique is also used to address the detection of citrus canker. Qin et al. (2009) developed a hyperspectral imaging approach in the spectral region between 450 and 930 nm for detecting Ruby Red grapefruits with cankerous, normal and other common peel diseases including greasy spot, insect damage, melanose, scab, and wind scar. An overall correct classification of 95.2 % for canker was obtained using their developed algorithm in the task of detecting and classifying canker lesions from sound peel and other diseases.

29.3.2.3 Melons

Automatic sweetness has been predicted for many fruits using a NIR spectroscopy technique (Hasegawa 2000). However, because sugar contents are unevenly distributed within a fruit, some parts of a fruit labeled sweet may taste insipid due to the lack of distribution information of sweetness. Tsuta et al. (2002) developed a universal method using the absorption band of the sugar in the NIR wavelength region for visualization of sugar content. On the basis of the designed imaging system, images at 846, 874, 902, and 930 nm of each half-cut sample were acquired with an exposure period of 3.7 s. Once the average absorbance at each wavelength was obtained, the second-derivative absorbance at 902 and 874 nm was calculated. Then, MLR analysis using this second-derivative absorbance was carried out to acquire the calibration curve for the sugar content of the imaging system, which is given below:

$$^{O}\mathrm{Brix} = 19.01 - 438.84d^2A_{902} + 70.32d^2A_{874}. \qquad (29.2)$$

A high correlation coefficient of 0.891 and SEC of 1.090 was obtained using the preceding function, which shows a good visualization ability of sugar content for the imaging system. From the visualized distribution map, an increment of sugar content could be seen from the rind to near the seeds, and the central upper area within the melon was found to be sweeter than the bottom area, although these phenomena were difficult to observe by with naked eye. The foregoing results show the potential of using NIR imaging for visualization of the distribution of sugar in melons.

29.3.2.4 Tomatoes

Tomatoes are a kind of health-stimulating fruit that are popular due to the rich antioxidant properties of their main compounds (Velioglu et al. 1998). Carotenes and phenolic compounds are the two main antioxidants in tomato. Because the ripening of tomatoes includes the breakdown of chlorophyll and the buildup of carotenes, their distribution within tomatoes is important for ripeness grading. Polder (2004) used hyperspectral imaging to determine the spatial distribution of the concentration of chlorophyll and carotenes in tomatoes for which reference compound concentrations were measured by HPLC. The resulting RMSEP values were 0.17, 0.25, 0.24, 0.31, and 0.29 for lycopene, lutein, b-carotene, chlorophyll-a, and chlorophyll-b, respectively. The results show the feasibility of determining the composition concentration in a space-preserving way using hyperspectral imaging. Moreover, after applying the established PLS model to the spectrum of each pixel, the distribution maps of individual compositions were generated, which can give a clear understanding of the spatial distribution of these compositions.

29.3.2.5 Mushrooms

As popular edible fungi with dietary and healthy benefits, white mushrooms (*Agaricus bisporus*) are valued for their white appearance, while a brown cap means low quality (Green et al. 2008), which is mainly caused by physical impact in the processes of picking, packaging, distribution, and storage. Gowen et al. (2008) demonstrated the potential of hyperspectral imaging for identifying vibration-induced damage on the mushroom surface. The hyperspectral image of each mushroom was analyzed by PCA, and the first several principal component score images were obtained. A data set comprised of 300 undamaged spectra and 300 vibration-damaged spectra was used to build a classifier for the separation of the spectra from sound and damaged tissues. These spectra were selected interactively at different points from elevation regions of normal or damaged mushroom, and the mean was then normalized. As a result, undamaged and damaged classes were well separated along PC1 in a score plot of PC1 versus PC2 for each spectrum.

29.3.2.6 Cucumbers

The quality measurement of cucumbers should be taken to reduce the economic loss of damage for the industry because cucumbers are prone to damage in the enlargement, harvest, transport, and processing stages. Ariana et al. (2006) designed a NIR hyperspectral imaging system in a spectral region of 900–1,700 nm for the detection of external bruises in pickling cucumbers. They found that the greatest reflectance difference between normal and bruised tissues was found in the region of 950–1,350 nm. When the first principal component was considered, the spectral region of 950–1,350 nm obtained the best classification accuracies for all days following mechanical stress. A good classification accuracy of 94.6 % was attained within 2 h of bruising, while the rate decreased over time to 74.6 % at day 6 after bruising. Moreover, to accelerate image acquisition and the process for the use of real-time application, fewer (two or three) wavelengths are required instead of full spectra. The researchers applied the calculation of the ratio or difference of two wavelengths followed by image segmentations using a threshold. Wavelengths at 988 and 1,085 nm with a threshold value of 0.79 were found to be the best for the ratio calculation, and their classification accuracy values were 81.8–92.7 % over a period of 0–6 days. On the other hand, the best wavelengths were 1,346 and 1,425 nm, and classification accuracy values of 83.6–92.7 % were obtained over a period of 0–6 days with a threshold value of 0.16.

29.3.3 Other Applications

Besides the potential for using a hyperspectral imaging technique for the quality evaluation of meats, fruits, and vegetables, the quality parameters of many other food products can be successfully determined on the basis of hyperspectral imaging, such as cereals, dairy products, maize, and nuts.

29.3.3.1 Wheat

Wheat is a widely consumed staple food around the world. Because the variety and grade of wheat are related to the market price, their identification is important for the wheat industry. Several factors affect the grading of wheat, such as physical and chemical properties, growing region and season, color, texture, nutritional content, and hardness/vitreousness. Hyperspectral imaging is considered an efficient and reliable tool for monitoring and classifying wheat samples. As early as 1998, a shortwave NIR imaging system (632–1,100 nm) was applied for the classification of wheat into color classes (Archibald et al. 1998). Shahin and Symons (2008) classified vitreous and nonvitreous wheat kernels using a NIR hyperspectral imaging technique. The detection of vitreous kernels is considered one of the most difficult quality indices for visual monitoring. The spectrum for each kernel was represented by the averaged reflectance spectra of kernel pixels. Savitzky–Golay smoothening and second derivative were used for spectral preprocessing. There are clear spectral differences in the preprocessed spectra between vitreous, starchy, piebald, and bleached kernels, which indicate the feasibility of building supervised classification models for the identification of vitreous wheat kernels with the aid of hyperspectral imaging.

29.3.3.2 Dairy Products

Nowadays, rapid and noninvasive alternatives are required for monitoring the quality of products during their production, and hyperspectral imaging can play an important role in the field of quality determination for dairy products Gowen et al. (2009). Burger and Geladi (2006) investigated the feasibility of using a hyperspectral imaging technique with a spectral range of 960–1,662 nm for the classification of 13 commercially produced cheeses containing varying amounts of protein, fat, and carbohydrate. Gowen et al. (2009) investigated the potential of hyperspectral imaging in the classification of cheese products. On the basis of its similarity to each mean spectrum, each pixel was classified as full fat or half fat. For comparison, an RGB image of the cheese samples studied was also acquired using a digital camera. A spectral angle mapping algorithm was used to classify each pixel of the hyperspectral image into one or more groups according to their spectral features. Finally, a visualization map was generated.

29.3.3.3 Maize

Maize kernel hardness, an important parameter for producers and processors in the grain trade, is generally a genetic expression but is also influenced by the environment and postharvest handling, such as drying, storage, handling, and processing. However, currently, most commonly used methods for assessing maize hardness are destructive. Williams et al. (2009) distinguished hard, intermediate, and soft maize kernels from inbred lines using a hyperspectral imaging technique. Following image preprocessing, PCA was used to effectively identify histological groups including glassy (hard) and floury (soft) endosperm. Two distinct clusters were obtained with the first three principal components calculated from the hyperspectral image data of the hyperspectral image data measured by the MatrixNIR camera and first two principal components of the hyperspectral images measured by the sisuChema camera, showing that a distinct difference between glassy and floury endosperm could be distinguished by a PCA process. Then a classification model was established using PLS-DA. The PLS-DA model had a RMSEP of 0.18 from the MatrixNIR image (12 kernels). The MatrixNIR image of the 24 kernels was further repeated, resulting in the same RMSEP. The repeatable results within the different data sets show that there is great potential for future classification uses with hyperspectral imaging for the accurate and noninvasive determination of maize hardness.

29.4 Conclusions

Hyperspectral imaging has already passed the phase of scientific curiosity and has developed significantly in the last two decades. It is now under dynamic investigation by researchers for more widespread utilization of this newly emerging technology in numerous fields. As a science-based automated food inspection technique, hyperspectral imaging can be used to reduce industrial dependence on human inspectors, improve product consistency and wholesomeness, increase production throughput, reduce production cost, and increase public confidence in the safety and quality of the food production and distribution system. Researchers have a responsibility to deploy this technology in various food science sectors to meet the urgent needs of the modern food industry for the nondestructive monitoring of food product quality. Therefore, significant opportunities and challenges will be presented to food technologists and food engineers in this research field. It is believed that hyperspectral imaging technology will play an important role in the food industry for real-time monitoring systems for food safety and quality control in the near future.

Acknowledgments The authors would like to acknowledge the financial support provided by the Irish Research Council for Science, Engineering and Technology dynamic investigations the Government of Ireland Postdoctoral Fellowship scheme.

References

Archibald DD, Thai CN, Dowell FE (1998) Development of shortwavelength near-infrared spectral imaging for grain color classification. In: Meyer GE, DeShazer JA (eds) Precision agriculture and biological quality. Proceedings of SPIE, Boston, MA, vol 3543, pp 189–198

Ariana DP, Lu R (2008) Quality evaluation of pickling cucumbers using hyperspectral reflectance and transmittance imaging- Part II. Performance of a prototype. Sens Instrum Food Qual 2:152–160

Ariana DP, Lu RF, Guyer DE (2006) Near-infrared hyperspectral reflectance imaging for detection of bruises on pickling cucumbers. Comput Electron Agric 53(1):60–70

Barbin DF, Elmasry G, Sun DW, Allen P (2012) Near-infrared hyperspectral imaging for grading and classification of pork. Meat Sci 90(1):259–268. doi:10.1016/j.meatsci.2011.07.011

Beucher S, Meyer F (1993) The morphological approach to segmentation: the watershed transformation. In: Dougherty ER (ed) Mathematical morphology in image processing. Marcel Dekker, New York, pp 433–481

Bremner HA, Brodersen K (2001) Exploration of the use of NIR reflectance spectroscopy to distinguish and measure attributes of conditioned and cooked shrimp (Pandalus borealis). Lebensm-Wiss Und-Technol-Food Sci Technol 34(8):533–541

Burger J, Geladi P (2005) Hyperspectral NIR image regression part 1: calibration and correction. J Chemometr 19(5–7):355–363

Burger J, Geladi P (2006) Hyperspectral NIR imaging for calibration and prediction: a comparison between image and spectrometer data for studying organic and biological samples. Analyst 131 (10):1152–1160

Cen HY, He Y (2007) Theory and application of near infrared reflectance spectroscopy in determination of food quality. Trends Food Sci Technol 18(2):72–83

Chao K, Yang CC, Chen YD, Kim MS, Chan DE (2007) Fast line-scan imaging system for broiler carcass inspection. Sens Instrum Food Qual 1:62–71

Chen X, Wu D, He Y, Liu S (2009) Detecting the quality of glycerol monolaurate: a method for using Fourier transform infrared spectroscopy with wavelet transform and modified uninformative variable elimination. Anal Chim Acta 638(1):16–22. doi:10.1016/j.aca.2009.02.002

Cluff K, Naganathan GK, Subbiah J, Lu R, Calkins CR, Samal A (2008) Optical scattering in beef steak to predict tenderness using hyperspectral imaging. Sens Instrum Food Qual 2:189–196

Cogdill RP, Dardenne P (2004) Least-squares support vector machines for chemometrics: an introduction and evaluation. J Near Infrared Spectrosc 12(2):93–100

Daugman JG (1980) Two-dimensional spectral-analysis of cortical receptive-field profiles. Vision Res 20(10):847–856

Daugman JG (1985) Uncertainty relation for resolution in space, spatial-frequency, and orientation optimized by two-dimensional visual cortical filters. J Opt Soc Am-Opt Image Sci Vis 2(7):1160–1169

Du CJ, Sun DW (2006) Learning techniques used in computer vision for food quality evaluation: a review. J Food Eng 72(1):39–55

ElMasry G, Wold JP (2008) High-speed assessment of fat and water content distribution in fish fillets using online imaging spectroscopy. J Agric Food Chem 56(17):7672–7677. doi:10.1021/Jf801074s

ElMasry G, Wang N, ElSayed A, Ngadi M (2007) Hyperspectral imaging for nondestructive determination of some quality attributes for strawberry. J Food Eng 81(1):98–107

ElMasry G, Wang N, Vigneault C, Qiao J, ElSayed A (2008) Early detection of apple bruises on different background colors using hyperspectral imaging. LWT- Food Sci Technol 41(2):337–345. doi:10.1016/j.lwt.2007.02.022

ElMasry G, Wang N, Vigneault C (2009) Detecting chilling injury in Red Delicious apple using hyperspectral imaging and neural networks. Postharvest Biol Technol 52(1):1–8

ElMasry G, Iqbal A, Sun DW, Allen P, Ward P (2011a) Quality classification of cooked, sliced turkey hams using NIR hyperspectral imaging system. J Food Eng 103(3):333–344

ElMasry G, Sun DW, Allen P (2011b) Non-destructive determination of water-holding capacity in fresh beef by using NIR hyperspectral imaging. Food Res Int 44(9):2624–2633. doi:10.1016/j.foodres.2011.05.001

Fathi M, Mohebbi M, Razavi SM (2009) Application of image analysis and artificial neural network to predict mass transfer kinetics and color changes of osmotically dehydrated kiwifruit. Food Bioprocess Technol. doi:10.1007/s11947-009-0222-y

Fletcher JT, Kong SG (2003) Principal component analysis for poultry tumor inspection using hyperspectral fluorescence imaging. In: Proceedings of the international joint conference on neural networks 2003, Portland, OR, vols 1–4, 1(1), pp 149–153

Gerlach RW, Kowalski BR, Wold HOA (1979) Partial least-squares path modeling with latent-variables. Anal Chimica Acta-Comput Tech Optim 3(4):417–421

Gomez-Sanchis J, Gomez-Chova L, Aleixos N, Camps-Valls G, Montesinos-Herrero C, Molto E, Blasco J (2008) Hyperspectral system for early detection of rottenness caused by Penicillium digitatum in mandarins. J Food Eng 89(1):80–86

Gowen AA, O'Donnell CP, Taghizadeh M, Cullen PJ, Frias JM, Downey G (2008) Hyperspectral imaging combined with principal component analysis for bruise damage detection on white mushrooms (Agaricus bisporus). J Chemometr 22(3–4):259–267

Gowen AA, Burger J, O'Callaghan D, O'Donnell CP (2009) Potential applications of hyperspectral imaging for quality control in dairy foods. In: (eds) 1st international workshop on computer image analysis in agriculture, Potsdam, Germany

Green JM, Grogan H, Eastwood DC, Burton KS (2008) Investigating genetic and environmental control of brown color development in the cultivated mushroom *Agaricus bisporus* infected with mushroom virus X. Int Soc Mushroom Sci 17:41

Haralick RM, Shapiro LG (1992) Computer and robot vision. Addison–Wesley, Boston

Hasegawa Y (2000) Merits and demerits of the automated sweetness sorting techniques. Fresh Food Syst 30:74–77

Huidobro A, Pastor A, Lopez-Caballero ME, Tejada M (2001) Washing effect on the quality index method (QIM) developed for raw gilthead seabream (Sparus aurata). Eur Food Res Technol 212(4):408–412

IEEE Standard 601.4-1990 (1990) IEEE standard glossary of image processing and pattern recognition terminology. IEEE Press, Los Alamitos

Kamruzzaman M, ElMasry G, Sun DW, Allen P (2011) Application of NIR hyperspectral imaging for discrimination of lamb muscles. J Food Eng 104(3):332–340

Kavdir I, Guyer DE (2008) Evaluation of different pattern recognition techniques for apple sorting. Biosyst Eng 99(2):211–219

Litwiller D (2005) CMOs vs. CCD: maturing technologies, maturing markets. Photon Spectra 39(8):54–61

Liu F, Song XD, Luo YP, Hu DC (2002) Adaptive thresholding based on variational background. Electron Lett 38(18):1017–1018

Liu YL, Chen YR, Kim MS, Chan DE, Lefcourt AM (2007) Development of simple algorithms for the detection of fecal contaminants on apples from visible/near infrared hyperspectral reflectance imaging. J Food Eng 81(2):412–418

Naganathan GK, Grimes LM, Subbiah J, Calkins CR, Samal A, Meyer GE (2008a) Partial least squares analysis of near-infrared hyperspectral images for beef tenderness prediction. Sens Instrum Food Qual Saf 2:178–188

Naganathan GK, Grimes LM, Subbiah J, Calkins CR, Samal A, Meyer GE (2008b) Visible/near-infrared hyperspectral imaging for beef tenderness prediction. Comput Electron Agric 64(2):225–233

Nagata M, Tallada JG, Kobayashi T (2006) Bruise detection using NIR hyperspectral imaging for strawberry (Fragaria × ananassa Duch.). Environ Control Biol 44(2):133–142

Noh HK, Lu RF (2007) Hyperspectral laser-induced fluorescence imaging for assessing apple fruit quality. Postharvest Biol Technol 43(2):193–201

Park B, Lawrence KC, Windham WR, Buhr RJ (2002) Hyperspectral imaging for detecting fecal and ingesta contamination on poultry carcasses. Trans ASAE 45(6):2017–2026

Park B, Kise M, Lawrence KC, Windham WR, Smith DP, Thai CN (2006a) Real-time multispectral imaging system for online poultry fecal inspection using UML. In: Chen Y-R, Meyer GE, Tu S-I (eds) Optics for natural resources, agriculture, and foods. Proceeding of SPIE, Boston, MA, pp 63810W-1–63810W-12

Park B, Lawrence KC, Windham WR, Smith DP (2006b) Performance of hyperspectral imaging system for poultry surface fecal contaminant detection. J Food Eng 75(3):340–348

Park B, Yoon SC, Lawrence KC, Windham WR (2007) Fisher linear discriminant analysis for improving fecal detection accuracy with hyperspectral images. Trans ASABE 50(6):2275–2283

Polder G (2004) Spectral imaging for measuring biochemicals in plant material. PhD Thesis, Delft University of Technology

Qiao J, Ngadi MO, Wang N, Gariepy C, Prasher SO (2007a) Pork quality and marbling level assessment using a hyperspectral imaging system. J Food Eng 83(1):10–16

Qiao J, Wang N, Ngadi MO, Gunenc A, Monroy M, Gariepy C, Prasher SO (2007b) Prediction of drip-loss, pH, and color for pork using a hyperspectral imaging technique. Meat Sci 76(1):1–8

Qin JW, Burks TF, Ritenour MA, Bonn WG (2009) Detection of citrus canker using hyperspectral reflectance imaging with spectral information divergence. J Food Eng 93(2):183–191. doi:10.1016/j.jfoodeng.2009.01.014

Rinnan A, van den Berg F, Engelsen SB (2009) Review of the most common pre-processing techniques for near-infrared spectra. Trac-Trend Anal Chem 28(10):1201–1222

Schaare PN, Fraser DG (2000) Comparison of reflectance, interactance and transmission modes of visible-near infrared spectroscopy for measuring internal properties of kiwifruit (Actinidia chinensis). Postharvest Biol Technol 20(2):175–184

Shahin MA, Symons SJ (2008) Detection of hard vitreous and starchy kernels in amber durum wheat samples using hyperspectral imaging. NIR News 19(5):16–18

Tsuta M, Sugiyama J, Sagara Y (2002) Near-infrared imaging spectroscopy based on sugar absorption band for melons. J Agric Food Chem 50(1):48–52

Velioglu YS, Mazza G, Gao L, Oomah BD (1998) Antioxidant activity and total phenolics in selected fruits, vegetables, and grain products. J Agric Food Chem 46(10):4113–4117

Williams P, Geladi P, Fox G, Manley M (2009) Maize kernel hardness classification by near infrared (NIR) hyperspectral imaging and multivariate data analysis. Anal Chim Acta 653 (2):121–130

Wold JP, Johansen IR, Haugholt KH, Tschudi J, Thielemann J, Segtnan VH, Narum B, Wold E (2006) Non-contact transflectance near infrared imaging for representative on-line sampling of dried salted coalfish (bacalao). J Near Infrared Spectrosc 14(1):59–66

Wu D, Feng S, He Y (2007) Infrared spectroscopy technique for the nondestructive measurement of fat content in milk powder. J Dairy Sci 90(8):3613–3619

Wu D, Feng L, He Y, Bao Y (2008a) Variety identification of Chinese cabbage seeds using visible and near-infrared spectroscopy. Trans ASABE 51(6):2193–2199

Wu D, Feng L, Zhang C, He Y (2008b) Early detection of Botrytis cinerea on eggplant leaves based on visible and near-infrared spectroscopy. Trans ASABE 51(3):1133–1139

Wu D, He Y, Feng S (2008c) Short-wave near-infrared spectroscopy analysis of major compounds in milk powder and wavelength assignment. Anal Chim Acta 610(2):232–242

Wu D, He Y, Feng SJ, Sun DW (2008d) Study on infrared spectroscopy technique for fast measurement of protein content in milk powder based on LS-SVM. J Food Eng 84(1):124–131

Wu D, Yang HQ, Chen XJ, He Y, Li XL (2008e) Application of image texture for the sorting of tea categories using multi-spectral imaging technique and support vector machine. J Food Eng 88 (4):474–483

Wu D, He Y, Nie PC, Cao F, Bao YD (2010) Hybrid variable selection in visible and near-infrared spectral analysis for non-invasive quality determination of grape juice. Anal Chim Acta 659 (1–2):229–237

Wu D, Chen XJ, Zhu XG, Guan XC, Wu GC (2011a) Uninformative variable elimination for improvement of successive projections algorithm on spectral multivariable selection with different calibration algorithms for the rapid and non-destructive determination of protein content in dried laver. Anal Methods 3(8):1790–1796

Wu D, Nie PC, Cuello J, He Y, Wang ZP, Wu HX (2011b) Application of visible and near infrared spectroscopy for rapid and non-invasive quantification of common adulterants in *Spirulina* powder. J Food Eng 102(3):278–286

Wu D, Nie PC, He Y, Bao YD (2011c) Determination of calcium content in powdered milk using near and mid-infrared spectroscopy with variable selection and chemometrics. Food Bioprocess Technol. doi:10.1007/s11947-010-0492-4

Yang CC, Chao KL, Kim MS (2009) Machine vision system for online inspection of freshly slaughtered chickens. Sens Instrum Food Qual Saf 3:70–80

Zhao JW, Vittayapadung S, Chen QS, Chaitep S, Chuaviroj R (2009) Nondestructive measurement of sugar content of apple using hyperspectral imaging technique. Maejo Int J Sci Technol 3 (1):130–142

Zou XB, Zhao JW, Povey MJW, Holmes M, Mao HP (2010) Variables selection methods in near-infrared spectroscopy. Anal Chim Acta 667(1–2):14–32

Chapter 30
Food Chain Safety Management Systems: The Impact of Good Practices

Raspor Peter, Ambrožič Mateja, and Jevšnik Mojca

30.1 Introduction

Quality management systems have reached a high level of professionalism in recent years and made a great impact on food systems. Food technology as a profession is responsible for the technical aspects of the development of food products, food processes, and the distribution of these products through the entire food chain to consumers. Since the ultimate target of these efforts is the satisfaction of consumer demands, it is essential to consider not only objective consumer needs (e.g. nutrition, safety, affordability) but subjective aspects of consumer satisfaction as well (e.g. sensorial properties and consumer attitudes). All this falls within the domain of consumer needs and wants, where some of the most difficult problems in fostering a rational food technology and nutrition system have arisen in recent years. Particularly important are the quantity and quality of food available to human populations, which plays a significant role in human evolution. Human existence is based on a knowledge of how to obtain enough food and how to get it right. The history of food safety is probably almost as old as human history itself. On the other hand, outbreaks of food pathogens and exaggerated concerns about food safety have become a permanent preoccupation of modern consumers. It is obvious that we must start to correlate the hard and soft sciences and share findings, experiences, and skills with consumers in an effort to engage all elements of the food supply chain.

R. Peter (✉) • A. Mateja
Chair of Biotechnology, Microbiology and Food Safety, Department of Food Science and Technology, Biotechnical Faculty, University of Ljubljana, Jamnikarjeva 101, SI-1000 Ljubljana, Slovenia
e-mail: peter.raspor@bf.uni-lj.si

J. Mojca
Department of Sanitary Engineering, Faculty of Health Sciences, University of Ljubljana, Poljanska 26a, SI-1000 Ljubljana, Slovenia

S. Yanniotis et al. (eds.), *Advances in Food Process Engineering Research and Applications*, Food Engineering Series, DOI 10.1007/978-1-4614-7906-2_30,

30.2 Food Supply Chain

Throughout the world, food production processes are fragmented. Food reaches consumers via supply chains that link many different types of organizations and stretch across multiple borders. The food supply chains encompass a sequence of players (Fig. 30.1), which add value to the product at each stage. A food supply chain is a network of food-related businesses involved in the creation and consumption of food products, where food products move from farm to table (Selvan 2008). All players in each food supply chain are linked by information, material and capital flows. One weak link can result in unsafe food that is dangerous to human health. Traceability is a very important element in a food supply chain to ensure product and process integrity, improve consumer trust and maintain quality standards (EC 178/2002; Trienekens and van der Vorst 2006). There are global and local supply chains, and each has its own features and rules, but in principle they must obey World Trade Organization (WTO) principles and the federal laws of countries where food is to be consumed.

30.2.1 Food Supply Chain Hazards

Quality and safety in a food supply chain are among the most important issues facing the food sector. The process of globalization, industrialization and centralization of

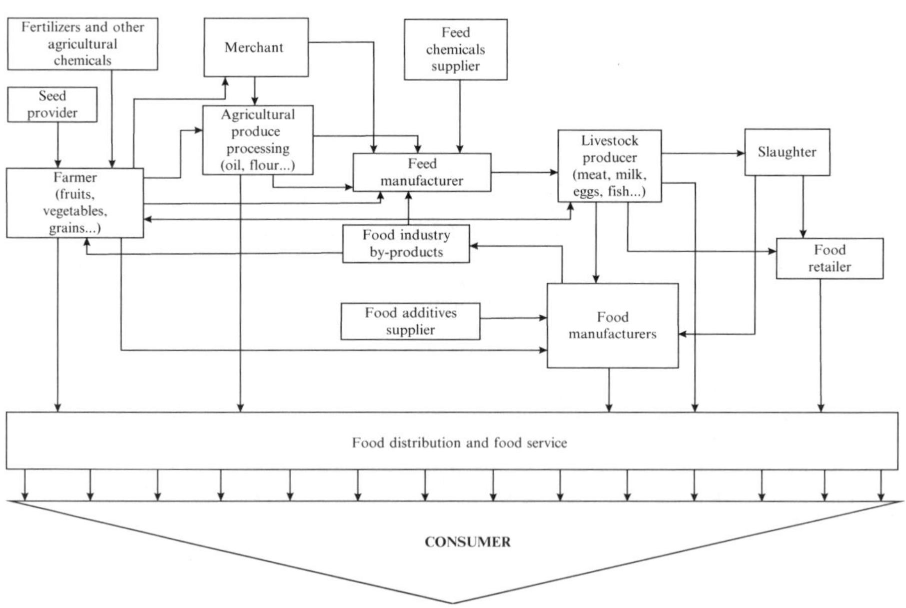

Fig. 30.1 Food supply chain complexity

food systems has changed the structure of those systems within a short period of time. A few decades ago the risks of food-borne illness were reduced by cooking and eating foods immediately after preparation or preserving them through drying, cooling or fermentation. Developments in food supply chains, like globalization, health and demographic trends, social situations (increased consumption of ready-to-eat foods on one hand and raw or minimally processed foods on the other) and environmental conditions (e.g. pollution, lack of drinking water) present new challenges for food safety. A globally integrated farm-to-table approach and new consumption patterns make it possible for a contaminated food product to be consumed by a large number of people worldwide in a short period of time. The public and private sectors require the implementation of coordinated systems such as a hazard analysis and critical control point (HACCP) system, but such systems are focused mainly on processing. Currently, more emphasis is being placed on identifying hazards at the preharvest stage to identify options for preventing hazards from entering the supply chain. But if hazards do enter into the food supply chain, they should be identified at an early stage before they have the chance to develop into real risks.

Food safety is concerned with the absence of contaminants or the presence, at acceptable and safe levels, of contaminants, adulterants, naturally occurring toxins or any other substances that could make food dangerous to human health. In food safety programs, each actor should be able to identify all hazards, analyse them, assess the likelihood of their occurrence and identify measures for their control. Food safety hazards can enter the food supply chain unexpectedly and anywhere during the growth, processing, storage, distribution and retail sale of food and at the end of the process during preparation activities in restaurants, cafes and catering establishments, as well as in residential kitchens. The occurrence of food safety hazards in food supply chains is connected with the presence of biological, chemical or physical agents in or condition of, food or feed with the potential to cause an adverse health effect (EC 178/2002; CAC 2003; EC 852/2004).

Food hazards are present at all stages in the food supply chain and should be identified and adequately controlled to ensure the safety and healthiness of food for consumers. However, in the case of the direct supply of small quantities of primary products, from the food business operator producing them to the final consumer or to a local retail establishment, it is appropriate to protect public health through national legislation, in particular because of the close relationship between producers and consumers (EC 852/2004).

Food safety hazards may be biological in origin, for example insects, rodents, pathogenic bacteria, parasites, prions and viruses (Lawley et al. 2008). The spectrum of food-borne infections has changed dramatically over time as well-established pathogens have been controlled or eliminated (Tauxe 2002) and new ones have emerged due to changed consumer eating habits and patterns. Microbial food safety is considered a significant public health issue but historically has focused mostly on the control of bacterial contamination. Lately, enteric viruses have been increasingly recognized as an important cause of food-borne disease, and control measures are being developed (Ambrožic et al. 2011).

We should be aware that food may be contaminated by various microorganisms and that they might increase at various stages from production to processing and, finally, in the last step of final preparation in the consumer's home (Tauxe 2002). Therefore, for all involved in a food supply chain, it is imperative to know and understand the mechanisms of microbiological hazards and ways to prevent or control them, which certainly applies to the final step of the food supply chain – the consumer (Raspor and Jevšnik 2008).

Hazards of a chemical origin in food are also a cause of great concern. Once a chemical contaminant is present in a food product, the food product cannot be decontaminated (Swanenburg et al. 2011). Appropriate measures implemented earlier (e.g. good quality raw materials) in the supply chain can prevent the presence of this type of hazard in later stages. Chemical contaminants (De Meulenaer 2006; Lawley et al. 2008) important in food safety are agricultural chemicals (e.g. pesticides, herbicides, fertilizer residues), veterinary drugs (e.-g. growth promoters, antibiotics), natural biological toxins (e.g. mycotoxins, fish toxins), environmental contaminants (e.g. mercury, lead), contaminants produced during processing (e.g. acrylamid), contaminants from food contact materials (e.-g. packaging materials, packaging inks) and cleaning and sanitizing chemicals and adulterants (e.g. melamine, illegal food dyes). Chemicals are also added to foods during processing in the form of food additives, flavourings and colours. Due to adaptation and resistance developed by (micro)organisms to chemicals, ever greater amounts of chemicals and new chemical compounds are used (Carvalho 2006).

When considering chemical hazards, food allergens should also be taken into account. In recent years, the problem of food allergies has increased. Food manufacturers have been encouraged to respond to this development by labelling foods very clearly. The control of allergens (e.g. cereals containing gluten, crustaceans, eggs, peanuts, soybeans, milk and sulphur dioxide) in food is now a rapidly developing aspect of food safety management (Lawley et al. 2008).

Physical hazards result from the inadvertent inclusion of harmful extraneous materials in the final product like machinery parts, wires, stones, wooden splinters, pens, paper clips and pencils. The majority of all reported incidents of illness or injury related to physical contaminants involves dental complaints and oral injury, but these incidents are nevertheless seldom life-threatening (Aladjadjiyan 2006).

Most conventionally produced food is extremely resource intensive. Industrialized production and processing methods use huge amounts of water and chemicals like herbicides, insecticides and fertilizers and produce tons of waste products that accumulate and pollute the land, water and air. That is why households with secure quantities of food focus mostly on food safety and health aspects (e.g. organic production of food), environmental conditions, geographic and social affinities (e.g. locally grown products, support of 'small' producers, fair trade considerations), religious affinity (e.g. halal, kosher food certificates) and animal welfare aspects. But we must be aware that novelties can incorporate new or even bring back old food safety hazards into the food supply chain without our awareness. This can be demonstrated by suggestions that the application of manure and reduced use of fungicides and antibiotics in organic farming could result in greater contamination of organic foods by microorganisms or microbial products (Williams 2002).

A currently hot topic in food safety is nanotechnology and its impact on food safety. Recently nanobiotechnology has also found application in food process technologies. Due to their small size, nanomaterials exhibit novel features that offer considerable opportunities for the development of innovative products and applications in the food sector to produce potentially safer and healthier products (Duncan 2011). Nanotechnology may be used in agriculture and food production (e.g. nanosensors for monitoring crop growth pesticide, fertilizer or vaccine delivery, animal and plant pathogen detection, targeted genetic engineering), food processing (e.g. encapsulation of flavour or odour enhancers, food texture or quality improvement, new gelation or viscosifying agents), quality control and testing (e.g. nanosensors), food packaging and storage (e.g. pathogen, gas or abuse sensors) and nutritional supplements (e.g. nutraceuticals with higher stability and bioavailability) (Duncan 2011). However, safety issues surrounding the use of nanotechnology in food have raised public and scientific concerns.

30.3 Food Safety Management Systems in Food Supply Chains

Food systems have been facing huge challenges due to several incidents of contaminated food (*E. coli* in alfalfa sprouts, the melamine incident), technology changes like irradiation, high hydrostatic pressure (HHP), nanotechnology, increasing operational complexity (globalization of food supply chains), frequently changing consumer needs, government regulations and dynamic market conditions. To build and maintain the trust of consumers with respect to food quality and safety, quality management systems (QMSs) are used to control the quality and safety of food products (Luning et al. 2006). At the same time, food systems are experiencing new possibilities due to fast technical and technological progress in the developed world such as novel plant products, processing without heat treatment, in situ control of products with biosensors, fast and precise detection of unwanted/undesired food-grade microorganisms, control of fluxes to reduce unwanted side reactions, intelligent packaging and polymer restructuring for increased shelf life and novel eatables, and, last but not least, the use of new approaches in food preservation and distribution, not to mention convenience foods and their effect on human health and well-being.

As a consequence, many countries and large international retailers have launched initiatives to enhance food quality. The increased demand for safer foods has resulted in the development and introduction of new food safety standards and regulations in an attempt to attain a higher level of food safety.

Companies around the world are using quality assurance systems to control, build and maintain the quality and safety of food products and the trust of consumers in food quality and food safety. Quality assurance is required at each step in the food production chain to ensure safe food and to show compliance with regulatory, retailer and customer requirements. In this development there has been a move from the previous end-of-line product inspection approach to an in-line inspection approach. Quality assurance is a modern term describing the control, evaluation and audit of a food processing system. It consists of the integration of all functions and processes within an organization to achieve continuous improvement in the quality of goods and services (Vasconcellos 2004).

The term food safety management system (FSMS) is composed of two concepts: food safety and management system. Food safety is defined according to the Codex Alimentarius (CAC 2003) as the assurance that food will not harm the consumer when it is prepared or eaten according to its intended use, which in reality is not always the case (Jevšnik et al. 2008). Quality management system refers to all activities that organizations use to direct, control and coordinate quality, including formulating a quality policy, setting quality objectives, quality planning, control, assurance and improvement (ISO 9000 2005). A food safety management system involves that part of the QMS specially focused on food safety (Raspor and Ambrožič 2011).

In recent decades many public and private food standards on food safety and quality have been established. The main purpose of private standards is to protect their business and that of public standards is to protect consumers (Trienekens and Zuurbier 2008). All standards have their origin in legislation and particularly in the Codex Alimentarius philosophy.

30.3.1 Public Standards Regulating Food Safety

At the national and international levels many government laws and regulations on the quality and safety of food have also been established through public standards like the Codex Alimentarius Commission, which focuses primarily on risks due to food hazards. On a global level, the Codex Alimentarius Commission, established by the Food and Agriculture Organization (FAO) and the World Health Organization (WHO) in 1961, has become the single most important international reference point for developments associated with food standards. The Codex Alimentarius Commission is committed to protecting the health of consumers, ensuring fair practices in the food trade and facilitating international trade in food (CAC 2006). Codex Alimentarius food standard issues range from specific raw and processed material characteristics to food hygiene, pesticide residues, contaminants and labelling to analysis and sampling methods (CAC 2006). Codex Alimentarius codes of practice are a collection of internationally adopted food standards presented in a uniform manner. The first developed Codex code of good practice was good hygiene practice, which was adopted in 1969 and still represents a solid foundation for ensuring food hygiene from farm to fork, highlighting the key hygiene controls at each stage (CAC 2003).

30.3.2 Private Standards Regulating Food Quality and Safety

Private standards such as International Organization for Standardization (ISO) standards, International Food Standard (IFS), British Retail Consortium (BRC) and Safe Quality Food (SQF) often consider food hazards as well as environmental, ethical, occupational health issues and other social responsibility issues. Private standards are driven by the food industry, retail buyers, buyer organizations, commodity groups, non-governmental organizations (NGOs) and others because the key factors driving private standards are brand protection, business improvement and efficiency and assisting in the response to consumer concerns (Henson and Reardon 2005; Fulponi 2006; Trienekens and Zuurbier 2008).

The major aims of private food standards are to (Vellema and Boselie 2003):

- Improve supplier standards and consistency and avoid product failure;
- Eliminate multiple audits of food suppliers/manufacturers through certification of their processes;
- Support consumer and retailer objectives by transferring their demands to parties upstream the chain; and
- Be able to provide concise information about production processes in case of food incidents.

Demands regarding private food safety standards are best represented by the Global Food Safety Initiative (GFSI) worldwide forum. This forum (GFSI 2011) is a global food network made up of approximately 400 retailers and manufacturers around the world. The rising demands of consumers, the increased liabilities of retailers and wholesalers, the increase in legal requirements and the globalization of product supplies all made it essential to develop a uniform quality assurance and food safety standard. Supplier audits have been a permanent feature of retailers' systems and procedures for many years. In the past they were performed by the quality assurance departments of the individual retailers and wholesalers. The GFSI was launched in May 2000 to harmonize international food safety standards and reduce the need for multiple supplier audits. The food safety schemes currently benchmarked by GFSI are divided into manufacturing schemes like BRC Global Standard Version 5, Dutch HACCP, ISO 22000, and IFS; primary production schemes like Global GAP; and Primary and Manufacturing schemes like PrimusGFS (GFSI 2011). The BRC Global Standard for Food Safety was the first standard to be approved by GFSI and was developed in 1998 by the food service industry to enable suppliers to be audited (BRC 2011). IFS Standard Food started in 2003 for auditing retailer and wholesaler branded food product suppliers (IFS 2011).

ISO was born in 1946 with the objective of facilitating the international coordination and unification of industrial standards. ISO is an NGO that forms a bridge between the public and private sectors. ISO is the world's largest standards development organization, ranging from standards for activities such as agriculture and construction, through mechanical engineering, to medical devices and the newest information technology developments (ISO 2011).

ISO standards are developed by technical committees (TCs). For efficiency they may establish subcommittees (SCs) and working groups (WGs). ISO TC 34 is responsible for the standardization of food products. The activity of TC 34 and its 17 SC covers standardization in the field of human and animal foodstuffs, covering the food chain from primary production to consumption, as well as animal and vegetable propagation materials, in particular, but not limited to, terminology, sampling, methods of test and analysis, product specifications, food and feed safety and quality management and requirements for packaging, storage and transportation. Till the end of 2011 781 standards fell under the responsibility of TC 34. The TCs for food products were established in 1947 (ISO 2011). In 2009, the subcommittee for the management systems for food safety was created (TC34/SC 17). Under its direct responsibility it published the ISO 22000 standards family, which was launched in September 2005. But as early as 1979 the TC for standardization in the field of quality management and quality assurance had already been established (ISO 2011).

EurepGAP started in 1997 as an initiative by retailers belonging to the Euro-Retailer Produce Working Group (EUREP), with British retailers in conjunction with supermarkets in continental Europe as the driving forces. They reacted to the growing concerns of consumers regarding product safety and environmental and labour standards and decided to harmonize their own often very different standards. EUREP started working on harmonized standards and procedures for the development of Good Agricultural Practices in conventional agriculture. In the 10 years following 1997, EurepGAP began to gain in global significance. In 2007, EurepGAP was renamed as GLOBALG.A.P. The GLOBALG.A.P. standard is primarily designed to reassure consumers about how food is produced on the farm by minimizing the detrimental environmental effects of farming operations, reducing the use of chemical inputs and ensuring a responsible approach to worker health and safety as well as animal welfare (GLOBALG.A.P. 2011)

Since the 1990s there has been an increase in food standards seven (Fig. 30.2). This can be interpreted by the fact that in the 1990s major world retailers like Walmart, Tesco and Carrefour entered international markets (Walmart 2011; Carrefour 2011; Tesco 2011). With these activities the food supply chains, especially retail chains, expanded. These activities required an increase in the number of suppliers that could satisfy the needs of customers worldwide, but they were obligated to produce equal-quality products regardless of the market for which they supplied food.

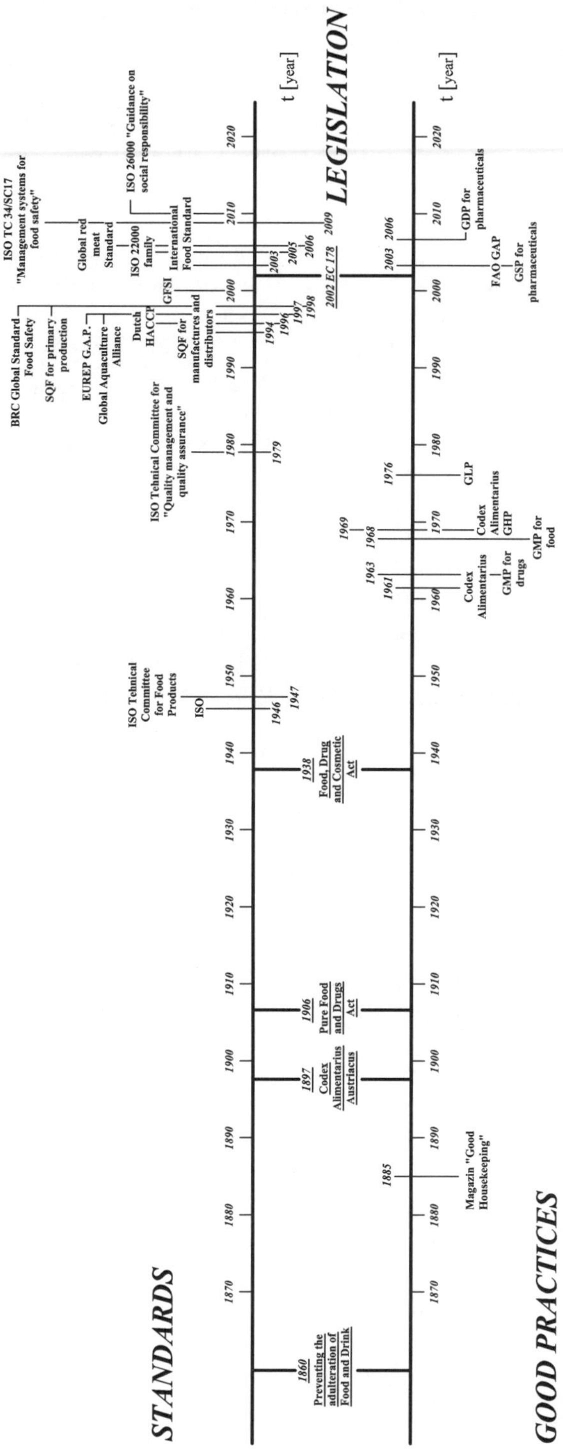

Fig. 30.2 Corresponding development of good practices, relevant standards and legislation to support quality and safety in food supply chain

30.4 Good Practices as Basics for Successful Food Safety Management Systems

Today's food industry and its sophisticated processing and distribution technologies produce a variety of foodstuffs available to the consumer in the form of various articles on the shelves of fast-growing commercial centres. With the development of food science and technology comes more complete knowledge of health risks, but the new interventions in technology and distribution have caused new risks. The high level of public health is one of the fundamental objectives of food legislation (EC 178/2002).

Good practices or prerequisite programs (ISO 22000 2005) are an essential element, but they are often neglected in the task of developing simple, effective FSMSs. Good practices are the basic conditions and activities that are necessary to maintain a hygienic environment throughout the food chain that is suitable for the production, handling and provision of safe end-products and food that is safe for human consumption (ISO 22000 2005).

Prior to the application of hazard analysis and critical control points (HACCPs) to any sector of the food chain, that sector should have in place prerequisite programmes (PRPs) (CAC 2003) such as good hygienic practices, appropriate codes of practice and appropriate food safety requirements (Raspor 2002, 2004). HACCP must be supported by a strong foundation of PRPs (Raspor 2008). PRPs or, even better, pre-requirements to HACCP may include (Sperber 1998) the following:

- Specifications for raw materials, finished products, and labelling;
- Supplier approval or certification, chemical control programmes;
- Audits and inspections;
- Product identification and retrieval procedures;
- Training;
- Water and air control, good manufacturing practices.

These PRPs to HACCP should be well established, fully operational and verified in order to facilitate the successful application and implementation of the HACCP system. Which PRP is needed depends on the segment of the food chain in which the organization operates and the type of organization (Raspor and Ambrožič 2011).

30.4.1 Historical View of Development of Good Practices

Humans' lives have been inseparably connected with the food they eat. From a historical point of view the challenge for humans has been to secure a sufficient supply of food to prevent hunger and starvation. All humans rely on food for sustenance and survival, but food has also formed and shaped culture and civilization. Several instructions on the manner of handling food are contained in the Old Testament. During the nineteenth century major changes in food production occurred. Due to industrialization and new distribution chains, which were connected to the rapid growth in urban populations together with a public health problem, food laws were created. Food laws were mostly the responsibility of local and state regulators; however, at that time knowledge and understanding of hygiene and the dangers of food adulterations was disregarded. The first comprehensive modern food law was the 1860 act for "Preventing the Adulteration of Food and Drink" in England (Lasztity et al. 2004), but it was voluntary and did not work. In 1875 it was replaced by a new act, which was mandatory. In the Austro-Hungarian empire between 1897 and 1911, a collection of standards and product description for a wide variety of foods was developed as the Codex Alimentarius Austriacus. The Codex Alimentarius Austriacus was not, strictly speaking, a collection of legally enforceable food standards. It lent its name to the present-day international Codex Alimentarius Commission (Randell 1995). However in the USA the first major federal consumer protection law with respect to food processing was the Pure Food and Drugs Act of 1906. The law prevented interstate and foreign commerce in misbranded or adulterated foods, drinks or drugs. The intent of the act was to prevent poisoning and consumer fraud. As more food products were manufactured in subsequent years, however, poor-quality food products and deceptive packaging continued to be produced due to loopholes in the law. In 1938 the Food Drug, and Cosmetic Act, which provides identity and quality standards for food, replaced the 1906 act (FDA 2011). After World War II, activity in international standardization started to grow intensively.

Tradition, practice and numerous technical and scientific advances helped shape the principles and techniques of how to achieve, in a given environment, acceptable food safety. Heterogeneous environmental conditions, a wealth of different materials, diversity of cultures and ways of work practical work helped shape the principles, some of which were later involved in legislation. Today food safety is managed through good practices at different levels of food production, distribution and consumption. The current maintenance of food safety in the food supply chain can be easily broken down because of the different kinds of barriers or simple misunderstandings among stakeholders including consumers (Raspor and Jevšnik 2008). HACCP represents the clearest example of this development (Raspor 2004). The previous quality control system was based on the finished product. The new food safety philosophy is based on the appropriateness of technological process in the chain through which food passes, which significantly reduces the risk of inadequate health of the final product (Sperber 2005a, b).

. Current good practices, parameters involved in food safety along the food supply chain, and consumer food safety dilemmas are presented. It has been shown that the current system of maintaining food safety in the food supply chain can easily break down because of the different kinds of barriers or simple misunderstandings and inconsistent terminology (Ambrožič et al. 2010) among stakeholders including consumers. As a result of the current situation, a new approach, called Good Nutritional Practice (GNP), has been proposed to balance the different food safety systems. It is shown how important it is to integrate actual food safety solutions within GNP, which includes consumers and is based on a model that covers subsystems from other relevant good practices (nine good practices in the food supply chain).

Food-borne diseases encompass a wide spectrum of illnesses which are a growing public health problem worldwide. Food-borne diseases not only significantly affect human health and well-being, but they also have economic consequences for individuals, families, communities, businesses and nations. According to the epidemiological data of WHO from 2010, food-borne diseases affect millions of people each year, in both developed and underdeveloped countries (Behrens et al. 2010). In the European Union (EU) one-third (36.4 %) of reported food-borne outbreaks are caused by improper handling of food in households, followed by restaurants, cafes, bars and hotels (20.6 %), schools and kindergartens (5.5 %) (EFSA 2011). *Salmonella* remains the most frequently detected causative agent in food-borne outbreaks reported in the EU. Developing countries bear the brunt of the problem due to the contamination of food and drinking water, which suggests to underlying major food safety problems.

Food safety is one of the highest public health priorities at national, community and international levels because eating and drinking are among the few things that are essential for everyone. It is a result of legislation for laying minimum requirements, official controls for checking food business operator's compliance and conscious food business operators (EC 178/2002). To achieve food safety for consumers, industry and government regulatory bodies must follow the recommendations of the United Nations Codex Alimentarius Commission, which is the global reference point for consumers, food producers and processors, national food control agencies and the international food trade (CAC 2006).

In terms of human health, safe food is a basic right of consumers. Assuring safe food is the most difficult task in preparation and distribution units, especially in small and medium-size companies (Walker and Jones 2002; Walker et al. 2003; Walczak and Reuter 2002; Sun and Ockerman 2005). One of the principal actions has been the development of HACCP (hazard analysis and critical control point)-based regulations or recommendations by federal agencies and the United Nations Codex Alimentarius Commission (Sperber 1998). To control and understand safety in the EU the *White Paper on Food Safety* is an important document that was published in January 2000 (EC 2000). The pursuit of a high level of protection of human life and health is one of the fundamental objectives of food legislation, as laid down in (EC) Regulation 178/2002. The principal objective of the new general and specific hygiene rules is to ensure a high level of consumer protection with regard to food safety (EC 852/2004). After that regulation 178/2002/EC and decision 97/579/EC were published, which exactly define the European Food Safety Authority. The use of HACCP principles at all levels of the food chain is, however, compulsory under EU Directive 93/43/EEC and Regulation 852/2004/EC (EU 1993; EC 852/2004). It is a responsibility of all concerned parties in the food chain to ensure food traceability and food safety by internal controls in all production phases. These safety assurance systems are based on the production process, the complexity of the product and human resources.

30.4.2 Introduction of Current Good Practices

In ancient times when food safety was the sole responsibility of the hunter/gatherer, the chain of responsibility was a very short one. Today, with important changes in lifestyles and demography and with globalization of food trade, we see the food supply growing ever more rapidly in size and diversity (Gorris 2005). To ensure farm-to-table food safety, it was necessary to establish a new concept for completely understanding food safety (Sperber 2005b). The HACCP system and its supporting programmes (good practices) represent the most intelligible example of this development (Raspor 2002).

Good practices are described in several different codes of practice designed by producer organizations, importers and retail consortia, and government bodies representing consumers. In general, good practice means the activity of quality assurance, which ensures that food products and food-related processes are consistent and controlled to assure quality procedures in food systems (Raspor 2008).

The development of good practices (Fig. 30.2) in the last 15 years has enabled the integration of all activities in the food supply chain, specific for each individual branch (Heggum 2001; Raspor 2004). Manufacturers started to think about the integrity control of individual stages and activities in the food supply chain. Positive experiences have developed, and it is known as Good Manufacturing Practice (GMP). From its first rules and principles the WHO set the course about meaning of enacting standard procedures dealing with personnel, building, equipment, documentation, production and quality control (Zschler 1989). The first published GMP in manufacture, processing packaging or holding for came out in 1963 for drugs (U. S. Federal Register 1963), but in 1968 a GMP for food processing facilities was finally proposed (FDA 2011). GMP connects all factors that assure the quality, safety and effectiveness of food, according to its specification and purpose. For example, the recent Regulation 2023/2006 defines GMP as an assurance of quality in terms of conformity with the rules at all production stages and for all types of food contact materials and articles intended to come into contact with food (Grob et al. 2009). Clearly set principles and the success of GMP soon laid the groundwork for developing many other good practices in the food supply chain.

Raspor (2008) pointed out three categories of good practices throughout a broad spectrum of the food industry. The first category of good practices is directly connected with food technology (i.e., Good Manufacturing Practice – GMP). GMP connects all factors that assure the quality, safety and effectiveness of food, according to its specification and purpose. Various people and entities are generally responsible for the handling, storage and distribution of products in different companies and countries (e.g. perishable products, cold chain storage). Distribution

(WHO 2006) and storage (WHO 2003) practices are also an important activity in integrated supply chain management.

Good Laboratory Practice (GLP) deals with the organization, process and conditions under which laboratory studies are planned, performed, monitored, recorded and reported. GLPs are intended to promote the quality and validity of test data. The formal, regulatory concept of Good Laboratory Practice originated in the USA because of concerns about the validity of non-clinical safety data submitted to the Food and Drug Administration (FDA) in the context of new drug applications. The inspection of studies and test facilities revealed instances of inadequate planning and incompetent execution of studies, insufficient documentation of methods and results, and even cases of fraud (e.g. replacing animals that had died during a study with new ones without documenting this fact; retrospectively changing raw data to "fit the result tables") (WHO 2009). GLP was published in 1976 for assuring the validity of studies (Baldeshwiler 2003).

Good Agricultural Practices (GAPs) are practices that address environmental, economic and social sustainability for processes on a farm and result in safe and quality food and non-food agricultural products; they were first presented to the FAO Committee on Agriculture in 2003 (FAO 2011). GAP codes, standards and regulations have been developed in recent years to codify agricultural practices at the farm level for a range of commodities. The objectives of GAP codes, standards and regulations are, to varying degrees (FAO 2011), as follows:

- Ensuring the safety and quality of produce in the food chain;
- Capturing new market advantages by modifying supply chain governance;
- Improving natural resource use, worker health and working conditions; and
- Creating new market opportunities for farmers and exporters in developing countries.

The principles and success of GMP have laid the foundation to develop many other good practices, which we present through steps of the food supply chain (Fig. 30.3).

In the context of food production, GAP is clearly the most important good practice for the next steps of the food supply chain in terms of hazards that make their way through the chain. Food processing includes several practices [GMP, GLP, Good Catering Practice (GCP), Good Retail Practice (GRP) and Good Hygiene Practice (GHP)] that are intertwined with each other.

GCP consists of practical procedures in catering. The guidelines for GCP concentrate on the essential steps needed to ensure that food served is always safe and wholesome. GRP consists of practical procedures and processes that ensure the right products are delivered to the right addressee within a satisfactory time period and under required conditions. A tracing system should enable any faulty product to be found, and there should be an effective recall procedure. GHP consists of practical procedures and processes that return the processing environment to its original condition (disinfection or sanitation programmes); keep building and equipment in efficient operation (maintenance programme); control cross contamination during manufacturing (usually related to people, surfaces, the air and the

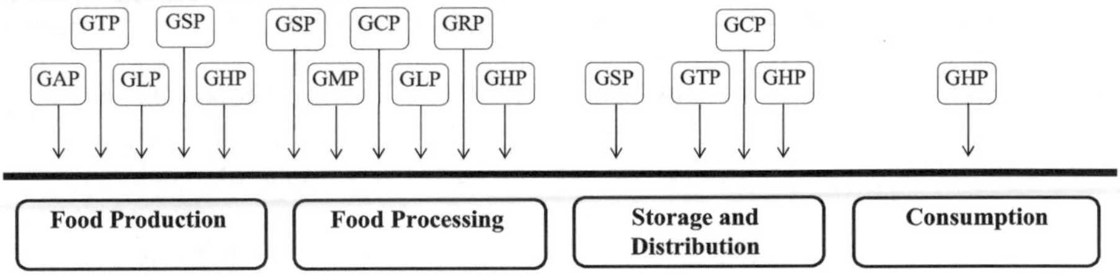

Fig. 30.3 Food supply chain elements with relevant good practices (*GAP* good agricultural practice, *GMP* good manufacturing practice, *GLP* good laboratory practice, *GHP* good hygiene practice, *GTP* good transport practice, *GSP* good storage practice, *GRP* good retail practice, *GCP* good catering practice)

segregation of raw and processed product) (Raspor 2008; Raspor and Jevšnik 2008).

The next stage in the food supply chain is storage and distribution. Good Storage Practice (GSP) consists of practical procedures and processes that ensure the appropriate handling of foods and address the implementation and control of product storage in accordance with a defined regime prior to use. Good Transport Practice (GTP) consists of practical procedures and processes that ensure a qualitative system governing the organization, implementation and control of transport of food products from producer to end user (Raspor 2008; Raspor and Jevšnik 2008).

The last step in the food supply chain is consumption. Consumers must also respect the principles of food safety at home. Good Housekeeping Practice (GHKP) relates to the selection of principles and techniques of food storage and food preparation at home (Raspor 2008; Raspor and Jevšnik 2008).

The second category of good practices (Raspor 2008) is indirectly connected with food issues (i.e. Good Research Practice, Good Educational Practice, Good Training Practice). If adequate training and appropriate education are not provided within human resources, then it will be unreasonable to expect to have professionals with highly developed skills or a high level of knowledge about relevant control and documentation (Raspor 2008; Raspor and Jevšnik 2008).

In addition, Raspor (2008) stressed that it is possible to create a third category, which in his view would be needed but currently does not exist. This practice would deal with all activities related to food in the home [i.e. Good Housekeeping Practice (GHKP)] (Raspor 2008). As early as 1885, Clark W. Bryan published the first edition of *Good Housekeeping* magazine with a mission to produce and ensure perfection as may be obtained in the household (Good Housekeeping Magazine 2011). Nowadays *Good Housekeeping* promotes safety in the home but also upholds a level of professionalism at work because effective housekeeping requires one to organize one's work and equipment, keep one's work area clean and clear and store things in their proper place.

The main purpose of all good practices in food safety is to provide consumers with safe, healthy and quality food. All good practices contain HACCP elements that compose the HACCP system as the main system in food practice today.

All practices are partial and not connected in a comprehensive system. GHKP is of the last importance for connecting food system chains into a food safety circle (Raspor 2004). According to many FBD, GHKP is still not applied in domestic food preparation.

In classic food chain strategy, all relevant activities are implemented for the benefit of human beings but locate them outside the system as consumers. The problem is that players in the food supply chain today behave self-sufficiently and do not understand their specific role in food safety integrity; holistic insight is missing. Based on the fact that good practices do not appear to be an integral part of the food safety circle, it is necessary to set the GNP as a way of providing the consumer with safe and quality food. The novel approach of GNP involves the last step in the food supply chain: the consumer and close link up all the relevant good practices. Global food safety will be achieved only when all links in the food chain, in both indoor and outdoor environments, properly oversee their own area and trust in the competence of both the previous and the following link in the food safety circle from farm to table. At the same time, the consumer must be aware of the potential risks, proper handling and preparation of food as these relate to a safe and balanced diet (Raspor and Jevšnik 2008).

The implementation and maintain age of all three sets of Good Practices will provide a secure global system of processes, procedures and documentation to assure that food is healthy and safe along the entire food supply chain.

30.5 Conclusions

Quality management systems will stay proactive in the development and implementation of new solutions in global and local food systems. Food professionals in the food supply chain are responsible for the technical aspects of development of food products, food processes and distribution of these products throughout the entire food chain. In the area of consumer needs, food safety management systems should develop relevant and user-friendly approaches to educating consumers about developments in food technology and nutrition and training them about novel paradigms serving safe and nutritious food in the context of their personal and professional needs. Clearly, we must start find ways to connect more profoundly the hard and soft sciences and share findings, experiences and skills with consumers, engaging all elements of the food supply chain. Currently, GNP offers an acceptable platform for implementing in actual practice along the food supply chain.

References

Aladjadjiyan A (2006) Physical hazards in the agri-food chain. In: Luning PA, Devlieghere F, Verhe R (eds) Safety in the agri-food chain. Wageningen Academic Publishers, Wageningen, pp 209–222

Ambrožič M, Jevšnik M, Raspor P (2010) Inconsistent terminology in food safety field: a permanent risk factor? J Food Nutr Res 49(4):186–194

Ambrožič M, Božič T, Jevšnik M, Raspor P (2011) Compliance of proposed Codex Alimentarius guidelines for virus management with principles of good practice. Acta Aliment 40(3):364–375

Baldeshwiler AM (2003) History of FDA good laboratory practices. Qual Assur J 7(3):157–161

Behrens JH, Barcellos MN, Frewer LJ, Nunes TS, Franco BDGM, Destro MT, Landgraf M (2010) Consumer purchase habits and views on food safety: a Brazilian study. Food Control 21(7):963–969

Carrefour (2011) Carrefour group: history. http://www.carrefour.com/cdc/group/history/. Accessed 10 Dec 2011

Carvalho FP (2006) Agriculture, pesticides, food security and food safety. Environ Sci Policy 9(7–8):685–692

Codex Alimentarius Commission (CAC) (2003) General principles of food hygiene CAC/RCP 1-1969- Rev 4. FAO/WHO, Rome

Codex Alimetarius Commission (CAC) (2006) Understanding the Codex Alimetarius, 3rd edn. FAO/WHO, Rome

De Meulenaer B (2006) Chemical hazards. In: Luning PA, Devlieghere F, Verhe R (eds) Safety in the agri-food chain. Wageningen Academic Publishers, Wageningen, pp 145–208

Duncan TV (2011) Applications of nanotechnology in food packaging and food safety: barrier materials, antimicrobials and sensors. J Colloid Interface Sci 363(1):1–24

EFSA (2011) The European Union summary report on trends and sources of zoonoses, zoonotic agents and food-borne outbreaks in 2009. EFSA J 9(3):2090

European Commission (EC) (2000) White paper on food safety, Brussels, 12 Jan 2000. http://europa.eu.int/comm/dgs/health_consumer/library/pub/pub06_en.pdf. Accessed 10 Dec 2011

European Union (EU) (1993) Council Directive 93/43/EEC of 14 June 1993 on the hygiene of foodstuffs. Off J EU Legis 175:1–11

U. S. Federal Register (1963) Part 133-drugs; current good manufacturing practice in manufacture, processing, packaging or holding, pp 6385–6387. www.cgmp.com/firstUScGMP.pdf. Accessed 10 Dec 2011

Food and Agriculture Organization of United Nations (FAO) (2011) Good agricultural practices. http://www.fao.org/prods/gap/. Accessed 10 Dec 2011

Fulponi L (2006) Private voluntary standards in the food system: the perspective of major food retailers in OECD countries. Food Policy 31(1):1–13

GFSI (2011) Global food safety initiative. http://www.mygfsi.com/. Accessed 10 Dec 2011

BRC Global Standards (2011) Food. http://www.brcglobalstandards.com/GlobalStandards/Standards/Food.aspx. Accessed 10 Dec 2011

GLOBALG.A.P. (2011) Global good agricultural practice. http://www.globalgap.org/cms/front_content.php?idcat=9. Accessed 10 Dec 2011

Good Housekeeping Magazine (2011) The history of the good housekeeping seal. http://www.goodhousekeeping.com/product-reviews/history/good-housekeeping-seal-history. Accessed 10 Dec 2011

Gorris LGM (2005) Food safety objective: an integral part of food chain management. Food Control 16(9):801–809

Grob K, Stocker J, Colwell R (2009) Assurance of compliance within the production chain of food contact materials by good manufacturing practice and documentation – Part 2: implementation by the compliance box; call for guidelines. Food Control 20(5):483–490

Heggum C (2001) Trends in hygiene management – the dairy sector example. Food Control 12(4):241–246

Henson S, Reardon T (2005) Private agri-food standards: implications for food policy and the agri-food system. Food Policy 30(3):241–253

International Featured Standards (IFS) (2011) Food. http://www.ifs-certification.com/index.php?page=home&content=public_content&desc=ifs_standards_food_5&language=english. Accessed 10 Dec 2011

International Organization for Standardization (ISO) (2011) International standards. http://www.iso.org/iso/home.html. Accessed 10 Dec 2011

ISO 22000 (2005) Food safety management systems – requirements for any organization in the food chain. ISO Central Secretariat, Geneva

ISO 9000 (2005) Quality management systems – fundamentals and vocabulary, 2nd edn. ISO Central Secretariat, Geneva

Jevšnik M, Hlebec V, Raspor P (2008) Food safety knowledge and practices among food handlers in Slovenia. Food Control 19(12):1107–1118

Lasztity R, Petro-Turza M, Földesi T (2004) History of food quality standards. In: Lasztity R (ed) Food quality and standards. Encyclopaedia of Life Support Systems (EOLSS), developed under the Auspices of the UNESCO. EOLSS, Oxford

Lawley R, Curtis L, Davis J (2008) The food safety hazard guidebook. RSC Publishing, Cambridge

Luning P, Marcelis W, van der Spiegel M (2006) Quality assurance systems and food safety. In: Luning PA, Devlieghere F, Verhe R (eds) Safety in the agri-food chain. Wageningen Academic Publishers, Wageningen, pp 249–302

Randell A (1995) Codex Alimentarius: how it all began. Food, nutrition and agriculture series, 13/14. www.fao.org/docrep/V7700T/v7700t09.htm. Accessed 10 Dec 2011

Raspor P (2002) Handbook for establishment and conducting HACCP system. Slovenian Institute of Quality, Biotechnical faculty, Ljubljana

Raspor P (2004) Current viewpoint on food safety. In: Gašperlin L, Žlender B (eds) 22nd Food Technology Days 2004 dedicated to prof. F. Bitenc. Biotechnical Faculty, Ljubljana, pp 1–14

Raspor P (2008) Total food chain safety: how good practices can contribute? Trends Food Sci Technol 19(1):405–412

Raspor P, Ambrožič M (2011) ISO 22000 food safety. In: Sun DW (ed) Handbook of food safety engineering. Wiley-Blackwell, Oxford, pp 786–816

Raspor P, Jevšnik M (2008) Good nutritional practice from producer to consumer. Crit Rev Food Sci Nutr 48(3):276–292

Regulation (EC) No 178/2002 of the European Parliament and of the Council of 28 Jan 2002 laying down the general principles and requirements of food law, establishing the European food safety authority and laying down procedures in matters of food safety. Off J Eur Commun Legis 31:1–24

Regulation (EC) No 852/2004 of the European Parliament and of the Council of 29 Apr 2004 on the hygiene of foodstuffs (OJ L 139, 30 Apr 2004). Off J Eur Commun Legis 139:1–54

Selvan NK (2008) Food supply chain: emerging perspective. In: Selvan NK (ed) Supply chain management in food industry. The Icfai University Press, Punjagutta, pp 3–12

Sperber WH (1998) Auditing and verification of food safety and HACCP. Food Control 9(2–3):157–162

Sperber WH (2005a) HACCP and transparency. Food Control 16(6):505–509

Sperber WH (2005b) HACCP does not work from Farm to Table. Food Control 16(6):511–514

Sun YM, Ockerman HW (2005) A review of the needs and current applications of hazard analysis and critical control point (HACCP) system in foodservice areas. Food Control 16(4):325–332

Swanenburg M, Rijsman VMC, Teeuw J, Mengelers MJB, Noordam MY, Schwarz-Bovee A (2011) De systematiek om tot indicatoren voor gevaren in de voedselketen te komen. Report 2001.05. Agricultural Economics Research Institute, Institute for Animal Science and Health,

Organization for Applied Scientific Research, State Institute for Quality Control of Agricultural Products, Wageningen
Tauxe RV (2002) Emerging food-borne pathogens. Int J Food Microbiol 78(1–2):31–41
Tesco (2011) Our history. http://www.tescoplc.com/about-tesco/our-history/. Accessed 10 Dec 2011
Trienekens J, van der Vorst J (2006) Traceability in food supply chains. In: Luning PA, Devlieghere F, Verhe R (eds) Safety in the agri-food chain. Wageningen Academic Publishers, Wageningen, pp 439–470
Trienekens J, Zuurbier P (2008) Quality standards in the food industry, developments and challenges. Int J Prod Econ 113(1):107–122
U. S. Food & Drug Administration (FDA) (2011) Current food good manufacturing practices. http://www.fda.gov/Food/GuidanceComplianceRegulatoryInformation/CurrentGoodManufacturingPracticesCGMPs/ucm110907.htm. Accessed 10 Dec 2011
Vasconcellos JA (2004) Quality assurance for the food industry – a practical approach. CRC Press, Boca Raton
Vellema S, Boselie D (2003) Matching performance, commitment and markets in fresh produce: an agenda for future research. In: Vellema S, Boselie D (eds) Cooperation and competence in global food chains: perspectives on food quality and safety. Shaker Publishing, Maastricht, pp 191–203
Walczak D, Reuter M (2002) Putting restaurant customers at risk: unsafe food handling as corporate violence. Hosp Manage 23(1):3–13
Walker E, Jones N (2002) An assessment of the value of documenting food safety in small and less developed catering businesses. Food Control 13(4–5):307–314
Walker E, Pritchard C, Forsythe S (2003) Hazard analysis critical control point and prerequisite programme implementation in small and medium size food businesses. Food Control 14(3):169–174
Walmart Corporate (2011) History timeline. http://walmartstores.com/aboutus/7603.aspx. Accessed 10 Dec 2011
Williams CM (2002) Nutritional quality of organic food: shades of grey or shades of green? Proc Nutr Soc 61:16–24
World Health Organization (WHO) (2003) WHO expert committee on specifications for pharmaceutical preparations, Technical report series no. 908. WHO, Geneva
World Health Organization (WHO) (2006) WHO expert committee on specifications for pharmaceutical preparations, Technical report series no. 937. WHO, Geneva
World Health Organization (WHO) (2009) WHO library cataloguing-in-publication data handbook: good laboratory practice (GLP): quality practices for regulated non-clinical research and development, 2nd edn. WHO, Geneva
Zschler R (1989) Good manufacturing practice (GMP) in the food industry. Zentralblat fur Bakteriologie Mikrobiologie und Hygiene 87(4–6):546–556

Part VI
Current and Future Issues

Chapter 31
Does Biofuel Production Threaten Food Security?

Walter E.L. Spiess

31.1 Introduction

The Sun is the source of all terrestrial energy. The incoming energy flux amounts to 174 PW; 89 PW are absorbed by the land mass and oceans, and an almost similar amount of 85 PW is reflected back into space (Fig. 31.1).

From the available energy flux of 89 PW, or 89,000 TW (TW = terawatt; 1 TW = 10exp 12 W) air movements, i.e. the wind, consume 870 TW, the movement of water 7.2 TW. Global human energy consumption amounts to 15 TW (Fig. 31.2). These considerations demonstrate that by far not all terrestrial energy sources are utilized.

The major share of energy consumption for maintaining our life in its various characteristics and activities is based on fossil energy sources. At the present time the largest source of energy consumed is oil, with almost one-third of the total consumption, followed by coal and natural gas. Renewable energy sources presently play only a minor role (Table 31.1).

Approximately 70 % of the consumed energy is used for residential, industrial and transport purposes. A minor part of fossil energy sources goes to the production of chemicals and other uses (Table 31.2). It is notable that 27 % of the aforementioned energy consumption is used for transport purposes.

At the beginning of the railway age, black coal was the dominant source of energy for transportation. Steam-powered engines, however, were inflexible and required huge backup systems. The invention of the Otto and Diesel engines allowed for a high flexibility and the use of those engines in almost any type of transportation system. The functioning of Otto and Diesel engines is based on the availability of combustible fuels. This fact explains that almost 80 % of worldwide

W.E.L. Spiess (✉)
Karlsruhe Institut für Technologie (KIT), Institut für Bio-und Lebensmitteltechnik,
Kaiserstr. 12, D-76131 Karlsruhe, Germany
e-mail: rose-walter.spiess@t-online.de

S. Yanniotis et al. (eds.), *Advances in Food Process Engineering Research and Applications*, Food Engineering Series, DOI 10.1007/978-1-4614-7906-2_31,

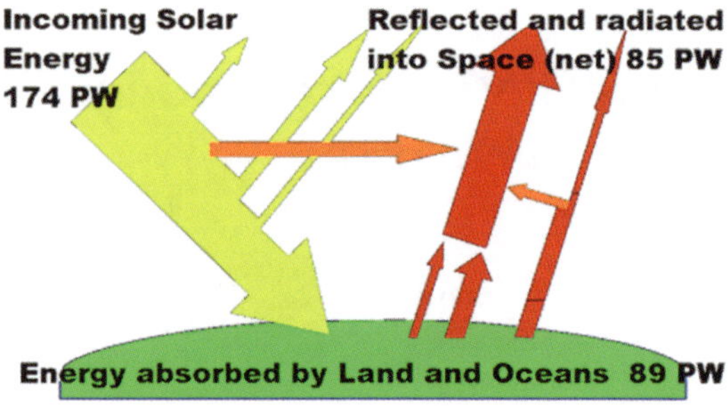

Fig. 31.1 Solar energy balance (2012)

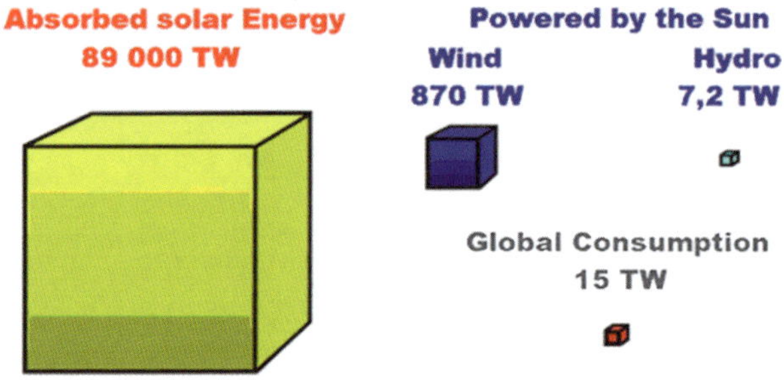

Fig. 31.2 Available energy sources and consumption (Solar Energy Balance 2012)

Table 31.1 Total primary world energy supply (2008) (1,426,652 PWh) (IEA Key World Energy Statistics 2010)

Source	Share (%)	Source	Share (%)
Oil	33.2	Renewable/waste	10
Coal/peat	27	Nuclear	5.8
Gas	21.1	Hydro and others	2.2/0.7

Table 31.2 World energy use by sector (2008) (total 98,022 PWh) (IEA Key World Energy Statistics 2010)

Sector	Share (%)	Sector	Share (%)
Industry	28	Residential and services	36
Transport	27	Non-energy use	9

petroleum production is used for transportation purposes (Table 31.3). Gasoline, with 46 %, accounts for the major share of consumption.

In 2008, total transport energy consumption amounted to 2,299.37 million tonnes of oil equivalent (Mtoe). The share of combustible biofuels was, at 23.22 Mtoe, around 1 %; the International Energy Agency (IEA) and the Organisation for Economic Co-operation and Development (OECD) expect that this share can be increased to 5–6 % by 2030 (IEA Key World Energy Statistics 2010; OFID Pamphlet 2012).

Table 31.3 Use of petroleum worldwide (IEA Key World Energy Statistics 2010)

Use	Share (%)	Use	Share (%)
Gasoline, various purposes	46	Heavy fuel oil	4
Diesel and related fuels	26	Asphalt	3
Various products/chemicals	11	Lubricants	1
Aviation fuel	9		

Table 31.4 Various types of biofuel

Solid	Wood, charcoal, dung, bagasse	
Liquid	Ethanol, fatty-acid-methyl-ester = FAME-bio-diesel	
Gaseous	Biogas (main compounds %)	
	Methane 50–70	Hydrogen 0–1
	Carbon dioxide 25–50	Hydrogen sulphide 0–3
	Nitrogen 0–10	

31.2 Biofuels, Relevance and Nature

Biofuels were certainly the first energy source exploited by humans. Wood and dried dung were, however, gradually replaced by fossil fuels, coal, petroleum and, more recently, natural gas because these energy sources were more convenient to handle and of higher energy density; their distinct disadvantage is, however, that during conversion into mechanical energy, considerable amounts of CO_2 are released.

By definition, biofuels are energy sources which are derived from renewable biomaterials such as plant materials, preferably which have a high carbohydrate and fat content and which lend themselves to fast crop rotation. Biofuels can also be produced from all types of organic waste, e.g. agricultural and food processing wastes and manure from various sources.

Biofuels can be classified as solid, liquid and gaseous (Table 31.4). Solid biofuels are wood, charcoal, dung and bagasse; liquid biofuels are mainly ethanol and fatty-acid-methyl-ester (FAME-Biodiesel); the most important gaseous biofuel is methane.

Excluding the huge amounts of wood and dung – which are still used in the developed but also in developing countries – the dominant biofuel is ethanol and bio-diesel.

Significant amounts of liquid biofuels entered the market in the early 1970s; the biofuel market grew steadily from this point on (Fig. 31.3). Present (2010) ethanol production amounts to approximately 70 billion L per annum; 80 % of this quantity is used as fuel for transport; 61 % of the ethanol production is sugarcane based, 39 % maize based, and the major producers are Brazil (sugarcane) and the USA (maize). Biodiesel production amounts to approximately 15 billion L per annum. A major global producer is Germany, with a share of more than 50 % (FAO 2012; IEA: World Energy Outlook 2008/2009). It must be expected that biofuels will

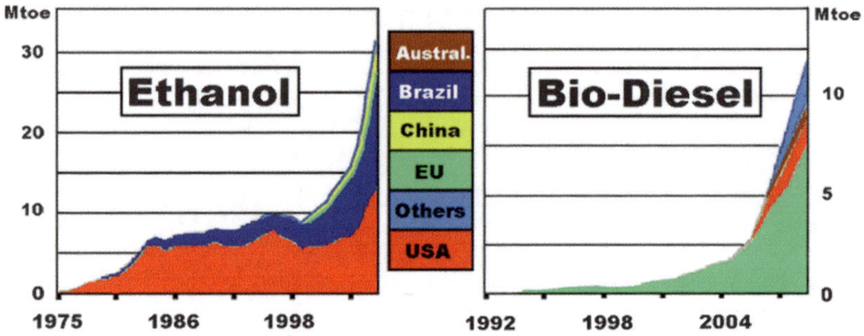

Fig. 31.3 World biofuel production (ethanol and bio-diesel) (OFID Pamphlet 2012)

gradually come to play a vital role in at least partially replacing petroleum as the energy source for transport; it is, however, also clear that in the foreseeable future fossil fuels will maintain their dominant role in the worldwide energy supply.

The search for new sustainable energy sources, i.e. energy sources with low or even no greenhouse gas (GHG) emissions, has sparked a worldwide interest in biofuels, despite the understanding that presently only in special cases does the production of biofuels have a positive impact on the reduction of GHGs. In particular, countries with a high agricultural surplus, e.g. the USA and Brazil, spurred the production of combustible biofuels by favourable tax policies and direct and indirect financial subsidies or other supportive measures like mandatory blending with fossil fuels. Worldwide efforts, sanctioned by national and international legal bodies and agreements, have also paved the way for the rise of biofuels in countries with deficient agricultural production.

The production of biofuels and the possible expansion of the production as requested by, amongst others, OECD countries and non-OECD (OECD 2007) countries have already had an impact on overall agricultural production and on the availability of agricultural produce for human consumption. It must be stated that the price of major raw materials suitable for biofuel and for food production, especially of cereals/roots and fruits/seeds containing high amounts of fat, increased sharply in recent years following a decline up till the early 1990s. This situation has without doubt increased the number of those who are not able to pay for their daily food demands and who are suffering from malnutrition and hunger (SIB 2010).

As previously mentioned, the main feedstock for the production of biofuels like ethanol and bio-diesel are agricultural products originally cultivated for human consumption, like maize and wheat but also root vegetables and soya or rapeseed, as well as fruit of the oil palm (Table 31.5). It is hoped that in the next decade those raw materials will be replaced by materials not suited for human consumption, such as wood, grass and tree nuts like jatropha. But the use of those materials will require new and more efficient conversion processes. Those processes are characterized as second-generation conversion processes/technologies and are based on pyrolysis, gasification or unconventional fermentation processes. In particular, if effective

Table 31.5 Major types of feedstock for biofuel production

Cereals	Maize, wheat, (Sorghum)
Root vegetable	Potato, cassava, sugar beet
Leguminous plants and Brassicaceae	Soya, rapeseed
Grass	Sugar cane, switchgrass
Tree and shrub fruit, wood	Palm tree, jatropha, willow, poplar

cellulase, ligninase and related enzymes (LMEs) are available to convert cellulose- and lignin-containing materials (wood and fibrous plant materials) into fuel, then the pressure on food-related biofuel feedstock will be relieved. Second-generation conversion of feedstock, such as forest or agricultural residues, wood waste and energy crops, however, is not yet ready for a larger industrial application, but pilot plants operate in most industrialized countries.

31.2.1 Biofuel Production from Food Feedstock: Consequences for Producers and Worldwide Food Supply

While the impact of biofuel production on global food supplies and availability is difficult to predict, it is without doubt closely related to the development of three major parameters: climate (with consequences for, amongst other things, agricultural production), population development (with consequences for, amongst other things, food demand) and energy demand and satisfaction (with consequences for, amongst other things, CO_2 emissions and, hence, climate).

A detailed discussion of the interrelation between biofuel production and food availability in the future must rely mainly on scenarios developed for those three major parameters. Various international working groups, including the International Institute for Applied Systems Analysis (IIASA), FAO and IEA, have devoted major efforts to the development of such scenarios; simplified elements of these scenarios, together with other publications, form the basis of the following considerations. The scenarios take into account the General Circulation Models (GCMs), the Agro-ecological Zone (AEZ) methodology, the World Food System Model and similar modelling approaches (OFID Pamphlet 2012).

The present discussion on energy demand follows, in broad terms and with some simplifications, the IEA World Energy Outlook Scenario (WEO). In particular, this projection takes into account that major OECD and non-OECD countries have passed legislation regarding biofuel production or have committed themselves to expanding biofuel production above present levels, e.g. the European Union

Table 31.6 Important mandatory targets regarding bio-fuel use by OECD and non-OECD countries (OECD 2007)

Country	Targets (MaTa)
Brazil	MaTa of 2 % bio-diesel in 2010; 5 % from 2013 on
Canada	MaTa at least 5 % biofuels of gasoline sold
China	Big cities/9 prov. MaTa E10 blending by 2012
EU-25	MaTa of 10 % for biofuels sales in 2020
India	MaTa 15 % ethanol and 2 % for biodiesel by 2015
Japan	MaTa 50 million L biofuels by 2011
Malaysia	MaTa of 5 % for bio-diesel in all diesel sales
New Zealand	MaTa of 3.4 % of total transport fuel sales by 2012
South Africa	MaTa of 4.5 % biofuels of liquid transport fuels
USA	Up to 28.35 billion L biofuels blended by 2012

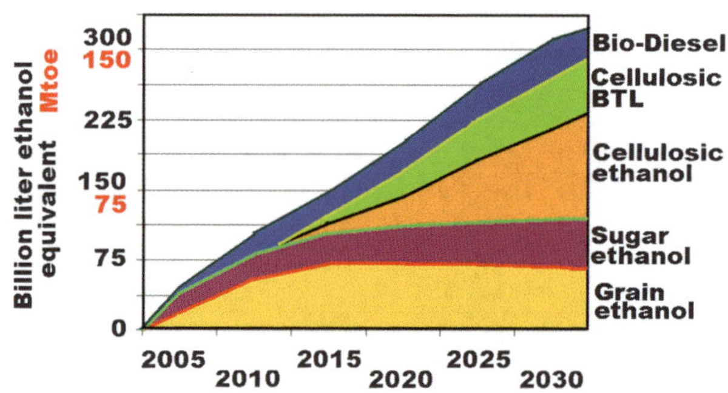

Fig. 31.4 World biofuel production by type (Office of Policy Analysis 2008)

(EU) has passed legislation setting a minimum target of 10 % for biofuels for transport energy consumption in 2020; other major countries have passed similar legislation (OECD 2007) (Table 31.6).

31.2.1.1 Underlying Scenarios

It is assumed that transport energy demand will develop as projected by the IEA in its WEO 2008 scenario assuming that mandatory and voluntary or indicative targets for biofuel use (as announced by the major OECD and non-OECD countries) will be implemented by 2020. This will result in approximately twice the biofuel consumption compared to 2008 (IEA Key World Energy Statistics 2010; OFID Pamphlet 2012; United Nations et al. 2012).

During the next decade ethanol produced from grains and sugar cane and bio-diesel produced from oil seeds will dominate the market. It is, however, expected that in about 20 years second-generation fuels from cellulosic materials and other non-food feedstock will have acquired increasing importance (Fig. 31.4).

At this point, palm oil and jatropha should be mentioned as being of special relevance. Up till now Europe has been the main producer of bio-diesel, with

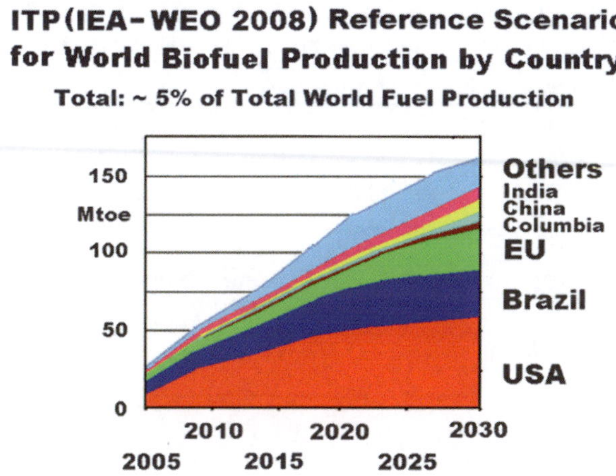

Fig. 31.5 World biofuel production by country (Office of Policy Analysis 2008)

production based mainly on rapeseed. With the USA, China, India, Brazil, Indonesia and Malaysia entering the arena, production will be based on different feedstocks, mainly palm oil, soybean, jatropha and, most probably in the more distant future, on algae.

Malaysia has already recognized the large potential of palm-oil-based bio-diesel production and passed legislation that all diesel fuel sold in Malaysia must contain 5 % palm oil.

According to the WEO scenario, the USA and Brazil will continue to be the major biofuel producers, closely followed by Europe. Interestingly, developing countries like Colombia, China and India will also contribute significantly to the total world production (Fig. 31.5).

31.2.2 Feedstock Production for Biofuels Versus Food Production

31.2.2.1 Biofuel Feedstock Demand

The yield obtained from various oil-containing plants or fruits depends very much on the variety, growing conditions and ultimately harvesting and post-harvest processing (Table 31.7). The presented figures represent average data; in individual cases the figures may vary considerably.

The largest raw material demand is required for ethanol production from sugar cane: for 1,000 L of fuel 13,000 T of raw material is needed (Table 31.8). The production of ethanol from sugar beet is equally demanding: 10,000 T of raw material are required for 1,000 L. If cereals are used for the production of ethanol, then the mass of the required raw materials is roughly one-third. This statement is also correct if biomass or cellulosic materials are used for the production of ethanol

Table 31.7 Average oil yield of important oil seeds and algae (FAO: World Agriculture: Towards 2012; FAO Natural Resources Management and Environment Department 2012)

Produce	Yield	Produce	Yield
Soybean	740	Palm oil	4,752
Sunflower	767	Algae	3,000
Rapeseed	954	Jatropha	3,000

Table 31.8 Feedstock requirements for 1,000 L ethanol alcohol or bio-diesel (average sugar/starch content, wet basis) (Vogelbusch 2012)

Feedstock requirements for 1,000 l alcohol (ethanol) (average sugar/starch content, wet basis)/bio-diesel			
	Amount	Biofuel production 2030	Raw material equivalent
Produce (product)	T/1,000/fuel	Billion litres	MMT
Maize (ethanol)	2,300	68	156
Wheat (ethanol)	2,800	68	190
Sugar beet (ethanol)	10,000	53	530
Sugar cane (ethanol)	13,000	53	689
Rapeseed (bio-diesel)	2,700	38	102
Biomass (to liquid) (bio-diesel)	2,000	57	114
Cellulosic materials (e.g. straw) (ethanol)	3,000	98	294

in a second-generation process. If the anticipated biofuel production in 2030 is considered, raw material equivalents between 690 million metric tonnes (MMT) (sugar cane) and 100 MMT in the case of rapeseed will be necessary.

31.2.2.2 Food Demand

According to international statistical data, population growth rates will decline in the years to come quite dramatically, whereas the cereal production growth rates will stabilize at a level of 1.2 % (Fig. 31.6). In other words the present world cereal production of 2,238 million tonnes will rise to close to 3,000 million tonnes in 2030.

The increased production of cereals will be absorbed by the growing world population of 8.3 billion people with a higher average daily caloric intake in 2030 compared with today (United Nations et al. 2012). It is estimated that in 2030 the average daily caloric intake will be 3,050 kcal/(capita · day) compared with today's 2,800 kcal/(capita · day). In developing countries the average will be somewhat lower at around 3,000 kcal/(capita · day).

By 2030 developing countries will produce only 86 % of the cereal needed to provide the daily requirements. That means that net imports will have to rise from current levels of 103 MMT to 265 MMT by 2030. The imbalances in production

Fig. 31.6 Population and cereal production growth rates (UN, Department of Economic and Social Affairs 2003; United Nations et al. 2012; Biofuels and Climate 2012)

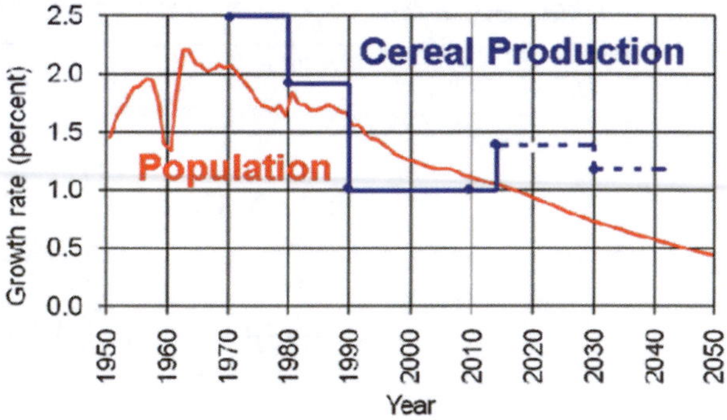

and consumption will also have a negative impact on the general price level, a fact which will mainly affect underprivileged segments of the population in developing countries (FAO: World Cereal Demand 2012; FAO: World Agriculture: Towards 2012; FAO Natural Resources Management and Environment Department 2012).

31.2.2.3 Land Use for Food, Feed and Biofuel Feedstock

The world land area is approximately 130,000,000 km^2, approximately one-third of which can be used for agricultural purposes, slightly more than 10 % as cropland. For major crops, less than 10 % is used, 5 % alone for the production of cereals (Table 31.9).

It is obvious that the impact of biofuel production on food security depends upon the crop grown and the land on which it is grown. Thus, if for example the land area presently used for agricultural production is not expanded, then, for example, 156 MMT of maize will not be available as a source of food. Given an average demand of 160 kg/(capita · year), approximately one billion people in addition to the already hungry people would be food insecure. This is certainly an alarming situation!

To obtain a more detailed understanding of the impact of biofuel production on food/feed supplies, three scenarios will be discussed: (a) use of existing arable land; (b) expansion of agricultural land and (c) use of marginal land and forests not otherwise suitable for crop production

Scenario 1: Use of Existing Arable Land. In the near future the land area used for agricultural production will not grow in a significant way. This means that sugar cane will be the major feedstock for biofuel production in the Southern Hemisphere, whereas cereals will dominate in the Northern Hemisphere as feedstock. It is expected that from 2015 on other, also non-conventional, feedstock like jatropha and algae will appear on the market produced mainly on newly cultivated areas, relieving the pressure of biofuel production on the availability of food/feed. The level of cereal prices will be quite high with a negative impact on the general supply situation (Berg Ch 2012).

Table 31.9 Cropland area in 2005 relative to world land area (FAOstat 2009)

Land type		km^2	
World land area		130,134,754	
Agricultural area		49,675,795	
Cropland		15,616,817	
Produce type	km^2	Produce type	km^2
Roots and tubers	33,801	Vegetables/melons	529,462
Cereals	6,877,512	Tree-nuts	82,154
Pulses	699,733	Oilcrops	2,520,130
Fibre crops	382,577	Fruit	523,412

Table 31.10 Expansion of agricultural land and use of marginal land and forest (Biomasse-vergasung 2012)

Scenarios 2 and 3: expansion of agricultural land and use of marginal land and forest			
Land requirements for alcohol (ethanol) (average sugar/starch content, wet basis)/bio-diesel production			
Produce (product)	Specific requirements (L/ha)	Biofuel production 2030 (billion L)	Required land resources (km^2)
Maize (ethanol)	1,960	68	346,938
Wheat (ethanol)	952	68	714,2879
Sugar beet (ethanol)	5,060	53	104,743
Sugar cane (ethanol)	4,550	53	116,483
Rapeseed (Bio-Diesel)	954	57	595,611

Scenario 2 and 3: Expansion of Agricultural Land and Use of Marginal Land and Forest. The additional land requirements for the production of biofuels, ethanol and bio-diesel seem quite demanding (Table 31.10). Thus, for example, the additional area required for the production of wheat for biofuel production is approximately 10 % larger than France; compared with the amount of land not used for agricultural purposes, the perspective changes, however.

In the case of Brazil as a major ethanol producer, it becomes obvious that in certain areas, especially in the Southern Hemisphere, huge reserves of unused land resources are available (Table 31.11). Besides the reserves in Brazil and other South American countries, huge reserves are also located in South Africa, e.g. in Mozambique (Table 31.12). In Asia, the reserves are limited; only China has the potential to expand its agricultural land if those areas can be sufficiently irrigated. In North America, the lack of water is also a limiting factor; in Europe land reserves are close to zero (Macedo 2009).

Those reserves can be utilized without touching virgin forest reserves. It is obvious that the land required for biofuel production is available and almost below 10 % of land reserves.

Besides expanding the land area used for crop production, crop production can be increased by higher cropping frequencies using higher yielding species and

Table 31.11 Land use in Brazil (total land area 8,500,000 km^2) (FAO Document 2012)

Land use (agricultural product)	Area (Mkm^2)	(Percentage of arable land)	Percentage of cultivated land
Forests	4.10		Unused resources: 0.77 Mkm^2
Arable land	3.40	100	
Pasture	2.00	59	
Cultivated land	0.63	19	100
Soybean	0.22	7	35
Maize	0.13	4	21
Sugar cane (total)	0.07	2	11
Sugar cane (ethanol)	0.04	1	6
Sugar cane (ethanol production 2030)	0.05	1.4	8

Table 31.12 Cultivated land area and gross reserves in million km^2 (Macedo 2009)

Continent	Total arable land	Land use 1990	Land use 2010	Land reserve
Latin America and Caribbean	10.59	1.90	2.17	8.42
Sub-Saharan Africa	10.09	2.13	2.55	7.54
South Asia	2.28	1.91	1.95	0.33
East Asia (excluding China)	1.84	0.88	1.03	0.81
North Africa and Near East	0.92	0.77	0.81	0.11
Total	25.72	7.59	8.51	17.21
Estimated demand of arable land for biofuel production				0.7–1.4

improving yields through improved agricultural methods (FAO: World Agriculture: Towards 2012).

More specifically, it is expected that agricultural production will be expanded through land expansion in South America and sub-Saharan Africa by 20 % (1.20 Mkm^2), through higher yields by 70 % (better soil management, genetically modified crops, organic agriculture), and through multiple cropping and shorter fallow periods by 10 %.

The expected overall effect of climate change on global food production by 2030 is likely to be small (FAO: World Agriculture: Towards 2012).

The Availability of Water. The availability of water is the crucial element in all efforts to expand agricultural production; without a minimum level of water supply any efforts to expand agricultural production will be in vain. The possibilities of extending present irrigation practices seem to be limited since water resources of sufficient quality have become scarce or too expensive to use, especially in developing countries and major agricultural countries, despite the fact that at the global level enough water is available. The problem will even increase in the future because of already ongoing climate change. More efficient use of rain in rain-fed agriculture requires more scientific attention amongst others more stress-tolerant varieties of crop plants must be bred, mainly through the use of genetic engineering.

Factors with an impact on soil, water, climate and crop must be put into more focus because they play an important role in optimizing the use of rainwater.

Presently, approximately 20 % of agricultural land is operated under irrigation production systems, providing 40 % of the world's food production. In 2000 in developing countries two million km^2 were irrigated; this area will be expanded to 2.42 million km^2 by 2030. Most of this expansion will occur in land-scarce areas where irrigation is already crucial, requiring considerable improvements in irrigation technologies (FAO: World Agriculture: Towards 2012).

31.2.3 Economy, CO_2 Reduction Goals and Food Security

According to the World Food System Model, a considerable increase in staple food prices must be expected. As predicted in the World Food System Model for 2020, a 30 % rise in prices is assumed against a situation without biofuel production. Especially alarming and of concern is the assumed dramatic increase in the prices of coarse grains, maize and sorghum. These cereals are of major importance for developing countries, especially in Africa, where they are the most important staple foods; high prices of coarse grains are therefore a distinct threat to food security in developing countries. Because of the large quantities of protein-rich by-products in first-generation biofuel production, especially in the case of bio-diesel production, the price of protein-rich feed materials may fall significantly below the reference level (no biofuel production) (Biofuels and Climate 2012).

The number of malnourished people worldwide continues to increase; the majority of food-insecure people live in Asia and the Pacific region, followed by sub-Saharan Africa.

If the CO_2 reduction goals and, in more practical terms, the goals set for the production of biofuel are maintained, the number of malnourished people will rise instead of falling, despite the fact that the land resources are available for both biofuel feedstock and food/feed production. Higher prices of staple foods will deny access to a nutritional and balanced diet to millions of people in developing countries. The only way to reduce the number of malnourished people in a significant way would be to cut biofuel production to zero. If biofuel production is maintained at its present level or expanded according to the recommendations of OECD countries and non-OECD countries, the number of malnourished people will grow dramatically. It is understood that even the introduction to the market of second-generation biofuels will not reduce the number of malnourished people below the present baseline (FAO 2011).

It is obvious that a reduction in traffic borne CO_2 emissions by replacing a large proportion of fossil fuels by biofuels may result in a transport fuel security approach to combustion energized transport. It is, however, also obvious that this development would expand considerably the number of hungry people in the world. It must be understood that, for now, biofuel production with the aim of achieving transport fuel security and food security cannot be dealt with as independent parameters; they are closely related. If biofuel production is increased as proposed, there will be a need for significant savings on food but also on the feed side (Dyson 1999).

Dramatic are however the reductions in food uses in developing countries, the quoted numbers underpin the statement that biofuel production reduces the nutritional basis, especially in poor developing countries. The chances of expanding profitable agricultural production into so far non-profitable areas are utilized for the production of second-generation biofuel feedstock may offer some means of alleviating the crisis, it may help in reversing migration to urban centres, especially in developing countries. The expectations of expanded feedstock production for biofuels should not be too high, however. Projections show that farmers in developed countries stand to profit the most, whereas farmers in developing countries see only marginal profit from development.

Improvements are possible in the case of small-hold farmers in developing countries, e.g. Africa. This could be because small-hold farmers would grow biofuel crops like jatropha (*Jatropha curcas*) or pongam (*Pongamia pinnata*) alongside food crops, which would help them to increase their cash income. Both trees are quite interesting. The collateral benefits from an expanded biofuel production could be that the farmers would earn additional cash for labour extensive production.

31.2.4 Are Their Any Real Benefits of Biofuel Production?

The underlying idea for biofuel production was and still is that the transportation-related part of fossil fuel use could be considerably reduced by replacing fossil fuels with biofuels, whose use would be emission neutral or even positive. The practice of biofuel production has, however, shown that considerable quantities of energy must be used for biofuel production, and only biofuels manufactured from biomass grown on degraded or abandoned agricultural lands incur little or no carbon debt. Starting with agricultural production including soil management, harvesting, transportation and processing, there are many steps in the chain leading to the final product which use energy. Projections developed on the basis of various scenarios do not reveal a clear picture; depending on the fertilizer's share of the balance, indications are that in the near future (2020–2030) the global carbon emission balance through the production of biofuels will be negative to neutral or slightly positive; only in the distant future (2030–2050) can a truly positive carbon balance be expected. The production of biofuels may also be connected with negative impacts on the environment through deforestation, loss of biodiversity and unfavourable interventions in the use of water resources (OFID Pamphlet 2012).

31.3 Concluding Remarks

As stated earlier, the impact of biofuel production on the availability of food will depend on various factors, mainly how fast second-generation biofuels appear on the market. The IEA is assuming that the contribution of second-generation biofuels to the share of biofuels will grow steadily starting in 2015/2020; it may then take 5–10 more years before it has a significant influence on biofuel production. It is anticipated that only from 2050 will the carbon debt from biofuels through second-generation biofuel production be entirely negative.

It is, however, evident that for now biofuel production will compete with food and feed production, with the effect that both food and biofuel prices will be interdependent. The level of both prices will depend on various factors, on market mechanisms, but ultimately on the political willingness of the developed countries to allow people in developing countries to develop a sustainable (agricultural) industry.

An important question is also whether export-oriented industrial production in developing countries will grow in the near future; with growing exports most developing countries will/should be able to afford, at least to a higher extent than presently, the imports of cereals required to provide food to their growing populations.

The prices and availability of food are also linked to patterns of food consumption, which are becoming more similar throughout the world, shifting towards higher-quality and more expensive foods such as meat and dairy products. Cereals will stay the most important staple food.

The change from plant-based proteins to more meat and milk/dairy products (Table 31.13), with all that implies, is more environmentally damaging than biofuel production.

The central question of this report of whether biofuel production threatens food security is difficult to answer. At the moment it certainly is having an impact as seen in the rise in coarse grain prices, and in the very near future this situation will continue. Beyond 2050 or even before, biofuel production will be based on non-food plant feedstock and will no longer interfere with food and feed production (IEA Energy Technology Essentials 2007).

An analysis of the presented data shows that the production of various types of first-generation biofuels will not be restricted by the availability of land resources and will not impede the production of sufficient food and feed to avoid food insecurity. There is certainly enough land available to produce sufficient food, feed and biofuels to meet the anticipated targets, including Millennium Development Goals 1 (Millenium Goals 2012). The existing food insecurity and the problems of the near and distant future can only be remedied by a new way of political thinking. The political classes in both developed and developing countries must assume their responsibility to think in a truly global way. It must be kept in mind that food security exists when all people, at all times, have physical and economic access to sufficient safe and nutritious food to meet their dietary needs

Table 31.13 Meat and milk/dairy products consumption in DC (Delgado 2012)

Year	Meat consumption (kg)	Milk/dairy products consumption (kg)
1966	10	28
1999	26	45
2030	37	66

and food preferences for a healthy and active life (FAO and Food and Agriculture Organization of the United Nations 1996).

If, in the current political climate, a target is set that around 10 % of the fuels used for transportation must be derived from renewable biological sources, then it must be expected that around 10 % of the global population will be food insecure, and this means that out of nine billion people 900 million people will be food insecure.

The authors of the study BIOFUELS and FOOD SECURITY arrived at the following conclusion:

"The mandates and targets together with substantial funding and subsidies to develop biofuels in the presumed interest of transport energy security should raise ethical and moral concerns regarding the failure of the international community to make decisive progress towards achieving world food security" (OFID Pamphlet 2012).

References

Berg Ch. and F.O. Licht World Fuel Ethanol, Analysis and Outlook, prepared for METI. http://www.meti.go.jp/report/downloadfiles/g30819b40j.pdf. Accessed 24 May 2012

Biofuels & Climate change – challenges to world food security in the 21st century. http://www.iiasa.ac.at/Admin/INF/feature_articles/Options/2009/biofuels_climate_change.html. Accessed 24 May 2012

Biomassevergasung. http://www.arhlive.de/ext-seite/index.php?url=+http%3A%2F%2Fde.wikipedia.org%2Fwiki%2FBiomassevergasung. Accessed 24 May 2012

Delgado CL (2012) IFPRI, rising consumption of meat and milk in developing countries has created a new food revolution. JN J Nutr 142(6). http://jn.nutrition.org/content/133/11/3907S.full. Accessed 24 May 2012

Dyson T., World food trends and prospects to 2025, Proc Natl Acad Sci USA 96:5929–5936, May 1999. http://www.pnas.org/content/96/11/5929.abstract . Accessed 24 May 2012

FAO 2011, The state of food security in the world. http://www.fao.org/publications/sofi/en/. Access 24 May 2012

FAO Document: how many people can the land support? Dimensions of need – an atlas of food and agriculture. http://www.fao.org/docrep/u8480e/U8480E0E.HTM. Accessed 24 May 2012

FAO Natural Resources Management and Environment Department: long-term scenarios of livestock-crop-land use interactions in developing. http://www.fao.org/docrep/W5146E/w5146e05.htm. Accessed 24 May 2012

FAO, Food and Agriculture Organization of the United Nations (1996) Declaration on world food security. World food summit, Rome, Italy. http://www.fao.org/docrep/003/w3613e/w3613e00.HTM. Accessed 24 May 2012

FAO: the state of food and agriculture 2008, part 1: biofuels: prospects, risks and opportunities services. ftp://ftp.fao.org/docrep/fao/011/i0100e/i0100e04.pdf. Accessed 24 May 2012. Licht, F.O.: world ethanol & biofuels report, Oct 2007 and May 2008. http://www.agra-net.com/portal2/puboptions.jsp?Option=archive&pubid=ag072. Accessed 24 May 2012

FAO: World Agriculture: Towards 2015/2030. Summary report. http://www.fao.org/docrep/004/y3557e/y3557e03.htm. Accessed 24 May 2012. FAO, world agriculture: towards 2030/2050, Interim report. http://www.fao.org/fileadmin/user_upload/esag/docs/Interim_report_AT2050web.pdf. Accessed 24 May 2012

FAO: World Cereal Demand in 2030. http://www.fao.org/docrep/004/y3557e/y3557e08.htm. Accessed 24 May 2012

FAOstat, Viridiplantae, 24 Jan 2009, cropland areas in 2005 relative to world land area. http://www-958.ibm.com/software/data/cognos/manyeyes/datasets/cropland-areas-in-2005-relative-to-w/versions/1. Accessed 24 May 2012

IEA Energy Technology Essentials ETE 02, Jan 2007, Biofuel production. http://www.iea.org/techno/essentials2.pdf. Accessed 24 May 2012

IEA: World Energy Outlook (2008/2009) http://www.worldenergyoutlook.org/. Accessed 24 May 2012

Macedo I.C., NIPE, Evaluation of GH gas emissions/mitigation in in ethanol production/utilization: current issues, in workshop: sugarcane ethanol sustainability CBTE, campinas, May 2009. http://www.bioetanol.org.br/hotsite/arquivo/editor/file/Workshop%20Sustentabilidade/CBTE%20Workshop%20-%20Isaias%20Macedo.pdf. Accessed 24 May 2012

IUFoST Scientific Information Bulletin (SIB) March 2010: impacts of biofuel production on food security by Lindsey Norgrove. http://iufost.org/sites/default/files/docs/IUF.SIB.Biofuel Production.pdf. Accessed 24 May 2012

Millenium Goals. www.un.org/millenniumgoals. Accessed 24 May 2012

OECD: a review of policy measures supporting production and use of bioenergy OECD TAD/CA/APM/WP (2007)24/FINAL. http://www.oecd.org/dataoecd/37/43/41037609.pdf. Accessed 24 May 2012

Office of Policy Analysis, Office of Policy and International Affairs, U. S. Department of Energy, Washington, DC 20585; Sept 15, 2008: Understanding the challenges to meeting the U.S. renewable fuel standard. http://www.ourenergypolicy.org/wp-content/uploads/2011/11/2008_09_USDOE_WorldBiofuelsProductionPotential.pdf. Accessed 24 May 2012

OFID PAMPHLET Series 38, biofuels and food security-IIASA. http://www.iiasa.ac.at/Research/LUC/Homepage-News-Highlights/OFID_IIASAPam_38_bio.pdf. Accessed 24 May 2012

Solar Energy Balance (2012) http://en.wikipedia.org/wiki/World_energy_consumption. Accessed 24 May 2012

UN, Department of Economic and Social Affairs, ESA/P/WP.187, 9 Dec 2003: population to 2300: http://www.un.org/esa/population/publications/longrange2/Long_range_report.pdf. Accessed 24 May 2012

United Nations, Department of Economic and Social Affairs, Population Division (ST/ESA/SER.A/236), world population to 2300 (page 18, A. world population). http://www.un.org/esa/population/publications/longrange2/WorldPop2300final.pdf. Accessed 24 May 2012

Vogelbusch: Planning a bioethanol plant. http://www.bioethanol.vogelbusch.com/en/faq.php. Accessed 24 May 2012

IEA key World Energy Statistics (2010) http://www.iea.org/textbase/nppdf/free/2010/key_stats_2010.pdf. Accessed 24 May 2012

Chapter 32
Academia-Industry Interaction in Innovation: Paradigm Shifts and Avenues for the Future

I. Sam Saguy

32.1 Introduction

The Europe 2020 strategy is focused on the European Union's (EU's) capacity to create millions of new jobs to replace those lost in the recent economic crisis and on the fact that future standards of living will depend on the ability to drive innovation in products, services, business models, and social processes. Innovation is at the heart of this strategy and has been identified as the best means of successfully tackling major societal challenges, such as population growth, land, energy and water resource scarcity, health, obesity, aging, and sustainability, to mention only a few, all of which are increasing in urgency. Innovation has been defined variously, but the most straightforward definition is the process of transforming an idea/invention into a good/service that consumers/customers are willing to purchase. It is important to note that to be defined as innovation, an idea or invention must be replicable at an economical cost and should satisfy specific consumer/customer needs. Innovation represents a significant driving force and a unique opportunity to address global economic pressures, unstable markets, accelerated exponential growth of scientific knowledge and technological complexity, and new consumer needs and expectations. Innovation is the application of ideas, technology, and processes in new ways to gain a competitive advantage and create value and plays a vital role in all facets of modern life (Saguy 2011a). Innovations can become commodities at an unprecedented speed, and consequently, continuous effort is required to sustain and nourish this process.

The mantra "innovate or die" is no longer sufficient. Open innovation (OI) and innovation partnerships could be the leitmotif for today's companies. OI has been defined as "a paradigm that assumes that firms can and should use external ideas as well as internal ideas, and internal and external paths to market, as the firms look to

I.S. Saguy (✉)
The Robert H. Smith Faculty of Agriculture, Food and Environment, The Hebrew University of Jerusalem, P.O. Box 12, Rehovot 76100, Israel
e-mail: sam.saguy@mail.huji.ac.il

S. Yanniotis et al. (eds.), *Advances in Food Process Engineering Research and Applications*, Food Engineering Series, DOI 10.1007/978-1-4614-7906-2_32,

advance their technology" (Chesbrough 2003). A more recent definition, "the use of purposive inflows and outflows of knowledge to accelerate internal innovation, and expand the markets for external use of innovation, respectively" (Chesbrough et al. 2006), highlights the fact that OI has become a widespread practice. OI is founded on the reality that, in a world of vastly distributed knowledge and accelerated rates of development, companies can no longer afford to rely on their own research; they must utilize outside sources and buy or license processes, technology, inventions, and solutions (Traitler 2009; Traitler and Saguy 2009).

OI has seen a massive expansion in recent years (Gassmann et al. 2010; Lindegaard 2010). The goal set back in 2000 by Procter & Gamble (P&G) for their OI model of "Connect + Develop" – that 50 % of their innovation be acquired from outside the company – has made significant inroads (Lafley and Charan 2008). Since 2000, OI has contributed dramatically to increasing the commercial success rate of P&G. For instance, from 2000 to 2008, the success rate of new P&G products increased from 15–20 % to 50–60 % (Lafley 2008). This significant increase has far-reaching ramifications in reducing new product development (NPD) costs and affecting the overall performance of the company and its status in the marketplace. It is therefore not surprising that OI has spread and mushroomed in many industries (e.g., pharmaceuticals, chemicals, biotechnology, drugs, software) and that the large food industry has followed suit (e.g., Erickson 2008; Fortuin and Omta 2009; Kuhn 2008; Saguy 2011a; Sarkar and Costa 2008; Traitler 2009; Traitler et al. 2011; Traitler and Saguy 2009), as have small and medium enterprises (SMEs; e.g., Kocher et al. 2011; van de Vrande et al. 2009).

However, despite OI's widespread applications, only 10 % of all companies are ready for it; another 30 % (called contenders) have seen the light and are struggling to make it work, and the remaining (60 %) companies (or pretenders) do not really know what OI is or why or how it could be relevant to them (Lindegaard 2010). SMEs and others firms operating in traditional sectors are struggling with its implementation due to their relatively low level of absorptive capacity and management challenges, which are perceived as unattainable (van de Vrande et al. 2009; Lindegaard 2010). The untapped potential and full adoption of OI is particularly relevant for the EU Food and Drink (F&D) sector – the EU's largest manufacturing sector, which employs some 4.4 million people generating 14 % of total manufacturing jobs. As elsewhere, the SMEs in this sector are especially struggling with OI implementation (Bianchi et al. 2010; Lee et al. 2010; Raymond and St-Pierre 2010), and although they comprise 99.1 % of the total 310,000 F&D companies, they generate only 49 % of the F&D turnover ((CIAA 2009); visited Nov. 14, 2010). This topic has attracted much attention and is under deliberation by various EU bodies, as well as the European Federation of Food Science & Technology (EFFoST) and the International Union of Food Science and Technology (IUFoST) (Saguy 2011a).

The overall objective of this chapter is to delineate the main elements of the strategy, vision, and changes required to enhance academia–industry interactions in innovation, namely: (1) identify key elements, and highlight academia–industry innovation paradigm shifts to meet mutual future challenges; (2) propose "pull" elements to enhance academia–industry collaboration; and (3) promote new approaches to encouraging partnerships, networking and innovative thinking, and implementation.

32.2 Academia–Industry Interactions in Innovation

Collaboration is a key piston in the engine that drives economic growth. Value creation is the ultimate goal of any partnership; without it, the concept holds no real merit for the partners. The five main principles of the Sharing Is Winning (SiW) model are as follows (Saguy 2011a; Traitler 2009; Traitler and Saguy 2009; Traitler et al. 2011): partner selection, cocreation of intellectual properties (IPs), joint creative problem-solving teams, implementing best practices, and sustainable and continuous processes affecting people, mindset, metrics, culture, and education. Ultimately, the overall objective of SiW is the alignment of the value chain with consumer-centric innovations. SiW extends the definition of OI in that it is a new avenue for collaboration in all areas of discovery and development, with external partners bringing competence, commitment, and speed to the relationship while also sharing the risk of innovation (Traitler et al. 2011). The SiW roadmap is founded on the OI principle of outside-in for cocreation with complementary partners through alliances, cooperation, and joint ventures and has proven very effective in a wide spectrum of industries and applications. However, the counter aspect of OI, inside-out, still faces steep slopes that are, in numerous cases, insurmountable. Many large food companies feel that they are not ready to share their IPs, and consequently, the inside-out route is seldom utilized. SMEs lag behind on both outside-in and inside-out approaches, and their utilization of SiW is limited or nonexistent due to a plethora of real or perceived issues, which fall into the following main groups (Braun and Hadwiger 2011; Massa and Testa 2008): barriers (e.g., limited resources, time, legal, cultural or emotional barriers, and barriers to trust), inadequate capabilities (e.g., knowledge, consumer understanding, external contacts, networks, education, training, language skills), control (e.g., reluctance to delegate authority or decision making), and monopoly (e.g., enforced by some large firms).

Despite the great potential of SiW, significant inroads still need to be made in collaborations between industry and academia. Academia's long history of working in isolation, its different value chains, and a general misunderstanding between parties furnish a partial explanation for academia's staying at arm's length. So-called ethical conflicts threaten academia by distracting it from teaching and basic research, undermining collegiality, encouraging secrecy, preventing or delaying publication, and devaluing the human component (Tyler 2009). Traditional collaborative conflicts between academia and industry revolve around issues that include confidentiality, publishing, IP rights, and ownership. The mindsets and research foci of the two institutions are also quite different: while universities are mainly centered on fundamental research (R), industry works primarily on development (D), with a typical D-to-R ratio in the food industry exceeding 4:1. Concerns about potential conflicts of interest arise when members of the academic community interact with industry (e.g., consultants, scientific advisors). These concerns are alarming, but SiW principles should serve as a platform for promoting the development of mutually beneficial collaborative relationships. Other factors, such as culture and funding, have also been identified as significant constraints stifling fruitful collaboration (Melese et al. 2009). It is important to note that recent studies clearly show that spending more on research and development (R&D) does not guarantee improved results. The most crucial factors identified in this respect were strategic alignment and a culture that supports innovation. It is worth noting that successful innovators are able to execute their distinct innovation programs when their activities are aligned with their overall business strategy. Innovators who had achieved this state of coherence were found, consistently and significantly, to outperform their rivals on several financial measures (Jaruzelski et al. 2011; Jaruzelski and Dehoff 2010). In addition, regardless of company size, strategy is also viewed as the most important best practice for NPD (Nicholas et al. 2011).

Another basic difference between academia and industry is their value chains. While industry is driven mainly by its bottom line and gaining full IP rights, academia is primarily motivated by the pursuit of basic science and knowledge dissemination, student education, publication, and, often, full IP rights as well. OI proliferation has proven that academic freedom is not affected and that biased company-sponsored research is quite rare. It is generally accepted, especially by EU countries, that university innovations have an underutilized and unrealized potential that continues to lie dormant. Consequently, sustained efforts to advance university innovations are needed. Building bridges between university researchers and businesses is critical for knowledge transfer, which is no longer an option but a must, and SiW can pave the way. However, even OI focuses on quite narrowly defined, short-term transactions, missing the opportunity to build much longer, trust-based relationships that can be used to engage diverse teams in tackling more diffuse and broadly framed challenges (Hagel et al. 2010). This issue might be due to the fact that OI has only recently seen extensive utilization, and it should therefore continue to adopt a larger scope for driving collaborations between academia and industry.

32.3 Discussion

For the purpose of implementing significant changes in academia–industry collaborations and the pursuit of innovation, SiW principles should be expanded and an overarching common vision developed. This game-changing vision should be based on new molding strategies (i.e., reshaping broader markets, industries, or social arenas; Hagel et al. 2010), which should be expanded to include academia as well. This new concept provides a fresh vision, and shaping strategies should address the broad scope of innovation and furnish industry and academia with opportunities to seek improved means and tools to develop platforms that will maximize mutual efforts, enable participation, promote learning and social responsibility, and lead to the creation of an innovation ecosystem. Academia should play a paramount proactive role in the conceptualization, deliberation, and design of this new vision and contribute to its content. The EU should also play a pivotal role in integrating all stakeholders and provide the leadership and support required to create rich seedbeds for innovation, learning, and collaboration. Continuous collaborations on this topic have already been addressed by the EU [e.g., European Technology Platforms (ETP 2010; ISEKI_Food 4 2010; Track_fast 2010; and European Commission Education & Training 2010; ISEKI 2010).

Academia–industry interaction toward innovation calls for several paradigm shifts, four of which are outlined in what follows.

32.3.1 Barrier Removal

The road from a discovery stemming from basic research to a commercial product, process, or service is long and rife with significant obstacles. Typically, a funding gap or "valley of death" (VoD) exists between basic research and commercialization (Markham et al. 2010). To simplify the VoD concept, the innovation sequence can be typically depicted in three stages: stage 1 is basic research, also termed pre-NPD, providing what is known as the front end (or fuzzy front end) of innovation; stage 2 includes the transformation from basic research outcomes into a potentially marketable product/service; stage 3 is commercialization and diffusion of a new product/service, translating projects into economic value. Government practices in R&D support may be the most significant contributor to the emergence of a VoD. Intervention at early stages of the innovation process can exacerbate the problem of underinvestment in intermediate-stage research. There are at least two ways in which this occurs: creation of a rift or "valley" in the innovation sequence by inflating the output of basic research above funding levels that will be invested at a later stage and altering the provision of funds at intermediate stages. Therefore, a government that is concerned with generating economic value from its basic R&D efforts should enhance its support of intermediate-stage research (Beard et al. 2009). Intimate involvement of the private sector is also needed to fund early-stage startups and incubators. These observations should be carefully considered by EU authorities when assessing whether and how to shift resources to enhance SME innovation. Even the remote possibility that EU funding (e.g., 7th Framework Programme) has contributed to proliferation of the VoD and has probably negatively affected innovation is shocking and warrants in-depth analysis.

The need for technology transfer (TT) or knowledge transfer (KT) can be highlighted with one typical initiative of TT and KT that includes European research organizations, industrial federations, and enterprises with the final goal of setting up a European Institute for Food Processing (i.e., HighTech Europe (2013), http://www.hightecheurope.eu/). This initiative facilitates the implementation of high-tech processing in the food industry by identifying, developing, and demonstrating potential and cost-efficient innovations via the ScienceCube approach [i.e., linking innovative scientific knowledge in biotechnology, nanotechnology, and information and communication technology (ICT) and the needs of the food industry] using scientific principles and basic food-processing operations, availability of high-tech pilot facilities, and connecting the regional KT. This ultimately results in interlinking regional KT, optimizing TT of the most recent R&D findings to industry, and providing Europe-wide suggestions for industrial needs in food processing.

To traverse the VoD successfully, academia should recognize its essential function by reaching out to industry and playing a proactive role. Conducting and excelling in basic and fundamental research is a prerequisite. Crossing the VoD by learning industry's needs and driving inventions at least past the pre-NPD stage, until the industry can pick them up, is also of paramount importance. The typical pre-NPD includes four steps: (1) affirming the technical viability of the invention as a product/service; (2) formalizing the product/service concept; (3) validating the concept with market research; and (4) developing a business case to gain commercial support, using consumer or market research (Markham et al. 2010). While initially the invention is the major driving force, consumer and market research and the business model take the lead at later stages. Therefore, a sustainable partnership is required. This partnership calls for both academia and industry to take a proactive stance and participate in each step of the innovation. Hence, it calls for a new mindset that supports and promotes a seamless free flow of knowledge, technology, and solutions across universities and industry boundaries, as outlined and promoted by OI and SiW. Some specific recommended changes for academia are as follows:

(1) *Applied research status* – enhancing the importance of applied research is imperative. Applied research plays an important role by significantly contributing to teaching quality, as it focuses on relevant topics. It can also contribute to and improve direct interactions between students and industry and create a magnet for resources that will further enhance collaborations. Ultimately, applied research can contribute to a university's reputation by offering numerous benefits (e.g., recruiting higher-quality faculty, increased research funding) and possible breakthrough advances in fundamental research.
(2) *Academia's new role* – the deep-rooted characteristics of the professor should be reassessed. He or she should play a proactive role in industry as well, motivated by the synergistic power of collaboration and driven by the overall goal of becoming a full member of an industrial team. This could require

devoting ample time to a particular industry and becoming an "organic" member of the industrial OI effort. The intimate presence of academicians in industry should create new possibilities, such as offering advanced industrial studies (Ph.D., M.Sc.). This approach could lead to significant outcomes, such as opening the door to, for example, industrial internships, fellowships, and advanced education. Last but not least, with the ballooning cost of equipment, access to sophisticated industrial laboratory equipment and resources is a huge and very significant benefit. Encouraging academic researchers to supervise joint theses carried out partially or entirely in industry is strongly recommended. These changes call for an in-depth assessment of course curricula, teaching methods, and learning modes. Institutes of higher education should also reconsider their standard "push" curricula, exposing students to codified information in a predetermined sequence of experiences, moving instead toward "platforms" designed to flexibly accommodate diverse needs and new and more efficient learning modes. Thus, several broad forces are addressed (e.g., increased uncertainty, growing abundance, intensified competition, increased emphasis on learning, role of consumers/customers) that are shaping the emergence and evolution of new roles for academia.

To address these new roles, academia should open its doors to several significant modifications. A recent book describes the power of "pull" (Hagel et al. 2010), which is based on three main elements: access (find, learn, and connect with people, products, and knowledge to address unanticipated needs), attract (people and resources one did not even know existed but are most relevant and valuable), and achieve (drawing out personal and institutions' full potential with less time and greater impact). This offers a very plausible approach to creating value and rapidly driving performance to new levels. It calls for the creation of environments that effectively integrate teams within a broader learning ecology so that performance improvements accelerate as more participants join in. Adapting this approach would allow us to move away from the old way of push to the new way of pull; it requires the creation of an innovation ecosystem that will go beyond single university boundaries by transforming learning and teaching methods, encourage the combined efforts and full participation of all the diverse shareholders, and, most importantly, make use of passionate human resources. Accomplishment of this new paradigm requires a new mindset that allows thoroughly revised or brand new curricula and teaching and learning methods, so that ultimately academia will reinvent and reinnovate its domain.

(3) *Industrial involvement* – the new paradigm calls for more intimate industrial involvement, mainly of its leadership and experts, transforming their role into a proactive one by teaching graduate courses, mentoring research, serving on university committees and boards, contributing to the strategic thinking of universities, and enhancing its overall contribution to social responsibility.

(4) *The fourth helix* – the three main axes listed previously have been described as a triple helix (Etzkowitz 2008), encompassing university–industry–government interactions as a crucial key to innovation in increasingly knowledge-based

societies. This triple helix intersection of relatively independent institutional spheres generates hybrid organizations such as TT or KT offices in universities, technology-based firms, and government research labs. Although this concept is very appealing in that it integrates the key players in the innovation ecosystem, it lacks a very significant fourth player, namely the private sector, which includes banks, private and venture capital (VC) funds, and angels: these are extremely important in jump-starting the innovation process and providing required and essential resources. This concept has been defined as the fourth helix (Saguy 2011b).

32.3.2 Revised IP Model

Innovation cannot exist without IP rights, and this creates what is known as a Gordian knot. Resolving this issue is paramount. The traditional role of the technology transfer office (TTO), acting as a broker between academia and industry by providing expertise and managing the commercialization process related to TT, patenting, licensing, and the creation of startup companies, should be modified. The TTO's main objectives should be reformulated to increase the likelihood of maximum impact. Focusing only on IP rights has frequently become an impassable and sometimes even crippling barrier to innovation success. This concern is even more salient in an OI ecosystem. To avoid stagnant situations, the complex IP issue requires special attention and new business models for cosharing. An example of IP management can be taken from the University of California at Berkeley's office of Intellectual Property and Industry Research Alliances (IPIRA), whose overview of the university's transactions was replaced with a more flexible approach, facilitating the achievement of an array of outcomes that match the mutual goals of industry and the university. This array opened up an entire relationship continuum. IPIRA supports Berkeley's research enterprise and its goal of deploying research results for social impact and public benefit. When universities elect to make academic discoveries proprietary by obtaining IP rights, and when they license those rights, they are demonstrating good stewardship (Mimura 2010). Hence, it is important to note that revenue generation from IP rights should be considered in the overall context of network collaborations, partnerships, social impact, and optimal accessibility. Multiple and diverse adaptable IP-management strategies are therefore required. Supporting and implementing this approach is a significant burden that rests on the shoulders of the academician and calls for leadership by personal example.

32.3.3 *Management's Role*

Management commitment and leadership are vital in nourishing, embracing, and facilitating OI and SiW principles, broadening the group of participating partners, and enabling and sustaining the innovation process. Management's foremost role in industry is to recognize that they are the "gatekeepers" of the innovation flow (Watzke and Saguy 2001) and must promote, leverage, and drive the required organizational changes and culture to increase the likelihood of success. The counterpart to management in academia should develop a strategy and sustain a culture that promotes collaborations and elevates and enhances the academic status of applied R&D. Note that academia needs to undergo significant changes beyond embracing and promoting applied research: it must also value the importance of the overall impact of its research and inventions through the lens of contributions to society and social responsibility. This calls for further in-depth consideration, and each university should develop its own mission to comply with this requirement.

Random collisions and interactions among innovation contributors are not an option. The new mantra has changed from "innovate or die" only a decade ago to "partner or perish" today – a new tune and way of life (Traitler 2009; Traitler et al. 2011). To thrive, management should institutionalize alliances and partnerships (i.e., outside-in and inside-out) to benefit from cross fertilization and synergy. Aligning university and industry in the codevelopment of sustainable innovation is not straightforward: it requires considerable management planning and commitment. Different cultures and mindsets are significant hurdles to overcome in obtaining a sustained codeveloped innovation process. Organization expediencies, cultures, and, especially, people's personalities are also significant factors that need to be considered. Moreover, such an alignment requires truly new thinking at both academia and industry levels. True changes do not come about through simple incremental developments, and conventional stepwise improvements will not suffice. The changes must be bold, requiring novel thinking and new leadership (Moskowitz and Saguy 2013). Openness cannot simply be wished for; it must be designed and engineered into the new system by its leaders. Academia and industry need a mutually shared vision, a coordinated thrust toward reforming the old systems, not only of relationships between industry and academia but, perhaps more importantly, of teaching, learning, and studying. Academic leaders must plan to bring students into the new corporate reality that has emerged in the last decade and will continue to develop in the foreseeable future. Perhaps the first task is to inculcate in students a recognition of the new reality, namely, that science, technology, knowledge, business, and social responsibility are all part of today's new world requirements (Moskowitz and Saguy 2013). Last but not least, to facilitate partnerships and improve the chances of a successful innovation outcome, industrial and academic management should be aligned so that all stakeholders can split the efforts and share the benefits.

32.3.4 Social Responsibility

For a business to create value for its shareholders over the long term, it must also bring value to society. Since its emergence, the notion of corporate social responsibility (CSR) has moved from ideology to reality: today it represents an important dimension in contemporary business practices and promotes OI (Jenkins 2009). The concept of created shared value (CSV) recognizes that societal needs, not just conventional economic ones, define markets. It also recognizes that social harm or weaknesses frequently create internal costs for firms – such as wasted energy or raw materials, costly accidents, and the need for remedial training to compensate for inadequacies in education. Addressing societal harms and constraints does not necessarily raise costs for firms because they can innovate through the use of new technologies, operating methods, and management approaches – and, as a result, increase their productivity and expand their markets (Porter and Kramer 2011).

Universities also have an important social responsibility and therefore should play a major role in maximizing the outcome of their research's impact while simultaneously considering CSV. This should include scientific merit as well as the overall contribution to society. Metrics for quantifying scientific contributions have been developed over the years (e.g., journal quality, impact factor, number of citations), as have measures of financial success (e.g., patents, licensing, royalties). However, the social impact associated with CSR and CSV is still vague. The development of suitable metrics to assess and evaluate research's overall contribution lies at our doorstep, and it is our responsibility to address this complex topic. Metrics cover a spectrum of important dimensions: on the one hand, they should continue to promote the high quality of fundamental research and reward scientific breakthroughs, enhanced OI, and partnerships. On the other, they should also facilitate and augment contributions to society. A genuine concern for society in all actions and decisions should become the norm and an integral part of the innovation process (Saguy 2011a).

32.4 Conclusions

Four paradigm shifts are recommended: barrier removal, revised IP model, management's role, and social responsibility, highlighting measurable and meaningful actions. They constitute a blueprint for jump-starting the process required to meet the innovation challenges facing academia and industry. Although significant changes are recommended, it is realized that, in addition to moving forward, a consensus needs to be created, requiring open deliberations and further discussions. To effectively cope with the accelerating development of science and technology, universities and industry need each other. This need is amplified by the quest for highly qualified human resources and the ever-increasing cost of research and equipment. The innovation process creates value and should also include implicit social responsibility. SiW offers a win-win approach to addressing these topics, but it requires additional input from all key players to galvanize the impetus of change. It requires passionate and committed leaders, academicians, human resources, and organization. Partnerships offer an opportunity to co-innovate the future. Integrating the whole innovation process should therefore take into consideration the social contribution. That which has been the norm is no longer sufficient; together we can make a difference, and we must not fail to try.

References

Beard TR, Ford GS, Koutsky TM, Spiwak LJ (2009) Avalley of death in the innovation sequence: an economic investigation. Res Eval 18(5):343–356

Bianchi M, Campodall'Orto S, Frattini F, Vercesi P (2010) Enabling open innovation in small-and medium-sized enterprises: how to find alternative applications for your technologies. RD Manag 40(4):414–431

Braun S, Hadwiger K (2011) Knowledge transfer from research to industry (SMEs): an example from the food sector. Trends Food Sci Technol 22(2011):S90–S96

Chesbrough HW (2003) Open innovation: the new imperative for creating and profiting from technology. Harvard Business School Press, Boston

Chesbrough H, Vanhaverbeke W, West J (2006) Open innovation: researching a new paradigm. Oxford University Press, Oxford

CIAA (2009) CIAA annual report. http://www.fooddrinkeurope.eu/documents/brochures/annual%20report%20CIAA%2009.pdf. Accessed 14 Nov 2010

Erickson P (2008) Partnering for innovation. Food Technol 62(1):32–37

Etzkowitz H (2008) The Triple Helix: university-industry-government innovation in action. Routledge, New York

European Technology Platforms (ETPs) 2010 http://etp.fooddrinkeurope.eu/asp/index.asp. Accessed 4 Aug 2010

European Commission Education & Training (2013) The lifelong learning programme: education and training opportunities for all. http://ec.europa.eu/education/lifelong-learning-programme/doc78_en.htm. Accessed 4 Aug 2010

Fortuin FTJM, Omta SWF (2009) Innovation drivers and barriers in food processing. Br Food J 111(8):839–851

Gassmann O, Enkel E, Chesbrough H (2010) The future of open innovation. RD Manag 40(3):213–221

Hagel JI III, Brown JS, Davison L (2010) The power of pull. How small moves, smartly made, can set big things in motion. Basic Books, New York

Hightech Europe (2013) The first European food processing network of excellence. http://hightecheurope.com/. Accessed 2 Aug 2010

ISEKI_Food 4 (2010) Towards the innovation of the food chain through innovation of education in food studies. http://www.iseki-food4.eu/. Accessed 4 Aug 2010

Jaruzelski B, Dehoff K (2010) The global innovation 1,000: how the top innovators keep winning. s + b Winter Issue 61. http://www.strategy-business.com/media/file/sb61_10408.pdf. Visited 4 April 2011

Jaruzelski B, Loehr J, Holman R (2011) Why culture is key. s + b Winter Issue 65. http://www.strategy-business.com/media/file/sb65-11404-Global-Innovation-1000-Why-Culture-Is-Key.pdf. Visited 3 Nov 2011

Jenkins H (2009) A 'business opportunity' model of corporate social responsibility for small-and medium-sized enterprises. Bus Ethics A Eur Rev 18(1):21–36

Kocher P-Y, Kaudela-Baum S, Wolf P (2011) Enhancing organisational innovation capability through systemic action research: a case of a Swiss SME in the food industry. Syst Pract Action Res 24:17–44

Kuhn ME (2008) Driving growth through open innovation. Food Technol 62(6):76–82

Lafley AG (2008) P&G's innovation culture. s + b Winter Issue 52. http://www.strategy-business.com/media/file/sb52_08304.pdf. Visited 8 June 2010

Lafley AG, Charan R (2008) The game changer. Crown Business, New York

Lee S, Park G, Yoon B, Park J (2010) Open innovation in SMEs-an intermediated network model. Res Policy 39:290–300

Lindegaard S (2010) The open innovation revolution essentials, roadblocks, and leadership skills. Wiley, Hoboken

Markham SK, Ward SJ, Aiman-Smith L, Kingon AI (2010) The valley of death as context for role theory in product innovation. J Prod Innov Manag 27(3):402–417

Massa S, Testa S (2008) Innovation and SMEs: misaligned perspectives and goals among entrepreneurs, academics, and policy makers. Technovation 28:393–407

Melese T, Lin SM, Chang JL, Cohen NH (2009) Open innovation networks between academia and industry: an imperative for breakthrough therapies. Nat Med 15:502–507

Mimura C (2010) Nuanced management of IP rights: shaping industry-university relationships to promote social impact. In: Dreyfuss RC, Zimmerman DL, First H (eds) Working within the boundaries of intellectual property for the knowledge society. Chapter 9. Oxford University, Oxford

Moskowitz HR, Saguy IS (2013) Reinventing the role of consumer research in today's open innovation ecosystem. Crit Rev Food Sci Nutr 53(7):682–693. (http://dx.doi.org/10.1080/10408398.2010.538093)

Nicholas J, Ledwith A, Perks H (2011) New product development practice in SME and large organizations: theory and practice. Eur J Innov Manag 14(2):227–251

Porter ME, Kramer MR (2011) The big idea: creating shared value. Harv Bus Rev 89(1–2):62–77

Raymond L, St-Pierre J (2010) R&D as a determinant of innovation in manufacturing SMEs: an attempt at empirical clarification. Technovation 30(1):48–56

Saguy IS (2011a) Academia and food industry paradigm shifts required for meeting innovation challenges. Trends Food Sci Technol 22:467–475

Saguy IS (2011b) Paradigm shifts in academia, food industry and government to accelerate innovation. Presented at innovation and technological linkage in foods: challenges and opportunities for the province of Córdoba, Cordoba, 27–28 Oct

Sarkar S, Costa AIA (2008) Dynamics of open innovation in the food industry. Trends Food Sci Technol 19:574–580

Track_fast (2013) Europe's food science and technology on a fast track. https://www.trackfast.eu/. Accessed 2 Aug 2010

Traitler H (2009) Innovation partnerships as a vehicle towards open innovation and open business. In: Moskowitz HR, Saguy IS, Straus T (eds) An integrated approach to new food product development. Chapter 5. CRC Press, Boca Raton

Traitler H, Saguy IS (2009) Creating successful innovation partnerships. Food Technol 63(3):22–35

Traitler H, Watzke HJ, Saguy IS (2011) Reinventing R&D in an open innovation ecosystem. J Food Sci 76(2):R62–R68

Tyler JE III (2009) Advancing university innovation: more must be expected—more must be done. Minn J Law Sci Technol 10(1):143–212

van de Vrande V, de Jong JPJ, Vanhaverbeke W, de Rochemont M (2009) Open innovation in SMEs: trends, motives and management challenges. Technovation 29(6–7):423–437

Watzke HJ, Saguy IS (2001) Innovating R&D innovation. Food Technol 55:174–188

ERRATUM TO:

Chapter 14
Aroma Encapsulation in Powder by Spray Drying, and Fluid Bed Agglomeration and Coating

Turchiuli Christelle and Dumoulin Elisabeth

S. Yanniotis et al. (eds.), *Advances in Food Process Engineering Research and Applications*, Food Engineering Series, DOI 10.1007/978-1-4614-7906-2, pp. 255–265,

DOI 10.1007/978-1-4614-7906-2_33

The publishers regrets that in the Table of Contents, Chapter 14 opening page, and the Footnotes of the print and online versions of this book, the names of the authors are written backwards. The correct order of their names should be "Christelle Turchiuli" and "Elisabeth Dumoulin".

The online version of the original chapter can be found at
http://dx.doi.org/10.1007/978-1-4614-7906-2_14

Index

A

Abadias, M., 398
Academia-industry interaction in innovation
  barrier removal
    fourth helix, 651–652
    industrial involvement, 651
    pre-NPD stages, 650
    research status, 650
    role of, 650, 651
    stages, 649
    technology/knowledge transfer, 650
    VoD concept, 649, 650
  concepts, 649
  definition, 645
  ethical conflicts, 647, 648
  management's role, 653
  open innovation
    applications, 646
    definition, 645, 646
    proliferation, 647
    significants, 646
  revised IP model, 652
  SiW model, 647
  social responsibility, 654
Acheson, K.J., 439
Achiwa, N., 396–398
Acrivos, A., 53
Agglomerated spray-dried powders, 262–263
Aguayo, E., 389
Ahn, H.-J., 385
Akbas, M.Y., 388, 390
Alemán, G.D., 379, 380
Alginate gums, 272–274
Allende, A., 393
Allergen protein polymerization
  fruit and vegetables, juices, 324–326
  peanut, 323–324
Almonacid, S., 197–206
Almonds, textural quality of
  activation energy, 532–533
  Arrhenius plots, of diffusion coefficient, 531–532
  bio-yield point, 533–534
  diffusion coefficient of water, 525–527
  equilibrium moisture content, 529, 530
  Fermi's distribution function, 534, 535
  force-distance relationship curves, 533
  material preparation, 523
  mechanical testing and properties, 524–525, 534
  moisture adsorption data, 525
  Monterey unpasteurized almonds
    moisture content *vs.* textural properties, 539, 541
    storage conditions, 539, 541
    temperature effects, 537, 539, 540
  nonpareil pasteurized almonds
    experimental moisture contents, 535, 536
    force-distance curves, 533
    fracture force values, 537, 538
    moisture adsorption curves, 529, 530
    percentage deviation, of predicted variables, 536, 537
    water adsorption curves, 530, 531
  predictive model, verification of
    flow diagram, 527, 528
    mean relative percentage deviation modulus, 528, 539, 542
    percentage of deviation, 528–529, 536, 537
    results, 535–542
  water adsorption curves, 530–531
  water migration, kinetics of, 524

S. Yanniotis et al. (eds.), *Advances in Food Process Engineering Research and Applications*, Food Engineering Series, DOI 10.1007/978-1-4614-7906-2,

Anandaramakrishnan, C., 164
Ananta, E., 344
Angell, C.A., 80, 81, 83–85, 87
An, G.J., 380, 381
An, J., 389
Area/plane scanning method, 584, 585
Ariana, D.P., 600
Aroma encapsulation
  applications, 454–455
  definition, 255
  efficiency, 256
  microencapsulation, 255–256
  powder properties
    agglomerated spray-dried powders in fluidised bed, 262–263
    agglomerate production, 260–261
    coated agglomerates, 263–264
    spray-dried aroma powders, 261–262
  preparation of powders
    emulsion and water, 258, 259
    fluidised-bed agglomeration and coating, 259–260
    products, 258
    spray-drying, 258–259
  rate of release, 257
Aroma microencapsulates, 241
Artés-Hernández, F., 391, 393
Aseptic processing and packaging, 143
Asteriadou, I.K., 209–224
Atmospheric freeze drying (AFD), 157–158
Augustin, W., 217
Autio, K., 348
Avalosse, T., 52

B
Backward difference method, 477
Baker, C.G.J., 167
Baker, R.A., 384, 386
Ballmer-Weber, B.K., 314
Banga, J.R., 486
Baranyi, J., 548
Baranyi model, 570, 571
Barbosa-Cánovas, G.V., 375–402
Bari, M.L., 396
Basaran, P., 400
Basbayraktar, V., 386
Batch coffee roaster
  with afterburner, 173–174
  afterburner bypassing, 192–193
  aroma concentration, 180
  bean temperature profiles, 178–179
  effects, 174–175
  effects in R Roasters, 188
  effects of operating and feed conditions, 186–187
  emission of reaction products, 180–181
  ending roasts, 178
  energy-saving methods, 193
  enthalpy balances, 185–186
  explosion hazards, 182
  green-bean load, 174
  heat transfer
    bean-hardware heat exchange, 183
    bean heating, 182
    efficiency, 182–183
    exothermic heating, 183
    rate of temperature variation, 183–184
    roasting time, 184–185
  high-speed roasting, 188
  kJ energy/kg bean feed, 186
  mass-flow rates, 177–178
  NO_X and CO, 176–177
  predrying of beans, 189
  recycling, 177
  roast darkness, 179–180, 189
  single-pass operation, 177
  SP roasting, 188
  stack-gas heat outflow reduction
    compact heat exchanger, 191–192
    fouling, 192
    heat-exchange system, 190–191
    volatile organic compounds and CO purification, 192
  volatile organic compounds
    in bean-cooler air, 189–190
    oxidation, 175–176
Baur, S., 388
Bayer, A.G., 44
Beltran, D., 388
Benzie, I., 434
Bermudez-Aguirre, D., 379, 400
Bialka, K.L., 390
Bibi, N., 386
Bigelow model, 569
Bintsis, T., 392
Bioactive component formulation
  carotenoids, 22–24
  droplet size *vs.* emulsifier addition, 19, 20
  homogenisation process, 19
  lipophilic components, 17–19
  microbial stable emulsions, 29–30
  microemulsions, 19
  phytosterols, 25–29

premix membrane emulsification, 20, 21
schematic representation, 18
in stable supersaturated solutions, 21, 22
Biocatalyst immobilisation
beer production, 248
dairy and meat fermentations, 249–250
wine and cider production, 249
Biochemical indicators, 483
Biofuel production
analysis of, 642, 643
benefits of, 641
classification, 631
definition, 631
energy source and consumption, 629, 630
feedstock *vs.* food production
agricultural land expansion, 638
arable land uses, 637
in Brazil, 638, 639
cereal production growth rates, 636–637
conversion process, 632, 633
CO_2 reduction goals, 640–641
cropland area, 637, 638
economic level and security, 640–641
gross reserves, 638, 639
marginal land and forest uses, 638
OECD and non-OECD countries, 633, 634
parameters, 633
raw material demands, 635–636
water availability, 639, 640
WEO 2008 scenario, 634–635
world food supply, 633–635
meat and milk/dairy products, 642, 643
petroleum uses, 629–631
solar energy balance, 629, 630
sources, 632
Biological fouling process, 210
Bioyield point, 533–534
Bird, M.R., 217
Blel, W., 217
Blue water, 9, 10
Boisseau, P., 386
Bostwick consistometer test, 512–514
Botulinum toxins, 72
Boukouvalas, C.J., 125
Bousquet, J., 316
Boynton, B.B., 384
Brandainão, T.R.S., 568–578
Britten, M., 113
Bryan, C.W., 621
Bublin, M., 316
Burger, J., 602
Butris, S., 385

C
Cal, K., 164
Callaghan, P.T., 501
Canned foods, thermal processing of, 483–485
Canning process, 142–143
Cao, S., 395
Carbopol solution, MRI-based rheological measurement
data processing procedure, 503
field of view, 504
GUI programs, 503
materials and methods, 502–504
neutralization, 503
no-flow and flow conditions, 504–505
repetition time, 504
rheological properties, 503–504
sample velocity profile image, 503
shear rate *vs.* apparent viscosity, log-log plot of, 504–506
tomography-based methods, 502
velocity sweep width, 504
Carboxymethyl chitosan, 70
Carob bean, 271–272
Carotenoid formulation
bioavailability assessment, 23–24
process for, 22–23
properties and nutritional effects, 22
Carrageenan, 274–276
Castell-Perez, E., 384
Cells and enzymes immobilisation. *See* Immobilisation technology
Cellulose gum, 279–280
Centralized processing, 144
Champagne, C.P., 247
Chapagain, A.K., 9
Chassagne, D., 445–455
Chemical heat pump system, 156–157
Chen, X.D., 164, 217
Chervin, C., 386
Chilled meals, 148–149
Chitosan
microbial polysaccharide research, 268, 280–281
in nanotechnology, 65, 69
polysaccharides, 233
Choicharoen, K., 161
Choi, Y., 493
Christelle, T., 255–264
Christian, G.K., 217
Chung, S.Y., 324, 326
Clark, J.P., 141–152
Clausius-Clapeyron equation, 521–522
Claussen, I.C., 158

Cleaning process, 210–211
  adhesion, 220
    AFM force, 221–222
    micromanipulation, 222–223
  classification of prototypes, 212–213
  diffusion-reaction removal
    brewery yeast films, 218–219
    factors, 217
    heat transfer coefficient *vs.* pressure drop, 217–218
    mechanisms, 216–217
    milk proteins, 217
  mechanical fluid removal
    core cleaning stage, 214, 216
    effects of velocities, Reynolds number, 215–216
    film removal stage, 216
    lab and pilot plant, time *vs.* wall shear stress, 214
    patch removal stage, 216
    weight of toothpaste in pipe, 214–215
Closed-loop HPD. *See* Heat-pump-assisted dryer (HPD)
Coacervation, encapsulation system, 237–238
Coated agglomerates, 263–264
Coffee roaster. *See* Batch coffee roaster
Cold plasma (CP)
  emerging food processing technology
    atmospheric-pressure devices, 360–361
    biological cells, impact, 361
    current applications and development, 363
    description, 360
    enzymes, impact, 362
    food constituents, impact, 362
    process-structure-function interactions, 362–363
    studying approaches, 363
  nonthermal emerging technologies
    advantage, 401
    electric field intensity, 399
    gaseous state, 398–399
    microbial inactivation, of fruit and vegetables, 400
    stable glow discharge, 399
Cole, P.A., 214
Collocation approach, 555
Comic-themed cereals, 149
Commercialization
  issues with limited success
    dry pasteurization, 146–147
    freeze drying, 146
    high hydrostatic pressure processing, 145
    irradiation, 144–145
    microwave and radio frequency heating, 145
    retort pouch foods, 145–146
  issues with null success
    fads, 149–150
    freeze concentration, 147
    meat analogs, 148
    pulsed electric fields, 148
    ready/chilled meals, 148–149
  issues with success
    aseptic processing and packaging, 143
    canning, 142–143
    centralized processing, 144
    individual quick freezing, 144
    membrane separations, 143
  principles of
    economics, 150
    market needs, 150–151
    right team, 151
    robust technology, 151–152
    timing of, 151
Complete heat transfer model, 475–477
Computational fluid dynamics (CFD) method, 168–169, 478–479
Contact-sorption drying, 163
Continuous solids, 122
Continuum formulation, 460
Controlled instantaneous pressure drop method, 162–163
Cooke, S., 321
CO_2 reduction, biofuel production, 640–641
Corthésy, B., 331
Costa, S.D, 385
Critzer, F.J., 363, 400
Crystallisation fouling process, 210
Curdlan, 276–277
Cyclic pressure drop method, 162–163

D

Daghigh, R., 157
Das, B.K., 389
Datta, A.K., 459–470, 473–478, 481
Daucus carota, 575–578
Debeaufort, F., 445–455
Deformable porous media framework, 465–466
Demirci, A., 390
Deng, S., 362, 400
Dens, E., 549
Devahastin, S., 161, 169

Dhall, A., 459–470
Ding, W.K., 270, 273
Disruption systems, 43–44
Donato, L., 439
Drip loss (DL), 577–578
Drying technologies
CFD, 168–169
contact-sorption drying, 163
developments, 154–155
finite-element mesh, 169
HPD
atmospheric freeze drying, 157–158
chemical system, 156–157
components, 155
continuous production line, 158
features, 156
limitations, 156
principle of, 155–156
solar dryers, 157
hybrid drying systems
infrared drying, 165
microwave freeze drying, 165
multistage drying options, 164–165
spray freeze drying, 166
industrial dryers
energy efficiency, 167
sustainability, 167–168
intermittent batch drying, 161–162
ISD, 160–161
PC drying, 166
selection of dryers, 155
Simprosys, 169
spray drying, 163–164
SSD
capital costs, 160
constant rate periods, 158–159
falling rate periods, 159
limitations, 160
quality, 159
recycling, 160
schematic representation, 158–159
types of, 158–159
swell drying, 162–163
Dry pasteurization, 146–147
Dube, M., 316
Dudak, F.2., 361
Dynamic approach. *See* Quality and safety assessment

E

E. coli O157:H7, 73
Ehlbeck, J., 361
Electrical pulsed energy (EPE)
assisted extraction applications
HVED method, 425
PEF, 421–425
induced effects, in foods
damage kinetics, 419, 420
electroporation, 416–417
power consumption, 420–421
supplementary effects, 417
synergy, of simultaneous electrical and thermal treatments, 419–420
techniques, for characterization, 417–419
laboratory experiments, 415
nanometer-sized biological membranes, 415
PEF and HVED treatment application, 416
Electrolyzed oxidizing water (EOW)
ability, 396
effect, 396
utilization, 397–398
Elisabete, M.C., 389
Elisabeth, D., 255–264
ElMasry, G., 595, 596, 598
Emerging food processing technology
HPP, 342–349
impact and potential of, 342
medical device technology, 341–342
opportunities, 341
PEF, 349–356
plasma treatment, 360–363
US, 356–360
Emin, M.A., 41–56
Emulsification, encapsulation system, 237
Encapsulating functional components, starch-based matrices
CFD simulation, flow characteristics, 51
dispersive mixing efficiency
capillary number, 53–54
capillary ratio, 55
cumulative droplet size distributions, 56
maximum shear stresses, 54–55
extrusion trials and droplet size, 51
lipophilic bioactive components, 50
numerical simulation, 52–53
rheological characteristics, 52
shear and elongational stresses, 50–51
Encapsulation system
active agent
controlled delivery, 230–231
easy handling, 231
protection of actives, 230
taste masking, 231
coacervation, 237–238
criteria for materials selection, 231–232
definition, 229

Encapsulation system (*cont.*)
emulsification, 237
extrusion methods, 236–237
fluidised bed coating techniques, 235, 236
immobilisation of cells and enzymes
adsorption, gel entrapment and covalent binding, 246
features of, 247
in fermentation processes, 248–250
gel-particle technologies, 246–247
lipids, 233
melt injection and extrusion processes, 235–236
microencapsulates, 238
aroma, 241
carotenoids, 245
colouring agents, 245
essential oils, 243–244
fish oils, 244
phenolic compounds, 241–243
probiotics, 239–241
vitamins, 244
molecular inclusion, 238
polymers, 233
polysaccharides, 232–233
proteins, 233
spray-chilling and-cooling techniques, 234–235
spray-drying techniques, 234
vacuum-and freeze-drying techniques, 235
Energy density/salt reduction, food structure
frozen low-fat ice cream, 431
gases, 430
release and perception, 431–432
sugar, 430–431
water, 430
Engel, R., 1–37
Enzymatic hydrolysis and in vitro digestion, food allergenicity
fruit and vegetables, 316
legumes, 317
milk, 316
peanut and peanut allergens, 316
rice, 317
wheat flour, 317
Epstein, N., 210
Equilibrium moisture content (EMC), of almonds, 529, 530
Erdoğdu, F., 473–486
Escalona, V.H., 391
EurepGAP, 614
Evapotranspiration, 5
Extrusion methods, encapsulation system, 236–237

F
Fan, X., 384, 385
Faridah, M.S., 386
Fat replacers, food allergenicity, 321
Feng, H., 394
Fermentation process
legumes, 319
shrimp, 320
whey proteins, 319
Fernandez-Diaz, M.D., 351
Fernandez-Gutierrez, S.A., 399
Finite difference method, 33, 474, 477–478, 481, 483
Finite element method, 52, 169, 477–478
Finite volume method, 477–478
First-order kinetic models, 570
Fluidised bed coating techniques, 235, 236
Fluidised bed Uni Glatt, 259–260
Flux balance analysis (FBA), 557–558
Foley, D.M., 384
Fonseca, J.M., 391, 394
Food allergenicity
enzymatic hydrolysis and in vitro digestion
fruit and vegetables, 316
legumes, 317
milk, 316
peanut and peanut allergens, 316
rice, 317
wheat flour, 317
fermentation process
legumes, 319
shrimp, 320
whey proteins, 319
gamma irradiation
egg proteins, 318
gliadin, 318
milk proteins, 318
protein aggregates, 317
Sebastiania jacobinensis bark lectin, 318
shrimp, 318–319
genetic modification
hen egg white allergen ovomucoid, 322
legumes, 321
peanut, 322
transgenic rice, 322
heat processing
eggs, 311
fish and seafood, 311, 313
fruits and vegetables, 314–315
legumes, 311
Maillard reaction products, 311
meat and meat products, 313

milk and milk proteins, 315
pasta, grain, and baked products, 315
peanuts and almonds, 313–314
high-pressure treatment and pulsed electric field
alpha amylase inhibitor, 327
bovine milk proteins, 329
fruits and vegetable, juices, and allergens, 327–328
peanut and apple allergens, 329–330
reactions, 326
rice, 327
sterilization, 327
tests, 325
Mal d1 structure, 291–292
materials and methods, 293
microparticulation technology, 321
no. of reviewed papers, 293, 311, 313
polymerization of allergen proteins
fruit and vegetables, juices, 324–326
peanut, 323–324
principle of, 291–292
pulsed ultraviolet light treatment
peanut, 322–323
soy, 323
refining process
peanut oil, 320
soy lecithin and soybean oils, 320
sunflower oil, 320
reviewed papers *vs.* tests, 293, 311–312
tests, 293–310
Food biopolymers, 132
Food dehydration. *See* Drying technologies; Water interactions
Food preservation process design
kinetic models
microbial kinetics, 490–491
physical transport models, 492–493
quality attributes, 491–492
retention of heat-sensitive vitamins, 492
maximum quality retention, 496
objectives, 489
process time, 494
reduction in microbial population, 493
thermal process optimization, 495–496
Food process control
MRI-based rheological measurement
bruising, 515–516
carbopol solution, 502–506
challenges, 515
direct/indirect composition measurements, 515
efficiency, 516
flow loop variations, 500, 501
MRI signal intensity, 500
vs. NMR, 500
tomato concentrate, 511–515
yogurt, 506–511
viscosity measurements, 499
Food process design
heat transfer coefficients, 133–135
mass transfer
coefficients, 136
operations, 135
phase equilibria, 135
size, 119
transfer equations, 119
transport property
bulk density variation, 126, 127
comparative figures, of porosity, 126, 128–129
dry solid material density, 125
Fick diffusion equation, 132
gases and liquids, 120–122
heat, 130–131
mass diffusivity, 132–133
measurement techniques, 122–124
rheological properties, of food, 129–130
typical values, of diffusivity, 133
Food safety management systems (FSMS)
concepts of, 612
food supply chain hazards
in biological origin, 609
in chemical and physical origin, 610
complexity, 608
developments, 608, 609
nanobiotechnology, 611
programs, 609
traceability, 608
good practices
classic strategy, 622
consumption, 621
description, 619
development of, 616–619
GAP, 620
GCP and GRP, 620
GHP, 620, 621
GLP, 620
GMP, 619
GSP and GTP, 621
HACCP/PRPs, 616
incidents of, 611
private standard regulations
development of, 614, 615
EurepGAP, 614
GFSI forum, 613
goals of, 613
ISO, 613, 614
public standard regulations, 612

Food safety management systems (FSMS) (*cont.*)
quality assurance, 612
technical and subjective aspects, 607
Food structure
designing complex, with tailored properties
encapsulation, of aroma compounds, 432–433
energy density/salt reduction, 430–432
fortification and encapsulation, 433–436
improved nutritional profile, 438
modulated digestion, 439–440
nutritional application, 438–439
pro-and prebiotics, 436–438
stabilization and bioavailability, of bioactive compounds, 432, 433
pharmaceutical products, 429
transport properties
bulk density with moisture content, dehydration, 126–127
continuous solids, 122
density, porosity, and shrinkage, 122–124
freeze-drying pressure, 129
granular materials, 122
pore formation and distribution, 126
porosity of extrudates, 126, 128
semiempirical model, 122
types of legumes, 126
Fortification and encapsulation, food structure
amorphous
carrier particles, 435–436
matrices, 436
bioavailability and stability, of micronutrients, 434
micronutrient deficiencies, 433
multiple-emulsion system, 435
vitamins, 434
zeaxanthine bioavailability, 434
Forward difference method, 477
Fouling process
antifouling surface, 220, 221
classification of mechanisms, 210
definition, 209
design for reduction, 211–212
predictive models, 210
Frameworks
challenges, 466–468
continuum formulation, 460
for food process modeling, 460, 462
large pore formulation, 460, 461
multiphase continuum equations, 460
porous media formulation
advantages, 466
deformable porous media, 465–466
description, 461, 463
rigid porous media, 463–464
small pores, 464–465
quality modeling, 468–469
safety prediction framework, 469–470
single-phase transport equations, 460
Freeze-drying techniques, 146, 235
Fridman, A., 360
Froehling, A., 341–365
Frozen foods, thermal conductivity, 130–131
Fruits and vegetables
allergen protein polymerization, 324–326
cold plasma, microbial inactivation, 400
enzymatic hydrolysis and in vitro digestion, 316
heat processing, food allergenicity, 314–315
high-pressure treatment and pulsed electric field, 327–328
hyperspectral imaging technology
apples, 598
citrus and oranges, 598–599
cucumbers, 600–601
melons, 599
mushrooms, 600
tomatoes, 600
Fryer, P.J., 209–224
Fustier, P., 247

G

Gamma irradiation, food allergenicity
egg proteins, 318
gliadin, 318
milk proteins, 318
protein aggregates, 317
Sebastiania jacobinensis bark lectin, 318
shrimp, 318–319
Garcia, A., 326
Gas-fired batch coffee roaster, 173–174
Gas-liquid equilibria, 135
Gaussian thermostat method, 96
Geeraerd, A.H., 548, 570
Geladi, P., 602
Gels, rheological behavior of, 506
Genome-scale metabolic networks, 551–552

Ghai, G., 197–206
Gharsallaoui, A., 451
Giannouli, P., 278
Gibbs-Helmholtz equation, 522
Gillham, C.R., 217
Gil, M.M., 568–578
Glass transition data
  DMA and DEA measurements, 80
  food components, 79
  relaxation times
    dielectric and mechanical, 88–89
    and fluidness, 81–87
    in food systems, 87
  scanning calorimetry, 80
Global food safety initiative (GFSI), 613
Gomes, C., 385
Gómez, P.L., 391
Gomez-Sanchis, J., 598
Gompertz model, 570, 571
Gonainçalves, E.M., 567–578
Gonzalez-Aguilar, G.A., 391
Good agricultural practices (GAP), 620
Good catering practice (GCP), 620
Good hygiene practice (GHP), 620, 621
Good laboratory practice (GLP), 620
Good manufacturing practice (GMP), 619
Good retail practice (GRP), 620
Good storage practice (GSP), 621
Good transport practice (GTP), 621
Gowen, A.A., 600, 602
Grandison, A.S., 383
Granular materials, 122
Green coffee bean batch, 174
Greenspan, L., 523
Green water, 10, 11
Grey water, 10, 11
Guar gum, 269–271
Guentzel, J.L., 398
Guggenheim-Anderson-de Boer (GAB) model, 521, 529
Gulati, T., 467
Gunes, G., 384, 385
Guo, B., 322

H

Hagenmaier, R.D., 384, 386
Halder, A., 475, 481
Hambleton, A., 445–455
Hammad, A.A., 386
Hanotel, L., 387
Harper, J.M., 486
Harrison, S.L., 381
Harvey, B., 446, 451
Hassenberg, K., 388
Hawkes, J.G., 492
Hawlader, M.N.A., 157
Hazard analysis and critical control points (HACCP), 616, 618, 619, 621
Heat exchangers, 134
Heat processing, food allergenicity
  eggs, 311
  fish and seafood, 311, 313
  fruits and vegetables, 314–315
  legumes, 311
  Maillard reaction products, 311
  meat and meat products, 313
  milk and milk proteins, 315
  pasta, grain, and baked products, 315
  peanuts and almonds, 313–314
Heat-pump-assisted dryer (HPD)
  atmospheric freeze drying, 157–158
  chemical system, 156–157
  components, 155
  continuous production line, 158
  features, 156
  limitations, 156
  principle of, 155–156
  solar dryers, 157
Heat transfer
  coefficients, 133–135
  in roasting chambers
    bean-hardware heat exchange, 183
    bean heating, 182
    efficiency, 182–183
    exothermic heating, 183
    rate of temperature variation, 183–184
    roasting time, 184–185
Heat transport properties, 130–131
Heldman, D.R., 489–496
Helm, R.M., 321
Hen egg white allergen ovomucoid (Gal d1), 322
Henningsson, M., 214
Herschel-Bulkley (H-B) parameters
  carbopol solution, 503, 504
  yogurt, 506, 507
High hydrostatic pressure processing, 145
High-pressure homogenizers
  conventional dairy homogenization process, 44–45
  conventional full stream and partial homogenization process, 48–49
  homogenized cream and standardization, 46

High-pressure homogenizers (*cont.*)
impact on process efficiency and product quality, 49–50
orifices and nozzles, 43–44
pressures, 42
simultaneous emulsification and mixing unit
distribution and intensity of turbulent kinetic energy, 46–47
globule aggregation and coalescence, 48
principal design of, 45–46
single-piston pump, 42–43
transient stresses, 43
valves, 43–44
High pressure processing (HPP)
biological cells, impact
affects, 343–344
internal and external disintegration, 346
isorate lines, 344
membranes, 345
p^H, 345
proteins and enzymes, 344–345
ribosomes, 345
spore inactivation, 345
current applications and development, 348
enzymes, impact, 346–347
food constituents, impact, 347
industrial scale, 342–343
needs and challenges, 348–349
nonthermal emerging technologies
application, 379–380
cause, 378
critical process factors, 381
enzymes, 381
Le Chatelier-Braun principle, 378
microbial inactivation, 379
process-structure function interactions, 347–348
High-pressure treatment and pulsed electric field, food allergenicity
alpha amylase inhibitor, 327
bovine milk proteins, 329
fruits and vegetable, juices, and allergens, 327–328
peanut and apple allergens, 329–330
reactions, 326
rice, 327
sterilization, 327
tests, 325
High-voltage electrical discharge (HVED) method, 425
Hinch, E.J., 53
Hoekstra, A.Y., 6, 7, 9, 10, 15
Hoffman, D.R., 311
Horrick, J.M., 384
Host, A., 315
Houska, M., 291–331
HPP. *See* High pressure processing (HPP)
Hsu, W.-Y., 385
Huang, T.-S., 395
Hunaefi, D., 375–402
Hybrid drying techniques
infrared drying, 165
microwave freeze drying, 165
multistage drying options, 164–165
spray freeze drying, 166
Hyperspectral imaging technology
dairy products, 602
fruit and vegetables
apples, 598
citrus and oranges, 598–599
cucumbers, 600–601
melons, 599
mushrooms, 600
tomatoes, 600
instrumentation
area detectors, 592
calibration of, 593
light sources, 591, 592
line-scan systems, 592–593
wavelength dispersion devices, 592
maize, 602
meat quality evaluation
beef, 594
chicken and poultry, 596, 597
fish, 595
lamb, 596–598
pork, 595
optical sensing technologies, 581
principles of
area/plane scanning, 584, 585
classes, 583
cube, 583–584
image acquisition modes, 585–586
line scanning, 584, 585
point scanning, 584, 585
single shot method, 585
processing methods
chemical images visualization, 590–591
correction of, 587
enhancement and preprocessing, 587
feature extraction, 588
model evaluation, 590
multivariate analysis, 589
optimal wavelength selection, 590
segmentation, 588

typical steps, 586
vs. spectroscopy, 581, 582
wheat, 601

I

Imaging spectroscopy. *See* Hyperspectral imaging technology
Immobilisation technology
adsorption, gel entrapment and covalent binding, 246
features of, 247
in fermentation processes
beer production, 248
dairy and meat fermentations, 249–250
wine and cider production, 249
gel-particle technologies, 246–247
Impinging stream dryers (ISD), 160–161
Individual quick freezing (IQF), 144
Industrial dryers
energy efficiency, 167
sustainability, 167–168
Intellectual property (IP) model, 652
Intermittent batch drying, 161–162
International Association for the Properties of Water and Steam (IAPWS), 198
International fresh-cut produce association (IFPA), 375
International organization for standardization (ISO), 613, 614
Islam, M.R., 157, 161
Isobe, S., 388
Issa-Zacharia, A., 396
Itoh, K., 397
Izumi, H., 398

J

Jaeger, H., 341–365
Jamaleddine, T.J., 168
Jangam, S.V., 153–170
Jang, J.H., 395
Jaya, D.S., 154
Joana, F., 389
Jousse, F., 481

K

Kamat, A.S., 385
Kamruzzaman, M., 596
Kananeh, A.B.., 220
Karbowiak, T., 445–455
Kasim, M.U., 394
Kato, T., 315
Kato, Y., 315
Kaufmann, S.F.M., 429–440
Keasling, J.D., 560
Keener, K.M., 390
Keskinen, L.A., 397
Khader, M.A., 390
Kim, B., 400
Kim, B.S., 384, 388
Kim, H.J., 391
Kim, J.-G., 389, 390
Kinetic compensation effect, 533
Klockow, P.A., 390
Knoerzer, K., 1–37
Knorr, D., 341–365
Köhler, K., 41–56
Koide, S., 397
Kokini, J.L., 64–75
Koseki, S., 388, 396, 397
Krasaekoopt, W., 247
Krokida, M., 119–136
Kucera, P., 291–331
Kudra, T., 154, 161, 163, 166, 167
Kuriakose, R., 164

L

Lag phase phenomenon, 561
Lamikanra, O., 392–394
Landuyt, A., 240
Large pore formulation, 460, 461
Latorre, M.E., 387
Lazou, A.E., 126
Lebovka, N.I., 352, 354, 415–426
Lee, N.Y., 386
Lequin, S., 445–455
Levine, H, 79, 80
Lian, W.C., 240
Liapis, A.I., 91–108
Li, B., 380, 381
Line scanning method, 584, 585
Lipophilic components, 17–19
Liquid crystal termography methods, 483
Liquid-liquid equilibria, 135
Li, S.-J., 164
Liu, W., 216
Liu, W.-G., 52
Lo, C.T., 278
Locust bean gum (LBG), 271–272
Lodinová-Zadniková, R., 331
Logist, F., 547–561

López, L., 385
Loveday, S.M., 111–117
Lo, Y.M., 267–284
Luecha, J., 64–75
Lu, R.F., 598
Luscher, C.M., 346
Lu, Z., 383, 385

M
Mackie, A.R, 293
Mackie, A.R., 293
Magnetic resonance imaging (MRI)-based rheological measurement
  bruising, 515–516
  carbopol solution
    data processing procedure, 503
    field of view, 504
    GUI programs, 503
    materials and methods, 502–504
    neutralization, 503
    no-flow and flow conditions, 504–505
    repetition time, 504
    rheological properties, 503–504
    sample velocity profile image, 503
    shear rate *vs.* apparent viscosity, log-log plot of, 504–506
    tomography-based methods, 502
    velocity sweep width, 504
  challenges, 515
  direct/indirect composition measurements, 515
  efficiency, 516
  flow loop variations, 500, 501
  MRI signal intensity, 500
  vs. NMR, 500
  tomato concentrate
    Bostwick consistometer test, 512–514
    gravity currents, 513–514
    ketchup Bostwick measurement, 514–515
    quality grades, 511–512
    shear viscosity *vs.* shear rate, 513
  yogurt
    comercially available products, 506
    flow behavior, 506–507
    no-flow and flow conditions, 508, 509
    pressure drop values, 509
    shear rate *vs.* apparent viscosity, log-log plot of, 507–508, 510
    thixotropic behavior, 508, 511
    wall slip velocity, 507
Magnetic resonance imaging (MRI), temperature measurement, 482–483
Mahadevan, R., 552
Maillard, M., 240
Mal d1 structure, 291–292
Malnutrition, 16–17
Manojlovic, V., 247
Manzocco, L., 391
Marco-Moles, R., 351
Maroulis, Z.B., 122, 125, 129
Marshall, R.G., 379
Martiáez-Sánchez, A., 389
Mass diffusivity, 132–133
Mass transfers
  coefficients, 136
  encapsulation, of aroma compounds application, 454–455
  food process design
    coefficients, 136
    operations, 135
  liquid-liquid interface, 450–451
  liquid-solid interface
    apparent diffusivity measurement, 452, 453
    infra-red signal, 452, 453
    Levich model, 452
    permeability, 453–454
  thermodynamics, of sorption
    aroma partitioning, food and air, 446
    wine active compounds, cork interactions, 447–450
Mass transport properties, 132–133
Mateja, A., 607–622
Mathematical modeling, food processing operations
  analytical solutions, 479
  aspects of, 473–475
  backward difference method, 477
  boundary conditions, 480
  bread baking process, 476–477
  canned foods, thermal processing of, 483–485
  complete heat transfer model, 475–477
  computational fluid dynamics method, 478–479
  description, 474
  experimental validation, 481
    biochemical indicators, 483
    liquid crystal termography methods, 483
    MRI, 482–483
    particle image velocimetry, 482

time-temperature indicators, 483
finite difference method, 477, 478
finite element method, 477–478
finite volume method, 477–478
forward difference method, 477
heat transfer coefficient, 481
numerical solutions, 477–478
observation and physics-based models, 473–474
optimization, 485–486
shrinkage problem, 481
thermal and physical properties, 480–481
Mathys, A., 345
Maxwell's equations, 32
M3B model, 93
McCarthy, K.L., 499–516
McCarthy, M.J., 499–516
McKellar, R.C., 550
Meat quality evaluation
heat processing, food allergenicity, 313
hyperspectral imaging technology
beef, 594
chicken and poultry, 596, 597
fish, 595
lamb, 596–598
pork, 595
Melamine, 73
Melt extrusion process, 235–236
Melt injection process, 235–236
Membrane separations, 143
Mercade-Prieto, R., 217
Metabolic networks
genome-scale, 551–552
mathematical model reduction, 555
optimal cell behavior, 556–558
pseudosteady-state assumption, 556
Meyer-Pittroff, R., 331
Michailidis, P.A., 122
Michel, M., 348
Microbial exopolysaccharides
curdlan, 276–277
xanthan gum, 277–279
Microbial growth phases
adaptation/lag phase, 561
exponential growth phase, 559–560
stationary phase, 560
Microbial polysaccharides, 267
Microbial research, systems biology approach
macroscopic level
description, 549–550
microbial dynamics, 550
vector representation, 551
mesoscopic level, 550
metabolic networks (*see* Metabolic networks)
microscopic and macroscopic levels
assumptions/simplifications, 553–554
multiple shooting and collocation, 554–555
single shooting approach, 554
microscopic level
extracellular dynamics, 553
intracellular dynamics, 552
Microbial thermal inactivation
Baranyi model, 570, 571
Bigelow model, 569
broth-based experiments, 574, 575
development of, 569
first-order kinetic models, 570
Gompertz model, 570, 571, 573
growth kinetic models, 569
in liquid medium, 573, 574
parameters, 573, 574
temperature dependence of, 571
Microcrystalline cellulose (MCC), 280
Microemulsions, 19
Microencapsulates
aroma, 241
carotenoids, 245
colouring agents, 245
essential oils, 243–244
fish oils, 244
phenolic compounds, 241–243
probiotics, 239–241
vitamins, 244
Microwave processing and utilisation
advantages, 30, 31
design, 31
model
computational methods, 33–34
electromagnetism, 32, 33
Fourier equation, 32
geometry, 31
material properties, 33
Maxwell's equations, 32
measured heating profile, 34
process conditions, 33
qualitative comparison, 34
quantitative model validation, 35
simulated heating profile, 34–36
temperature distributions, 35–37
validation, 34
and radio frequency heating, 145
Miller, F.A., 567–578
Mills, E.N.C., 293

Mini cookies, 149
Mintier, A.M., 384
Moisan, M., 361
Moisture diffusivities, food systems, 132–133
Mojca, J., 607–622
Molecular dynamics (MD), 92–93
Moles, C., 555
Mondal, A., 476
Mondry, V.A., 179
Monterey unpasteurized almonds
 moisture content *vs.* textural properties, 539, 541
 storage conditions, 539, 541
 temperature effects, 537, 539, 540
Moon, B., 267–284
Moreau, M., 361
Morris, E.R., 278
Mujumdar, A.S., 153–170
Müller, U., 320
Mullineux, G., 510
Multiphysics simulation. *See* Microwave processing and utilisation
Multiple shooting approach, 554–555
Multivariate classification, 589

N

Naganathan, G,K., 594
Nanoemulsions, 71
Nanoencapsulation
 bread crumb, electron microscopic image of, 70
 high-amylose corn starch, 70
 zein, 70–71
Nanotechnology
 microfluidic devices
  botulinum toxins, 72
  food adulteration, 73
  melamine, 73
  miniaturized chips, 71
  zein-based microdevices, 72
 microorganisms and pathogens, detection of, 73–74
 nanoemulsions, 71
 nanoencapsulation, 69
  bread crumb, electron microscopic image of, 70
  high-amylose corn starch, 70
  zein, 70–71
 nanostructured materials
  food nanoparticles, 64–65
  food protein and synthetic nanotubes, 65–66
  nanodots/quantum dots, 66–68
  polymer-nanoclay nanocomposite technology, 69
Nanotubes
 halloysite, 66
 α-lactalbumin, 65
 self-assembly process, 65
 SWNTs, 66
Navier-Stokes analog of Darcy equation, 461
Nedovic, V.A., 236, 246
Newtonian fluids, 129–130
Next-generation predictive models. *See* Top-down systems biology approach
Niamnuy, C., 169
Nicolaï, B.M., 570
Nicolai, T., 439
Niemira, B.A., 400
Nimmol, C., 161
Noh, H.K., 598
Nonfrozen foods, thermal diffusivity, 131
NonNewtonian fluids, 129
Nonpareil pasteurized almonds
 experimental moisture contents, 535, 536
 force-distance curves, 533
 fracture force values, 537, 538
 moisture adsorption curves, 529, 530
 percentage deviation, of predicted variables, 536, 537
 water adsorption curves, 530, 531
Nonstarch polysaccharides, 267
 chitosan, 280–281
 microbial exopolysaccharides
  curdlan, 276–277
  xanthan gum, 277–279
 microcrystalline cellulose, 280
 microencapsulation, 267–268
 probiotics, 268
 rheological properties
  conformational and structural characteristics, 282
  NMR, 283
  surface morphology, 282–283
  viscosity and dynamic modulus, 281
 seaweed extracts-derived gums
  alginate, 272–274
  carrageenan, 274–276
 seed-derived gums
  guar gum, 269–271

locust bean gum, 271–272
sodium carboxymethyl cellulose, 279–280
Nonthermal emerging technologies
AcEW (*see* Electrolyzed oxidizing water (EOW))
cold plasma
advantage, 401
electric field intensity, 399
gaseous state, 398–399
microbial inactivation, of fruit and vegetables, 400
stable glow discharge, 399
EOW
ability, 396
effect, 396
utilization, 397–398
fresh-cut plant products
cutting, 376
physiological changes, 377
high capital investment, 378
HPP
application, 379–380
cause, 378
critical process factors, 381
enzymes, 381
Le Chatelier-Braun principle, 378
microbial inactivation, 379
IFPA, 375
irradiation
cold process, 382
FDA, 383
low-dose application, 383–387
mechanism, 382
microorganisms, 382
ozone
application, 387–389
assessment, 390
gaseous treatments, 390
low level, 390
shelf-life extension, of fresh-cut salad mix products, 392
thermodynamically ideal condition, 390
use, 387
PEF
enzyme inactivation mechanisms, 382
membrane permeability, 381
technical drawbacks, 382
tissue-softening effect, 382
shelf-life extension, of fresh-cut products
application, 401–402
ultrasound
application, 394–395
bacterial inactivation effect, 394
investigation, 394
treatment, 395
UV
application, 391–393
Lambert-Beer law, 393
low-dose, 393–394
mathematical model, 393
microbial inactivation, 392
treatment, 393
utilization, 392
Norgaard, A., 315
Norton, I., 431
Nuclear magnetic resonance (NMR)
direct/indirect composition measurements, 515
vs. MRI, 500
nonstarch polysaccharides, rheological properties, 283
Nunes, T.C.F., 386

O
Oikonomopoulou, V.P., 126
Oil-in-water (o/w) emulsions. *See* Bioactive component formulation
Okos, M.R., 493
Ölmez, H., 388
Olsson, E.E.M., 476
Open innovation (OI)
applications, 646
definition, 645, 646
proliferation, 647
significants, 646
Ozdemir, M., 390
Ozone, nonthermal emerging technologies
application, 387–389
assessment, 390
gaseous treatments, 390
low level, 390
shelf-life extension, of fresh-cut salad mix products, 392
thermodynamically ideal condition, 390
use, 387

P
Palabiyik, I., 214
Palazoğlu, T.K., 477
Palekar, M.P., 384
Palou, E., 379
Palzer, S., 429–440
Park, C.-M., 397

Park, E.J., 397, 398
Particle image velocimetry (PIV), 482
Pattern recognition. *See* Multivariate classification
Peleg, M., 84
Perera, N., 380
Perni, S., 363, 400
Perre, P., 169
Perrier, C., 331
Peter, R., 607–622
Petroleum uses, biofuel production, 629–631
Pflug, I.J., 493
Phromraksa, P., 319
Physics-based models, 473–474
Phytoglycogen, nanotechnology in, 65
Phytosterol formulation
  investigation of, 28
  properties and nutritional effects, 25–26
  stabilisation of, 26–27
  volume density distribution, 28, 29
Pillai, S, 384
Pin,C., 548
Place prerequisite programmes (PRP), 616
Plasma treatments. *See* Cold plasma (CP)
Point scanning method, 584, 585
Polder, G., 600
Poly(D-L-lactide-co-glycolide) (PLGA), 71
Polymer-nanoclay nanocomposite technology, 69
Porous media formulation
  advantages, 466
  deformable porous media, 465–466
  description, 461, 463
  rigid porous media, 463–464
  small pores
    liquid phase transport, 464
    vapor phase transport, 465
Porter, C.J.H., 440
Powders
  preparation
    emulsion and water, 258, 259
    fluidised-bed agglomeration and coating, 259–260
    products, 258
    spray-drying, 258–259
  properties
    agglomerated spray-dried powders in fluidised bed, 262–263
    agglomerate production, 260–261
    coated agglomerates, 263–264
    spray-dried aroma powders, 261–262
Prakash, A., 385
Pramanik, J., 560
Precipitation fouling process, 210
Predictive microbiology, 547–549
Premix membrane emulsification, 20, 21
Probiotic microencapsulates, 239–241
Probiotics, 268
Process analytical technology (PAT), 516
Pruss, A., 202
Puértolas, E., 423
Pulse combustion (PC) drying, 166
Pulsed electric field (PEF)
  biosuspensions
    disruption, of microorganisms, 423
    E. coli, 424
    efficiency, 423–424
    S. cerevisiae, 424
    selectivity, 425
  emerging food processing technology
    biological cells, impact, 350
    design and optimization, 355
    electric field intensities, 349
    electroporation mechanism, 349
    enzymes, impact, 350–351
    food constituents, impact, 351–352
    high-intensity treatment, 354–355
    low-intensity treatment, 353–354
    process-structure-function interactions, 352–353
    undesired effects, 356
  and high-pressure treatment, food allergenicity
    alpha amylase inhibitor, 327
    bovine milk proteins, 329
    fruits and vegetable, juices, and allergens, 327–328
    peanut and apple allergens, 329–330
    reactions, 326
    rice, 327
    sterilization, 327
    tests, 325
  nonthermal emerging technologies
    enzyme inactivation mechanisms, 382
    membrane permeability, 381
    technical drawbacks, 382
    tissue-softening effect, 382
  vegetable and fruit tissues
    dielectric breakage models, 422
    efficiency, 422
    electroporation, 423
    investigated materials, 421
    treatment, 421, 422
Pulsed ultraviolet light treatment, food allergenicity

peanut, 322–323
soy, 323
Pushbroom method. *See* Line scanning method

Q

Qin, J.W., 599
Quality and safety assessment
degradation of
Daucus carota sample, 575, 576
drip loss, 577–578
TA.HDi texture analyser, 575
texture, 576–577
frozen stored vegetables, 568, 571–572
goals, 567
microbial thermal inactivation
Baranyi model, 570, 571
Bigelow model, 569
Broth-based experiments, 574, 575
development of, 569
first-order kinetic models, 570
Gompertz model, 570, 571, 573
growth kinetic models, 569
in liquid medium, 573, 574
parameters, 573, 574
temperature dependence of, 571
time-varying temperature conditions, 567, 568
Quality by Design (QbD) approach, 516
Quality of food, 439
Quantum dots (QD)
with bread sample labeling, 68
fluorescent microscopic images, 66, 67
multifunctional quantum dots, 67, 68
pathogenic bacteria, 66

R

Rabiey, L., 113
Ragni, L., 362
Rahman, M.S., 131
Rahman, S.M.A., 163
Ramaswamy, H., 379
Ramsden, L., 278
Rao, M.A., 111–117
Ray, M.B., 168
Reaction fouling process, 210
Ready meals, 148–149
Refining process, food allergenicity
peanut oil, 320
soy lecithin and soybean oils, 320
sunflower oil, 320
Reineke, K., 341–365
Relaxation times, glass transition data
dielectric and mechanical
food components with high solid contents, 88
skim milk-maltodextrin solid systems, 89
sticky point, 89
and fluidness
Arrhenius relationship, 85
DEA and DMA, 86
Fermi model, 84
fragility concept, 81, 83, 85
glass-forming materials, plot, 86
non-crystalline solids, 81
viscosity *vs.* temperature, 82, 83
vitrification, 82
VTF relationship, 82
WLF equation, 83, 84
in food systems, 87
Retort pouch foods, 145–146
Rheological properties
fluid and semifluid foods, 129–130
nonstarch polysaccharides
conformational and structural characteristics, 282
NMR, 283
surface morphology, 282–283
viscosity and dynamic modulus, 281
Richard, O.A., 392, 394
Rico, D., 388
Rigid pressure-tight module, 199, 200
Robbins, P.T., 209–224
Rodriguez-Fernandez, M., 555
Roos, Y.H., 79–89
Ruiz-Beviá, F., 522
Rushing, J.W., 391, 394

S

Sagalowicz, L., 435
Saguy, I.S., 485, 486, 645–655
Sahu, K.C., 214
Sampson, H.A., 321
Samuelsson, E.G., 315
Saravacos, G.D., 119–136
Scheibenzuber, M., 325, 327
Schenk, M., 391
Schlueter, O., 341–365
Schmidt, H.M., 385
Schoessler, K., 341–365
Schubert, H., 1–37, 41–56
Schuchmann, H.P., 1–37, 41–56
Schuetz, R., 560
Schwartzberg, H., 173–194

Seaweed extracts-derived gums
alginate, 272–274
carrageenan, 274–276
Seed-derived gums
guar gum, 269–271
locust bean gum, 271–272
Selcuk, M., 400
Selma, M.V., 388, 389
Setinova, I., 291–331
Seymour, I.J., 394, 395
Shahin, M.A., 601
Shah, N.P., 270, 273
Sharing is Winning (SiW) model, 647
Sheehan, P., 99
Silva, C.L.M., 567–578
Silveira, A.C., 389
Simmons, M.J.H., 510
Simulation models
amylose molecules, 95
atomistic representation, 94, 95
Gaussian thermostat method, 96
M3B model, 93
Morse potential, 94
polysaccharides, 93
realistic pore structures, 95
shifted dihedral functions, 94
Singh, H., 111–117
Singh, N., 390
Singh, R.P., 521–543
Single-pass (SP) roasters, 177
Single shooting approach, 554
Single shot method, 585
Single-walled carbon nanotubes (SWNT), 66
Skandamis, P.N., 550
Skim milk-maltodextrin solid systems, 89
Slade, L., 79, 80
Sletten, G., 311
Smelt, J.P., 344
Smetanska, L., 375–402
Sodium carboxymethyl cellulose (CMC), 279–280
Soft-center cookies, 149
Sokhansanj, S., 154
Solar-assisted heat pump dryers, 157
Solar energy, biofuel production, 629, 630
Solidification fouling process, 210
Sollohub, K., 164
Soma, P.K., 267–284
Somkuti, J., 329
Sozer, N., 64–75
Spiess, W.E.L., 629–643
Spray-chilling techniques, 234–235
Spray-cooling techniques, 234–235
Spray-dried aroma powders, 261–262
Spray dryer Niro Minor, 258–259
Spray drying (SD) techniques, 163–164, 234
Stang, M., 44
Starch
amylose, 93
fabrication of nanoparticles, 64
nanofillers, 64
viscosity, 64
Static gravimetric method, 523
Steffe, J.F., 507
St. John's gum, 271–272
Stock Pilot Rotor PR-900-0 BV-VA, 199–201
Stoichiometric matrix, 551, 552
Stopforth, J.D., 398
Strickland, W., 392
Sui, Q., 351
Sun, D.-W., 581–603
Superheated steam drying (SSD) system
capital costs, 160
constant rate periods, 158–159
falling rate periods, 159
limitations, 160
quality, 159
recycling, 160
schematic representation, 158–159
types of, 158–159
Swell drying, 162–163
Symons, S.J., 601

T

Taghavi, S.M., 216
Taitano, L.Z., 521–543
Tanabe, S., 331
Teixeira, A., 197–206
Teixeira, A.A., 474, 486, 495
Thermal conductivity, 130–131
Thermal diffusivity, 131
Thermal processing with flexible packages
detectors, 198
flexible retortable trays with burst, 197–198
International Association for the Properties of Water and Steam, 198
model-predicted internal pressure profiles, 203, 205
temperature and pressure with distilled water, 203–204
test module for head space pressure
mathematical model, 202–203
rigid pressure-tight module, 199, 200

Stock Pilot Rotor PR-900-0 BV-VA, 199–201
wireless instrumentation, 199, 201
UNIFAC model, 204, 206
Thermodynamics, of sorption
gaseous SO_2 enthalpy, 449–450
H_2O and SO_2 isotherms, 447–448
hydrated cork powder, 449–450
isotherms, of SO_2, 448–449
sulphur dioxide and water enthalpy, 448, 449
Thymoquinone, 71
Ting, E.Y., 379
Tomato concentrate, MRI-based rheological measurement
Bostwick consistometer test, 512–514
gravity currents, 513–514
ketchup Bostwick measurement, 514–515
quality grades, 511–512
shear viscosity *vs.* shear rate, 513
Top-down systems biology approach
individual cell dynamics, 548
macroscopic predictive microbial growth models, 548
microbial growth phases
adaptation/lag phase, 561
exponential growth phase, 559–560
stationary phase, 560
microbial population heterogeneity, 549
microbial research
macroscopic level, 549–551
mesoscopic level, 550
metabolic networks (*see* Metabolic networks)
microscopic and macroscopic levels, 553–555
microscopic level, 550, 552–553
primary and secondary models, 548
Torres, J.A., 379
Transport phenomena, mathematical modeling of. *See* Mathematical modeling, food processing operations
Transport properties
food structure
bulk density with moisture content, dehydration, 126–127
continuous solids, 122
density, porosity, and shrinkage, 122–124
freeze-drying pressure, 129
granular materials, 122
pore formation and distribution, 126
porosity of extrudates, 126, 128
semiempirical model, 122
types of legumes, 126
gases and liquids
equations, 120
mass diffusivity, 121–122
viscosity and thermal conductivity, 121
heat transfer coefficients, 133–135
mass diffusivity, 132–133
mass transfer coefficients, 136
mass transfer operations, 135
phase equilibria, 135
rheological properties, 129–130
thermal conductivity, 130–131
thermal diffusivity, 131
Trigo, M., 386
Tsuta, M., 599

U
Ultrasound (US)
emerging food processing technology
biological cells, impact, 357
bubble collapse, 356
current applications and development, 359–360
defination, 356
enzymes, impact, 357
food constituents, impact, 358
overview, 360
process-structure-function interactions, 358–359
nonthermal emerging technologies
application, 394–395
bacterial inactivation effect, 394
investigation, 394
treatment, 395
UNIversal Functional Activity Coefficient (UNIFAC) model, 204, 206

V
Vacuum-drying techniques, 235
Valley of death (VoD) concept, 649, 650
Van den Einde, R.M., 52
Van Derlinden, E., 547–561
Van Impe, J.F., 547–561, 570
Vapor-liquid equilibria, 135
Variable pressure drop drying, 162–163
Varjonen, E., 315
Velazquez, G., 379
Vercammen, D., 547–561
Vicente, A.A., 394
Villota, R., 492

Virtual water
- application of, 7–9
- blue water, 9, 10
- definitions, 6–5
- flow of, 11, 12
- green water, 10, 11
- grey water, 10, 11
- import and export, 11
- modified concept of, 13–15
- water footprint, 6, 7, 12–13

Viscosity
- Newtonian fluids, 129–130
- nonNewtonian fluids, 129

Voilley, A., 445–455
Vorobiev, E., 415–426

W

Wagner, W., 202
Wall slip, 507
Walzel, P., 164
Wang, H., 388, 397
Wang, J.-C., 91–108
Wan, J., 361
Wanninge, L.A., 486
Watanabe, M., 317
Water interactions
- food dehydration, 92
- molecular dynamics, 92
- results and discussion
  - amylose, 100
  - average interaction potential energy, 105–106
  - concave meniscus, 102
  - cylindrical disks, 97
  - energetics, 98, 99, 105
  - enthalpic effect, 104
  - food macromolecules, 101
  - Gibbs free energy, 106
  - lattice, 97
  - lower-density polysaccharide systems, 104
  - MD modeling and simulation approach, 103
  - pore openings, 97, 98
  - porous structure, 96–97
  - thermodynamic analysis, 107
  - upper regions, 101–103
  - water-macromolecule interactions, 100
  - water-water interactions, 99
- system formulation and simulation
  - amylose molecules, 95
  - atomistic representation, 94, 95
  - Gaussian thermostat method, 96
  - M3B model, 93
  - Morse potential, 94
  - polysaccharides, 93
  - realistic pore structures, 95
  - shifted dihedral functions, 94

Water-macromolecule interactions. *See* Water interactions
Water requirements
- availability, 4
- earth's water, 4–5
- facts, 3, 4
- malnutrition, 16–17
- scarce resource, 15–16
- substantial developments, 2
- technical innovations, 2, 3
- virtual water (*see* Virtual water)

Water-water interactions. *See* Water interactions
Whiskbroom method. *See* Point scanning method
Wielinga, W.C., 270
Williams-Landel-Ferry (WLF) equation, 83, 84
Williams, M.L., 82, 83
Williams, P.D., 267–284, 602
Wine active compounds, of sorption
- gaseous SO_2 enthalpy, 449–450
- H_2O and SO_2 isotherms, 447–448
- hydrated cork powder, 449–450
- isotherms, of SO_2, 448–449
- sulphur dioxide and water enthalpy, 448, 449

Wolbang, G.M., 380
Wold, J.P., 595
Woo, M.W., 165
Wu, D., 581–603
Wu, Z., 168

X

Xanthan gum, 277–279
Xu, C., 395

Y

Yamamoto, S., 327
Yang, W., 394
Yaun, B.R., 393
Yield stress, 500, 504, 506
Yogurt, MRI-based rheological measurement
- comercially available products, 506

flow behavior, 506–507
no-flow and flow conditions, 508, 509
pressure drop values, 509
shear rate *vs*. apparent viscosity, log-log plot of, 507–508, 510
thixotropic behavior, 508
curve fitting results, 511
isothermal conditions, 510
wall slip velocity, 507
Yoo, J.Y., 217
Yoon, W.B., 506–508, 510
Yuk, H.-G., 388

Z
Zambre, S.S., 390
Zeece, M., 347
Zehnder, A.J.B., 3
Zhang, H.H., 379, 380
Zhang, L., 379, 380, 384, 388
Zhang, M., 165
Zhao, T.W., 598
Zhou, B., 394, 395
Zhuang, H., 389
Zogzas, N.P., 121
Zong, W., 379–381

Printed by Printforce, the Netherlands